MW00987865

SAP® Ariba®

Rachith Srinivas, Matthew Cauthen

SAP® Ariba®

Editor Will Jobst
Acquisitions Editor Emily Nicholls
Copyeditor Julie McNamee
Cover Design Graham Geary
Photo Credit Shutterstock.com: 54556654/© swinner
Layout Design Vera Brauner
Production Hannah Lane
Typesetting III-satz, Husby (Germany)
Printed and bound in the United States of America, on paper from sustainable sources

ISBN 978-1-4932-2044-1

3rd edition 2021

Library of Congress Cataloging-in-Publication Data
Names: Srinivas, Rachith, author. | Cauthen, Matthew, author. | Ashlock, Justin. SAP Ariba,
Title: SAP Ariba / Rachith Srinivas, Matthew Cauthen.
Description: 3rd edition. | Bonn, Germany ; Boston : Rheinwerk Publishing, [2021] | Earlier edition authored by Justin Ashlock and Rachith Srinivas.
Identifiers: LCCN 2020048629 (print) | LCCN 2020048630 (ebook) | ISBN 9781493220441 (hardcover) | ISBN 9781493220458 (ebook)
Subjects: LCSH: SAP ERP. | Accounts payable--Software. | Electronic funds transfers--Software.
Classification: LCC HF5679 .A84 2021 (print) | LCC HF5679 (ebook) | DDC 658.15/244--dc23
LC record available at https://lccn.loc.gov/2020048629
LC ebook record available at https://lccn.loc.gov/2020048630

Contents at a Glance

1 Introduction to SAP Ariba 27
2 Supplier Collaboration 53
3 Supplier Lifecycle and Performance 93
4 Supplier Risk Management 171
5 Sourcing 199
6 Contract Management 279
7 Guided Buying 327
8 Operational Procurement 375
9 Invoice Management 477
10 Spend Analysis 529
11 SAP Ariba Integration 569
12 Conclusion 673

Dear Reader,

After a disruptive 2020, 2021 will be a year of *re-*. *Re*starting our processes with coworkers, clients, and collaborators. Revising our expectations of a workday, a workweek, a fiscal year. Rethinking our priorities, our workplaces, our measures for success.

Here, at SAP PRESS, each new edition offers us the possibility to reimagine a book, so it seems fitting that this revision is publishing at the start of 2021. In this new edition, we have a ground-up revision of this book, reorienting around the goal of making this the comprehensive SAP Ariba guide.

Refined and redesigned, this book is an example of that process! *SAP Ariba*, led by Rachith Srinivas and Matthew Cauthen with contributors Justin Ashlock and Juan Barrera, this is a book for our times.

What did you think about *SAP Ariba*? Your comments and suggestions are the most useful tools to help us make our books the best they can be. Please feel free to contact me and share any praise or criticism you may have.

Thank you for purchasing a book from SAP PRESS!

Will Jobst
Editor, SAP PRESS

willj@rheinwerk-publishing.com
www.sap-press.com
Rheinwerk Publishing · Boston, MA

Contents

Foreword 17
Preface 19

1 Introduction to SAP Ariba 27

1.1 Procurement: From On-Premise Solutions to the Cloud 28

1.2 SAP Ariba 31

1.2.1 SAP Ariba Process Areas 34
1.2.2 SAP Ariba Mobile App Features 34

1.3 Cloud Solutions at a Glance 35

1.4 Cloud, On-Premise, and Hybrid Models 36

1.5 Hybrid Deployments 39

1.5.1 Strategic Sourcing 39
1.5.2 MRO, Indirect, and Services Procurement 40
1.5.3 Direct Procurement 41
1.5.4 Accounts Payable and Invoicing 43

1.6 Cloud Implementation Model 44

1.6.1 SAP Activate 45
1.6.2 Time-to-Value Acceleration 49
1.6.3 Enhancement and Modification Limitations 51

1.7 Summary 52

2 Supplier Collaboration 53

2.1 What Is Supplier Collaboration? 54

2.1.1 Supplier Collaboration Strategies 54
2.1.2 Ariba Network 54

2.2 Implementing Ariba Network 56

2.2.1 Strategy 57
2.2.2 Design and Build 58
2.2.3 Supplier Onboarding 60
2.2.4 Network Growth 62
2.2.5 Resources 63

2.2.6 Change Management ... 63
2.2.7 Customer Long-Term Supplier Enablement Program ... 65

2.3 Onboarding Suppliers ... 66
2.3.1 Supplier Information Portal ... 67
2.3.2 Uploading Suppliers ... 68
2.3.3 Assigning and Monitoring Tasks ... 69
2.3.4 Buyer Tasks ... 70
2.3.5 Monitoring Suppliers ... 71

2.4 Enabling Suppliers ... 72
2.4.1 Configurations ... 73
2.4.2 Supplier Enablement Notifications ... 76
2.4.3 Supplier Enablement Reports ... 76
2.4.4 Supplier Integration ... 78

2.5 Becoming a Supplier on the Sales Side ... 81

2.6 Buying with Ariba Network ... 85

2.7 Purchase Order and Invoice Automation with Ariba Network ... 86

2.8 Summary ... 92

3 Supplier Lifecycle and Performance 93

3.1 What Is Supplier Lifecycle Management? ... 93
3.1.1 Supplier Management Strategies ... 93
3.1.2 SAP Ariba Portfolio ... 95

3.2 Implementing SAP Ariba Supplier Lifecycle and Performance ... 96
3.2.1 Application at a Glance ... 96
3.2.2 Monitoring Your Supplier Management Activities ... 99
3.2.3 Using the Supplier 360° View ... 99
3.2.4 Supplier Information Process Flow ... 105
3.2.5 Supplier Request Process ... 105
3.2.6 Supplier Registration ... 112
3.2.7 Supplier Qualification Process ... 120
3.2.8 Preferred Supplier Management Process ... 128
3.2.9 Modular Questionnaire Process ... 131
3.2.10 Scoring in Modular Questionnaires ... 134
3.2.11 Approving, Denying, or Requesting Additional Information during Modular Questionnaire Approvals ... 135
3.2.12 Configuring Supplier Certification Management Using Modular Questionnaires ... 135

3.2.13 Configuring SAP Ariba Supplier Lifecycle and Performance 136
3.2.14 Creating and Configuring Supplier Request Templates 138
3.2.15 Configuring the Supplier Registration Template 145
3.2.16 Configuring Workflow Approval Tasks 148
3.2.17 Configuring the Supplier Qualification Template 150
3.2.18 Configuring the Supplier Disqualification Template 152
3.2.19 Configuring Preferred Supplier Management Project Template 154
3.2.20 Configuring Modular Questionnaire Templates 156
3.2.21 Configuring Banking Questions in Supplier Management Questionnaires 158
3.2.22 Configuring Tax Information Questions 159
3.2.23 Loading Master Data 160
3.2.24 Configuring Analytical Reports 161
3.2.25 Responding as a Supplier 166

3.3 Summary 169

4 Supplier Risk Management 171

4.1 What Is Supplier Risk Management? 171
4.1.1 Supplier Risk Management Strategies 171
4.1.2 SAP Ariba Portfolio 172

4.2 Implementing SAP Ariba Supplier Risk Management 173
4.2.1 Application at a Glance 173
4.2.2 Planning the Implementation of SAP Ariba Supplier Risk 183
4.2.3 Configuring SAP Ariba Supplier Risk 185
4.2.4 Configuring Risk Assessment Project Templates 191
4.2.5 Configuring Issues Management Project Templates 192

4.3 Managing Supplier Risks 194
4.3.1 Monitoring Supply Risk 194
4.3.2 Creating an Engagement Risk Project 195

4.4 Summary 198

5 Sourcing 199

5.1 What Is Sourcing? 199
5.1.1 Sourcing Strategies 200
5.1.2 SAP Ariba Portfolio 201

5.2 SAP Ariba Sourcing 201
5.2.1 Application at a Glance 203
5.2.2 Planning the Implementation 217
5.2.3 Configuring SAP Ariba Sourcing 221
5.2.4 Creating and Configuring Sourcing Request Templates 221
5.2.5 Creating and Configuring Sourcing Project Templates 226
5.2.6 Creating and Configuring Request For Event Templates 233
5.2.7 Configuring Sourcing Library 237
5.2.8 Configuring Approval Tasks in Templates 238
5.2.9 Sourcing as a Buyer 241
5.2.10 Sourcing as a Supplier 242

5.3 SAP Ariba Discovery 242
5.3.1 Application at a Glance 243
5.3.2 Recommended Sourcing Scenarios for SAP Ariba Discovery 244
5.3.3 Using SAP Ariba Discovery 246

5.4 SAP Ariba Strategic Sourcing for Product Sourcing 249
5.4.1 Application at a Glance 251
5.4.2 Planning Your Implementation 258
5.4.3 Creating and Configuring Simple Request For Event Templates for Materials 258
5.4.4 Configuring Parameters for Product Sourcing 260
5.4.5 Sourcing as a Buyer 262
5.4.6 Sourcing as a Supplier 264

5.5 Guided Sourcing 265
5.5.1 Configuring Guided Sourcing 276
5.5.2 Creating a Guided Sourcing Event from Template 276

5.6 Summary 278

6 Contract Management 279

6.1 What Is Contract Management? 280
6.1.1 Contract Management Strategies for Indirect Procurement 280
6.1.2 SAP Ariba Contracts 282
6.1.3 Planning Your Implementation 284

6.2 Configuring SAP Ariba Contracts 286
6.2.1 Creating Project Templates 287
6.2.2 Contract Authoring Process 297
6.2.3 Configuring Workflow Approval Tasks 300

6.2.4 Create Custom Fields Required on Contract Requests and Contract Workspaces 304
6.2.5 Loading Master Data 307
6.2.6 Enabling Electronic Signatures 308
6.2.7 Enabling Reports on the Contracts Dashboard 310

6.3 Creating Contracts 311
6.3.1 Contract Creation 312
6.3.2 Contract Execution and Consumption 313

6.4 Consuming Contracts 317

6.5 Amending Contracts 317

6.6 Using Contracts with Other Applications 320
6.6.1 Using Contracts in SAP Ariba Buying and Invoicing 321
6.6.2 Using Contract Compliance in SAP Ariba Buying and Invoicing with SAP Ariba Contracts 323
6.6.3 Using Contracts in the SAP ERP Backend 323
6.6.4 Using SAP Ariba Contracts in SAP S/4HANA Cloud 324
6.6.5 Using SAP Ariba Contracts in SAP Ariba Fieldglass 325
6.6.6 Contract Application Programming Interface 326

6.7 Summary 326

7 Guided Buying 327

7.1 What Is Guiding Buying? 327
7.1.1 Buying Strategies 328
7.1.2 Guided Buying Capability 330
7.1.3 Planning Your Implementation 331

7.2 Configuring the Guided Buying Capability 333
7.2.1 Approval Flows 334
7.2.2 Spot Buy 335
7.2.3 Images 335
7.2.4 Tiles 338
7.2.5 Landing Pages 338
7.2.6 Home Page 340
7.2.7 Users and Groups 341
7.2.8 Filtering for Purchasing Units and Job Functions 342
7.2.9 Categories 342
7.2.10 Catalogs 343
7.2.11 Suppliers, Preferred Suppliers, and Supplier Management 344
7.2.12 Forms 344

7.2.13 Purchasing Policies 347
7.2.14 Help Community 349
7.2.15 Tactical Sourcing 353
7.2.16 Mobile Solution 356

7.3 Using the Guided Buying Capability 357
7.3.1 The Guided Buying User Experience 358
7.3.2 Creating Requests 361
7.3.3 Purchase Orders 365
7.3.4 Receiving 365
7.3.5 Simple Non-PO Invoices 368
7.3.6 Accessing Forms 370
7.3.7 Initiating a Tactical Sourcing Request 372

7.4 Summary 374

8 Operational Procurement 375

8.1 SAP Ariba Buying and Invoicing, SAP Ariba Buying, and SAP Ariba Catalog 375
8.1.1 SAP Ariba Buying and SAP Ariba Invoicing and Buying 377
8.1.2 Services Procurement with SAP Ariba Buying and SAP Ariba Buying and Invoicing 385

8.2 Configuring SAP Ariba Buying and SAP Ariba Buying and Invoicing 389
8.2.1 Procurement Master Data 389
8.2.2 Catalog and Non-Catalog Requisitioning 393
8.2.3 Spot Buy 393
8.2.4 Total Landed Cost: Taxes, Allowances, and Charges 396
8.2.5 Requisition Approval Process 399
8.2.6 Receiving 402
8.2.7 Purchase Order-Based Invoicing 402
8.2.8 Non-PO Invoicing 404
8.2.9 Contract Invoicing 405
8.2.10 Services Procurement 405

8.3 Using Advanced Buying and Invoicing Functions 409
8.3.1 Uploading Requisitions 409
8.3.2 Creating Planned Service Orders 410
8.3.3 Creating Service Entry Sheets 412
8.3.4 Evaluated Receipt Settlement 413

8.4 Implementing SAP Ariba Buying and Invoicing and SAP Ariba Buying 414
8.4.1 SAP Ariba Buying and Invoicing and SAP Ariba Buying Projects 415
8.4.2 SAP Ariba Buying Deployment 424
8.4.3 Standalone Implementations 444
8.4.4 Integrated Implementations 445
8.4.5 Defining Project Resources and Timelines 450
8.4.6 SAP Ariba Buying, Multi-ERP Edition 458

8.5 SAP Fieldglass Vendor Management System 461

8.6 Implementing the SAP Fieldglass Vendor Management System 467
8.6.1 Planning Your Implementation 468
8.6.2 Defining Project Resources, Phases, and Timelines 470

8.7 Summary 476

9 Invoice Management 477

9.1 What Is Invoice Management? 477
9.1.1 Invoice Management Strategies 478
9.1.2 SAP Ariba Portfolio 478

9.2 SAP Ariba Invoice Management 479
9.2.1 Electronic Invoice Management at a Glance 481
9.2.2 Integrated Implementations 487
9.2.3 Data Sources and Solution Landscape Inventory 487
9.2.4 Configuring Ariba Network Transaction Rules for Invoicing 488
9.2.5 Configuring Invoice Approvals in SAP Ariba Invoice Management 492
9.2.6 Configuring Invoice Exceptions 494

9.3 SAP Ariba Discount Management 496
9.3.1 SAP Ariba Discount Management at a Glance 496
9.3.2 Planning Your Implementation 498
9.3.3 Integration Scenarios 499
9.3.4 Configuring Early Payment Requests 499
9.3.5 Configuring Scheduled Payment Extraction and Vendor Master Setup 501

9.4 SAP Pay 502
9.4.1 SAP Pay at a Glance 503
9.4.2 Planning Your Implementation 510
9.4.3 Configuring SAP Pay 516
9.4.4 Creating and Managing Payments 518

9.5 Summary 528

10 Spend Analysis 529

10.1 What Is Spend Analysis? 529
10.1.1 Spend Analysis Strategies 531
10.1.2 SAP Ariba Spend Analysis 532

10.2 Planning Your Implementation 535
10.2.1 Prepare and Kickoff Phases 536
10.2.2 Data Collection 541
10.2.3 Data Validation 541
10.2.4 Data Enrichment 542
10.2.5 Deployment 543

10.3 Configuring SAP Ariba Spend Analysis 544
10.3.1 Configuring File Validation 544
10.3.2 Setting Opportunity Search Date Ranges 545
10.3.3 Configuring the Star Schema Export 546
10.3.4 Configuring Non-English UNSPSC Code Display 546
10.3.5 Integrating SAP Ariba Spend Analysis and SAP Ariba Data Enrichment Services 547
10.3.6 Enrichment Change Request Setting 548
10.3.7 Managing Data Access Control 551

10.4 Integrating SAP Ariba Spend Analysis with SAP Analytics Cloud and On-Premise Data Sources 553

10.5 Mining Procurement Operations for Data 553
10.5.1 Data Types 554
10.5.2 SAP Ariba Spend Analysis: Areas for Analysis 554
10.5.3 Work Areas 555
10.5.4 Key Reports and Corresponding Impact Areas 557
10.5.5 Key Data Sources and Options for Importing into SAP Ariba Spend Analysis 563
10.5.6 Creating Sourcing Requests from a Spend Analysis Report 566

10.6 Summary 567

11 SAP Ariba Integration 569

11.1 SAP Ariba Integration Projects and Connectivity Options 569

11.2 Ariba Network Purchase Order and Invoice Automation with Purchasing Integration Options 574
11.2.1 Conducting Transactions with Ariba Network Integrated with SAP S/4HANA 575

11.2.2 Ariba Network Customer and Supplier Interaction 578
11.2.3 Confirming and Order Fulfillment 579
11.2.4 Confirming Goods Receipts 582
11.2.5 Invoice Creation 583

11.3 SAP Ariba Sourcing Integration with SAP S/4HANA 586
11.3.1 Key Process Steps 586
11.3.2 Conducting Transactions with SAP Ariba Sourcing Integrated with SAP S/4HANA 587
11.3.3 Creating a New Request for Quotation in SAP S/4HANA 588
11.3.4 R for Quotation in SAP Ariba Sourcing 591
11.3.5 Supplier Processing of Request for Proposals 595
11.3.6 Bid Processing in SAP Ariba 597
11.3.7 Creating Follow-On Documents for Requests for Proposals 599

11.4 SAP Ariba Supply Chain Collaboration for Buyers 600
11.4.1 SAP Ariba Supply Chain Collaboration for Buyers Capabilities 603
11.4.2 Supplier Record in SAP ERP 606
11.4.3 Supplier and Buyer Collaboration in the Ariba Network 607
11.4.4 Scheduling Agreements in SAP Ariba Supply Chain Collaboration for Buyers 608
11.4.5 Scheduling Agreement Release Collaboration 608
11.4.6 Contract Manufacturing Collaboration 612
11.4.7 Order Collaboration Process for Direct Materials 614
11.4.8 Purchase Order Line Item Views 614
11.4.9 Invoice Enhancements for Self-Billing in the Ariba Network 615
11.4.10 Consignment Collaboration 615
11.4.11 Conducting Transactions in SAP Ariba Supply Chain Collaboration for Buyers 616

11.5 SAP Ariba Cloud Integration Gateway 635
11.5.1 SAP Cloud Platform Integration and SAP Ariba Cloud Integration Gateway 638
11.5.2 Backend Configuration 642
11.5.3 General Configuration 646
11.5.4 Ariba Network Integration 650
11.5.5 Additional Integration Options for Older Versions of SAP ERP 654
11.5.6 Integration Pointers for SAP ERP and SAP Business Suite 656
11.5.7 SAP Master Data Integration in SAP ERP 658
11.5.8 SAP Ariba Sourcing Integration 658
11.5.9 SAP Ariba Procurement Integration in SAP ERP 662
11.5.10 SAP Ariba Cloud Integration Gateway Extensions 670
11.5.11 Migrating to SAP Ariba Cloud Integration Gateway 670

11.6 Summary 672

12 Conclusion 673

12.1 Summary 673

12.2 The Future of Procurement Solutions 676

12.2.1 Procure-to-Pay and Order-to-Cash Processes 676

12.2.2 In-Memory Computing, Real-Time Analytics, Machine Learning, and Decision-Making 677

12.2.3 Big Data 677

12.2.4 Blockchain 677

12.2.5 Consumer-Grade User Experiences 678

12.2.6 SAP's Intelligent Enterprise 678

12.2.7 Conclusion 679

The Authors 681

Index 683

Foreword

More than fifteen of my twenty years in technology have been spent in procurement. I've seen the profession move from back-office to strategic, and to the advent of the modern CPO. I've seen the aspiration of our practice grow from savings and quality to prediction and purpose-environment, sustainability, humanity, and fair trade. And I've seen the technology itself evolve through on-premise and custom, multi-tenant and cloud, to a hyper-scaled, distributed, networked ecosystem. More significantly, that networked ecosystem of buying and selling trading partners around the globe, with next-generation technology is actually becoming abstracted from the software itself—which is orchestrating extremely complex and for-fit global business processes at enterprise scale—and it is working itself towards autonomous.

I'm honored to provide the foreword for the third edition of this book about SAP Ariba. Considering the start we've had to the 2020s, with the COVID-19 pandemic and how our procurement space and world at large is in the midst of perpetual change, and thinking about what I've seen my team go through and accomplish together, this is a rather unique edition to be writing the forward for. We all are managing through the new normal and with entirely new challenges. As I write this, the SAP Procurement Services team that I'm proud to lead has delivered essentially 100% of our work remotely since April 1st. In late February, when I made my last travel for business to the SAP Procurement Technology Advisory Board meeting, I never imagined that it would be my last for 2020.

The global COVID-19 pandemic of 2020 forced a step-change in digital transformation for all. For all reaches of the globe, for all generations and demographics, as families and communities, and within our daily lives, our education and work, our B2B and our B2C, we are all being driven to continuous digital transformation. SAP customers are managing their supply chains and ensuring business continuity, while reducing effort and expenditure as technology has evolved from enabler, to value creator.

I strongly believe that the COVID crisis of 2020 will eventually end; however, I do not believe that we will ever see a "return to normal". Chaos, change, and disruption are now truly normal. Fortunately, in times of chaos, disruption and change, business—which, at the end of the day is buying and selling—is a necessary and stabilizing global factor. CPOs and other leaders that are digitizing and thoughtfully controlling their supply chains to levels and with precision never possible before by taking advantage of contingent labor markets; strategically redistributing sources of supply; leveraging payment terms and discounts, while doing so sustainably and humanely, are pioneering our normal every day.

This book comprehensively describes SAP Ariba procurement solutions and the SAP procurement network in today's complex corporate environment, with in-depth chapters on implementing and integrating the solutions within enterprise procurement processes.

Thank you, I hope you enjoy the book and find it valuable in your own continual, digital transformation.

Jason Jablecki
SVP, Global Head of Procurement Services
SAP Services Intelligent Delivery Group

Preface

Procurement gets scant attention in some companies and great focus in others, but it has always been a necessary function of just about every company and government.

Corporate procurement departments focus on efficiencies, supplier optimization and partnerships, and, above all, cost savings. Governmental procurement departments focus on some of the same elements as corporations, but they also spend much of their time trying to optimize their procurement practices to avoid the impression of impropriety. Until the 1980s and 1990s, most procurement was conducted using paper. An organization had to identify a need for a service or product and then identify a supplier, either directly or through a bidding process. After the supplier was in place, a purchase document (i.e., a contract or purchase order) was issued to the supplier. The supplier fulfilled this order, with the buying organization confirming receipt, whereupon the supplier was within its rights to submit an invoice. The buying organization then processed the invoice and paid the supplier. This historical process lives on, albeit largely improved and enhanced in the digital age, and particularly within the SAP portfolio of procurement solutions.

One of the challenges within paper-based procurement, in addition to the unwieldy forms and drawn out procurement times, is understanding spend and financial positions. Drawn out and time-consuming procurement processes, especially during a crisis such as the COVID-19 pandemic, could be detrimental to organizations such as pharmaceutical companies, hospitals, government health departments, and their end users. When computing started to offer electronic means for supporting parts of the procurement process, the goal of having real-time financial postings and the ability to process reams of data for reporting became tangible. Although the generation of solutions available in the 1990s through the mid-2000s streamlined the paper-based, cumbersome procurement processes and focused on email correspondence and web interfaces, they still required installation, configuration, and maintenance at the customer's physical location. In the cross-company areas of exchanges and markets, locating these procurement systems and processes behind a corporate firewall limits collaboration with the supplier and restricts the ability to acquire up-to-date, Internet-based product and pricing information. On-premise solutions are thus at a disadvantage for truly integrated procurement, before even considering the maintenance side of the equation. From a maintenance point of view, it's preferable to have all the customers running on one platform in the cloud, versus supporting multiple versions at different customer sites, as SAP and other software vendors have to do for on-premise implementations.

The future for SAP's procurement solutions is in the *cloud*, an industry shorthand for Internet-based, multitenanted solutions accessed via a customer's browser with an Internet connection. As the leading provider of software and solutions supporting business processes, SAP has defined a wide portfolio of procurement solutions in the cloud, which was enhanced with the acquisitions of SAP Ariba, with its holistic portfolio of procurement solutions, as well as focused solutions in SAP Fieldglass for contingent labor management, and SAP Concur for travel and expense.

Ariba began as a focused solutions provider for Internet-based procurement in 1996. Originally, Ariba offered many of its solutions as on-premise software, similar to SAP Supplier Relationship Management (SAP SRM) and other on-premise SAP solutions. However, by the mid-2000s, Ariba began charting a clear direction toward the cloud, buttressed immensely by what was then called the Ariba Supplier Network (now Ariba Network), along with a suite of cloud solutions. It should be noted that some of SAP Ariba's large customers are still transitioning from their on-premise implementations of SAP Ariba, such as SAP Ariba Buying. The focus of this book, however, is on the current and, at times, future solutions that SAP Ariba has on offer.

In mid-2012, SAP announced the acquisition of Ariba. In March 2014, SAP also acquired Fieldglass, provider of the market-leading Cloud Vendor Management System, to further augment contingent and statement of work labor-lifecycle capabilities. SAP Fieldglass Vendor Management System has cross-functionality applications with both human resources and procurement in order to manage contingent labor and contractors, both from a requisitioning standpoint as well as a human resource-focused one.

Like a reflecting pool mirroring a grand building, SAP continues to maintain a portfolio of on-premise procurement solutions that support similar procurement processes in the cloud portfolio, with one exception: Ariba Network. This network is the core collaboration engine for customers and suppliers and has a transaction run rate of $3 trillion with more than four million companies currently—including more than half of Fortune 2000, in 190 countries transacting in 176 currencies. Ariba Network underpins SAP Ariba solutions and can be integrated with on-premise procurement projects based on SAP S/4HANA, as well as SAP ERP and third-party enterprise resource planning (ERP) platforms.

This kind of global reach of the Ariba Network, coupled with the end-to-end source-to-pay and external workforce cloud procurement solutions from SAP Ariba and SAP Fieldglass, allows customers to quickly and effectively respond to rapidly changing environments. During the COVID-19 pandemic in March 2020, companies around the world were facing unprecedented times and supply chain disruptions. SAP stepped up to help by offering SAP Ariba Discovery for free. With SAP Ariba Discovery, any buyer can post immediate sourcing needs and any of the more than four million suppliers on Ariba Network can respond, without incurring fees until December 31st, 2020. In addition, get up and running quickly with other SAP cloud solutions such as SAP Ariba Contracts (with 30% savings on a one-year contract) to allows supplies to flow in through signing contracts electronically with suppliers; SAP Ariba Sourcing (two free user licenses to create up to two request for [RFx] events per month) to help find competitive suppliers and reduce disruptions; and new SAP Ariba reports to provide insights on supplier disruptions. Finally, the SAP Fieldglass External Talent Marketplace is available for free in the United States until December 31st, 2020, which helps you quickly find temporary labor.

A cloud procurement solution doesn't imply that no work is required to roll out this solution at your company. Rather, the focus, project participants, and project phases change. The focus shifts to the business and its business processes up front, rather than information technology (IT) landscape issues and software install. The project participants change, as IT's role shifts toward network and integration topics. The business engages earlier and drives the business case definition and resulting key performance indicators.

With systems readily available on day one, you can leverage more agile and iterative approaches for setting up parts of the system to the business user's requirements. Supplier collaboration and adoption rates are not an afterthought in these cloud solutions. Business models for SAP's procurement solutions in the cloud acknowledge the importance of supplier participation. As this is a subscription model, much emphasis and focus from SAP Ariba is also placed on the successful usage of the SAP Ariba system, as the time quickly arrives when the customer will decide whether to continue their subscription. This in turn influences the phases. Rather than traditional kickoff, blueprint, realization, test, and go-live phases, and variations thereof, some of the design and realization phases are blended together. Not all the requirements need to be defined before configuration in the system can occur, nor can all the configuration be done directly by the project team. As this is a cloud environment, core configuration is handled by the SAP Ariba shared services organization, as this group is the sole group able to access certain configuration areas of the solution.

This book covers SAP Ariba's solutions in detail, providing a deep dive into functionality, as well as "day-in-the-life" scenarios showing an end user how it all comes together in the system, as well as laying out project implementation structures and guidelines for running successful implementations in each of the areas. Finally, we cover integration topics when cross-system integration is required. This book will enhance your understanding of what is possible with SAP Ariba and provide you with a springboard for launching your SAP Ariba procurement projects.

Target Audience

This book should be of interest to anyone working with SAP Ariba or considering rolling out procurement solutions in a cloud environment. Business users, procurement managers, and consultants are the primary beneficiaries. This book assumes a familiarity with and interest in procurement solutions and terminology, as well as project management and delivery topics around these solutions. For the integration topics, a more advanced understanding of middleware and technical configuration skills is helpful. Finally, critical reasoning skills in understanding just where these solutions will provide your organization the most value individually and in concert will serve you in this journey.

Objective

This book's focus is on implementing SAP's cloud procurement portfolio, principally SAP Ariba. SAP Ariba solutions cover a vast area of collaboration and processes between your company and its suppliers. While it's not possible to cover every permutation of configuration in these solutions, this book will provide an in-depth look at each of the solution areas, how to use and consume them as a business user, and how to furnish project-specific guidelines, methodologies, required roles, and timelines.

Structure of This Book

This book is organized into 12 chapters. The first half will cover what SAP Ariba previously called "upstream" procurement solutions, that is, solutions used to identify, vet, and onboard suppliers; create and manage sourcing events; and award orders and contracts. Beginning with Chapter 7, this book will turn its focus on to the "downstream" parts of the procurement process. By chapter, this book covers the following:

- **Chapter 1: Introduction to SAP Ariba**
  SAP has made sizable investments in its procurement solution portfolio. This chapter introduces the various solutions offered by SAP Ariba and in the context of the shift from on-premise procurement solutions to procurement in a cloud-based environment, as well as hybrid approaches to realizing the best options from both the digital core area, which can sometimes be on premise within an organization, and the cloud-based procurement solutions from SAP Ariba.

- **Chapter 2: Supplier Collaboration**
  The supplier collaboration chapter provides an overview of Ariba Network, including the processes for onboarding suppliers, enablement, and becoming a supplier on the sales side of Ariba Network. This chapter also details a purchase order/invoice exchange on Ariba Network, both supplier-side and buyer-side.

- **Chapter 3: Supplier Lifecycle and Performance**
  This chapter outlines the SAP Ariba solutions for understanding and managing supplier relationships, with a focus on a new product called SAP Ariba Supplier Lifecycle and Performance, which replaces SAP Ariba Supplier Information and Performance Management. This chapter details SAP Ariba Supplier Lifecycle and Performance's functionality and implementation, as well as how it's used.

- **Chapter 4: Supplier Risk Management**
  This chapter outlines the effective risk management solution from SAP, which simplifies risk management and allows businesses to make smart, informed, and timely decisions during their procurement process. This solution is natively integrated to the SAP Ariba Supplier Lifecycle and Performance solution, providing a comprehensive solution for onboarding and managing suppliers while continuously identifying and mitigating risk in the company's supply base. This chapter details SAP Ariba Supplier Risk's functionality and implementation, as well as how it's used.
- **Chapter 5: Sourcing**
  SAP Ariba Sourcing enables you to develop a sourcing strategy and negotiate, source, and manage the resulting contracts and supplier data. An augmenting solution is SAP Ariba Discovery, which suppliers across Ariba Network can register with to access any new requests for information (RFIs), requests for proposals (RFPs), and requests for quotes (RFQs), collectively known as RFx. This chapter outlines SAP Ariba Sourcing and SAP Ariba Discovery as solutions, as well as SAP Ariba Sourcing for direct procurement, and highlights their key functionalities and how to implement and use them.
- **Chapter 6: Contract Management**
  SAP Ariba Contracts provides a powerful platform to efficiently and effectively manage procurement and sales contracts, license agreements, internal agreements, and various other kinds of contracts. SAP Ariba Contracts also automates and accelerates the entire contract lifecycle process, including the contract signature process by enabling digital signatures; standardizes the contract creation, authoring, and maintenance processes; and strengthens operational, contractual, and regulatory compliance. This chapter reviews the SAP Ariba Contracts solution and how to implement and use these features.
- **Chapter 7: Guided Buying**
  The guided buying capability in SAP Ariba is a new entrant since the last edition of this book. Guided buying enables a user to enjoy a consumer-grade procurement experience, leveraging the collaboration capabilities of Ariba Network. This chapter provides an overview of the guided buying functionality, as well as how to implement and use it.
- **Chapter 8: Operational Procurement**
  After a source of supply has been identified and an agreement, such as a contract, has been put in place, it's time to go shopping. This chapter outlines SAP Ariba Buying and SAP Ariba Catalog. It connects the dots between the solutions and shows how to implement and consume them.
- **Chapter 9: Invoice Management**
  This chapter covers SAP Ariba's solutions in the area of accounts payable, which have been streamlined down to SAP Ariba Invoice Management and SAP Ariba Pay. This chapter will outline the solutions' functionalities, business benefits, usage, and implementation approaches.

- **Chapter 10: Spend Analysis**
  This chapter outlines SAP Ariba Spend Analysis capabilities and shows how to implement it to get a closer look at procurement operations at large as well as learn where to make changes. Depending on the SAP Ariba solutions you've implemented, you may also run reports on other areas of procurement, such as sourcing events, contracts, purchase orders, requisitions, and suppliers. However, SAP Spend Analysis is the main reporting areas featuring enhanced enrichment of aggregated spend data, including machine learning capabilities.
- **Chapter 11: SAP Ariba Integration**
  SAP Ariba can integrate with SAP ERP, SAP S/4HANA, and a multitude of third-party ERP environments. The latest integration approach and future go-to platform is the SAP Ariba Cloud Integration Gateway. This chapter provides an overview of the integration options, as well as a detailed view into SAP Ariba Cloud Integration Gateway and integration with SAP S/4HANA.
- **Chapter 12: Conclusion**
  This chapter provides a synopsis of the preceding chapters, discusses procurement in general, and makes some predictions for the future of procurement solutions.

References and Resources

In the course of writing this book, we used the following resources as references, which you may want to refer to for continued learning:

- **SAP Ariba Knowledge and Connect**
  http://knowledge.ariba.com and *http://connect.ariba.com*
- **SAP product documentation, downloads, service, and Rapid-Deployment Solution (RDS) content (integration)**
  http://service.sap.com

Acknowledgments

We would like to thank SAP PRESS for providing us with the opportunity to write and share this book. To all of our readers, we hope you find this book helpful on your journey through SAP Ariba.

Writing several chapters in this book, while juggling a fulltime job and personal life, was definitely challenging—yet very fulfilling. This could not have been possible without the support of my family and guidance from my many colleagues and friends within SAP Ariba and outside of SAP.

In particular, I would like to thank our editors, Will Jobst and Emily Nicholls of SAP PRESS; my coauthor, Senior Consultant, Matthew Cauthen; as well as the contributing authors: my manager and Vice President of Procurement Delivery Services, Justin

Ashlock, and Head of Services Enablement, North Americas, SAP Ariba, Juan Barrera. I would like to thank our leaders for their support and guidance: President, Intelligent Spend and Business Networks, SAP, John Wookey; Head of Procurement Services, SAP, Jason Jablecki; and Global Vice President, Functional Deployment & Network Services, SAP Ariba, Kenneth Psnipechous. Thanks also to my mentor Britt Freund, senior lecturer of supply chain management and assistant dean for McCombs School of Business, UT Austin, for his valuable inputs for all things supply chain. I would also like to thank our product management colleagues, Ashley Vandenhoek and Michael Waugh, as well as my other colleagues for their valuable inputs: Lauren Beck, Muthukumar Nagappan, JohnRoss Rao, Mangesh Thele, Tom Duculan, Lauren Shoup, Joshua Semler, Shreya Shetty, Tina Price, Ganesh Kotti, Harinath Malvathu, Firat Kasru, Jesse Clark, Lia Sorce, Amit Chokshi, Ram Ranganathan, Gustavo Cassuriaga, Andrew Gambelin, and Saurabh Sethi.

Finally, I could not have written this book without the unwavering support and love throughout my life and career from my family: my mom, Padma, and dad, Dr. R. Srinivasa; my two boys, Viren and Krish; my wife, Nita; my sister, Rachana; and twin brother, Dr. Rakshith Srinivas.

Rachith Srinivas
Dallas, TX
October 2020

I would like to thank my family for providing the support and encouragement needed to see this through from beginning to end. An undertaking like this could only have been done with their support and encouragement, of which I am forever grateful.

At SAP PRESS, I would like to thank Will Jobst and Emily Nicholls. Their patience and support were instrumental in getting this edition released, my coauthor, Rachith Srinivas, and contributing authors Justin Ashlock and Juan Barrera, as well as the leadership at SAP Ariba for all the support, guidance, and opportunity that made this book possible.

I would also like to thank my colleagues for sharing their time and expertise. All have provided essential insight, and without them this book could not have been written: Maria Barcarse, Abhishek Agrawal, Ashwani Kumar, Emily Szelestey, Awanendra Singh, Bob Dudas, Elizabeth Smith-Gales, Gourav Kumar, and Rita Gates.

Matthew Cauthen
Atlanta, GA
December 2020

Chapter 1
Introduction to SAP Ariba

In this chapter, we'll provide an overview of SAP Ariba in the context of SAP's procurement solution portfolio as well as describe SAP-recommended deployment and implementation strategies for integrating SAP Ariba with SAP's digital core, SAP S/4HANA.

SAP was founded in 1972 with the vision of providing integrated applications supporting business processes in real time. Procurement was an early business process supported in the first SAP ERP applications. Practically all business and government SAP customers procure items and services to support their day-to-day operations and to create and deliver goods and services to the marketplace or, in the case of the government, to their citizens. Fast-forward almost a half of a century, and procurement is still a core area supported within the SAP portfolio of products and solutions. This procurement area has changed a great deal, as with all areas of computing, over this time span.

An early prophecy of this computing revolution was Moore's Law, which predicted the doubling of transistor circuits on computer chips every year. This law was articulated in 1965 and has held true so far. However, transistors weren't the only area in computing evolving at a breakneck pace. Areas such as memory/storage, data throughput capabilities, and algorithmic-based software have shown even greater leaps in capabilities. No other industry has realized this type of better-faster-cheaper growth curve over the last half century than the technology industry. The products in the SAP ERP suite have taken advantage of and driven this trend by offering ever faster and more integrated business applications, including the reunification of both online analytical processing (OLAP) and online transaction processing (OLTP) in SAP's latest ERP production iteration, SAP S/4HANA.

In addition to the exponential growth and expansion of capabilities, the Internet has changed the way computing applications and interactions are leveraged by both customer and supplier. No longer are perpetual licenses and large, up-front investments in both hardware and software necessary to acquire enterprise-level applications. Now, customers can purchase SAP solutions, including SAP Ariba and SAP S/4HANA Cloud, as subscriptions and consume these solutions via web browsers on smartphones, tablets, and desktop computers.

At the forefront of procurement in the cloud are the cloud-based solution offerings from SAP Ariba. In this chapter, we'll introduce the various solutions offered by SAP Ariba, as well as the recommended implementation approaches for integrating the digital core of SAP S/4HANA with SAP Ariba for procurement scenarios and various levels of organizational complexity. We'll walk you through the main considerations in the procurement solutions space in Section 1.1; SAP Ariba, in Section 1.2; cloud, on-premise, and hybrid models in Section 1.3 and Section 1.4; hybrid deployments in Section 1.5; and implementation methodologies for SAP's cloud-based services and solutions in Section 1.6. In the following chapters, you'll gain focused insights into SAP Ariba's individual solution areas and applications, as well as accelerated project management approaches for each solution as we highlight key objectives and considerations for successful implementations and provide step-by-step instructions for configuration and use.

1.1 Procurement: From On-Premise Solutions to the Cloud

Starting with paper-based processes, procurement has always been about buyers understanding an organization's needs and interacting with suppliers to obtain the required materials and services. Today, core business processes such as procurement are moving to cloud-based solutions at an increasing rate.

From the start of procurement, processes and systems were needed to optimize efficiencies and address potential conflicts of interest, also known as the principal-agent problem, which we'll discuss further in Section 1.2. Prior to the Internet, computer-based procurement systems relied on an on-premise model for the system, as information technology (IT) was largely managed on servers and mainframes within the four walls of the company's headquarters. SAP R/3 and follow-on versions of SAP ERP brought this approach into real time, allowing finance and buying organizations to view orders and their corresponding liabilities/financial implications as they occurred, instead of after the fact during a month-end closing or reconciliation processes. SAP S/4HANA, the latest iteration in SAP's line of ERP solutions, provides not just up-to-date views into these areas but also predictive modeling, machine learning, and dashboards presenting this information in consumable and actionable form (e.g., with tiles that alert users to overages and other issues).

As the ERP platform, SAP S/4HANA represents the "digital core" for an enterprise, from which customer engagement, manufacturing and supply chain, people management, and network and spend management can be conducted, as shown in Figure 1.1. Connected to network and spend management are a host of spend management solutions in the cloud (shown in Figure 1.2).

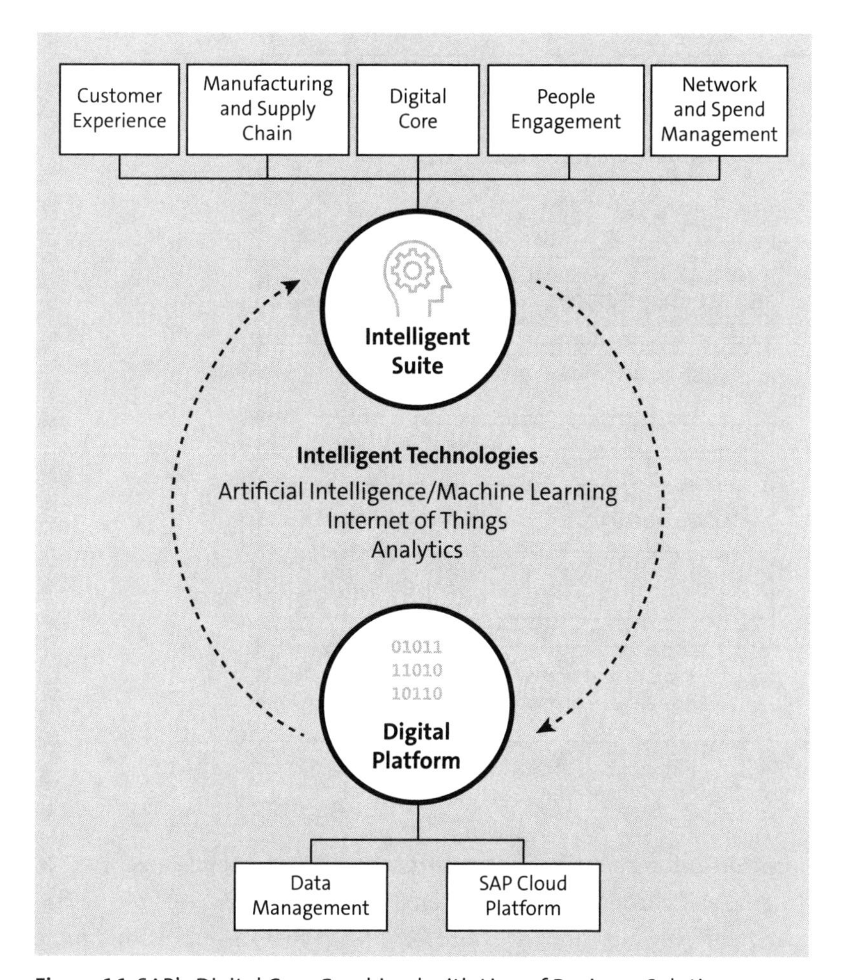

Figure 1.1 SAP's Digital Core Combined with Line of Business Solutions

The solutions comprise SAP S/4HANA and SAP Ariba for direct procurement activities, with the direct procurement activities originating in SAP S/4HANA, generated in modules and processes, such as production planning (PP), sales and distribution (SD), and materials management (MM). After the demand is generated, SAP Ariba can be used to support collaboration with suppliers for purchase order (PO) and invoice collaboration, using any of these solutions: SAP Ariba Supply Chain Collaboration for Buyers, SAP Ariba Sourcing, SAP Ariba Supplier Lifecycle and Performance and SAP Ariba Supplier Risk, and SAP Ariba Discovery, via Ariba Network. Content management can be handled via SAP Ariba Catalog and the guided buying capability.

As shown in Figure 1.2, the further spend categories supported and augmented by cloud solutions comprise indirect procurement and maintenance, repair, and operations (MRO), supported by SAP Ariba and SAP S/4HANA; flexible workforce management for contingent labor, supported by SAP Fieldglass; and travel procurement, supported by SAP Concur.

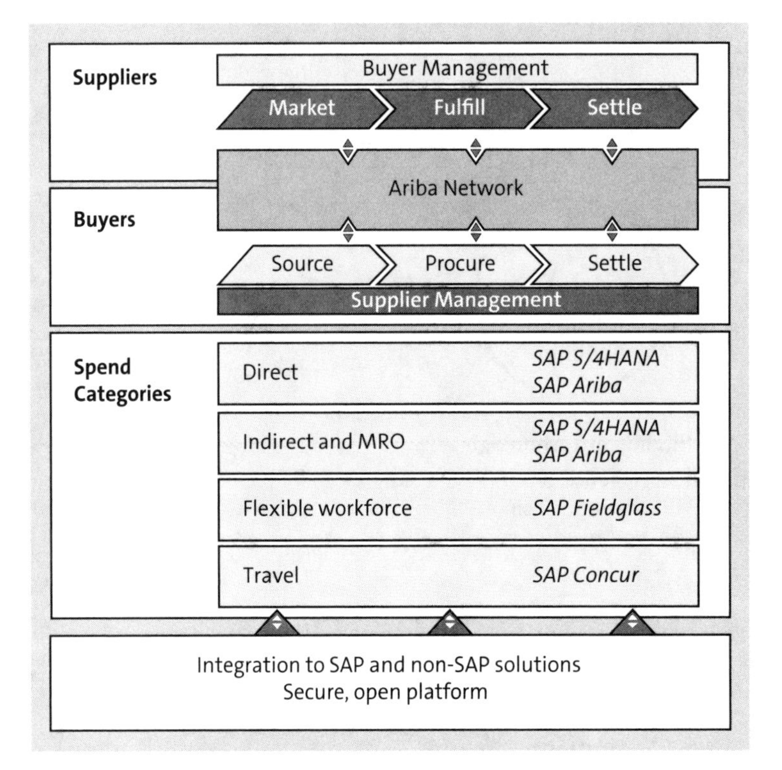

Figure 1.2 SAP Line of Business Procurement Solutions

Why does SAP provide a cloud solution to support each of these spend categories? In today's ever changing and challenging business environment, exacerbated by the current COVID-19 pandemic, organizations are transforming to be lean, nimble, and more adaptive to change. These organizations are reluctant to make huge up-front investments in on-premise solutions, IT infrastructures, and so on. Instead, they're choosing cloud- and subscription-based solutions, with very little up-front costs, zero maintenance costs, and a pay-as-you-use model. SAP recognizes this shift.

Procurement lends itself nicely to the cloud model, as continuous interaction between suppliers and customers is required for successful procurement operations. On-premise software can address much of the process from behind a customer's firewall, but for interacting with a supplier at the system level, phone calls and emails are often required to supplement the lack of capabilities for real-time, in-system interaction with the supplier.

Organizations are loath to open up their integrated ERP systems, which are usually ensconced behind firewalls for protection, even to controlled access by suppliers. The protocols for access are often so unwieldly and unique to the organization that interacting with an organization could be prohibitively complex and expensive for the supplier. Collaboration at this level can add to the costs of procurement, rather than lowering costs. Cost-saving pressures grow on IT departments tasked with building what are essentially standard processes in their in-house environments on an individual basis. In traditional, on-premise software implementations, all the servers, installation, configuration, customization, and administration are maintained by the customer. Connecting a supplier to this type of environment for document exchange is equally cumbersome using electronic data interchange (EDI), as establishing EDI linkages requires individual, point-to-point mapping of the transmission for each supplier. Several iterations may be required to set up a single linkage.

Another continual benchmark for corporate procurement is what users experience in their online shopping experiences at home. If a corporate system is clunky and difficult to use, these shortcomings are made more glaring by the seamless shopping experiences available online today on many sites. Seamlessness is most pronounced in procurement systems—a financial application or warehouse management system isn't something you would normally access on a day-to-day basis online at home. Almost everyone, at some point, has bought something online for personal use. In the highly competitive online shopping arena, online retailers are continuously setting the bar higher to stay relevant to consumers.

Procurement, at its core, involves tight interaction with entities outside the four walls of an organization, that is, with suppliers. While not all the linkages between ERP and procurement can be decoupled, especially in the finance and PP areas of the process, better integration is already available between SAP Ariba and cloud-based procurement solutions, with SAP S/4HANA running as a true cloud platform, whether hosted or on premise. This integration of procurement solutions between SAP ERP isn't limited to SAP Ariba. SAP Concur for travel and expenses and SAP Fieldglass for contingent labor also integrate in a similar fashion to their dependent modules in SAP S/4HANA, creating an integrated, holistic cloud platform for all different types of procurement.

In the next section, we'll cover SAP Ariba solutions in further detail, beginning with how SAP Ariba addresses key challenges in procurement and how procurement processes are supported by these SAP Ariba toolsets.

1.2 SAP Ariba

SAP Ariba offers a comprehensive array of tools and solutions to cover core procurement activities and challenges. As with any process, procurement's complexity arises in the steps and details. The main considerations and challenges for the procurement process include the following:

- **Principal-agent problem**
  Also known as the agency problem, the principal-agent problem refers to a conflict of interest that can occur when an agent acts on behalf of an organization. Procurement transactions typically operate as zero-sum games, whereby a supplier either wins the deal and does business with the purchasing company or doesn't. Under this

pressure to win, suppliers are tempted to create an "edge" wherever possible to win the bid. Thus, the purchasing agent in this interaction holds a powerful position, as the purchasing agent can influence or even decide the outcome of this situation in favor of a particular supplier. The purchasing agent's interests may not always align with the principal's in practice. As a result, a conflict of interest scenario may arise, whereby the agent colludes with the supplier, accepting bribes or less overt forms of payment in kind to steer the deal to the favored supplier. To counter this conflict of interest, organizations can seek to establish systems of checks and balances in procurement operations and systems. SAP Ariba provides a rigorous set of controls, processes, and workflows in all the solution areas to avoid the principal-agent problem.

- **Demand planning**
  Reacting to needs that have arisen and which the organization needs filled yesterday puts the organization in a tougher negotiating position vis-à-vis suppliers and creates inefficiencies because divisions and actors in the organization must wait for supplies and services to arrive. Some form of demand planning, based on known targets, goals, historical needs, and external factors, can help alleviate the last-minute, reactive nature of tactical procurement. SAP Ariba Supply Chain Collaboration for Buyers and the extensive reporting capabilities available in SAP Ariba in general allow buyers to proactively formulate an understanding of procurement needs and subsequently create procurement strategies to address these needs.
- **Suppliers**
  After a demand has been defined and/or raised, suitable suppliers need to be identified, and a procurement plan must be put in place. How do you define the group of eligible suppliers, create urgency and competition around your requirements, and arrive at the best price, product, and supplier? SAP Ariba Discovery allows you to identify new suppliers in Ariba Network's suppliers, which in turn enriches the sourcing events and activities provided in SAP Ariba Sourcing. Finally, using SAP Ariba Supplier Lifecycle and Performance, your organization can iteratively improve supplier interactions and management processes.
- **Purchase agreements**
  After negotiating with potential suppliers and selecting the most suitable one, a contract or PO must be defined for the particular item/service being procured. This agreement must clearly define payment terms and other items, while being usable at an operations level to allow for the consumption of the contract. SAP Ariba Contracts provides a platform for creating and managing contracts, serving as a lynchpin connecting the various procurement solutions available in SAP Ariba and in the digital core of SAP S/4HANA.
- **Consumption of goods**
  With an order/contract in place, the item or service must be made ready for consumption/deployment by your organization. How do employees find and consume

this item within the proper approvals and processes? The key is to make the "finding" portion of the process intuitive for your employees, while still conforming to company approval processes and financial regulations. SAP Ariba Buying and SAP Ariba Catalog provide a platform for expert and casual users alike to find and consume the items and services vetted and negotiated by the organization's procurement department.

- **Invoicing**
  As a company consumes items following the previous steps, the supplier begins to submit bills for the consumed items, unless the agreement required payment up front. Either way, a company requires an approach to process invoices: a process that matches invoices to actual requests, thus ensuring that payments made to suppliers reflect the agreement's terms and the amounts actually ordered and delivered by the supplier. SAP Ariba Invoice Management and SAP Pay provide a seamless solution for managing the entry and intake of invoices, ensuring correct invoices up front, and enabling you to obtain further discounts and payment efficiencies during the payment portion of the procurement process.
- **Analysis**
  After payment is verified, this process can start again—ideally smarter the next time, leveraging the knowledge gleaned from the preceding procurement cycle. In this area, organized data and analytics, as well as the processing power to perform timely analysis on large swaths of transactions and other data, are needed. SAP Ariba Spend Analysis helps procurement and finance departments gain this enhanced understanding of their procurement operations to ensure that continuous improvements can be made.

SAP Ariba addresses these areas with a host of solutions and services, as shown in Figure 1.3.

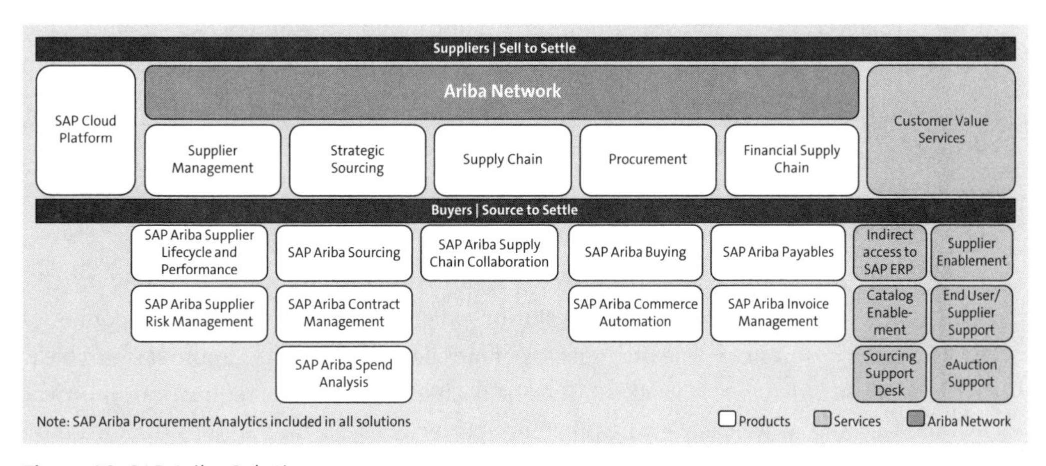

Figure 1.3 SAP Ariba Solutions

1.2.1 SAP Ariba Process Areas

The process areas in SAP Ariba span the full sourcing and procure-to-pay process. The main process areas include the following:

- **Sourcing (finding a deal)**
  SAP Ariba solutions for this include SAP Ariba Sourcing, SAP Ariba Discovery, SAP Ariba Spend Analysis, SAP Ariba Supplier Lifecycle and Performance, and SAP Ariba Supplier Risk.
- **Signing the deal**
  SAP Ariba Contracts supports this area.
- **Broadcasting the deal**
  SAP Ariba Catalog provides a content repository for users to search and select products and services curated by the procurement department.
- **Buying items and services**
  SAP Ariba Buying supports the procure-to-order and procure-to-pay processes for items and services.
- **Being invoiced**
  SAP Ariba Buying and Invoicing and SAP Ariba Contract Invoicing manages interactions between suppliers and buying organizations as the supplier submits invoices, thus ensuring that the supplier submits an invoice in the form that is most easily processed by the receiving organization via extensive business rule options and workflows.
- **Paying invoices**
  SAP Pay supports payments and additional discount negotiations after an invoice has been submitted.
- **Collaborating with suppliers**
  Ariba Network comprises of more than four million customers and suppliers and manages more than $3.2 trillion in transactions annually. Ariba Network serves as a platform to underpin and augment all the other SAP Ariba solutions on offer, facilitating supplier/buyer collaboration at every step of the procurement process.

1.2.2 SAP Ariba Mobile App Features

Mobility and intuitive user interfaces (UIs) also deserve mention. Beginning in 2015, SAP has made access to SAP Ariba solutions via mobile devices a core component of the product's strategy. The first releases of mobile apps supported approvals and basic lookups in SAP Ariba. Now, the SAP Ariba mobile app supports requisition, approval, and tracking, as well as viewing previous approvals. These mobile apps run on a multitude of devices, such as phones and tablets.

In the SAP Ariba Supplier app, suppliers can do the following:

- View POs and invoice activity.
- Confirm POs.
- Pin POs requiring further follow up.

Users of SAP Ariba Supply Chain Collaboration for Buyers can use the SAP Ariba Supply Chain mobile app. Users can monitor key supply chain procurement documents, such as scheduling agreement POs and other complex PO types supported in SAP Ariba Supply Chain Collaboration for Buyers, confirmations, and advanced shipping notifications (ASNs), as well as monitor the availability of supply and receive alerts for unconfirmed items. The app includes full item details, drilldown capabilities, and search capabilities as well as communication options to call or email suppliers to resolve identified issues in real time.

As of SAP Ariba 14s, the UI has been completely refreshed, beginning with SAP Ariba mobile 1.5 for iOS under an initiative called the *Total User Experience*. The SAP Ariba Total User Experience represents a complete paradigm shift in the meaning of user experience (UX)—not just for SAP Ariba customers but also for the industry as a whole. SAP is delivering that shift through three fundamental tenets:

- **Work anywhere**
  Users should be able to access SAP Ariba applications anytime and anywhere.
- **Learn in-context**
  Users can access in-context guides and community expertise from within the application.
- **Modern UX principles**
  Users can benefit from the adoption of SAP Fiori–based user interaction paradigms.

With this initiative, SAP Ariba incorporates modern UX principles utilizing SAP Fiori-based user interaction paradigms for a new and improved visual design for all SAP Ariba on-demand solutions.

1.3 Cloud Solutions at a Glance

For cloud computing and services, an Internet connection and a subscription are all that is required. With the advent of cloud computing, both sides of the equation, integration with corporate environments and supplier interaction, have been simplified. In addition, the expensive infrastructure supporting a company-specific instance of software is no longer needed. Procurement also benefits from the network effect: The more suppliers you can access and understand, the better the competition around your purchasing becomes, which leads to finding better suppliers, relationships, and pricing.

The more activity and potential customers that suppliers see in a particular network, the more likely they are to join the network, leading to a virtuous cycle. The most obvious example of an offline supplier network of this type is a traditional marketplace, where buyers and sellers meet to exchange goods.

Point-to-point networks, a contradiction in terms, were often the only option for on-premise solutions. In this scenario, you would laboriously set up each relationship and linkage by supplier, then broadcast requirements to individual connections, and conduct purchasing transactions with the winning supplier. Networks and online marketplaces are a key argument for cloud solutions in procurement, as you'll reap the benefits of supplier interaction and competition while avoiding the lengthy setup and maintenance requirements of an EDI or other on-premise approach. Ariba Network is a powerful example of a transaction platform in the cloud, and one of the largest of its kind, with more than $3 trillion in transactions already flowing on an annual basis. Suppliers can set up their transmission methods in the network, and new customers transmitting purchasing documents to the supplier won't need to worry about additional setup steps for transmission methods—these connections are already all set up and maintained in the network.

Support also is less burdensome in a cloud solution. If something fails to transmit, a supplier needs to register, or an auction needs focused support, the network host can support and resolve any issues not on the internal IT department. If you only use a specific supplier monthly, or even annually, a supplier record can quickly become out of date. If a new order is placed, the supplier record may first require updating via phone or email. Worse yet, the PO could be issued to an outdated address, and the order may not be fulfilled in a timely manner. These manual interventions and updating cycles are minimized on the customer side with a network model. When a supplier updates its record on Ariba Network, the supplier records held by buyers on the network also receive the update.

1.4 Cloud, On-Premise, and Hybrid Models

With the acquisitions of SAP Ariba and SAP Fieldglass, SAP offers cloud procurement solution counterparts to all its on-premise solutions, covering all spend types and stages of the procure-to-pay process, as shown in Table 1.1.

Stage	On-Premise Solutions	Cloud Solutions
Spend analysis	SAP S/4HANA apps for spend performance	SAP Ariba Spend Analysis

Table 1.1 Cloud and On-Premise Solutions Addressing All Spend Types

Stage	On-Premise Solutions	Cloud Solutions
Sourcing	SAP S/4HANA Sourcing and Procurement and SAP Supplier Relationship Management (SAP SRM)	SAP Ariba Sourcing and SAP Ariba Discovery
Contract management	SAP S/4HANA for Legal Content and contract functionality (and SAP SRM)	SAP Ariba Contracts
Operational procurement	SAP S/4HANA Sourcing and Procurement, MM, and SAP SRM	SAP Ariba Buying, SAP Ariba Catalog, Spot Buy capability, guided buying capability, and SAP Fieldglass Vendor Management System
Invoice management	SAP Invoice Management by OpenText	SAP Ariba Invoice Management and SAP Pay
Supplier information and performance management	SAP S/4HANA for supplier management apps, SAP Supplier Lifecycle Management	SAP Ariba Supplier Lifecycle and Performance and SAP Ariba Supplier Risk
Mobile procurement	SAP Fiori	SAP Ariba mobile
Supplier collaboration	N/A	Ariba Network

Table 1.1 Cloud and On-Premise Solutions Addressing All Spend Types (Cont.)

The main solution covering supplier collaboration for both on-premise and cloud, however, is Ariba Network. While an on-premise solution for collaboration exists, supplier self-services in SAP SRM can't provide the same level of collaboration as Ariba Network.

Several integration options are available for connecting your on-premise SAP ERP and/ or SAP SRM solutions to Ariba Network and for interacting with suppliers via the network, while still leveraging your on-premise investments. As shown in Figure 1.4, the SAP S/4HANA digital core can connect to Ariba Network and other applications via the SAP Ariba Cloud Integration Gateway.

The digital core ERP doesn't necessarily need to be on-premise. While most companies have existing SAP ERP implementations in place that house core processes and functions connected to procurement and typically reside on-premise, if you're starting fresh, you could house all your operations and processes in SAP S/4HANA Cloud, enterprise edition, which would integrate with SAP Ariba, SAP Concur, and SAP Fieldglass cloud solutions. In this scenario, you would be following a pure cloud model.

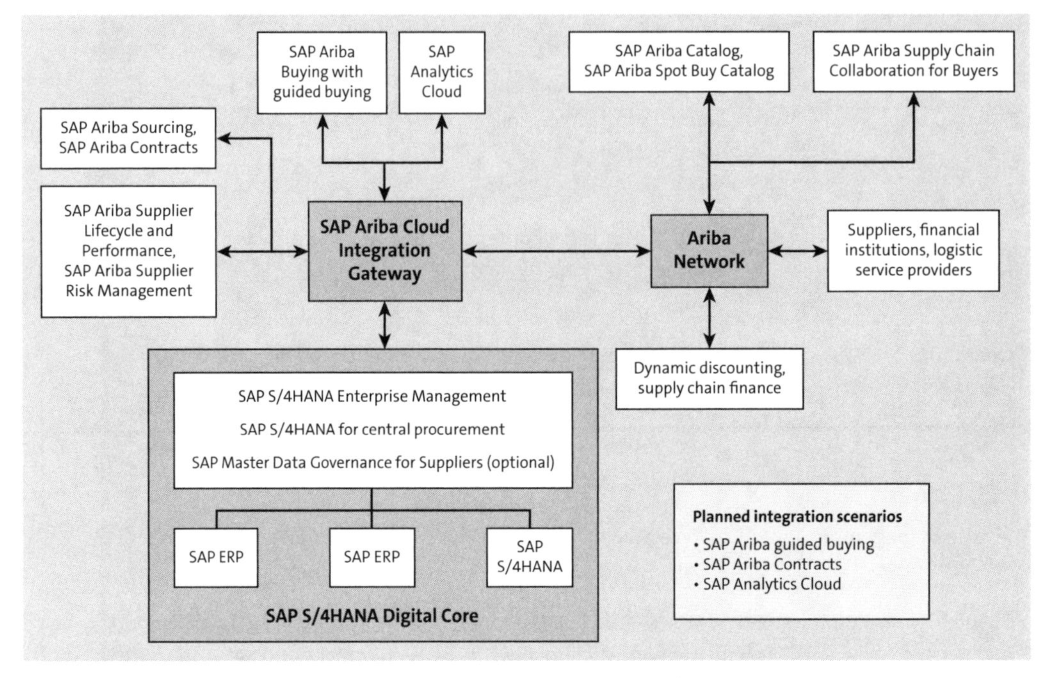

Figure 1.4 Procurement Reference IT Landscape for SAP S/4HANA and SAP Ariba

On-demand SAP Ariba provides new and enhanced features that are automatically deployed and driven by partner requests. The reasons for going this route and for leveraging the cloud in general center around automatic upgrades, standardized processes, accelerated innovation, and technology foundation advantages. More specifically, these include the following:

- **Features and functions**
  Delivered on a regular release schedule.
- **Lower total cost of ownership**
  Reduced maintenance, infrastructure, and support costs.
- **Customization and configuration**
  Replace heavy customization with standardized processes.
- **Security**
  Data stored in the cloud in a secure hosting environment with SAS 70 and WebTrust.

As a cloud user, not only can you upgrade to the latest release much more easily, but the innovation in a cloud solution is also more robust and continuous. For example, new releases for SAP S/4HANA are released annually (the date and month comprise the release number, so for the 2018 September release, the version is called "1809"). For SAP S/4HANA Cloud, enterprise edition, you'll receive quarterly updates. In the cloud, if the system shows issues with certain navigation areas, those issues can be identified and addressed for all the system's users much more quickly than in the on-premise world.

On-demand SAP Ariba provides new and enhanced features that are automatically deployed and driven by partner requests.

As enticing as a cloud model is, most companies will need to face the realities of their existing investments and infrastructure when evaluating a pure cloud model. Some public sector enterprises or large corporate customers simply can't move to a pure cloud model in one step or even multiple steps, given the complexity of their operations and the expense of moving off of existing investments in IT. Public sector and strategic industries may have further regulations that preclude them from conducting some areas of their business in a cloud-based environment. These organizations may even be inclined to stick with a purely on-premise approach, especially if regulations, high levels of customization, complexity, and/or existing investments in on-premise solutions complicate moving to the cloud at this point. In general, you can mix and match solutions from both of SAP's procurement solution portfolios, SAP S/4HANA and SAP Ariba, based on requirements, landscape/existing environments, and regulatory requirements.

Even in cases where regulations preclude the use of the cloud for certain transactions and processes, cloud solutions can still be leveraged in several ways. These approaches typically leverage a hybrid approach to cloud solutions. Some areas, such as contingent workforce management/vendor management systems (VMSs) with SAP Fieldglass, SAP Ariba Catalog, Ariba Network, and SAP Ariba Buying and Invoicing, all provide ample integration options with SAP S/4HANA and SAP SRM to augment existing investments. While Chapter 11sdiscusses integration approaches in more detail, in the next section, we'll outline SAP's recommended approaches for hybrid deployments of both the digital core and SAP Ariba.

1.5 Hybrid Deployments

Choosing the right procurement path forward when faced with this large and varied set of options can be a bit daunting at first. In acknowledgment, SAP provides a set of recommend approaches for typical procurement scenarios. For source-to-pay processes, we begin with strategic sourcing, indirect, MRO, and services, then discuss direct procurement, ending with accounts payable and invoicing.

1.5.1 Strategic Sourcing

Strategic sourcing lends itself well to a cloud model. A buyer needs to go beyond the firewall and interact directly with suppliers, which puts on-premise systems at an immediate disadvantage. Many of these suppliers may not be onboarded to an internal system yet. The buyer thus benefits from a preexisting platform that can be leveraged to design sourcing strategies and to publish requests for pricing, quotes, information, and bid responses. Auctions and other, more advanced bidding activities are easier to conduct from a robust platform, where support for complex activities is centralized with the platform provider. As a result, an organization's resources aren't tied up when trying to conduct sensitive bidding activities in real time.

The sourcing process typically generates a purchasing document at the end, such as a contract and/or a PO, that confirms the supplier and the price. After a fruitful bidding process, the conversion to a contract should be relatively seamless, which is the overarching goal and mission of SAP Ariba Strategic Sourcing and is also the recommended approach from SAP. As shown in the following list, SAP Ariba Strategic Sourcing leverages the cloud for project/category management, sourcing, contracts, and catalogs:

- **Project/category management (SAP Ariba Sourcing)**
  - Sourcing pipeline
  - Category management
  - Savings tracking
- **Sourcing (SAP Ariba Sourcing)**
  - Bidding: request for information, (RFI), request for quotation (RFQ), and request for proposal (RFP)
  - Auctions
  - Category best practices
- **Contract (SAP Ariba Contracts)**
  - Contract authoring
  - Workflow approval
  - Contract repository
- **Operational contract (SAP S/4HANA)**
  - Contract release
  - Contract update
  - Contract distribution
- **Contract monitoring (SAP S/4HANA)**
  - Contract consumption (this can occur as source of supply default/option or be selected via a catalog item with contract assigned in SAP Ariba Catalogs)
  - Resulting contract distributed to the digital core in SAP S/4HANA for further consumption and purchasing alignment in activities driven in SAP S/4HANA, such as plan-driven direct procurement

1.5.2 MRO, Indirect, and Services Procurement

The guided buying capability in SAP Ariba is SAP Ariba's overarching procurement entry point for both casual and specialized users. The guided buying capability represents a consumer-grade guided buying experience, incorporating negotiated contract pricing, company policies, and a spot buying marketplace for longtail spend, enabling simple, effortless employee requisitioning via SAP Ariba Buying and SAP Ariba Catalog. The process then maps to SAP S/4HANA for PO execution and inventory management, returning back up to Ariba Network for PO and invoice collaboration, and then back again to SAP S/4HANA for accounts payable processing, as shown here:

- **Guided buying (SAP Ariba Buying)**
  - One place to go
  - Consumer-grade user
  - Policy and price compliance
- **Catalog (SAP Ariba Catalog)**
  - Managed content
  - Spot Buy marketplace
  - Contract compliance
- **PO or reservation (SAP S/4HANA)**
  - PO execution
  - Inventory reservation
  - Inventory transfer
- **Supplier collaboration (Ariba Network)**
  - PO confirmation, PO acknowledgement, ASN, service entry sheet
  - Smart invoice
  - Payment status
- **Goods receipt and invoice receipt (SAP S/4HANA)**
  - Goods receipt
  - Invoice receipt
  - Invoice workflow

The guided buying capability is a key nexus point of SAP Ariba, providing a diverse assortment of end users an easy and simple UX combined with rich catalog content. Catalog content is managed by the supplier and requires little to no maintenance from the buying organization. SAP Ariba processes in the cloud are deeply integrated with core processes in SAP S/4HANA, including operational transactions, inventory management, and plan-driven procurement, making the guided buying capability a single point of entry for a variety of procurement processes following a procurement approach recommended by SAP.

1.5.3 Direct Procurement

ERP-based direct procurement processes typically originate in SAP S/4HANA. During material requirements planning (MRP), a trigger for purchasing occurs when requirements are generated that are either unsourced, not stocked or available in your internal warehouses, or both. These requirements are created during PP and can't be ignored—your organization needs these items to deliver goods or services to the marketplace in a timely, committed fashion. Requirements planning takes place in the context of production and business planning, with the resulting requisitions documents in SAP S/4HANA. To source and procure these requisitions, SAP recommends that all supplier collaboration for direct materials occur over Ariba Network. With SAP Ariba Supply Chain Collaboration for Buyers, Ariba Network includes capabilities to support the more complex types of ordering found in direct procurement, such as scheduling agreements, forecasts, and contract manufacturing (outsourced assembly), as shown in the following list:

- **MRP (SAP S/4HANA)**
  - MRP run
  - Supply plan
  - Minimum/maximum replenishment
- **Requisition (SAP S/4HANA)**
  - MRP generates requisitions
  - Schedule lines
  - Scheduling agreement
- **Source determination (SAP S/4HANA)**
  - Auto source of supply
  - Source lists
  - Quota arrangements
- **PO (SAP S/4HANA)**
  - PO creation
  - Price compliance to
  - Outline agreements/purchase info records (PIRs)
- **Supplier collaboration (Ariba Network)**
  - Forecast/commit
  - Schedules, purchase order confirmation (POC), purchase order acknowledgment (POA), ASN
  - Consignment, contract manufacturing
  - Smart invoicing

Plant maintenance or in project management procurement activities, as shown in the following list, require dynamic sourcing and content management:

- **Plant maintenance (PM)/project system (PS) (SAP S/4HANA)**
  - Work order/project
  - Component planning
  - Materials and services

- **Managed catalog (SAP Ariba Catalog)**
  - Supplier-managed content
  - Contracted prices
  - Spot Buy marketplace
- **PO or reservation (SAP S/4HANA)**
  - PO execution
  - Inventory reservation
  - Inventory transfer
- **Supplier collaboration (Ariba Network)**
  - POC, POA, ASN, service entry sheet
  - Smart invoice
  - Payment status
- **Goods receipt/invoice receipt (SAP S/4HANA)**
  - Goods receipt
  - Invoice receipt
  - Invoice workflow

These transactions may originate in SAP S/4HANA but are then sourced by leveraging SAP Ariba Catalog, which hosts large MRO supplier catalogs and can apply contract pricing to these items for plant maintenance requesters. After the PO is created in SAP S/4HANA, the PO and invoice collaboration process is supported via Ariba Network as in the other scenarios, ultimately returning goods receipt and invoice processing to SAP S/4HANA.

1.5.4 Accounts Payable and Invoicing

SAP Ariba provides an array of solutions for accounts payable that goes well beyond the simple submission and processing of invoice documents. SAP Ariba enables suppliers of all sorts to create and submit electronic invoices via Ariba Network, via email-based smart invoices, or by leveraging direct business-to-business (B2B) connections brokered through Ariba Network. SAP Ariba invoicing includes more than a hundred configurable invoice rules to prevent erroneous or nonconforming invoices from ever being submitted into the system, as well as to allow invoices based on contracts (rate sheets) to reduce invoice errors. This control drives price compliance and tax regulatory compliance, ultimately permitting buyers and accounts payable to generate clean, error-free, mostly touchless invoices in SAP S/4HANA, all to help drive automation, reduce costs, and avoid headaches while processing and paying invoices efficiently.

In addition, SAP Ariba also provides dynamic and self-service discounting for suppliers and supply chain financing in SAP Pay, as shown in the following list:

- **Smart invoicing and contract invoicing (Ariba Network)**
  - Electronic invoicing
  - Invoice rules
  - Invoice against contract
  - Invoice two-way match
  - Rate sheet
- **Invoice receipt and approval (SAP S/4HANA)**
  - Invoice receipt
  - Invoice workflow
- **Self-service, discounting, supply chain finance (SAP Pay)**
  - Invoice status
  - Dynamic discounting
  - Supply chain finance
- **Settlement (SAP S/4HANA)**
  - Settlement with accounts payable
  - Remittance
- **Payment execution (SAP Pay)**
  - Payment

SAP Ariba augments frontend collaboration with suppliers from an efficiency and discount standpoint via the inclusion of SAP Pay, while still relying on SAP S/4HANA as the digital core for receiving the final invoice, updating the core financials, and creating payment runs.

SAP's cloud strategy spans its entire solution portfolio, especially in procurement, where SAP has invested more than $12 billion in cloud solution acquisitions in the form of SAP Ariba, SAP Concur, and SAP Fieldglass. When assessing any SAP solution area, especially procurement, you should understand your options in the cloud, your overall strategy, and the SAP road map and investment plans from a product standpoint.

1.6 Cloud Implementation Model

Managing a cloud implementation project is different from managing an on-premise project. With less hardware, fewer process permutations, and a different architecture to consider, more focus can be placed on solving business requirements and issues. In the following sections, we'll cover different implementation models for cloud projects.

1.6.1 SAP Activate

The overall implementation methodology for SAP cloud projects is SAP Activate. SAP Activate is a prescriptive and predictable methodology that is lean and fast, while at the same time incorporating the iterative and agile approach when appropriate, such as in configuration and testing. As a result, consultants and project team members can lead with best practices and simultaneously involve business subject matter experts (SMEs) in the configuration and testing cycles to ensure that the solution suits their needs.

SAP Activate's methodology, tools, and best practices include the following:

- Modular agile methodology (replacing ASAP and SAP Launch)
- Guided configuration and prescriptive content
- Ready-to-run scenarios
- Migration content
- Integration content

SAP Activate replaces previous versions of SAP's project implementation methodologies: ASAP and SAP Launch. SAP Activate relies heavily on agile approaches to project implementation, focusing on iterative sprints versus waterfall approaches and fine-tuning the project as you go to fulfill requirements as succinctly and efficiently as possible. Previously, a project team would laboriously define a full blueprint and design before the realization phase, but now, SAP Activate pushes forward the realization portion in the form of sprints much sooner, which allows you to show stakeholders and project participants up front what things will look like as the solution is implemented. For cloud-based implementations, the environment is already live and already supporting other customers, thus facilitating an agile approach such as SAP Activate. As shown in Figure 1.5, SAP Activate leverages four phases: prepare, explore, realize, and deploy.

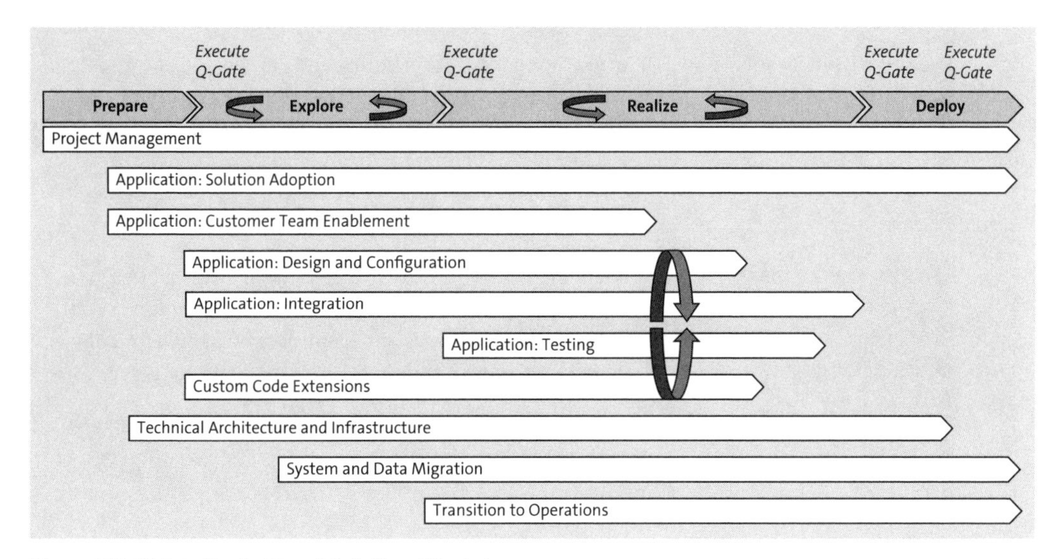

Figure 1.5 SAP Activate Cloud Solution Workstreams

Figure 1.6 shows a closer look at the SAP Activate phases.

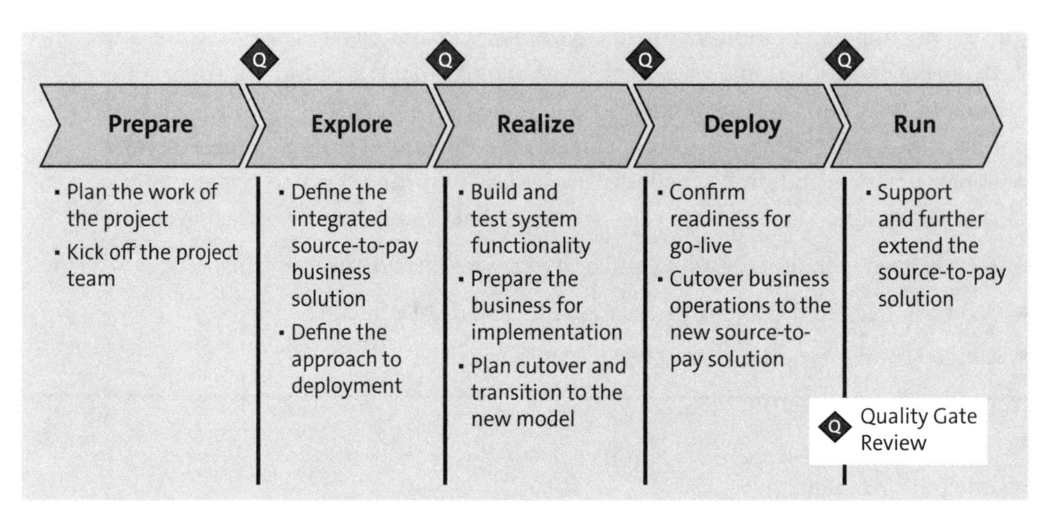

Figure 1.6 SAP Activate Phases

The main components of the methodology are called *workstreams*. Workstreams are collections of tasks required to achieve one or many deliverables and can span many phases, as shown in Table 1.2.

Workstream	Description
Project management	Covers planning, scheduling, governance, controlling, and monitoring the execution of the project.
Solution design	Covers the validation of scope, identification of detailed business process requirements, fit-gap analysis, and functional design of the solution.
Solution configuration	Covers the configuration, setup, and unit testing of the system (without custom development) to fulfill the customer's requirements per solution design. Items that can be configured include, but aren't limited to, forms, workflows, user permission/security, screen layouts, reports, master data setup, notifications, and so on.
Solution walkthrough	Covers the demonstration of the configured/developed solution to the customer project team, after each iteration cycle, for customer acceptance and identification of adjustments needed for the next iteration.
Integration preparation	Covers identification of integration requirements, integration points, integration approach, and integration solution design.

Table 1.2 SAP Activate Workstreams

Workstream	Description
Customer team enablement	Covers the enablement of the customer project team to work on the project effectively. This includes standard product orientation to prepare the customer for product requirements and design discussions, as well as key user and admin training to prepare the customer for test case development and test execution.
Data migration	Covers the discovery, planning, and execution of moving legacy data to the new system and the archiving of legacy data.
Integration setup	Covers the setup of the integration environment and middleware between the solution and any external systems.
Solution testing	Covers test strategy, planning, test case development, and the execution of user acceptance tests, integration tests, performance tests, and system tests.
Solution adoption	Covers value management, organization change management, and end-user training.
Support readiness	Covers the establishment of the helpdesk process, incident management process, post go-live change management process, and user-related operations standards and processes.
System management (not applicable to public cloud)	Covers the solution landscape, deployment concept, system architecture, technical system design, environment (development, testing, production, failover) setup, and technology operations standards and processes.
Custom extensions management	Covers the design, development, and deployment of system functionality that the standard product can't provide and which therefore needs to be custom developed.
Cutover management	Covers planning and execution of activities to cut the system over into production, including a heightened support period after cutover.

Table 1.2 SAP Activate Workstreams (Cont.)

The ASAP methodology was previously the go-to methodology for SAP project delivery, and SAP Activate maps quite closely with this methodology, as shown in Table 1.3.

SAP Activate Workstream	ASAP Equivalent
Project management	Project management Application lifecycle management (project standards)

Table 1.3 Workstream Alignment: SAP Activate and ASAP

SAP Activate Workstream	ASAP Equivalent
Solution design	Business process management
Solution configuration	
Solution walkthrough	
Integration preparation	Technical solution management (integration design)
Customer team enablement	Training (project team)
Data migration	Data migration Data archiving
Integration setup	Technical solution management (environment setup)
Solution testing	Test management
Solution adoption	Value management
	Organization change management Training (end user)
Support readiness	Technical solution management (helpdesk process, incident management, change control management) Application lifecycle management (operational standards for process and people)
System management (not applicable to public cloud)	Technical solution management (solution landscape/deployment concept, dev/Q, failover environment setup) Application lifecycle management (operational standards for technology)
Custom extensions management	Business process management (custom enhancements)
Cutover management	Cutover management

Table 1.3 Workstream Alignment: SAP Activate and ASAP (Cont.)

Workstreams are grouped into milestones called *quality gates*, for example, project verification and solution acceptance quality gates. Quality gates confirm that all stakeholders in the implementation project agree that specific deliverables meet requirements and consequently that the project can continue. Workstreams that show an ellipsis indicate that other services/streams can be connected, depending on the customer's business requirements. The phases, streams, and quality gates establish the sequence of implementation activities, as shown earlier in Figure 1.5.

Depending on the solution, SAP Ariba leverages aspects of the SAP Activate approach, but currently, the phases and streams can be somewhat different. This difference is due in part to variable project requirements for each solution. For example, to be successful, SAP Ariba Buying implementations require significant supplier onboarding, along with supplier and internal user adoption. Typically, supplier enablement is an afterthought for an on-premise project and usually the sole responsibility of the customer. In a procurement cloud, where subscription users and supplier participation are paramount and part of the payment model, these streams are fundamental to a project's success. So, in comparison to SAP Activate, you'll see more emphasis and separate workstreams set up for supplier and internal knowledge transfer in a focused SAP Ariba project methodology for SAP Ariba Buying. As some phases blend together in the SAP Ariba approach, especially the design and configuration phases, quality gate approaches become less frequent, unlike in on-premise projects, which have separate, defined phases.

1.6.2 Time-to-Value Acceleration

Combining phases is one of the many ways that cloud solutions seek to quickly deliver value after the implementation decision has been made. Because cloud-based solutions don't depend on the installation of software, architecting, or design of your landscape, nor the baseline configuration setup of your system (the realm arrives ready out of the box), this phase of the project is more condensed than an on-premise project. With all projects, but with cloud projects in particular, implementation can proceed quickly if both you the customer and the cloud solution provider agree on the urgency of getting the system into production. The SAP Fieldglass subscription model typically only applies after suppliers begin submitting invoices via their network, while SAP Ariba begins assessing subscription fees upon signature. In either instance, you can cancel the subscription at any time, so a strong emphasis on customer satisfaction and providing value exists from day one, not to mention going live in a timely manner to realize these objectives and key performance indicators (KPIs). Whereas software firms selling a perpetual license will invest the most in the sales cycle to ensure a successful sale, the subscription-based model drives a more balanced investment approach that looks beyond the sales cycle in the form of close alignment, customer engagement, and continual design improvements to drive customer satisfaction and renewal.

Another big difference is the nature of a multitenant cloud environment versus on-premise software deployments. The customer and consulting resources can't access all the areas of the SAP Ariba realm without going through authorized shared services resources with access to these sensitive areas of the overall systems. Certain changes and modifications to the system could upset the other customer systems running in the cloud, and, as such, changes must be carefully shepherded by shared services during the project and in production. Shared services have two main impacts on a project: realization phase and SAP Ariba shared services coordination.

Realization Phase

The realization phase of a traditional on-premise project can comprise a significant portion of a project's resources, time, and, ultimately, costs. Tight harmonization of technology deployment, landscape architecting, configuration, and development to simply realize the requirements and business processes defined in a blueprint are required during this phase. In a cloud deployment project, most of these areas are either already in place (technology and landscape) or at least partially in place (configuration and development). In addition, you're limited in what can be configured and developed in a cloud environment, thus simplifying decision trees and design processes. In a multitenant cloud environment, a customer/project can't completely subvert a business process, such as procurement, which would be possible when using an off-the-shelf piece of software and changing core code. (However, this customization method is never recommended by on-premise software vendors because your warranty would be invalidated.) The realization phase in a cloud deployment is thus focused more on business processes, asking, "What do you want this system to support in your business processes?" This phase focuses less on the technical minutiae of enabling the environment and code. All these considerations translate into a different type of realization phase.

SAP Ariba Shared Services

In an on-premise project, while the resources can be on-site or off-site, you generally have the option to colocate the hardware, the software, and the team at the customer site. In an SAP Ariba on-demand project, the SAP Ariba shared services team is a core part of the overall delivery and, ultimately, acts as the gatekeeper to the multitenant cloud environment hosting the solution. For a basic project, no on-site consulting resources are even required—SAP Ariba shared services can work with the customer to design and deliver the entire project remotely. Table 1.4 outlines the differences in role responsibilities between SAP Ariba shared services and on-site consulting.

SAP Ariba Shared Services	SAP Ariba/Partner Consulting Services
Only works on tactical activities directly related to setting up the SAP Ariba applications.	Performs strategic planning and management, as well as activities related to holistic business solution and change management.
Primarily guides and supports customer activities. Only executes activities that the customer can't perform, such as documenting and building certain configurations into the SAP Ariba applications.	Can execute any project activities, including high-bandwidth heavy lifting.

Table 1.4 SAP Ariba Shared Services and SAP Ariba or Partner Consulting Services

SAP Ariba Shared Services	SAP Ariba/Partner Consulting Services
Provides expert information on the configuration and functionality of SAP Ariba applications and generic best practices.	Analyzes customer-specific business challenges and provides expert advice on both holistic business solutions and how the SAP Ariba applications can advance them.
Works mostly or all remotely.	On-site and remote.
Shared across multiple clients.	Usually dedicated to one client at a time.

Table 1.4 SAP Ariba Shared Services and SAP Ariba or Partner Consulting Services (Cont.)

1.6.3 Enhancement and Modification Limitations

As a multitenant solution in the cloud, SAP Ariba on-demand and other SAP cloud solutions, such as SAP Concur and SAP Fieldglass, don't offer the flexibility for modification or for process-changing customizations to the same degree as in an on-premise environment, where you could reach directly into the code of the application either via user exits or outright modification. As mentioned earlier in Section 1.6.2, the latter has never been a preferred approach nor recommended on projects. Modifications essentially void the warranty on that area of your software, or the entire piece of software altogether. If you call a software provider with a support issue, and you're running heavily modified software, the provision of support received from your software provider may be severely restricted, as the software provider didn't build these modifications and isn't obligated, in standard contracts, to support code that the provider didn't write. When the time to upgrade the release version of your on-premise software arrives, if you've heavily modified the software, a simple upgrade may become intensely complex, requiring significant in-house development, testing, and revising to bring your system into the next release. Given the complexity of upgrading for some customers with modified systems, older releases of software are kept in production well beyond their end-of-lifecycle date on extended maintenance or no maintenance at all. If the system goes down, you'll need to address the issue with limited support from the provider, which isn't the preferred risk scenario for the IT department, executives, or others. In general, cloud updates and upgrades roll out on a quarterly basis, whereas an on-premise software customer base typically upgrades every two to three years. Some customers go well beyond that average, running decades-old legacy systems and reimplementing, rather than upgrading, when supporting the legacy system finally becomes unbearable.

For customizations, user exits in the software permit a developer to take action and add functionality not available in the standard version of the software. These types of customizations can be quite complex and drive significant behavior changes at runtime. If you log a support ticket, you still can obtain software from the rest of the package, as the software provider has created this user exit to allow for additional customizations in the software and can quickly determine whether the behavior/error is a bug in the provider's code or due to the user exit code. If your user exit code is found to be the source of the error, you'll need to address this error. If the code issue is found in the software itself, SAP typically will fix the code and issue a note to address the discrepancy, which you would then apply to your system. During a software upgrade, adjustments and testing may be required for the user exits to migrate your customizations to the new version or release, but an upgrade is less arduous in this case than with modified software.

In an on-demand environment, to some degree, only one codebase exists, which means you can neither modify the software (typically a good thing) nor customize the solution to the same degree via user exits. However, you do have significant customization latitude via the configuration, and some limited user exit-type code access points for interplay with the main codebase of the solution. Not all customizations are switch- or table-driven in SAP Ariba or in other procurement cloud solutions, but the nature of the cloud model restricts the options for customization and modification. This trade-off instills discipline in following best practices and standards driven by the centralization of the system. In exchange for significantly reduced up-front investment, ease of exit, no capitalization/assetization of investment, reduced maintenance requirements, accelerated implementation times, ease of updates/upgrades, and the general peace of mind that you won't be stuck in a dead-end system and have to reimplement, you just won't be able to completely change the solution, as you would in an on-premise implementation.

1.7 Summary

Solutions for procurement have evolved dramatically since the early days of computing. We're now at another inflection point, where whole procurement process areas are moving to cloud-based solutions and where real-time analysis is possible in the transaction system via in-memory database solutions such as SAP HANA. Business networks, mobile, big data, blockchain, artificial intelligence, and the Internet of Things (IoT) are also core themes and trends for procurement. With SAP Ariba solutions for cloud, with its rich background and install base for SAP ERP and SAP S/4HANA environments, and with SAP SRM procurement suites, SAP covers both on-premise and cloud-based implementations with complete end-to-end solution portfolios. In this book, we'll outline SAP Ariba's procurement cloud functionality, supported processes, and implementation approaches and discuss optimal integration strategies for deploying these SAP solutions in your organization, beginning with supplier collaboration in the next chapter.

Chapter 2
Supplier Collaboration

Ariba Network is SAP's core platform for enabling supplier collaboration. Collaborating via Ariba Network enables buyers and suppliers to interact in a streamlined environment with a complete toolset to facilitate transactional and strategic relationships.

Ariba Network is one of the largest Business-to-Business (B2B) networks in both size (millions of buyers and suppliers) and volume (over 3.2 trillion USD annual run rate as of 2020). As a B2B network, Ariba Network emulates the user experience you have today as a consumer. Online, you may leverage a host of sites to shop. After you've established an account with the site, you can reuse your payment and shipping information, revisit older purchases, set up recurring purchases, and conduct a host of other transactions digitally. More retail transactions are moving online for the simple reason that online retailing has made the shopping experience more efficient and cheaper, offering a much wider selection of goods. More recently, the Covid-19 pandemic has further accelerated this transition, with many consumers moving a large portion of their spending online and out of the brick-and-mortar retail experience, out of necessity. Leading online retailers can now boast millions of items and machine learning–based recommendations, not to mention financing in the form of store credit cards and loyalty rewards programs. Along these lines, leading companies look to manage their digital business networks more efficiently to engage with their trading partners.

In this chapter, you'll see the manner in which the supplier enablement process works and the tools and insights required to smoothly transition to the Ariba Network. While you're probably familiar with supplier enablement steps in general, this chapter will provide you with an understanding of supplier onboarding, the required permissions for the application, the supplier enablement process, portal and task configuration, supplier monitoring, and reporting. You'll also see, at a process level, how the network supports buyers and suppliers transact on Ariba Network integrated with a buyer's SAP ERP system.

In the following sections, we'll discuss the overall implementation approach for Ariba Network, which entails onboarding your suppliers to build out your own network on Ariba Network. In this chapter, we'll outline the implementation strategy and processes for onboarding suppliers using Ariba Network, the enablement process required to add suppliers into Ariba Network in bulk, and the steps for self-registering as a supplier on the sales side of Ariba Network. To start, in the next section, we'll create a common understanding of supplier collaboration as well as define strategies and put Ariba Network in context.

2.1 What Is Supplier Collaboration?

Supplier collaboration has many levels and can refer to simple interactions and meetings, all the way out to co-innovation between supplier and customer, where bills of materials are exchanged and honed. At a high level, supplier collaboration is the process by which customers and suppliers develop shared understandings and capabilities together, with the goal of cutting costs, improving processes, and even co-innovating and creating joint go-to-market strategies.

2.1.1 Supplier Collaboration Strategies

Before collaborating with a supplier, you, as the customer, would first define a strategy for the overall supplier relationship. If this relationship is purely transactional, with the goal of getting the best price, then the collaboration will need to take place on this level, with the customer obtaining and comparing pricing with the supplier's prices, potentially taking apart products to understand the cost of components and the corresponding margins or even dissecting the product in the lab. In short, in this low-trust relationship, each side tries to game their way to an economic advantage.

On the other end of the supplier collaboration spectrum, in strategic supplier relationships, a customer needs a true partner and isn't concerned with the price so much as the overall partnership and other facets of the relationship. These suppliers can move the needle on the customer's business in both directions. First, by co-innovating with a customer, a strategic supplier can make the customer's products even better and provide further competitive advantages in the marketplace. If a strategic supplier falls short on quality or execution, your business may also suffer a negative impact to a material degree. A strategic supplier relationship is a higher trust and dependency relationship. The good news is that Ariba Network manages both types of supplier relationships, as well as others in between.

2.1.2 Ariba Network

With Ariba Network, SAP Ariba delivers a connection-centric approach where buyers and suppliers can establish trading relationships in the cloud. Suppliers and buyers on Ariba Network manage their procurement collaboration and customer relationships through a single unified user interface (UI). The unified UI enables accurate and timely seller and buyer information, providing a single company profile in Ariba Network. This ease-of-access profile screen enables a unified cloud solution, providing greater efficiency through a many-to-many supplier-to-buyer relationship approach.

SAP Ariba is the largest business network of its kind, supporting sourcing to contract management, purchasing, or invoicing. SAP Ariba is the single place for you to obtain relevant leads of buyers who are ready to buy; for suppliers to get their product catalogs in front of requisitioners and submit proposals to buyers; and for negotiating contracts and performing collaborative redlining, receiving orders, and submitting invoices. Ariba Network isn't a private network for just one customer and one supplier, but for multiple customers and suppliers, all collaborating on one platform. This platform for suppliers, called Ariba Network for sellers, helps sales, marketing, and e-commerce professionals find buyers by putting products and services in front of end users—the hundreds or thousands of employees at each business making buying decisions every day. In turn, Ariba Network helps suppliers accelerate the sales cycle and helps close deals faster, leading to improved performance, improved customer retention, and suppliers receiving payment faster.

The sheer scale of Ariba Network is difficult to fathom. Ariba Network in 2018, in the preceding edition of this book, had almost 3 million companies conducting more than $1.5 trillion in transactions comprising around 250 million documents on Ariba Network. As of 2020, Ariba Network supports 4.62 million suppliers alone, with more than $3.21 trillion supported via 604 million transactions. Further fast facts for Ariba Network include the following:

- 4.62 million companies, are in the network, with a new one joining every five seconds.
- $3.21 trillion in annual commerce supports 190 countries and 176 currencies.
- 604 million documents include 597,000 timesheets, processed annually.
- 70% of the Fortune Top 50 companies are suppliers.
- 200,000 catalogs are hosted on Ariba Network.
- 461,000 leads are sent annually to sellers.
- 26 million users make use of the network.
- 670,000 suppliers joined in 2018 alone.

As a seller or buyer, joining Ariba Network and collaborating results in multiple benefits. A large supplier or buyer can immediately realize more sales and purchasing volumes through tighter integration with the purchase and invoice automation initiatives of the counterparty. An individual or a smaller company will have the ability to transact with many of the largest companies in the world. Some features and concepts such as dynamic discount management help you gain better insights into the deployment of cash flow. For example, with dynamic discount management in SAP Pay, you can negotiate early payments through Ariba Network and avoid taking on external capital to bridge the payment cycles.

The traditional option is called a *bridge loan* or *factoring*. A bridge loan is typically a short-term loan at a higher interest rate than a long-term one, which allows you to get from point A to point B liquidity events. These loans are obtained using some form of collateral or potentially a credit line from the bank. Factoring involves "selling" your unpaid invoices to a finance provider. For example, if you're a start-up that needs cash prior to one of your customers paying an invoice, you can factor your invoices with a credit provider. Many of these credit providers can charge up to 10% interest rates for the privilege of borrowing, making this form of short-term credit akin to going to a payday lending company to keep finances going until a customer pays an invoice.

Leveraging dynamic discount management allows you as the customer to act as a credit provider for a reduced fee to your suppliers. As a result, you'll accelerate and buttress your internal investments, thanks to enhanced visibility into cash flow opportunities. The supplier benefits by not having to pay onerous interest rates for short-term cash flow. Ariba Network will provide you, as a customer or a supplier, with many other commercial opportunities, such as participating in sourcing events, contract management lifecycles, transacting purchase orders (POs), shipment notifications, invoices, and payment statutes.

2.2 Implementing Ariba Network

To implement the SAP Ariba supplier enablement automation process for a large number of suppliers, you'll need the SAP Ariba supplier enablement team. You'll also need to be familiar with the SAP Ariba supplier enablement methodology to understand the resources, roles, phases, and timelines for successful enablement. Let's reinforce the concept of supplier enablement by highlighting the fact that enablement is always a function of the business, not just a single project. Supplier enablement isn't a finite activity that will take just a few months to implement; you'll need to see this program as ongoing to ensure that your supply chain needs are satisfied and that you're implementing best practices. As you learn from the process, you'll continue improving going forward. As shown in Figure 2.1, the implementation process breaks down into four main phases with some interim phases: strategy, design and build, supplier onboarding, and network growth.

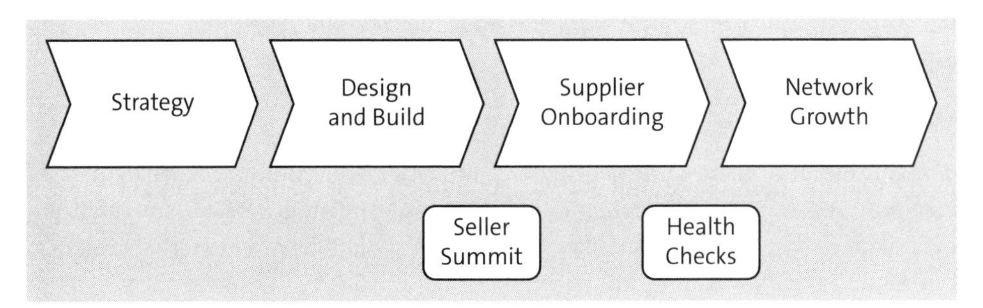

Figure 2.1 Supplier Enablement Process

To implement Ariba Network, you'll need the following roles:

- SAP Ariba supplier network lead
- Catalog enablement lead
- The necessary additional project roles and responsibilities if implementing SAP Ariba Invoice Management, SAP Ariba Discount Management, or SAP Pay

In addition to completing the four main phases shown in Figure 2.1, you'll also need to consider the resources necessary for the project, your strategy for change management, and your long-term strategy for supplier enablement.

2.2.1 Strategy

The strategy phase is where you'll analyze the way things currently stand with your supplier base and company spend data so you can develop a flight plan and envision a final product with which to move forward. During this phase, the flight plan and enablement prioritizations will take place. At this stage, your goal is visibility and insight into where the spend and volume reside, from an invoice and PO standpoint, so you can best determine for which suppliers to enable purchase and invoice automation using SAP Ariba Buying or to enable invoicing against contract catalogs using SAP Ariba Catalog.

Normally, depending on your supplier universe, a flight plan can have a two- to five-year supplier enablement wave plan with a clear set of goals. The final time frame will depend on how quickly you want to proceed; for each wave of suppliers you enable, you'll have a projected timing, starting by sending trading relationship requests to the relevant suppliers up to the go-live, when you start transacting with them.

The following roles will support and help meet your business goals and objectives:

- **SAP Ariba roles**
  - Understand your customer objectives, conduct an analysis on your customer spend, rationalize spend by commodity and suppliers, and develop a supplier flight plan.
  - Recommend and work with you to prioritize suppliers based on transactional volumes and values, existing Ariba Network suppliers, and different geographical regions.
- **Customer roles**
  - Provide the list of main commodities being procured by the company and the spend data for the analysis.
  - Review and provide feedback on the flight plan.
  - Provide refreshed and current accounts payables transaction data with supplier rationalization and duplicate filtering for strategy analysis to ensure the correct and active suppliers are targeted and to avoid unproductive effort.
  - Ensure that the accounts payables transaction data doesn't include account-related and/or sensitive information such as salaries and benefits.

2.2.2 Design and Build

After the strategy is set in place, you'll enter the design and build phase, when you'll conduct requirements-gathering workshops and start talking about supplier communications and communication plans about the order of events ahead required to bring the supplier to a state where it can transact. The objective of this phase is to develop the processes, infrastructure, and materials necessary to enable, educate, test, and support suppliers.

The following roles will be taken up by the SAP Ariba team and the customer in this phase:

- **SAP Ariba roles**
  - Conduct transaction requirements-gathering workshops.
  - Support with supplier communication and education materials.
  - Create an education portal infrastructure.
  - Assist with the buyer network account configuration.
- **Customer roles**
  - Participate in workshops to identify requirements.
  - Appoint an internal team to own the supplier enablement program.
  - Build the compliance message.
  - Validate supplier communications and education materials.
  - Collect supplier data and provide finalized supplier lists.
  - Designate an SAP Ariba account administrator.
  - Configure the Ariba Network account.

The supplier information portal infrastructure and the supplier accounts will need to be configured. After you've signed off on the business requirements, the SAP Ariba team can go ahead and build all the required materials. Input from the customer during this time ensures that the SAP Ariba team has the appropriate counterparts and the right resources working with them. Your feedback is crucial not only for supplier enablement but also for all other live SAP Ariba solutions and ensures that your attention is focused on the right areas.

Compliance Message

The next step is to build a compliance message. A *compliance message* is used for your internal stakeholders and customers as well as external suppliers. Part of change management, discussed in greater detail in Section 2.2.6, centers on establishing the expectations and usage of the tool. At this point, you'll need to think about supplier compliance because expectations need to be shared with the supplier for a successful program. Creating a compliance message should be an initiative pushed through the customer team while highlighting the importance of the business factors supporting the message. The supplier needs to know that this strategic move, perhaps a new direction for the business, affects both companies and that any deviation can detract from the expected business improvements for both companies. Giving a candid yet gentle message to your suppliers is critical. Normally, a stronger business compliance message directly correlates to a more successful campaign.

Be sure to articulate a clear direction, communication plan, and expectations. SAP Ariba will provide for communications, starting with requirements gathering on the supplier side. When creating blueprints for the PO or invoice automation application, or for catalogs in SAP Ariba Buying or SAP Ariba Catalog, you'll want to reflect the processes and rules defined in these applications in Ariba Network. Suppliers will have access to these rules, and you'll also communicate these rules during supplier summits. Suppliers can also go into their accounts and see what the buyer's transaction rules are (e.g., can non-PO invoices be sent, etc.). The network side isn't as complex as the application side, but the same principles apply, and both the network and the application should be in alignment.

Project Notification Letter

SAP Ariba follows the same communication and education plan on every deployment, coordinating the same critical high-level course of events. The project notification letter is the first message to suppliers from the customer team. This message introduces suppliers to the customer's campaign, articulating business objectives and what to expect in the steps ahead. The supplier will receive an email from SAP Ariba with information and instructions. Make sure this email also includes your compliance message. The supplier should be well aware of the program and be ready to receive the letter from SAP Ariba, thus preventing the dreaded question, "Who are you?" when a supplier enablement specialist from SAP Ariba calls. The purpose of the project notification is more to introduce the initiative than to detail its pricing aspects. At this point, you're introducing the supplier to the supplier enablement process and instructing the supplier to follow simple links to understand more about the SAP Ariba process as well as your own processes.

Pricing

SAP invests hundreds of millions of dollars in Ariba Network annually. There are costs for certain levels of supplier accounts on Ariba Network. Some suppliers are wary of paying anything for an account, and these suppliers can select the "light account" option. Light account suppliers can leverage the core functionality of the SAP Supplier Portal and Mobile Supplier app to receive POs and submit invoices. If light account suppliers want to access further functionality in the supplier portal, they can upgrade their account.

Supplier Pre-Enablement Summit Meeting

Both the customer and SAP Ariba teams/consulting partner teams will host a supplier pre-enablement summit meeting. Summit content is fully customizable to communicate your initiatives to your suppliers. The summit's purpose is to align your expectations with suppliers and present the mutual benefits for both companies. For example, perhaps your suppliers could speed up payment processes by submitting their invoices faster or eliminate paper and manual errors by sending documents over the network. Normally, suppliers who are already in Ariba Network will want to work with you in Ariba Network, so they don't have to support multiple communication methods.

Again, at this point, the supplier information portal and education materials must be ready. The supplier information portal is an Ariba Network–hosted, customer-managed web page, specifically dedicated to sharing customer-specific information. This portal is the primary supplier self-service method for communicating documents and training materials, such as the account configuration guide, PO management guide, invoicing guide, electronic data interchange (EDI) guidelines, commerce XML (cXML) guidelines, and Catalog Interchange Format (CIF) catalog training guide.

All education materials are created after Ariba Network requirements have been gathered. These education materials need to provide links to other documentation (e.g., SAP Ariba standard documentation/help, supplier membership page, etc.).

2.2.3 Supplier Onboarding

The objective of the supplier onboarding phase is to enable suppliers on Ariba Network to transact electronically through three steps: registration, education, and testing. You'll also need to implement health checks with regular project reviews, weekly status calls, process evaluations, and strategy refreshes, along with customer education. The final step of this phase will be supplier deployment.

First, you'll need to define a list of supplier candidates for onboarding, which involves three main steps: identification, invitation, and integration. Identifying potential network suppliers is the first step. From SAP Ariba's standpoint, the more the merrier. However, from your organization's standpoint, prioritizing suppliers using some simple questions will probably be necessary, such as the following:

- Is this supplier on Ariba Network already? These suppliers are typically the easiest to bring into your individual network because they're already transacting with other customers on Ariba Network.
- Is this supplier high volume, high spend, or both?
- How integrated is this supplier with our organization? Does this supplier have existing EDI linkages with our procurement processes for receiving POs and/or submitting invoices?

- Is the macro environment conducive for this supplier to join Ariba Network as one of my suppliers? Some suppliers may be working under regulations and country dynamics that make it more difficult for them to join a network.

After you've determined your suppliers and prioritized them, you can invite them to register on Ariba Network, if necessary.

Registration

During registration, you should determine your communication protocol and remember that, when you receive a PO, you're receiving a financially committed document. The following roles are required for this step:

- **SAP Ariba activities**
  - Obtain finalized supplier lists from customers in a vendor upload file.
  - Upload vendor files to Ariba Network.
  - Create Ariba Network accounts by sending relationship request letters.
  - Assign supplier enablement tasks.
  - Track supplier registration status/task completion.
  - Provide status reports.
- **Customer activities**
  - Provide finalized lists of suppliers with required data in the vendor upload file.
  - Assign enablement tasks to the supplier. These are the tasks the supplier must complete to begin transacting with the customer. For example, the supplier can configure electronic invoice and routing settings on Ariba Network, defining how their POs and invoices can be submitted, including via cXML and EDI.
  - Follow up on supplier escalations for the following:
    - Invalid supplier data
    - Nonresponsive suppliers
    - Noncompliant suppliers

Education

During the education step, the objective is to provide educational materials and communicate business requirements to suppliers. The activities required in this step are as follows:

- **SAP Ariba activities**
  - Make supplier education materials available as content on the supplier portal.
  - Conduct a supplier summit.
  - Record webinars to be posted on the portal for supplier viewing.
  - Perform supplier follow-up.

- **Customer activities**
  - Participate in supplier education meetings.
  - Review and approve supplier education material content.
  - Drive compliance messaging with suppliers, as well as within the organization.

Testing

The objective of testing is to confirm connectivity and the successful transmission of business documents with a representative subset of suppliers participating in customer testing sessions. The activities required during this step are as follows:

- **SAP Ariba activities**
  - Conduct end-to-end testing with a pilot group of suppliers.
  - Provide testing support to assist with pilot, integrated, and punchout catalog suppliers testing. *Punchout* catalogs refer to catalogs where the user accesses a catalog hosted by the supplier, effectively "punching out" of the purchasing application to the catalog.
- **Customer activities**
  - Determine an end-to-end testing plan.
  - Create/validate test transactions to suppliers.

Supplier Deployment

During supplier deployment, the following areas are covered:

- Migration of all production-ready suppliers to the production environment.
- Communication of go-live to suppliers and internally within the organization. This communication to suppliers will go out a week or two before go-live, when production orders can be expected.
- Post-go-live production monitoring and support.

2.2.4 Network Growth

After the SAP Ariba solution is live and running, transitioning to a network growth team is important to continue with supplier registration, education, and testing. The activities required during this step are:

- **SAP Ariba activities**
  - Provide a supplier enablement playbook to the customer.
  - Provide continuous supplier registration, education, and testing support.
  - Give strategy refresh support.
  - Assess network utilization.
  - Attain successful business cases.

  - Monitor suppliers for transactions.
  - Report progress and associated actions.
- **Customer activities**
  - Provide continuous involvement in supplier registration, education, and testing.
  - Maintain revisions of supplier education and communication materials.
  - Drive supplier compliance.
  - Maintain internal and external change management.

2.2.5 Resources

In resourcing the project, SAP Ariba will normally provide you with a program manager for the solution being deployed, as well as a customer enablement executive. Additionally, you'll have your own supplier enablement team, which itself will have a supplier enablement lead and a network growth manager working with multiple supplier managers (the supplier integration lead, an electronic supplier integration manager, and a catalog knowledge expert) behind the scenes.

Your organization will also require a program sponsor; finance, procurement, and legal subject matter experts (SMEs) to provide ad hoc support; one or two supplier enablement leads; an IT team to work on the integrations in scope; a change management lead; an Ariba Network account administrator; a catalog knowledge expert; and a functional team. The involvement needed from the IT team and change management depends on the complexity of the rollout plan for the program, the number of countries in scope, PO or invoice automation, and how employees are to be trained. The Ariba Network account administrator and catalog knowledge expert will transition into ad hoc support after the SAP Ariba solution goes live. The typical timeline for a supplier enablement project is 12–16 weeks from inception to the first wave of onboarding.

2.2.6 Change Management

The planning and execution of several change management activities are needed to support the supplier enablement process to increase awareness and to increase the probability of a successful migration path and adoption of Ariba Network. Supplier enablement requires a considerable initial amount of interaction and communication to help implement new technology and processes. Ideally, you'll start planning the change management activities as soon as the scope of the supplier enablement is determined in the supplier enablement wave plan. The following steps should be taken to ensure smooth change management for your project:

- Schedule an Ariba Network overview workshop for stakeholders to increase awareness of enablement and the benefits of using the network.
- Ensure executive support of compliance requirements.

- Ensure all divisions/business units sign off on suppliers targeted for enablement.
- Create memos for spend owners with instructions/FAQs to communicate with suppliers in advance of go-live.
- Build a solid migration path, which will result in higher compliance from suppliers.
- Ensure unified messaging so that suppliers hear the same message from all company personnel.
- Ensure all stakeholders are aware of what is going on with the project and what is expected from suppliers (buyers, commodity managers, business power users, etc.). IT and supplier enablement plans need to be aligned.
- Reduce delays:
  - Delays related to integration with customer ERPs or other systems on Ariba Network projects can occur if not properly scoped. While the latest release of SAP's ERP, SAP S/4HANA, includes native integration with Ariba Network and should have fewer integration items to complete, some data alignment and potential process integration areas will exist, depending on the scope of the Ariba Network implementation.
  - Aligned project plans also help reduce the impact of IT delays.
  - Be aware that the more manual integration required, the higher the likelihood of IT delays.
- Understand the SAP ERP release schedule to avoid any potential downtime or system freezes due to fiscal year or quarter closings that can impact the enablement activities.
- Be confident that you can answer the following questions:
  - Is the IT work appropriately staffed and scheduled?
  - Do certain milestones need to be completed on certain key dates?
  - Does a holiday "code freeze" policy exist that may impact end-of-year enablement efforts?

Supplier education depends on completed requirements and any network customizations; supplier testing depends on technology and test environment availability.

Depending on the approach, allocate a minimum of a part-time employee for every 500 suppliers during the data collection time period, which can last one to three months, depending on the quality of your supplier profile data. The main factors to consider are as follows:

- Supplier profile data doesn't exist in any one system and can be found in the contact folders for the procurement buyers. This data is needed from order processing and invoice processing.

- Data collection takes time and effort, which aren't always available from understaffed procurement departments. Dedicated resources may be needed. For example, you may need to hire temporary staff.
- Clean supplier data is essential. You'll need to clean up vendor record database, get a DUNS number, and confirm supplier contact information (name, address, email, and fax), among other things.

Tip

Start early in your efforts to clean up your supplier records!

2.2.7 Customer Long-Term Supplier Enablement Program

For your long-term supplier enablement program, SAP Ariba recommends that you consider the following key elements:

- How will this be supported long term (to be outlined in the post-SAP Ariba statement of work)?
- What resources are needed to support this process?
- How will roles within our organization change?
- Will there be adequate resource allocation and bandwidth?
- Who will own the program?
- Do we have a sustainable training plan? Do we have program/process experts?
- What is our plan for internal change management and education across departments?
- What is our compliance plan?
- What will the escalation path be for noncompliant suppliers?
- How will we track and manage internal compliance?
- Have we ensured that proper reporting is in place for tracking an end-to-end process?
- Have we created metrics tracking for program success, such as a scorecard? Have we created an escalation plan for noncompliant suppliers to appropriate staff and defined actions needed?

Ariba Network provides a seller collaboration console as a new way to enhance the seller experience with greater exposure, convenience, and control to "consumerize" business commerce. SAP Ariba brings all aspects of buyer/seller collaboration under one umbrella. Prior to the introduction of the seller collaboration console, supplier users had multiple IDs to Ariba Network and separate SAP Ariba profiles. Now, through the seller collaboration console, you can use your Ariba Network user ID and access all your SAP Ariba information and customer relationships in one place via a shared profile and document repository.

The seller collaboration console includes the following:

- Centralized administration, a single user ID for all SAP Ariba Commerce Cloud seller solutions, common terms of use, and a common SAP Ariba profile
- Leads through SAP Ariba Discovery
- Proposals through SAP Ariba Sourcing
- Contracts through SAP Ariba Contracts
- Orders and invoices through Ariba Network

In the next section, we'll detail the process for onboarding suppliers to Ariba Network.

2.3 Onboarding Suppliers

Onboarding suppliers is a customer-oriented initiative that can be done through an assisted- or self-service method. If your company has a large number of suppliers, the recommended onboarding procedure is to use Ariba Network methodology outlined in this section, which also details the steps to follow. Ariba Network itself keeps all your information in one place for easy access to better support the supplier enablement process and so you can track suppliers as they go through the enablement process.

First, you'll need a required minimal set of permissions to access and perform supplier enablement activities in your buyer realm, as summarized in Table 2.1.

Permission	Description
Supplier enablement program administrator	This permission includes all the other permissions available for supplier enablement automation in SAP Ariba. Configure supplier enablement automation, upload vendors, start supplier enablement, manage supplier enablement tasks, and access vendor data export reports.
Supplier enablement configuration	This permission allows you to configure supplier invitation letters, tasks, and activities for supplier enablement.
Supplier enablement task management	This permission allows you to manage and monitor tasks for supplier enablement and edit vendor details (vendor name, preferred language, email, contact information, address, vendor comments, enablement status).
Supplier enablement report administration	This permission allows you to access reporting and supplier enablement reports: ▪ Supplier Enablement Task Status ▪ Supplier Enablement Status

Table 2.1 Permissions Needed for Supplier Enablement

After you've assigned the appropriate set of security accesses displayed in the **Onboarding Suppliers** to the users administering your SAP Ariba realm, as outlined in Table 2.1, you, the buyer, will need to compete these steps to onboard suppliers:

1. Assign specific activities to the suppliers.
2. Evaluate how many suppliers are ready to transact on Ariba Network.
3. Identify suppliers that haven't taken action on their tasks.
4. Evaluate the overall status of the supplier enablement.
5. Review preconfigured activities and related tasks.
6. Configure and customize email notifications sent to suppliers.
7. Set up a supplier information portal with specific custom content.
8. Upload multiple vendors for supplier enablement.
9. Group vendors by type of trading document exchange, assign the appropriate predefined activities, and start the enablement.
10. Gather missing contact information or rectify incorrect contact details online.
11. Resend failed email or fax trading relationship request letters.
12. Complete, defer, and track pending tasks.
13. Review the supplier enablement status.
14. Rely on system-generated reminders and notifications when a task is overdue.
15. Access all pending and escalated tasks.

The next section will review the supplier information portal in more detail.

2.3.1 Supplier Information Portal

To allow suppliers easy access to your supplier enablement process, you can create and enable the supplier information portal. In the portal, you can upload any PDF, Word, Excel, or PowerPoint documents that you deem necessary to explain and support business processes and protocols that suppliers should follow when transacting with your company via Ariba Network. For example, you can give suppliers extra information concerning invoices or specific financials processes, or you can provide the cXML or EDI formats necessary to transact with you through Ariba Network. You can even consider uploading a timeline displaying the expected testing phase and when the production orders will go out.

The portal is housed in Ariba Network, which you can access next to the supplier relationship session. To change its content, we recommend sending your changes or documents to the SAP Ariba supplier enablement team and having them make the changes on your behalf, which you can approve before being published. The portal is accessible by suppliers and is for informational purposes only, but suppliers won't need to go through the portal to interact with your organization.

After a supplier receives an invitation to join Ariba Network, they typically ask what they have to do, and the answer is attend supplier summits. Large customers with large supplier bases typically have supplier enablement plans with defined supplier summits.

Tip

Ariba Network certifications can make a supplier an almost touchless candidate to start a trading relationship immediately.

Suppliers also have the ability track any new trading relationships with their customers. As a supplier, you must have an Ariba Network supplier account to access the SAP Ariba supplier information portal, or you can use the secure link from the trading relationship request letter. A link to the supplier information portal is available on the following pages:

- Enablement tasks
- Task details
- Customer relationships
- Trading relationship request letters

2.3.2 Uploading Suppliers

You can upload a set of new suppliers in Ariba Network using a comma-separated value (CSV) file or zipped archive of CSV files. The first step is to go to the **Upload Vendors** tab in the **Supplier Enablement** section, click on the **Upload** button, and then **Upload Vendors**. The **Upload Vendors** page will appear, and by clicking **Download Template**, you can download a copy of the CSV file format you'll need to use.

As shown in Figure 2.2, for uploading suppliers with a CSV file, the following syntax rules apply:

- The file must be saved in CSV format.
- String values in a CSV file can have commas, but quotes must be used around the string.

The following fields are required in the CSV file:

- Vendor name
- Vendor ID (required only if the tax ID isn't included in the CSV file)
- Tax ID (required only if the vendor ID isn't included in the CSV file)
- Vendor city or postal code
- Vendor country code

- Vendor province/state/region (required only if the vendor city or vendor postal code is in a US state)
- System ID (required only for multi-ERP buying organizations)
- A buyer Ariba Network ID required only for third parties managing supplier enablement on behalf of a buying organization using a provider account

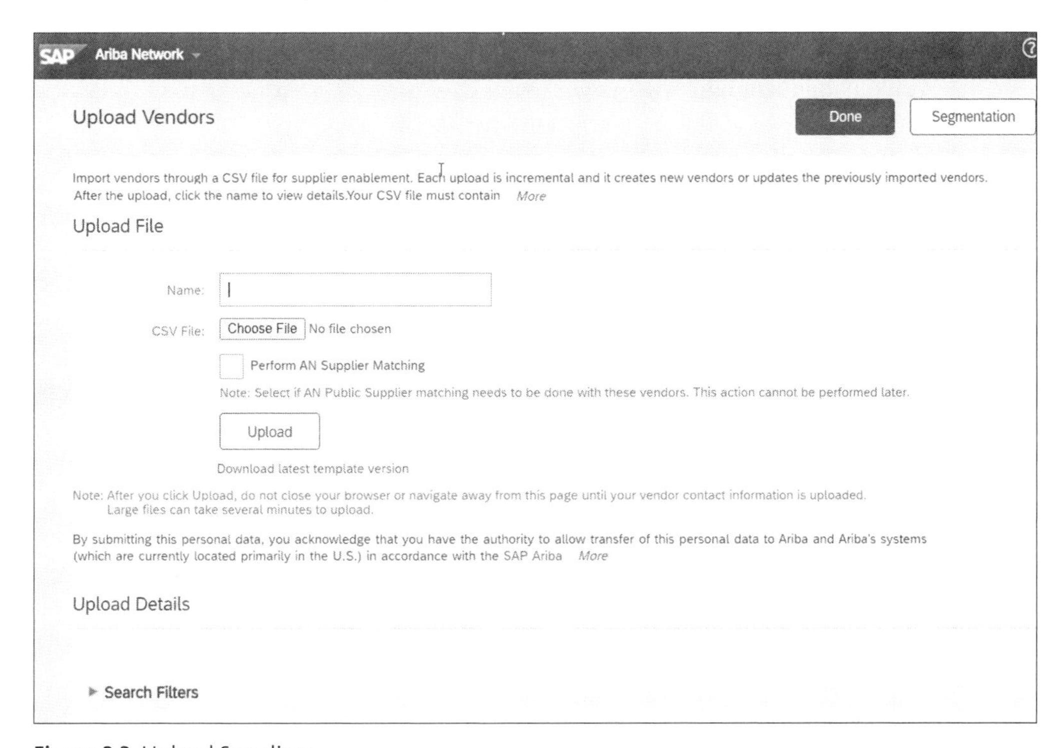

Figure 2.2 Upload Suppliers

The template also provides basic details on the available fields, and the file must specify the file encoding in the first row.

As a buying organization, you can leave optional fields blank or just delete the columns you don't use. You can upload up to 100,000 records at a time.

2.3.3 Assigning and Monitoring Tasks

After the suppliers are loaded, you can assign tasks to them. You'll notice that most of the tasks will be automatically assigned by the system based on your previous supplier network configuration activities and tasks that you entered in the screen shown in Figure 2.2.

Figure 2.3 outlines the enablement area for suppliers. Suppliers will appear in the **Monitor** area only in the following cases:

- **Case 1**
  Vendors are invited through quick enablement methods (inviting suppliers through POs, payment proposals, carbon copy [CC] invoices, requests for quotation [RFQs], or invoices through the SAP Ariba invoice conversion services add-on). These vendors are automatically added to the account activity and appear for monitoring in the supplier enablement automation process under the **Monitor** area.
- **Case 2**
  You can manually add vendors and assign activities. After assigning the vendor, the activity becomes visible under the **Monitor** area for tracking.
- **Case 3**
  After the completion of an assigned task (task completed), the vendor won't appear in the **Monitor** area because no tracking activity occurs, and the task was completed successfully. However, you can still see the task completed under the **Tasks** area, as shown in Figure 2.3.

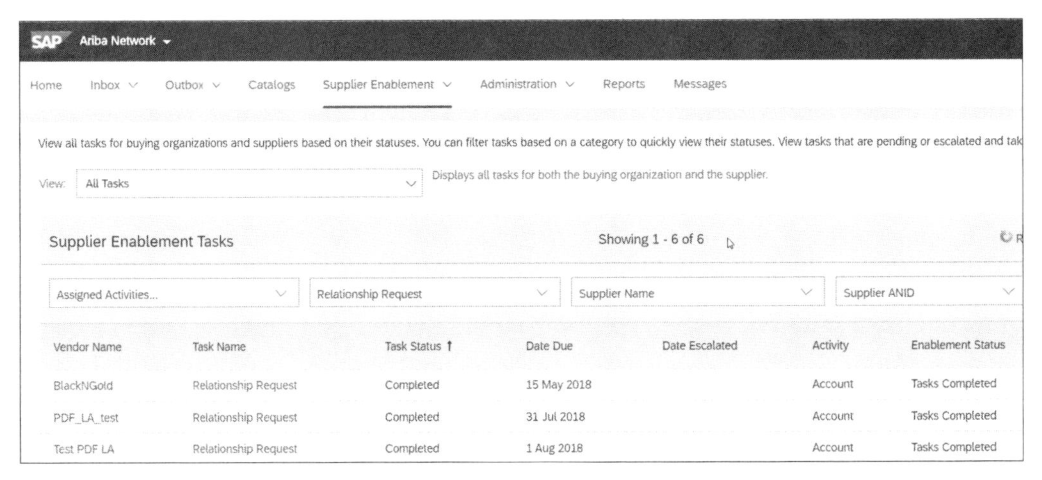

Figure 2.3 Task Status: Completed

A supplier will have a *green status* when their account is active and its tasks are completed on time, even if some tasks are still pending; *yellow status*, when the supplier has an overdue task; and *red status*, when the tasks failed or have been escalated.

2.3.4 Buyer Tasks

Buyers must complete several tasks to send a PO. If the buyer doesn't perform specific tasks, then the supplier won't be able to complete their own tasks. Buyer tasks can be viewed under the **Tasks** tab, as shown in Figure 2.4.

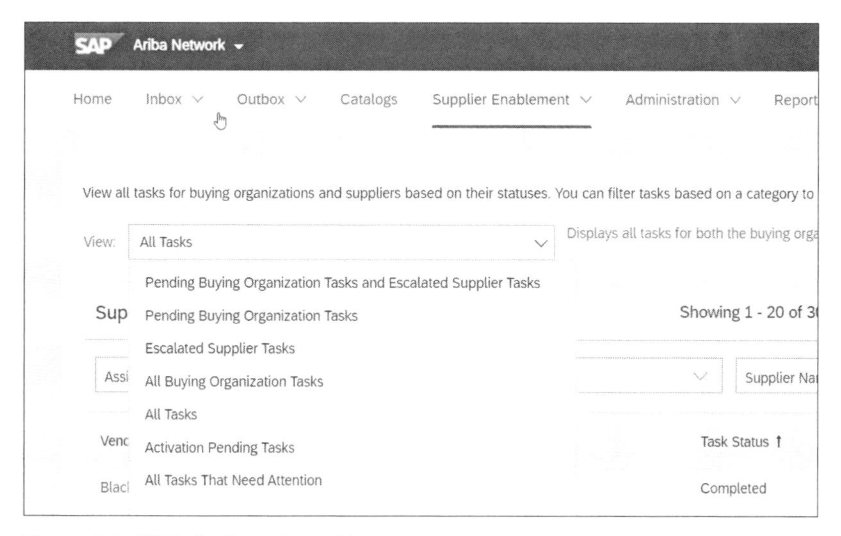

Figure 2.4 All Tasks Dropdown Menu

The number of pending tasks with actions required to complete will be highlighted in red in the supplier's view screen.

2.3.5 Monitoring Suppliers

You can track your supplier enablement status for all your vendors by selecting the **Monitor** option in the **Supplier Enablement** dropdown menu. You can view one supplier at a time or all suppliers at a glance, and retrieve their statuses based on the descriptions displayed at the bottom of the monitoring screen, as shown in Figure 2.5. Here, you can monitor all the tasks for one supplier at once, which is important to help you understand when a process is being held up and you need to follow up with the supplier.

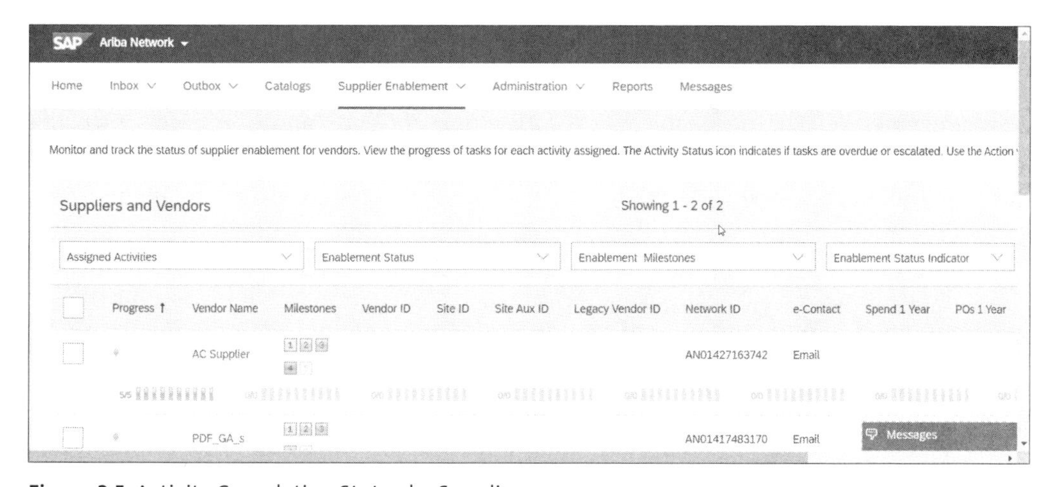

Figure 2.5 Activity Completion Status by Supplier

After suppliers have been onboarded, they'll need to be transformed into active participants on Ariba Network and begin realizing value for both themselves and their customers. The next section looks at enablement approaches for suppliers.

2.4 Enabling Suppliers

Supplier enablement is the process of configuring and enabling an unlimited number of suppliers in Ariba Network, which enables your organization to transact electronically with those suppliers. The enablement methodology encompasses the design and development of processes, infrastructure, and materials necessary to support the enablement, education, testing, and support of suppliers in Ariba Network. Complete enablement lasts for the life of the contract and involves SAP Ariba supplier enablement services from an SAP Ariba shared organization designed to execute and support the enablement tasks. The SAP Ariba shared services can be deployed based on the project timeline and the execution phases of the program. The key drivers of the deployment are the numbers of countries in scope, the number of languages, the buyer's back-office enterprise systems, the type of technical adaptors used, and the transaction types:

- PO (SAP ERP to Ariba Network)
- Change order (SAP ERP to Ariba Network)
- Order confirmation/acknowledgement (Ariba Network to SAP ERP)
- Advanced shipping notice (Ariba Network to SAP ERP)
- Invoice (Ariba Network to SAP ERP)
- Invoice status update (SAP ERP to Ariba Network)
- SAP ERP–initiated invoice or CC invoice (SAP ERP to Ariba Network)
- Payment proposal request (SAP ERP to Ariba Network)
- Pay me now request (Ariba Network to SAP ERP)

The standard method is for buyers to ask their suppliers to join Ariba Network to work together and find mechanisms to work more efficiently and effectively on all the shared aspects of business commerce: from sourcing proposals, contracts, and orders, to invoices and payments, in a way that can save time, reduce error handling, and save money and resources. More than likely, if a supplier receives an order through the network, the supplier will use the same mechanism to send the invoice, which is where suppliers will find more value in the process.

The supplier enablement process is also a cycle, as shown in Figure 2.6. Supplier enablement allows you as a buyer to establish trading relationships with suppliers and configure the necessary information to start transacting electronic documents through the Ariba Network.

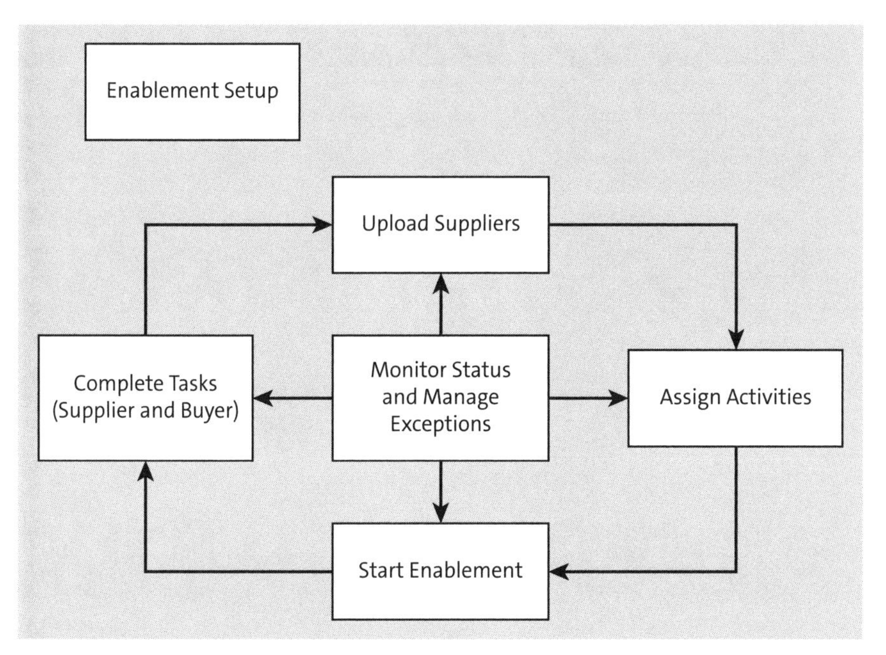

Figure 2.6 Supplier Enablement Process/Cycle

The SAP Ariba supplier enablement automation process requires three initial steps: configure the trading relationship request letter, add your contact information in the contacts screen, and, finally, set up your company's supplier information portal. You can skip the supplier information portal if your company already has a supplier-facing website where you post and communicate instructions to your suppliers.

In the following sections, we'll cover all the steps necessary for supplier enablement, from configurations and uploading suppliers, to assigning and monitoring tasks, to notifications and reports.

2.4.1 Configurations

The SAP Ariba supplier enablement automation process starts by drafting the trading relationship request letter or quick enablement letter. From the **Supplier Enablement** dropdown, click the **Configure** option and then the **Letter Content** tab, as shown in Figure 2.7. Under the **Letter Content** tab, you'll define the content of the letters to send to the suppliers to be enabled in Ariba Network.

When a supplier is invited to join Ariba Network through quick enablement or through the SAP Ariba supplier enablement automation process, an invitation letter is sent to the supplier. These letters are called *trading relationship request letters* and are sent by fax or email. To create or edit a letter, click on the **Actions** hyperlink associated with the letter you want to modify.

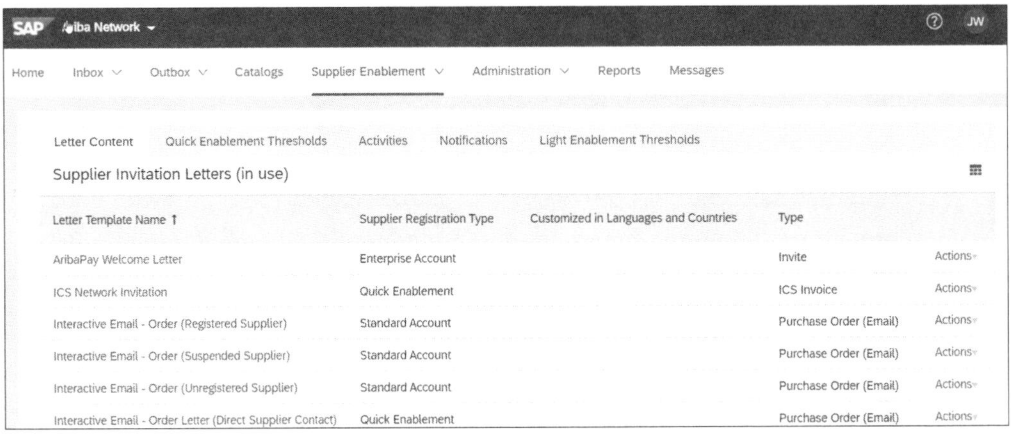

Figure 2.7 Supplier Enablement: Configure Letter Content

Different types of trading relationship letters can be sent to the suppliers. You'll determine exactly which letter to send based on the business process you're enabling for your organization in Ariba Network. If you aren't sure which option to choose, you should contact your SAP Ariba customer executive to help you determine how to enact an effective supplier enablement strategy. You can also invite your suppliers to use Ariba Network by using quick enablement and selecting any of the following: **Purchase Order**, **ICS Invoice** (for invoices sent from an invoice conversion service provider), **Payment Proposal**, **Request for Quotation**, or **CC Invoice** (for invoices sent from the SAP Ariba procurement solution or the SAP ERP system).

You have the option of configuring the thresholds for POs and SAP Ariba invoice conversion services add-on invitation letters sent to quick-enabled suppliers. For example, if you set the invoice threshold to "5," when Ariba Network receives the fifth invoice from a given supplier, an invitation letter will be sent to the supplier. The supplier can use the link in the invitation to register on Ariba Network and view the statuses of invoices.

After a trading relationship letter is sent to the supplier, your Ariba Network configuration needs to determine the list of tasks you expect the supplier to perform for the business process to be considered fully configured and tested. For example, to enable the PO process, select **Purchase Order** in the **Assign Activities** screen, as shown in Figure 2.8.

Thus, a supplier must complete this task when a PO is sent to the supplier via Ariba Network. This example brings us to another topic: Some tasks, like sending the PO, will be automatically completed when the PO passes through Ariba Network, while other tasks will have to be manually completed and closed by the task owner after the requested action has been completed. For example, the manual task **Buying Organization is Ready to Send Orders** is required, but it can only be completed and closed after you've communicated and coordinated all the required logistics between you and the supplier to send the supplier a PO.

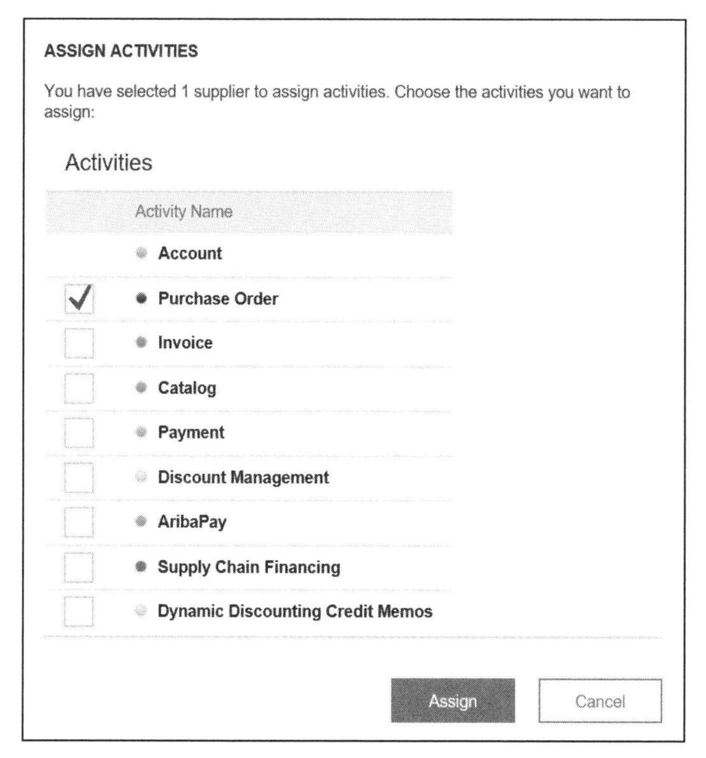

Figure 2.8 Assigning Activities to a Supplier

You can also choose to send notifications to up to three email addresses if a task is overdue or escalated, as shown in Figure 2.9. If you're enabling a large number of suppliers, you can use the **Reports** tab, or select **Tasks** under the **Supplier Enablement** tab, to see all the suppliers with **Failed** or **Overdue** statuses.

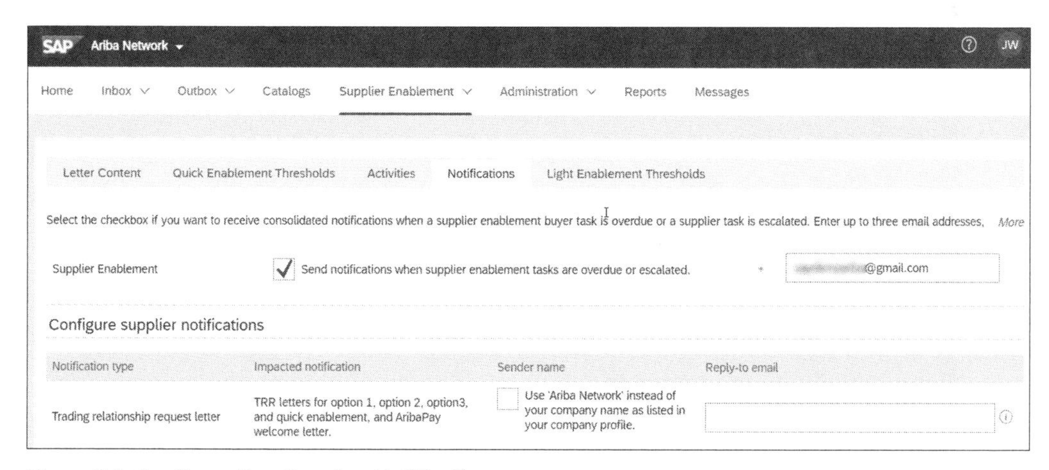

Figure 2.9 Configuration Overdue Notification

2.4.2 Supplier Enablement Notifications

The buyer will receive notifications when a supplier with an active supplier enablement status has overdue, escalated, or failed tasks in its profile that haven't been deferred. To complete your overdue tasks, follow these steps:

1. Log in to your Ariba Network buyer account (*http://buyer.ariba.com*).
2. Click the **Supplier Enablement** tab, and then click **Tasks**.
3. Review your task list, and complete the overdue and escalated tasks.

System-generated reminders will be sent to the supplier account administrator when a task is overdue. Tasks will escalate if the task isn't complete according to preconfigured time frames for the task. Reminders can be sent by filtering **Private Suppliers** and clicking **Send Invitation** in the **Actions** dropdown menu.

Note

Reminders aren't sent if a supplier fails to accept the terms of use.

2.4.3 Supplier Enablement Reports

The profile details also include changes made online to the existing/saved data. To create a report, log in to your Ariba Network account, click on the **Reports** tab on the dashboard home page, and follow these steps:

1. Click **Create**.
2. Enter a unique **Title** to help you and other users easily identify this report from a list.
3. Enter a **Description**.
4. Choose a value from the **Time zone** list.
5. Choose a **Language**.
6. Choose a **Report type**.
7. Click **Next**.
8. Click either **Manual Report** or **Scheduled Report** (Ariba Network automatically generates scheduled reports using the frequency that you specify).

If you choose **Scheduled Report**, Ariba Network refreshes the page and displays the **Scheduling** section. In this section, you should do the following:

1. Select a date range.
2. Choose a value from the **Automatically Run** list to set the report generation frequency.
3. Enter up to three email addresses, each separated by a single comma. Ariba Network sends a notification to these addresses when the report status is **Processed**.
4. Click **Next**.

5. Select the parameters for the report.
6. Click **Submit**.

Ariba Network saves the created manual report under the **Reports** tab for you to run at your convenience. To run a previously created manual report, follow these steps:

1. Log in to your account at *http://buyer.ariba.com*.
2. Click the **Reports** tab on the dashboard home page. You can view and download the report when the status is **Processed**.
3. To view and download the report, select the report template, and click **Download**.
4. Click on **Open**. Your browser starts Microsoft Excel and displays the report. Save this report.

Three reports are available to provide buyers information regarding supplier enablement:

- **Supplier Enablement Status Report**
  This report contains detailed information about suppliers on Ariba Network assigned to supplier enablement activities. This report includes information on the enablement status, activity status, supplier enablement attributes (e.g., wave, supplier address), and the relationship status of the supplier. The report also displays information about suppliers with an **On-Hold** status, including the reason for this status change.
- **Supplier Enablement Task Status Report**
  This report contains detailed information about supplier task statuses, including information about the supplier's status, vendor details, activities, and tasks and their statuses. This report helps when creating aggregated summary reports about tasks and the overall enablement status.
- **Vendor Data Export Report**
  This report, shown in Figure 2.10, contains all vendor profiles.

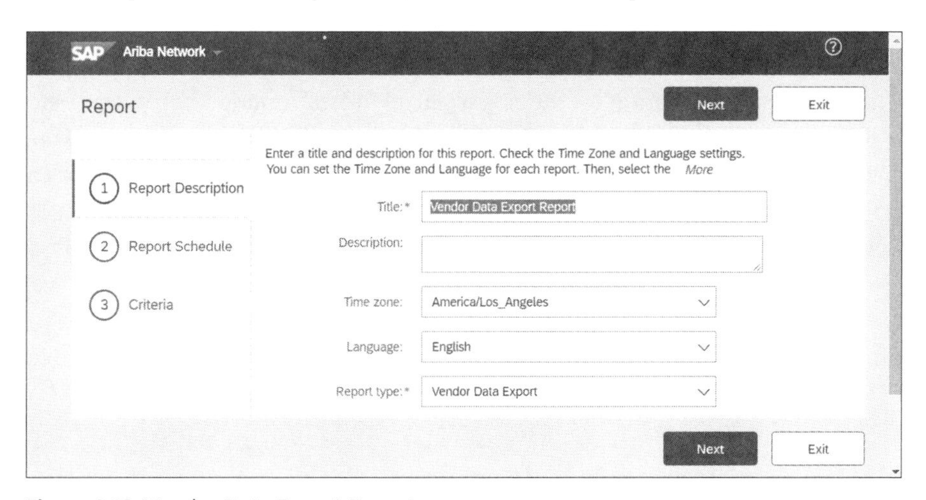

Figure 2.10 Vendor Data Export Report

2.4.4 Supplier Integration

Integration design is something that starts right after project kickoff with the requirements-gathering workbook. With this workbook, the SAP Ariba team or the consulting partners will try to retrieve as much information as possible to blueprint what the network side will entail. The blueprint is then leveraged as the road map for building out the guides and configuration. These guides will teach suppliers how to set up their accounts on their network, including EDI and cXML specifications detailing the technical requirements suppliers will need to fulfill to send you your company's specific data through the network.

Supplier Network ID

The supplier network ID is an important component automatically created by SAP Ariba, either when the supplier self-registers or when supplier information is loaded from SAP ERP. The supplier network ID serves as the conduit between Ariba Network and SAP ERP. Therefore, to determine the correct approach, you'll need to determine whether SAP ERP will update SAP Ariba, or vice versa: The supplier must have the freedom to update its profile information in Ariba Network, your company must be able to update the vendor master records in SAP ERP, and both records will need to be aligned.

At this point in the integration process, you should work with the SAP Ariba supplier enablement lead or the SAP Ariba supplier integration lead to understand the approach and the data flow between all applications before you start preparing any data. Something to keep in mind is that your existing SAP ERP vendor maintenance doesn't go away and usually continues to be updated through your existing vendor management process. For more on supplier lifecycle performance and risk management, see Chapter 3.

Remittance Information

Remittance information is another critical component in most integrations. When the invoicing component is enabled, the remittance becomes a cornerstone of the invoicing process. All invoices should have banking information, and many suppliers will have their banking information on the invoice. Banking information becomes essential to speeding up and streamlining the payment process: Having accurate banking information is critical and will save you a lot of process time. Instead of waiting for an invoice exception to be generated due to an incorrect remittance, you can look ahead and validate the banking information by matching the supplier's banking information with the information you have in the supplier master record in your SAP ERP instance, ideally before the invoice arrives in Ariba Network. Even if all the information can't be validated or matched, validating banking information before an invoice is processed in Ariba Network is a good approach.

When SAP Ariba downstream solutions are enabled, the supplier's banking information may not always contain remittance information. However, you can require the supplier to submit a remittance ID to ensure that this information is up to date. A remittance ID on the invoice allows you to verify that the address is linked to the supplier. To determine where to send payments to, some customers use the vendor ID, company code, purchasing organization, business unit, and several other descriptions to perform a match back in SAP ERP. In this case, SAP Ariba won't capture or determine the appropriate remittance information on the network because these tasks will take place through SAP ERP.

SA Ariba Supplier Mobile App

The SAP Ariba Supplier mobile app allows suppliers to manage their activities and collaboration on Ariba Network via their mobile devices. The app supports most phones and tablets running iOS 7.x or higher or Android 4.4x or above. The app allows for login, document creation, notifications, order/invoice sharing via PDF (with markup/comments to help resolve questions internally or with the customer), new leads review and response from the app, and invoice status. The app is available in 24 languages. For more information or to download the app, go to *www.ariba.com/ariba-network/ariba-network-for-suppliers/fulfillment-on-ariba-network/sap-ariba-supplier-mobile-app*.

Supplier Integration

The integration part of the deployment is where you'll connect your SAP ERP system to Ariba Network using middleware or a communication mechanism that can utilize cXML over standard HTTPS post, single sign-on (SSO), or a reply-response communication. You can be a buyer in Ariba Network, and the network will allow you to have many supplier relationships. For older versions of SAP ERP, such as SAP ECC, SAP's integration solutions, as shown in Figure 2.11 and Table 2.2, include SAP NetWeaver or the SAP Business Suite Add-On, which allows communication to take place via cXML.

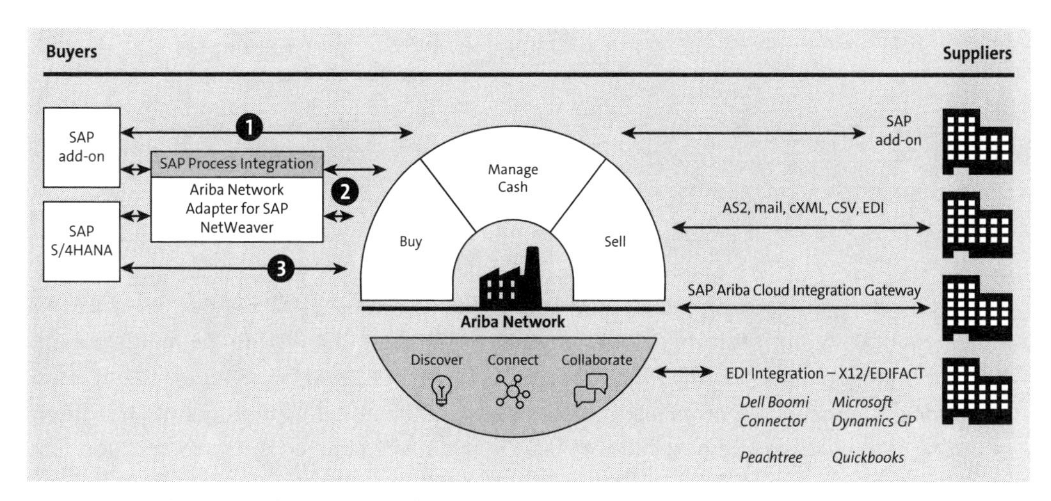

Figure 2.11 Ariba Network Integration for SAP ECC

SAP Ariba Buyer-Side Integration	SAP ERP	SAP S/4HANA	
	On-premise	On-premise	Cloud
SAP Business Suite Add-On ❶	Yes		
SAP Cloud Integration solutions ❷	Yes	Yes	
Native integration for SAP S/4HANA ❸		Yes	Yes

Table 2.2 Ariba Network Integration

In addition to the previous approaches to integration with Ariba Network, you can use the SAP Ariba Cloud Integration Gateway, which is the go-to integration platform for future integration approaches with SAP Ariba, including Ariba Network. SAP S/4HANA has native integration with Ariba Network for commerce automation; however, as with all things ERP, there may be some configuration required to support your preferred processes and customizations in your environment.

On the sales side, suppliers can specify how they want to receive transactions via a specific order routing method: cXML, EDI, ERP inbox, fax, email, or using the Ariba Network interface to create manual invoices. Figure 2.12 shows the main integrated documents for a seller connecting its SAP ERP to Ariba Network.

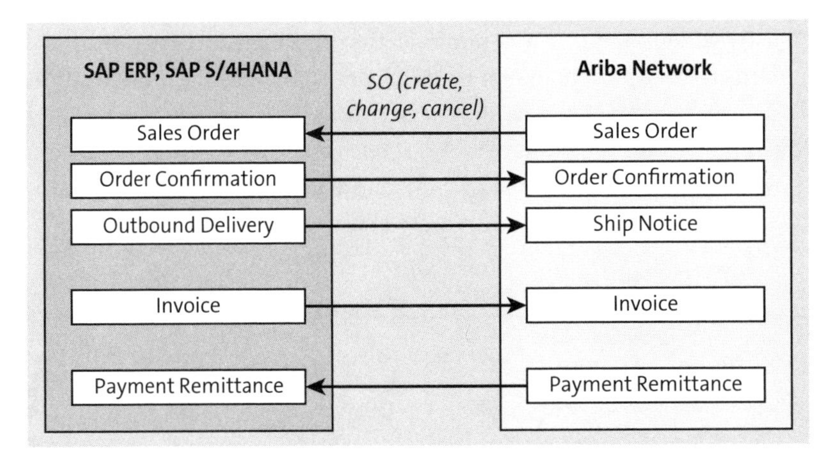

Figure 2.12 SAP Ariba Integration: Seller

You'll need to determine what, how, and when you want the SAP ERP system and Ariba Network to communicate. All information will be captured during the requirements-gathering sessions, utilizing the supplier enablement transactional requirement workbook, where the SAP Ariba team will tell you how the integration works and the different permutations and possible scenarios that can be managed by the integration. The integration team will provide you with the standard mappings for the integrations; conduct several mapping sessions; and review the cXML, files, and data. Note that for SAP S/4HANA integrations, the integration between SAP S/4HANA and Ariba Network is native. In other words, predefined integration and field mappings between SAP S/4HANA and Ariba Network can be activated via guided configuration in SAP Activate, best practices in SAP S/4HANA, and account settings in Ariba Network.

The integration process will take the following approach:

- Utilize the enablement workbook to document transaction enablement requirements. The workbook will be a living document throughout the project.
- Complete a requirements-gathering questionnaire.
- Schedule working sessions/discussions to understand and document transaction enablement and supplier enablement requirements.
- Provide opportunities to discuss the implications of selecting certain configurations.
- Document specific questions, concerns, and follow-ups.
- Gather information needed to build supplier education materials and cXML/EDI technical specifications.
- Understand any customization requirements, and so on.

The enablement workbook will have the following main sections:

- Project objectives
- Supplier registrations
- Communication/education/support
- Testing
- PO requirements
- Invoice requirements
- Supplier country-based invoice rules
- Customer-specific processes, fields, and functionality that aren't part of the core SAP Ariba (also known as *extrinsics*).
- Supplier groups

The more technical your supplier gatherings are and the more granular you get, the less reworking you'll have to do later. You should always have someone technical from your integration team engaged to help you understand how the two systems communicate and to start the mapping and documentation. These types of sessions can be quite technical.

2.5 Becoming a Supplier on the Sales Side

Ariba Network delivers the ability for sellers to connect and collaborate electronically with their customers, helping them optimize their business processes and collaborate effectively with their customers. Some benefits include the following:

- **Simplified usability**
  Buyer profile visibility allows for better alignment and status feedback during the spot quote process and improved password reset capability for occasional users.
- **Business collaboration and growth**
  Align with your customers' business requirements better with new extended buyer profiles, enabling multiple buyer sold-to and bill-to addresses, and creating comparisons between business configurations of different buyers.
- **Ordering, invoicing, and payment enhancements**
  Start invoice creation by converting a shipping notification or a goods receipt into an invoice. Based on your customer's use of Ariba Network capabilities, goods scheduling information can now be made available to you in Ariba Network.

Tip

By obtaining the appropriate SAP Ariba certifications, you can become a more attractive supplier in a customer's supply chain.

- **Globalization**
  New business rules to support tax invoicing requirements in Mexico, Chile, Colombia, Brazil, and France to drive comprehensive compliance.

Tip

The customer can segregate and filter suppliers by location, region, and country. Be sure your information is as up to date as possible prior to beginning these steps.

- **Supplier fees**
  Suppliers can start with a free standard supplier subscription (the light account), mentioned earlier in Section 2.2.2, and pay fees only if the supplier leverages further tools and functionality, or they can determine the type of SAP Ariba supplier membership program tier suitable for their organization. To determine whether you'll need to pay any fees based on the anticipated transaction level, and to compare costs and benefits, Ariba Network has a value calculator, as shown in Figure 2.13.

Tip

You must have your remittance information on hand, and the best way to accomplish this task is by having your SAP Ariba supplier network profile and customer supplemental profile updated.

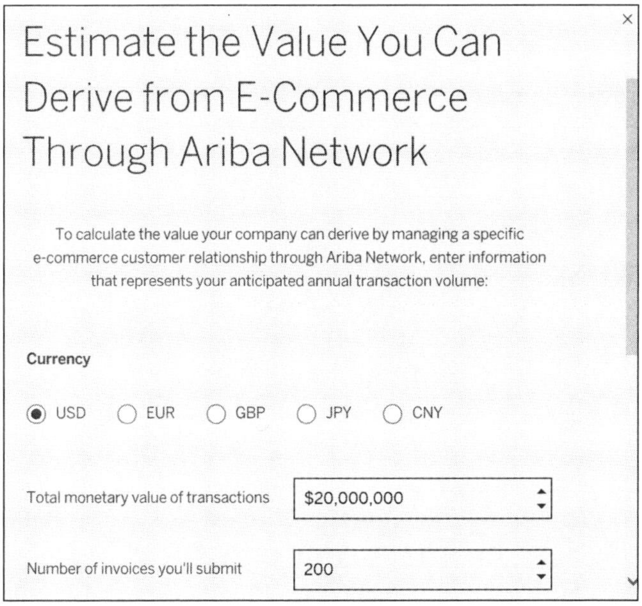

Estimate the Value You Can Derive from E-Commerce Through Ariba Network

To calculate the value your company can derive by managing a specific e-commerce customer relationship through Ariba Network, enter information that represents your anticipated annual transaction volume:

Currency

USD EUR GBP JPY CNY

Total monetary value of transactions $20,000,000

Number of invoices you'll submit 200

Figure 2.13 Seller Value Calculator

The value calculator is available at *https://supplier.ariba.com* under the **Learn More** section. As shown in Figure 2.13, the value calculator estimates the benefits of doing business through Ariba Network in terms of three components:

- **Bottom-line Benefits for Sellers**
  Most potential benefits center on reducing processing costs. Through automation, you can more efficiently respond to POs and submit invoices, thus minimizing lost and mishandled documents. When you provide a catalog or have electronic transactions with your customer via Ariba Network, orders, shipment notifications, confirmations, and invoices are more accurate and are paid on time, which helps reduce your days sales outstanding (DSO) and improve your capital position.
- **Top-Line Benefits**
  Through network connections and automation, you can boost sales by making it easier for customers to routinely buy from you, which helps customers with compliance and accommodates their procurement needs, increasing the likelihood they will stick with you.
- **Additional Benefits**
  Your SAP Ariba subscription includes value-added features designed to provide an even better purchasing experience, including, for example, licensing, data services, and sales and marketing activities. Figure 2.14 shows further benefits.

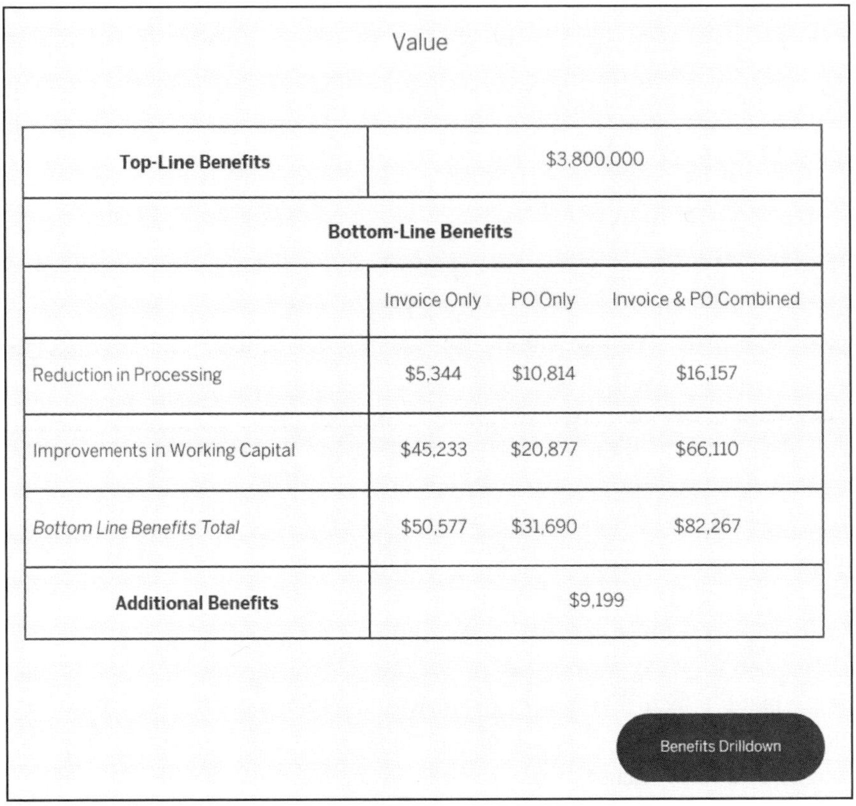

Value

Top-Line Benefits	$3,800,000		
Bottom-Line Benefits			
	Invoice Only	PO Only	Invoice & PO Combined
Reduction in Processing	$5,344	$10,814	$16,157
Improvements in Working Capital	$45,233	$20,877	$66,110
Bottom Line Benefits Total	$50,577	$31,690	$82,267
Additional Benefits	$9,199		

Benefits Drilldown

Figure 2.14 Benefits Drilldown: Supplier

Supplier costs are made up of two components: a subscription fee and network transaction service fees. The maximum annual transaction costs to a supplier, if the supplier is transacting billions of dollars over Ariba Network with you, are capped currently at $20,000, as shown in Figure 2.15.

At any time, you can email and discuss your results with an Ariba Network specialist at *commerceassistance@ariba.com*.

To configure an account, you'll just need to follow a few simple steps. First, go to *https://supplier.ariba.com*. SAP Ariba Buying also has a remittance update transaction where the SAP ERP system can update Ariba Network and SAP Ariba Buying on the status of scheduled payments. This transaction allows SAP ERP to send the payment details (i.e., Automated Clearing House [ACH] confirmation, or check details) to Ariba Network and SAP Ariba Buying, so that suppliers and buyers can see which payments have been made without having to log in to SAP ERP. This transaction also allows you to send updates about whether a payment was canceled or rejected in SAP ERP.

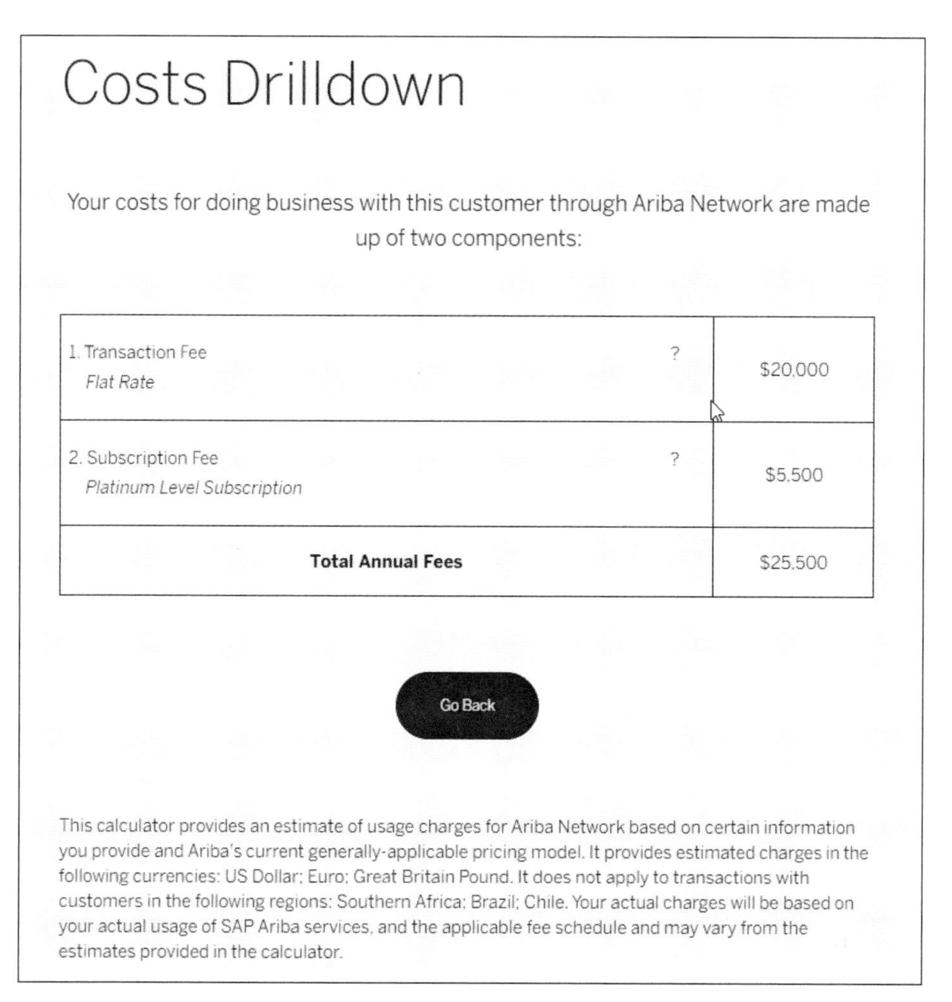

Figure 2.15 Cost Drilldown for Calculator

Other onboarding steps and tools include supplier registration in Ariba Network and profile updates to allow you to establish a trading relationship with the buyers. Once registered, you can sign in and determine the type of goods and services your company provides.

2.6 Buying with Ariba Network

You can transact on Ariba Network using all SAP Ariba Cloud solutions, running the sourcing, purchasing, and invoicing in the various SAP Ariba solutions and tying them together with Ariba Network. Or, you can integrate Ariba Network with an ERP system and use a hybrid approach—generating the purchasing documents in SAP S/4HANA or an older version of SAP ERP and exchanging these documents on Ariba Network, also called *purchase order and invoice automation*. Figure 2.16 shows the integration between the various SAP Ariba and non-SAP Ariba components in further detail.

In the next section, we'll focus on automating the PO and invoice exchange between supplier and customer via Ariba Network.

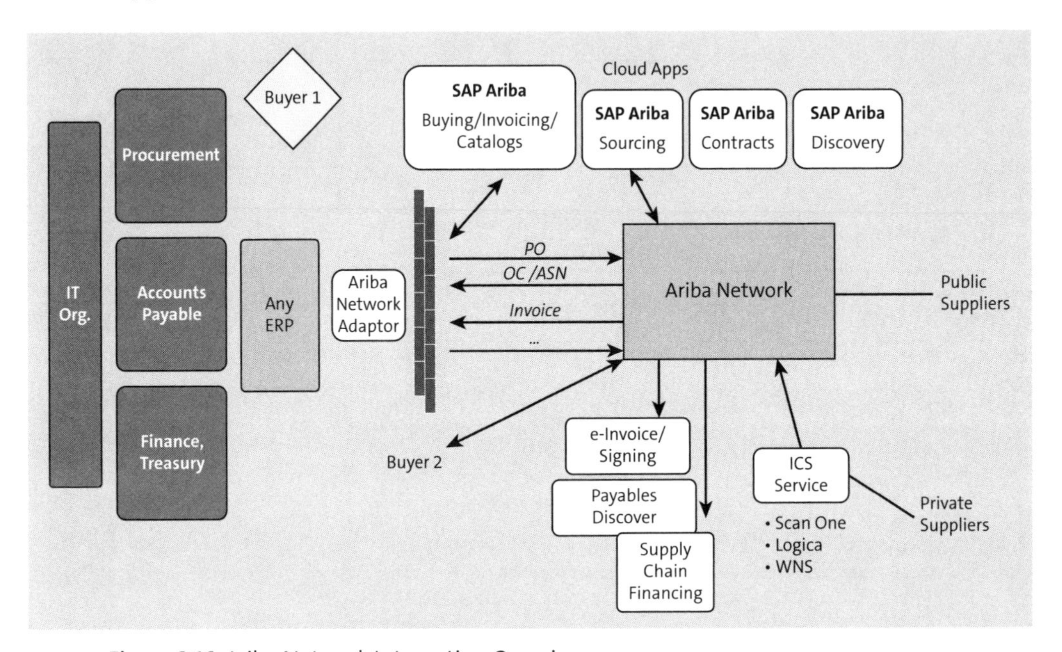

Figure 2.16 Ariba Network Integration Overview

2.7 Purchase Order and Invoice Automation with Ariba Network

Many collaboration processes are available on Ariba Network for suppliers and buyers. The most often-used processes are PO and invoice automation, which allow you to transmit POs to suppliers leveraging the network so that the supplier confirms, processes, and fulfills the PO and then submits an invoice back to the supplier. All these steps are mediated in Ariba Network, ensuring transaction integrity and flow.

In this scenario, a buyer creates a PO in SAP ERP, as shown in Figure 2.17.

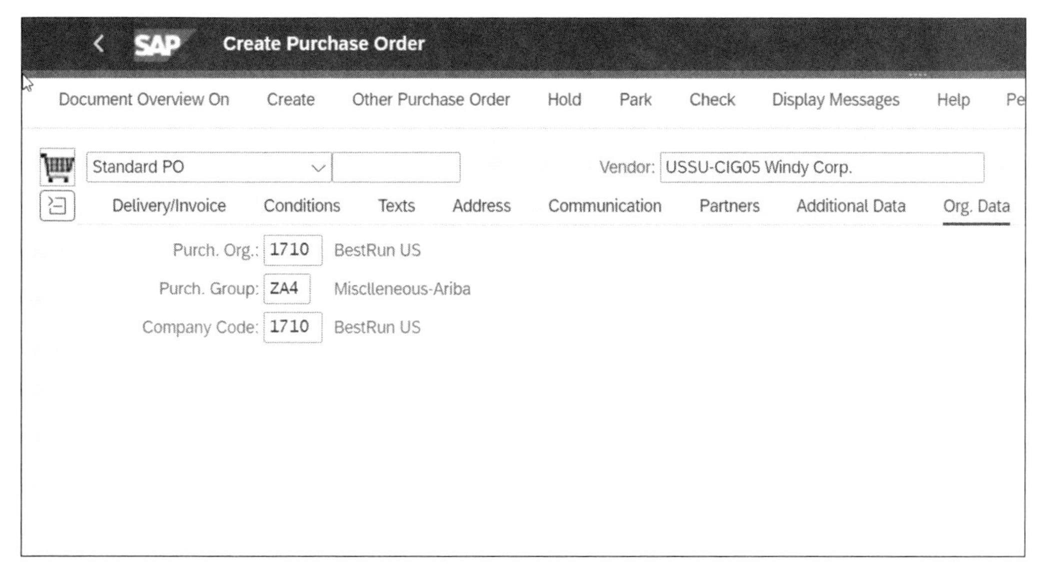

Figure 2.17 Creating a PO in SAP S/4HANA

Next, the buyer completes the line item information and issues the PO, as shown in Figure 2.18.

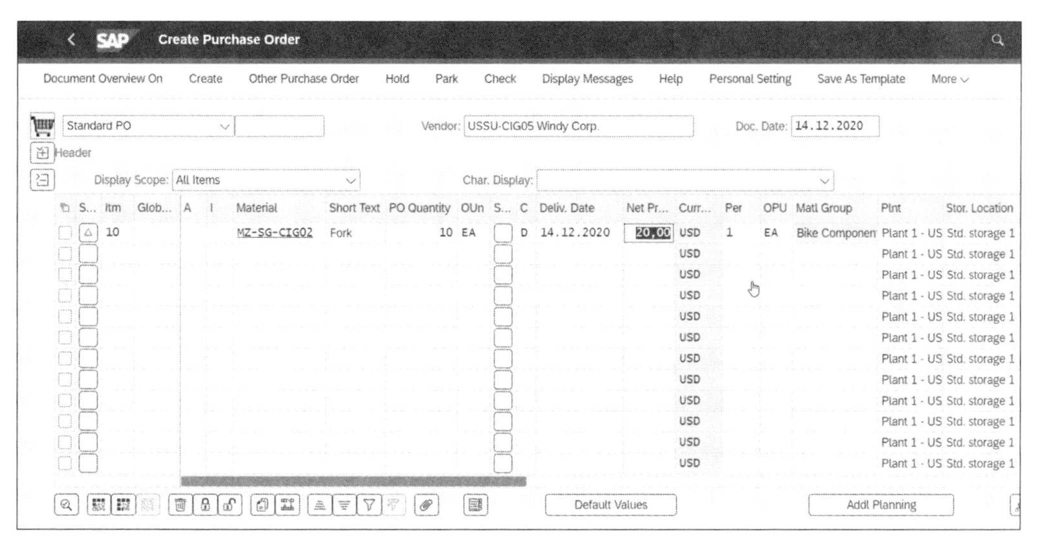

Figure 2.18 Purchase Order Line Item Complete

Once issued, the PO is transmitted to Ariba Network, and the receiving supplier logs into their account, as shown in Figure 2.19.

The supplier reviews the new PO by clicking on the PO number in the **Order Detail** tab, as shown in Figure 2.20.

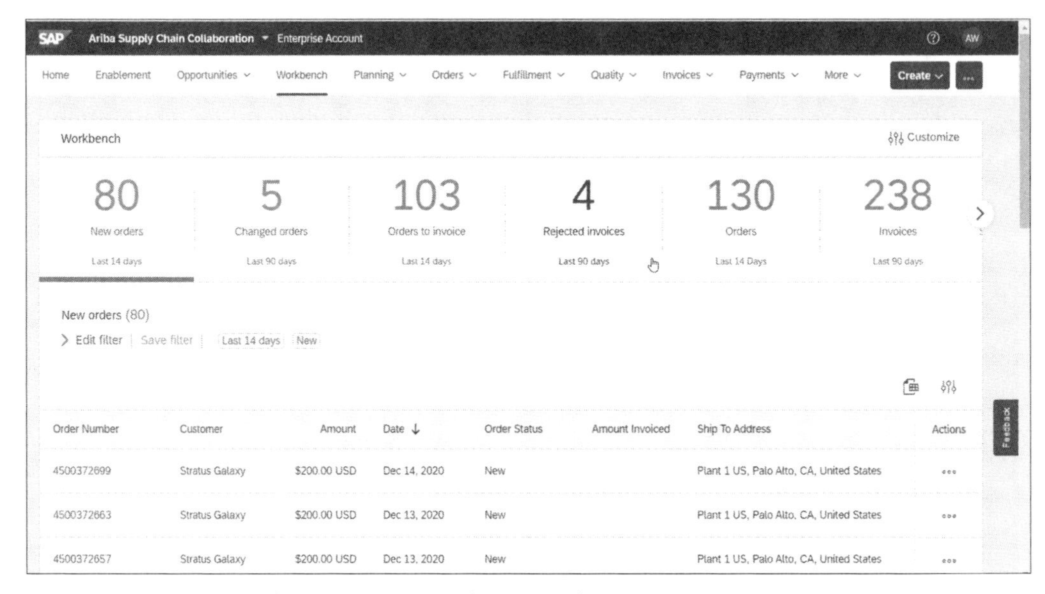

Figure 2.19 Supplier Login Home: Ariba Network

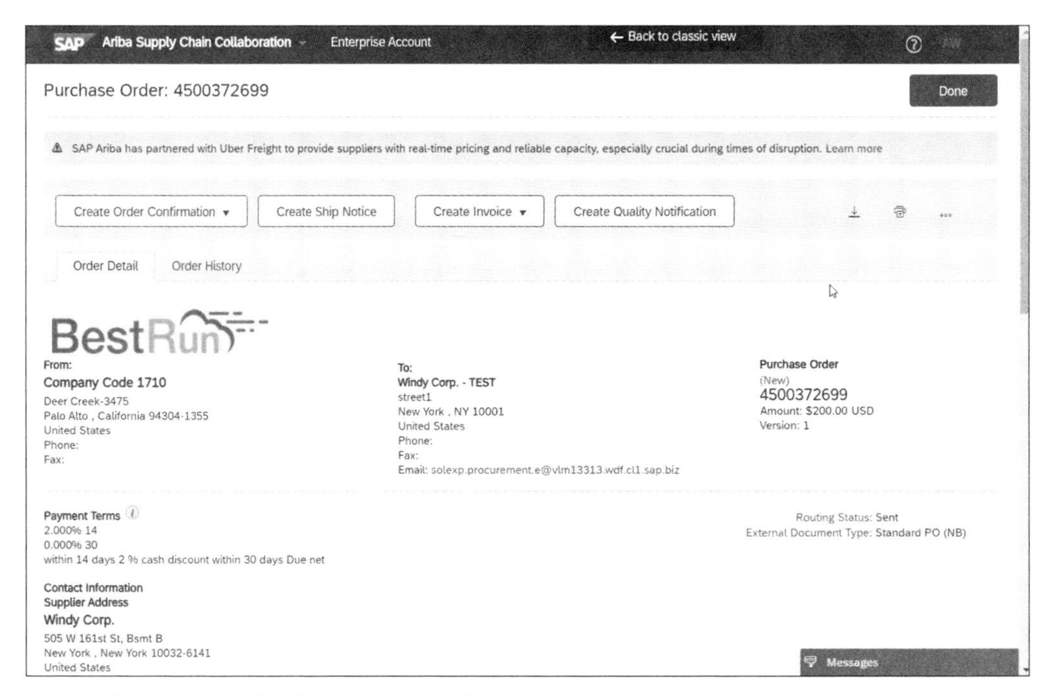

Figure 2.20 Reviewing the PO: Supplier

Within the PO, the supplier can create a confirmation as well as a shipping notice. As shown in Figure 2.21, the supplier can confirm the total or a partial quantity or reject the order entirely.

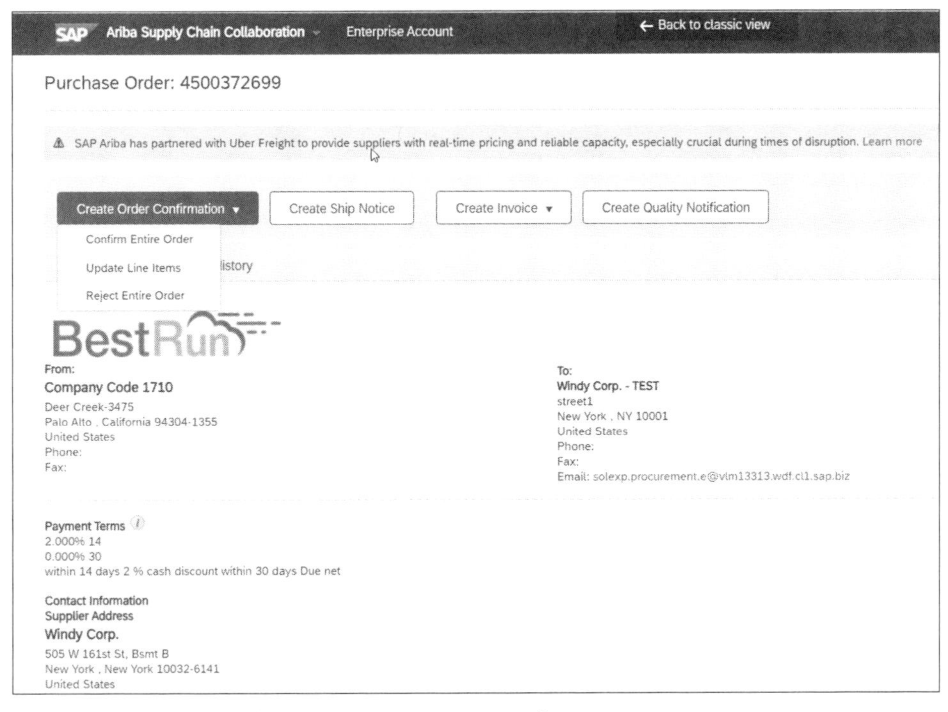

Figure 2.21 Confirm, Update Line Items, or Reject Order

In the details screen for the confirmation, the supplier can further specify shipping costs, taxes, and shipping/delivery date, as shown in Figure 2.22.

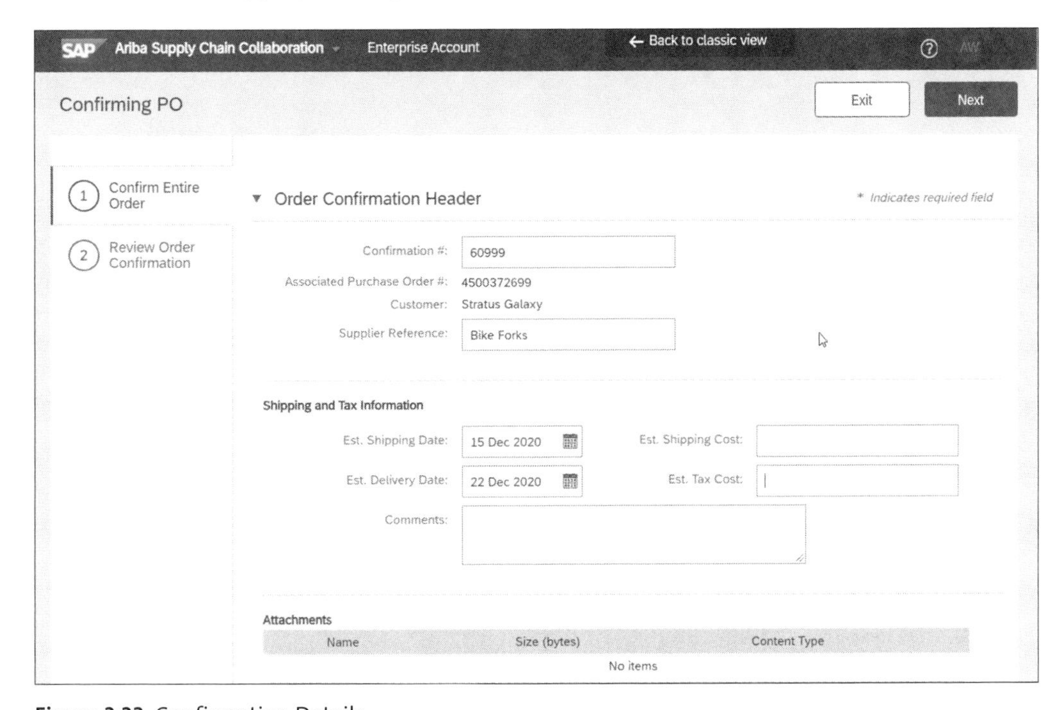

Figure 2.22 Confirmation Details

Once submitted, the PO confirmation is sent back to the customer's SAP ERP system. Next, the supplier creates a shipping notification, as shown in Figure 2.23.

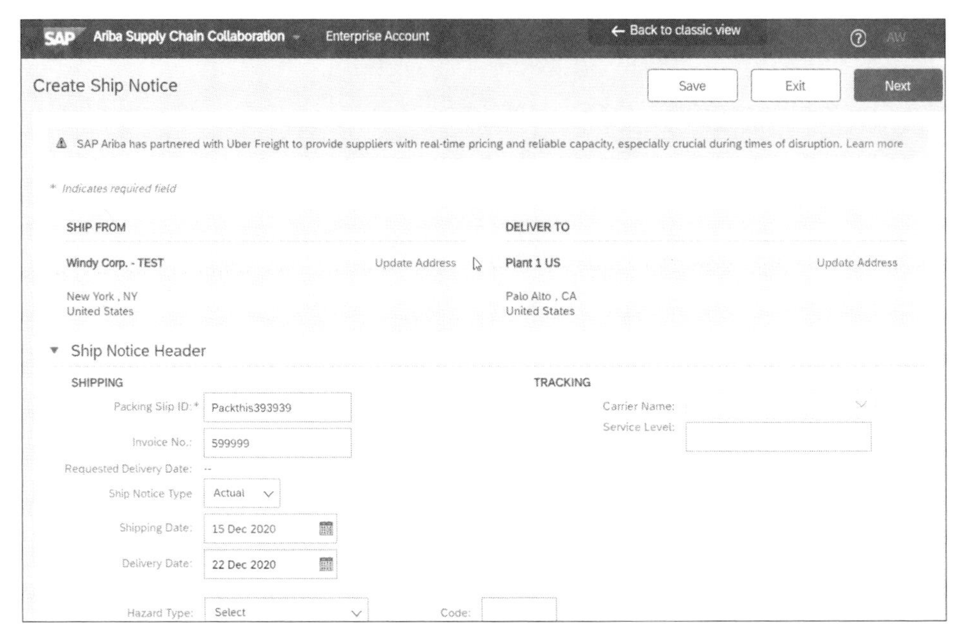

Figure 2.23 Create Ship Notice

The ship notice contains the shipping, tracking, and delivery information, as well as lines for additional items the supplier needs to add to this order, as shown in Figure 2.24.

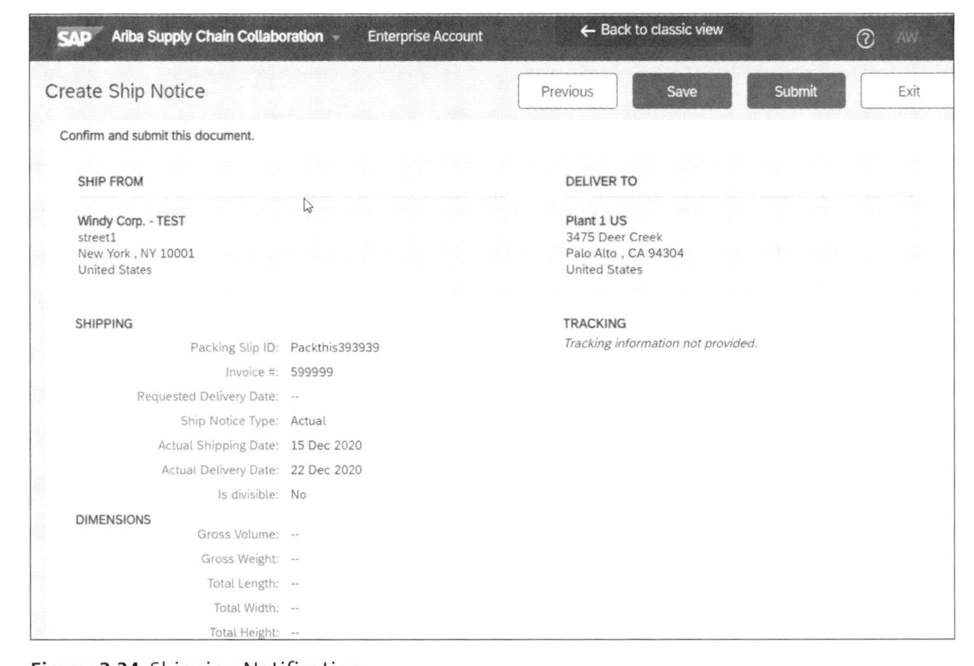

Figure 2.24 Shipping Notification

After the ship notice has been created, the supplier can invoice for this order if no goods receipt is required, as shown in Figure 2.25. Note that the supplier can create a standard invoice or a credit memo by selecting the option of the same name from the dropdown list.

SAP Ariba Supply Chain Collaboration Enterprise Account
← Back to classic view
Create Invoice
Update Save Exit Next
Invoice Header
* Indicates required field
Add to Header
Summary
Purchase Order: 4500372699
Invoice #:* 60944993
Invoice Date:* 14 Dec 2020
Service Description:
Supplier Tax ID:
Remit To UKBANK
LON , CO
United States
Bill To: Company Code 1710
Palo Alto , California
United States
Subtotal: $200.00 USD
Total Tax: $0.00 USD
Total Gross Amount: $200.00 USD
Total Net Amount: $200.00 USD
Amount Due: $200.00 USD
View/Edit Addresses
Tax
Header level tax
Line level tax
Remove
Feedback

Figure 2.25 Creating the Invoice

The supplier creates an invoice, as shown in Figure 2.26, and confirms the submission, as shown in Figure 2.27.

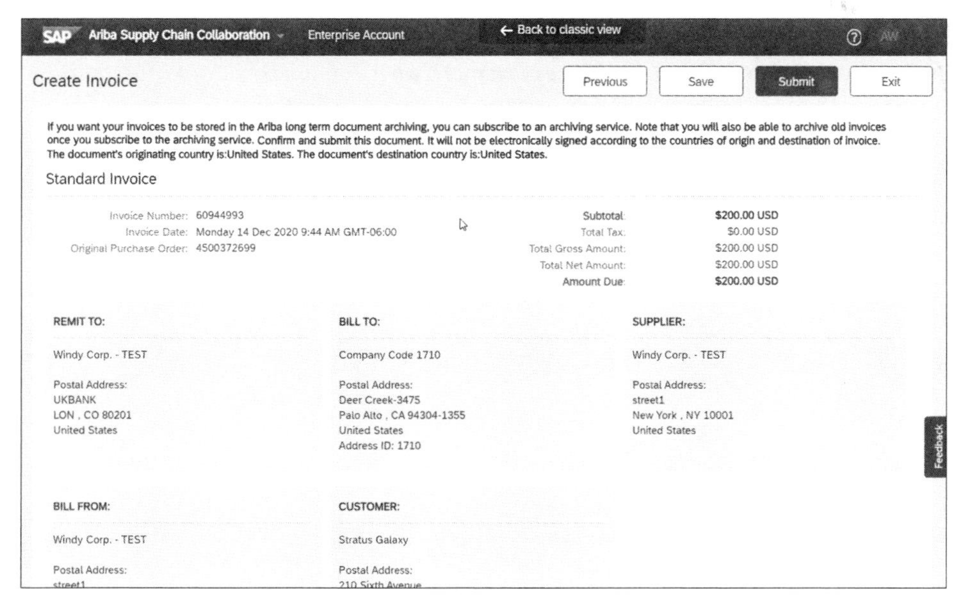

Figure 2.26 Creating Invoice Details

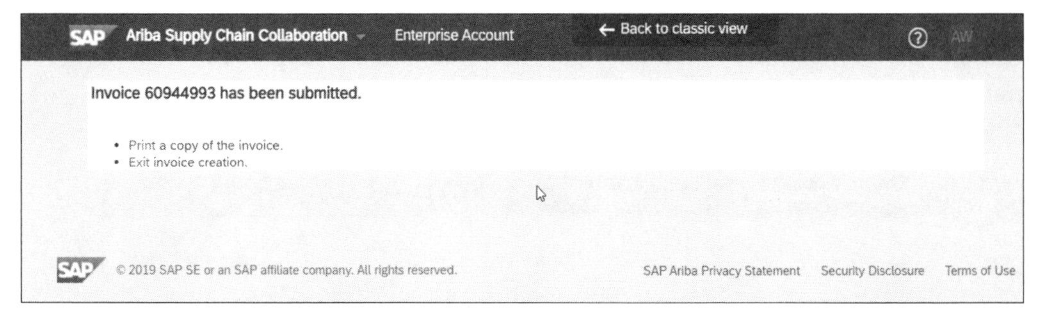

Figure 2.27 Invoice Submitted

The invoice, as with the confirmation and ship notice, transfers to the customer's SAP ERP and can be viewed and processed for payment.

2.8 Summary

In this chapter, we introduced you to Ariba Network—a 21st-century tool for supplier collaboration. SAP Ariba provides options for moving suppliers to an electronic process by leveraging proven methodologies either through a services team that handles supplier enablement for you or through the self-service tool. These tools help you quickly target and enroll suppliers to support your procurement business objectives, thus ensuring an effective and efficient trading partner collaboration. Ariba Network is the largest network of its kind, with millions of trading partners and trillions of dollars in transactions. After you've set up Ariba Network and begun leveraging its supplier networks and groups to derive tremendous savings and efficiencies, you'll likely want to further refine your supplier mix and expand your understanding of your existing suppliers. Along these lines, in the next chapter, we'll explore the SAP Ariba solutions designed for further understanding and managing your suppliers.

Chapter 3
Supplier Lifecycle and Performance

SAP Ariba Supplier Management is the only end-to-end solution portfolio that lets you manage supplier information, lifecycle, performance, and risk, all in one place.

Supplier management is probably one of the most ignored disciplines. Historically, procurement organizations have concentrated more on other procurement functions, such as sourcing and contract management. In an economy where globalization has exponentially increased the number of potential suppliers, scaling supplier performance management can be very tricky. Typically, processes for onboarding, qualifying, and segmenting suppliers aren't integrated to other procurement solutions. In addition, managing vendor master data from multiple sources and a lack of visibility in your supply base can add to the complexity of supplier management.

Now, with SAP Ariba Supplier Lifecycle and Performance, you have a comprehensive supplier management solution that can transform your supply management processes and help you better manage your supply base. In this chapter, we'll provide detailed insights on these two solutions and describe implementation approaches to get these solutions up and running in your enterprise.

3.1 What Is Supplier Lifecycle Management?

Supplier management is a broad term. In today's business landscape, supplier management isn't just a set of administrative tasks to load and maintain supplier records. Companies have started to take a more strategic approach in managing their supply base. Supplier management has transformed into supplier lifecycle management. In simple terms, supplier lifecycle management is the process of managing a supplier throughout its lifecycle from initial registration and onboarding to qualification to performance management and development.

3.1.1 Supplier Management Strategies

The basic process of engaging with new suppliers and managing supplier data has evolved from simple supplier information management to more comprehensive supplier lifecycle management. Technology can enhance the supplier onboarding process with the right checks and balances, supplier self-services, and approval processes.

Let's look at some strategies to enhance supplier management processes.

Cross-Functional Collaboration

As shown in Figure 3.1, onboarding a new supplier must be a collaborative effort between procurement, accounts payable, and perhaps other corporate entities such as treasury or risk management.

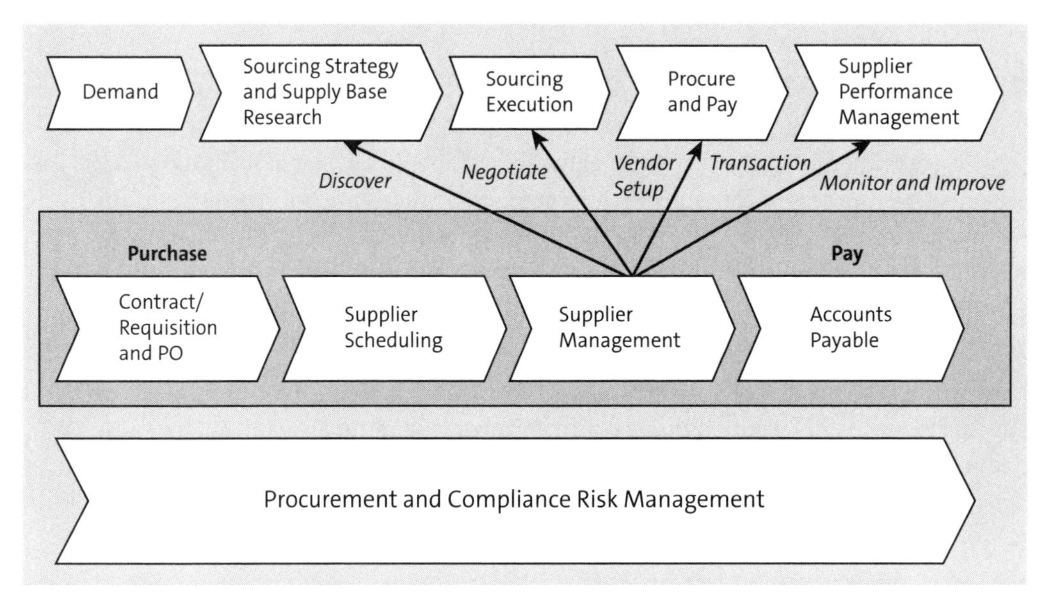

Figure 3.1 Supplier Collaboration

The onboarding process should include supplier checklists based on the supplier's category, geographic scope, and other criteria, such as anticipated spend. Supplier registration and onboarding must be reviewed and approved by procurement based on category strategy, by risk management and treasury based on compliance, and by other corporate entities as needed. Adding new suppliers should not just be a function of one entity (e.g., accounts payable).

Supply Base Strategy

The fewer the suppliers a company has, the greater the opportunities for spend reduction and lower procurement costs. In addition, having fewer suppliers reduces risk exposure and makes automating processes easier. Requests to add a new supplier could be vetted to determine whether the new supplier is truly necessary or a preferred supplier could meet the need instead.

Risk Management

Supplier risk management is a proactive process of identifying and mitigating risks cost effectively across a buyer organization's supply chain. With a global supply chain, performing thorough risk-related due diligence research before onboarding new suppliers into your supply base is critical.

An effective risk management strategy helps an organization evaluate its risk profile and define which risk factors must be managed in its supply base, including how suppliers should be assessed. Typically, risk assessment criteria include financial, human resource, environmental, supply chain disruption, performance, and relationship factors.

Defining Procure-to-Pay Channels

Best-in-class organizations have a clear, well-defined process for requesting, buying, and paying suppliers for the goods and services being procured. These organizations define buy channels by spend category and guide their requisitioners through these channels. To maintain this type of rigor, new suppliers should be onboarded and set up within these defined channels, which includes additional enablement processes, such as catalog setup, purchasing cards (p-card) setup, and so on.

Support of Working Capital Objectives

In today's highly competitive market, managing and enhancing working capital performance is crucial. Working capital objectives can be addressed early on during the supplier onboarding process by negotiating payment terms based on your organization's working capital strategy, by setting up p-card usage, and by defining dynamic discounting processes.

These days in business, global and ever-changing supply chains have become increasingly complex and hard to manage. Managing your organization's supply chain is critical for your company to survive. An efficient supplier information and lifecycle management solution, along with a comprehensive supplier risk management solution, can considerably help your organization keep the business running, achieve business goals, and remain competitive.

3.1.2 SAP Ariba Portfolio

As shown in Figure 3.2, the SAP Ariba Supplier Management portfolio consists of SAP Ariba Supplier Lifecycle and Performance and SAP Ariba Supplier Risk.

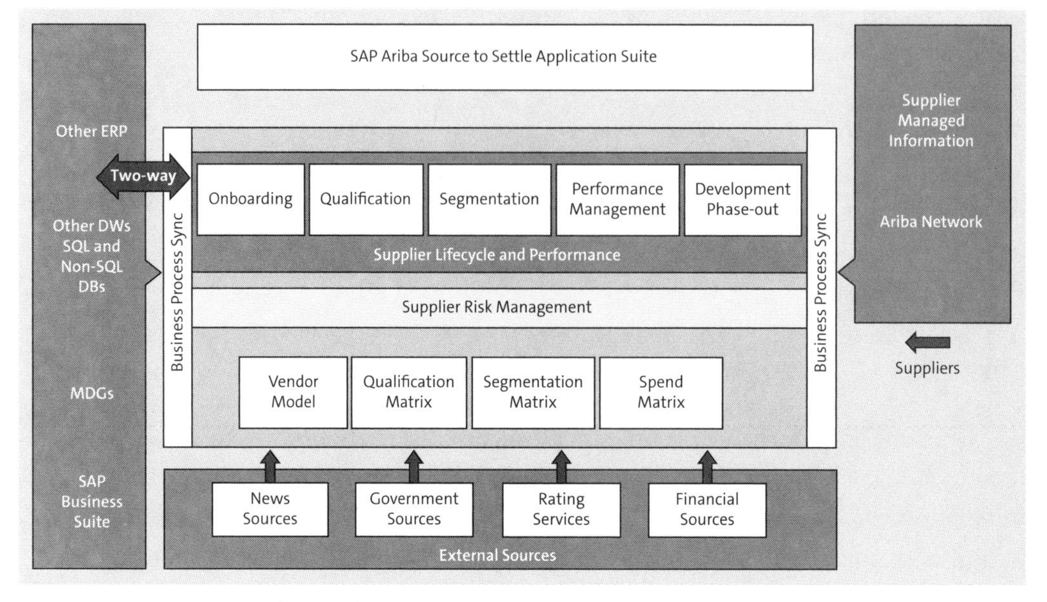

Figure 3.2 SAP Ariba Supplier Management Portfolio

3.2 Implementing SAP Ariba Supplier Lifecycle and Performance

SAP Ariba Supplier Lifecycle and Performance, a reincarnation of SAP Ariba's older solution, SAP Ariba Supplier Information and Performance Management, is a supplier information management system. The vendor model in SAP Ariba Supplier Lifecycle and Performance is designed around the SAP vendor model, which consists of the following:

- General information
- Company code information
- Purchasing organization information

3.2.1 Application at a Glance

SAP Ariba Supplier Lifecycle and Performance provides comprehensive tools to help you onboard, qualify, segment, and manage supplier performance more effectively. As shown in Figure 3.3, SAP Ariba Supplier Lifecycle and Performance is integrated into your procurement processes and lets you drive spending to your preferred suppliers and to scale compliance for your entire supply base using an array of key capabilities, such as the following:

- A unified vendor data model in the cloud provides a single accurate supplier record.
- Supplier self-service in the cloud via Ariba Network makes it easy for suppliers to maintain their own information.

- A flexible matrix for supplier qualification and segmentation lets you manage suppliers based on specific parameters.
- Full integration with other SAP Ariba procurement applications supports speed and consistency throughout the entire procurement process.
- A comprehensive 360-degree view aggregates business processes and data from multiple SAP Ariba solutions and presents a single view of the supplier.
- Syncing with your SAP ERP backend is bidirectional (two-way).

SAP Ariba Supplier Lifecycle and Performance has a new supplier management dashboard. The **SUPPLIER MANAGEMENT** tab shows all the activities and suppliers that you manage directly. You can also search for suppliers under the **SUPPLIER MANAGEMENT** tab.

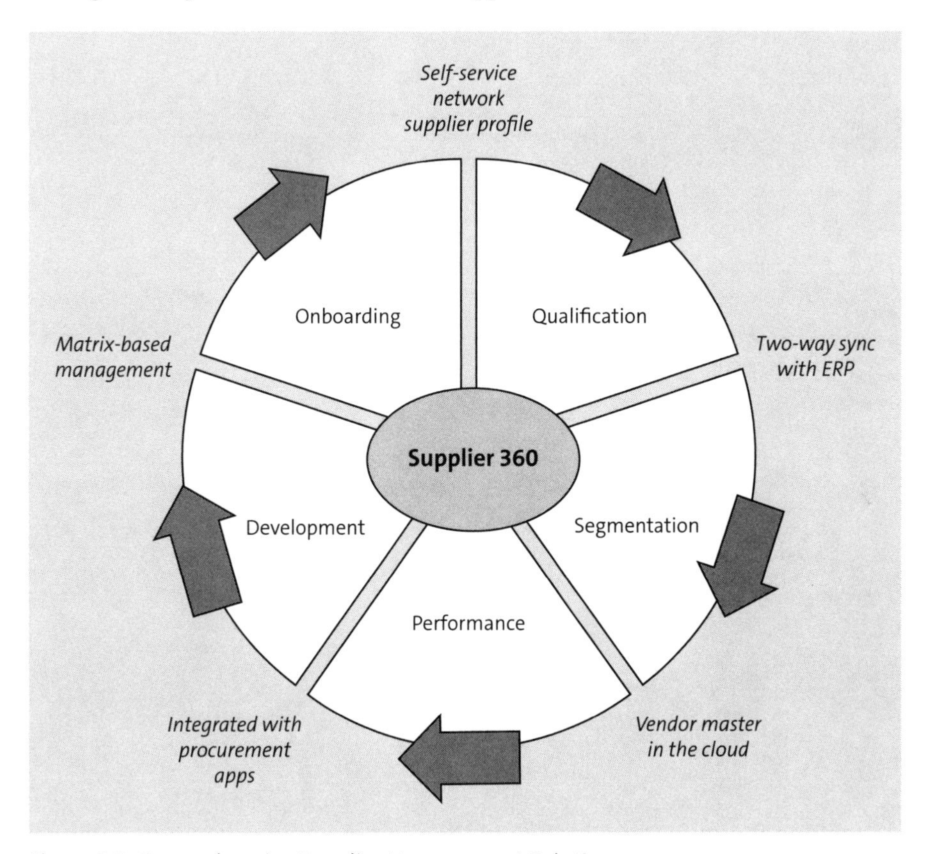

Figure 3.3 Comprehensive Supplier Management Solution

Under the **SUPPLIER MANAGEMENT** tab, as shown in Figure 3.4, you can perform the following activities:

- Search for suppliers.
- Click any supplier's name to see their Supplier 360° View, which includes all their profile information. If you're a category or supplier manager assigned to the supplier, you can manage registration, qualification, and other processes from this tab.

- Monitor your supplier management activities, such as the supplier requests you've created, the registrations you've started, and so forth.
- Monitor the suppliers you manage, including their qualifications and preferred status levels.

Figure 3.4 Supplier Management Dashboard

In the following section, we'll look at how your organization can use SAP Ariba Supplier Lifecycle and Performance to monitor and manage the lifecycle of a supplier.

3.2.2 Monitoring Your Supplier Management Activities

The **My Activities** area under the **SUPPLIER MANAGEMENT** tab shows information about all the supplier processes you've started so that you can monitor their progress. You'll only see this area if you have permission to start a supplier process (e.g., requesting a new supplier or inviting a supplier to register) and have started at least one process. The area shows a tile for each process type with the number of processes you've started.

Click a tile to see a list of processes with the following information:

- The supplier name, which you can click to open the **Supplier 360° View**.
- The current status of the process. If the process is still in approval, the **Status** column shows the number of days the process has been in the approval stage and the names of the outstanding approvers so that you can monitor the approval's progress and send reminders as needed. Click the filter icon to filter the activity list by status.
- Commodities and regions associated for qualification, disqualification, and preferred supplier processes.
- A **View** button, which you can click to open the process on the **Supplier 360° View** to review questionnaires, reissue invitations, and complete tasks.

Note

The **My Activities** area only shows supplier processes that you've initiated manually and that require approval. If a supplier request uses auto-approvals, your request won't appear in the **My Activities** area. In addition, the **My Activities** area doesn't show registrations that were initiated through automatic registration or mass invitation.

3.2.3 Using the Supplier 360° View

The **Supplier 360° View** shows all the information your company has about a supplier. If you're a category or supplier manager assigned to the supplier, you can also manage the supplier from this view. Open a **Supplier 360° View** by clicking the supplier's name on the **SUPPLIER MANAGEMENT** tab or in the search results. As shown in Figure 3.5, the **Supplier 360° View** includes the supplier name, supplier ID: **SM Vendor ID** (SAP Ariba internally generated ID), **ERP Vendor ID**, **Ariba Network ID**, and links to sections of supplier information, such as **Summary**, **Contacts**, **ERP data**, **Public profile**, **Certificates**, **Activity log**, **Registration**, and **Qualifications**. Here you can see:

- **Summary**
  This section (Figure 3.5) displays information about the **Primary Contact**: **Name** and **Email**; **Location**: **Address**; **Status**: **Supplier Creation Date**, **Registration Status**, **Qualification Status**, and **ERP Integration Status**. This section also contains a subsection

that displays the **Origin of the Supplier**: **Requested by**, **Approved by**, and other details from the supplier request.

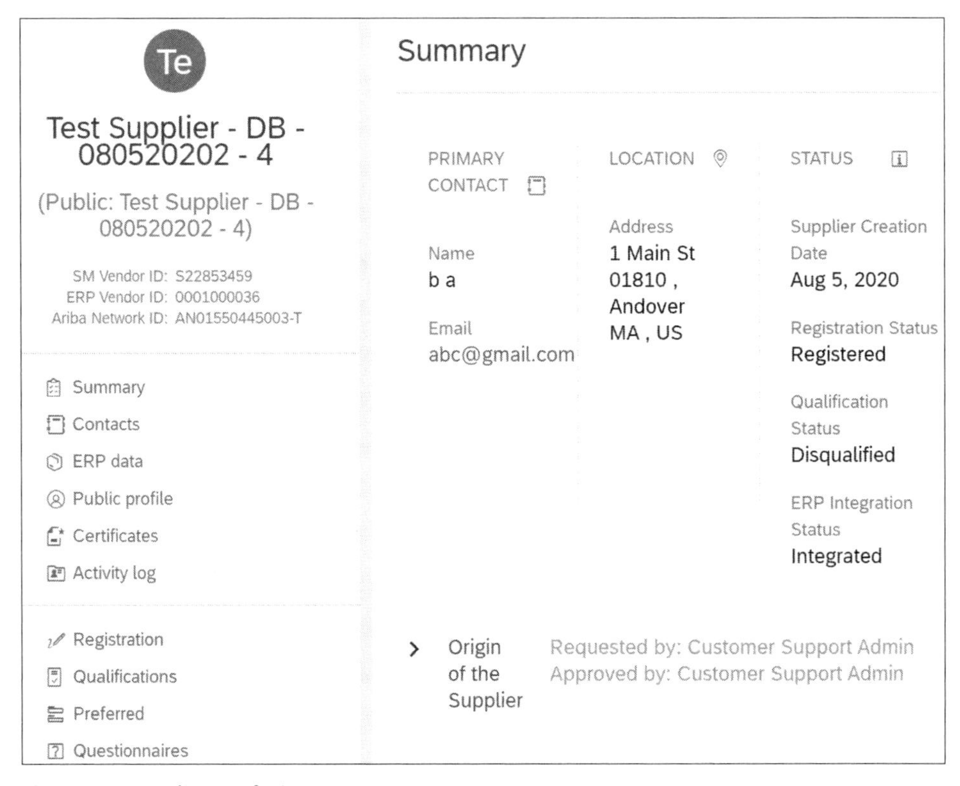

Figure 3.5 Supplier 360° View: Summary

- **Contacts**
  This section (Figure 3.6) displays the **Supplier Manager** assigned to manage this supplier. In addition, all supplier contacts will be listed in this section. From this section, the supplier manager can edit existing supplier contact information or add/delete supplier contacts.

- **ERP Data**
  This section (Figure 3.7) shows the detailed profile information from your company's integrated SAP ERP system. This includes **General**, **Company code**, and **Purchasing data**. **Synced to ERP** status and the **Resync now** button to resync the supplier with the ERP backend.

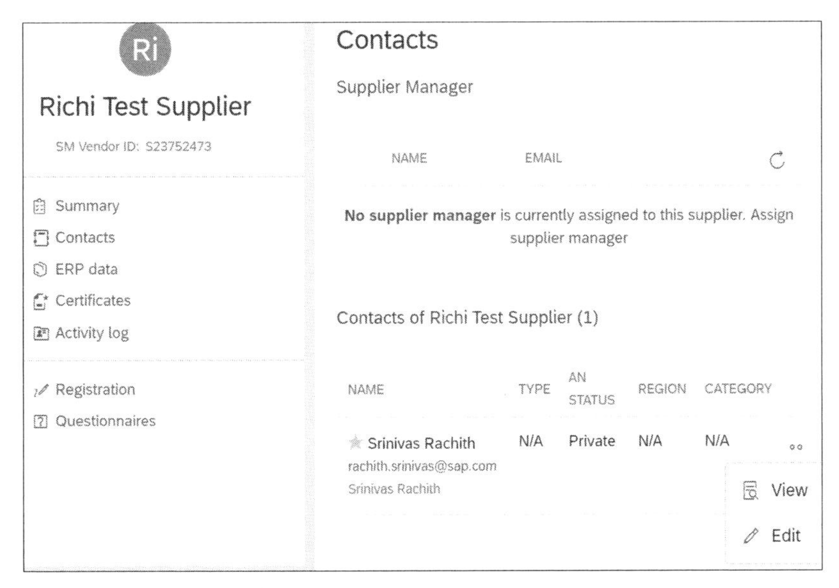

Figure 3.6 Supplier 360° View: Contacts

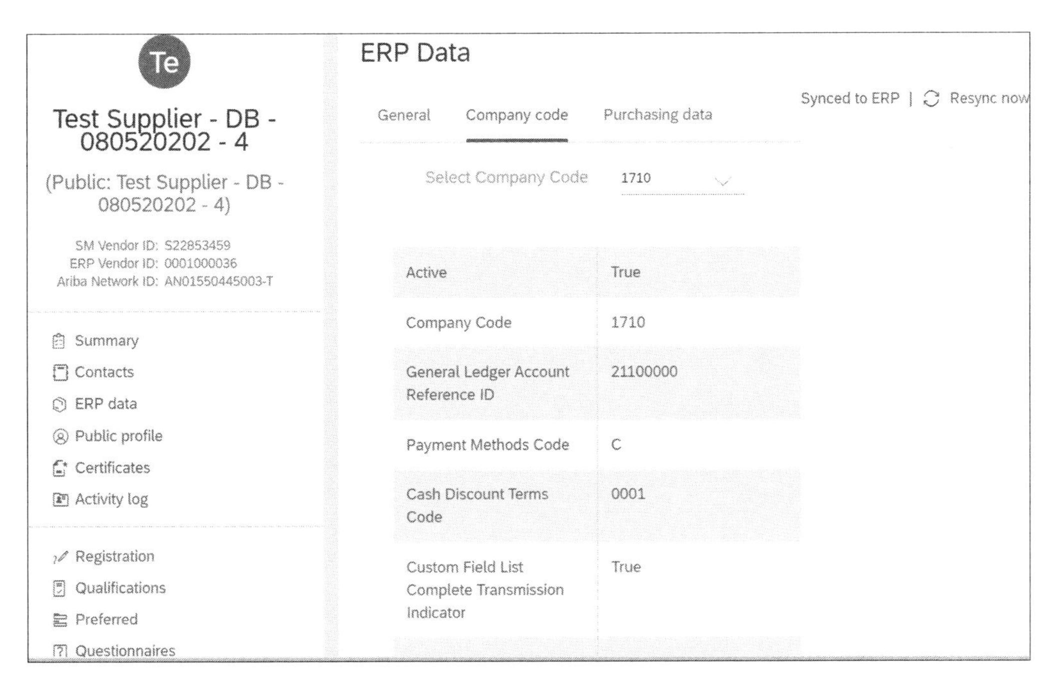

Figure 3.7 Supplier 360° View: ERP Data

- **Certificates**
  If the certificate management feature (Figure 3.8) is enabled in your site, this section displays certificates that the supplier has submitted for certificate-related modular

questionnaires along with expiration dates and any qualification status that corresponds to a certificate's combination of commodity, region, and department. Certificates for which notifications of upcoming expiration have been sent are highlighted in yellow, while expired certificates are highlighted in red. You can click on the certificate to view the modular questionnaire source and other certificate details: **Certificate Number** and **Certificate Issuer**; validity dates; and **Category**, **Region**, and **Business Unit** for which the certificate is applicable.

Figure 3.8 Supplier 360° View: Certificates

- **Public Profile**
  This tab shows information from the supplier's linked Ariba Network account.
- **Registration**
  SAP Ariba Supplier Management solutions include this tile where category or supplier managers can access suppliers in the registration process. On clicking the **Registration** link in the **Supplier 360° View** (Figure 3.9), the registration page appears (Figure 3.10). Supplier and category managers can invite suppliers to register or start internal registration, complete any registration to-do tasks, or approve or deny registrations.

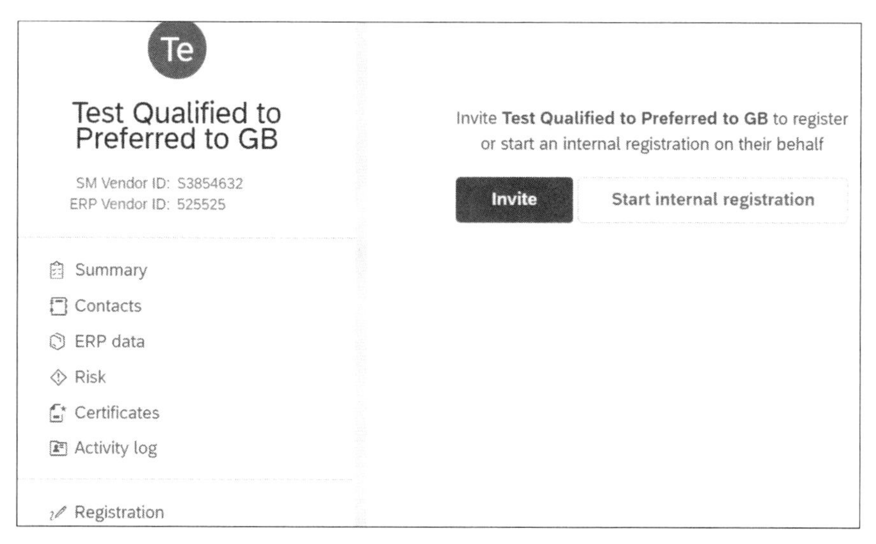

Figure 3.9 Supplier 360° View: Registration Invite

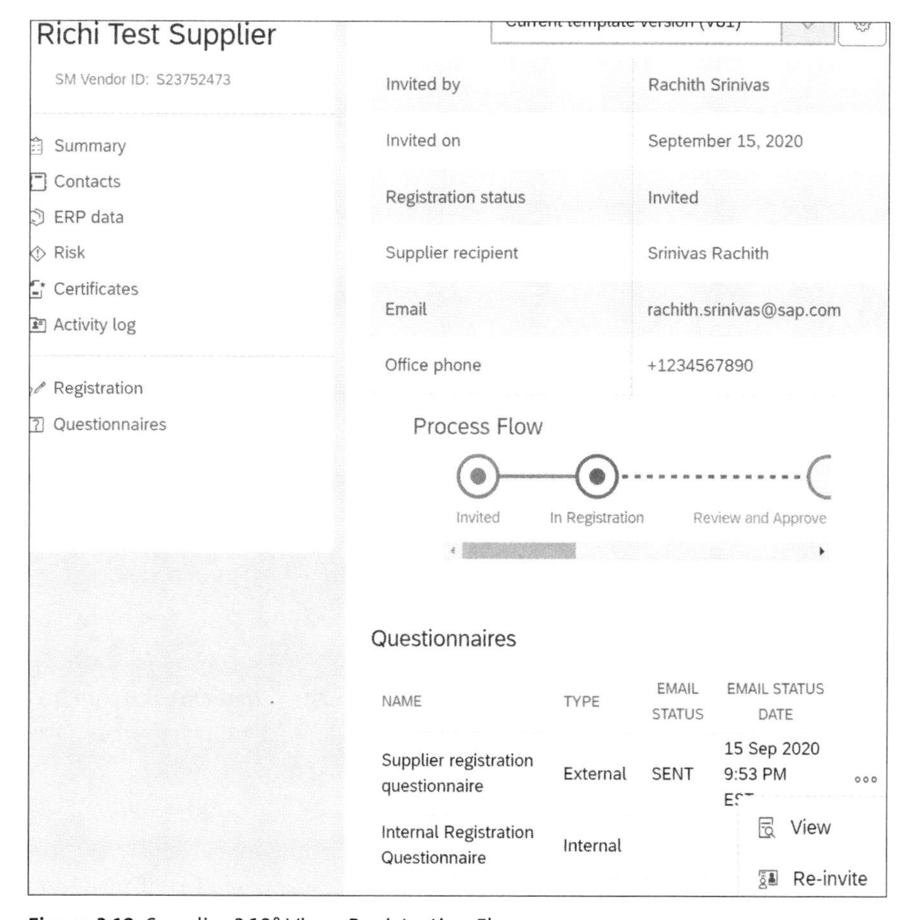

Figure 3.10 Supplier 360° View: Registration Flow

- **Qualification**
  SAP Ariba Supplier Management solutions include this tile where category or supplier managers can qualify a supplier for specific commodities, regions, and business units by assembling qualification questionnaires. On clicking the **Qualifications** link in the **Supplier 360° View**, the **Qualification** page appears. Supplier or category managers can complete any qualification to-do tasks, approve or deny qualifications, disqualify a supplier, or invite a supplier for a new qualification.
- **Questionnaires**
  This link in the **Supplier 360° View** is only available in solutions where the modular questionnaire or certificate management feature has been enabled. On clicking this link, the **Questionnaires** page appears (Figure 3.11). Category or supplier managers can review and approve modular questionnaires submitted by suppliers, which can be certificate-related, and complete any related to-do tasks.

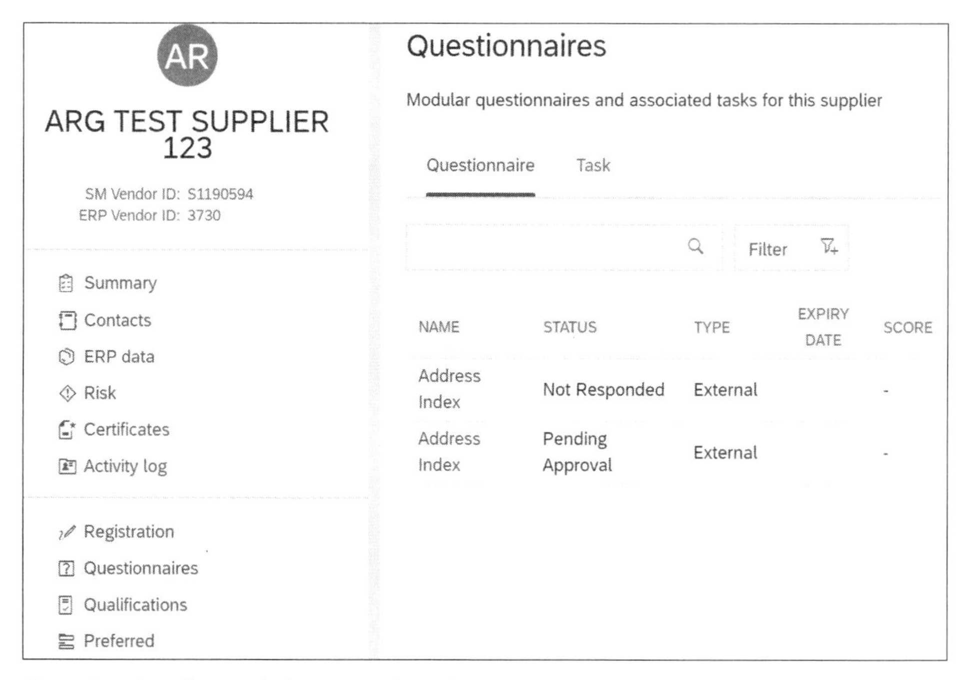

Figure 3.11 Supplier 360° View: Questionnaires

- **Risk**
  This link is only available in solutions that include SAP Ariba Supplier Risk. From this tile, users in the supplier risk user group can see detailed risk information about a supplier. For more information about using this tile, Chapter 4.
- **ACTIVITY LOG**
  This new link and page are part of a new feature added to the **Supplier 360° View** profile. This log shows all activities conducted in this project (supplier record), including automated activities. Search and filtering for a specific type of activity is allowed.

3.2.4 Supplier Information Process Flow

As shown in Figure 3.12, managing a supplier in SAP Ariba Supplier Lifecycle and Performance starts with the creation of a supplier request, completing the supplier registration process, qualifying the supplier using the supplier qualification process (optional), and, finally, setting up a supplier as a preferred supplier (optional).

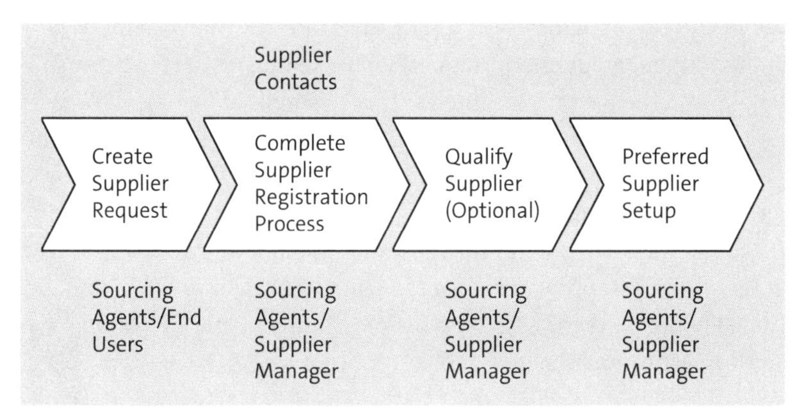

Figure 3.12 Supplier Lifecycle Process

After a supplier is created in SAP Ariba Supplier Lifecycle and Performance on the cloud, the supplier contact can maintain the supplier's own information by logging in to the SAP Ariba Supplier Lifecycle and Performance supplier portal on Ariba Network.

Information during the registration and qualification process is collected through a series of questionnaires. When a supplier contact submits the initial registration, qualification questionnaires, or updates to the questionnaires, the approval workflows can be set up so that any supplier information modified by the supplier is first reviewed and approved by the buyer organization before committing changes to the supplier record in SAP Ariba Supplier Lifecycle and Performance.

In the following section, we'll look at these key processes in SAP Ariba Supplier Lifecycle and Performance.

3.2.5 Supplier Request Process

The supplier request process allows internal users to request new suppliers. When a supplier request is approved, the supplier is created in your site.

Users in your company, such as procurement agents, functional buyers, sourcing or category managers, and other users looking to onboard a new supplier, can request a new supplier by filling out and submitting a supplier request form.

The internal supplier request template determines the following:

- The information buyers need to provide when submitting the request (e.g., the supplier's W9 to ensure duplicate suppliers aren't created)

- Whether the request is automatically approved on submission or must be manually approved
- The approvers who will be required to approve the request, if the request isn't set for automatic approval

Users can create supplier requests from the **Create** menu on the dashboard and monitor their progress in the **My Activities** area under the **SUPPLIER MANAGEMENT** tab. When a user submits a supplier request, the request is checked against the existing suppliers in the database, and the user is presented with any potential matches. The user then has the option of canceling the request (to avoid duplicate suppliers) or continuing with the request. Notifications let approvers know that a request requires approval. Approvers can also see outstanding requests using the **Manage My Tasks** area on the dashboard. Notifications also let the requester know when a request has been approved or denied and displays the final approver so that the requester can have further discussions about the denial. Approvers can also see a list of potential duplicate suppliers so that they can deny duplicate requests.

When a supplier request is approved, either automatically or manually, the supplier is created in your site's supplier database. Supplier request information is displayed in the **Overview** tab of the **Supplier 360° View** page, in the **Origin of Supplier** area.

Depending on your site's configuration, a supplier might be automatically invited to register at this point, or a category or supplier manager can manually issue a registration invitation. If your site is integrated with SAP ERP or SAP S/4HANA, your site can be configured to integrate the new supplier after the supplier request is approved, either automatically or through manual action, by a user with the right permissions.

Creating a Supplier Request

When you want to work with a supplier not already in your company's vendor database, you'll create a supplier request. Creating a supplier request starts your company's process for registering and managing the supplier.

Internal users must belong to one of the following system groups to create a supplier request:

- Supplier request manager
- Supplier management operations administrator

When you submit a new supplier request, SAP Ariba searches the database to see whether any existing suppliers have the same information. Standard questionnaires in the supplier request form, such as supplier name, address, and DUNS number, in addition to custom questionnaires configured to be included in the duplicate check, are all factors used to determine whether the request might create a duplicate supplier. The duplicate check also provides a match score. SAP Ariba uses fuzzy or approximate matching for supplier names and addresses, so these matches don't have to be exact. If SAP Ariba identifies existing suppliers that might match your request, you can either cancel the request or ignore and submit the request.

Note

SAP Ariba allows you to include custom questionnaires to be included as part of the duplicate check by setting new questionnaire attribute **Enable Duplicate Check**. This attribute is available when the **Custom Field-Based Deduplication** feature is enabled in your site by SAP Ariba support.

Now let's create a supplier request:

1. On the dashboard, click **Create • Supplier Request**, as shown in Figure 3.13.

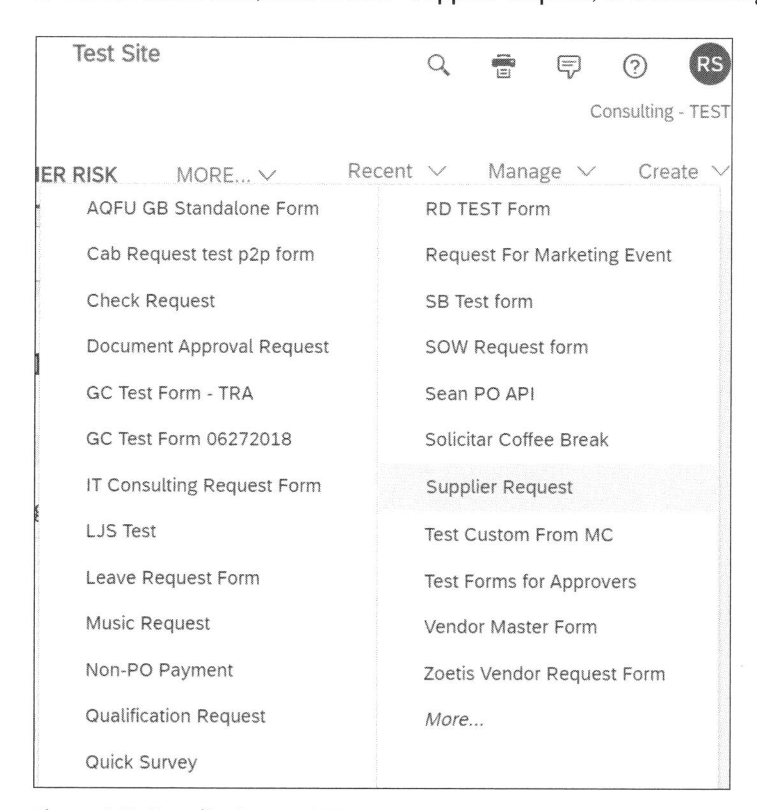

Figure 3.13 Supplier Request Menu

2. Enter information about the supplier that you want to add to the event in the supplier request form, shown in Figure 3.14.

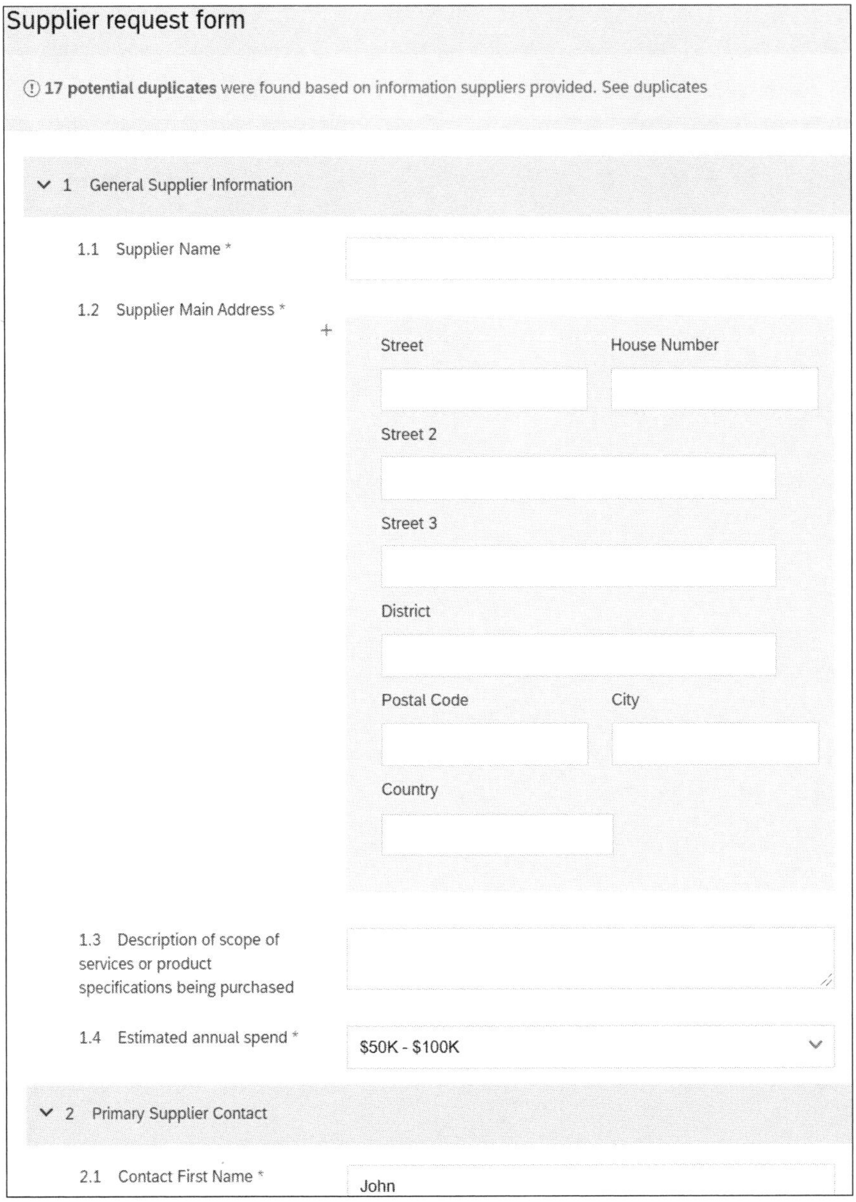

Supplier request form

ⓘ **17 potential duplicates** were found based on information suppliers provided. See duplicates

1 General Supplier Information

1.1 Supplier Name *

1.2 Supplier Main Address *

Street
House Number
Street 2
Street 3
District
Postal Code
City
Country

1.3 Description of scope of services or product specifications being purchased

1.4 Estimated annual spend *
$50K - $100K

2 Primary Supplier Contact

2.1 Contact First Name *
John

Figure 3.14 Supplier Request Form

3. Click **Submit**.
4. If SAP Ariba identifies any existing suppliers that match the information in your request, perform one of the following actions (see Figure 3.15):
   - If you agree with the suggested match, click **Cancel request**.
   - If you don't agree with the suggested match, click **Ignore and submit request**.

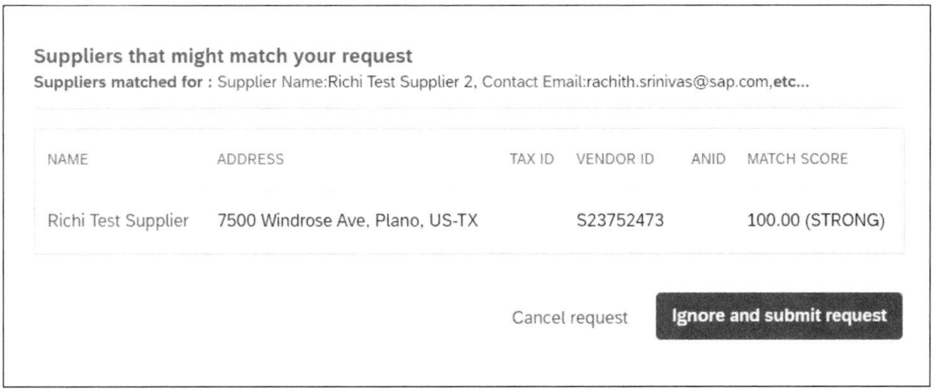

Suppliers that might match your request

Suppliers matched for : Supplier Name:Richi Test Supplier 2, Contact Email:rachith.srinivas@sap.com,**etc...**

NAME	ADDRESS	TAX ID	VENDOR ID	ANID	MATCH SCORE
Richi Test Supplier	7500 Windrose Ave, Plano, US-TX		S23752473		100.00 (STRONG)

Cancel request | **Ignore and submit request**

Figure 3.15 Duplicate Check with Score

After you've submitted the supplier request form, the request is sent for approval. You'll receive a notification when the request is approved or denied. If your request is approved, the new supplier is added to the supplier database. If the request is denied, it isn't.

Tip

The supplier request form should be designed to capture all the information that will be required by approvers to validate the business need for creating a new supplier and to perform additional duplicate checks within the organization before approving the request.

Supplier Self-Registration Request Process

The supplier self-registration request process allows suppliers who want to do business with your company to introduce themselves to you. When a supplier self-registration request is approved, the supplier is created in your site. The supplier self-registration request is only available in customer sites with supplier self-registration enabled. As part of self-registration enablement, SAP Ariba customer support provides your company with a self-registration URL specific to your site. Your company might provide this URL to potential new suppliers through specific outreach programs by publishing the URL on your corporate website or through other means. Suppliers self-register by clicking the URL and filling out and submitting the self-registration request form.

Note

Note the misnomer—although this process is called "self-registration," the form submitted is a request, the first step in working with a supplier. The next step in the supplier lifecycle process, supplier registration (described in more detail in Section 3.2.6), is a separate process that occurs after the supplier request is complete.

The supplier self-registration request template determines the following:

- What information suppliers need to provide when submitting the request
- Which category or supplier managers approve the request and whether other stakeholders at your company must also approve the request

The rest of the process is similar to the supplier request process defined earlier in this section.

Managing Supplier Requests

You can monitor, and, in some cases, edit, the requests you created under the **SUPPLIER MANAGEMENT** tab. To see the supplier requests that you've created, click the **Supplier Request** tile in the **My Activities** area under the **SUPPLIER MANAGEMENT** tab. You'll only see this area if you've created at least one supplier request and if your site doesn't use automatic approval for supplier requests. If a request hasn't yet been approved by the first approver in the approval flow, you can edit it. Click **View** to open the request, then click **Edit** to open the request form, make your changes, and click **Save**. After the first approver in the approval flow has approved or denied the request, the request is in an active approval process, and you can no longer edit the request.

Each supplier request approver has three choices:

- **Approve**
  The approval process moves to the next approver. If you're the last approver, the supplier request is approved. Supplier record is created in the SAP Ariba Supplier Lifecycle and Performance database. At this time, the supplier can be invited to register. SAP Ariba Supplier Lifecycle and Performance can be configured so that as soon as the supplier request is fully approved (all approvers have approved), the invitation for supplier registration is sent out automatically.
- **Deny**
  If one of the approvers clicks **Deny**, the supplier request is denied, and the requester is sent a notification with the reason for the denial.
- **Request additional info**
  Any of the approvers during the supplier request approval process can ask for additional information before deciding to approve/deny the request.

Approving or Denying Supplier Requests

If you're in the approval flow for a supplier request as shown in Figure 3.16, you can approve or deny it. Approving a supplier request creates the supplier in your site. Approvers can edit supplier requests to add missing information, correct mistakes, or supply information that was designed to be added by approvers before approving the request and creating the supplier.

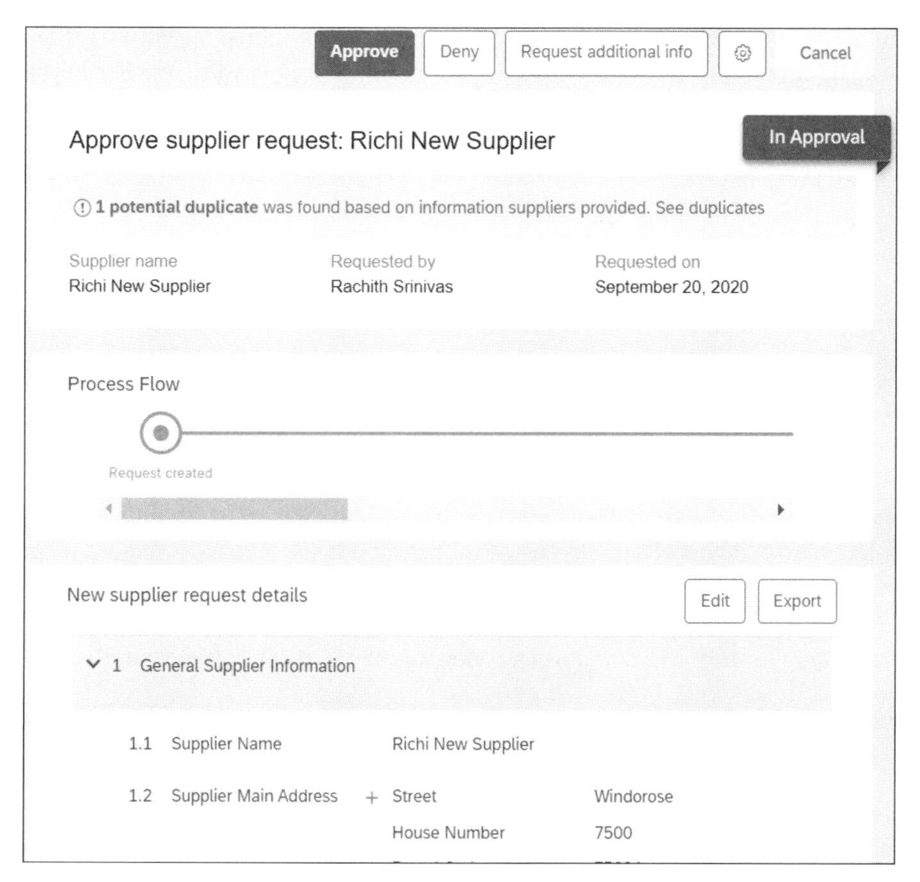

Figure 3.16 Supplier Request in Approval

To access an approval request and approve/deny the request, follow these steps:

1. Perform one of the following actions:
   - Click the link in the notification email to open the request.
   - On the dashboard, click **Manage • My Tasks**. Locate the supplier request you want to work with, click the associated approval task, and choose **Approve** or **Deny**.
2. The supplier request approval page opens. Review the information in the supplier request.
3. (Optional) Click **Edit** to edit the supplier request, and then click **Save**.
4. Perform one of the following actions:
   - Click **Approve** to approve the request.
   - Click **Deny** to deny the request. Enter explanatory comments to tell the requester the reason for the denial, and click **Confirm Denial**.
   - To request additional information, click **Request additional info**, enter a comment for the requester explaining what information you want, and click **Confirm request additional info**.

If you approved the supplier request and your site isn't configured to require validation in an integrated SAP Master Data Governance, Supplier system, the supplier will be created in your site with the **Not Invited** status, and the supplier contact specified in the request will be added as the primary contact. The supplier now has a **Supplier 360° View**, and its **Overview** tile shows information from the supplier request. If you denied the request, the supplier won't be created. A notification to the requester includes your comments explaining your denial. If you requested additional information, the supplier isn't created. The requester must edit and resubmit the supplier request, at which point, the request reenters the approval flow.

If your site is configured to require validation of approved supplier requests in an integrated SAP Master Data Governance, Supplier system, the supplier request shows a new automatic step, **MDGS Validation**, and the supplier is in **In External Approval** status. If SAP Master Data Governance, Supplier matches the request to an existing supplier, the request is automatically denied, and you'll be redirected to the existing **Supplier 360° View** profile. If SAP Master Data Governance, Supplier doesn't match the request to an existing supplier, the supplier is created in your site with the **Not Invited** status, and the supplier contact specified in the request is added as the primary contact. The supplier now has a **Supplier 360° View**, and its **Overview** tile shows information from the supplier request.

Resubmitting a Supplier Request

When additional information is requested, SAP Ariba Supplier Lifecycle and Performance sends a notification. The requester will follow this procedure to provide additional information:

1. Perform one of the following actions:
   - Click the link in the notification email to reopen the request.
   - In the **My Activities** area under the **SUPPLIER MANAGEMENT** tab, click the **Supplier Requests** tile, locate the request you need to resubmit, and click **View**.
2. Click **Edit**.
3. Make the requested changes to the supplier request.
4. Click **Save**.

3.2.6 Supplier Registration

After a new supplier is created, the supplier contact is invited to register. The registration process typically solicits detailed profile information from the supplier, for example, what commodities they supply, the regions in which they operate, whether they have certifications in broad categories such as diversity, what standards they adhere to or agree with, and so forth. Depending on how you've set up the process on your site, the process can include one or more external (supplier-facing) registration questionnaires. The supplier registration process can also include internal questionnaires that allow internal stakeholders to answer questions about your company's processes and add internally important information, such as cost centers, purchasing organizations, and so on, to the supplier's profile. Depending on the site's configuration and the particular supplier, registration invitations can be sent in the following ways:

- Automatically when the supplier request is approved
- Manually by a category or supplier manager
- Automatically whenever an uninvited supplier is invited to a sourcing event and your site requires that the supplier has registration status of at least **Invited**, if not higher, to participate in the event
- Through a mass invitation wave, which a customer administrator can create to onboard a specific set of suppliers at one time

The supplier registration process can include an initial registration phase and an update phase. If the process includes an update phase, suppliers will complete the initial registration and then can update their answers on an ongoing basis. Approvals can be set up for both the initial registration and for the subsequent updates submitted by the supplier contact. Internal users such as supplier managers or category managers can approve or deny the registration and updates.

The supplier registration template determines several factors, including the following:

- The number of external (supplier-facing) registration questionnaires and internal questionnaires (if any), and their content
- The approvers for the initial registration and updates
- Additional tasks that are part of the registration process, such as a to-do task to complete an internal questionnaire
- Whether you allow updates to the registration questionnaire after the initial registration, and, if so, which category or supplier managers and other stakeholders approve the updates
- Whether denial of an internal questionnaire update affects the supplier's registration status

Users with the appropriate permissions can invite suppliers to register by sending one or more questionnaires. These users can also approve or deny the supplier's answers or request more information from a supplier. They can also fill out internal questionnaires, complete other to-do tasks, and monitor the approval flow and status on the **Registration** tile on the **Supplier 360° View** profile under the **SUPPLIER MANAGEMENT** tab of the dashboard.

Suppliers must register with Ariba Network to access the registration questionnaires. The supplier's profile is automatically shared between Ariba Network and SAP Ariba Supplier Management solutions. Notifications are configured in SAP Ariba Supplier Lifecycle and Performance to perform the following activities:

- Let the supplier know that it has been invited to register. The email includes a link to create an account on Ariba Network and access the registration questionnaires.
- Let the user who created the original supplier request know that registration has started.
- Let approvers know that they need to approve a new or updated supplier registration.
- Let internal stakeholders know if they need to complete other tasks.
- Let the supplier know that the registration has been approved, denied, or requires more information. If a registration invitation includes multiple questionnaires, the supplier receives a separate notification for each questionnaire.

Supplier Registration Update Process

The supplier registration process can be set up so that it supports two different approval flows: one for the initial registration and one for registration updates. Suppliers therefore have a mechanism for notifying you of changing circumstances. In addition, a new timing rule in the registration template allows you to set a time interval to request an update from the supplier. Plus, a new notification email reminds suppliers to submit the updates.

Category managers and other stakeholders can approve or deny the updates. If your registration process supports updates, after all the tasks in the initial registration are completed, and the supplier is in **Registered** status, the previously closed questionnaire automatically reopens for 365 days. The supplier can navigate to the questionnaire on Ariba Network and submit updates at any time.

When a supplier submits an update, the registration status of the supplier is unchanged (**Registered**). A separate registration update status will be set to **Pending Approval**. The registration update status isn't displayed in the **Supplier 360° View**, but the questionnaire's detail page displays the **Last Updated** field, which shows the date when the last update was approved, as shown in Figure 3.17.

Figure 3.17 Supplier Registration Update

If the registration update isn't set to auto-approve, internal users assigned as approvers must approve or deny the registration update. If approved, these updates are added to the supplier profile. If the updates are denied, the supplier profile isn't updated with the proposed changes. In either case, the supplier remains in the **Registered** registration status but will have an **Approved** or **Denied** registration update status.

Note

SAP Ariba provides a Supplier Data application programming interface (API) that can be used to update the supplier external questionnaires (**Registration**, **Qualification**, and **Modular**) and internal questionnaires.

Synchronization with the SAP Backend

If your site is integrated with SAP ERP or SAP S/4HANA, your site's integration settings determine whether a new supplier is automatically or manually synchronized with your ERP vendor master. Options include after the request is approved, on inviting the supplier for registering, and after the registration is approved.

In the initial registration workflow, as shown in Figure 3.18, integration is set to synchronize the supplier automatically with the ERP vendor master after registration is approved.

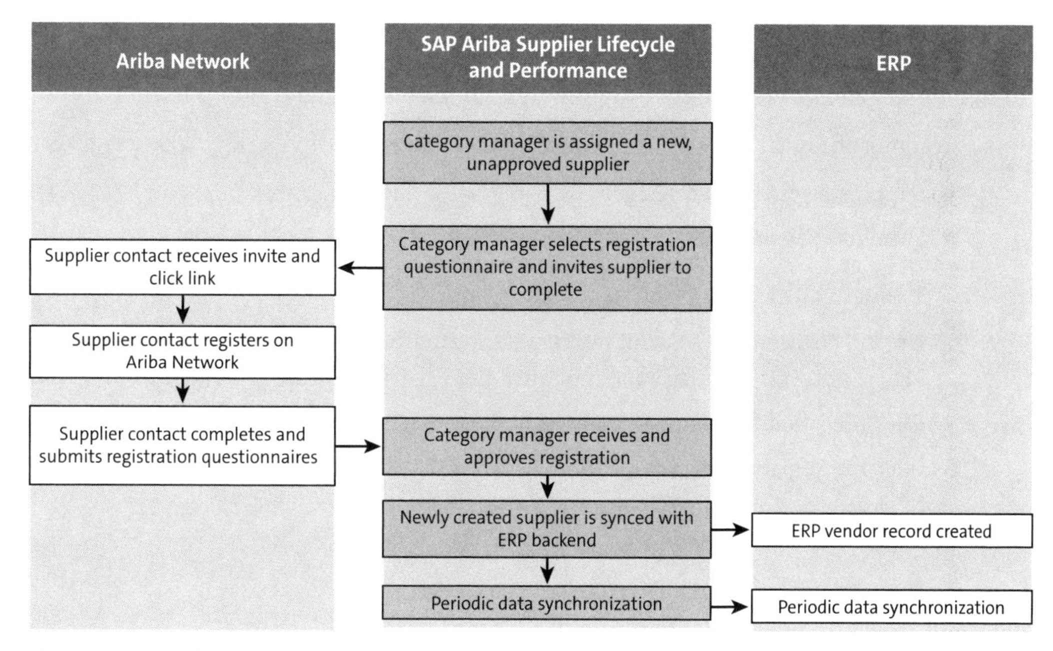

Figure 3.18 Supplier Synchronization with ERP

Registration Page

The **Registration** page in the **Supplier 360° View** profile is where you'll invite the suppliers to register as well as manage their registration and registration updates. If your site uses an internal registration questionnaire, users with the appropriate permissions can complete or update the internal questionnaires from this page.

You can perform the following tasks on the **Registration** page:

- Invite or re-invite a new supplier to register.
- Track the progress of the registration process on the status graph. On the graph, the approval nodes show the approver (either a user or a user group). Hover over approval tasks in the graph to see a list of possible approvers. If you hover over other tasks in the graph, such as to-do tasks, you can see their status as well.
- View the supplier's answers to the registration questionnaire in the **Questionnaires** table.
- See all the tasks associated with the registration in the **Tasks** table, and complete them if you're a task owner. After the registration is finally approved, you can click **View** for a task to see its status graph.
- Approve or deny the registration, if you're in the approval flow.

Inviting Suppliers to Register

Inviting a supplier to register involves sending the registration questionnaire to a supplier contact. The questionnaire is designed to gather basic profile information about the supplier for assessing the supplier's capabilities and fit.

You must belong to one of the following system groups to invite suppliers to register:

- Supplier registration manager
- Supplier management operations administrator

To invite suppliers to register, follow these steps:

1. Under the **SUPPLIER MANAGEMENT** tab, locate the supplier you want to invite, for example, by filtering for suppliers with the registration status **Not Invited**.
2. Click the supplier's name to open its **Supplier 360° View** profile.
3. Click the **Registration** link.
4. Click **Invite,** as shown earlier in Figure 3.9.

5. Check the questionnaire or questionnaires you want to send to the supplier, as shown in Figure 3.19.
6. Choose the supplier contact who will receive the questionnaire. You can add a new supplier contact at this point if necessary.
7. Click **Send**. One or more notification emails, one per registration questionnaire, will be sent out to the supplier contact you selected.

The supplier's registration status is now set to **Invited**, as shown in Figure 3.20.

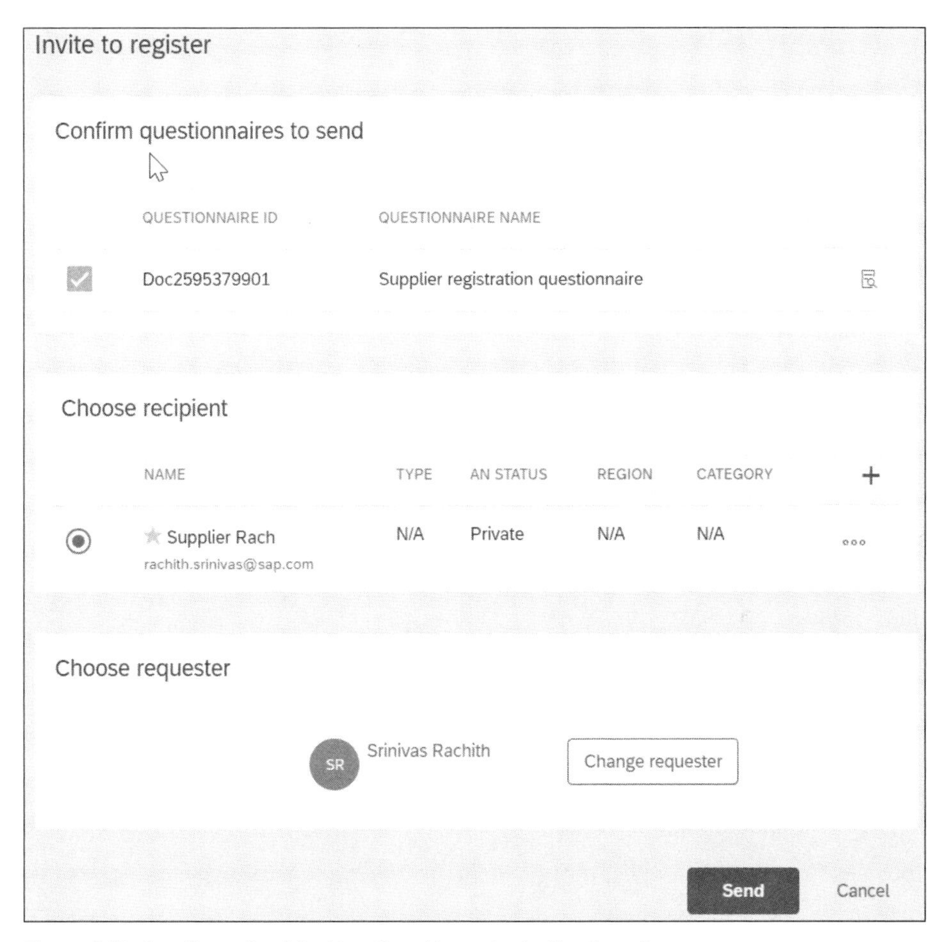

Figure 3.19 Sending a Registration Questionnaire to the Supplier

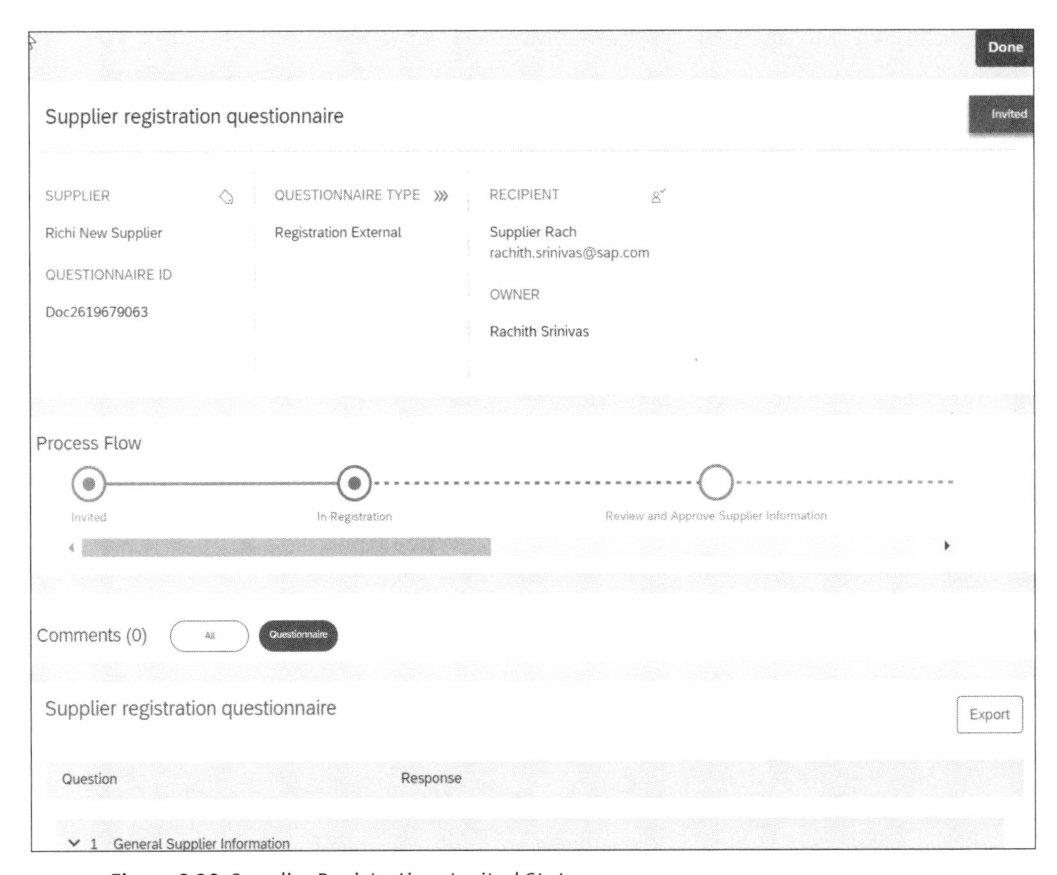

Figure 3.20 Supplier Registration: Invited Status

For each questionnaire, an invitation email is sent to the supplier contact you selected. The supplier contact can use the link in the email to create an Ariba Network account and to submit the filled registration questionnaires. In the **Supplier 360° View** profile, the **Questionnaires** table on the **Registration** page shows the questionnaires you just sent, along with the **Email Delivery Status** and **Email Delivery Status Date**. To view the questionnaire, choose **Action • View**.

If you need to re-invite a supplier to register, in the **Questionnaires** table, choose **Action • Reinvite**.

While inviting a supplier to register, if you're working with a supplier contact other than the primary contact created during the supplier request process, then you can use the **Add User** link in the **Invite to Register** screen to create a new supplier contact. On inviting a supplier to register, an email will be sent out to the email address provided for the supplier contact.

Note

All notification emails are customizable to include customer-centric messaging. SAP Ariba now allows you to set a template-level custom sender name for supplier registration invitations.

After the supplier contact receives the supplier registration request email, he can click on the link to register the supplier on Ariba Network. After creating a new account on Ariba Network, the supplier contact can access the supplier registration questionnaire and other questionnaires, such as supplier qualification questionnaires and event questionnaires (e.g., requests for proposals [RFPs], auctions, etc.) for which the supplier contact is invited.

The supplier contact can click on the supplier registration questionnaire to open the questionnaire and respond to the questions. The questionnaire can be saved and submitted at a later time. All required questions must be completed before submitting the response.

After the supplier submits a response to the supplier registration questionnaire, the supplier manager for this supplier receives an email that the supplier has completed and submitted the questionnaire. An email is also sent to the first approver in the chain of approvers to approve and accept the supplier registration response.

Approving, Denying, or Requesting Additional Information during Supplier Registration Approvals

Suppliers who have submitted registration questionnaires for the first time are in **Pending Approval** status. If your registration process is set up to allow updates, suppliers who have submitted updates to the registration questionnaire are in the **Registered** registration status and have a separate **Pending Approval** registration update status. Separate approvals exist for initial registrations and registration updates.

For registration updates, or if you've requested more information from the supplier, the approval task page shows answers in both the latest version and the previous version so that you can compare them. After the supplier has made more than one update to the questionnaire, the columns for the latest and previous versions include version numbers for the current and previous updates. The initial registration is always version 1.

Note

SAP Ariba creates a new version of the supplier registration questionnaire for each time the supplier updates the questionnaire and submits a response. Supplier managers can compare versions before approving changes submitted by the supplier.

For new registrations, if you're the final approver, the supplier is now in **Registered** or **Registration Denied** status and is notified of that status by email. If you're not the final approver, the supplier remains in "**Pending Approval**" status until the final approver takes action.

For registration updates, the supplier always remains in registration status **Registered** but will have its registration status updated to **Approved** or **Denied**. If you're the final approver and you approve the update, the supplier's most recent answers are added to its profile, and the registration questionnaire remains open so that the supplier contact can make future updates. If you're the final approver and deny the update, the supplier's most recent answers are discarded, and the registration questionnaire closes.

Note

Version histories of the questionnaires now have a label for supplier manager and approvers to clearly differentiate versions based on whether they were approved, denied, or held over for requests of additional information.

After the supplier is registered, the supplier qualification process can begin. In the next section, we'll discuss the supplier qualification and disqualification processes.

3.2.7 Supplier Qualification Process

After a supplier is registered, SAP Ariba Supplier Lifecycle and Performance provides a new feature that allows supplier managers or category managers to further qualify suppliers to deliver goods or services for a specific commodity, region, or department (or business unit) combination, if necessary.

Supplier administrators can configure multiple qualification questionnaires, one for every combination of a specific commodity, region, or department, to solicit information from the supplier to qualify that supplier. Based on the commodity, region, or department selected for the supplier, specific questionnaires are available for the project owner to send out to the supplier. The supplier can respond to these questionnaires. Qualification approvals set up by the administrator will route responses from the supplier to selected approvers. When a supplier either has been disqualified or had a qualification expire, they are eligible for requalification. Qualification and requalification use the same process; however, a requalification is always based on the same commodities, regions, and departments (business units) as the previous disqualification or expired qualification.

Similar to supplier registration, and all other event processes in SAP Ariba Strategic Sourcing Suite, the supplier qualification process is also based on a template that can be configured by an internal user with the template creator role.

The supplier qualification template provides the following:

- The questionnaire segments for various commodity, region, and department combinations that specifies which segments are required.

- Which category or supplier managers approve the qualification or requalification. In sites with buyer category assignments, approvers are automatically assigned based on the qualification or requalification commodities, regions, and departments.
- The content of the questionnaire segments.
- Whether other stakeholders at your company must also approve the qualification or requalification.
- Whether internal questionnaires or other tasks or documents are part of the qualification or requalification process.
- Whether the qualification expires after a specific date, and, if so, whether there is a waiting period before the supplier can be requalified.

The category or supplier managers assigned to the supplier or the owner of the supplier can start a new qualification or a requalification process by sending out the questionnaire for a specific commodity/region/department combination to the supplier contact. On receiving a response to the questionnaire, approvers in the approval chain for supplier qualification can review the supplier's answers, request more information, approve/deny the qualification or requalification, and monitor the approval flow and status in the **Qualification** page on the **Supplier 360° View** profile under the **SUPPLIER MANAGEMENT** tab of the dashboard.

The supplier answers the qualification questionnaire in its Ariba Network for Suppliers profile. The supplier's profile information is automatically shared between Ariba Network and SAP Ariba Supplier Management solutions.

Notifications in SAP Ariba Supplier Lifecycle and Performance for supplier qualification are configured to perform the following tasks:

- Let suppliers know that they've been invited to qualify for specific commodities, regions, and departments and provide links to the questionnaire.
- Let approvers know that they need to approve a supplier qualification or requalification.
- Let supplier managers or category managers know when one of their qualifications has expired or is eligible for requalification.
- Let the supplier know that the qualification or requalification has been approved, denied, or requires more information.

At any point after the supplier qualification or requalification has been approved, category or supplier managers can do the following:

- Start new qualifications for additional combinations of commodities, regions, and departments.
- Evaluate the supplier for a preferred status level (if your company uses preferred supplier levels).
- Disqualify the supplier for commodities, regions, and departments for which it was previously qualified (see Figure 3.21).

Figure 3.21 Supplier Qualification: Add New or Disqualify

Note

In solutions that include SAP Ariba Sourcing, event templates can be configured to only allow suppliers that are qualified for the event's specific commodity, region, and department to participate in the event.

Using the Supplier Qualification Page

The **Qualification** page in a **Supplier 360° View** profile is where you'll qualify, disqualify, or requalify the supplier. The **Qualification** tile only appears on the **Supplier 360° View** profiles of registered suppliers.

You can perform the following tasks in the **Qualification** tile:

- Start a qualification, disqualification, or requalification.
- See all the tasks associated with a qualification, disqualification, or requalification in the **Tasks** table and complete a task if you're the owner of the task. Click **View** for a task to see its status graph. On the graph, the approval nodes show the users or user groups in the approval process.
- Review the supplier's responses to qualification or requalification questionnaires or review internal users' responses to disqualification questionnaires in the **Questionnaires** table.
- See the expiration dates for qualifications.

- Approve or deny the qualification, disqualification, or requalification during the approval flow.
- Re-invite a supplier to fill out the qualification questionnaire.

Creating a New Supplier Qualification Process

Starting a new supplier qualification involves assembling a questionnaire based on specific commodities, regions, and departments and then sending this questionnaire to a supplier contact. The qualification process ensures that the supplier meets your company's standards for those commodities, regions, and departments. For more information on setting up a qualification questionnaire, refer to Section 3.2.17.

In the supplier **Qualification** page, click the **+** icon, shown earlier in Figure 3.21. In the screen that appears, select the commodity, region, and department for which you want to start a qualification process, and then select the qualification end date (Figure 3.22).

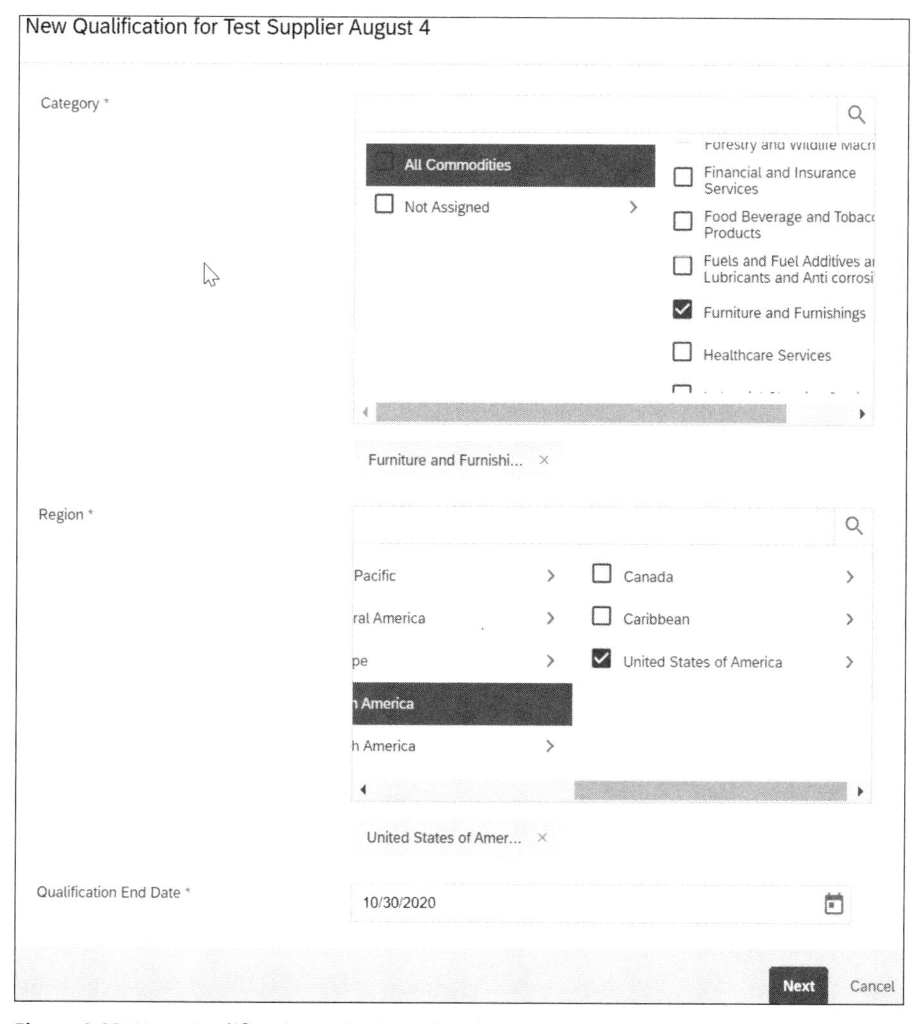

Figure 3.22 New Qualification Criteria and End Date

Note

Department-standard fields can be used to define any organizational unit for which you want to qualify suppliers.

Note

If your qualification of the supplier doesn't require all three dimensions, then set the value of the dimension that you want to ignore to **All**.

After you select a commodity (category), region, and department, and click **Next**, you're taken to the next screen where all configured qualification questionnaires for that combination are displayed. Questionnaires that aren't set as required by the supplier administrator can be deselected.

The following steps must be completed by the internal user to send qualification questionnaires out to suppliers:

1. Select the necessary qualification questionnaires, if not already selected, and deselect the optional questionnaires that don't need responses from your suppliers.
2. Select the supplier contact with whom you're working to send the notification out to that supplier contact. If the supplier contact you need isn't on the list of contacts, click **Add User** to add a new supplier contact.
3. Repeat these steps to send the qualification request to multiple supplier contacts.

The selected supplier contacts will receive an email notification for each qualification questionnaire selected. The supplier contact can log in to Ariba Network by clicking the link in the email and then respond to the questionnaire. All required questions must be completed before submitting the response.

Reviewing and Approving Supplier Qualification Responses

Category managers, supplier managers, or the supplier qualification request owner can review the status of each qualification request from the **Supplier Qualification** page, as shown earlier in Figure 3.21. By clicking **Action • View** for each qualification request that has been sent out to the supplier, you can view the supplier's response to the qualification questionnaire.

Approving, Denying, or Requesting Additional Information during Supplier Qualification Approval

Suppliers who have submitted qualification questionnaires are in the **Pending Qualification Approval** status.

Follow these steps to approve or deny a qualification or to request more information from the supplier:

1. Perform one of the following actions:
   - Click the link in the notification email to open the qualification.
   - Under the **SUPPLIER MANAGEMENT** tab, click the name of the supplier, and then click the **Qualification** page.
2. Review the supplier's answers by performing the following steps:
   - In the **Questionnaires** table, for each submitted questionnaire, click on the **View** icon.
   - After reviewing the supplier's answers, click **Done** to return to the **Qualification** page.
3. In the **Tasks** table, for the **Qualification** approval task, click **Approve/Deny**.
4. If you're approving an update, to display only the questions with updated answers, click **Updated**. To display all answers, click **All**.
5. (Optional) If you're approving an update, to compare the current answers with versions of the questionnaire that the supplier submitted before the previous version, click the version number in the **Previous Version** column, and then choose the version. The **Previous Version** column displays answers from the selected version. The **Latest Version** column continues to show answers from the current update.
6. In the top-right corner of the page, perform one of the following actions:
   - To approve the registration, click **Approve**.
   - To deny the registration, click **Deny**, enter an explanatory comment for the supplier, and click **Confirm denial**.
   - To request additional information, click **Request additional info**, enter a comment for the supplier explaining what information you want, and click **Confirm request additional info**.

Each qualification questionnaire's status is displayed in **Qualifications table.** If a questionnaire is approved by all approvers, the status changes to **Approved**. If one of the approvers rejects the questionnaire, the status is set to **Rejected**. If all qualification questionnaires have been approved, the **Qualification Status** on the **Summary** page in the **Supplier 360° View** is set to **Approved**.

Disqualification of a Supplier

Disqualification signals that a previously qualified supplier no longer meets your company's standards. Disqualifying a supplier involves filling out a form.

Note

You must be a member of the supplier qualification manager or supplier management operations administrator groups to disqualify a supplier. Based on how your SAP Ariba Supplier Lifecycle and Performance is configured, you may only be able to disqualify the suppliers to which you're assigned.

Supplier disqualification has an approval process that can be enabled. Upon approving a supplier disqualification, the supplier can be immediately disqualified, or SAP Ariba Supplier Lifecycle and Performance provides a phaseout period during which the supplier is set to **Restricted** status.

Supplier disqualifications are always based on an existing qualification. The qualification is for specific commodities, regions, and departments. A supplier can have multiple separate disqualification projects, each for a different combination of commodity, region, and department. You can disqualify a supplier for the same combination of commodity, region, and department *or* for subsets of the qualified commodities, regions, and department. For example, if a supplier has a qualification for IT in North America, you can disqualify the supplier for IT hardware in Texas. These partial disqualifications are nested under the original qualification on the **Qualifications** page. You can disqualify the supplier for additional commodities and regions from the original qualification. Because commodities, regions, and departments are hierarchical, with higher and lower levels, disqualifying a supplier for a commodity or region disqualifies them for the lower levels of the hierarchy tree. For example, disqualifying a supplier for the United States disqualifies that supplier for all 50 states.

Users can only start a manual disqualification for a supplier based on an existing qualification; however, data imports can disqualify suppliers that haven't previously been qualified. A supplier's disqualifications and associated questionnaires are displayed on the **Qualification** page. Only approvers can see unapproved questionnaires. After a questionnaire is approved, any user who has permission to view the **Qualification** page can see the answers.

To disqualify a supplier, follow these steps:

1. Under the **SUPPLIER MANAGEMENT** tab, locate the supplier you want to disqualify.
2. Click the supplier's name to open the **Supplier 360° View** profile.
3. Click the **Qualification** tile.
4. On the row of the qualification you want to fully or partially disqualify, choose **Action • Disqualify**, as shown in Figure 3.23.
5. Enter information in the disqualification form.
6. Click **Submit**.

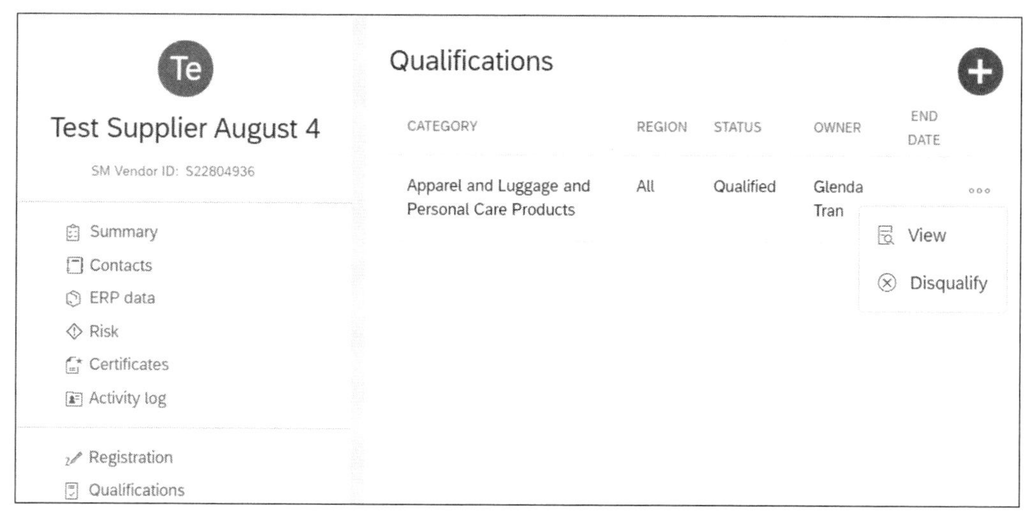

Figure 3.23 Disqualify Qualification

After the disqualification is finally approved, the supplier either has **Disqualified** status immediately or has **Restricted** status for a phaseout period before automatically moving to **Disqualified** status. When a supplier is in **Disqualified** status, that supplier can no longer participate in new events that use qualification as criteria, although disqualification doesn't remove it from any existing events. Disqualification also removes any related preferred status levels for the disqualified commodities and regions.

Requalifying a Supplier

Requalifying a disqualified supplier or an expired qualification involves relaunching the original qualification process. Supplier requalification is always based on the commodities, regions, and (optionally) departments of the original qualifications. To be requalified, the supplier must fill out and submit the qualification questionnaire again.

To requalify a supplier, follow these steps:

1. Under the **SUPPLIER MANAGEMENT** tab, locate the supplier that you want to requalify.
2. Click the supplier's name to open the **Supplier 360° View** profile.
3. Click the **Qualification** tile.
4. On the row with the disqualification or expired qualification, choose **Action • Requalify**.
5. Check the questionnaires you want to send to the supplier. The list of available questionnaires is based on the commodities, regions, and departments in the original qualification. You can't uncheck required questionnaires.
6. Choose the supplier contact who will receive the questionnaire. If you want to send the questionnaire to a new contact, you can add a new user.
7. Click **Send Qualification**.

If the requalification is based on a disqualification, the supplier remains in the **Disqualified** status until the requalification has been finally approved. At that point, the status moves to **Qualified**. If the requalification is based on an expired qualification, the supplier is now in the **Qualification Started** status. A notification email is sent to the supplier contact selected. The supplier contact can use the link in the email to log in to Ariba Network and then complete and submit the qualification questionnaire.

3.2.8 Preferred Supplier Management Process

After a supplier is qualified for a specific commodity, region, and (optionally) department combination, a category or supplier manager can designate the supplier as a preferred supplier for that combination.

Note

Preferred supplier status is usually based on a combination of commodity and region. If a business unit user matrix is enabled, then the preferred supplier status is based on a combination of commodity, region, and department (business unit).

The preferred supplier management template determines the following:

- What information users add to the form
- Whether the preferred status request requires approval and, if so, which category or supplier managers must approve it
- Whether other stakeholders at your company must also approve the status change

Up to five levels of preferred supplier statuses are allowed.

Internal users with the appropriate permissions can request a category status change for a specific combination of commodities, regions, and departments in the **Preferred** page in the **Supplier 360° View** profiles. Suppliers can also gain preferred levels via data import. This feature will be covered in Section 3.2.23. The preferred data import doesn't create a corresponding preferred supplier management project. However, a user with the appropriate permissions can always request a change for an existing preferred category status. In this case, the action of editing a preferred category status creates a preferred supplier management project based on the supplier's previous data import defined preferred levels, and the user who requests the change is the explicit project owner.

Qualifying a supplier for a combination of commodity, region, and department is a prerequisite for a user to request a preferred category level for that same combination of commodity, region, and department. However, data imports can also designate a supplier without a previous qualification as preferred.

As with qualifications, a supplier can have multiple separate preferred supplier management projects, each for a different combination of commodity, region, and department.

Users can always request changes to existing preferred category levels to upgrade, downgrade, or remove a supplier's preferred status. A preferred supplier management project closes after its final task is completed. When a user requests a preferred category status change on a supplier that was previously designated as preferred using a project (rather than via data import), the original preferred supplier management project reopens, and all its tasks start again.

Note

Disqualifying a supplier for a commodity, region, and department combination automatically removes any corresponding preferred status without reopening any associated preferred supplier management project.

If the process includes an approval flow, requesting a new preferred supplier status level or change to an existing preferred supplier status level triggers the workflow process.

Notifications in SAP Ariba Supplier Lifecycle and Performance for preferred supplier management are configured to perform the following tasks:

- Notify approver of a preferred status change request approval.
- Notify requesting user when preferred status change request is approved or denied.

To request preferred status for a supplier, follow these steps:

1. Under the **SUPPLIER MANAGEMENT** tab, locate the supplier for which you want to request preferred category status.
2. Click the supplier's name to open the **Supplier 360° View**.
3. Click the **Overview** tile.
4. In the **Category Status** area, perform one of the following actions:
   - To create a new request for preferred category status (for a commodity, region, and department), click **+ Add**.
   - To request a change in preferred category status (for a commodity, region, and department for which the supplier already has a preferred status), locate the status, and click **Action • Edit**.
5. Enter information in the form.
6. Click **Submit**.

Approving or Denying a Preferred Supplier Category Status Request

If your preferred supplier management process requires an approval flow, the request can be approved or rejected. To approve or deny a request for preferred status for a supplier, follow these steps:

1. From the **My Tasks** area, click the task name, and choose **Action • View Task Details**. Alternatively, you can select **Action • Approve** or **Action • Deny** to approve or deny the task without viewing the task details and document. An approval task page will open. The left pane contains a link to the preferred supplier category status request. The right pane contains buttons to complete the task.
2. Click **Deny** or **Approve**.
3. Perform one of the following actions:
   - If you clicked **Deny** in the previous step, enter a message to the user who created the request.
   - If you clicked **Approve** in the previous step, enter a message and click **OK**.

Removing a Preferred Supplier Category Status

If a supplier is no longer preferred for a specific commodity, region, and department combination, you can remove that preferred status by deleting it.

> **Note**
>
> You must be a member of the preferred supplier manager or supplier management operations administrator group to delete a preferred supplier category status.

You can't delete a preferred supplier category status that is still in approval. You can only delete a preferred supplier category status that has already been approved.

Deleting a preferred supplier category status doesn't require approval. If a supplier is disqualified for a commodity/region/department combination, any preferred status for that commodity/region/department combination is automatically removed.

To delete a preferred supplier category status, follow these steps:

1. Under the **SUPPLIER MANAGEMENT** tab, locate the supplier for which you want to request preferred supplier category status.
2. Click the supplier's name to open the **Supplier 360° View** .
3. Click the **Preferred** link.
4. Locate the preferred status you want to delete and click on **••• Delete** (as in Figure 3.24).
5. To confirm that you want to delete the preferred category status, click **Delete**.

After you delete a preferred supplier category status, the supplier is no longer eligible for sourcing events that use preferred statuses as criteria. You can request a new preferred supplier category status for the same commodity/region/department combination at any time.

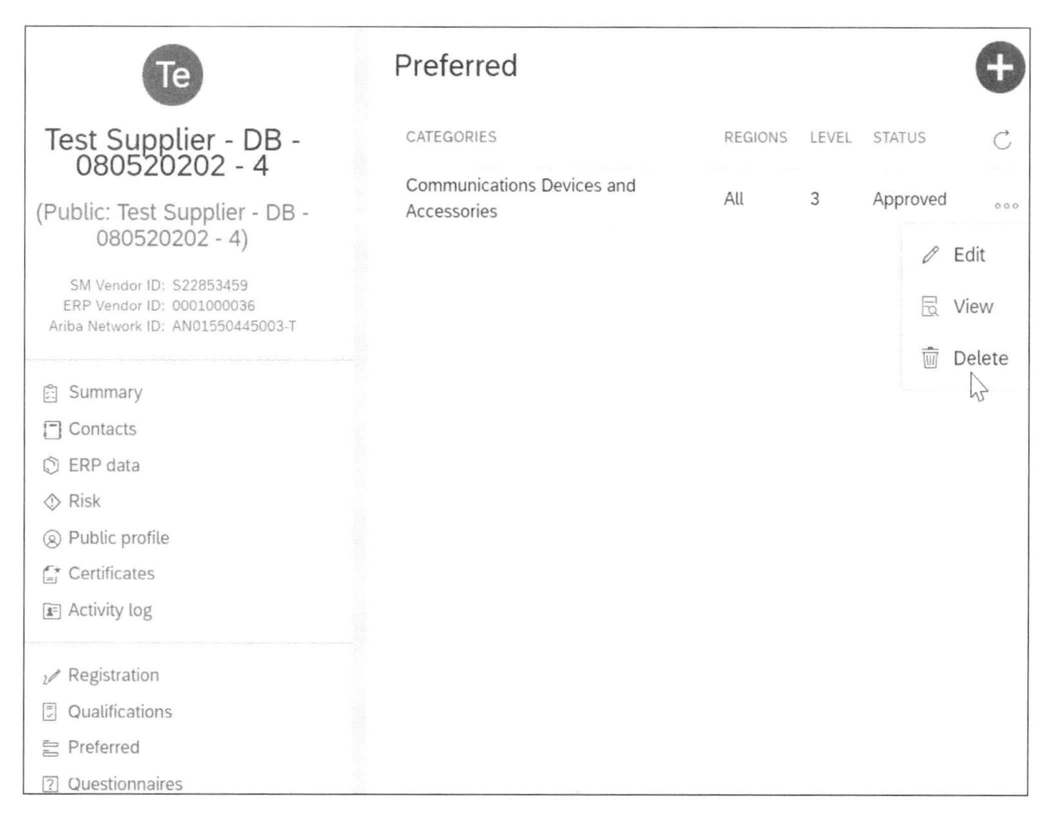

Figure 3.24 Deleting Preferred Status

3.2.9 Modular Questionnaire Process

The *modular questionnaire* is a new feature in SAP Ariba Supplier Lifecycle and Performance that provides a flexible framework for buyers to create standalone questionnaires, which can be applied for multiple commodities, regions, and departments, and recur with (optional) expiration schedules. These questionnaires can have their own approval flow.

This feature allows buyers to collect information once for multiple combinations without having suppliers repeat their responses, thus improving the supplier experience. Unlike registration projects, which collect basic profile information in one project, modular supplier management questionnaire projects are designed to collect specific, limited sets of information, such as a certificate or set of related certificates. A supplier can have any number of modular questionnaire projects, which exist as standalone projects and aren't currently linked to any other supplier management process.

The modular questionnaire includes the following capabilities:

- A modular questionnaire can have its own status and recurs. For example, the questionnaire requesting updated information from the supplier can be sent out to the supplier every two years or every five years.

- Expiration dates and automatic notifications are sent to the buyer and the supplier when a questionnaire is about to expire.
- Questionnaires can be set as open or closed. Open questionnaires are always open for suppliers to update.
- Modular supplier management questionnaire projects support the use of two phases, similar to the supplier registration process, to apply separate approvals and to-do tasks. One phase is for **First Time** submission of the modular questionnaire by the supplier, and the second is for **Updates** to the modular questionnaire by the supplier. (For more information, Section 3.2.20.)

Statuses supported in the modular questionnaire are **Not started**, **Pending Submission**, **Pending Approval**, **Pending Resubmit**, **Approved**, **Denied**, **Expiring**, and **Expired**.

After the modular questionnaire template is created, users with the supplier management modular questionnaire manager role can create modular questionnaires to send to a supplier. To send a modular questionnaire to a supplier, follow these steps:

1. Log in as a user with the supplier management modular questionnaire manager role.
2. Click on **Manage • SM Modular Questionnaires**, as shown in Figure 3.25.

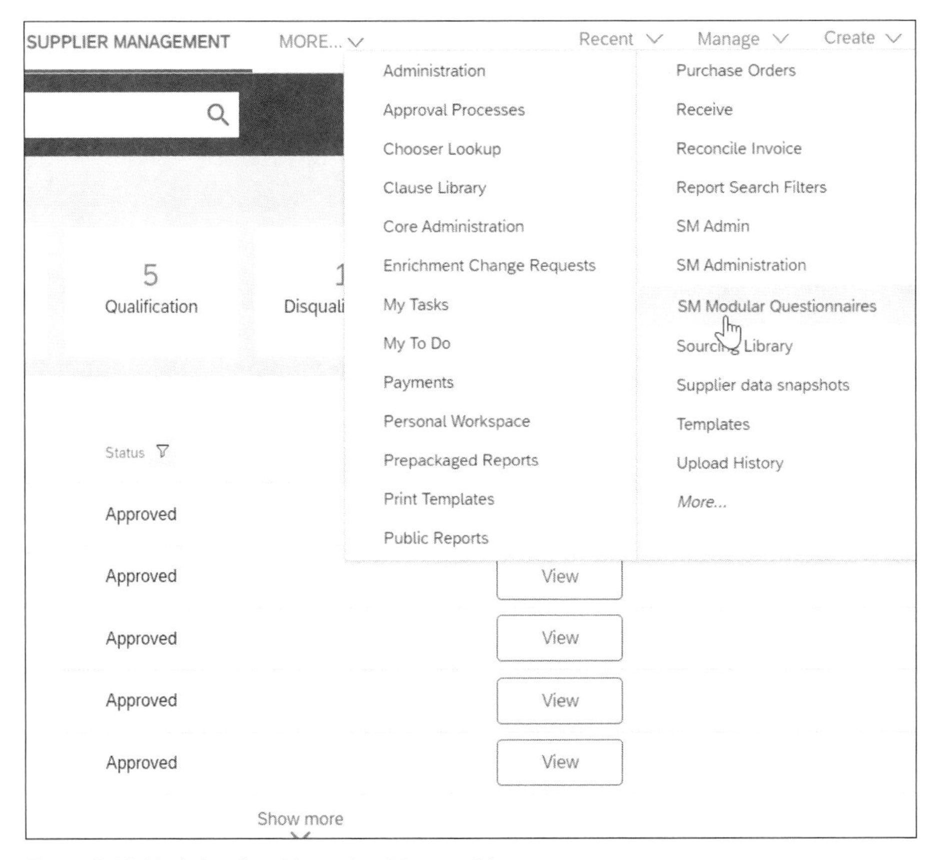

Figure 3.25 Modular Questionnaire: Manage Menu

3. Select from the list of active modular questionnaire templates.
4. After one or more templates are selected, click on **Select Suppliers to Send**.
5. Select one or more suppliers to send the modular questionnaires selected.
6. Click **Continue**. SAP Ariba Supplier Lifecycle and Performance will check to see if any selected suppliers lack contacts.
7. Click **Send** to send the modular questionnaires out to the suppliers selected, as shown in Figure 3.26.

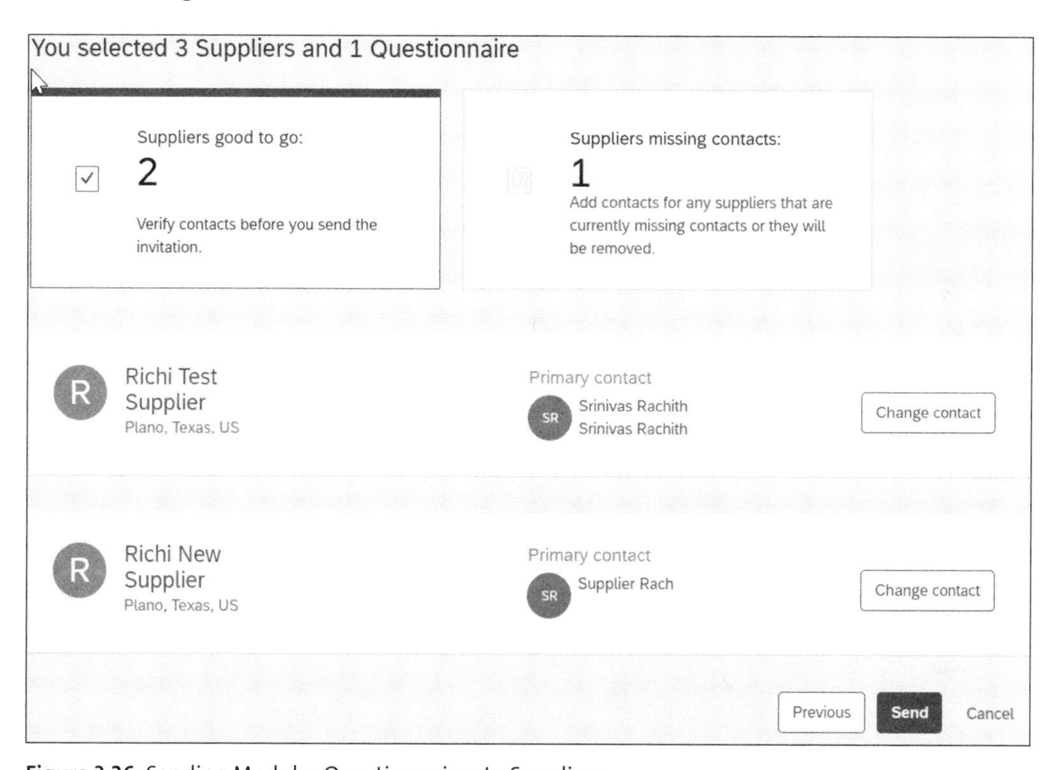

Figure 3.26 Sending Modular Questionnaires to Suppliers

The supplier contact receives an email requesting a response to the modular questionnaire. The contact can click the link in the email to log into Ariba Network to complete and submit the modular questionnaire.

In the new **Supplier 360° View**, a new link—**Questionnaires**—allows the supplier manager to view all modular questionnaires sent out to the supplier and the status of each questionnaire, as shown in Figure 3.27.

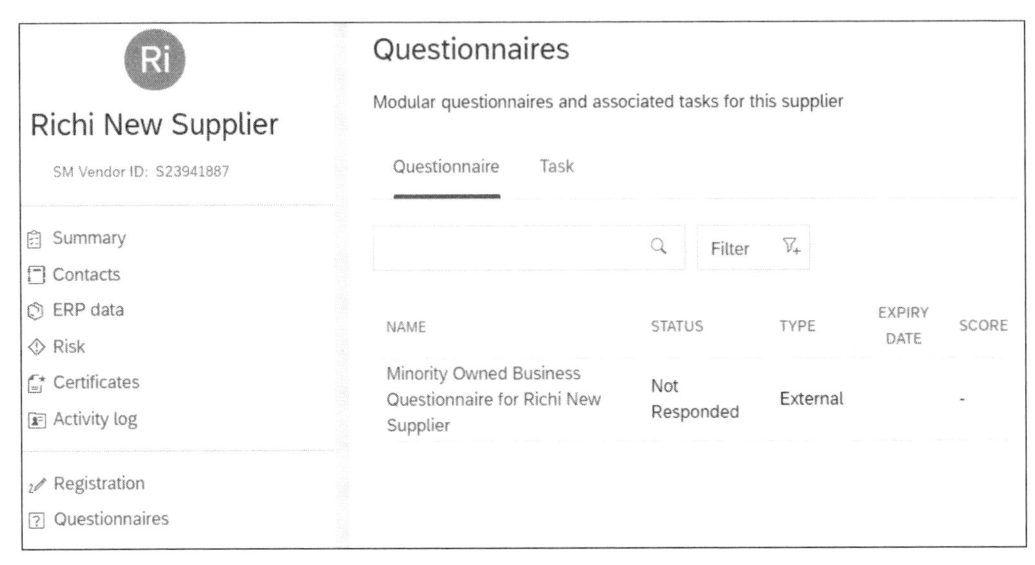

Figure 3.27 Supplier 360° View: Questionnaires

If the supplier hasn't responded, the supplier registration manager can change the supplier contact (as in Figure 3.28).

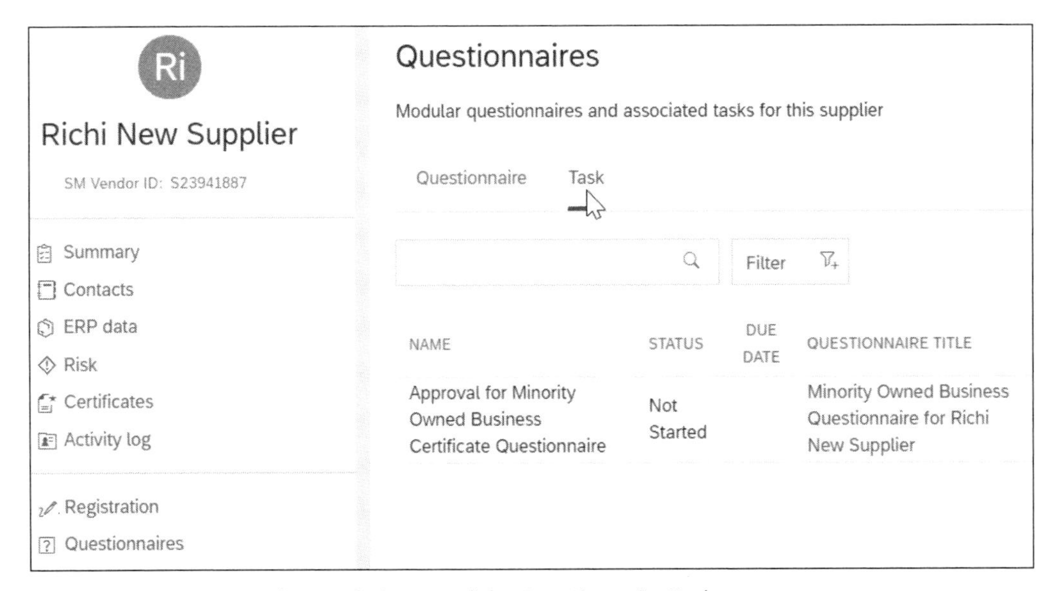

Figure 3.28 Supplier 360° View: Modular Questionnaire Tasks

3.2.10 Scoring in Modular Questionnaires

A new feature allows modular questionnaires to be set up with scoring. Based on the supplier's response, a score is calculated for each question and section. An overall score for the entire questionnaire is calculated. When you open the questionnaire details page, the overall score is displayed at the top of the page. Individual scores for each question and section are also displayed.

Note

Quantifiable question types, such as multiple-choice, yes/no questions, and numeric (e.g., numbers, dates, or %), answers can be scored; however, free text answers can't be scored.

3.2.11 Approving, Denying, or Requesting Additional Information during Modular Questionnaire Approvals

After a supplier submits a response to a modular questionnaire, the approver in the buyer organization receives an email to approve the responses submitted by the supplier. Approvers can access all modular questionnaire tasks related to a supplier from the **Supplier 360° View** profile in the **Questionnaire** page. Approval tasks will have an **Approve/Deny** button next to the task to take the approver to the task. The approver can review the responses submitted by the supplier and perform one of the following actions in the top-right corner of the page:

- To approve the registration, click **Approve**.
- To deny the registration, click **Deny**, enter an explanatory comment for the supplier, and click **Confirm denial**.
- To request additional information, click **Request additional info**, enter a comment for the supplier explaining what information you want, and click **Confirm request additional info**.

After a modular supplier questionnaire project is finally approved, if either the same or a different internal user sends the same modular questionnaire to the same supplier again using **Manage SM Modular Questionnaires**, the existing project reopens, and its tasks will restart. The explicit project owner remains the original user who sent the questionnaire.

3.2.12 Configuring Supplier Certification Management Using Modular Questionnaires

Supplier certificates can be managed using dedicated modular questionnaires with their own approval flow and expiration dates. Modular questionnaires can have one or more certificate type questions and other questions of different types in the same questionnaire. Certificates collected in these projects display on the **Certificate** page of the **Supplier 360° View** profile.

You can create different modular supplier management questionnaire project templates to collect different certificates or sets of certificates for different commodity, region, and department combinations. Each questionnaire has its own template, and each template also has its own approval flow, which you can leverage to route specific certificates to the relevant approvers.

The supplier's primary supplier manager or owner can be set up to receive notifications of upcoming or elapsed expirations of the certificate or the modular questionnaire in which it's housed. Modular questionnaires can be set up such that the status changes to **Expiring** or **Expired** when the corresponding certificate in the questionnaire changes to **Expiring** or **Expired** status.

Now that you know how SAP Ariba Supplier Lifecycle and Performance can be used for managing supplier information, in the next section, we'll take a look at how to plan for a successful SAP Ariba Supplier Lifecycle and Performance implementation.

Planning the Implementation

Implementation planning for SAP Ariba Supplier Lifecycle and Performance is similar to implementation planning for SAP Ariba Contracts, discussed in Chapter 6.

3.2.13 Configuring SAP Ariba Supplier Lifecycle and Performance

Within SAP Ariba Supplier Lifecycle and Performance and other SAP Ariba applications, objects such as supplier requests, registrations, qualifications, sourcing events, contract workspaces, and so on are called *projects*. To create a project, you first need a project template to be created.

Users with the customer administrator role can access the **Administration** section (**Manage • Administration**). Users with the template creator role can access the **Templates** section (**Manage • Templates**), as shown in Figure 3.29.

Figure 3.29 Accessing Templates from the Manage Menu

SAP Ariba Supplier Lifecycle and Performance uses project templates with special characteristics to define the supplier management processes. These templates can be found in the **Documents** tab of the template project within the site.

The following project templates must be used and modified to configure the supplier management processes on the customer side:

- Supplier request project template
- Supplier self-registration request template
- Supplier registration project template
- Supplier qualification project template
- Supplier disqualification project template

Only one project template for each type of document can exist in an SAP Ariba Supplier Lifecycle and Performance site. These project templates must be edited and customized to cater to your requirements. Modular questionnaires are also based on project templates. Multiple modular questionnaire templates can be created for each type of modular questionnaire.

Supplier management project templates use survey documents to create forms and questionnaires. Some types of projects involve questionnaires that are targeted to specific commodity, region, and department combinations. Supplier management processes center around forms or questionnaires to solicit information about a supplier. Therefore, each supplier management project template includes a default survey document, which forms the basis of that form or questionnaire. Projects can use additional internal questionnaires to supplement the form or questionnaire information.

The following types of forms or questionnaires in these project templates are available:

- **Forms**
  Used to solicit basic information with a limited number of questions. The content in this type of survey document isn't numbered. The default survey document in the supplier request project uses this type of a document. All the content is added to this default survey document without additional content documents.
- **Simple questionnaires**
  Used to solicit larger sets of information. The content in this type of survey document is numbered. The default survey document in the supplier registration project template is this type of a document, an externally facing questionnaire. All the content is added to this default survey document without additional content documents.
- **Internal forms or questionnaires**
  Used in some supplier management projects to supplement information provided by suppliers in the externally facing questionnaires with internal information that will be entered by internal category managers or other internal users. Multiple forms can be created with the supplier management projects and set as an internal form or internal questionnaire. A purchasing organization to which a supplier will

be extended is an example of information that only internal users can provide about a supplier.

- **Dynamic qualification questionnaires**
  Used only in qualification projects where the default qualification questionnaire is blank. Content is dynamically assembled based on multiple content document segments configured for each combination of commodity/region/department (business unit).

In the following section, we'll look at steps required to configure the supplier management project templates required to onboard and maintain a supplier throughout the supplier lifecycle.

3.2.14 Creating and Configuring Supplier Request Templates

As mentioned in the previous section, only one supplier request project template will be loaded in the SAP Ariba Supplier Lifecycle and Performance solution per site. This project template can be found in the **Templates** section. The **Supplier Request Project Template** folder contains the following items:

- Supplier request template
- Supplier self-registration request template

Click on the supplier request template, and click **Open** to customize the standard template based on your requirements.

If only two tabs appear (**Overview** and **Conditions**), then click **Action • Display • Full View**. The following tabs should now be visible: **Overview**, **Documents**, **Tasks**, **Team**, **Conditions**, **Advanced Options**, and **History**.

Designing the Supplier Request Template

To change the standard supplier request template, click **Action • New Version** in the **Overview** tab of the supplier request template. A new version of the template is created.

In the following section, we'll look at configuring each tab within a supplier request template. The **History** tab is used for tracking changes made in the template. This tab isn't configurable and therefore won't be covered in this section.

Overview Tab

This tab provides an overview of the template properties such as **Name**, **Description**, **ID**, **Owner**, **Base Language**, **Access Control**, and **Conditions**. To change these properties, click **Actions • Edit • Properties**.

The following properties can be changed:

- **Owner**
  Person who owns the template. If other owners need access to maintain the template, these users can be added to the project owner group in the **Team** tab.

- **Access Control**
  Allows a level of access provided to supplier requests created using this template. By default, the access control **Owner/Administrator** is set if this field is left empty. Usually the access control **Private to Team Members** is selected.
- **Conditions**
  Allows business rules to be defined to conditionally show/hide this template to end users for selection based on certain criteria. For instance, if a supplier request template is meant for usage in North America only, a condition on **Region = NA** can be set in the **Conditions** field on the **Overview** tab of the supplier request template. In this case, this template will only be available when a user selects **Region = NA** while creating a new supplier request.

Documents Tab

This tab provides a container to create different types of documents, as shown in Figure 3.30. You can also organize the documents created into folders. Documents and folders created in this supplier request template will be automatically loaded in all supplier request projects created from this template.

The main document for a supplier request is the supplier request form. This document is a survey type form designed to collect all the necessary information required by the supplier request approvers to achieve the following goals:

- Understand the business need for creating this new supplier.
- Check for duplicate suppliers. Typically, the supplier request form could ask the requester to upload the supplier's W9 (or W8) form, which can be used to check for duplicate suppliers in SAP Ariba.

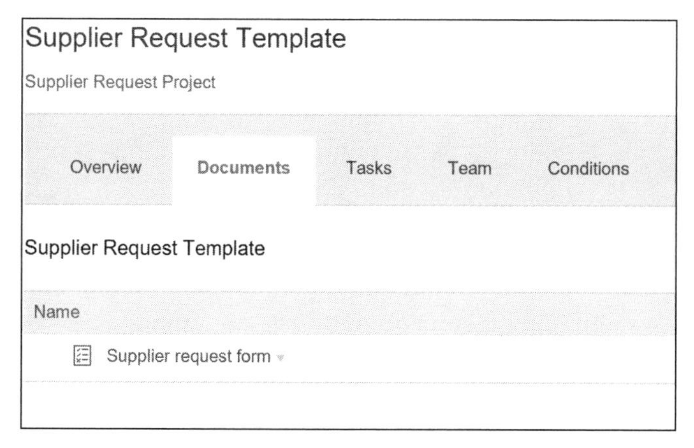

Figure 3.30 Supplier Request Form under the Documents Tab of the Supplier Request Template

Supplier Request Form

At a minimum, a supplier request form should solicit key information about the supplier whom the requester is trying to onboard, as shown in Table 3.1.

Description	Type
Supplier Name	Text field
Supplier Address	Address field
Supplier Contact First Name	Text field
Supplier Contact Last Name	Text field
Supplier Contact Email Address	Text field

Table 3.1 Required Questions in the Supplier Request Form

Additional content can be configured in the supplier request form. Typically, the following questions are included on the supplier request form:

- Information on the commodities (goods and services) that the supplier can provide
- Information on the regions where the supplier operates
- Tax-related information, such as W9 or W8 forms, so that existing suppliers with the same information can be identified as duplicates

Adding Content to the Supplier Request Template

To add content to the supplier request template, on the **Content** page, click the **Add** button at the bottom of the questions table, as shown in Figure 3.31.

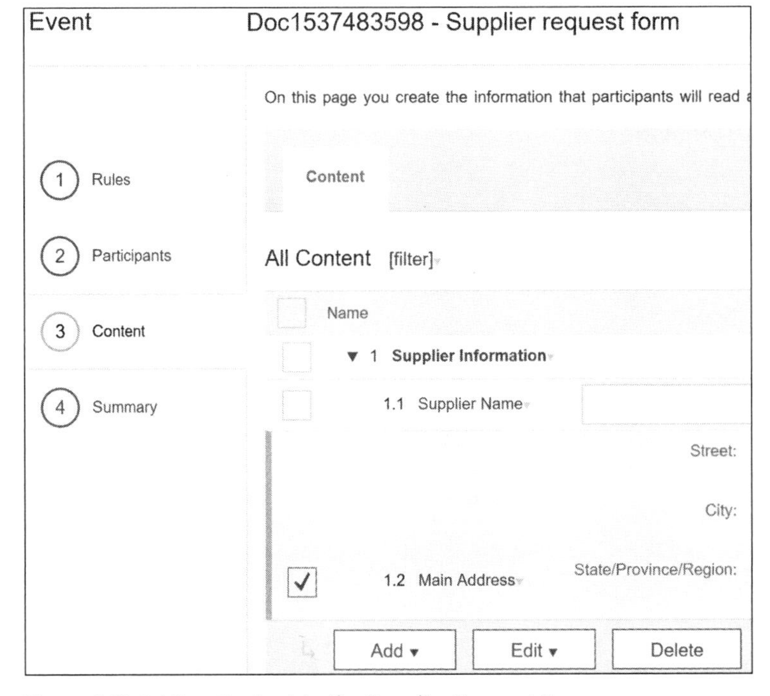

Figure 3.31 Adding Content to the Supplier Request Form

The following types of content can be added:

- **Section**
  A section can contain one or more types of questions.
- **Table Section**
  A section with a tabular data entry.
- **Repeatable Section Parent**
  A section with questions with a plus sign (+) associated with the section to allow for multiple section entries.
- **Question**
  Is a type of content that is used to ask specific questions and gather supplier responses.
- **Requirement**
  Is a type of content to provide specific requirements for a product or service that the customer is looking to procure.
- **Attachment from Desktop – Attachment**
  This type is used to upload any attachments that provides additional details to the suppliers.
- **Attachment from Library**
  Attachment from the sourcing library.
- **Content from Library**
  Content defined in the sourcing library.

Note

The **Repeatable Section Parent** option for a question is used to solicit information where more than one response may occur, for example, banking information.

Questions

Questions are created within sections. Several kinds of questions can be added to the content of a form or questionnaire:

- Text (single line limited)
- Text (single line)
- Whole number
- Decimal number
- Date
- Money
- Yes/no
- Attachment
- Certificate
- Address
- Percentage
- Quantity
- Commodity
- Region
- Department
- Supplier

Questions can be configured to be **Required by Owner or Participant** or **Optional**. Validation patterns can be set on each question to ensure that a response to the question follows a particular format. In addition, a question can be mapped to an SAP Ariba supplier field. If you've enabled synchronization between SAP Ariba Supplier Lifecycle and Performance and your SAP ERP backend, all values in the SAP Ariba supplier fields will be pushed to SAP ERP.

The visibility and editability of a question can be based on responses to previous questions. This option can be configured by using the **Visibility Conditions** field and the **Editability Questions** field and by creating a condition on a previous question. Project-level conditions, discussed in more detail later in this section, can be used as well. For now, in this section, let's look at how to set a questionnaire-level condition on a question.

Enabling Questionnaire-Level Conditions

Follow these steps to create a questionnaire-level condition:

1. Click on a supplier request form in the **Documents** tab.
2. Click **Actions • Edit**.
3. Click on the **Content** link.
4. Click on a question that you want to conditionalize, and then click **Edit**.
5. In the question, scroll down to the **Visibility Conditions** field, click **None**, and then click the down arrow button.
6. Click **Create Condition**.
7. Select a previous question.
8. Based on the type of previous question selected, enter or select a value for the question. (Note: This is an EQUAL TO expression.)
9. Click **OK**.

Now, the question will only be visible to an end user if the condition is satisfied, that is, if the answer to the previous question satisfies the condition. These steps can be repeated for the editability condition as well.

> **Note**
> You can also use project conditions defined in the **Conditions** tab to set the visibility and editability conditions of a question.

After the content of the supplier request form is completed, the administrator can use the **Tasks** tab to define the supplier request process and approval task, which we'll discuss next.

Note

These new question types, bank account and tax, can be used to collect banking and tax information. See Section 3.2.21 for more information on banking and tax information.

Tasks Tab

SAP Ariba Supplier Lifecycle and Performance allows administrators to define a standard process with phases and tasks for creating and approving supplier requests, similar to other project templates. For a supplier request project template, the following types of tasks can be created:

- **To-do tasks**
  Requires the task owner to perform some activity. Document folders can be linked to to-do tasks.
- **Review and approval tasks**
  Routes the documents associated with these tasks to reviewers or approvers. Reviewers can leave comments while reviewing the document. Approvers will need to approve/deny the document. Document folders can be linked to review or approval tasks. All documents within the document folder will be part of the review or approval process. (For more details on configuring approval tasks, Section 3.2.16.)

Supplier requests created from a supplier request template will automatically inherit the tasks defined in the template. Supplier requesters can choose to add additional tasks to the request process or delete optional tasks that are attached to the template.

Team Tab

Collaboration throughout the supplier request process is key for ensuring a faster and less error-prone execution of supplier requests. SAP Ariba Supplier Lifecycle and Performance allows collaborators or team members to be added manually or automatically to the supplier request in the **Team** tab. Project team groups, such as supplier managers, should be created first, before adding team members to these project team groups. The project owner team group is created automatically by the system. Other project team groups must be created manually.

Team members who will be added to the project team group can be individual users or system/custom user groups. Each project team group is defined a role that allows team member **View Only** or **View/Edit** access to the project.

Similar to tasks, project team groups and team members can be set up by the administrator at the template level. Supplier requests created from the template will automatically inherit the project team groups and team members, which is useful to enforce compliance.

Our example configuration shown in Figure 3.32 allows template administrators to restrict end users from making changes to project groups added at the template level. The template administrator can also associate a system group in the **Roles** setting. This setting assigns the underlying roles in the system group to all users added as members to this group.

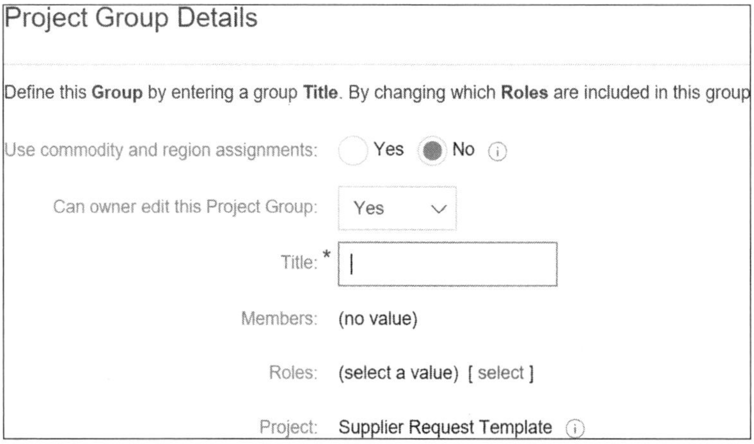

Figure 3.32 Project Team Group Details

SAP Ariba Supplier Lifecycle and Performance introduced a new feature, called *user matrix configuration*, that allows you to set users/groups to team member groups based on commodity/region/department (business unit) combinations. This user matrix is loaded by an administrator in the master data **Import/Export** area in SAP Ariba Supplier Lifecycle and Performance.

Conditions Tab

As mentioned earlier in this section, two kinds of conditions can be defined in a supplier request project:

- Within a questionnaire
- At the project level

Project-level conditions are defined in the **Conditions** tab. These project-level conditions can be used to set visibility and editability conditions on questions and thus set visibility condition on tasks and documents in the supplier request project.

Creating a Project-Level Condition

Follow these steps to create a project-level condition:

1. Within the **Conditions** tab, click **Add Condition**.
2. Provide a name for the condition.
3. Click on the expression **All are true**.
4. Choose **Condition • Field Match**.

5. Select a project field (click **More Fields** to see the whole list).
6. Select a value.
7. Click the inverted triangle icon next to the **All are true** expression, and change to **Any are true** or **None is true** if needed.
8. Click **Ok**.

After the supplier request template is fully configured, the template must be published before it can be used to create supplier requests. To publish a template, one of the template owners must click on the **Overview** tab and then click **Action • Publish**.

After a template is published, the status of the template changes from **Draft** to **Active**. The template is now available to users and will appear in the list of templates for the user to choose from when creating a new supplier request.

Supplier Self-Registration Template Design Restrictions

There are some restrictions when designing the supplier self-registration template for suppliers to submit a supplier request:

- Use the template's default survey document for the supplier self-registration request form or questionnaire, and create questions directly in the survey document. The supplier self-registration project template doesn't support any additional documents.
- You can add approval tasks in addition to the template's default approval task on the request survey document and chain them together as predecessors, but don't add any other type of task to the template. The supplier self-registration project template only supports approval tasks on its single survey document.

3.2.15 Configuring the Supplier Registration Template

After a supplier request is approved, the next step in the supplier lifecycle process is to get the supplier registered in SAP Ariba Supplier Lifecycle and Performance and Ariba Network so that the supplier can collaborate with the buyer organization on various documents, such as sourcing events, contracts, purchase orders (POs), and invoicing documents, as well as other documents from SAP Pay and SAP Ariba Discount Management. To register a supplier in SAP Ariba Supplier Lifecycle and Performance, the supplier registration template must be configured and published.

Designing the Content of Supplier Registration Questionnaires

As mentioned earlier, SAP Ariba Supplier Lifecycle and Performance comes with a default supplier registration project template that can be found in the **Documents** tab of the **Templates** section. This template can be accessed by users with the template creator role.

Open the template and create a new version of the template before customizing the standard template according to your own requirements. A supplier registration template has the same tabs as a supplier request template.

Documents Tab

A supplier registration project has a single supplier registration questionnaire.

Note

You should use the standard supplier registration questionnaire provided by SAP Ariba Supplier Lifecycle and Performance. Don't make a copy of this questionnaire because it's tied to the supplier registration and update approval tasks that can be enabled as needed.

Content for supplier registration, such as prequalification questionnaires, and vendor management-related questionnaires, should be added to the supplier registration questionnaire's **Content** section.

Note

We recommend adding all content related to supplier registration within the default supplier registration questionnaire provided. Additional external questionnaires can be created with associated to-do tasks and set as predecessors for the supplier registration questionnaire approval task. However, this approach is tricky and not recommended.

To add content, open the supplier registration questionnaire, and go to the **Content** section. Adding sections and questions to a supplier registration template is similar to the process for adding content to supplier request templates, discussed earlier in Section 3.2.14.

Tasks Tab

SAP Ariba Supplier Lifecycle and Performance allows administrators to define a standard supplier registration process with phases and tasks. The task could be a to-do task or a workflow approval task. Two important phases that can be configured are the **New Registration** phase and the **Registration Update** phase, as shown in Figure 3.33.

Tasks created within the new registration phase will be triggered when the supplier registers to Ariba Network and submits the completed supplier registration questionnaire for the *first* time. Typically, the supplier registration approval task (approving the information submitted by the supplier) will be the only task configured during this phase. In addition, if additional internal questionnaires need to be completed before the registration can be approved, then to-do tasks can be created.

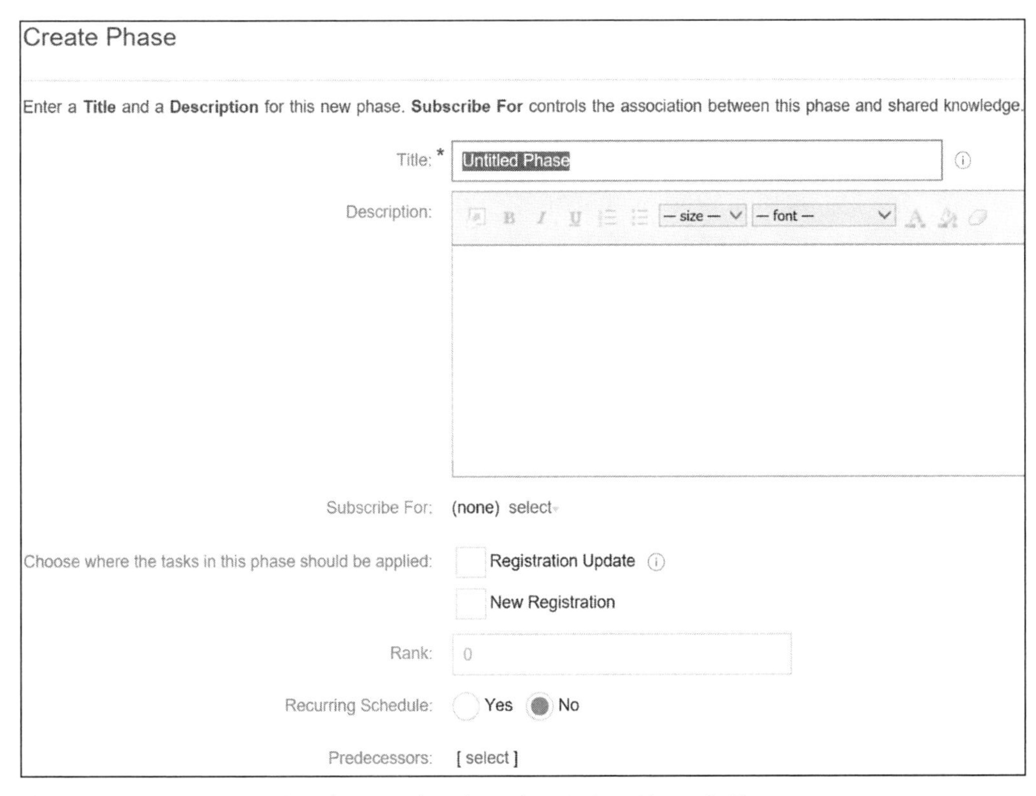

Figure 3.33 New Registration Phase and Registration Update Phase Options

Tasks created within the registration update phase will be triggered when the supplier updates the registration questionnaire and submits changes. Typically, the supplier registration update approval task (approving the updated information submitted by the supplier) will be the only task configured during this phase. In addition, if additional internal questionnaires need to be reviewed before the registration update can be approved, then to-do tasks can be created.

The following types of tasks can be created:

- **To-do tasks**
  Requires the task owner to perform some activity. Document folders can be linked to to-do tasks.
- **Review and approval task**
  Routes the documents that are associated with these tasks to reviewers or approvers. Reviewers can leave comments while reviewing the document. Approvers will need to approve/deny the document. Document folders can be linked to review or approval tasks. All documents within the document folder will be part of the review or approval process.

Conditions Tab

Similar to supplier request templates, this tab is used to create project conditions. See Section 3.2.14 for more information on configuring project-level conditions.

3.2.16 Configuring Workflow Approval Tasks

SAP Ariba Supplier Lifecycle and Performance provides approval workflow tasks can be configured during the supplier request, supplier registration, and supplier qualification and disqualification processes.

The workflow task can be defined in each template, is highly customizable, and can be set as a required task. It can also be associated to a specific document or to a folder within the **Documents** tab, as shown in Figure 3.34. When assigned to a folder, all documents within the folder will be included in the approval workflow process.

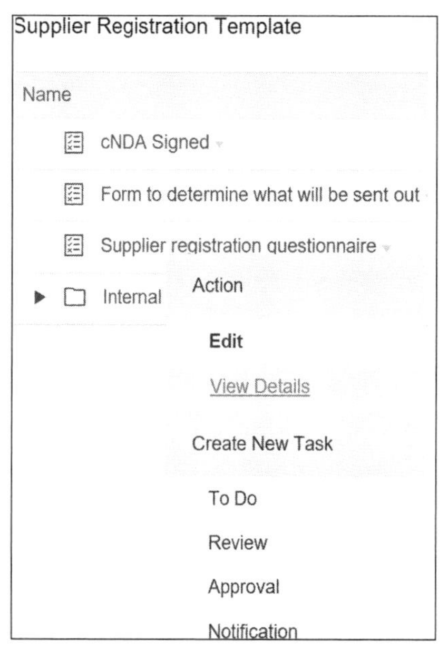

Figure 3.34 Creating an Approval Task on a Supplier Registration Questionnaire

On triggering the workflow task, the document or documents within the folder will be routed for approval, and an email notification will be sent out to the approvers in the approval flow.

Approval Flow

SAP Ariba Supplier Lifecycle and Performance provides a flexible approval workflow task that can be customized based on your approval process. As shown in Figure 3.35, the following approval flows are allowed:

- **Parallel**
  Multiple approvers/ groups of approvers will receive the approval email simultaneously, allowing access for approvers to take action on the approval task simultaneously.
- **Serial**
  Each approver/group within the approval flow will receive an approval email in a serial order. An approver/group will receive an approval request email only after the previous approver/group has approved the documents in the previous approval step.
- **Custom**
  Customers can define a more complex workflow with a combination of parallel and serial approvals. Customers with a more complex approval flow can use this feature to design their approval process.

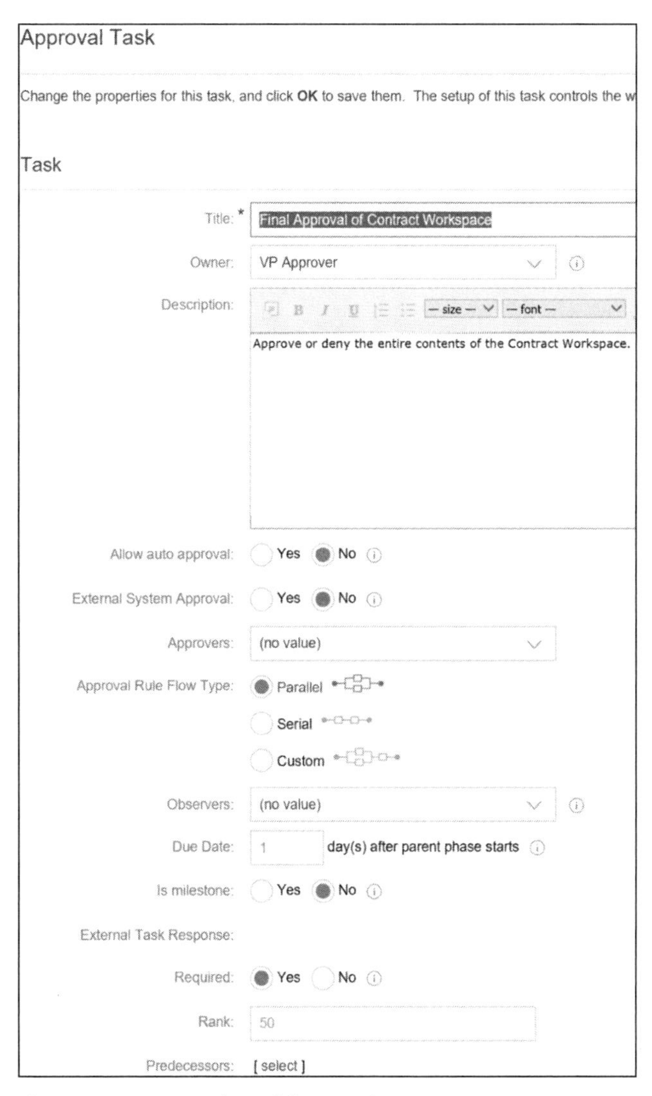

Figure 3.35 Approval Workflow Task

Approvers will be required to review and approve the document or documents within the folder that is associated with the approval task. Observers can be added to the approval workflow as well. Observers don't have to take any action throughout the process but will have access to review the status of the approval and review the documents within the approval task.

Adding Approvers to Custom Approval Flow

Custom approval flows provide a flexible option in SAP Ariba Supplier Lifecycle and Performance to define a customized and complex approval process. On selecting **Custom** as the flow, parallel and serial approval steps can be added to the approval process.

The approval flow can be customized based on your supplier onboarding and vendor management requirements. Individual users, project groups, or system groups can be added as approvers for each step in the approval flow.

3.2.17 Configuring the Supplier Qualification Template

After the supplier is registered, your organization can qualify the supplier for any combination of commodity, region, and department (business unit).

As mentioned earlier, SAP Ariba Supplier Lifecycle and Performance is loaded with a supplier qualification template project, which can be found in the **Templates** section. This template can be accessed by users with the template creator role. Open the template and create a new version of the template before customizing the standard template according to your requirements. The supplier qualification template has the same tabs as the supplier registration template.

Documents Tab

The supplier qualification project template must always include an external (supplier-facing) qualification questionnaire with an approval task. The project template can also include an internal questionnaire. Supplier qualifications are always based on a combination of commodities, regions, and departments (if your site has the supplier management business unit matrix enhancement enabled).

The external qualification questionnaire is dynamic and generated from two different types of template documents:

- An empty survey document, which functions as a vehicle for serving the qualification content. Approval tasks and other tasks are always on this empty survey document, which serves as the default supplier qualification questionnaire.
- Content documents, which define questionnaire segments for specific combinations of commodity, region, and department.

Users can start a new qualification in two ways, and the supplier qualification project template determines which method is used in your site:

- **Static application page**
  By default, users who start a supplier qualification project are presented with a static application page, where they can specify the commodities and regions for qualification. Submitting these answers creates the qualification project. This page doesn't allow users to specify departments or expiration information for the qualification.
- **Prequalification questionnaire**
  You can set up a prequalification questionnaire, and users who start qualifications are presented with the prequalification questionnaire rather than with the default static application page. If your company uses departments in qualifications, you must set up a prequalification questionnaire. You can also use prequalification questionnaires to specify expiration dates and requalification eligibility dates. If you don't use prequalification questionnaires but want to allow users to set expiration and requalification eligibility dates, you must set up an internal questionnaire instead.

Tip

We recommended using prequalification questionnaires, which can replace the default static application page. Up to five questions can be set up in this prequalification questionnaire: three for commodity, region, and department filters, and two for expiration date fields.

Tasks Tab

By default, SAP Ariba Supplier Lifecycle and Performance includes an approval task for approving the supplier qualification questionnaire in the supplier qualification template. The project owner of the supplier qualification project will be added as the approver, which can be modified according to your requirements. To-do tasks can also be added based on your requirements.

Team Tab

By default, the **Team** tab has a project owner group. The `aribasystem` user and the template creator system group are added as members of this group.

When designing a supplier qualification template, the following considerations must be kept in mind:

- This template must contain one external-facing survey questionnaire document that is *empty* as well as content documents with content specific to the combination of commodity/region/department that you want include. Don't add any other type of document.

- Never add tasks other than the to-do and approval tasks. No other type of task is permitted.
- The supplier qualification project template supports one external questionnaire survey document that should be left empty. You'll define the content of the external questionnaire in multiple content documents. Adding content to the external questionnaire survey document itself causes the questionnaire to fail with errors.
- The supplier qualification project template supports multiple internal questionnaire survey documents. You'll define the content of internal questionnaires in the survey documents themselves. Don't use content documents.
- Don't apply project-level visibility conditions based on commodity, region, or department to the content documents that define qualification questionnaire segments. When a user starts a qualification, only the questionnaire segments that apply to the qualification's commodities, regions, and departments will be displayed.

Some things to keep in mind when designing the qualification template:

- You must create a to-do task for each internal registration questionnaire survey document you add. This task ensures that the correct users can edit the survey and fill it out.
- If you want to qualify suppliers based on department as well as commodity and region, you must use a prequalification questionnaire.
- If you want qualifications in your site to expire, you must create a question with a date answer and map it to `project.ExpirationDate`. If you want to allow a category or supplier manager to establish a waiting period before an expired qualification can be requalified, add a second date question, and map it to `project.RequalificationEligibilityDate`. Depending on how you set up your supplier qualification project template, you can add these questions in the prequalification questionnaire (if you use one) or in an internal questionnaire (if you don't use the prequalification questionnaire).

3.2.18 Configuring the Supplier Disqualification Template

Suppliers qualified for a commodity, region, and department (business unit) can now be disqualified. A supplier disqualification project template that is available by default in SAP Ariba Supplier Lifecycle and Performance in the **Templates** section can be used to design the supplier disqualification process.

Deploying your solution includes a supplier disqualification project template with the default configuration described in this section, which includes an empty disqualification questionnaire survey document and an approval task. You must edit the template to define your company's specific processes.

Documents Tab

An empty internal questionnaire survey document is loaded by default in the **Supplier Qualification Template** project. The name of this survey document is the **Supplier disqualification questionnaire**, as shown in Figure 3.36.

Figure 3.36 Supplier Disqualification Questionnaire under the Documents Tab

Tasks Tab

The supplier disqualification template comes with a default approval task for approving the supplier disqualification questionnaire. By default, the owner of the supplier disqualification project is set as the approver, as shown in Figure 3.37. You can change the owner as needed.

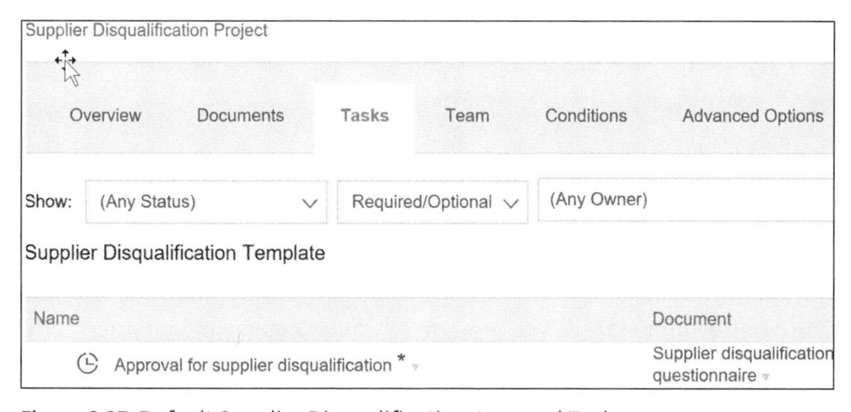

Figure 3.37 Default Supplier Disqualification Approval Task

To-do tasks can also be added according to your requirements.

Team Tab

By default, the **Team** tab has a project owner group. The `aribasystem` user and template creator system group are added as members of this group.

Keep in mind the following restrictions while designing the supplier disqualification template:

- This template must contain one internal questionnaire type survey with an associated approval task.
- Never add tasks other than the to-do and approval tasks. No other type of task is permitted.
- As the disqualification of a supplier is always tied to a commodity, region, and departments (if the business units matrix is enabled), the supplier disqualification questionnaire must include questions of the following types:
  - Commodity, mapped to `matrix.Categories`
  - Regions, mapped to `matrix.Regions`
  - Departments, mapped to `matrix.Departments`

Note

If you create a date question and map it to `project.DisqualificationDate`, after the questionnaire is approved, the supplier will be in **Restricted** status for a phaseout period until the date is reached and then will automatically be disqualified on that date. This mapped question is the only way to achieve a restricted qualification phase for the supplier before disqualification.

Note

If you create a date question and map it to `project.DisqualifiedUntilDate`, after the questionnaire is approved, the supplier can't be requalified for the same commodities, regions, and departments until the date is reached.

3.2.19 Configuring Preferred Supplier Management Project Template

After a supplier is qualified for a combination of commodity, region, and department, the preferred supplier management process can be used to set this supplier as a preferred supplier for this combination. The preferred supplier management project template can be edited to define this process in your site.

Deploying your solution includes a preferred supplier management project template with the following default configuration that includes an external (supplier-facing) category status questionnaire with an approval task. You must edit this template to define your company's specific processes.

Documents Tab

An internal questionnaire is loaded by default in the **Preferred Supplier Management Template** project. The name of this default survey document is **Category status**, as shown in Figure 3.38.

Figure 3.38 Category Status, Default Survey Document in the Documents Tab

Tasks Tab

The preferred supplier project template comes with a default approval task for approving the category status change request questionnaire. By default, the owner of the preferred supplier project is set as the approver. You can change the owner as needed. To-do tasks can also be added according to your requirements.

Team Tab

By default, the **Team** tab has a project owner group. The `aribasystem` user and template creator system group are added as members of this group.

Keep in mind the following restrictions while designing the preferred supplier management template:

- The preferred supplier management project template only supports one survey document for one internal questionnaire. Don't add any other type of document to the template, and don't add any more survey documents besides the template's default disqualification questionnaire.
- Never add tasks other than the to-do tasks and approval tasks. No other type of task is permitted.
- Preferred status for a supplier is always tied to a commodity, region, and departments (if the business units matrix is enabled). So, the category status change questionnaire must include questions of the following type:
  - Commodity, mapped to `matrix.Categories`
  - Regions, mapped to `matrix.Regions`
  - Departments, mapped to `matrix.Departments`

  These settings ensure that the preferred category status designations use the same commodity, region, and department data that is used for qualifications and disqualifications.

Note

For a question that prompts users to enter the preferred supplier level, set **Acceptable Values = Master Data Value** and **Type of Master Data = Supplier Preferred Level**.

3.2.20 Configuring Modular Questionnaire Templates

Modular templates can be found in the **Templates** section under the **SM Modular Template**. Modular questionnaire templates are of project type **SM Modular Questionnaire**, and each template must contain one external (supplier-facing) questionnaire with an approval task. In each modular questionnaire template, you'll set a questionnaire type and specify the commodities, regions, and departments to which this questionnaire applies. You can also specify an expiration schedule for the questionnaire. This schedule is specific to the questionnaire itself and operates independently of the expirations of any of the certificate questions it might contain.

Before creating a modular questionnaire template, the modular questionnaire type master data must be loaded through the import process. See Section 3.2.23 for more information on loading master data.

Unlike other supplier management projects, multiple modular supplier management templates can be created in your site, one for each questionnaire. To create a modular questionnaire template, follow these steps:

1. Click **Manage • Templates**.
2. Click **Action • Create • Template**.
3. Select **SM Modular Questionnaire** as the **Project Type**.
4. Select one of the questionnaire types that were loaded using master data import, and select a commodity, region, and department.
5. Click **Done** to create a new modular questionnaire template.

After the modular questionnaire is created, the draft template is now ready for you to load content with the questions you want the supplier to respond to and for setting up the approval task, as needed per your requirements.

Documents Tab

An external (supplier-facing) survey document of type questionnaire must be created. All content must be created within this survey.

Tasks Tab

Modular supplier management questionnaire projects support the use of two phases, similar to the supplier registration process, to apply separate approvals and to-do tasks. One phase is for first-time submission of the modular questionnaire by the supplier, and the other phase is for updates to the modular questionnaire by supplier. Questionnaire updates are automatically approved unless the project template includes an approval task in the update phase.

Note

To add phases and tasks to the modular questionnaire project template, you must have the template creator and supplier management modular questionnaire manager roles.

Each modular supplier management questionnaire project template requires at least one approval task. If you use phases, the approval task must be in the new questionnaire phase. If phases aren't used, then the tasks will only be triggered the first time the supplier responds and submits the modular questionnaire or when an internal user resends the modular questionnaire requesting for updates. If the supplier submits updates without an internal user resending the questionnaire, and no update phase is set up, then no tasks are triggered.

If approval and to-do tasks are required for updates to the modular questionnaire, then phases must be created. Modular supplier management project templates support phases with special new questionnaire and questionnaire update settings to control the order in which the phases start and whether the tasks in the phase are one-time-only (for new questionnaires) or should recur (for every questionnaire update). The new questionnaire phase starts one time, that is, immediately when the supplier is invited to fill out the questionnaire. The questionnaire update phase starts again every time a supplier updates the modular questionnaire.

Within the new questionnaire and questionnaire update phases, you can add separate approval and to-do tasks to the same questionnaire survey document to define the workflows for new and updated questionnaires. The following types of tasks can be created in each phase:

- **To-do tasks**
  Requires the task owner to perform some activity. Document folders can be linked to to-do tasks.
- **Review and approval tasks**
  Routes the documents that are associated with these tasks to reviewers or approvers. Reviewers can leave comments while reviewing the document. Approvers will need to approve/deny the document. Document folders can be linked to review or approval tasks. All documents within the document folder will be part of the review or approval process. (For more details on configuring approval tasks, Section 3.2.16.)

Keep in mind the following considerations while setting up modular questionnaire phases and their tasks:

- You can only add two phases to modular supplier management questionnaire projects: one with the **New Questionnaire** setting and one with the **Questionnaire Update** setting. These phases don't use the **Subscribe For**, **Rank**, or **Predecessor** settings. The new questionnaire phase automatically precedes the questionnaire update phase.

- Make sure that all the template tasks are inside either the new questionnaire phase or the update questionnaire phase. If you use these phases in a modular supplier management project template, adding tasks outside these two phases isn't supported.
- Make sure that the tasks you specify as predecessors are within the same phase. Don't make tasks in one phase predecessors of tasks in a different phase.
- You can't apply conditions to modular supplier management questionnaire tasks or phases themselves.

To create a new questionnaire or questionnaire update phase, follow these steps:

1. Under the **Tasks** tab of the **Modular Questionnaire Project Template**, click **Action • Create • Phase**.
2. Select only either the **New Questionnaire** option (for a new questionnaire phase) *or* the **Questionnaire Update** option (for a questionnaire update phase).
3. Make sure that **Recurring Schedule** is set to **No** so that the tasks in the phase are used only once, when the questionnaire is first submitted.
4. Click **OK**.

3.2.21 Configuring Banking Questions in Supplier Management Questionnaires

SAP Ariba has a special question type that allows you to easily gather a supplier's banking information. This question is also designed to be used within a repeatable section so that you can gather information about any number of bank accounts that the supplier wants to provide to the buyer organization.

Bank account question types include a set of predefined fields to collect bank information that is automatically mapped to the corresponding `vendor.bankInfos` field. A new enhancement to the supplier bank ID management automatically generates a unique, sequential, numerical bank ID. Optionally, a country-specific prefix can be used with the automatically generated ID.

Note

The enhancement to ERP data synchronization with the external supplier management questionnaire, which pushes updates from an integrated ERP system to all external questionnaires that aren't in approval, must be enabled for the automatically generated bank ID to work properly.

Keep in mind the following consideration when adding a bank account question to an external (supplier-facing) questionnaire:

- The IBAN number and account number are always masked so that only users in the sensitive data access group can see the full numbers. By default, bank account questions use a masking pattern of (.*).(4), which masks all but the last four digits of every IBAN or account number. You can edit or remove this masking pattern, but if you remove it entirely, the default masking pattern of (.*).(4) will still apply.

Bank account questions and responses are automatically displayed in the **ERP Data** page of the **Supplier 360° View** profile. Some of the bank account fields are validated based on default syntax validations. A new data import task allows you to define your own country-specific syntax validations so that they match the syntax requirements of your ERP system.

3.2.22 Configuring Tax Information Questions

A new feature in SAP Ariba Supplier Lifecycle and Performance introduces a new answer type—tax—for questions in supplier management questionnaires. The tax component question includes a set of customer-configured, country-specific fields for tax ID details. Tax questions automatically include an initial field for **Country**. After the supplier chooses a country, the question displays the tax ID fields for that country, which are defined in your site via a metadata file import. For more information on data imports for supplier management, Section 3.2.23. The comma-separated values (CSV) that defines your tax ID fields also allows you to specify which tax ID fields show for each country, including validation patterns, sample values that display in validation error messages, and mappings to different tax number fields in the vendor database.

Keep in mind the following considerations when adding a tax question to a supplier management questionnaire:

- By default, tax questions use a masking pattern of (.*).(4), which masks all but the last four digits of every tax ID. You can edit or remove this masking pattern, but if you remove it entirely, the default masking pattern of (.*).(4) will still apply.
- Tax questions use the field mapping `vendor.taxExt` (in single questions) or `vendor.taxExt[$index]` (in questions in repeatable sections). This mapping stores answers in the supplier database fields that correspond to different tax IDs as defined in the `TaxCode` column of your tax metadata CSV file.

Because suppliers can have multiple tax IDs in one or more countries, you can include tax questions in repeatable sections in external (supplier-facing) questionnaires. You can also include single tax questions in internal questionnaires, but the use of repeatable sections isn't supported in internal questionnaires. The SAP Ariba supplier database can store multiple tax IDs for each supplier, and all tax information in mapped tax questions is synchronized to integrated ERP systems.

> **Note**
>
> In the SAP Ariba supplier database, the **Country** and **Tax ID** fields in tax questions are encrypted.

Note

If you flag a supplier as inactive in your site for more than 30 days, SAP Ariba Supplier Lifecycle and Performance solution automatically deletes all data in bank account, tax, and supplier contact database fields. To reactivate the supplier, this information must be recollected.

3.2.23 Loading Master Data

Master data imports are crucial for enabling supplier management processes in SAP Ariba Supplier Lifecycle and Performance. Supplier master data is imported in the **SM Administrator** or the **Data Import/Export** area under **Administration**. Users with one of the following roles (members of the following system user groups) can import supplier master data:

- Supplier management ERP administrator
- Supplier management operations administrator
- Supplier risk manager
- Customer administrator
- Supplier/customer manager and supplier management operations administrator

The following master data can be loaded using **SM Administrator** or **Data Import/Export**:

- Questionnaire type (*SMQuestionnaireType.csv*)
- Certificate types (*CertificateType.csv*)
- Country-specific tax ID fields in the tax question type (*SapTaxMetadata.csv*)
- Suppliers
- Supplier contacts
- Supplier qualification data (a list of supplier qualifications by commodity code, region, and department in a CSV file)
- Preferred supplier list data (a list of preferred suppliers by commodity code, region, and department in a CSV file)
- Supplier factory data (information about supplier factories in a CSV file)
- Purchasing organization data (the purchasing organizations associated with the supplier, including defaults, in a CSV file)
- Custom display names (for custom labels for registration and qualification statuses in a CSV file)
- Primary supplier manager (the names of primary internal contacts for suppliers in a CSV file)

- User matrix (user assignments to commodities, regions, and supplier management project groups in a CSV file)
- Supplier risk data (risk data for suppliers in SAP Ariba Supplier Risk)
- Risk exposure score configuration data for SAP Ariba Supplier Risk (used to import the CSV file exported from the risk scoring configuration workbook)

Importing Master Data

To import master data, click the **Import** tab on the **Data Import/Export** screen. Four choices are available when importing a file:

- **Load**
  Creates and modifies objects in the database using values in the data file. If an object in the data file doesn't already exist in the database, the object is created. If an object in the data file already exists in database, it's modified using the value in the data file.
- **Create**
  Creates new objects in the database using values in the data file. If an object in the data file already exists in the database, the object isn't modified.
- **Update Only**
  Modifies existing objects only in the database using values in the data file. If an object in the data file doesn't already exist in the database, the object won't be created. If you don't want to modify a particular object, don't include the object in the data file. This operation can cause deactivated objects to be reactivated.
- **Deactivate**
  Deactivates objects in the database based on objects in the data file. If you don't want to deactivate a particular object, don't include the object in the data file.

Select one these choices, and then click the **Browse** button to choose the master data object file you updated and saved on your desktop. Click **Run** to execute the import process. If any failures arise, a new tab will display errors. Correct these errors and reupload the file.

> **Note**
>
> To import master data, we recommend you first export all the values for the master data from the **Export** tab. Update the exported CSV file with new data or with updates to existing data, and import the modified file using the **Import** tab.

3.2.24 Configuring Analytical Reports

SAP Ariba has extended the prepackaged reports to include **Supplier Lifecycle Reports**. These reports can be accessed by clicking on **Manage-Prepackaged Reports**, as shown in Figure 3.39.

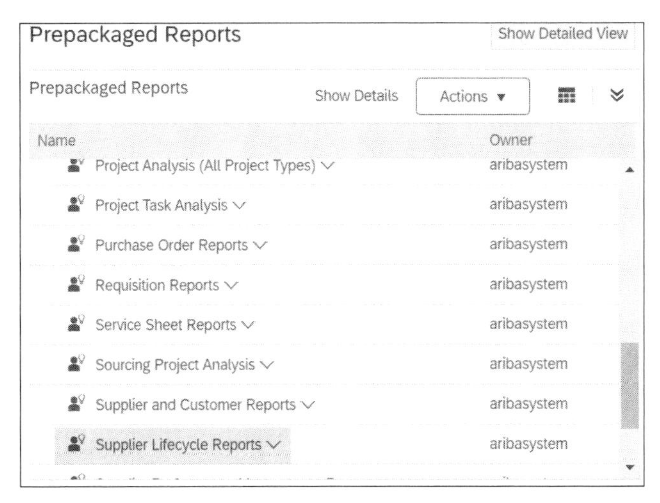

Figure 3.39 New Prepackaged Supplier Lifecycle Reports

The following prepackaged reports are now available for SAP Ariba Supplier Lifecyle Management, as shown in Figure 3.40.

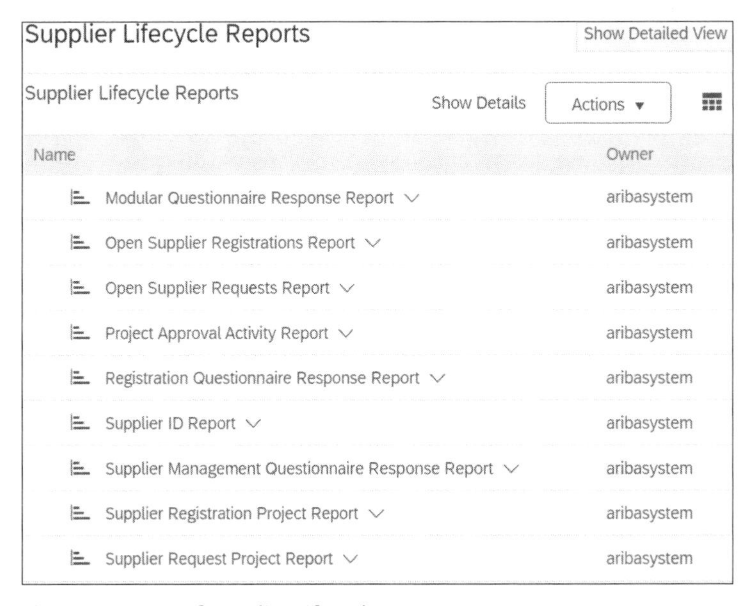

Figure 3.40 List of Supplier Lifecycle Reports

Open Supplier Registration Report

This report can be used to get a list of suppliers that haven't responded to the supplier registration invitation, to get a list of those that haven't yet submitted their responses, or to identify bottlenecks in the supplier registration approval process, as shown in Figure 3.41. The report displays supplier registration projects that are still waiting on the supplier to respond, pending approvals, or pending resubmission from the supplier.

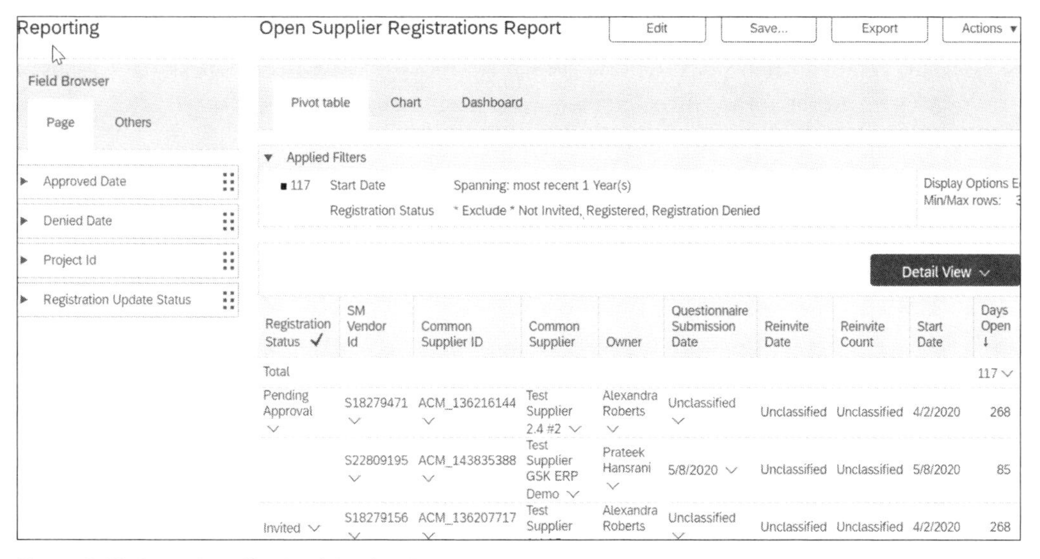

Figure 3.41 Open Supplier Registration Report

Open Supplier Requests Report

This report can be used to list all the supplier request projects that haven't yet been approved or denied, as shown in Figure 3.42. The report shows the number of days the supplier request project has been open.

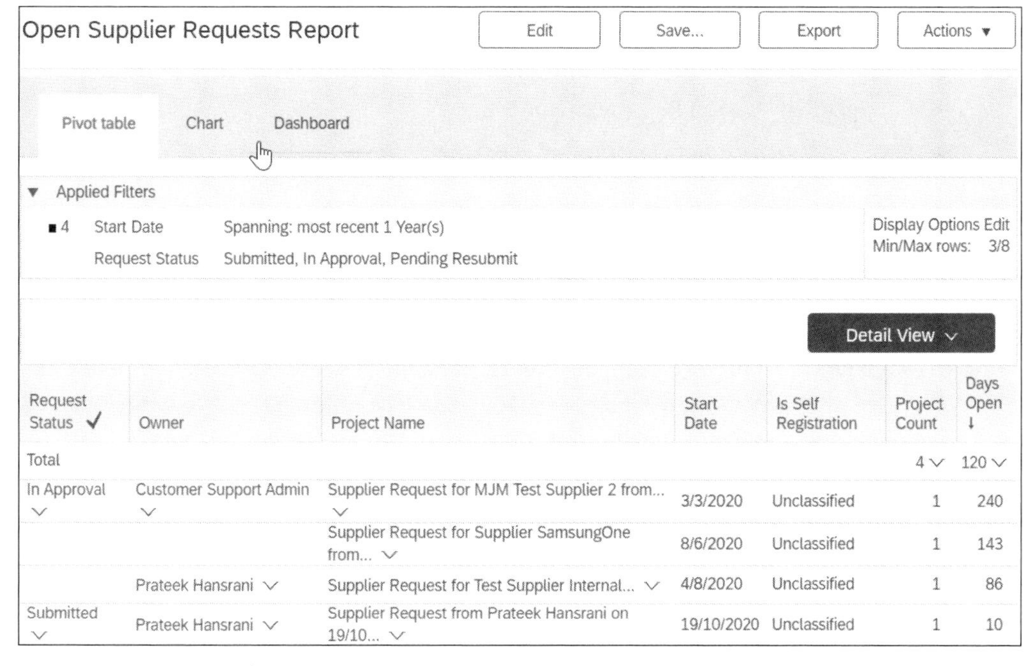

Figure 3.42 Open Supplier Requests Report

Project Approval Activity Report

This report displays information on approval tasks related to supplier management projects, including approvers, approval comments, and the number of days the approval has been open (see Figure 3.43).

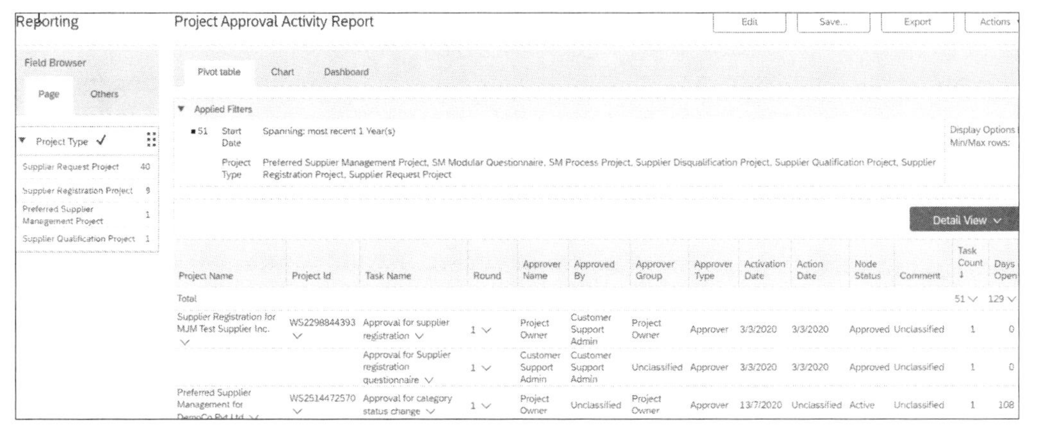

Figure 3.43 Project Approval Activity Report

SM Questionnaire Project Response Report

SAP Ariba now has a new reporting app—SM Project Questionnaire Response—and a set of new prepackaged reports for reporting on supplier responses to supplier management questionnaires:

- Modular Questionnaire Response Report
- Supplier Registration Response Report
- Supplier Management Questionnaire Response Report

A user in the analyst user group and senior analyst user group can copy a prepackaged report, as shown in Figure 3.44, and create a new report that can then be modified per requirements.

In addition, users in the analyst user group and senior analyst user group can create a new analytical report by clicking on **Create – Analytical Report**, as shown in Figure 3.45. The user can pick the source tables or main, second, and third facts. Users can add project fields to the report and define the pivot layout, as shown in Figure 3.46.

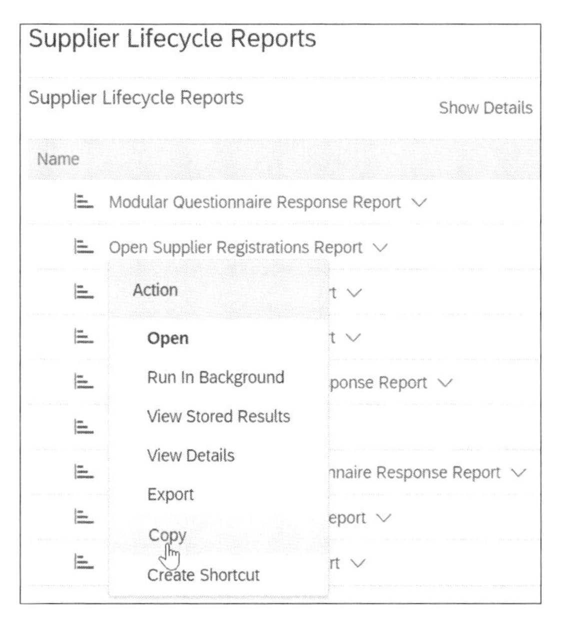

Figure 3.44 Copying Prepackaged Reports

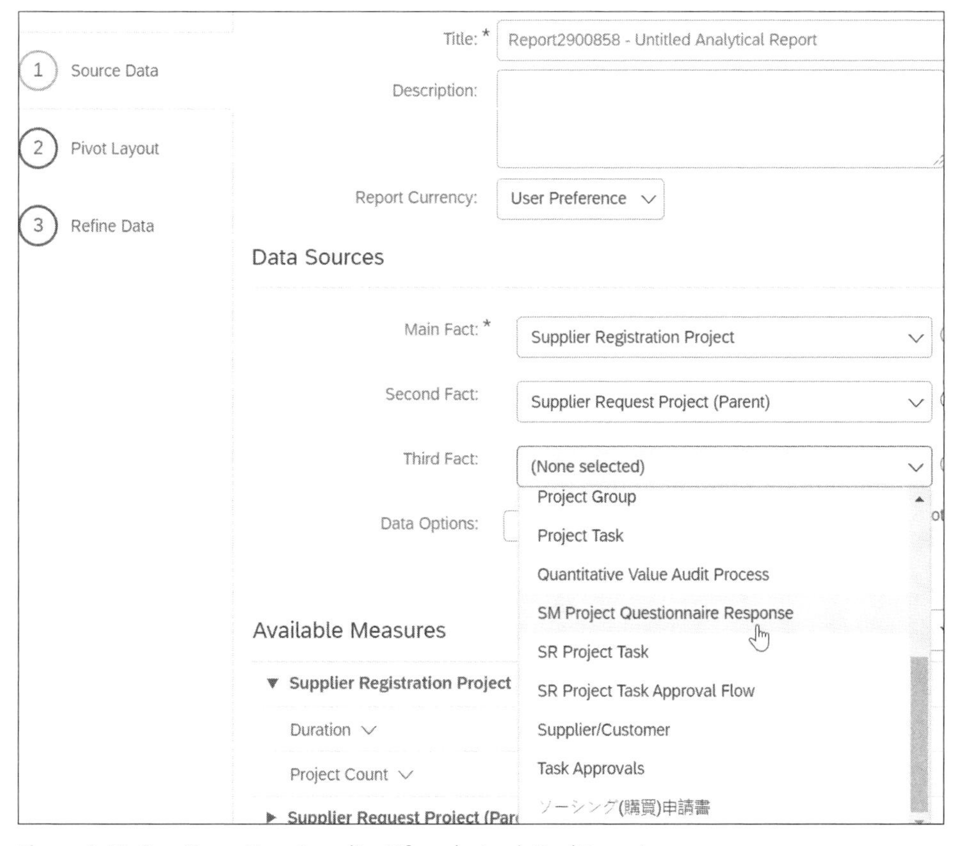

Figure 3.45 Creating a New Supplier Lifecycle Analytical Report

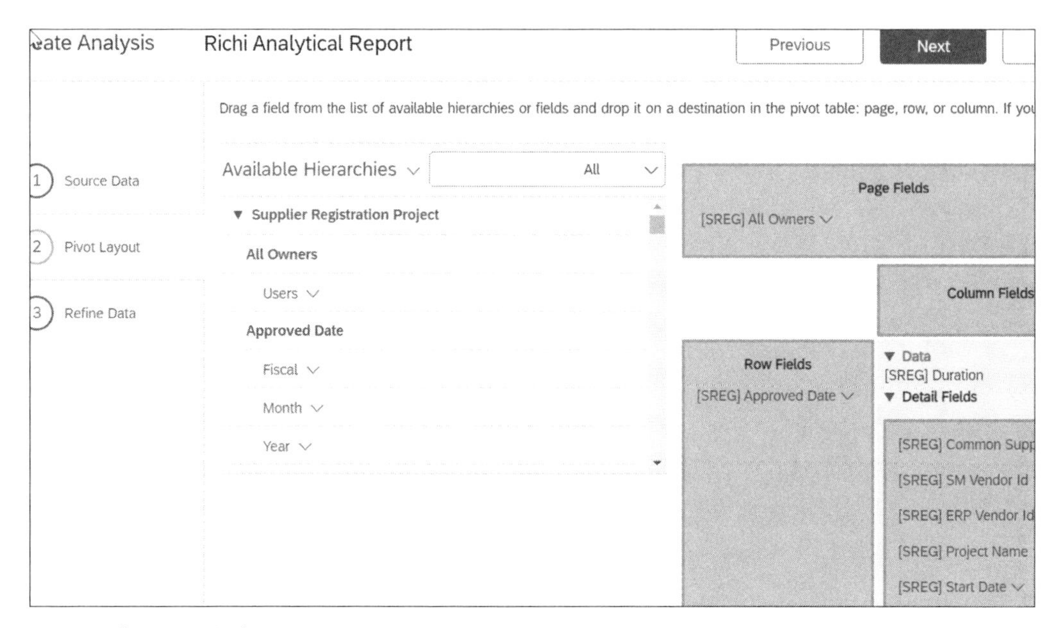

Figure 3.46 Pivot Layout

Finally, users can refine data by adding filters, such as date range or owners, in the **Refine Data** section of the report.

3.2.25 Responding as a Supplier

Suppliers can respond to SAP Ariba Supplier Lifecycle and Performance registration requests, qualification requests, modular questionnaires or other event requests such as SAP Ariba Sourcing event participation request by creating an account on Ariba Network.

During the new supplier onboarding process, after a new supplier request is approved, and a supplier registration invitation is sent to the supplier primary contact, the primary contact receives an email, as shown in Figure 3.47. The supplier can access Ariba Network from the link provided in the email to register the new supplier on Ariba Network and respond to the supplier registration questionnaires and other questionnaires that follow in the supplier lifecycle process, respond to sourcing events, and negotiate contracts with the buying organization.

The supplier primary contact can click on the link to register the supplier on Ariba Network. Once registered, the supplier primary contacts and other supplier contacts associated with this supplier on Ariba Network can access all supplier registration and qualification questionnaires from the SAP Supplier Portal on Ariba Network, as shown in Figure 3.48.

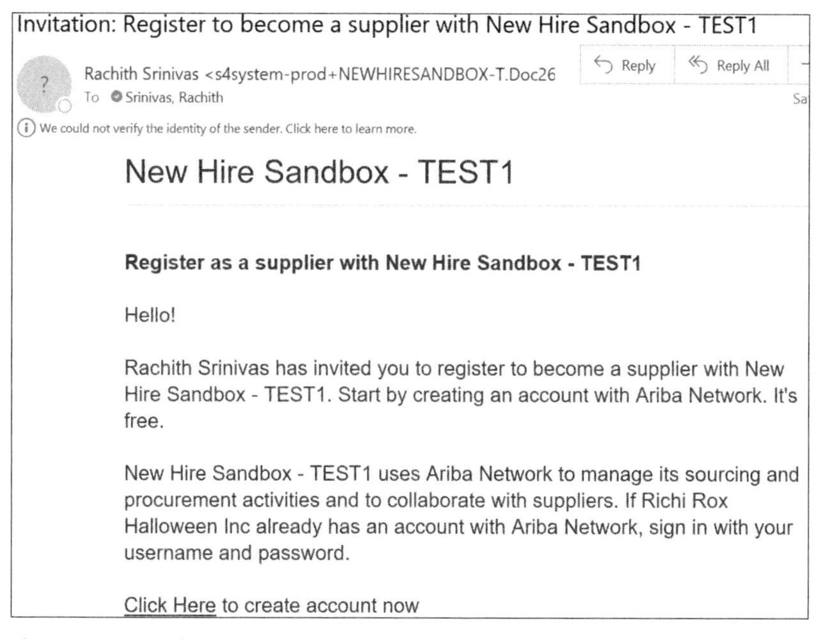

Figure 3.47 Supplier Registration Email Invitation

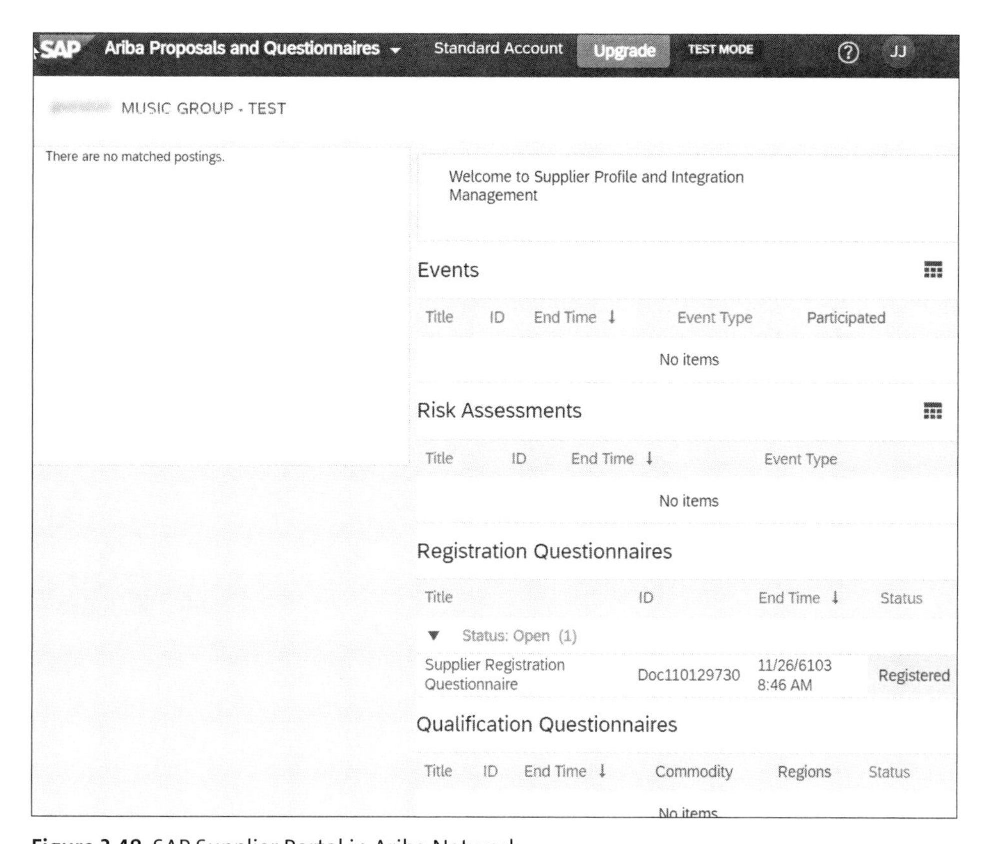

Figure 3.48 SAP Supplier Portal in Ariba Network

They can click on each questionnaire, accept the invitation to respond, and submit their responses to the registration and qualification questions. Required questions in the questionnaire must be answered by the supplier, without which, the response won't be allowed to be submitted. The supplier contacts can choose to revise their response and resubmit, as shown in Figure 3.49.

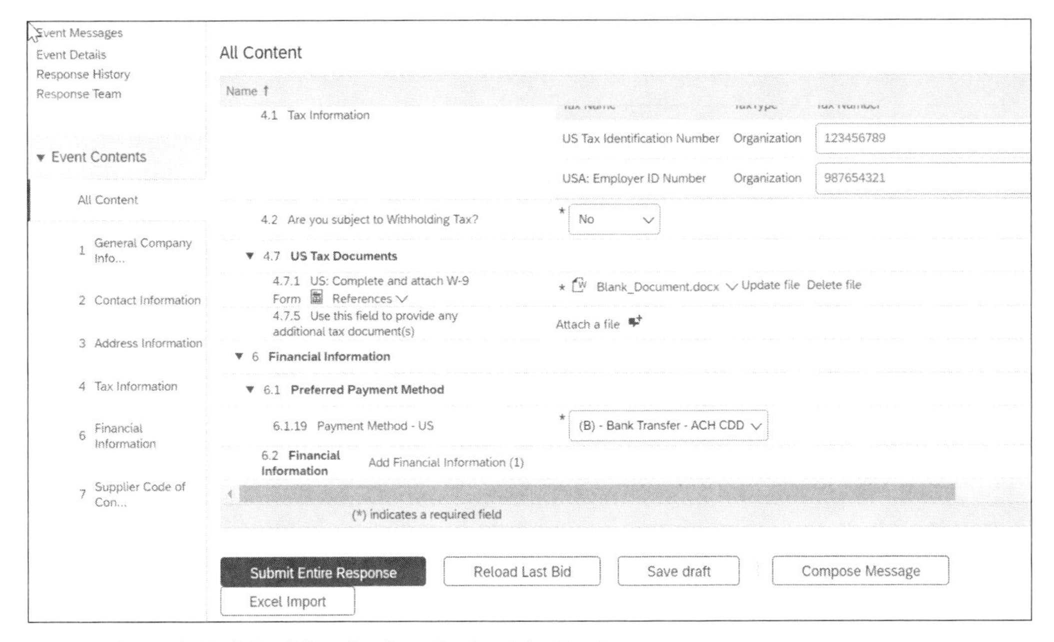

Figure 3.49 Submitting the Supplier Registration Response

After the supplier submits responses, as shown in Figure 3.50, the supplier can monitor the status of the questionnaires on Ariba Network. After the responses are approved by the buying organization, the questionnaire status on Ariba Network is changed to **Registered** or **Qualified**, as shown previously in Figure 3.49.

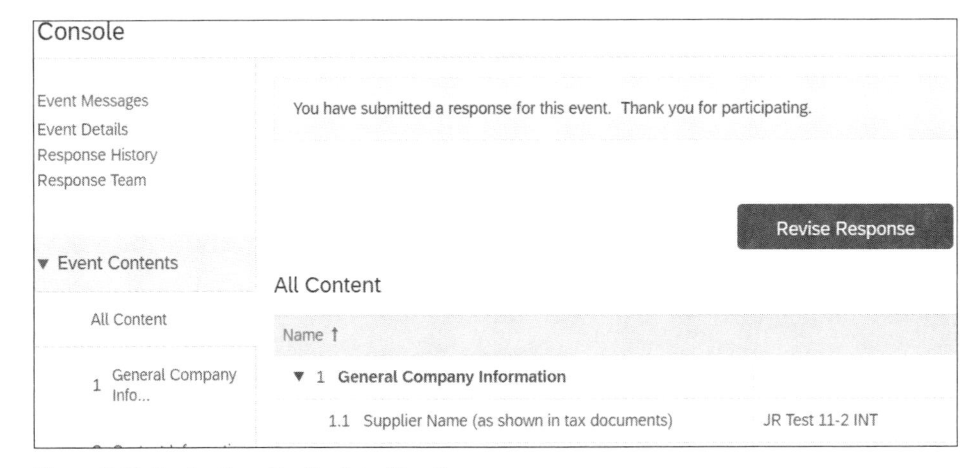

Figure 3.50 Revise Supplier Registration Response

In addition, SAP Ariba Supplier Lifecycle and Performance allows the buying organization to set expirations on the registration and qualification questionnaires. On expiration, the system sends an email to the supplier contact requesting that the supplier resubmit responses to the registration and qualification questionnaires.

Suppliers can also communicate with the buying organization during the registration and qualification process via the **Event Messages** shown in Figure 3.51.

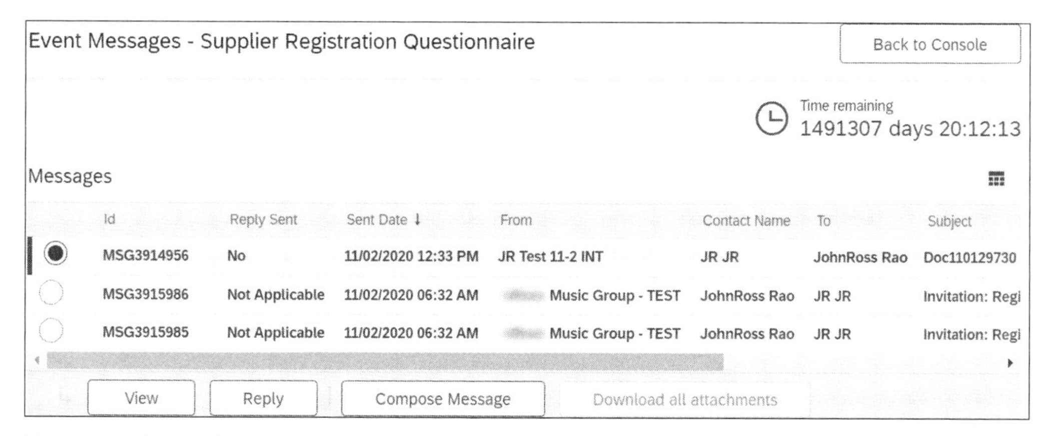

Figure 3.51 Event Messages in Supplier Registration and Qualification

SAP Ariba Supplier Lifecycle and Performance sends email notifications to the supplier contacts throughout the supplier registration and qualification process to notify supplier contacts to take necessary action.

Finally, like the supplier registration questionnaire, qualification questionnaires and modular questionnaires—if any—will be made available to the supplier in the SAP Supplier Portal page on Ariba Network.

Access to other transactional documents such as purchase orders and invoices was covered in Chapter 2.

3.3 Summary

Our goal in this chapter was to describe the benefits of implementing SAP Ariba Supplier Lifecycle and Performance and SAP Ariba Supplier Risk. The features defined in this chapter on SAP Ariba Supplier Lifecycle and Performance, such as supplier request, supplier registration, supplier qualification and preferred supplier setup, modular questionnaires, and supplier certificate management, can be easily deployed in a multi-ERP environment to efficiently manage the full supplier lifecycle from initial onboarding to collecting supplier information to setting the supplier to preferred status. In addition, the features defined in this chapter on SAP Ariba Supplier Risk, such as risk score calculation, engagement request and engagement risk assessment projects, and issues management projects, can be easily deployed as a standalone solution or with other SAP Ariba solutions to provide a comprehensive solution to ensure that risk-related due diligence research is an integral part of your procurement process, thus preventing supply chain disruptions.

Chapter 4
Supplier Risk Management

SAP Ariba Supplier Risk Management is the only end-to-end solution portfolio that lets you manage supplier information, lifecycle, performance, and risk, all in one place.

Avoiding damage and reputation from disruption to an organization's supply chain is becoming more important than ever. The growing pace of business, rapid economic changes, and increased regulatory requirements across the globe call for supplier risk management as an essential part of the procurement process. However, supplier risk management can be tough to implement in a business because risk-related due diligence research tends to be sporadic, costly, and time-consuming, making staying informed about your supply base and about market signals almost impossible.

Now, with SAP Ariba Supplier Lifecycle and Performance and SAP Ariba Supplier Risk, you have a comprehensive supplier management solution that can transform your supply management processes, help you better manage your supply base, and avoid risk. In this chapter, we'll provide detailed insights into SAP Ariba Supply Risk and describe implementation approaches to get these solutions up and running in your enterprise.

4.1 What Is Supplier Risk Management?

Supplier management is a broad term. In today's business landscape, supplier management isn't just a set of administrative tasks to load and maintain supplier records. Companies have started to take a more strategic approach to managing their supply base. Supplier management has transformed into supplier lifecycle management. In simple terms, supplier management is the process of managing a supplier throughout its lifecycle from initial registration and onboarding to qualification to performance management and development.

4.1.1 Supplier Risk Management Strategies

Some of the supplier management strategies described in Chapter 3, Section 3.1.1, such as the supply base strategy and cross-functional collaboration, can help with mitigating supplier risk by engaging risk management during the process of onboarding new suppliers across the enterprise. In addition, let's look at other strategies to enhance supplier risk management processes.

Enterprise-Wide Supplier Information Management Capital System

Using a comprehensive solution, such as SAP Ariba Supplier Lifecycle and Performance as described in Chapter 3, as a single repository to capture and maintain information of every supplier in the enterprise allows for effectively consolidating supplier data across the organization and providing visibility and real-time information updates about a supplier. This enables supply chain transparency, proactive compliance, and ongoing performance review of suppliers, resulting in reduced supplier risk.

Risk Management

Supplier risk management is a proactive process of identifying and mitigating risks cost effectively across a buyer organization's supply chain. This includes segmenting suppliers into risk categories and continuously accessing and updating risk profiles.

An effective risk management strategy helps an organization evaluate its risk profile and define risk factors that must be managed in its supply base, including how suppliers should be assessed. Typically, risk assessment criteria include financial, human resource, environmental, supply chain disruption, performance, and relationship factors.

4.1.2 SAP Ariba Portfolio

SAP Ariba Supplier Risk is SAP's cloud-based supplier risk management solution. As shown in Figure 4.1, the SAP Ariba Supplier Management portfolio consists of SAP Ariba Supplier Lifecycle and Performance and SAP Ariba Supplier Risk.

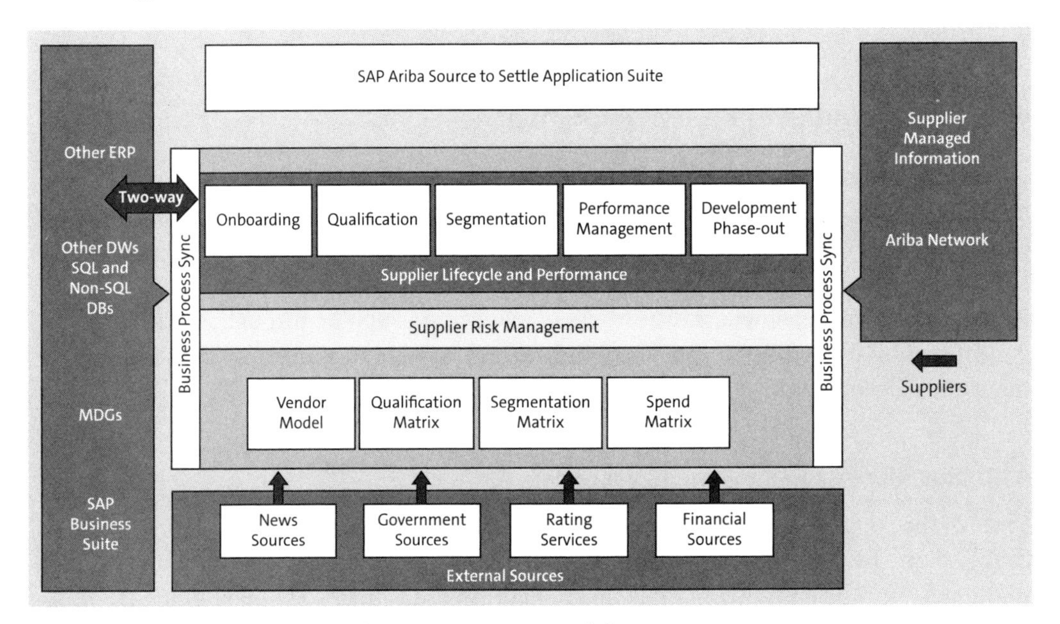

Figure 4.1 SAP Ariba Supplier Management Portfolio

4.2 Implementing SAP Ariba Supplier Risk Management

SAP Ariba Supplier Risk is an effective risk management solution from SAP, which simplifies risk management so businesses make smart, informed, and timely decisions during their procurement process. This solution is natively integrated to the SAP Ariba Supplier Lifecycle and Performance solution to provide a comprehensive solution for onboarding and managing suppliers while continuously identifying and mitigating risk in the company's supply base.

Most buyer organizations must keep an eye on the following risk categories:

- **Regulatory and legal compliance**
  Sanctions and watch lists, bribery and corruption, legal, IT security, fraud, anticompetitive behavior, corporate crime.
- **Environmental and social**
  Human rights, labor issues, health and safety, environmental issues, conflict minerals, unethical practices, decertification.
- **Financial**
  Bankruptcy, insolvency, mergers and acquisitions, divestiture, credit rating downgrade, downsizing, liquidation, tax issues.
- **Operational**
  Natural disasters and accidents, plant disruption or shutdown, labor issues, product issues, project delays, labor issues, and product issues.

SAP Ariba Supplier Risk comes with automated integration to data feeds from multiple public data sources (for more information, see section "Enrichment of Supplier Risk Data") that are continuously scanning for incidents across the globe. Incident types are mapped to the preceding risk categories.

Category and supplier managers can use this risk data to do the following:

- Decide which suppliers to approve for procurement, sourcing, and other activities.
- Segment suppliers by risk levels and risk category.
- Identify problem suppliers and initiate risk assessment and due diligence activities.

In the next section, let's look at how SAP Ariba Supplier Risk can help alleviate the pain of managing supplier risk.

4.2.1 Application at a Glance

SAP Ariba Supplier Risk provides a platform for buyers in any organization to make safer and data-driven decisions during their purchasing process. With this tool, you can keep your buyers well informed about your suppliers and ensure that risk-related due diligence research is an integral and natural part of your procurement process, thus preventing supply chain disruptions.

SAP Ariba Supplier Risk is a market-leading supplier risk solution that allows you to tailor risk views and alerts to your business. You can also segment suppliers based on your risk exposure. This solution provides a comprehensive risk profile for each of your suppliers, enabling you to make more timely, contextual, and accurate business decisions.

Integrated to your procurement processes, this tool helps reduce risks, avoid damage to your revenue or reputation, and improve collaboration with your trading partners with several key capabilities:

- **Integrated and unified vendor model**
  Provides a single accurate supplier record.
- **Risk monitoring**
  Continuously updates monitoring throughout the supplier's lifecycle, from onboarding to qualification.
- **Supplier segmentation**
  Groups suppliers based on risk exposure and risk categories for focused due diligence.
- **Tailored alerts**
  Provides customizable views and risk alerts per your business's risk appetite and need.
- **Integrated to SAP Ariba Supplier Lifecycle and Performance**
  Allows integration for supplier segmentation and qualification based on supplier's category and risk exposure (see Chapter 3).
- **Integrated to other SAP Ariba solutions**
  Allows integrations to other solutions as well, such as SAP Ariba Contracts Management.
- **Supplier 360° View**
  Provides an integrated and comprehensive 360-degree view of the supplier, aggregating data from various processes, including supplier risk profiling.

SAP Ariba Supplier Risk Dashboard

SAP Ariba Supplier Risk allows internal users who have access to the **Supplier Risk** dashboard to follow suppliers, as shown in Figure 4.2.

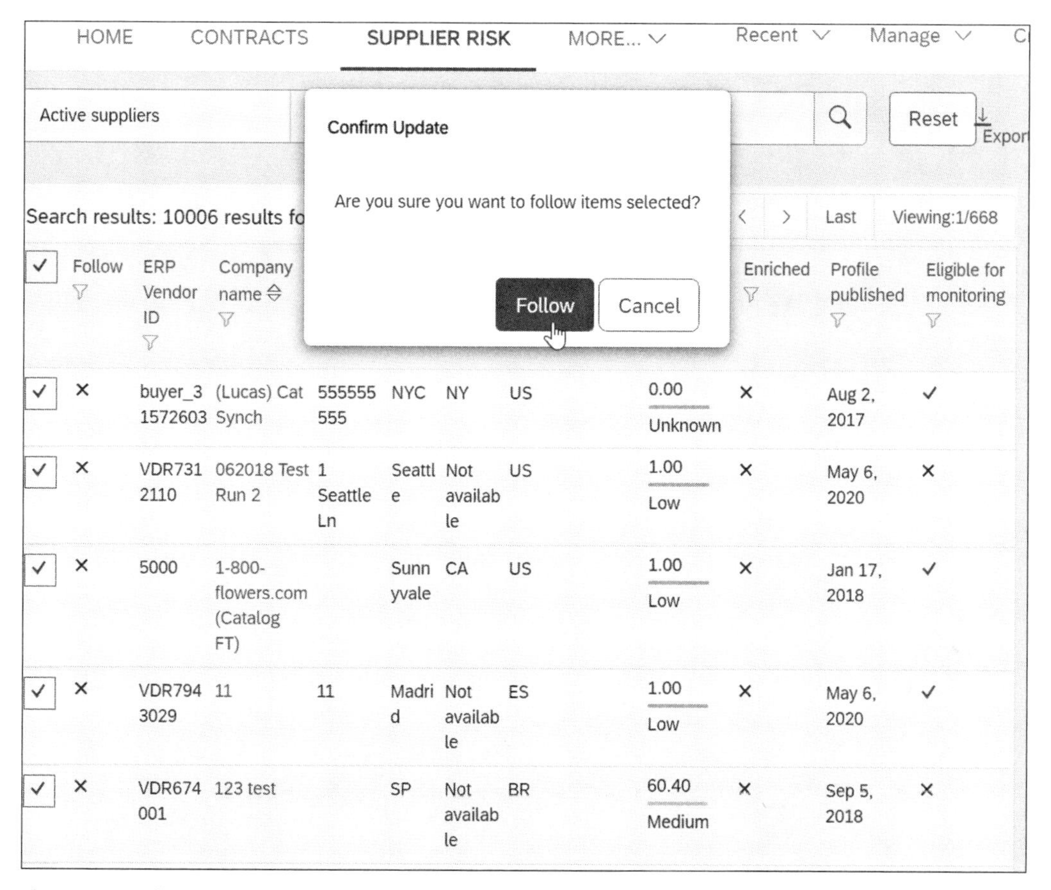

Figure 4.2 Follow Suppliers

On the dashboard, the **Risk summary** tile shows the risk ratings of the suppliers you follow (see Figure 4.3). The **By risk category** tile displays the number of suppliers in each of the high, medium, and low risk levels, by category. The **Alert feed** tile displays the latest alerts for the suppliers you follow. Finally, SAP Ariba Supplier Risk has a geospatial map that shows the number of suppliers you're following in each region. You can filter by specific risk types, risk levels (high, medium, low), or industry, as shown in Figure 4.3.

At the bottom of the screen, there is a table with the suppliers within your filter criteria. The table shows the country risk score and the risk exposure for each supplier, as shown in Figure 4.4.

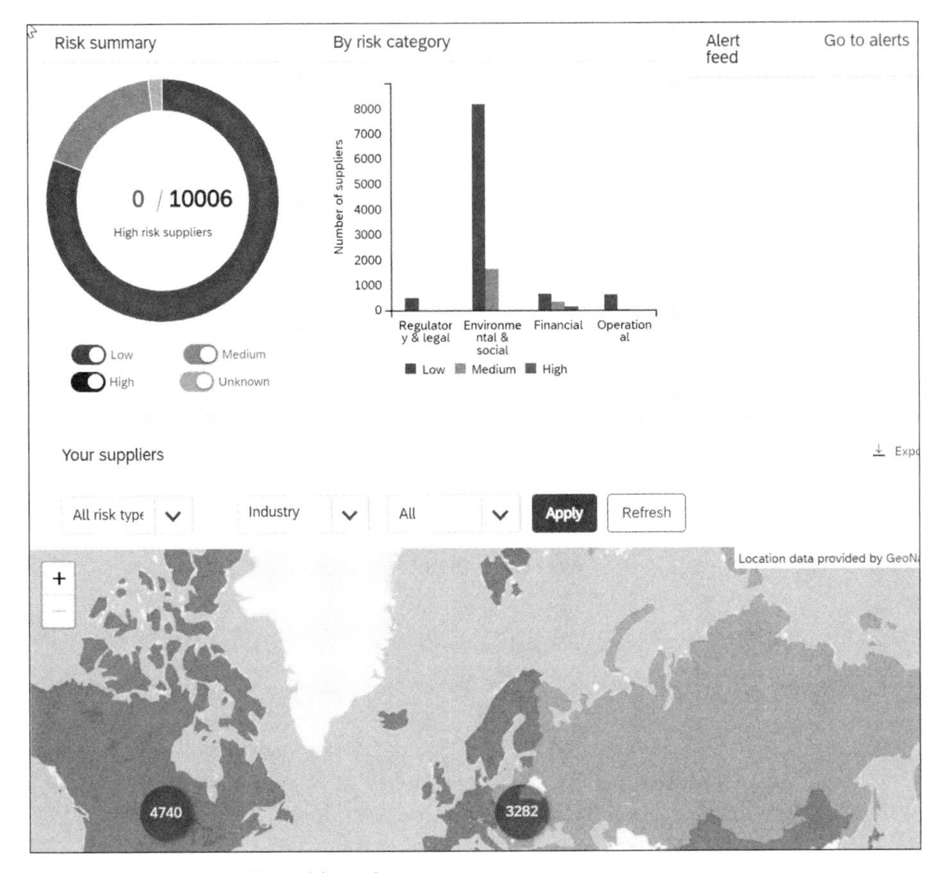

Figure 4.3 Supplier Risk Dashboard

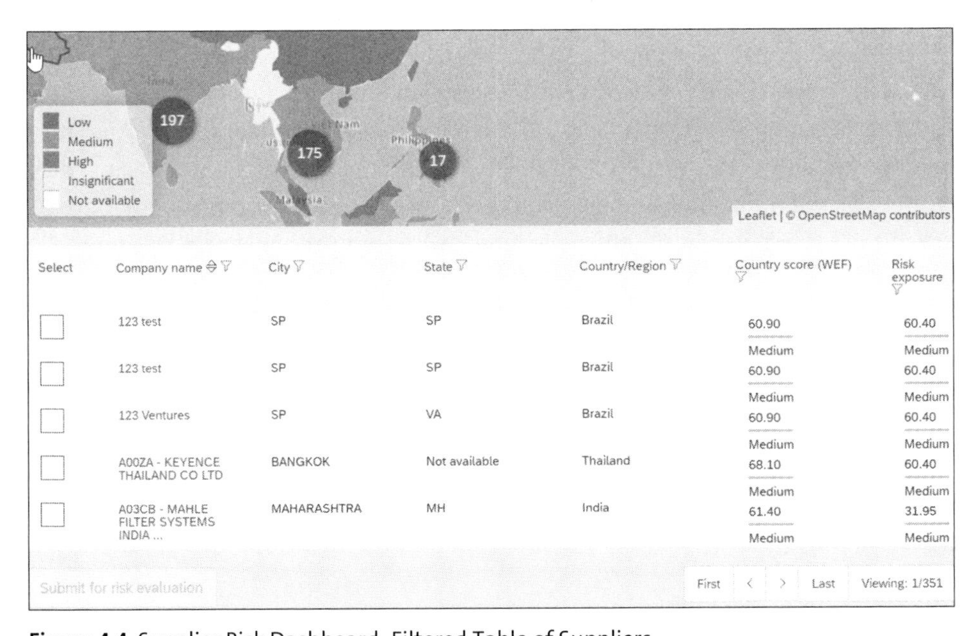

Figure 4.4 Supplier Risk Dashboard: Filtered Table of Suppliers

In addition, clicking on the **Medium** zone, for example, of the **Risk summary** tile at the top shows the count of suppliers you're following that are in the medium risk exposure level (see Figure 4.5).

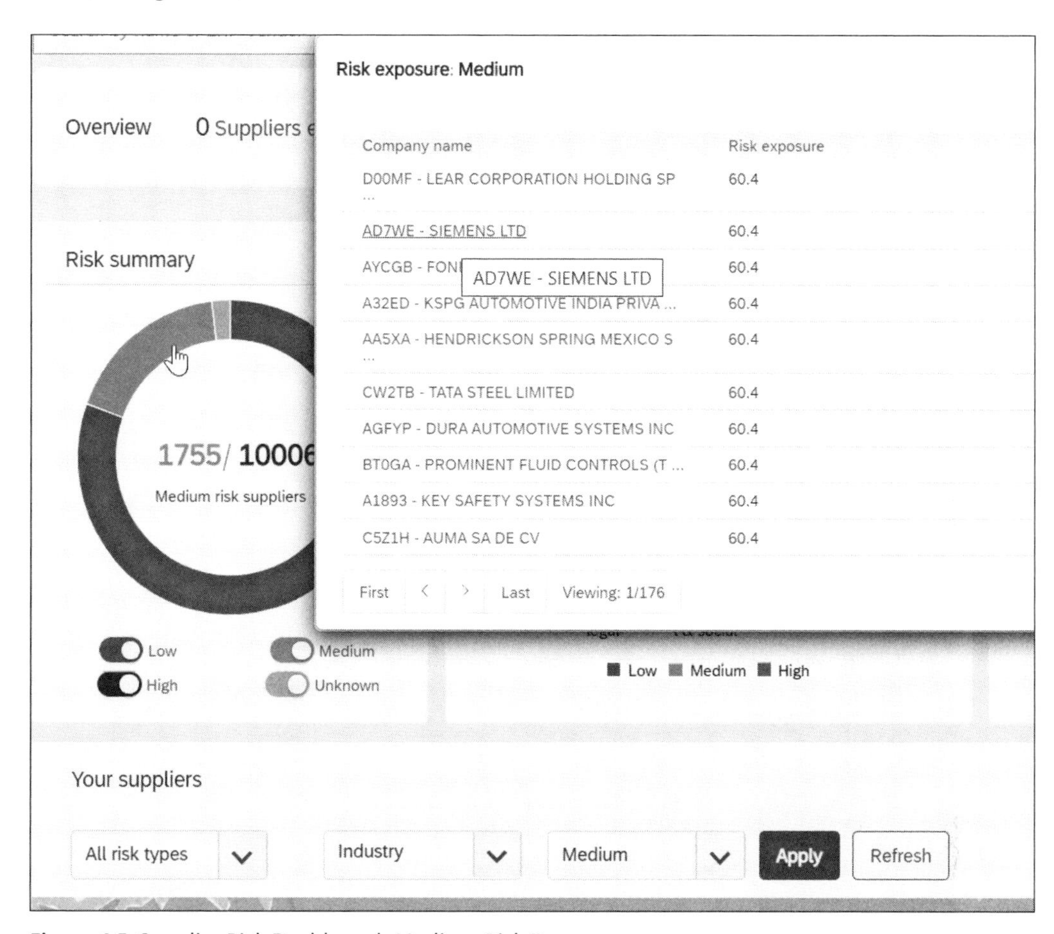

Figure 4.5 Supplier Risk Dashboard: Medium Risk Exposure

If you click on one of the supplier names on the list shown in Figure 4.5, you'll be taken to the **Supplier 360° View** for that supplier. If you click the **Risk** link on the left side of the **Supplier 360° View** page, the **Risk exposure** page for this supplier will open with details on risks identified over a period of time. There are additional tabs for **Risk incidents**, **Enriched corporate info**, and **Engagement requests** for this supplier, as shown in Figure 4.6.

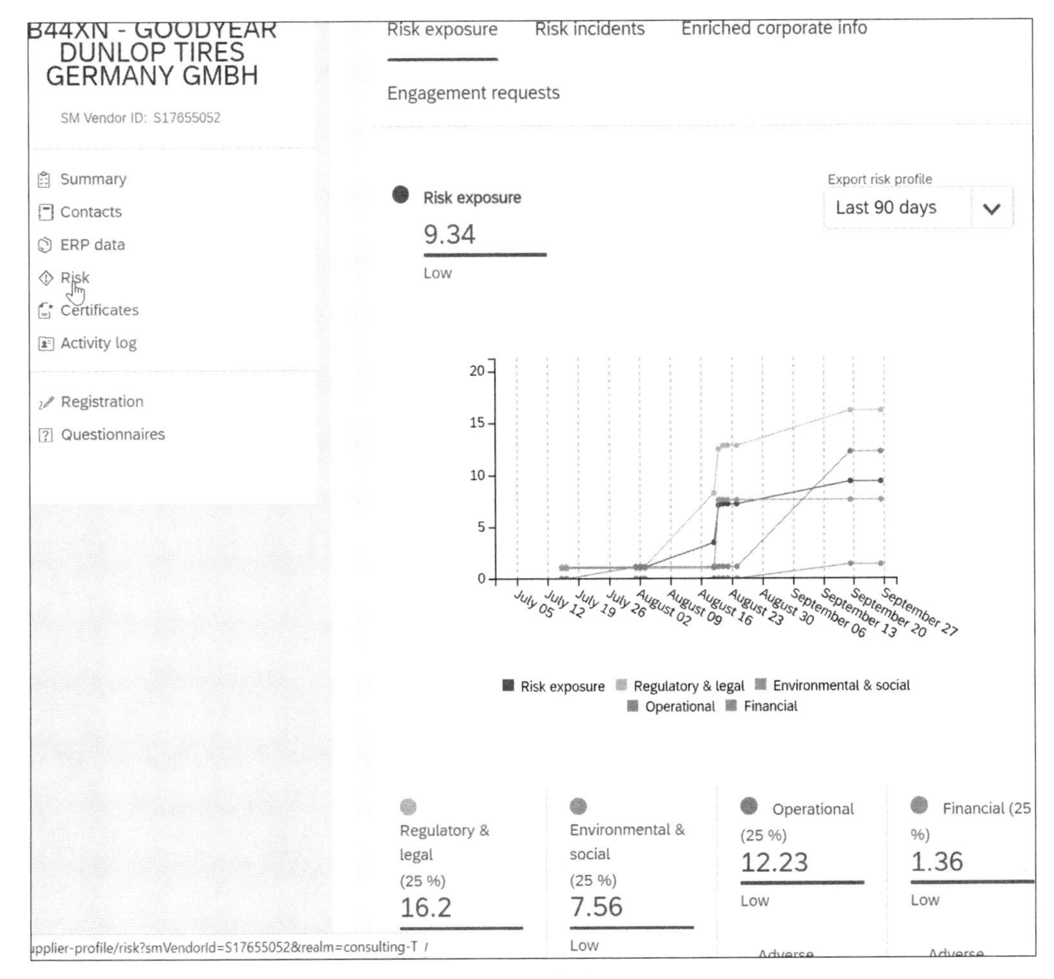

Figure 4.6 Supplier Risk Page in Supplier 360° View

These tabs allow you to do the following:

- **Risk incidents**
  Review all the incidents by incident type for this supplier. You can also filter by a specific incident type, as shown in Figure 4.7. Clicking on each incident will take you to the actual source report.
- **Enriched corporate info**
  View information on a company's parentage structure, diversity indicators, number of litigations, and so on.
- **Engagement requests**
  View details on all the engagement projects with a count of total projects, projects in progress, and projects completed. In addition, issues created for the engagement projects are displayed.

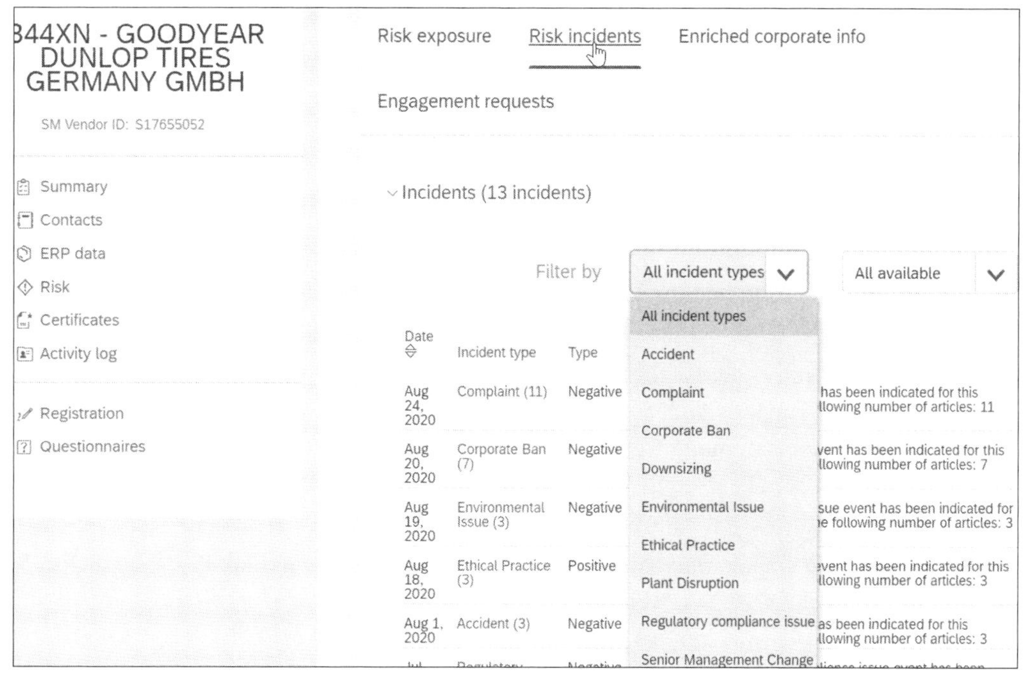

Figure 4.7 Supplier Risk Incidents in the Supplier 360° View

SAP Ariba Supplier Risk Process

The SAP Ariba Supplier Risk module is integrated with public databases out of the box to monitor for incidents related to the following risk categories: legal and regulatory compliance, environmental and social, financial, and operational. Other categories and databases can be configured based on a customer's existing license to other databases. Risk scores can be configured for specific risk categories. The supplier risk score is periodically updated based on the enrichment of supplier risk data from the configured public database sources.

The following is a typical process flow:

1. An incident in one of the risk categories is reported by one of the standard or custom licensed database sources. The risk score is updated for the specific supplier based on the risk score configurations for that category or risk.
2. A supplier risk engagement requester who reviews this incident in the **Supplier Risk** dashboard can create an engagement project to assess the risk in more detail and take any necessary mitigation steps. The requester can also email this risk to internal users from the user interface.
3. Risk governance users can pick up these engagement projects and gather additional evidence related to the risk. They can create a risk issue at the engagement level and work on controlling the risk.

4. If the risk can't be controlled by the internal organization through, for instance, a contract, the supplier can be asked to create a risk issue at the control level and work on remediating the risk.
5. Resolution is then reviewed by the buyer organization, and the engagement project is either approved or rejected based on the review.

Enrichment of Supplier Risk Data

The SAP Ariba Supplier Risk base solution includes enrichment of the supplier data and monitoring from the following data sources related to supplier risk:

- **Enrichment**
  A unified supplier database with more than 245 million records, including data such as supplier parentage (information on parent company), number of suits, liens, bankruptcy indicators, diversity indicators, and so on.
- **Financial**
  Data-following sources, such as RapidRatings, Bureau van Dijk (BVD), and Dun & Bradstreet (D&B). This will impact the financial risk scoring for the supplier.
- **Country risk/forced labor**
  Through the World Economic Forum, the Made in a Free World database includes country-specific competitive rankings, risk management rankings, and information on forced labor (e.g., industry forced labor score and category forced labor score).
- **New alerts**
  Semantic Vision provides real-time alerts on both positive and negative events by scanning more than 600,000 global and regional sites, expert blogs, government websites, and private databases.
- **Natural disaster alerts**
  Global Disaster and Alert Coordination System (GDACS) provides disaster-related information. This is a joint initiative of the United Nations and European Commission to disseminate such information in order to improve coordination of international relief efforts.
- **Compliance monitoring**
  ScreenIQ by Exiger is an optional add-on feature that can be enabled to monitor sanctions and watch lists, anti-corruption and bribery, and regulatory and compliance violations.

Note

SAP Ariba Supplier Risk allows customers to add additional third-party databases for which they have an existing license. After these third-party databases are added, incidents reported on suppliers from these databases will be automatically fed into SAP Ariba Supplier Risk and included in the overall risk scoring.

Risk Assessment Process

Risk assessment projects provide a process for evaluating the risk of engaging with a supplier. Based on the evaluation, the buyer organization can determine whether or not to engage with this supplier. If they decide to engage, then these evaluations can determine the level of monitoring required. For example, some engagements, such as consulting engagements, may involve access to confidential information or company networks or facilities, thus requiring stringent risk assessments.

A risk assessment project typically includes four stages:

1. **Requesting the engagement and inherent risk assessment**
   A user in your company who wants to engage with a supplier or other third party requests a new engagement risk assessment by creating an engagement request and filling out the engagement request form.
2. **Sending detailed engagement-level risk assessments**
   After the engagement request is approved, the person responsible for reviewing the inherent engagement risk sends risk assessments to internal stakeholders and (if applicable) to suppliers or other third parties.
3. **Responding to risk assessments**
   Recipients are notified of the assessments they need to fill out. Internal stakeholders fill out their risk assessments on the **Engagement** page in SAP Ariba Supplier Risk.
4. **Evaluating and approving risk assessments**
   Depending on how the buyer organization's assessments are set up, a residual risk score might be calculated for each assessment based on the submitted answers.

The risk assessment project template functions in much the same way as project templates in other SAP Ariba solutions. However, the way users experience projects created from this template is quite different. Most users who participate in risk assessment projects don't see the classic project interface with its tabs for documents, tasks, team, and so forth. Instead, they work in the **Supplier 360° View** and in the **Supplier Risk** dashboard for an individual supplier, where task owners can send engagement-level risk assessments to various stakeholders, review answers, and complete tasks. Project owners don't manage the project's team, upload additional documents, or perform other activities associated with the classic project interface in other types of SAP Ariba projects.

> **Note**
> Users who are members of the supplier risk engagement governance analyst group will see an **Advanced View** link on the **Engagement** page. This link will take these users to a classic view of the project.

Engagement Risk Issues Management Process

Issue management is the process by which internal users, governance experts, and other stakeholders at your company raise, analyze, and resolve issues related to supplier or third-party engagement risk assessments. The nature and severity of an issue, and whether or not a satisfactory resolution has concluded, are some of the factors that approvers of engagement risk assessment projects take into account when approving or denying an engagement.

At any time between when the request is submitted and the engagement is completed or canceled, the requester and person responsible for review can create issues to highlight potential problems or concerns with the engagement as a whole and then track and resolve them. Requesters and people responsible for the review of these assessments can also create issues related to specific engagement-level risk assessments at any time between when the first assessments are sent out and when the engagement is completed. Each issue is a separate project with its own workflow and approvals embedded within the supplier risk assessment project.

The issue management process provides a process for gathering all the relevant information about an issue and involving relevant experts and other stakeholders in its analysis and resolution. The process includes five stages:

1. **Issue creation**
   A user becomes aware of a potential issue with a proposed supplier or third-party engagement with an assessment project in progress—either with the entire engagement or with a specific engagement-level risk assessment—and creates an issue with **Draft** status. The user who created the issue might fill out most or all the information in the **Issue Details** area, including specifying the assignee, or might leave most of the issue's fields blank. The **Comments** area isn't yet available during issue creation.
2. **Issue definition**
   The issue assignee (if one exists at this point) and owners of various issue definition tasks edit the issue to provide more detailed information, add comments, and complete their assigned tasks. The issue then moves from **Draft** to **In Progress** status.
3. **Issue analysis**
   The assignee (if one exists at this point) and owners of various issue analysis tasks review the issue details, edit the issue to update or add information if necessary, add comments, and complete their assigned tasks. These users might or might not propose resolutions at this stage. If the issue hasn't yet been assigned, they can also specify a user who can resolve the issue as the assignee at this point.
4. **Issue resolution**
   The assignee and owners of various issue resolution tasks review the issue information, edit it to propose or finalize its resolution, and complete their assigned tasks. If the fields of the **Inherent Issue Document** area haven't yet been filled out, they are finalized at this point.
5. **Issue resolution acceptance**
   Task owners complete any other assigned tasks related to issue resolution acceptance, and the approvers assigned to the issue review the resolution and finally

approve it. The issue then moves from **In Progress** status to either **Resolved** or **Request Denied** status.

4.2.2 Planning the Implementation of SAP Ariba Supplier Risk

As with SAP Ariba Supplier Lifecycle and Performance, SAP Ariba Supplier Risk is implemented using the SAP Activate methodology for the cloud. To implement SAP Ariba Supplier Risk, the following resources will be required on the SAP Ariba side:

- SAP Ariba project manager
- SAP Ariba Supplier Risk functional lead
- SAP Ariba shared services lead
- SAP Ariba technical lead

The following resources will be required on the customer side:

- Project sponsor
- Project manager
- Functional lead
- Subject matter experts (SMEs)
- System administrator for SAP Ariba solutions
- Support lead

Prepare Phase

Activities in this phase are similar to the activities described in the prepare phase of an SAP Ariba Supplier Lifecycle and Performance implementation.

Explore Phase

Activities in this phase are similar to the activities described in the prepare phase of an SAP Ariba Supplier Lifecycle and Performance implementation.

After the kickoff is complete, the blueprinting meetings and breakout sessions will be conducted by the SAP Ariba Supplier Risk functional lead. The objective of this phase is to document the detailed requirements and to design the future-state with SAP Ariba Supplier Risk. For SAP Ariba Supplier Risk, the following key areas will be covered:

- Supplier risk score calculation configurations
- Risk assessment processes
- Engagement risk issues management processes
- Master data requirements
  - Master data to be loaded
  - Master data refresh frequency and master data strategy

Assumptions during the Prepare and Explore Phases for Deploying SAP Ariba Supplier Risk

Let's look at some of the assumptions during the prepare and explore phases:

- **Risk scoring and customization enablement (optional)**
  Also known as risk taxonomy and scoring, this optional feature may be configured during the deployment. Services to deploy this feature include guidance and training to walk through risk score capabilities while outlining relevant fields and their impact on scores. The product supports an out-of-the-box model and provides you options to configure the scoring model, if needed, as outlined in the risk score.
- **Engagement risk enablement (optional)**
  Deployment of this feature includes guidance and training to walk through the engagement risk template and the creation of assessments, which can be multiple documents (surveys) used to facilitate the engagement risk workflow.
- **Access to multiple sites**
  Access is provided to the following sites: test, production, and (as applicable and provided by SAP in its discretion) development. The applicable sites will remain available for the subscription term.
- **Reviewing customer data**
  After SAP Ariba receives the completed data collection documents with your information, SAP Ariba will first review the information and identify any potential gaps and errors. If any errors arise, SAP Ariba will notify you for any corrections.

As part of this phase, the SAP Ariba technical lead will train your users on the SAP Ariba Supplier Risk interface and functionality, including training sessions demonstrating site navigation, alert setup, and administrative capabilities and user management settings for news alerts, engagement risk setup (optional), and risk score configuration (optional).

Realize Phase

Activities in this phase are similar to the activities described in the realize phase for an SAP Ariba Supplier Lifecycle and Performance implementation.

The SAP Ariba Supplier Risk functional lead works with the SAP Ariba shared services lead to perform the following activities:

- Configure supplier risk score calculation.
- Register third-party providers.
- Configure the risk assessment project template.
- Configure the issues management project template.
- Load master data such as suppliers, supplier risk data, and other necessary master data according to the enablement workbook defined in the explore phase.

Note

In a typical SAP Ariba Supplier Risk implementation, the scope for deployment by the SAP Ariba technical lead is defined in the deployment descriptor risk assessment template:

- **One engagement risk request template**
  - Up to 20 questions
  - 5 tasks
  - 5 conditions
- **Two risk assessment surveys (internal or external)**
  - Up to 20 questions

Deploy Phase

Activities in this phase are similar to the activities described in the deploy phase for an SAP Ariba Supplier Lifecycle and Performance implementation. After user acceptance testing (UAT) is completed, you're ready to deploy SAP Ariba Supplier Risk in your production environment.

4.2.3 Configuring SAP Ariba Supplier Risk

SAP Ariba Supplier Risk provides detailed data on potential supplier risk in a number of areas, including financial, regulatory, and legal risk.

Category and supplier managers use this risk data to do the following:

- Choose which suppliers to approve for procurement, sourcing, and other activities.
- Segment suppliers by risk levels and risk category.
- Identify problem suppliers and initiate risk assessment and due diligence activities.

Setting up SAP Ariba Supplier Risk in a customer realm involves configuring risk scoring, making sure users have permission to work with risk data, and importing suppliers into the database so that they are associated with the risk data.

In the next section, we'll look at how to configure SAP Ariba Supplier Risk.

Configuring Risk Exposure

SAP Ariba Supplier Risk allows customers to configure the risk exposure based on their company's risk appetite. Supplier risk manager can do the following:

- **Enable/disable default data sources**
  Enable or disable the default data sources that provided by SAP Ariba, as shown in Figure 4.8.

Figure 4.8 Risk Exposure: Data Sources

- **Define risk category weights**
  Configure weights and risk thresholds to risk categories based on customer's risk priorities (Figure 4.9).

Figure 4.9 Risk Exposure: Category Weight Configuration

- **Define the risk score for standard and custom fields**
  For all the standard fields and risk categories, configure weights, threshold order (safer to riskier and riskier to safer), and assign less than and greater than scores (see Figure 4.10). Based on data feeds from these standard databases, risk scores will be assigned per configuration. Custom fields can be configured for internal users to assign custom risk scores that can be included in the overall score calculation.

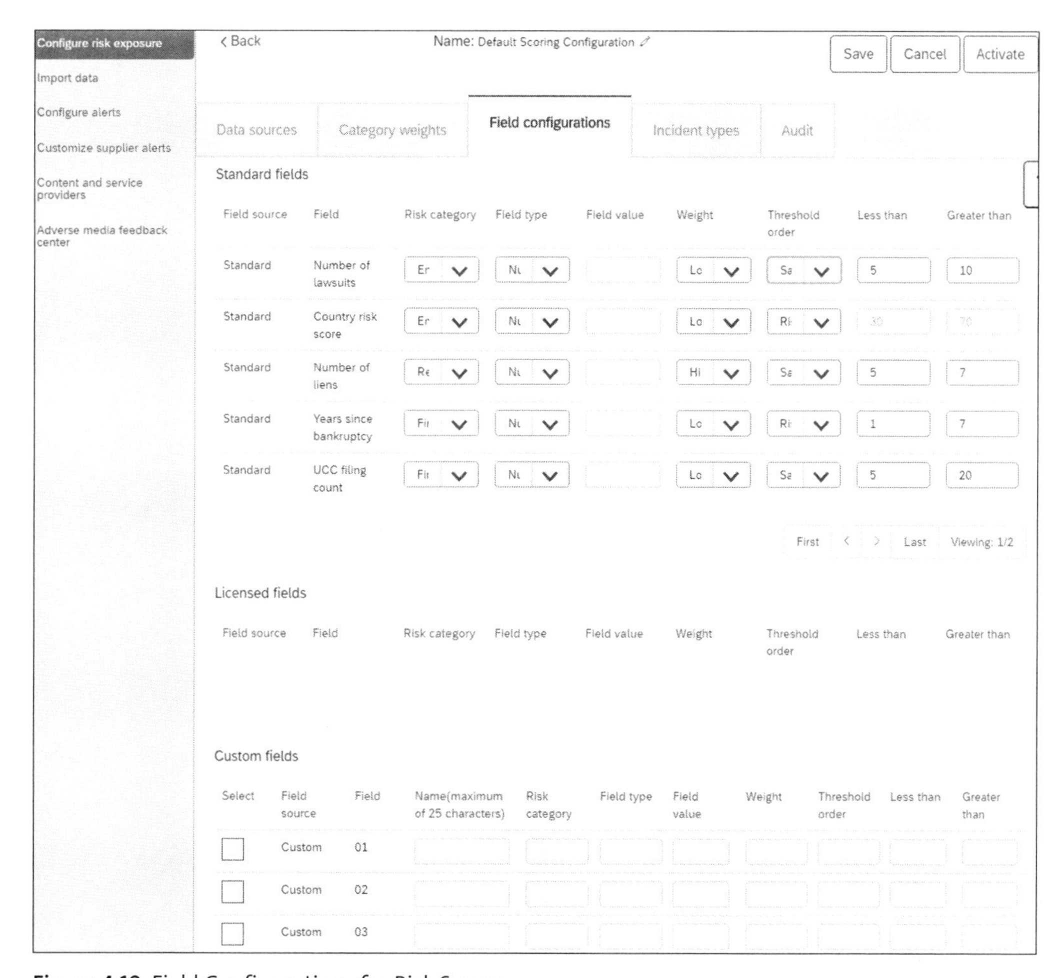

Figure 4.10 Field Configurations for Risk Scores

Managing Supplier Risk Users

Users who need access to supplier risk data in the customer realm must be added to the appropriate supplier risk system user group. Table 4.1 lists system user groups used in SAP Ariba Supplier Risk.

Group Name	Access
Supplier risk manager	▪ View supplier risk information under the **Supplier Risk** tab on the dashboard and in **Supplier 360° Reports**. ▪ Perform tasks in the **SM Admin** area, including importing and exporting supplier-related data. ▪ Download the risk scoring configuration workbook, and run risk metrics reports. ▪ Register licenses for third-party providers of supplier risk data.
Supplier risk user	▪ View supplier risk information under the **Supplier Risk** tab on the dashboard and in **Supplier 360° Reports** as well as receive and manage alerts for followed suppliers.
Supplier risk engagement requester	▪ Create engagement requests, and fill out the internal risk assessments.
Supplier risk engagement analyst	▪ Fill out internal risk assessments, and run compliance report updates.
Supplier risk engagement expert	▪ Fill out internal risk assessments.
Supplier risk engagement governance analyst	▪ Send out or skip risk assessments for an engagement request, fill out internal risk assessments, and specify ad hoc approvers for engagement requests and risk assessments with no defined approval flow.
Supplier management operations administrator	▪ Import and export data in the **SM Admin** area.

Table 4.1 System User Groups in SAP Ariba Supplier Risk

Importing Supplier Master Data

See Chapter 3, Section 3.2.22, for information on loading supplier master data.

> **Note**
>
> To enrich supplier data, the supplier import loads should include the country code or DUNS number.

Configuring Alerts

SAP Ariba Supplier Risk allows customers to configure the severity of incident types per their risk appetite.

To configure the incident alerts, follow these steps:

1. Log in as a supplier risk manager.

2. Click on the **Setting** button on the **Supplier Risk** dashboard.
3. Select **Configure alerts** on the left side of the screen.
4. Change the severity for the different incident types, as shown in Figure 4.11. Select **Ignore** to inactivate incident type alerts.

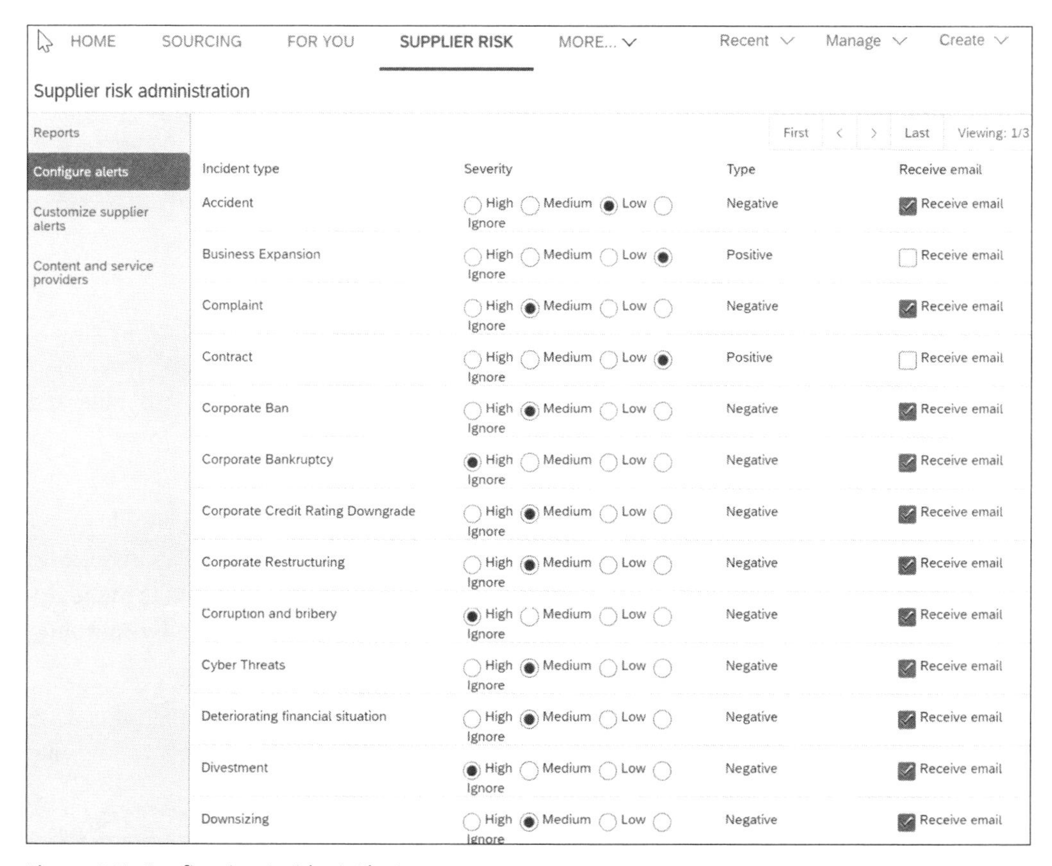

Figure 4.11 Configuring Incident Alerts

Configuring Risk Scoring

SAP Ariba Supplier Risk calculates potential risk scores for each supplier. A supplier's risk score is a numerical value (from 1 to 5) that indicates the supplier's risk level with 5 being the riskiest and 1 being the least risky. Users in the buyer organization can take these scores into consideration when making decisions about the company's relationships with its suppliers.

Risk scores can be based on a number of different factors, including the following:

- News items about the supplier
- Corporate information about the supplier
- Geographical data on natural disasters
- Legal, regulatory, and environmental compliance information about the supplier

- Risk data associated to the supplier's country profile
- Structured risk information based on the supplier's corporate hierarchy
- Internal buyer organization rating based on spend volume and strategic or preferred supplier status

Note

The buyer organization can define the risk score calculation based on the organization's appetite, using an Excel spreadsheet workbook to define weights and priorities to different components of the risk score.

Workbook to Configure Risk Score

The risk score configuration workbook is a Microsoft Excel workbook that you can use to configure the risk scoring model to meet your company's policies and thresholds and to define any custom fields you'll use. You can also maintain your supplier and supplier risk data in the workbook.

Workbook to Specify the Data Sources Used in Supplier Risk Score Calculation

SAP Ariba Supplier Risk uses data from several different sources to calculate risk scores for suppliers. You can enable or disable these sources on the **Data Sources** tab of the risk score configuration workbook. Data sources can be enabled or disabled by checking/unchecking them on the data source worksheet.

After this workbook is filled out, export the Excel file to a comma-separated value (CSV) file, and import the CSV file using the supplier management administration data import.

Note

To import supplier and supplier risk data in the supplier management administration area, you must be a member of the supplier risk manager system group.

Registering Third-Party Licenses

SAP Ariba Supplier Risk receives data from several third-party data sources to identify risk. To receive such data, the customer must subscribe to the third-party providers, and the license must be imported into SAP Ariba Supplier Risk.

To import third-party licenses, follow these steps:

1. Log in as a supplier risk manager.
2. From the **Supplier Risk** dashboard, click on the **Settings** icon.
3. Select **Content and service providers** on the left side of the page.

4. Click on the available providers to view status **Enabled**, as shown in Figure 4.12.
5. Click the **Edit** icon or click **Request**, and then enter your credentials.
6. To register, click **Continue**.

Figure 4.12 Checking the Status of Available Service Providers

4.2.4 Configuring Risk Assessment Project Templates

SAP Ariba Supplier Risk is loaded with *one* default risk assessment template—the engagement risk assessment project template. This template can be accessed by users with the template creator role.

Open the template and create a new version of the template before customizing the standard template according to your requirements.

Documents Tab

The risk assessment process is initiated by an engagement request, which solicits information about the inherent risks of engaging with a supplier or third party and starts the evaluation process.

The following default documents are available in the **Documents** tab:

- **Forms and questionnaires**
  The risk assessment template project includes a default survey document that forms the basis of the engagement request form. You can add other survey documents to the template to create additional engagement-level risk assessment questionnaires for suppliers and internal stakeholders.

- **Grading and scoring**
  Target grades, pre-grades, and weights can be set up for questions in template survey documents. Risk assessment projects use these grades and weights to calculate residual risk scores for engagement-level risk assessment questionnaires.

Tasks Tab

Under this tab, you can create four phases, one for each stage of the risk assessment process. To-do tasks and approval tasks can be set up for the engagement risk assessment process. You can add an associated to-do task for each engagement-level risk assessment survey document (but not for engagement request survey documents). These to-do tasks ensure that external assessments are sent to suppliers and that internal recipients can edit these internal assessments to submit answers.

The risk assessment project template includes a default approval task on the default engagement request survey document. You can add approval tasks to additional survey documents and organize them into phases to define the order in which project questionnaires are sent, evaluated, and approved.

Make sure that the approval task on the engagement request survey document is the first task to trigger in the entire project. This positioning is what defines the survey as the form that users fill out after choosing **Create Engagement Request** on the dashboard and defines the other template survey documents as additional internal or external risk assessments.

> **Note**
>
> The template survey document with the first approval task is always treated as the engagement request. If you want to move straight from the request to the engagement-level risk assessment stage without a manual approval, set the task to auto-approve but maintain its position.

Team Tab

The default project owner group is available in the **Team** tab. The creator of the risk assessment project will be automatically added to this project.

4.2.5 Configuring Issues Management Project Templates

The issue management project template defines the process by which internal users in the buyer organization can raise, analyze, and resolve issues related to engagement risk assessment projects and their assessment questionnaires.

The project template must always contain one survey document with an associated approval task and a specific configuration of phases and tasks, which we'll describe in this section.

Documents Tab

The issue management project template only supports one survey document for the **Inherent Issue Document**. This default survey document is included in the template, which includes default questions. The following default questions include specialized supplier field mappings that define their functions in the issue management workflow:

- **Title**
  This is mapped to `project.Title`.
- **Issue Description**
  This is mapped to `project.IssueDescription`.
- **Issue Severity**
  This is mapped to `project.IssueSeverity`.
- **Issue Probability**
  This is mapped to `project.IssueProbability`.
- **Issue Assignee**
  This is mapped to `project.Assignee`.
- **Due Date**
  This is mapped to `project.IssueDueDate`.
- **Issue Type**
  This is mapped to `project.IssueType`.
- **Resolution Type**
  This has possible answers of **Unspecified**, **None**, **Remediate**, **No Action**, and **Defer**.
- **Resolution Description**
  This is used to provide details on the actions taken to resolve/mitigate the risk.
- **Mitigation Plan**
  This is an attachment field type.

These questions with the field mappings are required for the proper functioning of the issue management workflow. You can edit the title or other supported settings of the questions, but they must be present in the issue management project template survey document.

Tasks Tab

The **Tasks** tab contains four phases by default, which are in the following specific order to perform the issues management workflow process:

1. **Issue definition phase**
   The first task in your issue management workflow must be in the issue definition phase and can't have any predecessors. This phase becomes active automatically when a user creates an issue. If you add another task in the issue definition phase, it must specify that first task as its predecessor, and so on, so that all the tasks in the phase are chained together as predecessors.

2. **Issue analysis phase**
   The first task in this phase must specify the last task in the issue definition phase as its predecessor.
3. **Issue resolution phase**
   The first task in this phase must specify the last task in the issue analysis phase as its predecessor.
4. **Issue resolution acceptance phase**
   The first task in this phase must specify the last task in the issue resolution phase as its predecessor, and the last task of this phase must not be set as a predecessor to any other task. A default approval task exists in this phase. By default, the project owner of the issues management project is set as the approver, but the owner can be changed.

This order must not be changed because each phase has to-do or approval tasks in a specific order with predecessors. Again, the order of these tasks should not be changed.

Note

Members of SAP Ariba Supplier Risk user groups can create issues and edit the issues that *they have created*. Only members of the SAP Ariba Supplier Risk engagement governance analyst group can edit issues *created by other users*.

4.3 Managing Supplier Risks

Here we'll take a look at the steps to manage supplier risk. We'll begin with monitoring supply risk, and then we'll move into engagement risk projects.

4.3.1 Monitoring Supply Risk

As discussed previously in Section 4.2.1, internal users such as supplier risk engagement users, category managers, or supplier managers with access to the **Supplier Risk** dashboard can follow suppliers that are under their purview and monitor their suppliers using the **Supplier Risk** dashboard. Clicking on the **Supplier 360° View** for each supplier provides risk scores based on the risk score configurations set up for their buyer organization.

In addition, SAP Ariba sends adverse media alerts daily. If one or more of the suppliers you follow are part of the alerts, you'll receive an email with the alerts, as shown in Figure 4.13. Buyer users can log in to **Supplier Risk** dashboard and drill down into additional information and risk scores and then take further actions such as creating an engagement risk assessment project or an engagement issue.

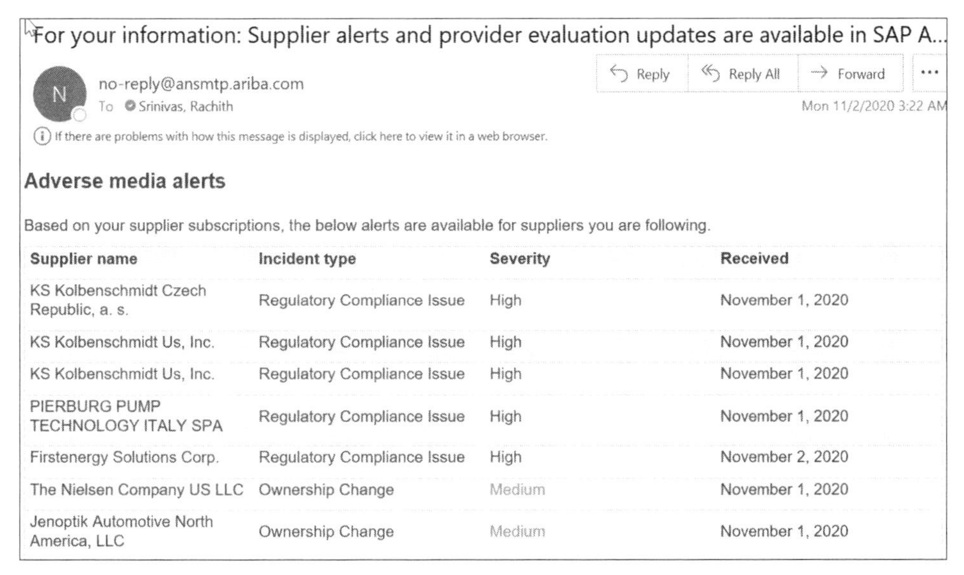
For your information: Supplier alerts and provider evaluation updates are available in SAP A...

no-reply@ansmtp.ariba.com
To Srinivas, Rachith
Reply | Reply All | Forward
Mon 11/2/2020 3:22 AM

If there are problems with how this message is displayed, click here to view it in a web browser.

Adverse media alerts

Based on your supplier subscriptions, the below alerts are available for suppliers you are following.

Supplier name	Incident type	Severity	Received
KS Kolbenschmidt Czech Republic, a. s.	Regulatory Compliance Issue	High	November 1, 2020
KS Kolbenschmidt Us, Inc.	Regulatory Compliance Issue	High	November 1, 2020
KS Kolbenschmidt Us, Inc.	Regulatory Compliance Issue	High	November 1, 2020
PIERBURG PUMP TECHNOLOGY ITALY SPA	Regulatory Compliance Issue	High	November 1, 2020
Firstenergy Solutions Corp.	Regulatory Compliance Issue	High	November 2, 2020
The Nielsen Company US LLC	Ownership Change	Medium	November 1, 2020
Jenoptik Automotive North America, LLC	Ownership Change	Medium	November 1, 2020

Figure 4.13 Adverse Media Alerts Email

4.3.2 Creating an Engagement Risk Project

Risk assessment projects provide a process for evaluating the risk of engaging with a supplier. Based on the evaluation, the buyer organization can determine whether or not to engage with this supplier. If you decide to engage, then these evaluations can determine the level of monitoring that would be required. For more information, review the "SAP Ariba Supplier Risk Process" subsection in Section 4.2.1.

To create an engagement risk project, the supplier risk engagement requester can click on **Create • Engagement Request**, as shown in Figure 4.14.

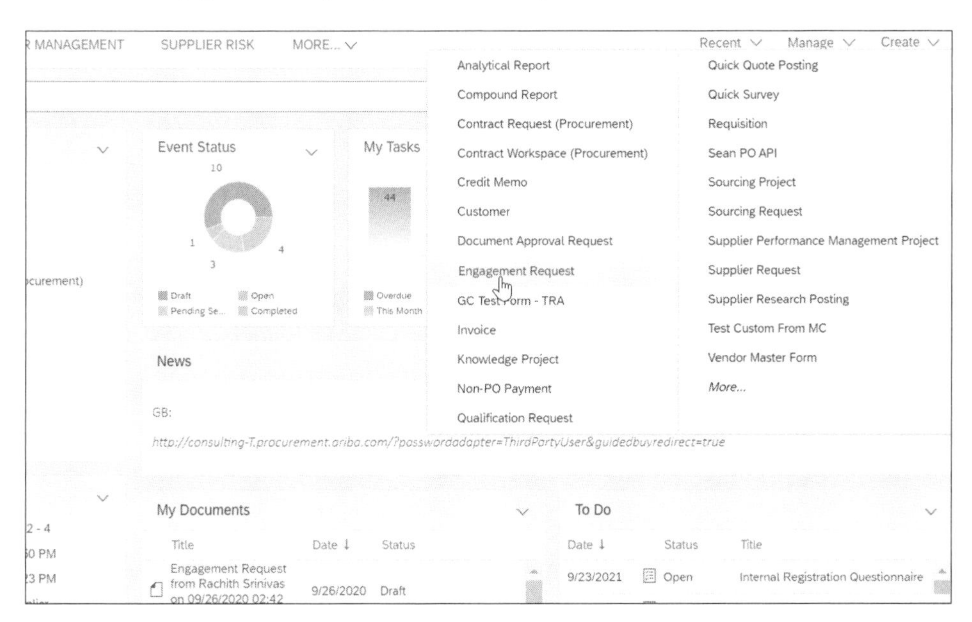

Figure 4.14 Creating an Engagement Request

The supplier risk engagement requester can then complete the request form shown in Figure 4.15.

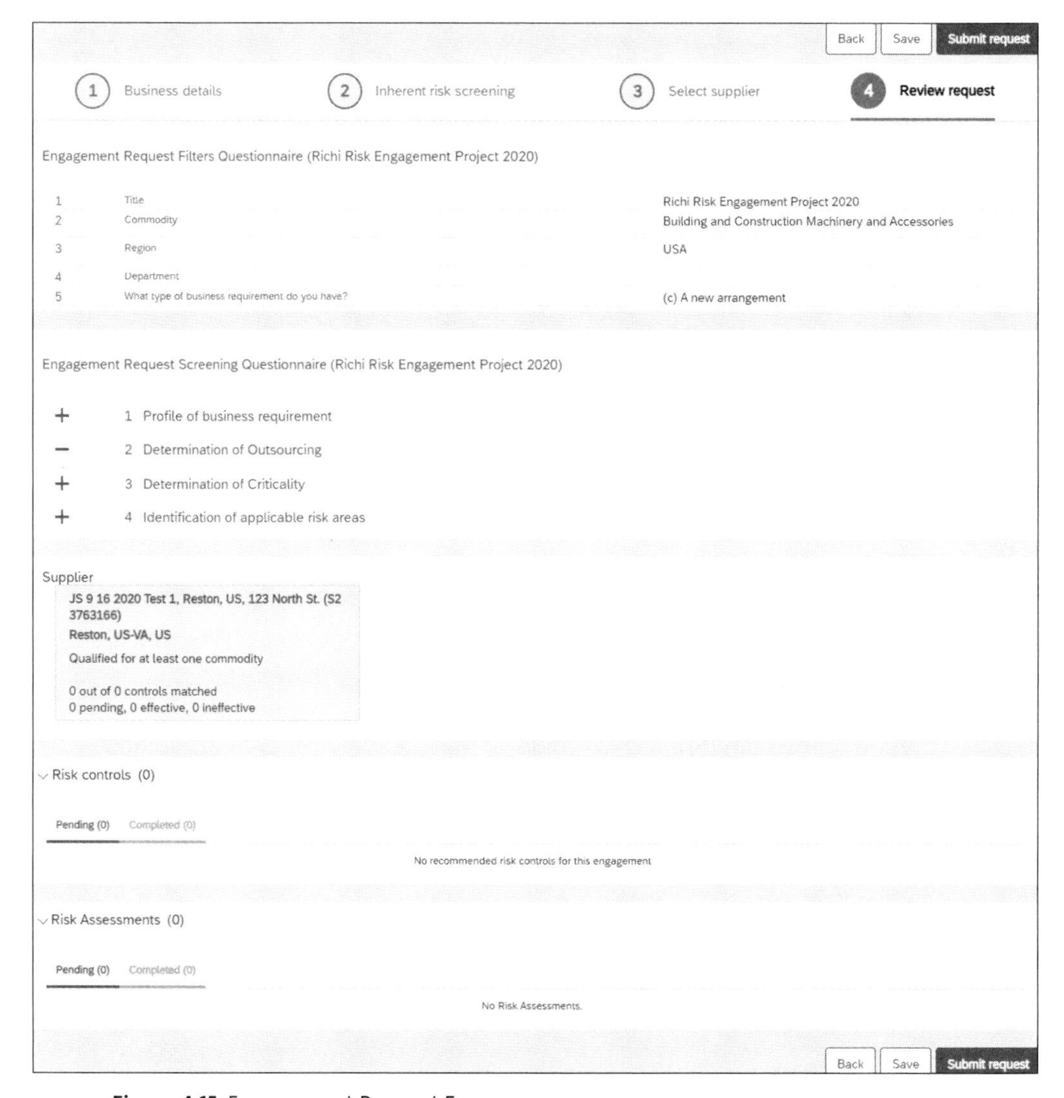

Figure 4.15 Engagement Request Form

After the risk assessment project is approved, the person responsible for reviewing the inherent risk sends the assessment out to the internal stakeholders. Assessments can be sent to suppliers as well, as shown in Figure 4.16.

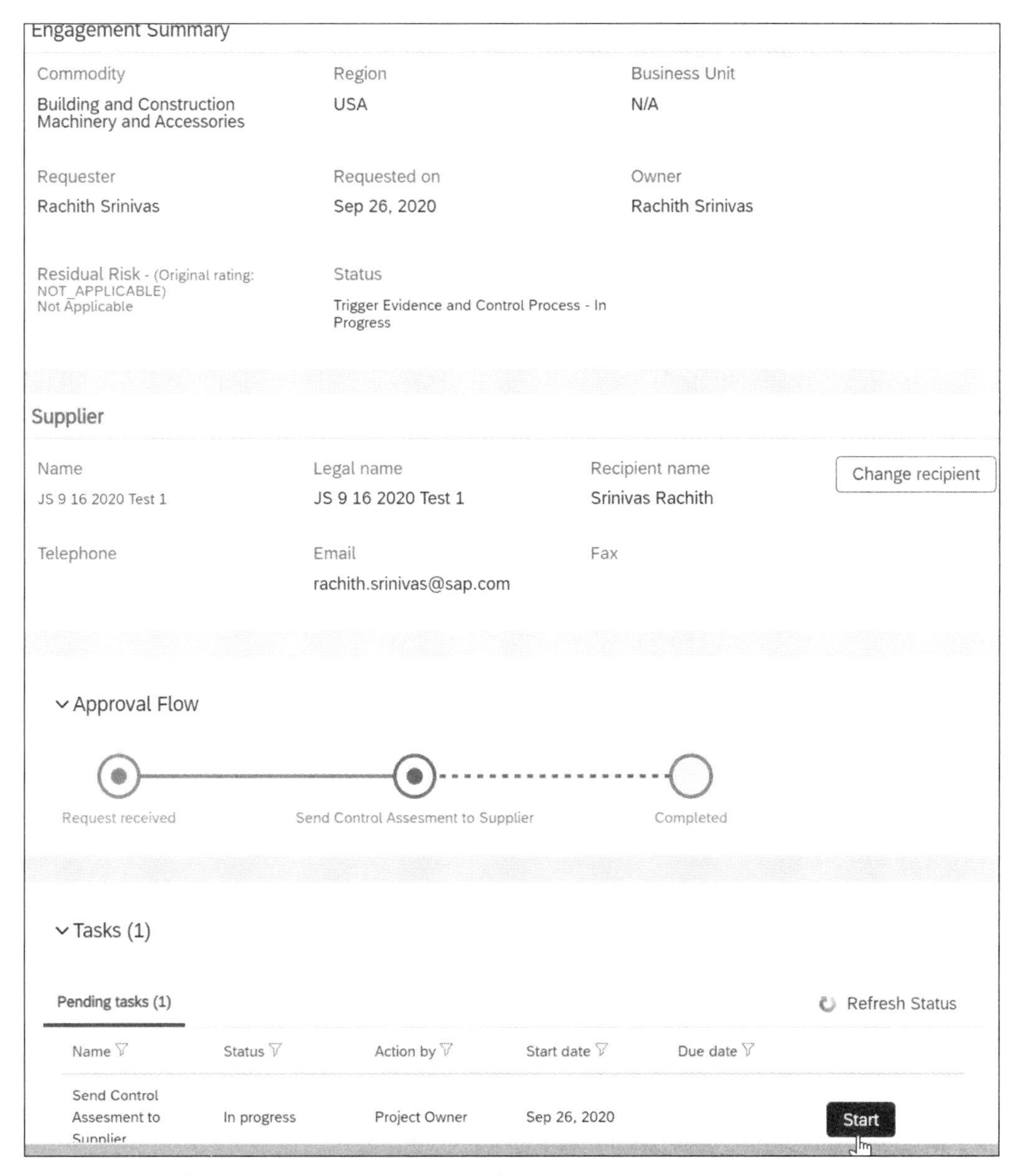

Figure 4.16 Sending a Risk Assessment to a Supplier

Notification emails are sent out by SAP Ariba to the recipients. Recipients click the link in the email to log in to SAP Ariba Supplier Risk and complete the risk assessments on the **Engagement** page. Based on the configuration, a residual risk score is calculated for each assessment according to the submitted answers.

4.4 Summary

Avoiding damage and reputation from disruption to an organization's supply chain is becoming more important than ever. Supplier risk management can be tough to implement in a business because risk-related due diligence research tends to be sporadic, costly, and time-consuming, making staying informed about your supply base and about market signals almost impossible. SAP Ariba Supplier Lifecycle and Performance and SAP Ariba Supplier Risk, you have a comprehensive supplier management solution that can transform your supply management processes, help you better manage your supply base, and avoid risk. In this chapter, we provided detailed insights into SAP Ariba Supply Risk and described implementation approaches to get these solutions up and running in your enterprise.

Chapter 5
Sourcing

SAP Ariba Sourcing provides a best-in-class sourcing and negotiation technology with access to the world's largest business network. With SAP Ariba Sourcing, your enterprise can enable sustainable savings, connect to one of the largest business networks in the world, and simplify sourcing processes across all spend categories: services, indirect materials, and direct materials.

Sourcing is more than just finding a supplier and negotiating a price. Sustainable savings requires a lot more effort through identifying cost-saving opportunities, defining stringent processes for supplier selection, and creating contracts for realizing the negotiated savings. This model requires evaluating the total cost of sourcing and not just the lowest purchase price, while reducing the time and effort required for completing the process.

In addition, effective supplier discovery and supplier management are critical for driving sustainable savings. With SAP Ariba Strategic Sourcing Suite, your enterprise can greatly realize cost savings, while improving the quality and performance of the supply chain. With the ever-expanding global supply chain, the increasingly competitive environment, and the soaring energy and commodity costs, achieving your full sourcing potential is almost impossible without the right strategic sourcing solution.

The SAP Ariba Strategic Sourcing Suite and the SAP Ariba Sourcing solution are the right solutions for enterprises of all sizes to drive toward achieving their full sourcing potential. The SAP Ariba Sourcing solution is the most widely used and comprehensive strategic sourcing tool. In this chapter, we'll provide detailed information on SAP Ariba Sourcing, including the implementation approach for deploying this solution in your enterprise.

5.1 What Is Sourcing?

Simply put, *sourcing* is the process of finding the right suppliers to provide goods or services at the right price. Strategic sourcing, on the other hand, is the process of continuously improving and evaluating the purchasing processes of a company. Strategic sourcing considers the lowest total cost and not just the lowest purchasing price.

5.1.1 Sourcing Strategies

In today's highly competitive business environment, getting ahead of the competition and having a solid positive cash flow to innovate and grow are crucial for every business. Procurement and procurement strategies, including a sound sourcing strategy, play important roles in achieving these goals. In addition, cost-saving strategies are more consistent and long lasting than sales increases.

A good sourcing strategy will help you perform the following tasks:

- **Identify opportunities**
  Analyze spend across departments, and work with department owners or project leads to understand their types of spend, the frequency, and the kinds of relationships with existing suppliers they have, as well as to identify opportunities for savings. Gain a clear understanding of the spend categories you're focused on.
- **Evaluate the supplier market**
  Perform secondary market research on the suppliers available for the spend category. Shortlist the supplier you plan to engage.
- **Send initial requests for information**
  Send out initial requests for information (RFIs) to determine whether a supplier meets basic requirements that are required by your organization and whether the supplier can provide the goods and services needed within the timeline you may set.
- **Prepare a sourcing plan**
  Understand the goals and key value drivers, such as cost reduction, process efficiency, fiscal controls and compliance, cash management, and so on for this sourcing initiative. Identify whether this initiative will have multiple requests for (RFx) event rounds, including an auction event. Look at the competitiveness of the marketplace to determine whether a negative impact may affect a strategic relationship with a key existing supplier.
- **Execute RFx events**
  Create and execute RFx rounds based on your sourcing plan. Invite incumbent suppliers and other new suppliers that you've identified during the market research phase. Add weights to ensure all criteria are considered, such as switching cost and other costs. Focus on total cost of ownership (TCO) rather than just cost reduction.
- **Award business to suppliers**
  Thoroughly analyze responses to your RFx events. Use grading and collaborate with peers on awarding business to one or more suppliers. Create and analyze different award scenarios, and, where possible, use optimization software to help you decide on the best supplier.
- **Create contracts**
  Create contracts with awarded suppliers, and ensure strong contract compliance to maximize cost savings. Engage with awarded suppliers to work together and build partnerships.

- **Continuously monitor progress**
  For sustainable sourcing savings, continuously monitor and improve your initiatives.

5.1.2 SAP Ariba Portfolio

SAP Ariba provides two solutions for implementing a best-in-class sourcing solution at your organization:

- SAP Ariba Sourcing
- SAP Ariba Strategic Sourcing Suite bundles together the following solutions:
  - SAP Ariba Sourcing
  - Guided sourcing (new)
  - SAP Ariba Contracts (see Chapter 6)
  - SAP Ariba Supplier Lifecycle and Performance (see Chapter 3)
  - SAP Ariba Sourcing, savings and pipeline tracking add-on
  - Bonus/penalty lookup table formulas
  - SAP Ariba's product sourcing functionality
  - SAP Ariba Discovery
  - SAP Ariba Spend Analysis (see Chapter 10)

In this chapter, we'll cover SAP Ariba Sourcing; SAP Ariba Sourcing, savings and pipeline tracking add-on; and the product sourcing functionality.

5.2 SAP Ariba Sourcing

A strategic approach to sourcing is essential for achieving your immediate needs as well as for sustaining enterprise-wide cost reductions. Effective supplier discovery and supplier information management are critical for the sourcing process to drive sustainable results, yet many organizations struggle to keep their supplier information current. As mentioned earlier, most enterprises strain to achieve their full sourcing potential without a comprehensive sourcing solution.

The SAP Ariba Sourcing solution, a software-as-a-service (SaaS) solution, is the most widely adopted and complete strategic sourcing offering in the marketplace. It's used by thousands of companies to create and implement competitive best-value agreements.

SAP Ariba Sourcing with is its industry leading online sourcing and negotiation technology and with the worlds largest business network consisting of over 800 million buyers and suppliers, companies of any size can drive fast and sustainable results throughout their sourcing process, including the following:

- **Strategy development**
  Identify savings opportunities, assess market dynamics, and develop an informed sourcing strategy.
- **Sourcing and negotiating**
  Structure the most appropriate process for simple to complex sourcing opportunities. Identify and qualify suppliers, negotiate best-value agreements, derive optimal award allocations, drive project collaboration, standardize processes, and manage knowledge.
- **Monitoring and managing suppliers and agreements**
  Quickly implement supplier agreements, track and realize savings, and manage supplier performance.

As shown in Figure 5.1, SAP Ariba Sourcing is part of SAP Ariba's Strategic Sourcing Suite with native integration to SAP Ariba Contracts and SAP Ariba Supplier Lifecycle and Performance solutions.

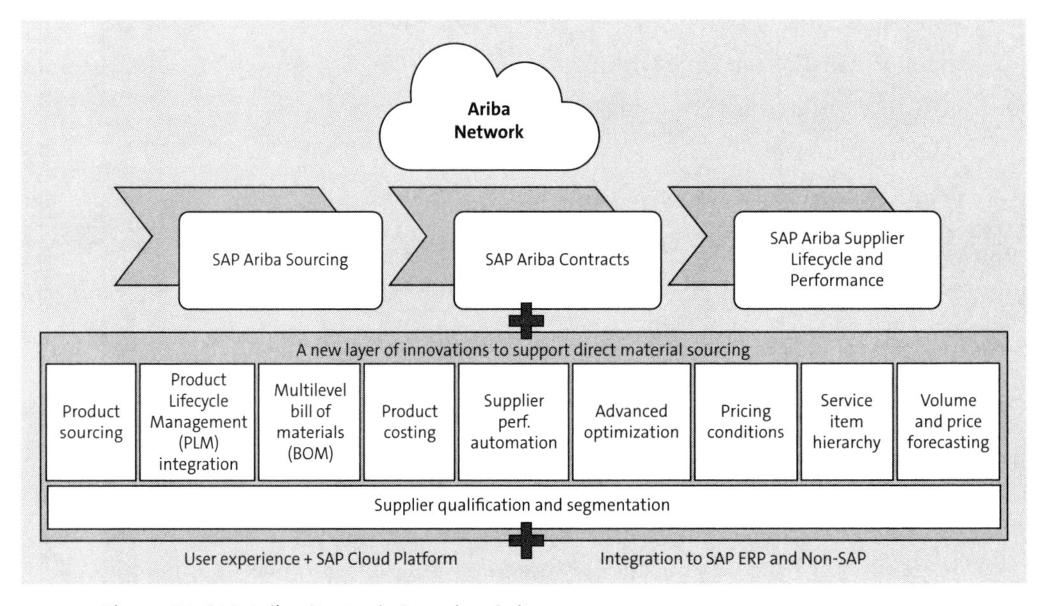

Figure 5.1 SAP Ariba Strategic Sourcing Suite

Standard integration to an SAP ERP backend is also available as an option. In addition to natively integrating to other SAP Ariba solutions, SAP Ariba Sourcing can also integrate with third-party systems that use web services and file transfer protocol (FTP) channels.

The SAP Ariba Sourcing solution features the following technical capabilities:

- RFx creation and management
  - A broad set of RFx types, including requests for information (RFIs), requests for proposals (RFPs), reverse auctions, and forward auctions

  - Integrated supplier discovery
  - Rapid RFx creation (patented competitive bidding and timing options)
  - Sealed envelope bidding, Dutch auctions, and total cost events
  - Matrix and tiered pricing
  - Bid optimization and decision support
  - Flexible supplier bidding options, including buyer and supplier bundles
  - Supplier response management
  - Team grading and collaborative scoring
  - Conditional content, table questions, and event prerequisites
  - Communications and messaging
  - Global, multilingual, and multicurrency capabilities
- Category management
  - Project management
  - Workflow and approval management
  - Document management
  - Knowledge management
  - Resource management
- Sourcing analysis and reporting
- Integration to third-party systems using web services and file channels
- Savings pipeline and tracking

The SAP Ariba Sourcing solution also offers the following community features:

- Integrated access to Ariba Network, the world's largest trading community for efficient and effective supplier discovery, qualification, risk assessment, and more competitive negotiations
- Unique peer benchmarking program with dedicated customer success teams
- Access to the SAP Ariba Exchange user community, a unique community designed to drive networking and to encourage the sharing and adoption of best practices

In the next section, we'll look at the features provided by the SAP Ariba Sourcing solution in more detail.

5.2.1 Application at a Glance

In this section, we'll look what SAP Ariba Sourcing can do. In Section 5.2.3, we'll look at how to configure these solutions.

The SAP Ariba Sourcing solution offers a dashboard that displays role-based content, as shown in Figure 5.2. Users can also personalize the dashboard, while administrators can enforce what content must appear on the dashboard. The dashboard includes the following:

- Each user has a personal calendar.
- Users can add data, such as watched sourcing projects, event statuses, announcements, to-do lists, and document folders, to their dashboards.
- Company news displays important information to users. This content can show data from RSS feeds. You can configure this news content for your sites.

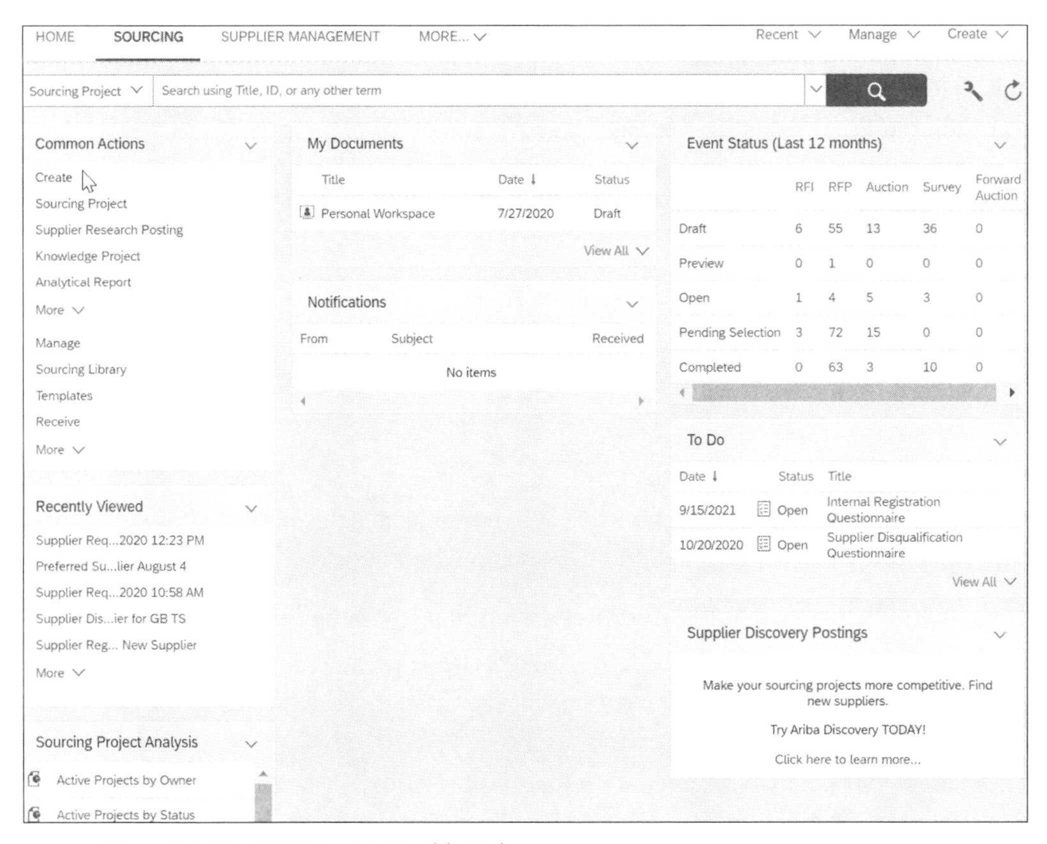

Figure 5.2 New SAP Sourcing Dashboard

In the following section, we'll look more closely at the features provided by the SAP Ariba Sourcing solution.

Sourcing Requests

The sourcing process typically starts with a *sourcing request*. End users in your enterprise looking to purchase goods and services can create a sourcing request and provide information about the goods and services they need. Sourcing managers or category managers may be required to approve the request before a sourcing project can be created. A typical process flow is shown in Figure 5.3.

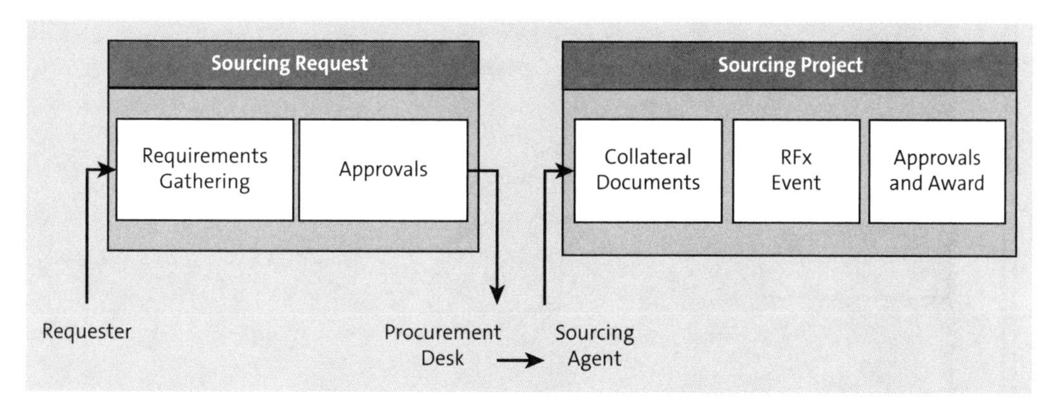

Figure 5.3 SAP Ariba Sourcing Process Overview

Any user with access to the SAP Ariba Sourcing application with the internal user role can create a sourcing request.

Let's look at the process of creating a sourcing request:

1. Log in to SAP Ariba Sourcing.
2. Click the **Create** button in the top-right corner, as shown in Figure 5.4, or select **Create** under the **Common Actions** menu on the left, as shown in Figure 5.5.

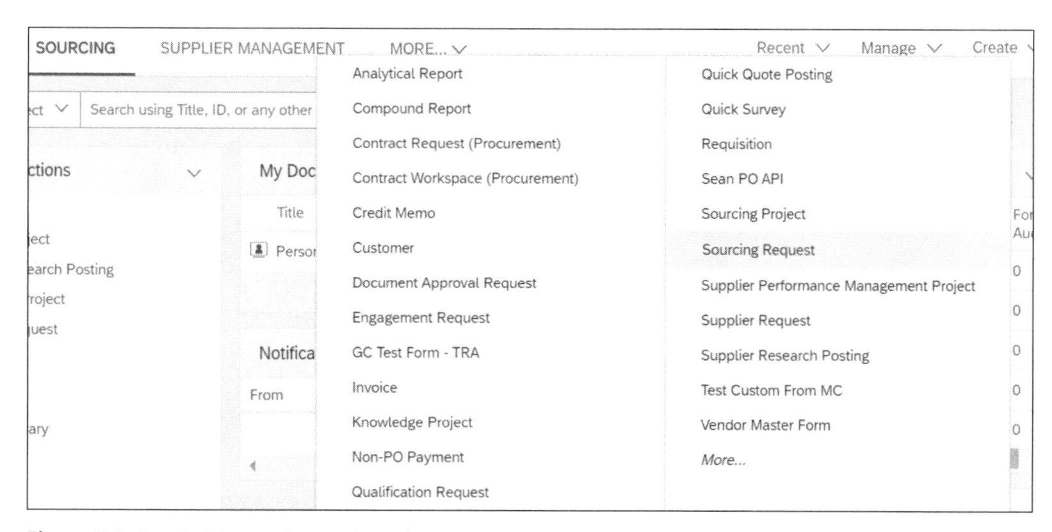

Figure 5.4 Create Menu: Create Sourcing Request

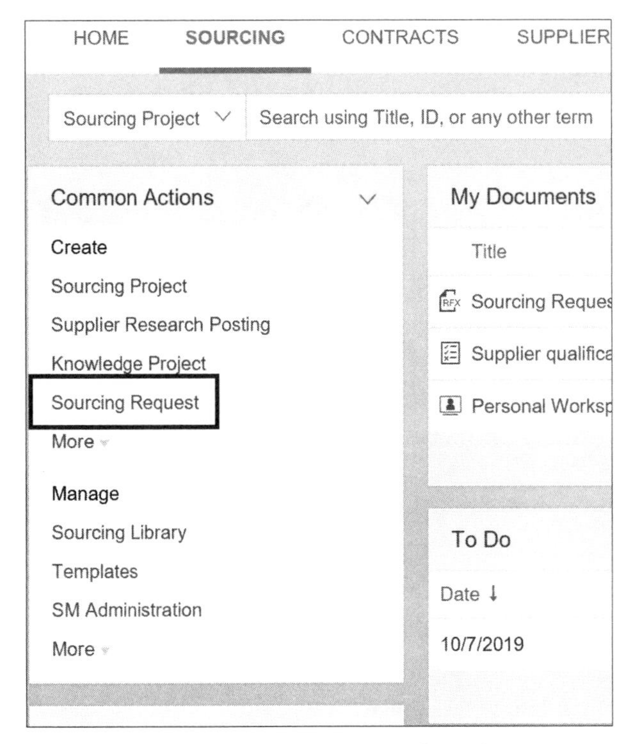

Figure 5.5 Creating a Sourcing Request from the Common Actions Menu

3. Click on **Sourcing Request**.
4. The sourcing request creation page will appear to the end user, who must complete the standard and custom fields to create a new sourcing request.

Sourcing templates, which appear at the bottom of the page, will be automatically suggested based on the values entered in the standard and custom fields, thus ensuring that end users select the right template.

A *sourcing request template* determines the following things:

- Visibility of standard and custom header fields
- Sourcing request item document
- Sourcing request process, including to-do and approval workflow tasks
- Team members who can access the sourcing request, that is, sourcing managers, sourcing agents, and so on

End users can provide information on the goods or services they want to procure in the *sourcing request item form* under the **Documents** tab, as shown in Figure 5.6. This document will be automatically available for end users if the template creator added the document to the sourcing request template.

Figure 5.6 Sourcing Request Item Form

To add items to a sourcing request item form:

1. Click the **Documents** tab in the sourcing request.
2. Click on the sourcing request items form, and click **Edit**.
3. In the **Content** page of the sourcing request items document, click **Add • Line Item**.

After the items are added and any other documents, such as specifications, are uploaded in the **Documents** tab, the end user can complete the **Prepare Sourcing Request** task under the **Tasks** tab, as shown in Figure 5.7.

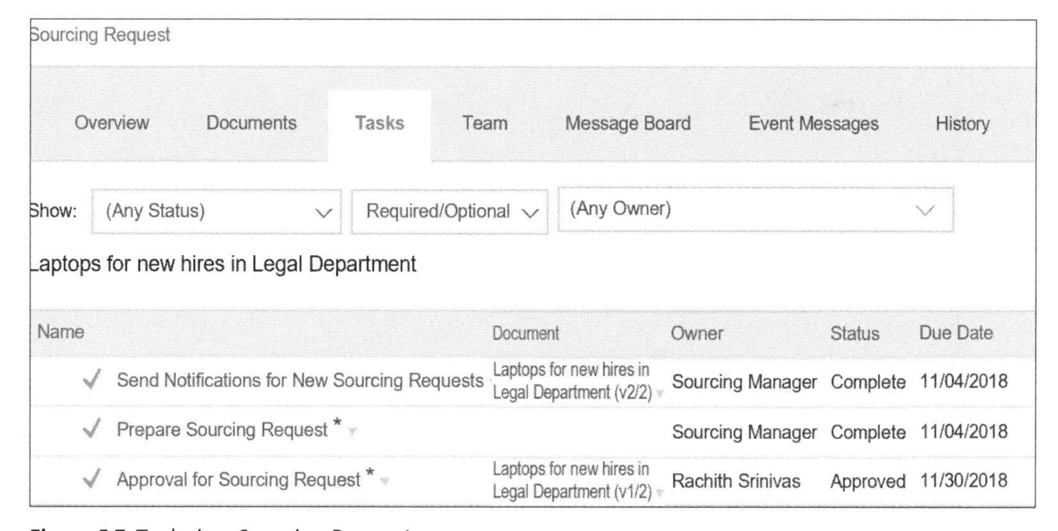

Figure 5.7 Tasks in a Sourcing Request

After the **Prepare Sourcing Request** task is completed, the end user can submit the **Approval for Sourcing Request**, as shown in Figure 5.8.

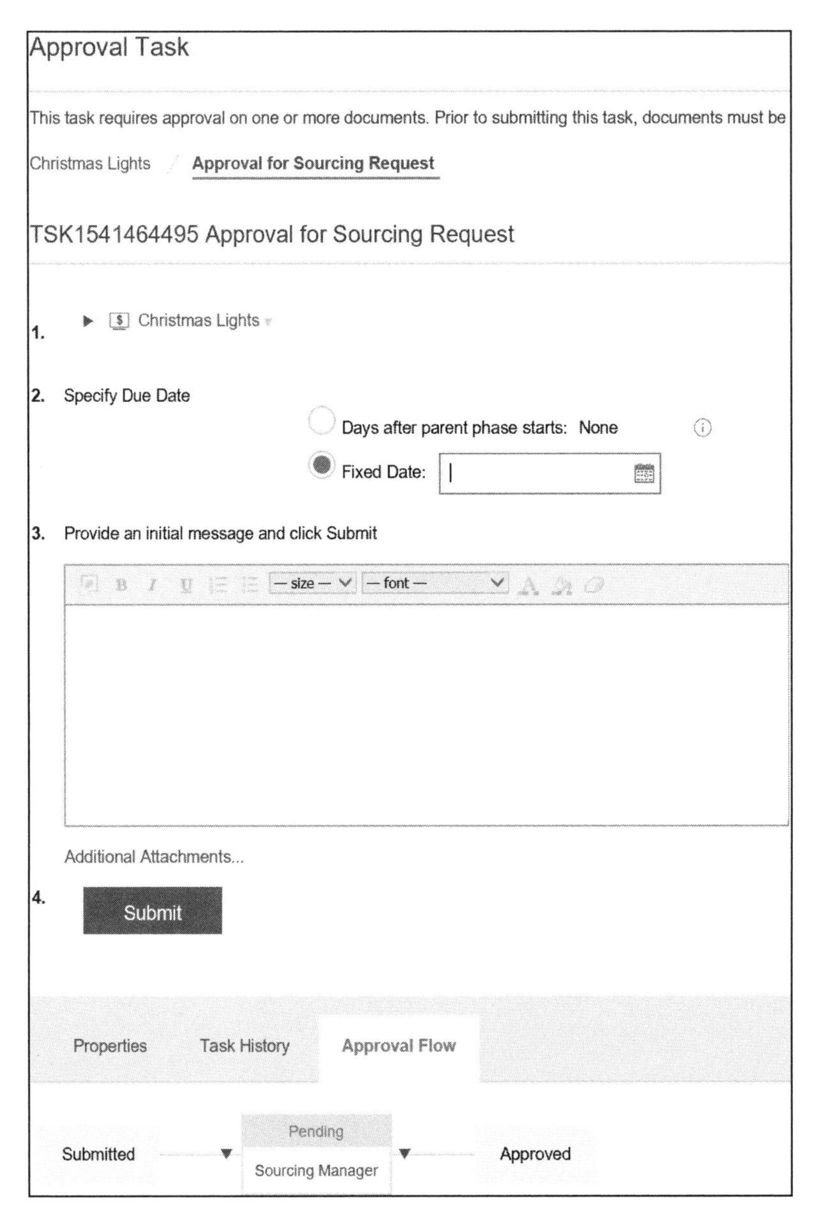

Figure 5.8 Submitting an Approval Task in a Sourcing Request

Approvers will receive an email or see the approval request in their **To-Do Tasks** channel on the dashboard.

As shown in Figure 5.9, a sourcing request approver has two choices:

- **Approve**
  The approval request moves to the next approver. If you're the final approver, the sourcing request is approved. After the approval flow is complete, the sourcing project link will appear in the **Documents** tab.

- **Deny**
  If one approver clicks **Deny**, the sourcing request is denied, and the requester is sent a notification with the reason for the denial.

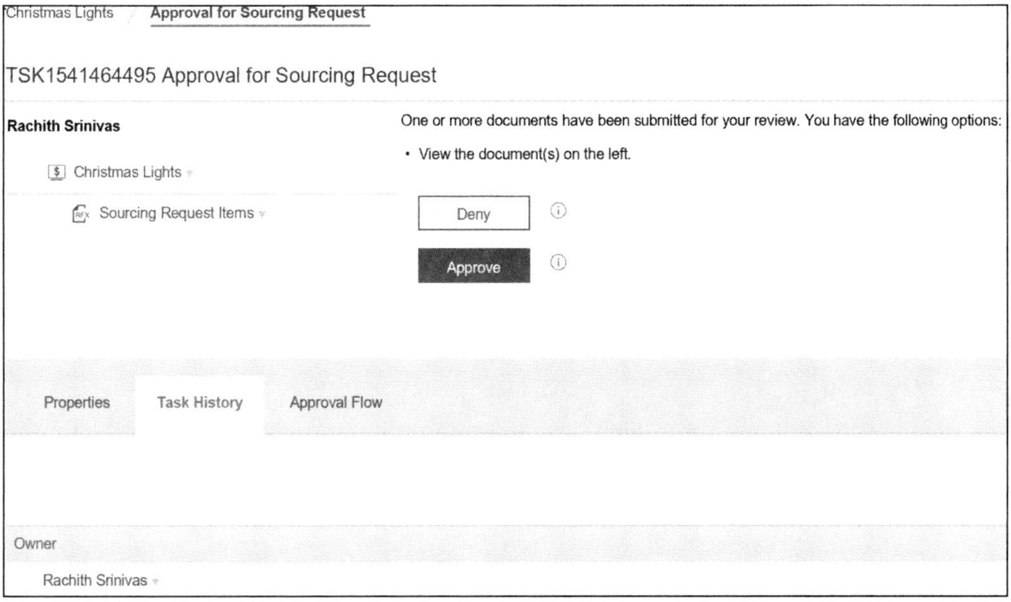

Figure 5.9 Approval Options for a Sourcing Items Form

Sourcing Projects

After the sourcing request is complete, the sourcing project link appears under the **Documents** tab. The sourcing agent assigned to handle the sourcing project can now create a new sourcing project from the sourcing request.

To create a new sourcing project, click the **Sourcing Project** link that appears in the **Documents** tab after the **Sourcing Request** is approved, as shown in Figure 5.10, to create a new sourcing project.

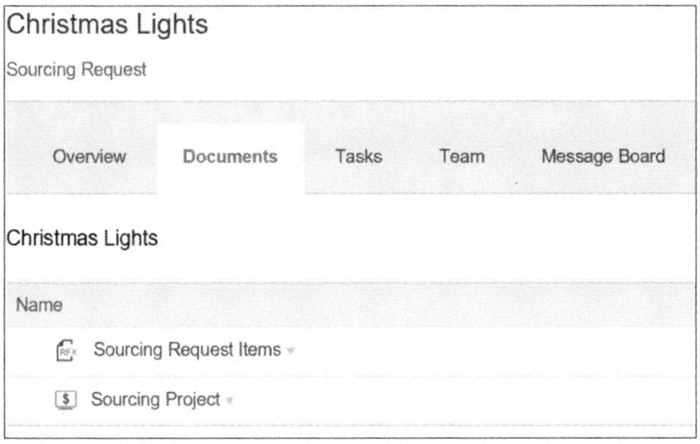

Figure 5.10 Sourcing Project Link in a Sourcing Request

The sourcing process can be managed using a sourcing project, which consists of tasks and phases, project teams, documents, milestones, dependencies, review and approval flows, subprojects, and follow-on projects. Like all projects in SAP Ariba, sourcing projects are created via custom-defined *sourcing project templates* that can be configured to capture and enforce best practice processes. We'll discuss how to configure sourcing project templates in more detail in Section 5.2.3.

Sourcing projects can have one or more related events. Examples of events include RFx, online auctions, sealed bids, e-negotiations, quick surveys, and quick projects. Automatic project configurations based on project attributes (standard and custom header fields) and project template questions can be used to define what tasks and events, including review and approval steps, must be completed by a sourcing agent to complete a sourcing project and award business to one or more suppliers.

Note

Sourcing projects are counted toward this solution's usage metrics in the month or year (as applicable) in which the project's start date occurs.

As mentioned earlier, each sourcing project is created from a project template and can contain one or more sourcing events (i.e., RFPs, RFIs, auctions, etc.). Each sourcing event has sourcing content or questionnaires for participants (supplier contacts) to fill out as part of their event response. This content can be configured within the sourcing project template, or the Sourcing Library repository can be used to build specific sourcing content for each category. The sourcing agent creating the sourcing event can pull content from the Sourcing Library as needed.

Savings forms are documents inside sourcing projects that allow buyers to manage their sourcing pipelines and track project savings, including estimated, negotiated, implemented, and actual savings. Savings forms are reportable and searchable.

Note

Savings and pipeline tracking require separate deployment services. This is an add-on feature as defined in the SAP Ariba Strategic Sourcing Suite bundle discussed previously.

Any user with the following roles can create a sourcing project:

- Sourcing agent
- Sourcing manager

To create a sourcing project, follow these steps:

1. Click the **Sourcing Project** link under the **Documents** tab of the sourcing request, or select **Create** from the **Common Actions** menu on the left. Click on **Sourcing Project**.
2. Select **Full Project** or **Quick Project**.
3. Select a sourcing project template.
4. Select event types to be included in the sourcing project (if a full project).
5. Click **Create**.

Note

If you're creating a sourcing project from a sourcing request, most of the content from the sourcing request (values entered in the standard and custom fields) will be copied over to the sourcing project.

As mentioned above there two types of sourcing projects exist:

- **Full sourcing project**
  Can contain initial tasks required to manage a strategic sourcing initiative, such as identify sourcing opportunities, develop sourcing strategy, and so on. This type of project also contains tasks for completing sourcing events (RFPs, RFIs, auctions) that were included as part of the sourcing project when the project was being created. Sourcing event documents appear under the **Documents** tab. If the savings and pipeline tracking add-on feature of the SAP Strategic Sourcing Suite has been enabled, then the savings and pipeline tracker form may also appear under the **Documents** tab. A full sourcing project can also contain multiple approval tasks, such as the sourcing participants approval task and the sourcing award approval task.
- **Quick sourcing project**
  Contains a single sourcing event, such as an RFI, RFP, or auction.

Sourcing Events

As part of a project (whether quick or full), SAP Ariba Sourcing allows you to create and run events in which you exchange business information with other companies. Depending on the type of information you want to collect, you'll create different types of events. All events are created from templates, which define the rules and the types of information (e.g., types of pricing terms) for the event. The event templates covered in this chapter are provided out of the box with the product. Only members of the template creator group or template creator team can modify a template.

In SAP Ariba Sourcing, an event follows a process from creation to awarding contracts to participants, as shown in Figure 5.11. An event has a status corresponding to each stage in the event process, which determines the actions a user can take.

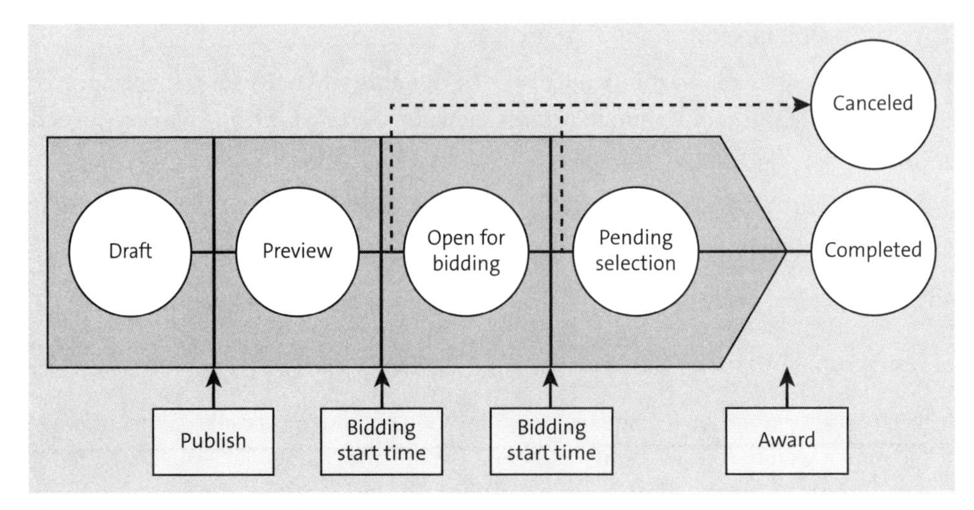

Figure 5.11 Sourcing Event Statuses

The status of an event, listed in Table 5.1, is visible in multiple places on the user interface (UI):

- On the upper right-hand corner of the event monitoring interface
- In the **My Documents** content item on the dashboard home page

Status	Description
Draft	The event is still being creating and hasn't been published.
Preview	When setting up the event, you can choose to have a period of time before the event opens up for bidding. You can optionally allow pre-bids in which suppliers can submit an initial bid or response.
Open	The event is open for participant responses. You can edit, cancel, or close the event.
Pending Selection	The event is closed for bidding. You can now select one or more participants to award business to. You can also reopen or edit the event.
Completed	Event is complete. It can't be reopened.
Canceled	At any point after publishing an event, you can choose to cancel the event. Canceling an event bypasses all the other statuses and immediately ends the event. The **Canceled** status indicates that you aborted the event. You can undo the cancellation of an event.

Table 5.1 Sourcing Event Status Definitions

Note

The dashboard home page queries your event database every six hours and displays the events you created over the past three months, six months, or one year, depending on the number of events you've created.

Note

The **My Documents** content item displays a maximum of 20 events (RFIs, RFPs, auctions, and surveys) and a maximum of 50 projects (sourcing and contracts).

A sourcing project can have multiple sourcing events that a sourcing agent or sourcing manager may need to include. A typical best practice is to run an RFI event with a larger set of participants (suppliers) and then run an RFP event with a short list of participants based on the previous RFI round. Finally, you can run an auction event for more competitive pricing savings with further short-listed participants before awarding your business to one or more participants.

Sourcing events can also be published to SAP Ariba Discovery so that suppliers on Ariba Network can ask to participate in your sourcing events. Native integration to Ariba Network (connected to more than 3.4 million companies) and to SAP Ariba Discovery is one major reason that SAP Ariba Sourcing is a leading strategic sourcing tool and delivers the best sourcing results. We'll discuss SAP Ariba Discovery in more detail in Section 5.3.

Sourcing Event Content

The content of a sourcing event varies based on the type of sourcing event. For instance, a typical RFI will usually have an initial set of questions to be completed by the participants. An RFP event will usually have further detailed questionnaires as well as the line items/lots for which you're seeking quotes from participants. Sourcing items can also come with standard terms that can be configured based on your requirements.

SAP Ariba Sourcing comes with numerous out-of-the-box templates with item content configuration that you can use or modify to meet your requirements. We'll discuss how to configure sourcing event templates in more detail in Section 5.2.6.

Sourcing Event Types

Depending on the type of information you want to collect and your sourcing solution, you can create the following types of events using SAP Ariba Sourcing:

- **Requests for information**
  This event type is used to send questions to participants, gather participant feedback,

and prequalify participants based on their responses. You can weigh and grade participant responses and create an overall score for each participant.

- **Requests for proposal**
  This type is used to create a questionnaire with sections, questions, requirements, or line items to collect pricing information. With this information, you can qualify participants, possibly for an auction. You can weigh and grade participant responses, create an overall score for each participant, and review price breakdowns and total costs.
- **Large-capacity RFP**
  To solicit pricing or other information for a large number of items (more than 2,000 items), you can create a large-capacity RFP. Large-capacity RFPs are noncompetitive RFPs that can contain up to 10,000 line items. Contents for large-capacity RFPs are accessed using Microsoft Excel spreadsheets and aren't directly visible in the UI.
- **Reverse auctions**
  You can use a reverse auction to create a competitive bidding event, based only on price, for line items or lots.
- **Bid transformation auction**
  You can use a bid transformation auction to create a competitive bidding event for line items and lots, while including factors other than price. This event format is the same as for a reverse auction.
- **Total cost auction**
  You can use a total cost auction to create a competitive bidding event for line items or lots, while including factors other than price. This event format is the same as for a reverse auction.
- **Index auction by amount**
  You can use an index auction by amount to create a competitive bidding event based on an index. Participants will bid and compete on an amount from a given index. This event format is the same as for a reverse auction.
- **Index auction by percentage**
  You use an index auction by percentage to create a competitive bidding event based on an index. Participants will bid and compete on a percentage from a given index.
- **Dutch auctions**
  You can use this auction type to create a Dutch-style competitive bidding event for line items or lots. You use a Dutch forward auction to sell things such as surplus inventory. You can invite prospective buyers to bid in the event.
- **Forward auctions**
  You can use a forward auction to sell things such as surplus inventory. You can invite prospective buyers to bid in the event. This event format is the same as for a reverse auction.

- **Forward auctions with bid transformation**
  You can use a forward auction with bid transformation to sell things such as surplus inventory. Bid transformation allows you to "transform" buyers' bids by adding cost terms that you can define. As a result, buyers will compete on your total costs instead of on their raw prices. This event format is the same as for a reverse auction.
- **Japanese auctions**
  Japanese auctions require suppliers to accept pricing at levels that automatically adjust at regular intervals. In a reverse Japanese auction, you're the buyer, and the price level falls at each configured interval; in a forward Japanese auction, you're the seller, and the price level rises at each interval. Participants can choose to accept price levels as they drop (or rise). By default, a participant who doesn't accept a price level for an item becomes inactive and is unable to accept any further price levels for the item.

  Bidding ends for an item when either the number of active participants drops to or below the configured minimum value or the target price is reached.

Communications during the Event Process

Several kinds of messages are used during the event process:

- **Event messages**
  The **Message** tab in the event monitoring interface stores all event messages. Suppliers and buyers can communicate using event messages; buyers can provide event-related information, and suppliers can ask questions and receive answers.
- **Private messages**
  Users can send messages to all suppliers or to other buyers on the event team.
- **Project messages**
  Project message boards facilitate communication between project team members.
- **Notifications**
  Customer sites generate a number of automatic notifications related to invitations, changes to events, event closings, awards, and so forth.

To help you monitor the event process, the following reporting capabilities are provided:

- Reporting on individual events during event monitoring
- Cross-event reporting
- RFI reporting
- Supplier activity reporting
- Audit log reporting
- Project and project task reporting
- Custom analytical reporting, including reporting across multiple fact tables

Application Programming Interfaces for Sourcing

Enterprises deploying SAP Ariba solutions usually have a complex enterprise landscape and, in many cases, a centralized reporting framework. While the SAP Ariba Reporting framework and SAP Ariba Spend Visibility can serve the need for consolidated spend reporting across an enterprise, many customers may be using a homegrown or other third-party management reporting application. In such cases or in other business scenarios, there is a definite need to be able to extend and integrate SAP Ariba solutions. SAP Ariba has tackled this need through SAP Ariba application programming interfaces (APIs).

SAP Ariba APIs allow enterprises to quickly extend, integrate, and optimize SAP Ariba applications and Ariba Network to meet unique domain- or region-specific business requirements. SAP Ariba APIs provide a simple, scalable, and secure way to build new or extended functionality using Representational State Transfer (REST) APIs (as shown in Figure 5.12). SAP Ariba also provides easy-to-use tools and developer resources for rapid prototyping and deployment of client programs to call the SAP Ariba APIs.

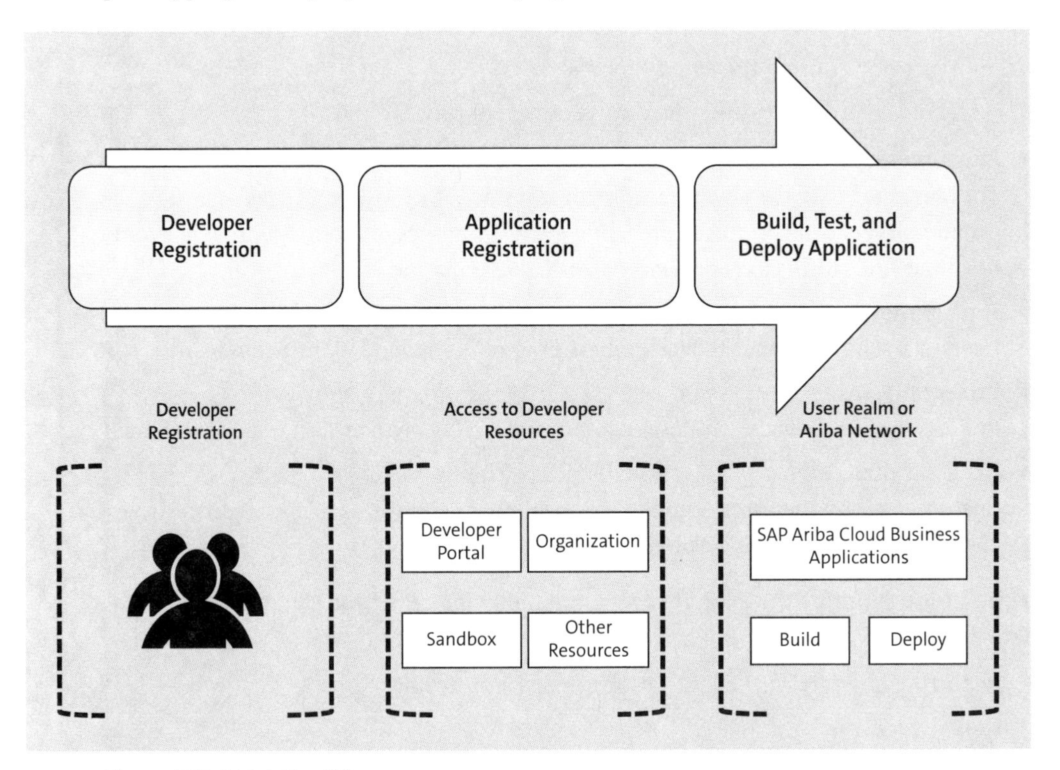

Figure 5.12 SAP Ariba API

The following are some of the APIs available for SAP Ariba Sourcing:

- **Operational Reporting for Strategic Sourcing API**
  Can be used to extract transactional sourcing data such as tasks that are due, events waiting on approvals, events scheduled to close the next day, and so on.

- **Strategic Sourcing API**
  Can be used to pull status information on file channel integration import and export events for both transactional data and master data in SAP Ariba Strategic Sourcing Suite solutions.
- **Project Management API**
  Can be used to add or delete team members in a project group that is part of a sourcing project or a contract workspace. This API can also retrieve a list of project groups and team members.
- **Event Management API**
  Can be used to publish or republish a sourcing event. The API provides options on how previous participant responses are handled (**Keep** or **Discard**) and if participants receive notifications for the modifications.
- **Surrogate Bid API**
  Can be used by buyers to submit surrogate bids on behalf of a supplier.
- Product Hierarchy Management API
  Can be used with the new product hierarchy feature to return a list of product questionnaires responded to by a supplier, along with associated sourcing events and line items.

Additional details on the APIs are out of scope for this book. To learn how APIs deliver the speed and flexibility you need to extend and integrate your SAP Ariba solutions and Ariba Network, visit *http://www.ariba.com/solutions/solutions-overview/platform-solutions/sap-ariba-apis*.

5.2.2 Planning the Implementation

The key to a successful implementation is planning and stakeholder ownership. The SAP Ariba Sourcing team must be engaged to implement this solution. Implementing SAP Ariba Sourcing follows the SAP Ariba on-demand deployment methodology, which is based on the SAP Activate methodology. In this section, we'll provide an overview of the implementation methodology, calling out the different roles played by the SAP Ariba team and the customer team throughout the process.

The SAP Activate methodology consists of four phases: prepare, explore, realize, and deploy.

To implement the SAP Ariba Sourcing solution, the following roles are required on the SAP Ariba side:

- SAP Ariba project manager
- SAP Ariba Sourcing functional lead
- SAP shared services lead
- SAP technical lead
- SAP integration lead

The following resources will be required on the customer side:

- Project sponsor
- Project manager
- Functional lead
- Subject matter experts (SMEs)

In the following section, we'll review the four phases in the SAP Ariba Sourcing deployment process that you must consider.

Prepare Phase

The prepare phase is the first implementation phase, usually after a discover phase that is completed by you and the SAP Ariba sales organization. The initial project scope, timeline, and budget are agreed upon during the discover phase. The purpose of this prepare phase is to confirm value drivers, goals and objectives, project scope, and success metrics for a successful implementation. In this phase, the project team is identified. The project plan, project governance framework, issues and risk plan, and project roles and responsibilities matrix are defined and finalized with the customer. After the sign-off to the project charter is completed, the project kickoff is conducted with all the project stakeholders, including the customer project sponsors, customer project team, and SAP Ariba team.

A typical deployment kickoff includes but isn't limited to the following information that will be presented to all resources involved with the project:

- **Section 1**
  Business goals, objectives, and success criteria (content usually owned by the customer and the SAP Ariba value realization lead, where applicable).
- **Section 2**
  Project methodology, overall timeline, and governance (content owned by the SAP Ariba project manager).
- **Section 3**
  Solution implementation framework, roles and responsibilities, and meeting schedules (content owned by the SAP Ariba Sourcing lead).

Explore Phase

In the explore phase, the requirements gathering and design workshops are held to clearly document your requirements and pain points and to design future processes and global templates for deploying SAP Ariba Sourcing in your procurement landscape. The purpose of this phase is to confirm requirements and prepare your technical infrastructure.

After the kickoff is complete, the requirements gathering and design breakout session will be conducted by the SAP Ariba Sourcing lead. The requirements gathering and design workshops generally start off with a solution overview demonstration, followed by initial sessions and breakout sessions. After the breakout sessions are complete, consolidated sessions can bring together the newly designed processes, organizational change decisions, and policies and procedures defined for the sourcing processes. The integration session looks at the processes in the SAP Ariba Sourcing solution, emphasizing the end-to-end processes across the enterprise landscape.

The following key areas should be covered during your breakout sessions:

- **Sourcing request blueprinting**
  - Custom fields required on the sourcing request form
  - Custom terms required on the sourcing items document
  - Tasks required to create and submit sourcing requests
  - Tasks required to approve requests and create sourcing projects
- **Sourcing project blueprinting**
  - Custom fields required on the sourcing project header
  - Type of RFx events required
  - Collaborators, tasks, conditions, and template questions required to initiate and execute a sourcing project, including the sourcing events associated with the sourcing project
  - Tasks required to approve supplier selection and awards during each sourcing event
- **Sourcing events blueprinting**
  - Content to be included in the sourcing event templates
  - Custom terms to be defined for the items in the sourcing events
  - Event rules to be defined
- **Master data**
  - Master data to be loaded
  - Master data refresh frequency and master data strategy
- **Reporting**
  - Analytical report requirements

For effective breakout sessions, we strongly recommend you involve SMEs who have complete knowledge of current processes. These experts should be able to articulate the requirements and current pain points, should help map out future processes, and should have decision-making authority to choose the right options during the design sessions.

The usual scope during deployment is shown in Table 5.2.

Configuration	Deployment Description
Master data	▪ Loading master data using the enablement workbook
Custom header fields	▪ Up to five custom header fields
Custom report	▪ One custom report
Sourcing process (one template deployed in standard implementation)	▪ Up to 50 tasks ▪ Up to 10 process configuration conditions ▪ Up to 20 related documents (attached to process or loaded to the Sourcing Library)
Form (one template deployed in standard implementation)	▪ Up to 50 fields/line items ▪ Up to 10 conditions
Sourcing event templates	▪ Up to 1 RFP template ▪ Up to 15 lines (sections, questions, lots, and items)
Savings form	▪ Up to five custom fields

Table 5.2 Scope of Deployment

Note

The scope per deployment we've described can vary based on what you've agreed upon with SAP Ariba sales. These changes should be defined in the statement of work (SOW).

Realize Phase

The purpose of this phase is to configure and test the solution. This phase begins immediately after you sign off on the business requirements document. Now, the SAP Ariba Sourcing team begins performing functional configuration in a test realm based on the documented requirements. The SAP Ariba Sourcing functional lead works with the shared services lead to perform the following tasks:

- Create sourcing templates:
  - Define the sourcing request process and approval workflow.
  - Define the sourcing project process and approval workflow.
  - Define sourcing event content configurations, such as line item/lot configuration for RFIs, RFPs, and auction events.
- Load master data, such as data about suppliers, regions, departments, and purchasing organizations, and other necessary master data as specified in the enablement workbook defined during the explore phase.

- Work with the SAP integration lead and customer resources to enable integration to enterprise resource planning (ERP) (backend) solutions such as SAP ERP (if in scope).

Deploy Phase

The purpose of this phase is to go live and transition the customer to SAP Ariba support. The deploy phase begins when you sign off on the user acceptance testing (UAT) developed during the realize phase. Now, you're ready to deploy the new SAP Ariba Sourcing solution in your production environment. The SAP Ariba Sourcing lead will work with the SAP shared services lead and the SAP Ariba technical lead to migrate all configurations from your test environment to your production environment.

5.2.3 Configuring SAP Ariba Sourcing

SAP Ariba's spend management solution, which includes SAP Ariba Sourcing and other SAP Ariba solutions, has multiple dashboards, one for each solution. Once logged in, a user will see the dashboard home page. The particular information displayed on the various dashboards and access to specific features and other dashboards will depend on a user's permissions and roles. For example, a user with the customer administrator role can access the **Administration** page (**Manage • Administration**). Users with the template creator role can access the **Templates** section (**Manage • Templates**).

SAP Ariba Sourcing uses project templates to define sourcing request and sourcing project processes. These templates can be found under the **Documents** tab of the template project within the site. SAP Ariba provides out-of-the-box templates that can be copied and modified according to your requirements. In addition, SAP Ariba also provides sourcing event templates that, for the most part, can be used without modification. In this section, we'll look at the configuration steps required to implement the SAP Ariba Sourcing solution.

5.2.4 Creating and Configuring Sourcing Request Templates

SAP Ariba provides several sourcing request templates out of the box. A best practice for creating customer-specific sourcing request templates is to copy one of the existing templates.

Copying a Sourcing Request Template

To copy a sourcing request template, begin by follow these steps:

1. Log in as a user with the template creator role.
2. In the template project, search for the **Sourcing Templates** folder.
3. Look for a sourcing request template.
4. As shown in Figure 5.13, click on **Sourcing Templates • Sourcing Request • Copy**.

5. Enter a title for the new sourcing request template, and click **OK**.

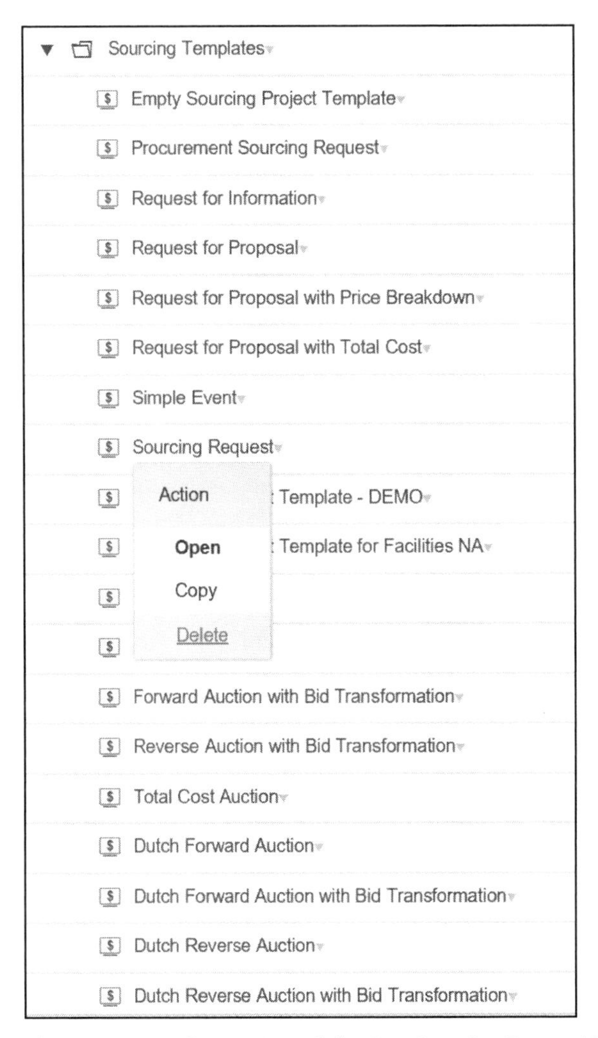

Figure 5.13 Copying an Out-of-the-Box Sourcing Request Template

The new sourcing request template is created as a draft. Configure the sourcing request template by maintaining the properties found under each of the tabs, which we'll cover next.

Overview Tab

The properties of the sourcing request template can be configured in this tab by clicking **Actions • Edit Properties**. You can also set template visibility conditions and access controls in this tab, as shown in Figure 5.14.

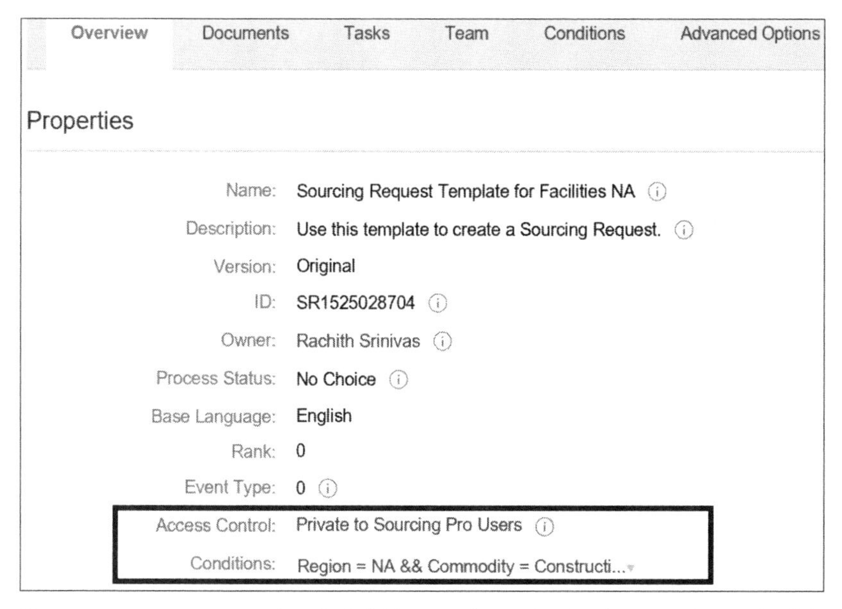

Figure 5.14 Access Control and Visibility Configuration in the Overview Tab of the Template

Based on the configuration shown in Figure 5.14, this template is only visible to end users when they select the **Region = NA** and **Commodity = Construction Services**. In addition, sourcing requests created from this template can be accessed by all SAP Ariba Sourcing Professional users. We'll discuss how to set these conditions when we cover the **Conditions** tab later in this section.

Note

A best practice is to allow access control to a small set of users. Usually, the access control is set to **Private to Team Members**. Users who need access to a particular project can be added under the **Team** tab.

Documents Tab

As shown in Figure 5.15, the following documents should be copied over from the out-of-the-box sourcing request template:

- Sourcing request items
- Sourcing project

When a sourcing request is created using this template, these documents will automatically be added in the **Not Created** status. Sourcing project documents are only visible when the **IsApproved** condition is satisfied. This condition and other conditions can be configured under the **Conditions** tab of this template.

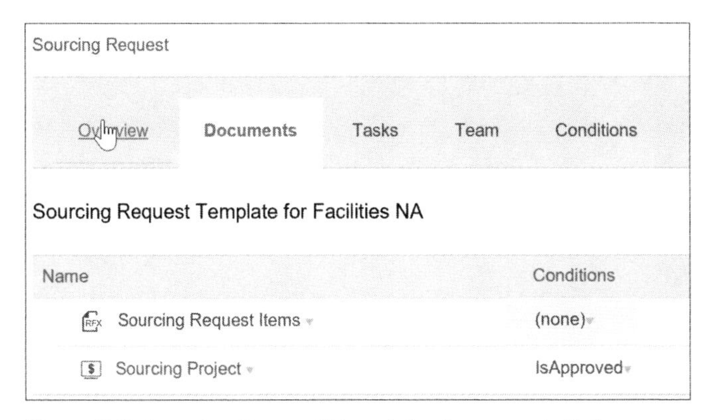

Figure 5.15 Sourcing Request Template: Documents Tab

Tasks Tab

In this tab, the sourcing request process can be configured as a set of required or optional tasks for completing the sourcing request process. The following types of tasks can be created:

- To-do tasks
- Approval tasks (Section 5.3.2 for more on configuring approval tasks)
- Review tasks
- Notification tasks

When you create a copy of an out-of-the-box sourcing request template, standard tasks from that template will be replicated in the copy, as shown in Figure 5.16.

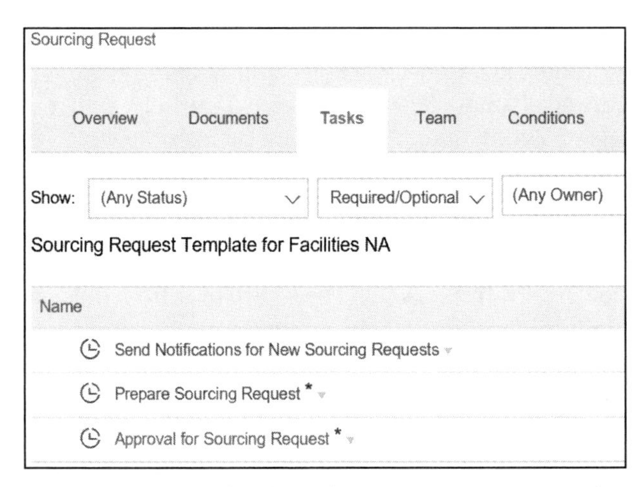

Figure 5.16 Standard Tasks in the Sourcing Request Template

These tasks can be modified, or new tasks can be added. Conditions on tasks can also be changed to meet your requirements. Tasks will only be enabled if the condition on the sourcing request is satisfied.

Conditions Tab

Conditions in SAP Ariba can be defined on project attribute values. As shown in Figure 5.17, the following conditions are already available out of the box:

- **IsApproved**
  Satisfied when a sourcing request created from this template is approved (status set to **Approved**).
- **IsExternalOrigin**
  Satisfied when a sourcing request is created from an RFQ sent from SAP ERP.

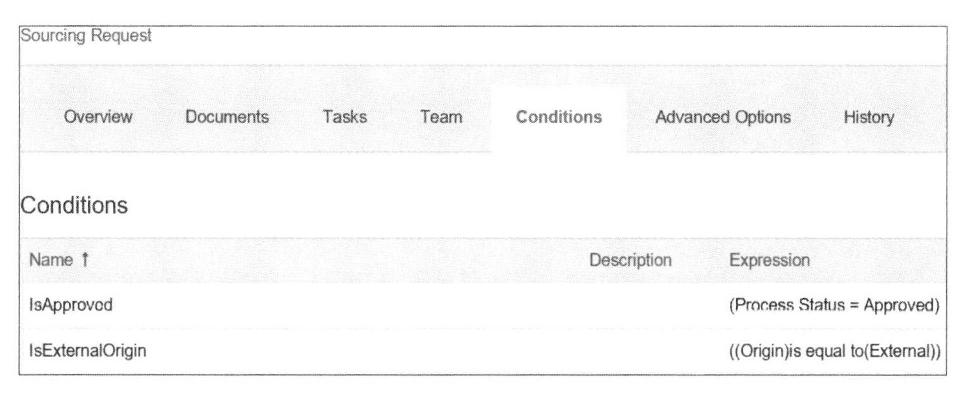

Figure 5.17 Standard Conditions in the Sourcing Request Template

As shown in Figure 5.18, additional conditions can be created and used when configuring templates by following these steps:

1. Click **Add Condition**.
2. Enter a name to describe the condition.
3. Select one of the following expressions:
   - **All Are True**
   - **Any Are True**
   - **None Is True**
4. You can further refine a condition by clicking on the inverted triangle icon next to an expression and selecting one of the following options:
   - **Field Match**
   - **Reference to Condition** (to refer to an existing condition)
   - **Subcondition** (to refer to a nested condition)
5. Select a field from the list of available fields, or select an existing condition.
6. From this list filtered by field or condition, select a comparison expression.
7. Select the value to compare the sourcing request against.
8. Click **OK** to create the new condition.

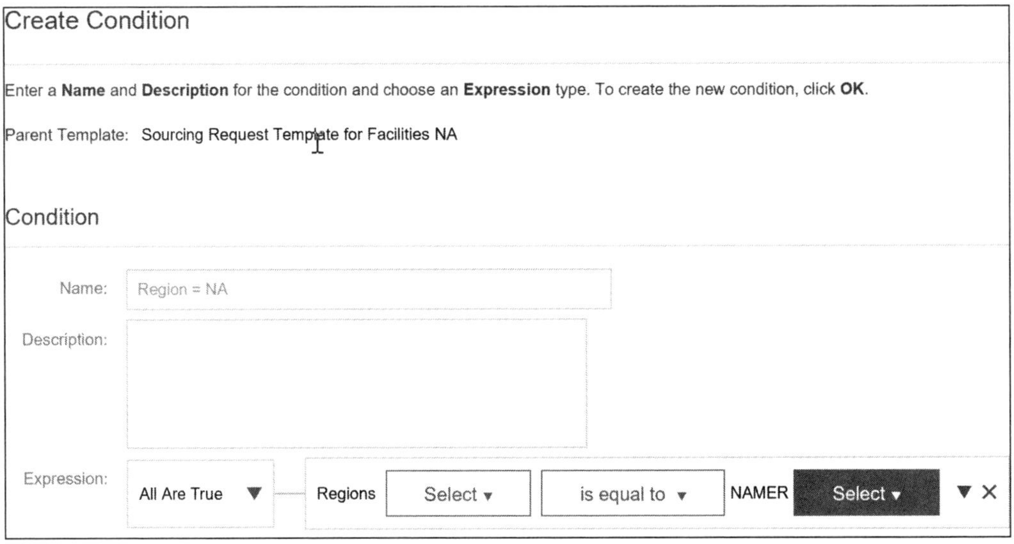

Create Condition

Enter a **Name** and **Description** for the condition and choose an **Expression** type. To create the new condition, click **OK**.

Parent Template: Sourcing Request Template for Facilities NA

Condition

Name: Region = NA

Description:

Expression: All Are True ▼ Regions Select ▾ is equal to ▾ NAMER Select ▾ ▼ ✕

Figure 5.18 Creating a New Condition

Conditions created in this tab can be used across the template to define the visibility (and, in some cases, the editability) of documents and tasks. These conditions can also be used to set the visibility of this template. For instance, a sourcing request template (facilities North America) can be made available to end users only when they select the **Commodity = Facilities** and **Region = NAMER**, under the **Overview** tab, as shown earlier in Figure 5.14.

5.2.5 Creating and Configuring Sourcing Project Templates

SAP Ariba provides several sourcing project templates out of the box. A best practice for creating customer-specific sourcing project templates is to copy one of the existing templates.

Copying a Sourcing Project Template

To copy a sourcing project template, begin by following these steps:

1. Log in as a user with the template creator role.
2. In the template project, search for the **Sourcing Templates** folder.
3. Look for an SAP Ariba Best Practice Sourcing project template.
4. Click on the template, and click **Copy**.

A new sourcing project template is created as a draft. Configure the sourcing project template by maintaining the properties found under each of the tabs, which we'll cover next.

Overview Tab

The properties of the sourcing project template can be configured in this tab by clicking **Actions • Edit Properties.** You can also set template visibility conditions and access controls in this tab.

If only two tabs appear (**Overview** and **Conditions**), then click **Actions • Display • Full View**. Now, all the tabs should be visible, including the **Overview**, **Documents**, **Tasks**, **Teams**, **Conditions**, **Advanced Options**, and **History** tabs.

Properties of the template can be modified to change access controls or visibility conditions. Access controls will be inherited by the sourcing projects created from this template. The **Conditions** field is used to control when this sourcing project is available for sourcing agents/sourcing managers to use when creating a sourcing project. In the configuration, for instance, this sourcing project template will only be available when the sourcing agent/sourcing manager selects **Region = North America** and **Commodity = Construction Services**.

Note

A best practice is to allow access control to a small set of users. Usually, the access control is set to **Private to Team Members**. Users who need access to a particular project can be added under the **Team** tab.

Documents Tab

As shown in Figure 5.19, the following documents should be copied over from the out-of-the-box sourcing project template:

- RFI template
- RFP template
- Auction template

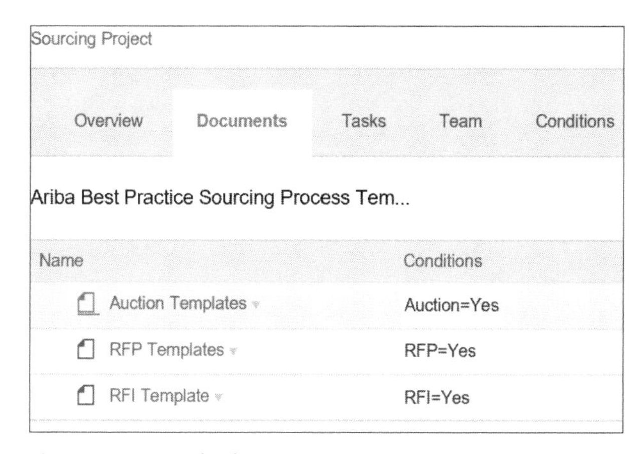

Figure 5.19 Standard Sourcing Project Documents

When a sourcing project is created using this template, these documents will automatically be added but in the **Not Created** status. The documents will only be visible if the conditions are satisfied. These conditions are driven by the responses provided to the template questions defined under the **Conditions** tab.

Tasks Tab

In this tab, the sourcing request process can be configured as a set of required or optional tasks for completing the sourcing request process. The following types of tasks can be created on sourcing events in the sourcing project (Section 5.3.2 for configuring approval tasks):

- **RFI sourcing event**
  - To-do
  - Review
  - Notification
  - Approval for publish
  - Approval for award
- **RFP sourcing event**
  - To-do
  - Review
  - Notification
  - Approval for publish
  - Review for team grading
  - Approval for award
- **Auction sourcing event**
  - To-do
  - Review
  - Notification
  - Approval for publish
  - Review for team grading
  - Approval for award

The following tasks are copied from the SAP Ariba Best Practices Sourcing project template:

- Phase 1: Identify Opportunity, Analyze Category (Offline Activity)
- Phase 2: Develop Strategy (Offline Activity)
- Phase 3a: Source and Negotiate (Online Activity) – RFI
- Phase 3b: Source and Negotiate (Online Activity) – RFP

- Phase 3c: Source and Negotiate (Online Activity) – Auctions
- Phase 4: Project Close-Out Phase

These tasks can be modified or deleted, and new tasks can be added. Conditions on tasks can also be changed according to your requirements.

Conditions Tab

Conditions in SAP Ariba can be defined on project attribute values. As shown in Figure 5.20, additional conditions can be created and used when configuring templates by following these steps:

1. Click **Add Condition**.
2. Enter a name to describe the condition.
3. Select one of the following expressions:
   - **All Are True**
   - **Any Are True**
   - **None Is True**
4. You can further refine a condition by clicking on the inverted triangle icon next to an expression and selecting one of the following options:
   - **Field Match**
   - **Reference to Condition** (to refer to an existing condition)
   - **Subcondition** (to refer to a nested condition)

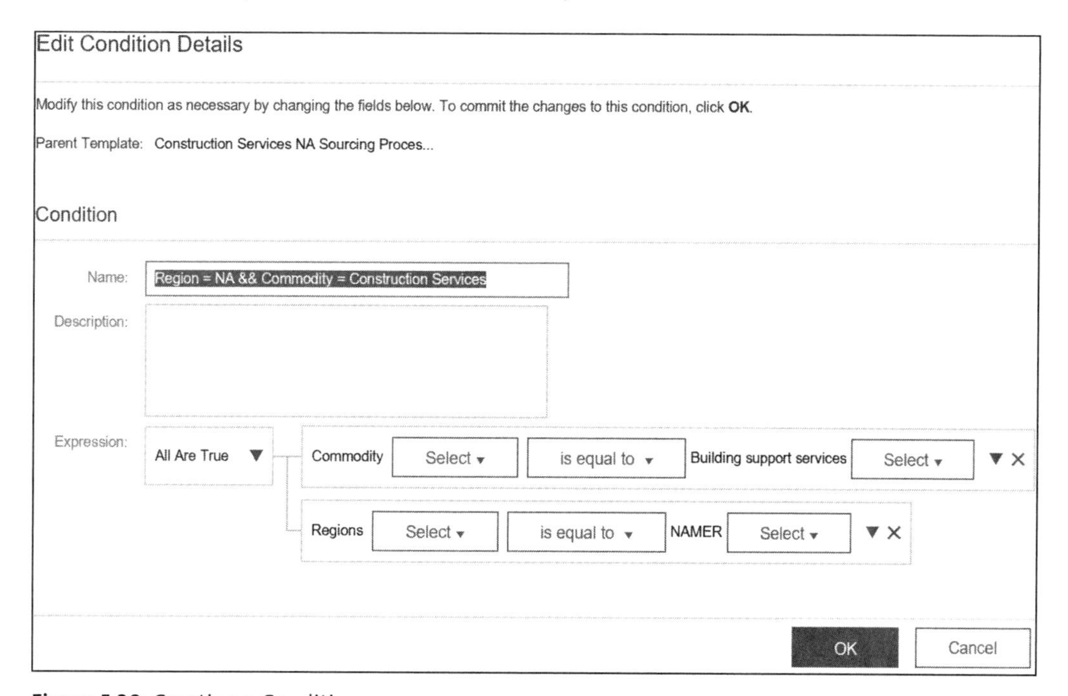

Figure 5.20 Creating a Condition

5. Select a field from the list of fields available, or select an existing condition.
6. From this list filtered by field or condition, select a comparison expression.
7. Select the value to compare the sourcing project against.
8. Click **OK** to create the new condition.

Conditions created in this tab can be used across the template to define the visibility (and, in some cases, the editability) of documents and tasks. These conditions can also be used to set the visibility of this template. For instance, a sourcing request template (Facilities NAMER) can only be made available only to end users when they select the **Commodity = Facilities** and **Region = NAMER** under the **Overview** tab, as shown earlier in Figure 5.14.

Template Questions

In addition to conditions, template questions can also be used to define visibility. Follow these steps to configure template questions:

1. Click on **Add Question** under the **Conditions** tab of the template.
2. Enter a question.
3. Enter answers and define conditions for each answer.
4. Set the default answer. Select the **Default must be changed** checkbox if necessary.
5. Click **OK**.

Figure 5.21 shows some template questions copied from the standard SAP Ariba Best Practice Sourcing project template.

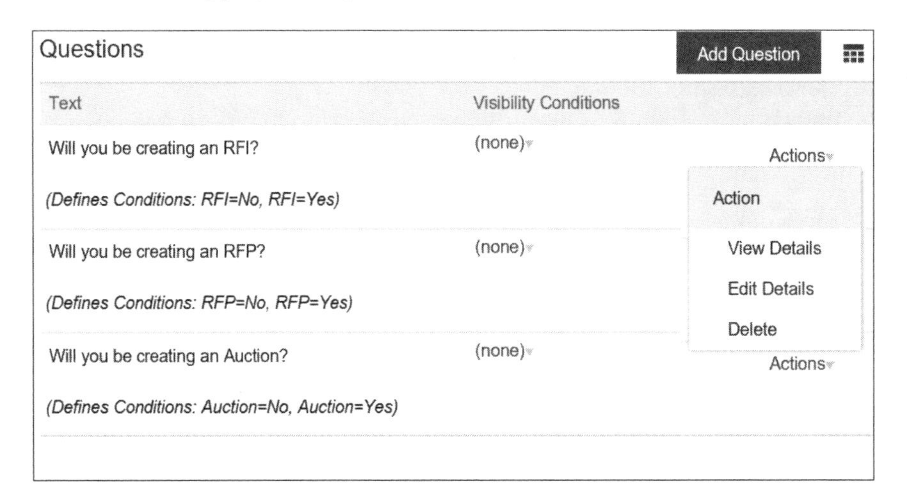

Figure 5.21 Standard Questions in a Sourcing Project Template

The sourcing project creator (sourcing agent/sourcing manager) must answer the template questions, shown in Figure 5.22, before using the sourcing project template to create a sourcing project.

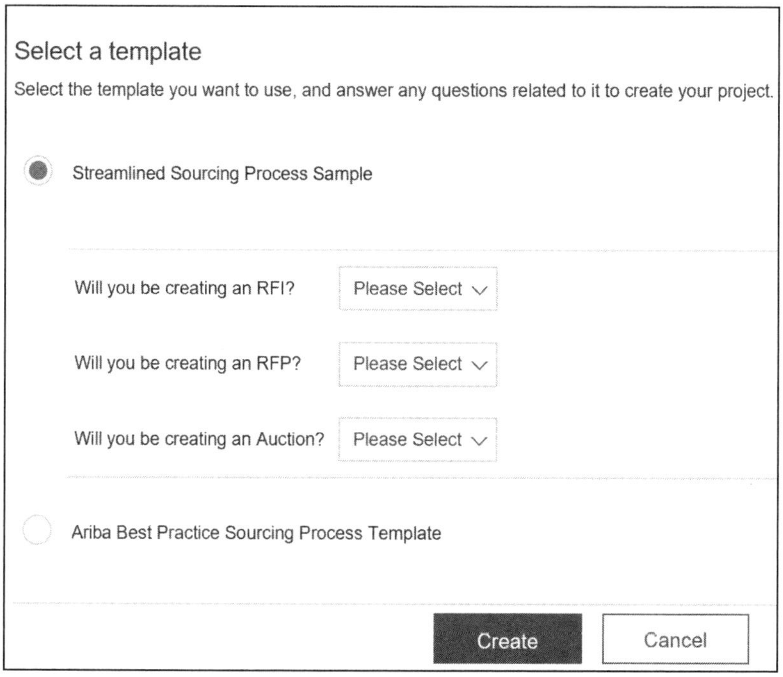

Figure 5.22 Standard Questions When Creating a Sourcing Project

Based on the responses to these template questions, sourcing event documents and associated tasks will be made visible in the created sourcing project.

Team Tab

SAP Ariba Sourcing allows administrators and template creators to automatically add team members to facilitate collaboration, approvals, and project administration. Template creators can also ensure that some project groups can't be deleted by end users in sourcing projects, which helps enforce business policies in terms of access to sourcing projects created. End users creating sourcing projects can also provide access to their projects to additional users by adding the individual users or user groups under the **Team** tab, as shown in Figure 5.23.

To add users or groups to the **Team** tab, you first need to create a project group and then define the role for that project group. Based on the role assigned to the project group, the users or groups added as members to this project group will get either read-only access or both read/write access to the document. With the read/write access, users can execute the project on behalf of the owner. In many cases, a project group may not be assigned any role. Such groups are usually used for approvals. The project group is added to the approval flow, and any team member added to this project group becomes an approver.

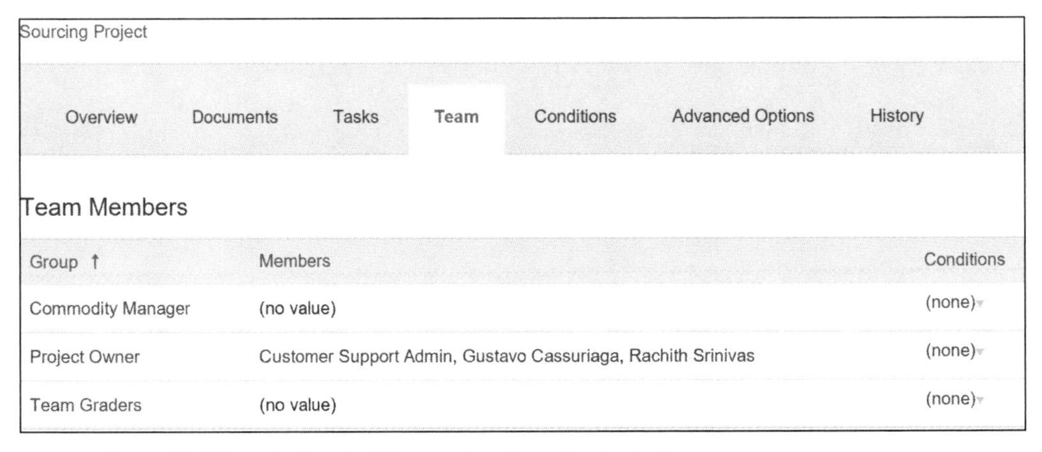

Figure 5.23 Standard Project Groups in a Sourcing Project

The project owner group is a default group. The creator of the sourcing project will be added automatically to this group. If you want to set additional users as project owners, then you can add these users as team members to this group. All tasks owned by the project owner group will now be owned by all the users in this group.

The team grader group is a special group in SAP Ariba Sourcing projects. Members of this group will be automatically added as graders to evaluate the responses to sourcing events during the sourcing project.

As shown in Figure 5.24, additional custom project groups can be added by following these steps:

1. Click **Actions • Edit** from the dropdown menu in the **Team** tab.
2. Click **Add Group**.
3. When creating a new project group in a template, select **Yes** from the **Can owner edit this Project Group** dropdown list if you want to allow the project owner to delete this group in the sourcing project.
4. Enter a **Title** to identify the project group.
5. Select a role using the system groups.
6. Click **OK**.

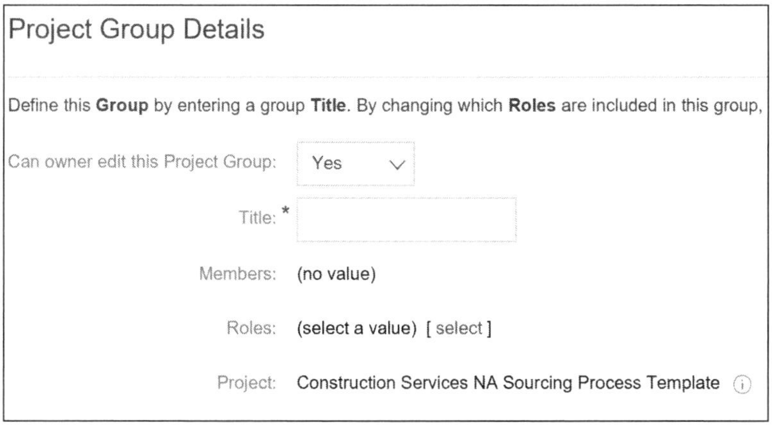

Project Group Details

Define this **Group** by entering a group **Title**. By changing which **Roles** are included in this group,

Can owner edit this Project Group: Yes

Title: *

Members: (no value)

Roles: (select a value) [select]

Project: Construction Services NA Sourcing Process Template

Figure 5.24 Creating a New Project Group

5.2.6 Creating and Configuring Request For Event Templates

SAP Ariba provides several RFx event templates out of the box. The RFx event templates are an event type. A best practice for creating customer-specific event templates is to copy one of the existing templates.

Adding an Event Template

To add an event template, begin by following these steps:

1. Log in as a user with the template creator role.
2. In the template project, search for the **Sourcing Templates** folder.
3. Look for an SAP Ariba Best Practice RFx template (e.g., **RFP Template**).
4. Click on the template, and then click **Copy**.

A new RFx event template is created as a draft. Some typical configuration steps usually used with an RFx event template include the following:

- Adding content
- Adding additional terms to line items

Adding Content

To add content to an RFx event template, begin by following these steps:

1. Click on an RFP event document under the **Documents** area, as shown in Figure 5.25.
2. Click on **Request for Proposal - MARKETING**.
3. Click on **Content**, as shown in Figure 5.26.

Request for Proposal

Sourcing Project

Overview Conditions

Properties Actions

Name: Request for Proposal

Description: Use this RFP to create a questionnaire with sections, questions, requirements, and/or line items to collect pricing information, and/or qualify suppliers, possibly for an auction. Buyers can weigh and grade supplier responses, and create an overall score for each supplier. Supplier responses are not revealed to other suppliers.

Version: v5 (editing)

ID: WS1481627317

Owner: aribasystem

Process Status:

Base Language: English

Rank: 0

Event Type: RFP

Access Control: Private To Team Members

Conditions: Match Project Type

Documents

Request for Proposal - MARKETING

Figure 5.25 RFx Event Template

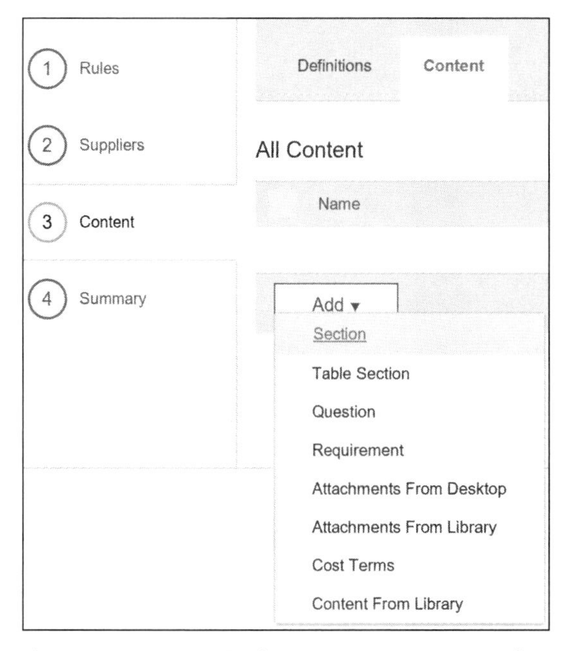

Figure 5.26 Content in the Sourcing Project Template

The RFx content you configure, such as questions and requirements, will always be asked for every RFx event created from this template and can be added to the template content. Content added to the template is automatically pulled into every single RFx event created from this template. End users creating RFx events can delete the content or add additional content on an ad hoc basis.

Modifying Item Definitions

The definition of an item can be modified in the template to add additional out-of-the-box terms, delete terms, or add new terms.

Adding a New Lot

In addition, new lots can be added so end users can bundle items into lots for suppliers to respond to.

Adding Product Questionnaires

SAP Ariba released a new feature that can be used to collect additional details on each line item. Previously, when questions were added to a sourcing event, they were associated with the event and not with specific line items. Now, questions can be added at the line item level as part of a new object called a *product questionnaire* as shown in Figure 5.27.

Now, you can add questions to a questionnaire that is specific to a product category so that when an item from this product category is added to an event as a line item, the product questionnaire is associated with that line item. In this way, questions can now also be made mandatory for specific line items.

Buyers can create product questionnaires in the following two ways:

- Add a product questionnaire to a product category from the **Category Attribute Management** page. Whenever an article from that category is added to a sourcing event, the product questionnaire is added automatically under the line item.
- Add a product questionnaire as a term. When adding a line item to a sourcing event, a user can add a product questionnaire to the line item as a term. You can add multiple questionnaires to a line item, provided each questionnaire is associated with a term that has the **Product Questionnaire** answer type. So, for example, you might have a **Certification Questionnaire** and an **Insurance Questionnaire** associated with a single line item.

Conditions are supported in product questionnaires. So, when adding answers to questions in a product questionnaire, if a supplier adds an answer to a question that triggers a conditional question, that question is then added to the questionnaire automatically.

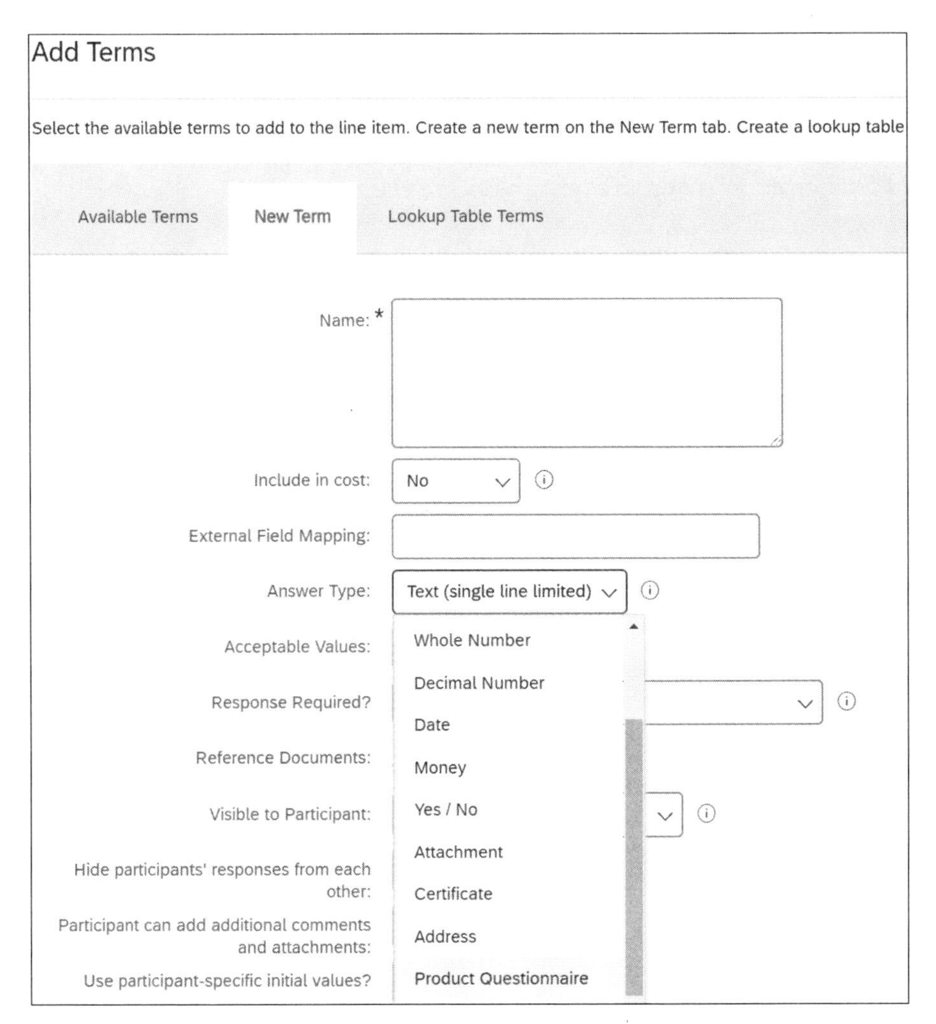

Figure 5.27 Product Questionnaire Term

When a supplier provides answers for questions in a product questionnaire, those answers are saved with the questions. So, if the buyer later adds the same product questionnaire to another sourcing event for the same supplier, the old product questionnaire answers from that supplier are included as default answers in the product questionnaire. Both buyer and supplier can choose to change the default value.

If a contract line items document (CLID) is created from a sourcing event that contains a product questionnaire, when you open the line items document on the contract workspace, the product questionnaire is listed as a term column heading. You can click on a link under the product questionnaire column heading to open a product questionnaire. The answers provided by the supplier are displayed by default, but the buyer can edit these if required.

Note

This feature also provides buyer organizations with a new Product Hierarchy Management API that can be used to return a list of product questionnaires responded to by a supplier along with associated sourcing events and line items.

5.2.7 Configuring Sourcing Library

Content that is usually repeated in every RFx event created by end users can be added to the template or maintained in the Sourcing Library so that end users don't have to build the content from scratch every time they create a sourcing event. Sourcing Library content can be accessed from the RFx event's **Content** section, as shown in Figure 5.28.

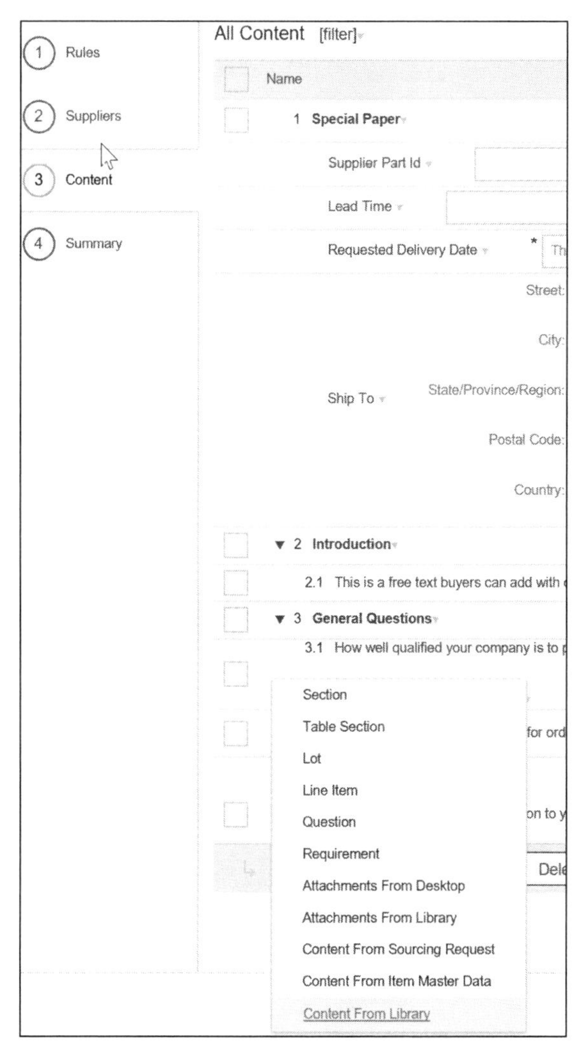

Figure 5.28 Adding Content from the Sourcing Library

The **Sourcing Library** screen, shown in Figure 5.29, can be accessed from **Manage • Sourcing Library**.

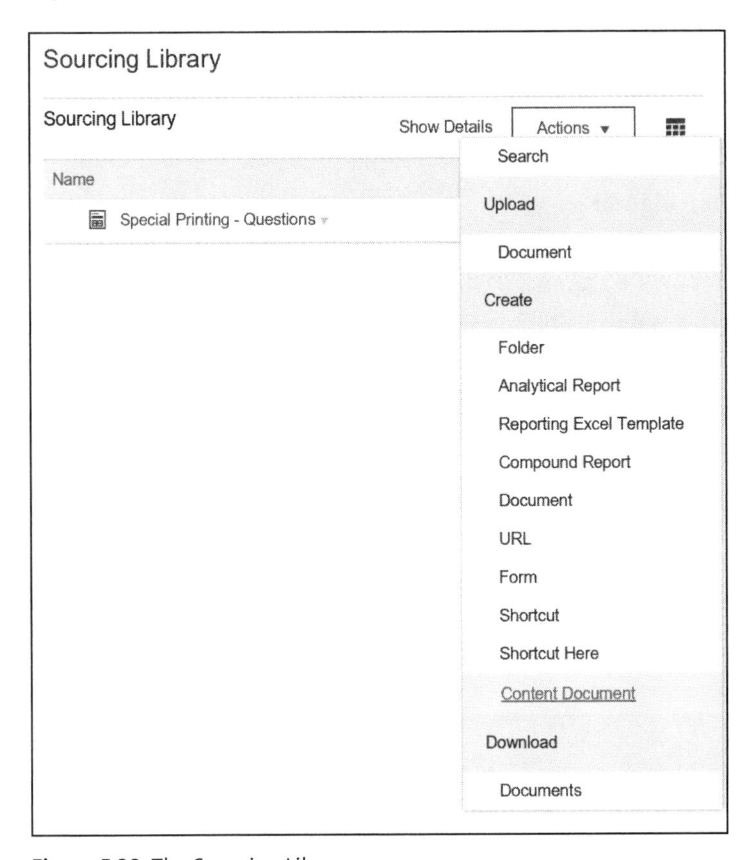

Figure 5.29 The Sourcing Library

5.2.8 Configuring Approval Tasks in Templates

A *workflow task* can be defined in a template, can be require, and is highly customizable. This task can be associated to a specific document or to a folder within the **Documents** tab. When assigned to a folder, all documents within the folder will be included in the approval workflow process.

Creating an Approval Task

To create an approval task, begin by following these steps:

1. Click on a document or folder under the **Documents** tab.

2. Select the approval task from the list of tasks available, as shown in Figure 5.30.
3. Configure the approval task, and click **OK**.

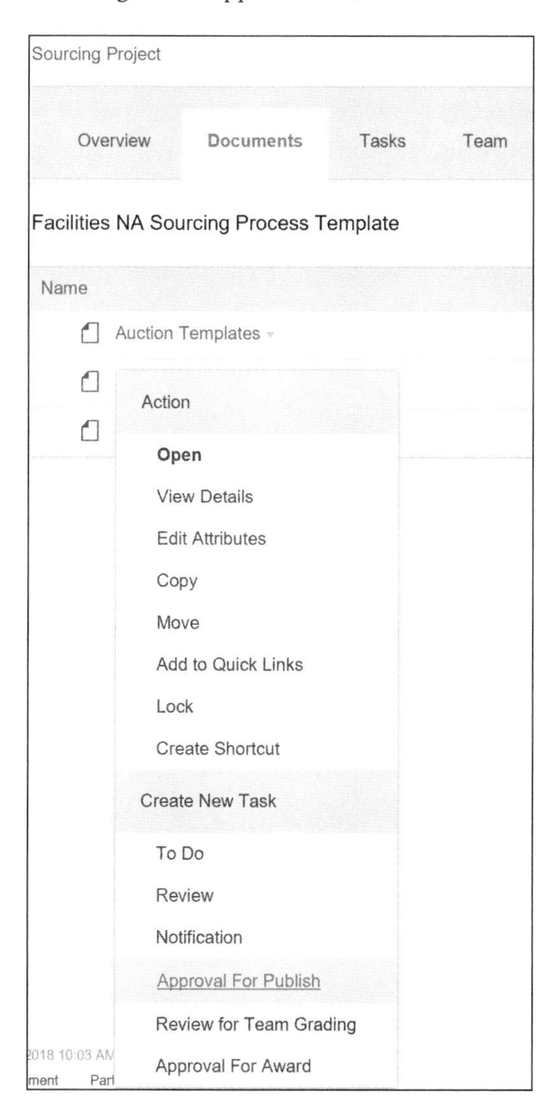

Figure 5.30 Creating an Approval Task on a Sourcing Document

After the workflow task is triggered, the documents within the folder—if the approval task was created on a folder—will be routed for approval, and an email will be generated and sent out to the approvers specified in the approval flow.

Configuring the Approval Task

SAP Ariba Sourcing provides a flexible approval workflow task that can be customized based on your approval process. The following approval flows are allowed:

- **Parallel**
  Multiple approvers/groups of approvers will receive the approval emails simultaneously, allowing access for approvers to take action on the approval task simultaneously.
- **Serial**
  Each approver/group within the approval flow will receive an approval email in a specified series. An approver/group will receive an approval request email only after the previous approver/group has approved the documents in the previous approval step.
- **Custom**
  You can define a more complex workflow with a combination of parallel and serial approvals.

Approvers will be required to review and approve the documents within the folder associated with the approval task. Observers can be added to the approval workflow as well. Observers aren't required to take any action throughout the process but will have access to review the status of the approval and review the documents within the approval task.

Adding Approvers to Custom Approval Flows

Custom approval flows add flexibility to SAP Ariba Sourcing to define a customized and complex approval process. If you select **Custom** as the flow, both parallel and serial approval steps can be added to the approval process, as shown in Figure 5.31. **User or Group – System and Project Group** can be added as approvers to each step, as shown in Figure 5.32.

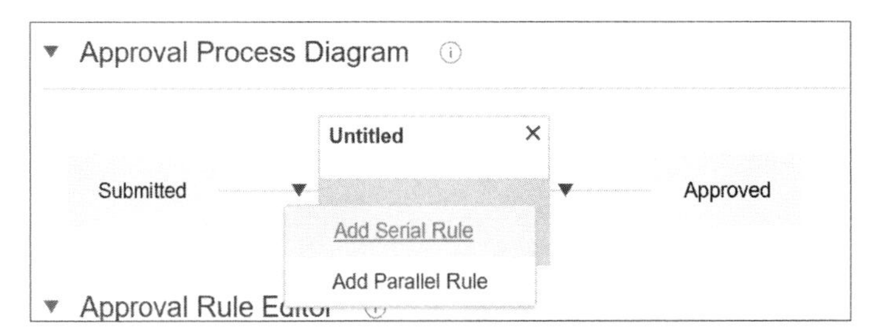

Figure 5.31 Custom Approval Flow: Serial or Parallel

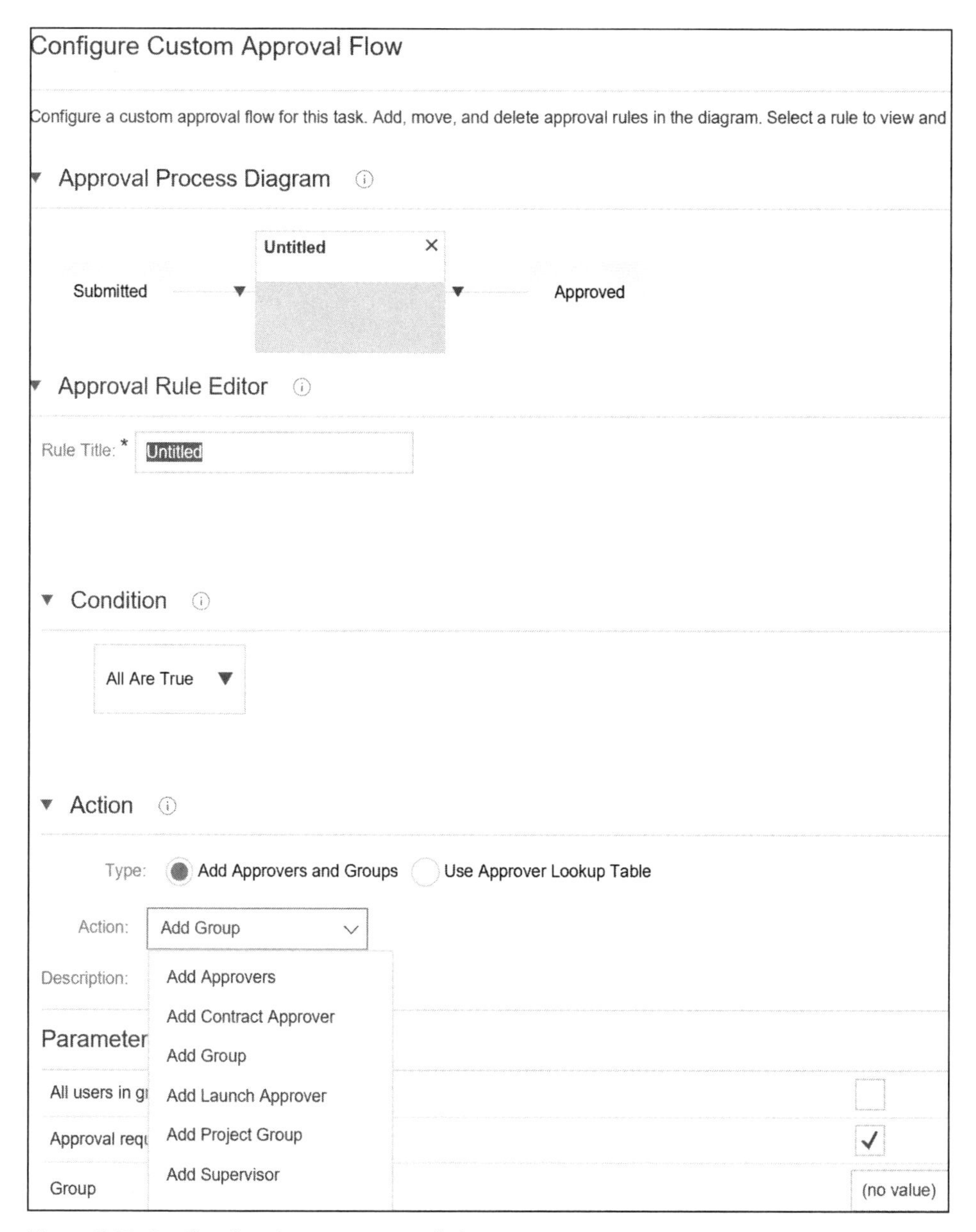

Figure 5.32 Configuring Custom Approval Flow

5.2.9 Sourcing as a Buyer

Buyers with the **Sourcing Manager** or **Sourcing Agent** can use SAP Ariba Sourcing to create sourcing projects or sourcing events. These users will approve sourcing requests created by other end users in the organization, create sourcing projects and sourcing events, and pull in incumbent suppliers and new suppliers into these sourcing events to competitively source goods or services. After running the sourcing events, based on the responses from the participating suppliers, the buyers can select one or more suppliers to award business.

Note

Awarding business during a sourcing event isn't an obligation but an intention to do business. Buyers can choose to send out the award information to the selected suppliers via an email. This email is configurable to include company-specific verbiage.

Note

Buyers must ensure that new suppliers invited to participate in the sourcing event are fully onboarded using SAP Ariba Supplier Lifecycle and Performance before awarding business.

5.2.10 Sourcing as a Supplier

Incumbent suppliers with existing relationships with the buyer's organization or new suppliers can be added to sourcing events by a buyer. The supplier contact added to the events will receive email notifications requesting the contact to participate in the sourcing event. The supplier contact can choose to either participate or decline participation. If the supplier contact decides to participate in the sourcing event, he should respond to the sourcing event by answering the questions and providing quotes for the goods or services that the buyer has included in the sourcing event. The supplier contact must log in to Ariba Network to respond to the sourcing events.

If the supplier is new without an Ariba Network account, then the supplier contact added to the sourcing event must first register the supplier to Ariba Network before accessing the sourcing event.

Buyers in an organization usually work with different supplier contacts for different sourcing events. The buyer first creates the supplier contact in the supplier profile in SAP Ariba Supplier Lifecycle and Performance before inviting the new supplier contact to participate in a sourcing event.

5.3 SAP Ariba Discovery

SAP Ariba Discovery is a service that allows buyers to find new suppliers and read profile information, feedback, and other information about them. SAP Ariba Discovery isn't technically included in other SAP Ariba solutions but is rather a separate service. In some circumstances, you can access SAP Ariba Discovery from within your solution.

SAP Ariba Discovery is currently free for buyers. Before you first access SAP Ariba Discovery, you must accept the online terms of use, and those terms will apply if you choose to proceed and use SAP Ariba Discovery.

SAP Ariba Discovery augments SAP Ariba Sourcing for suppliers with the following functionalities:

- **Searching for suppliers**
  An intelligent matching tool to identify suppliers that meet buyers' needs in a category, pulled from a global database of millions of suppliers in 140 countries. Supplier profiles contain enriched information with community insights.
- **Engaging with suppliers**
  The ability to invite newly identified suppliers to qualification events or RFx. After a supplier has responded to a posting and has passed the initial qualification rules set, you can move the supplier further along in the qualification process or invite the supplier directly into an RFx or auction.
- **Onboarding suppliers**
  If you've awarded the business, you have the ability to onboard a supplier quite easily. The tool includes ongoing self-maintenance of a supplier's profile information with a unified supplier portal—a one-stop shop for suppliers to update their information and collaborate on RFx, contracts, and so on.

In the following sections, we'll cover how to incorporate SAP Ariba Discovery into your sourcing process and provide some recommended sourcing scenarios that make good use of SAP Ariba Discovery. As with sourcing in general, scenarios will benefit from SAP Ariba Discovery, such as a fragmented market where you lack familiarity with suppliers but want to invite as many as possible to your RFx or auction to achieve further price reductions. Conversely, in a situation where the supplier enjoys a monopoly for a particular good or service, SAP Ariba Discovery may not be able to provide many more options.

5.3.1 Application at a Glance

SAP Ariba Discovery allows you to quickly and easily find new suppliers, conduct market research, and retrieve quotes. You can access SAP Ariba Discovery at different points in your SAP Ariba Sourcing solution. Four primary use cases define SAP Ariba Discovery:

- **Supplier research posting**
  In this use case, you'll find and evaluate new suppliers before a sourcing event. A common way to access SAP Ariba Discovery is by using the supplier research posting link to find new suppliers and conduct market research before creating a sourcing event. You don't need to own or create a sourcing event to conduct a supplier research posting.

- **Quick quote posting**
  In this use case, you'll obtain quick quotes for use with the Spot Buy capability. The quick quote feature also retrieves fast responses for spot buys and noncontract items.
- **Posting as part of a sourcing event**
  After you've selected one of the links in SAP Ariba Sourcing, you'll be taken to SAP Ariba Discovery to begin your posting. Clicking on **Post Now!** will launch a three-step process for creating a posting on SAP Ariba Discovery.
- **Inviting additional suppliers**
  In this use case, you can also invite additional suppliers to SAP Ariba Sourcing by adding suppliers from your database to those automatically matched to your posting by SAP Ariba Discovery.

After you're satisfied with the posting content, click **Publish**. Your posting will go live on SAP Ariba Discovery within 12 hours, and an email notification will be automatically sent to suppliers that match your posting.

After you've created a posting, you'll want to review responses, answer questions, and possibly edit your posting. For convenience, SAP Ariba Discovery allows you to manage postings from several areas, such as the following:

- Within an SAP Ariba sourcing event
- From the SAP Ariba Sourcing dashboard
- From SAP Ariba Discovery

You can then award the posting and import suppliers into SAP Ariba Sourcing. When you close the project by awarding the project, you can save newly discovered suppliers on SAP Ariba Discovery or import them into your sourcing database.

5.3.2 Recommended Sourcing Scenarios for SAP Ariba Discovery

Finding the right balance of suppliers for each event in sourcing is challenging. Inviting too few suppliers leads to suboptimal outcomes in savings, while inviting too many can quickly degenerate into confusion, as unqualified suppliers overshadow and underbid potentially qualified ones. SAP Ariba Discovery enables you to add additional qualified suppliers to sourcing events that otherwise would lack a sufficient number. Table 5.3 shows where having more suppliers adds the most savings and where the marginal utility of having an additional supplier starts to taper off.

Managing quote solicitation and processing entails real effort and, thus, expense, on the part of your purchasing organization, so at some point, adding more suppliers can actually reduce the savings realized.

The optimal number of suppliers in a sourcing event can vary by material or by the service being procured. Market constraints can be geographic or due to the consolidation of suppliers (market share) or the complexity and strategic significance of the item or service. In a market where only a few providers operate, having a few suppliers participate in a bidding exercise may be all that is possible. For more commoditized markets with many suppliers and wide variances in pricing and quality, having a large number of suppliers participate may provide significant cost savings.

As shown in Table 5.3, you can use SAP Ariba Discovery for complex sourcing, strategic sourcing, and spot buying. The timing of when to create postings in SAP Ariba Discovery varies by scenario. For complex sourcing, you would leverage SAP Ariba Discovery during the RFI template process, whereas, during a complex sourcing event, SAP Ariba Discovery should be used earlier to identify additions to the targeted supply base via research. Supplier profiles can reveal much about the applicability of a supplier to a sourcing event.

Project	Complex Sourcing	Strategic Sourcing	Spot Buying
Description	The roles: ■ Sourcing event manager ■ Supply material research Main uses: ■ Large spend of $3 to $50 million ■ Multiple teams and events	The roles: ■ Strategic sourcing manager Main uses: ■ Large spend of $100,000 to $5 million ■ Single strategy owner and event	The roles: ■ Maintenance, repair, and operations (MRO) buyer ■ Plant manager ■ Procurement agent Main uses: ■ Competitive bidding for $5,000 to $100,000 ■ Quick Spot Buy, no sourcing event
SAP Ariba products	SAP Ariba Sourcing: ■ Use full projects ■ Launch postings in RFI template ■ Link RFI to RFP or auction event	SAP Ariba Sourcing: ■ Launch postings early in the sourcing process with independent supplier market ■ Research posting	SAP Ariba Sourcing: ■ Create quick quote posting from the **Create** menu SAP Ariba Buying and Invoicing: ■ Without a requisition: create a quick quote posting from the **Create** menu ■ With a requisition: create a noncatalog item first, then create a posting

Table 5.3 How to Use SAP Ariba Discovery

A supplier profile in SAP Ariba Discovery contains the following information:

- Company information
- Product and service categories, ship-to and service locations, and industries
- Diversity, quality, and green classifications
- Transacting relationships
- SAP Ariba-ready certification
- Customer references and ratings
- Dun & Bradstreet (D&B) credit scores

For tactical Spot Buying, posting at the time of purchase (as soon as the need arises) is still preferable to not activating SAP Ariba Discovery at all.

Planning Your Implementation

SAP Ariba Discovery is part of the SAP Ariba Strategic Sourcing Suite, and no additional planning is required for this module. SAP Ariba Discovery is automatically enabled in the production realm and can be used from within SAP Ariba Sourcing.

Configuring SAP Ariba Discovery

SAP Ariba Discovery doesn't require any additional configurations and is automatically enabled in the production realm.

5.3.3 Using SAP Ariba Discovery

Buyers and sellers can use SAP Ariba Discovery to find new sources of supply or to find new leads, respectively.

As a Buyer

As a buyer, the buying organization can use SAP Ariba Discovery to find new sources of supply by following three simple steps (as seen in Figure 5.33):

1. **Create a posting.**

   The buyer creates a supplier discovery post describing what the organization is looking for. SAP Ariba Discovery automatically sends the posting to high-quality sellers who are a good match to be a source of supply for the product or services that the buying organization is looking for.

   The buyer can create the supplier discovery posting directly in SAP Ariba Discovery or via a sourcing event such as an auction or RFx event.

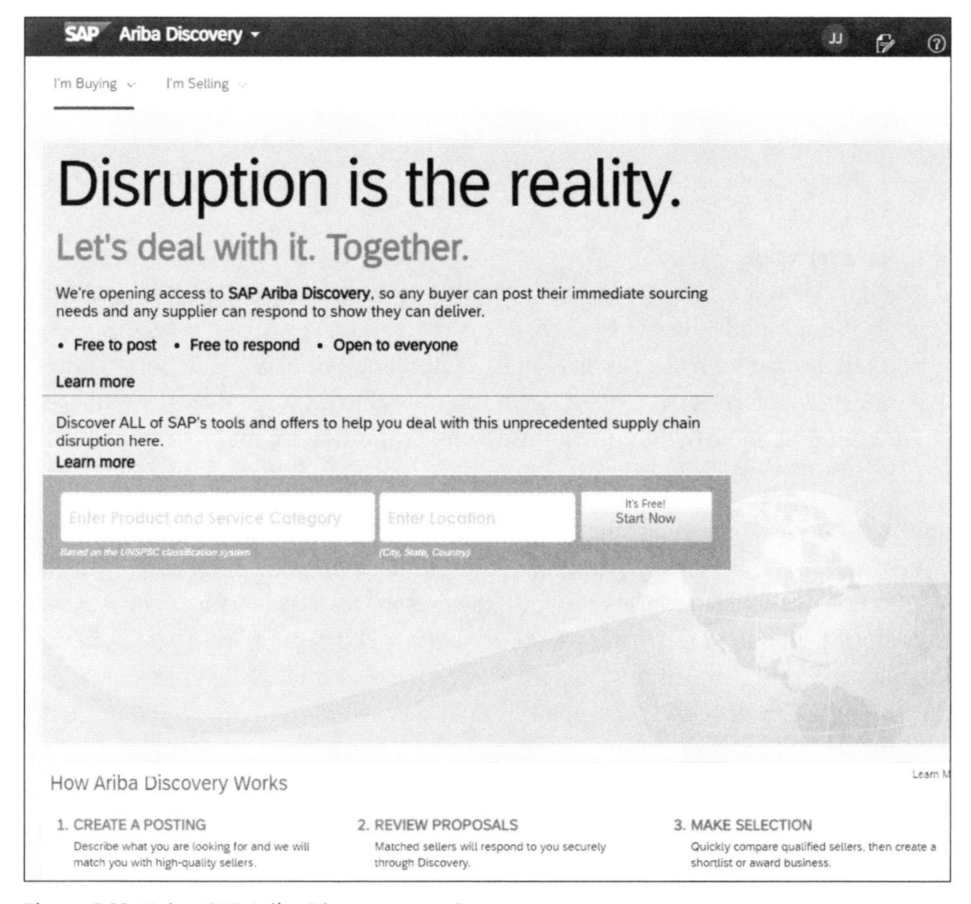

Figure 5.33 Using SAP Ariba Discovery as a Buyer

Creating a Supplier Discovery Posting

Buyers can log in directly to SAP Ariba Discovery and click on **Posting**. Buyers can then create a posting by clicking on **Create Posting**.

Buyers can log in to SAP Ariba Sourcing and click **Create – Quick Quote Posting or Supplier Research Posting** on the dashboard to access SAP Ariba Discovery.

From within a sourcing event, in the event monitoring page, click the **Discovery Supplier** tab. Click the link at the top of the page to access SAP Ariba Discovery.

2. **Review proposals.**
   Matched suppliers who are interested in participating will respond to the posting with quotes by logging in to SAP Ariba Discovery and responding to matched leads. The buyer accesses SAP Ariba Discovery as described earlier and clicks on **Postings** in SAP Ariba Discovery. The buyer can see all the postings he has created. By clicking

into an individual posting, the buyer can access responses to the posting from the participating sellers.

If the supplier discovery posting was created from within a sourcing event in SAP Ariba Sourcing, the buyer can directly access the responses to the posting from the **Discovery Supplier** tab within the SAP Ariba Sourcing event by clicking on **Action – View Responses**.

3. **Make a selection.**
   The buyer can review responses to individual postings as described previously and short-list or award business to a seller. The buyer can also review the seller's profile and ratings on SAP Ariba Discovery before deciding on the winning seller. After a winning seller is selected, a notification is sent to the winning seller. The other sellers who participated in the post will also be sent a notification that another seller has been awarded business.

Buyers can also manage their supplier discovery postings by accessing the postings on SAP Ariba Discovery and editing the post. Changes made to the posting are republished. Notifications are sent to sellers to resubmit their responses based on the changes, if they have already submitted responses. The buyer can also choose to delete a posting as necessary by the buying organization before the posting is awarded. Sellers participating in the posting will be notified via an email.

As a Seller

As a seller, the selling organization can drastically expand its reach and build new customer relationships using SAP Ariba Discovery by following three simple steps (as seen in Figure 5.34):

1. **Register.**
   Register your selling organization for free on SAP Ariba Discovery. Enter information on commodities that your organization can supply and the geographical locations that are covered.
2. **Receive leads.**
   Suppliers with matching commodities and territories included in a buyer's supplier discovery posting will automatically receive an email with the new lead.
3. **Discover new relationships.**
   Selling organization can choose to respond to new leads, giving them an opportunity to build new relationships and grow their business. Based on the responses from the sellers, the buying organization decides which seller best suits their need and awards business to the selling organization.

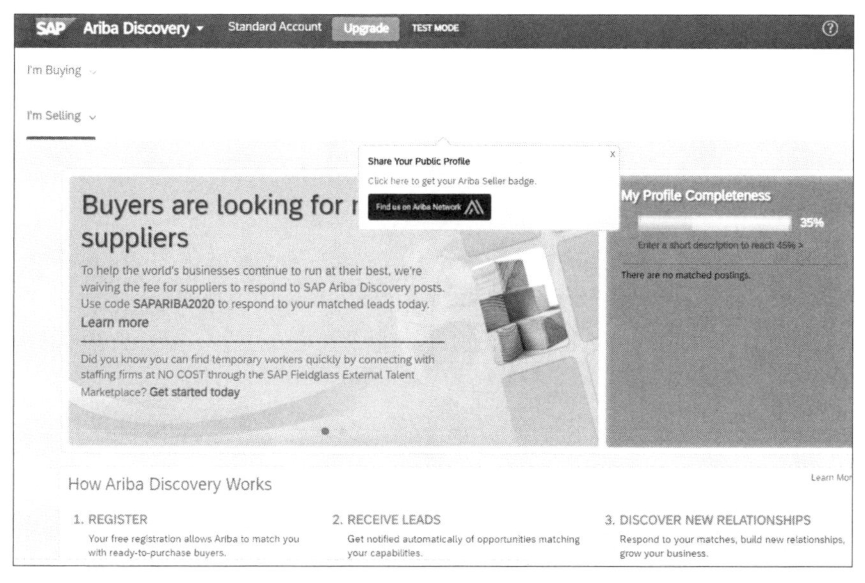

Figure 5.34 Using SAP Ariba Discovery as a Seller

5.4 SAP Ariba Strategic Sourcing for Product Sourcing

As mentioned earlier in Section 5.1.2 on the overall SAP Ariba portfolio, and as shown in Figure 5.35, SAP Ariba's product sourcing functionality is part of the SAP Ariba Strategic Sourcing Suite.

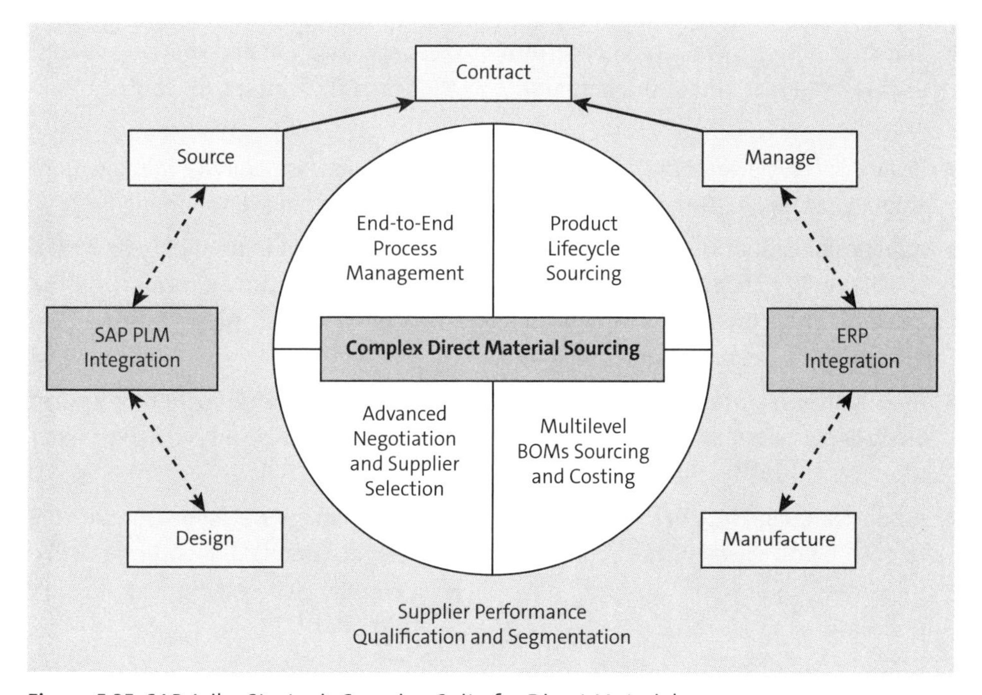

Figure 5.35 SAP Ariba Strategic Sourcing Suite for Direct Materials

The SAP Ariba Strategic Sourcing Suite is a comprehensive solution that works for all spend categories. The product sourcing functionality manages bills of materials (BOMs) from product lifecycle management (PLM) systems and manages the entire source-to-contract process from sourcing, to contract management, to contract compliance integration. The SAP Ariba Strategic Sourcing Suite enables you to do the following:

- Source directly from a BOM and integrate to your PLM and SAP ERP systems.
- Upload a BOM from your PLM systems and source directly from it. All items from your BOM will be priced and have suppliers assigned.
- Source items from complex BOMs while integrating to your ERP system to ensure your purchase orders are generated directly from those contracted terms.
- Ensure a selection of authorized suppliers compliant with regulatory requirements.
- View supplier qualification criteria at a detailed level.
- Export pricing information to an external system, which can then be used to create purchasing information records (PIRs) or analytical reports.

The SAP Ariba Strategic Sourcing Suite integrates seamlessly with ERP, PLM, and material master (MM) solutions from SAP. You can also model product costing for complex, multilevel BOMs, which can help you understand final product costs and negotiate better savings—all while managing suppliers at the item or plant level to enhance supplier qualification and performance. This integration enables you to get products to market faster by integrating sourcing and product design teams.

The SAP Ariba Strategic Sourcing Suite has the following built-in functionalities for procuring direct goods:

- End-to-end process management, from product specification and sourcing to contract management that's integrated with SAP ERP and that minimizes contract leakage
- Product lifecycle sourcing with integration to PLM, MM, and vendor management solutions from SAP for seamless data sharing and demand-driven sourcing
- Advanced negotiation and supplier selection functions, including support for additional event types, extended events, advanced scoring, and award optimization and analysis, which increase materials savings across more of your supply base and enables highly granular supplier selection
- Product costing of multilevel BOMs for materials and services, including what-if analysis for better volume pricing and more accurate predictions of final product costs
- Product sourcing supplier qualification, segmentation, and performance monitoring to ensure you select authorized, optimal suppliers; accurately assess their quality, timeliness, and responsiveness; and avoid noncompliance or supply chain disruptions

5.4.1 Application at a Glance

The SAP Ariba Strategic Sourcing Suite comes with dashboards for each solution included in the suite. A dashboard is also available for product sourcing. The information and tiles displayed on your dashboard depend on what group you belong to:

- If you're a member of the materials viewer group, you'll see information about the materials assigned to you.
- If you're a member of the materials manager group, you'll see information about BOMs. You'll also see information about materials, such as materials for which owners need to be assigned.

In the product sourcing dashboard, you can configure tiles or filter information from a selected tile. Based on the search filter entered, the material or the BOM will be displayed. More BOM details can be viewed by clicking on the BOM links displayed below in the table, as shown in Figure 5.36.

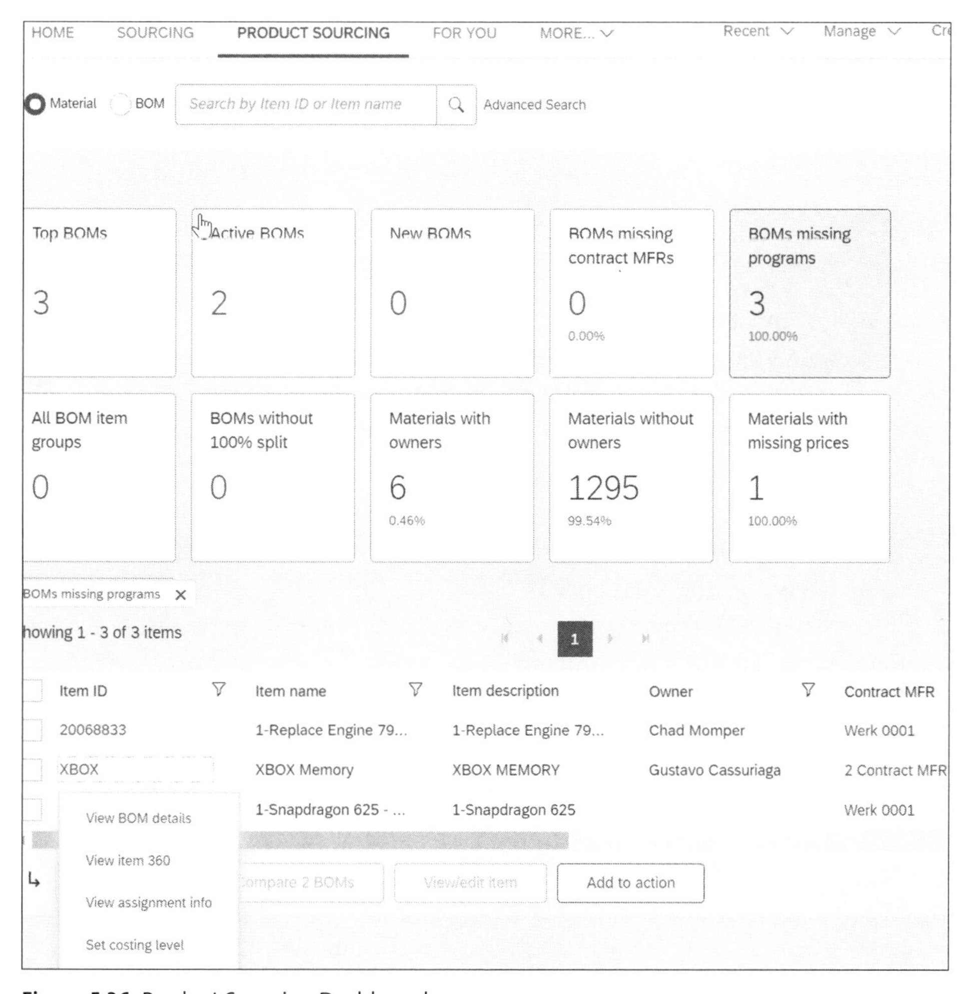

Figure 5.36 Product Sourcing Dashboard

In the following section, we'll look at some of the features available in the product sourcing functionality.

Action Tiles in Product Sourcing Dashboard

The following tiles can be added to the product sourcing dashboard:

- **Top BOMs**
  Displays the number of BOMs frequently used.
- **Active BOMs**
  Displays the number of active BOMs.
- **New BOMs**
  Displays the number of newly created BOMs.
- **BOMs Missing Contract MFRs**
  Displays the number of BOMs with unassigned contract manufacturers.
- **BOMs Missing Programs**
  Displays the total number of BOMs and percent without programs.
- **BOMs without 100% Split**
  Displays the total number of BOMs and percent of BOMs with splits that don't total 100%.
- **Materials with Owners**
  Displays the total number of materials and percent of materials with assigned owners.
- **Materials without Owners**
  Displays the total number of materials and percent of materials without owners assigned. This action tile shows the number of materials without owners. This tile displays a bar chart with the overall number materials that lack owners, with a bar for each part type and the number of materials without owners for that part type.
- **Materials with Missing Prices**
  Display the total number of materials and the percent of materials without pricing.
- **Materials with missing AML Splits**
  Displays the total number of materials and the percent of materials with missing Approved Manufacturer List (AML) splits.

Clicking any of the tiles displays the BOMs related to the tiles in the bottom table. Users can click the BOMs in the table to take further action, such as **View BOM Details**, **View Item 360**, **View Assignment Info**, **Set Costing Level**, and **Unset Costing Level**.

Product Sourcing Tile Visibility

The action tiles are visible to users based on their role. Table 5.4 shows a list of action tiles and the relevant user group that activates the various tiles.

Action Tile	Visible to Members of Group
■ BOMs missing contract manufacturers ■ BOMs missing programs ■ BOM item groups without 100% split	■ Materials manager
■ Materials with owners ■ Materials without owners ■ Materials with missing pricing ■ Materials with missing AML splits ■ Materials with missing lead time	■ Materials manager ■ Materials viewer

Table 5.4 List of Action Tiles and the Relevant User Group That Activates the Various Tiles

Item 360 View

Users can access the **Item 360** view by doing the following:

1. Log in to SAP Ariba, and click on the **Product Sourcing** dashboard.
2. Click on the interested tile, such as **Top BOMs**.
3. Click on one of the BOMs displayed in the table.
4. Select **View Item 360**.

Here you can see the 360° view (Figure 5.37), the activity log (Figure 5.38), and the pricing split (Figure 5.39).

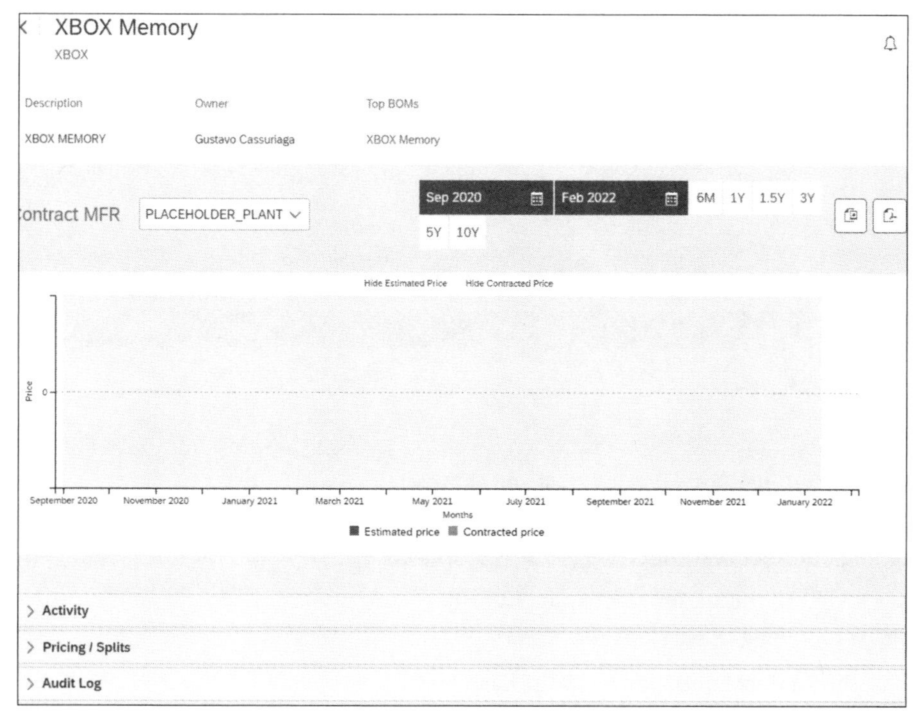

Figure 5.37 Item 360° View

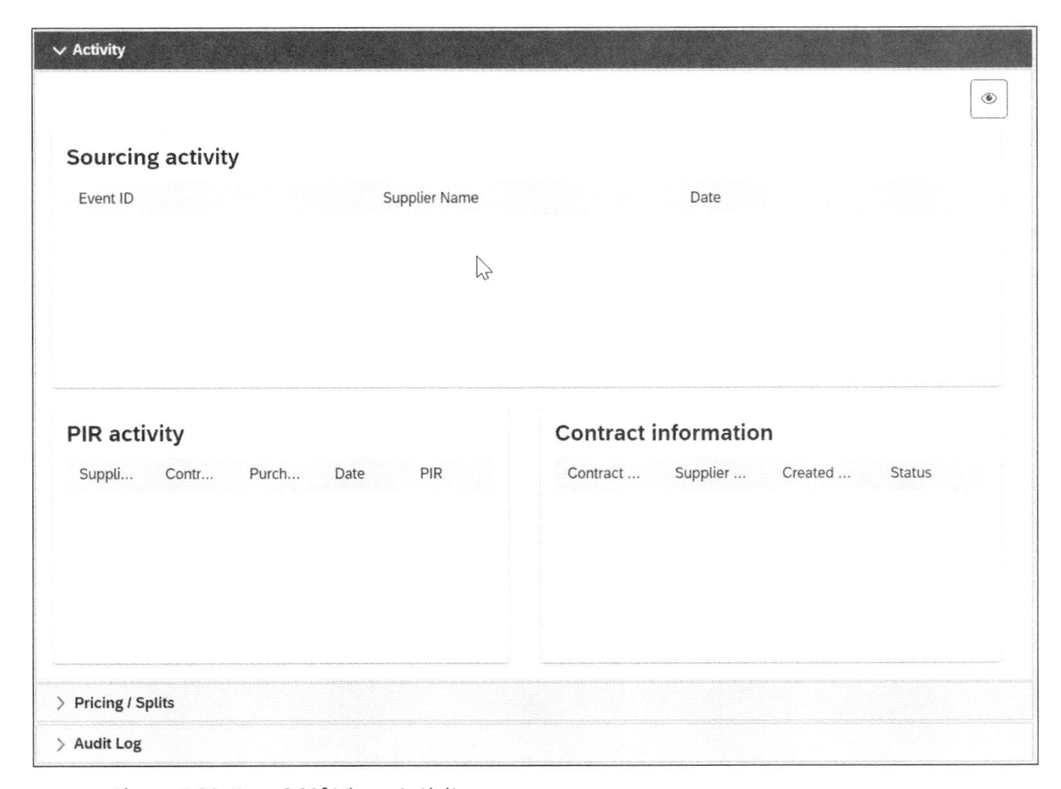

Figure 5.38 Item 360° View: Activity

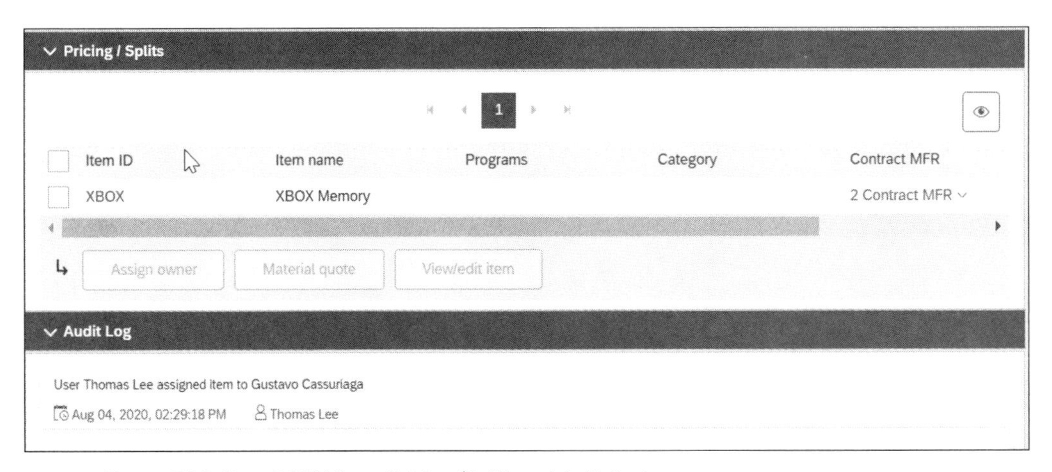

Figure 5.39 Item 360° View: Pricing/Split and Activity Log

Assigning Contract Manufacturers to Bills of Material

A user added to the materials manager group can assign contract manufacturers to BOMs by following these steps:

1. Click the **Product Sourcing** tab on the dashboard.

2. Select the BOMs to which you want to assign contract manufacturers by doing one of the following:
   - Check a BOM directly on the dashboard.
   - Click a BOM name, click **View details**, and select a BOM.
   - Click the **BOMs missing contract MFRs** action tile to view a list of BOMs without the contract manufacturers, and select a BOM.
3. Click **Assign**, and then click **Contract MFR**.
4. In the popup, choose the manufacturer you want to assign to the selected BOMs, and click **Assign**.

Assigning Programs to BOMs

This is similar to assigning contract manufacturers to BOMs, but you'll search for BOMs, click **Assign**, and select the owner from the list of users.

Assigning Owners to Materials

This is similar to assigning contract manufacturers to BOMs, but you'll search for materials without owners, click **Assign** to assign an owner, and then click **Program** to select the program to assign to the BOM (as in Figure 5.40).

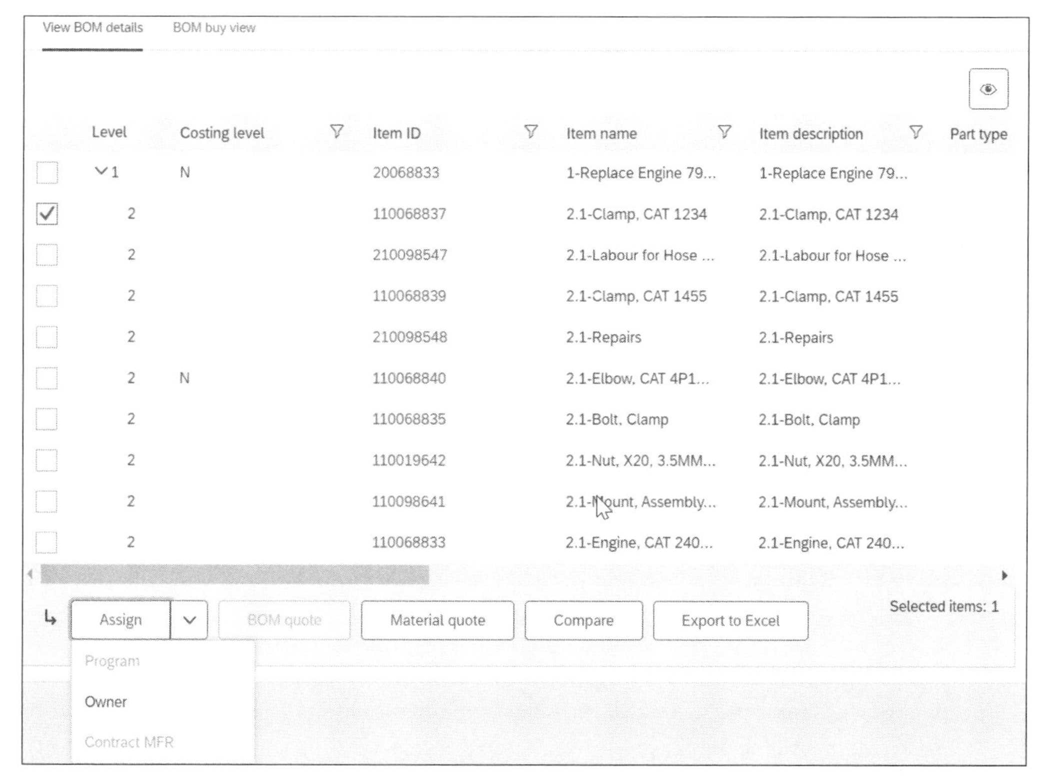

Figure 5.40 Product Sourcing: Assign Options

Setting Costing Levels

Costing levels can be set manually to items or assemblies. When a costing level is set, the cost is rolled up for the item but not for any child items below. If not set, the cost is rolled up for the item and any child items below it. Costing levels for an assembly are usually set when the price of the assembly already includes the costs of its child items.

To set costing levels, follow these steps:

1. Click the **Product Sourcing** dashboard.
2. Click on one of the tiles, such as **Active BOMs**.
3. In the table, click the BOM name of the materials for which you want to a set the costing level.
4. Click **Set costing level**. SAP Ariba displays the **Set Costing Level** confirmation screen.
5. Click **Set** (as shown in Figure 5.41).

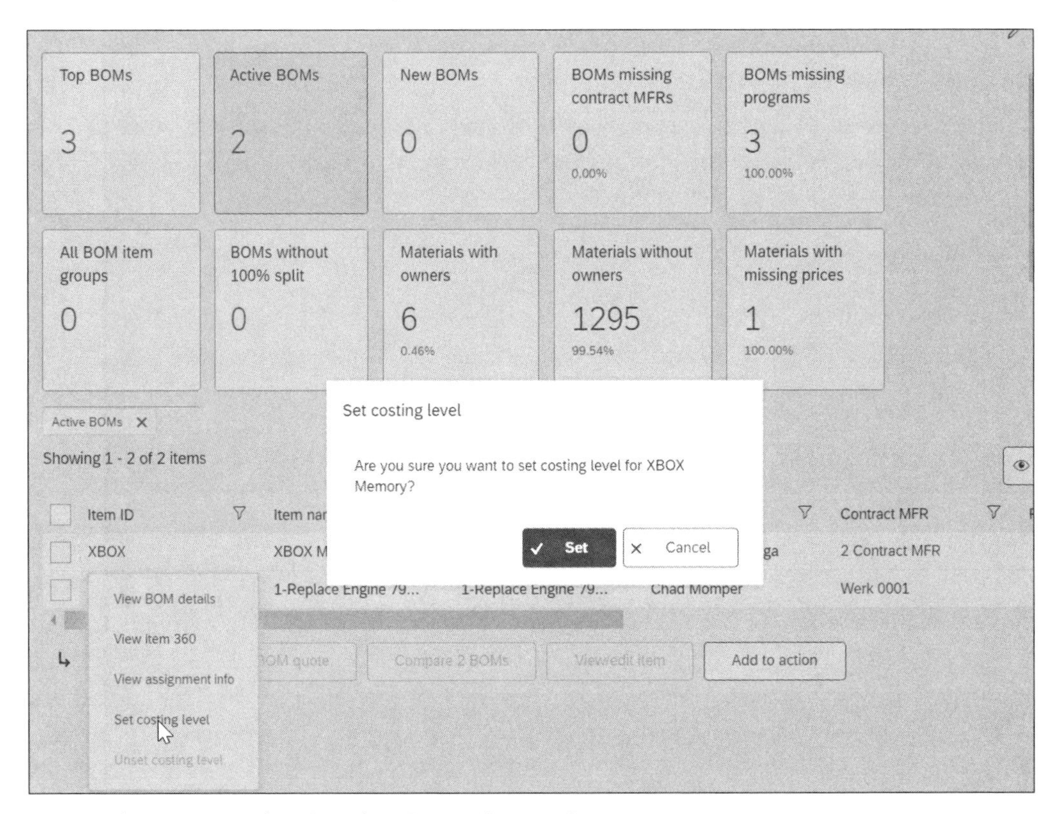

Figure 5.41 Product Sourcing: Set Costing Level

Default Plant

Buyers who don't use plants or only have one plant can use a default plant to store prices in the pricing database without having to select a contract manufacturer (plant). The default plant can be enabled to appear on UIs. In sites integrated with SAP ERP, when you create an item, the default plant is added to the assembly when the data comes from SAP ERP. The default plant is used for every assembly.

When a default plant is configured to be hidden from the UI, note the following considerations:

- The default plant is still present in the system and is added to the assembly when data is sent from SAP ERP.
- The **Contract MFR**, **Contract MFR ID**, and **Assign Contract MFR** fields, as well as the **BOMs missing contract MFRs** tile, aren't shown.

Note

Only members of the materials manager group can see the default plant on the dashboard.

Note

A default plant can only be configured when initially enabling the realm, and one and only one default plant can exist, which can't be changed. The default plant name is DEFAULT PLANT, and the ID is 9999. The name can't be changed.

Simple Request For Events for Materials

The sourcing event creation process requires multiple steps and may be too time-consuming for buyers that need to create basic, simple events. Simple events also often don't use the full event functionality available in SAP Ariba Sourcing. Now, buyers can create simple RFx events for materials that don't require complex functionality.

Complex sourcing scenarios on SAP ERP integrated systems and partial costing often require pricing conditions more complex than only cost terms, so simple RFx for materials supports cost terms that have time components to describe how costs are impacted in the future. The bidding steps for suppliers responding to simple RFx events for materials has also been simplified. Simple RFx events for materials enable suppliers to specify how long their quotes are valid for; SAP Ariba automatically expires these quotes and notifies the material owner.

Simple RFx events for materials offer buyers the following benefits:

- Ability to efficiently create simple RFx events in minutes
- Ability to capture pricing information for products and items on an ongoing basis
- A single, vertically scrolling event creation page
- Display for all users simple predefined RFx templates for buyers to choose from during the creation process

- Minimal fields required to create simple RFx events:
  - **Title**
  - **Select Template**

Simple RFx events for materials offer suppliers the following benefits:

- Suppliers can skip the select lots step when all items are required.
- Suppliers can skip the bidder agreement step when bidder agreements aren't required.

Note

The simple RFx feature is disabled by default. To enable this feature, you must log a service request ticket with SAP Ariba. This feature is only available for the SAP Ariba Strategic Sourcing Suite where product sourcing with BOMs and materials integrated with SAP ERP have been enabled.

Pricing Conditions in Simple Request For Events

Simple RFx events for materials enable you to enter validity periods for pricing. Specifying validity periods enables you to collect different prices during different validity periods. You can map validity periods on pricing conditions to external SAP ERP systems. Event owners can specify pricing condition settings as part of the term's attribute definition. Event owners can set the validity periods. You can also export a Microsoft Excel file containing the pricing conditions for each validity period and item and supplier combination. To download the pricing comparison Microsoft Excel file, click **Download Pricing Comparison**.

The pricing conditions functionality has the following limitations:

- Support for pricing conditions is limited to SAP ERP integrated events.
- Validity periods are available on contracts and RFPs, but not for auctions.
- Validity periods are only available on money-type terms.

5.4.2 Planning Your Implementation

Planning the implementation of the product sourcing functionality is the same as planning for SAP Ariba Sourcing, which we discussed in detail earlier in Section 5.2.

5.4.3 Creating and Configuring Simple Request For Event Templates for Materials

SAP Ariba provides several simple RFx event templates out of the box. A best practice for creating customer-specific RFx event templates is to copy one of the existing templates.

Copying a Request for Event Request Template

To copy an RFx event request template, begin by following these steps:

1. Log in as a user with the template creator role.
2. In the template project, search for the **Sourcing Templates** folder.
3. Look for and click on a simple RFx event template.

The template creator can now give a name to the newly created RFx simple event template. The copied template is tied to the same event type as the out-of-the-box template. Modify the newly created event based on your requirements, for example, by adding new terms.

Adding a New Term

New terms can be added to the line item within SAP Ariba Sourcing. The following section provides information on adding a new term:

1. Go to the **Content** tab of the template.
2. Edit the line item.
3. In the **Item Terms** table, click **Add • Term**.
4. From the **Source Type: Global** dropdown list under the **Available Terms** tab, choose the term (e.g., **Lead Time** term).

Enabling Pricing Condition

The bidding rules section under the **Overview** tab of the simple RFx event template contains a rule to enable pricing conditions for this template.

To enable the **Allow Pricing Conditions** rule, select **Yes** to allow event creators to configure validity periods. This rule must be enabled to use pricing conditions, validity periods, and volume scales in RFP sourcing events. Select **No** to disable validity periods.

When the **Allow Pricing Conditions** rule is enabled, the following rules appear:

- **Validity Period Type**
  This field is required. This rule determines the type of period for which you want to collect pricing. The following options are available for this rule:
  - **Monthly**
  - **Quarterly**
  - **Annually**
  - **Buyer Defined**
  - **Perpetual**
  - **Supplier Defined**

- **Start Date**
  This is the date on which the validity periods begin. The validity period automatically ends after the duration time expires.
- **Number of Periods**
  Enter the number of validity periods.
- **Scales**
  Enter the volume tiers for which you want to collect pricing.
- **Enable period quantity**
  Choose **Yes** to display the **Period Quantity** field on the pricing condition page. The period quantity value will either be a calculated field based on BOM volumes or a buyer-entered value.
- **Suppliers can view period quantity**
  This rule appears when you choose **Yes** for the **Enable period quantity** rule. Choose **Yes** to show suppliers the **Period Quantity** field. The **Period Quantity** field appears as a noneditable field for suppliers.

The **Add Term to Pricing Conditions Rule** in the **Edit Term** page can be used to add terms to pricing conditions. Choose **Yes** to include the term when collecting the price by time and volume information.

When the **Add Term to Pricing Conditions** rule is enabled, a **View Pricing Conditions** link appears in the **Item Terms** section on the **Add Item** and **Edit Item** pages. You can click **View Pricing Conditions** to view the volume scale, validity period, and cost terms.

The pricing condition functionality supports the following data types:

- Text
- Short strings
- Whole numbers
- Money
- Extended price

The new template must be published to be available for end users. Under the **Overview** tab for the template, click **Action • Publish**.

5.4.4 Configuring Parameters for Product Sourcing

SAP Ariba's product sourcing functionality includes some features that are controlled by site-level parameters that you can enable in SAP Ariba or product sourcing using manager-level parameters that members of the materials administrator group can configure. Let's look at some of the site-level parameters that can be configured.

Site Level Parameters Set by SAP Ariba

The `Application.ACM.AcceptPricing.ValidationRule` parameter specifies who can accept pricing for materials. By default, only a material's owner can accept (approve) pricing for a material in an event. This parameter can be configured to allow additional users to accept pricing for materials.

This parameter has three valid values:

- `MATERIAL OWNERS ONLY`
  Only the material owner can accept pricing.
- `PROJECT OWNERS`
  Material owners who have project owner capabilities can accept pricing for the material.
- `NO VALIDATION`
  Users with project owner capabilities can accept material pricing, regardless of whether the user is a material owner or not.

Product Sourcing Manager Parameters

These parameters can be configured by users in the materials administrator group. There parameters can be found under **Administration • Product Sourcing Manager**.

The following parameters can be configured:

- `BOM_UPLOAD_SERVICE.ENABLE - Default: TrueBOM_UPLOAD_SERVICE.ABORT_ON_FAILURE - Default: False.`
  For a payload with multiple BOMS:
  - `True`: Stops further processing if one of the BOM uploads fails.
  - `False`: System continues to process the payload after encountering the error. At the end of the upload, a partial success status shows for the upload.
- `DEFAULT_PLANT.ENABLED - Default: False.DEFAULT_PLANT.UI_DISPLAY - Default: False.`
- `DEFAULT_SUPPLIER - Default: False.`
  Determines if a default supplier can be selected for materials that don't have a supplier in the pricing database for product sourcing.
- `EXCEL.DISPLAY_PRICING_IN_MONTHS - Default: 18.`
  Specifies how many months of pricing, AML splits, and item group (BOM) splits are exported in Excel.
- `PIR_DEFAULTS.DefaultExternalSystem`
  SAP Ariba Strategic Sourcing Suite can send PIR data to only one external system. If your installation has multiple external systems configured on the **Master Data Manager • External System Configuration** page, set this parameter to the ID of the external system to use for PIR integration. When SAP Ariba Strategic Sourcing Suite sends

PIR data to an external system that is an SAP system, each material must have a valid SAP plant ID. You can set this parameter to a valid SAP plant ID to be used as the default plant ID.

- `TOGGLE.ITEM_360_DEMO`
  Provides enhancements to the 360° view of materials, which you can see after clicking on **View Material Details**. Enhancements include a price trend graph, information about the AML suppliers, sourcing events for the material, PIR activities, information about contracts for the material, and the BOMs that include the material.
- `UI.BOMS_WITH_MISSING_BOM_SPLIT_MONTH_INTERVAL - Default:18.`
  This count is the number of BOMs that don't have item group allocations that total 100% for each month in the next *n* months (including the current month).
- `UI.DISPLAY_PRICING_IN_MONTHS - Default:18.`
  Specifies how many months of pricing, AML splits, and item group (BOM) splits are displayed (exported) in the UI.
- `UI.MATERIAL_MISSING_AML_SPLIT_MONTH_INTERVAL - Default:18.`
  Specifies the number of months used to determine materials with a missing AML splits count. This count is the number of materials that don't have approved manufacturer allocations with prices that total 100% for each month in the next *n* months, including the current month.
- `UI.MATERIAL_MISSING_PRICE_MONTH_INTERVAL - Default:18.`
  Specifies the number of months used to determine materials with missing prices count. This count is the number of materials that don't have a price for each month in the next *n* months (including the current month).

5.4.5 Sourcing as a Buyer

Like sourcing for indirect material and services, buyers can get quotes for product materials by creating a sourcing project or a simple RFx project.

Creating Quotes for Materials Using a Sourcing Project

Category managers, commodity managers, sourcing agents, or procurement agents can create a quote using the following steps:

1. Click the **Product Sourcing** tab on the dashboard.
2. Complete one of the following actions to display materials:
   - Click one of the material action tiles. The data table automatically filters to display materials according to the action tile you selected.
   - Enter a material name or ID in the search box, and click the **Search** icon.

- Click **Advanced Search**. On the advanced search page, enter one or more material IDs, separated by spaces.

3. Check the materials for which you want to request supplier quotes.
4. Click **Get Quote**, and then click the following option:
   - **Using Sourcing Project**
     The **Create Sourcing Project** page displays if you choose **Using Sourcing Project** where you can create a sourcing event for materials.
5. The buyer enters information as needed and clicks **Publish**.
6. SAP Ariba sends invitation emails to the invited suppliers.
7. Invited suppliers sign in and submit their responses.

Creating Quotes for Materials Using a Simple Request For Event

Simple RFx events for materials only require that buyers enter information on the **Create Quote Request** page. Category managers, commodity managers, sourcing agents, or procurement agents can create a quote using the following steps:

1. Click on the **Product Sourcing** dashboard.
2. Search for the material for which you want to create a quote.
3. Click on the material, and go to the **Item 360°** view.
4. In the **Pricing/Splits** section, select the material, and click the **Material Quote** button, as shown in Figure 5.42.
5. Select **Material Quote Using Simple RFX**, as shown in Figure 5.43.
6. The buyer enters information as needed and clicks **Publish**.
7. SAP Ariba sends invitation emails to the invited suppliers.
8. Invited suppliers sign in and submit their responses.
9. After the simple RFx event for materials closes, the buyer chooses the best suppliers as well as the item combinations for the most acceptable supplier pricing schemes.

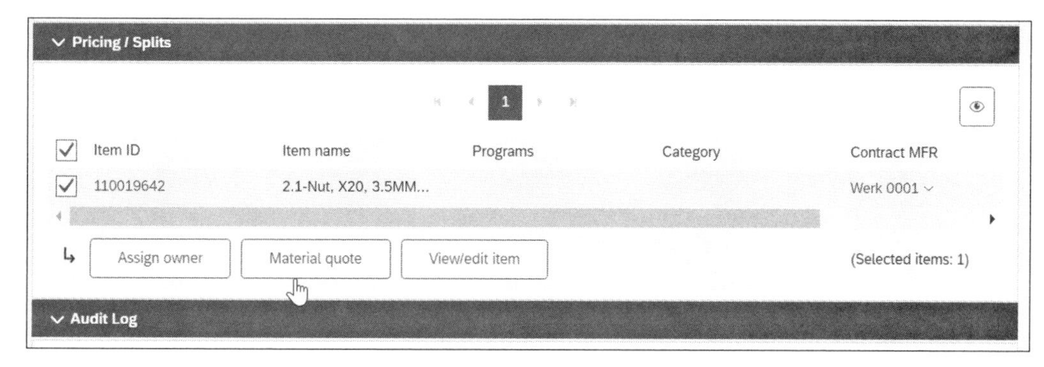

Figure 5.42 Product Sourcing: Creating a Material Quote

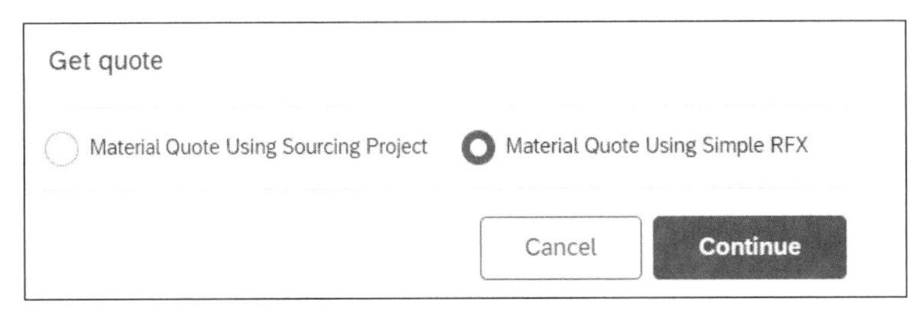

Figure 5.43 Product Sourcing: Creating Material Quote Using Simple RFX

5.4.6 Sourcing as a Supplier

The supplier contacts invited to product sourcing events receive invitation emails asking suppliers to submit quotes. The supplier contact clicks the link in the email to register the supplier on Ariba Network. The supplier contact responds to the product sourcing event using the following steps:

1. On Ariba Network, go to **Ariba Proposals and Questionnaires** dashboard (see Figure 5.44).

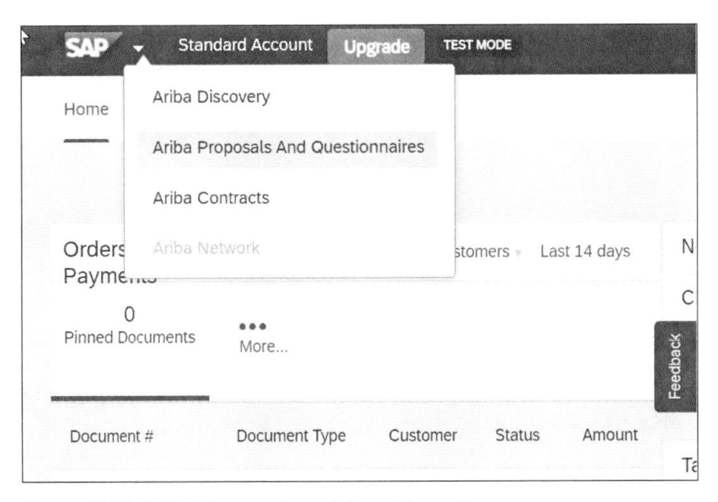

Figure 5.44 Ariba Proposals and Questionnaires

2. The **Events** section displays the product sourcing material quotes events for which the supplier is invited.
3. Click on the specific event to open it.
4. Click on **Intend to Respond**, and select **Yes**.
5. Select the **Material Lot**, and respond with a quote.
6. Click **Submit** to submit the response.

5.5 Guided Sourcing

Guided sourcing is a new feature released by SAP Ariba to further enhance the user experience (UX) for users creating and executing sourcing events. Guided sourcing is very intuitive, which increases user adoption and reduces the time required to create an RFI or RFP event by almost 50%. It also comes with its own dashboard called **FOR YOU**. Guided sourcing events are monitored from this dashboard, as shown in Figure 5.45.

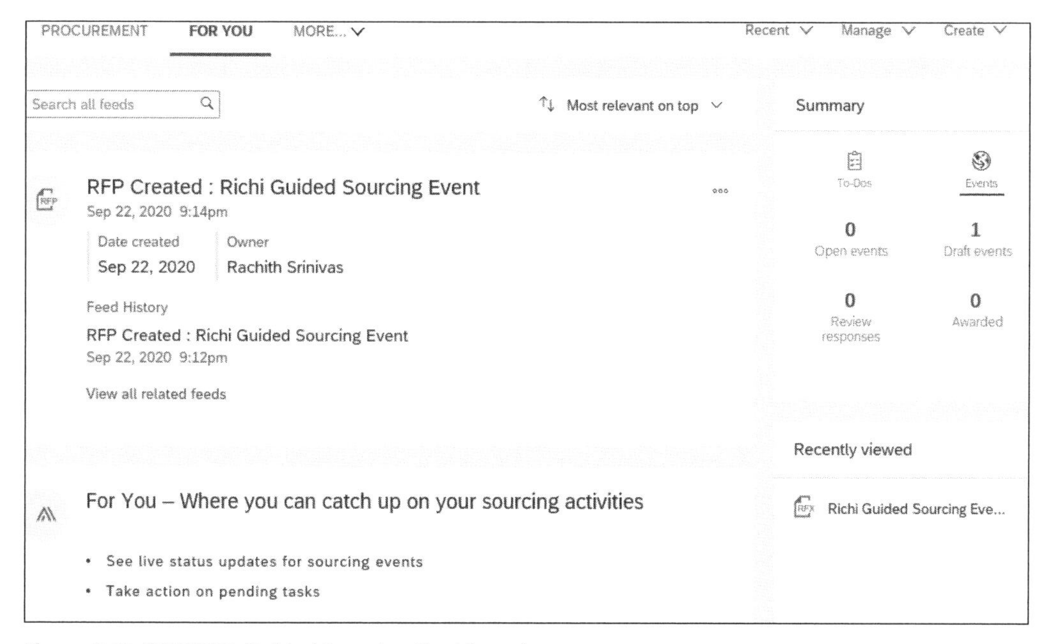

Figure 5.45 FOR YOU: Guided Sourcing Dashboard

Users with category buyer permissions can create new guided sourcing events.

> **Note**
>
> Users can only create RFI and RFP events using guided sourcing at this time. Auctions or full sourcing project can't be created.

Guided sourcing supports the following existing features in SAP Ariba Sourcing:

- RFI and RFP events
- Line items, lots (item lots), and sections
- Event content, such as questions, prerequisite questions, requirements, and attachments
- Importing line items and sections using smart Excel and simplified Excel
- Participant-specific initial values for terms
- **Approval for Publish** and **Approval for Award** tasks

- Event monitoring
- Preconfigured award scenarios: best bids, best savings, best bid with limited number of suppliers
- Manual award scenarios
- Event message boards
- Event notification emails
- Guided sourcing events available in SAP Ariba Sourcing reporting facts
- Email bidding
- Multicurrency events
- Creating a contract from a guided sourcing award

Guided sourcing events can be created from a guided sourcing template or from a smart import from Excel; they can also be copied from a previous guided sourcing event.

To create a guided sourcing event using template, follow these steps:

1. Log in to SAP Ariba Sourcing.
2. Click **Create • RFx**, as shown in Figure 5.46.
3. Select **Create from template**.
4. Enter the event **Name**, and select the **Event type** as **RFI|RFP**.
5. Enter event information such as **Commodity**, **Regions**, **Departments**, **Currency**, and **Baseline Spend**.
6. Select a guided sourcing **Template**.
7. Click the **Create** button on the top-right corner of the screen, as shown in Figure 5.47.

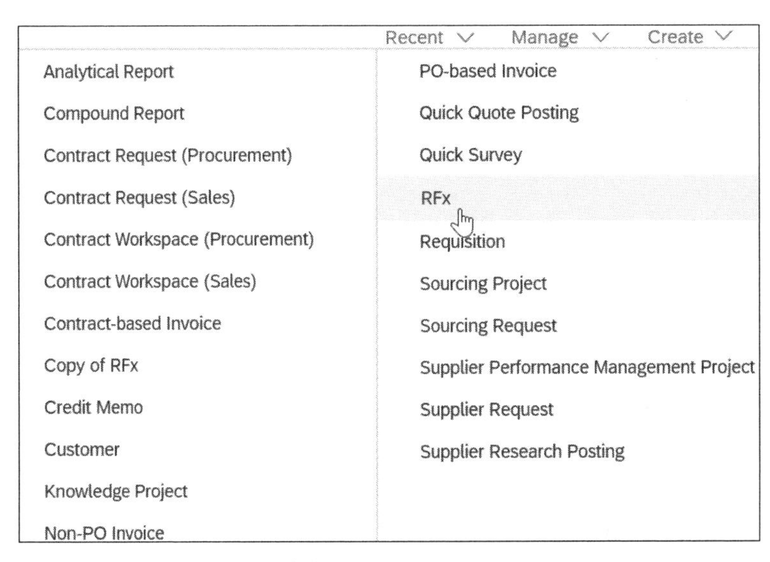

Figure 5.46 Creating a Guided Sourcing Event

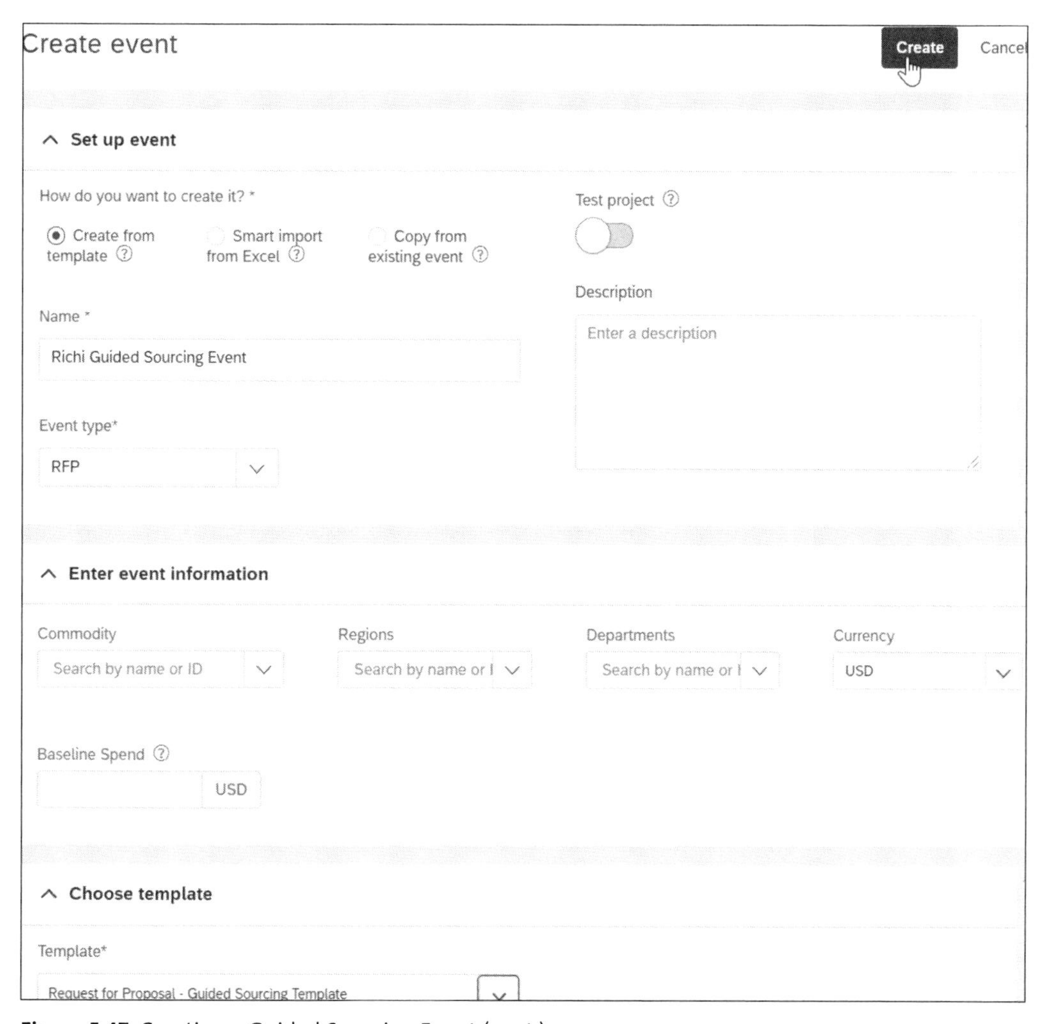

Figure 5.47 Creating a Guided Sourcing Event (cont.)

The guided sourcing event is created. This is a streamlined event with all the information in one page. Guided sourcing guides the user through creation to award in a very user-friendly UI that allows for a quicker event setup, as shown in Figure 5.48.

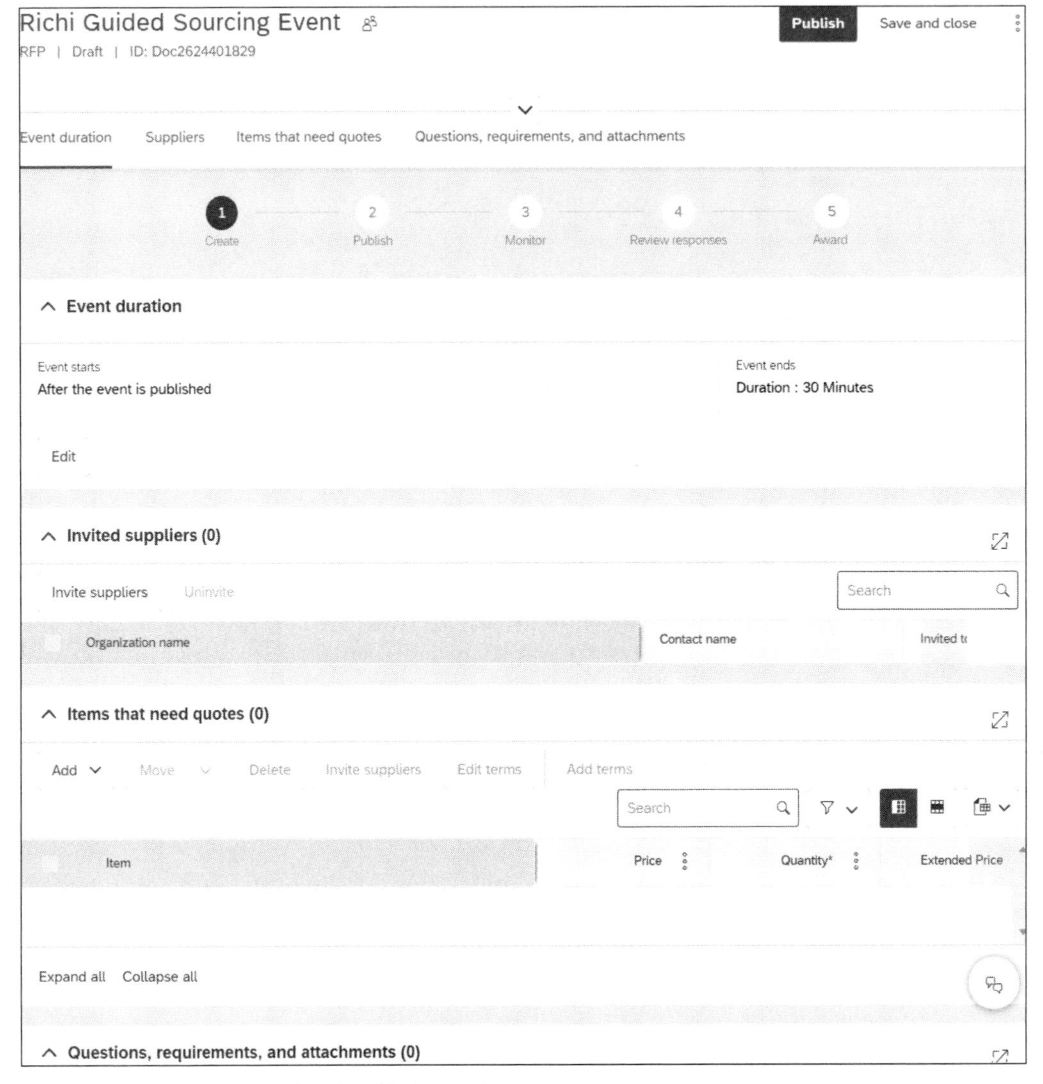

Figure 5.48 Streamlined Guided Sourcing Event

Guided sourcing includes a new feature called *SAP Enable Now*. This feature provides guided tours to walk the user through the process flow for key tasks. It also provides contextual help tiles with useful how-to guides for the user creating the guided sourcing event. Users spend less time looking for documentation and more time creating events, thus reducing the time required to create the event. Clicking the **?** icon opens SAP Enable Now with contextual help, as shown in Figure 5.49.

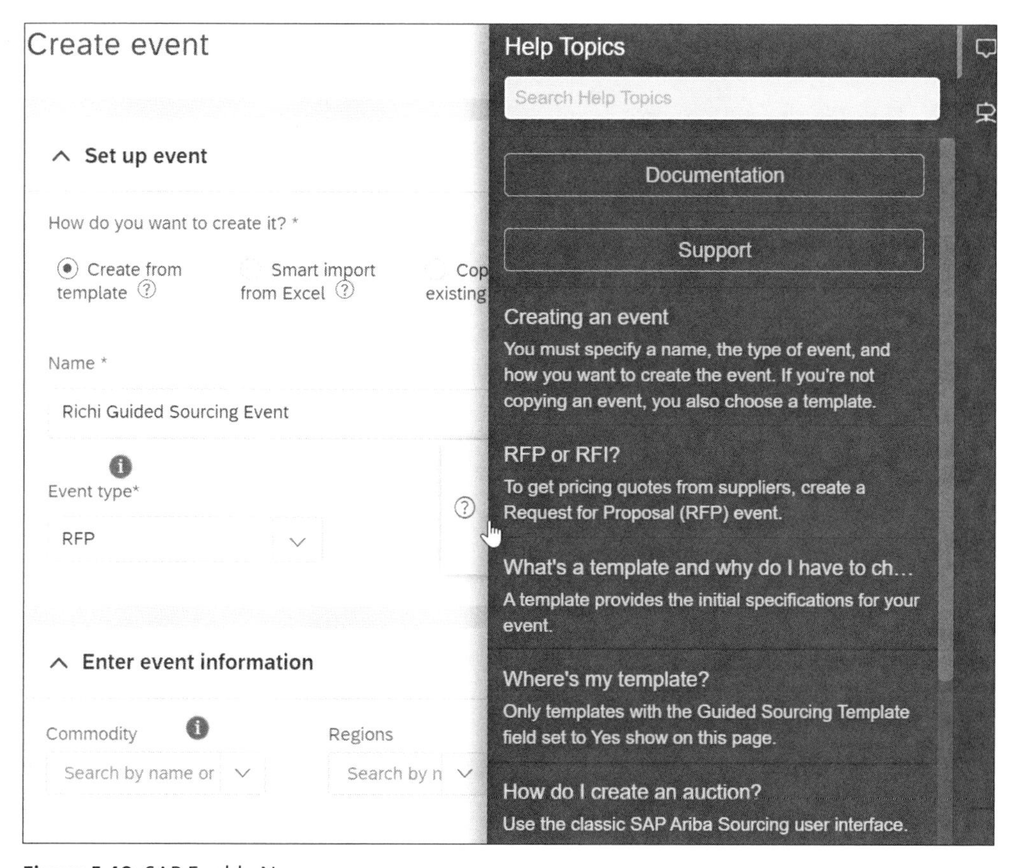

Figure 5.49 SAP Enable Now

Invite one or more suppliers by clicking the **Invite suppliers** button under the **Invited suppliers** section, as show in Figure 5.50. Suppliers can be selected by filtering based on **Qualification status** or **Preferred level**. In addition, if the right supplier isn't found, users can choose to create a new supplier by clicking **Create • New Supplier** or choose to add a new contact to an existing supplier by clicking **Create • New Contact**, as shown in Figure 5.51. If the user clicks on **Create • New Supplier**, the user is taken to the Supplier Request form in SAP Ariba Supplier Lifecycle and Performance (see Figure 5.51). See Chapter 3 for more information on creating a new supplier.

Figure 5.50 Guided Sourcing: Inviting Suppliers

Figure 5.51 Guided Sourcing: Selecting Supplier or Creating New

User can then add sections, lots, or items to the **Items that need quotes** section of the page. A popup page is displayed to fill in information about the item to be added, as shown in Figure 5.52.

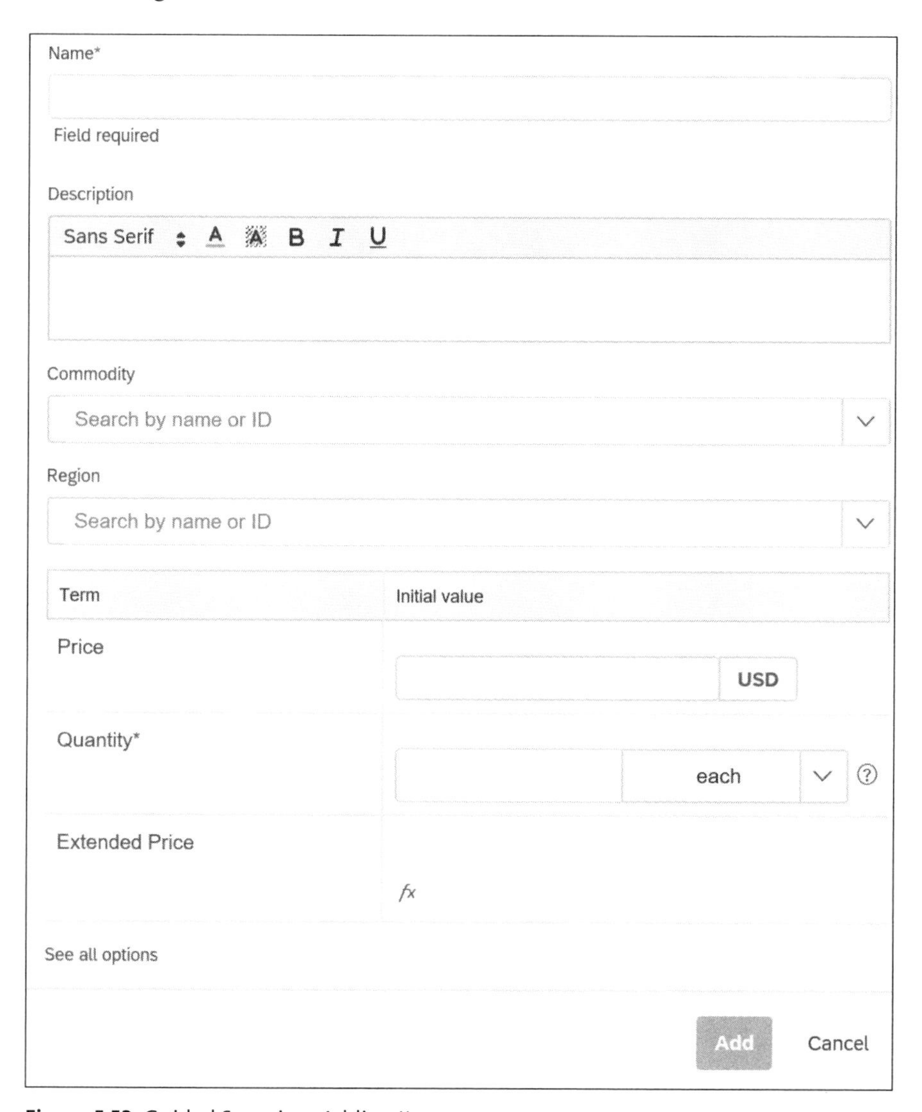

Figure 5.52 Guided Sourcing: Adding Items

In the next section, users can add questions, requirements, and attachments, as shown in Figure 5.53, to gather additional information from the suppliers.

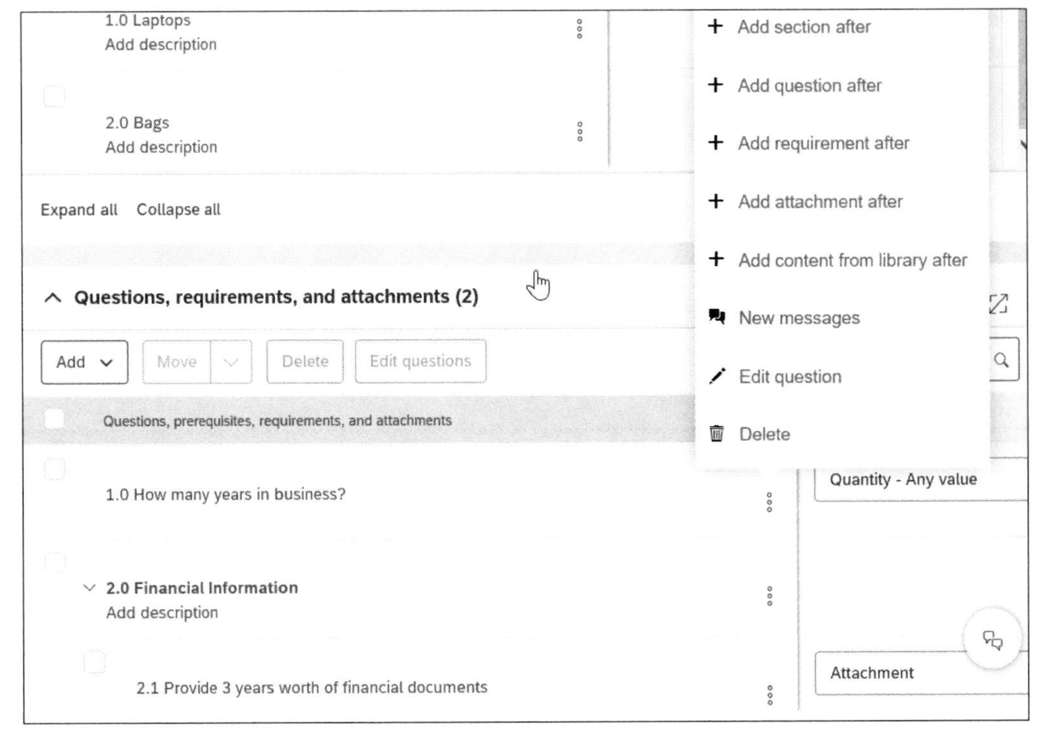

Figure 5.53 Guided Sourcing: Adding Questions

When the event is ready, click the **Publish** button on the top of the screen to publish the event to invited suppliers (see Figure 5.54).

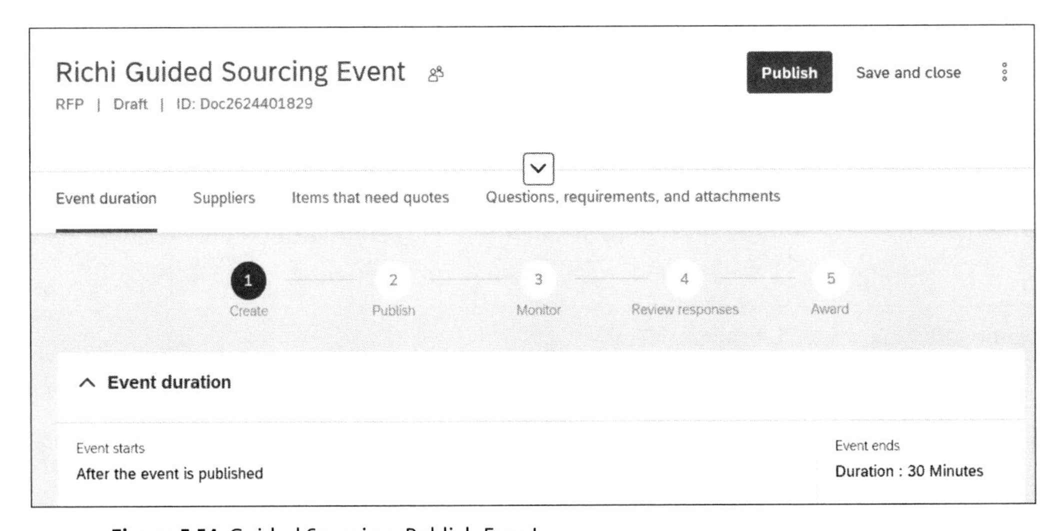

Figure 5.54 Guided Sourcing: Publish Event

After the event is published, the user can monitor the event as displayed in Figure 5.55.

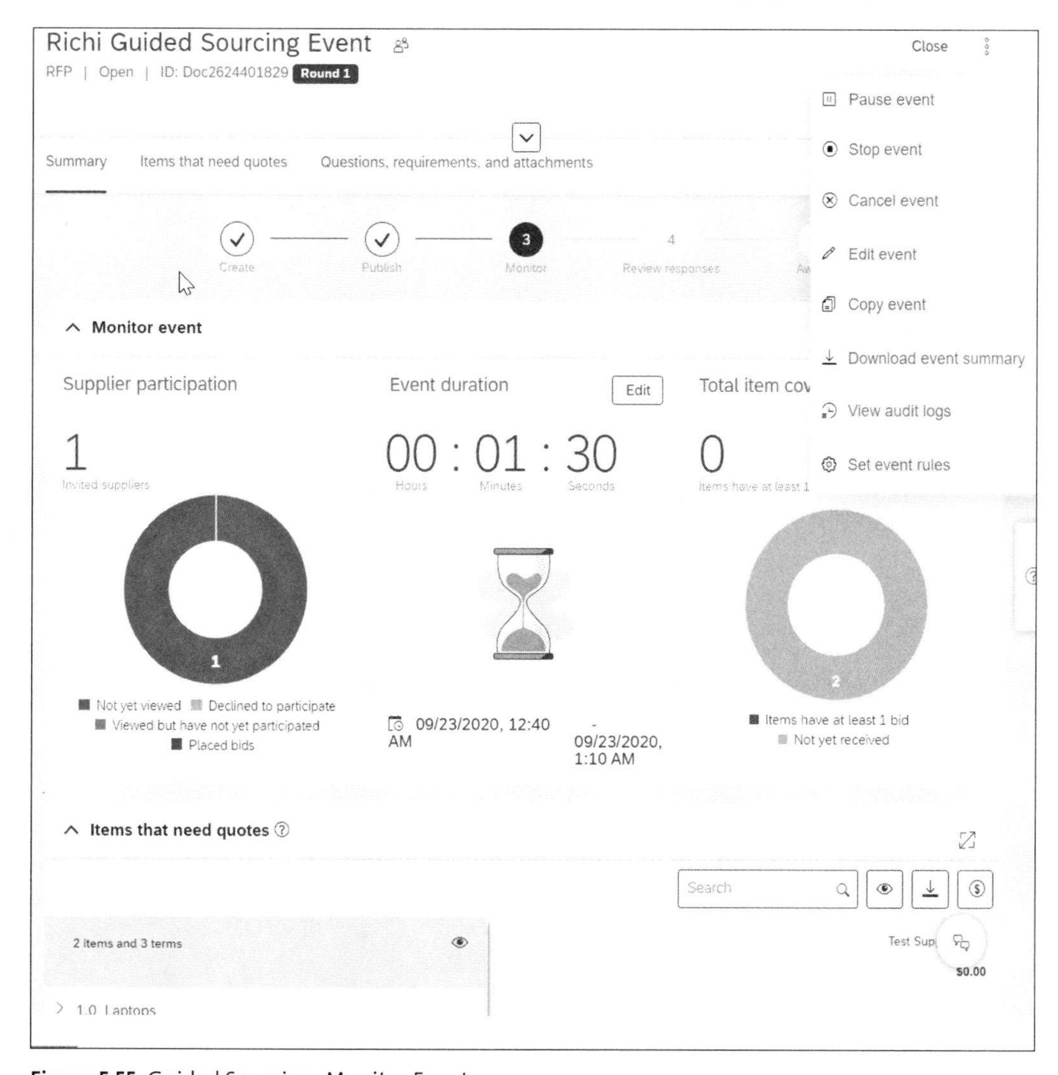

Figure 5.55 Guided Sourcing: Monitor Event

The buyer can extend the duration of the event by clicking the **Edit** button in the **Event duration** section.

Suppliers invited to the sourcing event will receive an email inviting them to participate. Suppliers can click on the link provided in the email and respond to the guided sourcing event invitation.

After responses are received and the event is closed, the owner of the event can review responses from the suppliers and award business by clicking the **Award** button at the top of the screen shown in Figure 5.56.

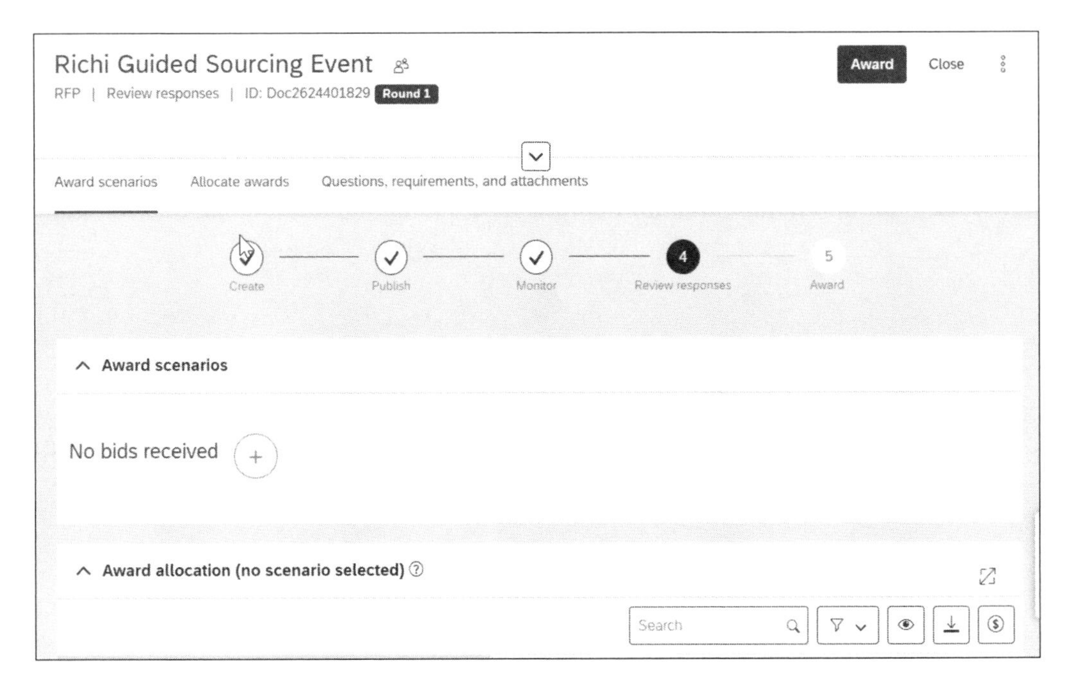

Figure 5.56 Guided Sourcing: Review Responses and Award

In addition, the guided sourcing events feature comes with enhanced item management that allows for enhanced searching and bulk edit capabilities while reducing the need to scroll in full screen mode.

Guided sourcing events also include intelligent event creation. An event can be created from unstructured data and an unstructured Excel import. Let's look at the process of creating a guided sourcing event using a template from an unstructured Excel import. Follow these steps:

1. Log in to SAP Ariba Sourcing.
2. Click the **Create • RFx**, as shown earlier in Figure 5.46.
3. Select **Create from Smart Excel**.
4. Enter the event **Name**, and select the **Event type** as **RFI|RFP**.
5. Enter event information such as **Commodity**, **Regions**, **Departments**, **Currency**, and **Baseline Spend**.
6. Select a guided sourcing **Template**.
7. Click the **Create** button on the top-right corner of the screen (Figure 5.57).
8. Drag a file or browse to upload the Excel file.

A common sourcing practice is to run multiple round events to zero in on the right suppliers. Previously, multiple rounds required separate sourcing events to be created. With the guided sourcing event feature, multiple rounds can be run within the same event (see Figure 5.58).

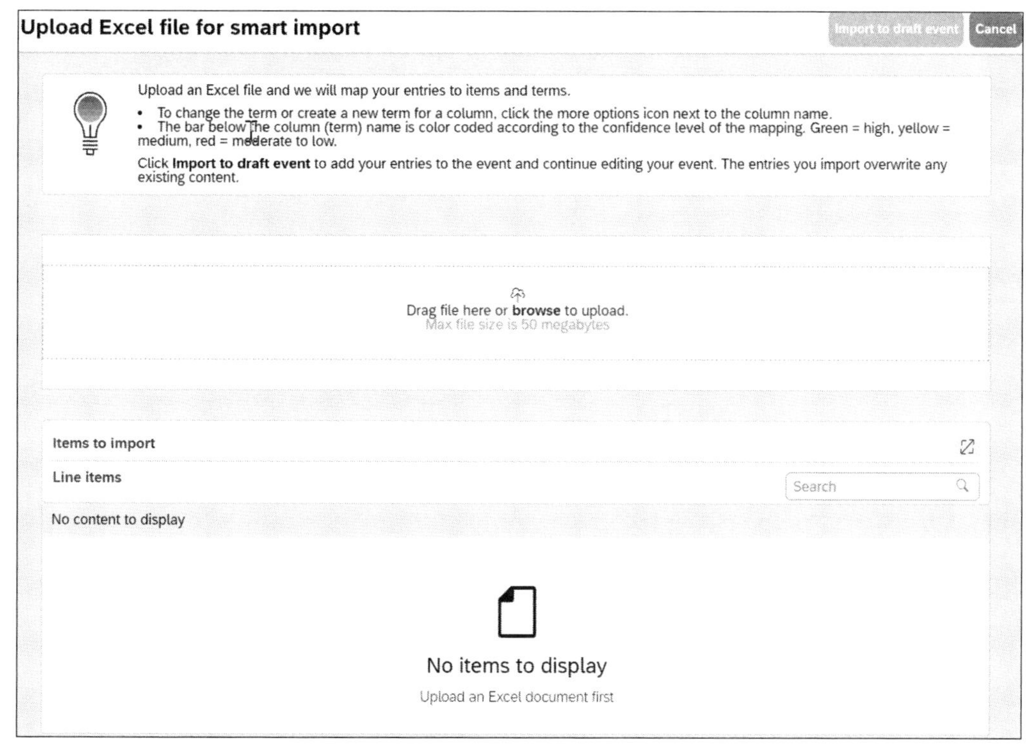

Figure 5.57 Guided Sourcing: Intelligent Event Creation

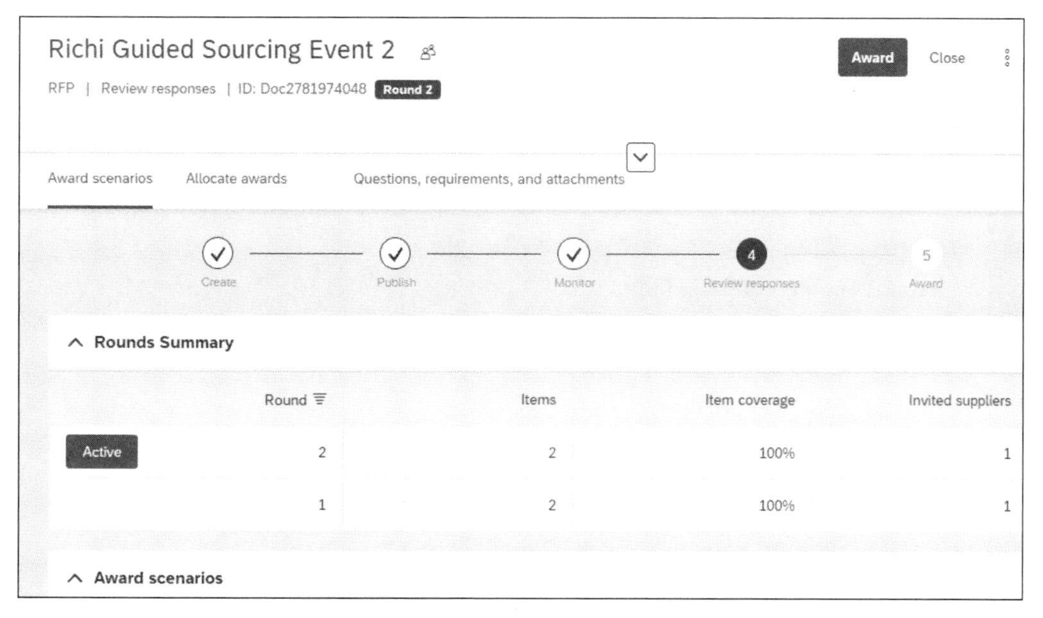

Figure 5.58 Guided Sourcing: Multiple Rounds

Planning Your Implementation

Guided sourcing implementation is part of the standard SAP Ariba Sourcing implementation project. See Section 5.2.2 for more information.

5.5.1 Configuring Guided Sourcing

Users who are members of the customer administrator group can enable guided sourcing by taking the following steps:

1. On the dashboard, click **Manage – Administration**.
2. Click **Event Manager – Enable Guided Sourcing**.
3. On the **Enable Guided Sourcing** page, click **Enable**.

After guided sourcing is enabled, the following templates will be installed and will be visible inside the **Template** area:

- **Request for Information – Guided Sourcing Template**
- **Request for Proposal – Guided Sourcing Template**

Members of the category buyer group are the only ones who can use the guided sourcing interface. To add users to this group, as a customer user admin or customer administrator, edit the category buyer group and add users who need to use the guided sourcing interface to the group.

5.5.2 Creating a Guided Sourcing Event from Template

As mentioned earlier, all events are created from a template. Templates control the rules and the information in the events created from the template. Guided Sourcing templates can be created by following these steps:

1. On the dashboard, click on **Manage – Templates**.
2. Click **Actions • Create • Template**.
3. Select the **Project Type for Template** as **Sourcing Project**.
4. In the **project** field, select **Quick Project**.
5. In the **Event type** field, select **RFI** or **RFP**.
6. Select **Guided Sourcing Template = Yes**.
7. Click **OK**.
8. In the **Properties** area, choose **Actions • Publish**.

A guided sourcing template will be created with the specific event template added to the **Documents** tab. Based on the **Event type** field, either the **Request for Proposal – Guided Sourcing** or **Request for Information – Guided Sourcing** event template will be added to the **Documents** tab.

In addition, the template creator can add Approval for Publish or Approval for Award tasks and content such as prequalification questions to the template.

Setting Event Rules

The guided sourcing template rules have been simplified, as shown in Table 5.5. These are the only rules that can be changed in a guided sourcing template.

Type	Rule	Default Value	Visibility or Availability in Events
Supplier eligibility criteria	Minimum supplier status for event participation	Not invited	Hidden
	Minimum supplier status for award eligibility	Not invited	Hidden
	Supplier qualification level	Event level	Hidden
Timing rules	Response start date	When the **Publish** button is clicked on the **Summary** page	Delegated (shown in the **Event** duration panel)
	Due date duration	30 minutes	Delegated (shown in the **Event** duration panel)
	Due date reminder	None	Hidden
Bidding rules	Allow surrogate bids when event is closed	No	Hidden
	Allow participants to submit bids by email	No	Delegated
	Enable multiround bidding	Yes	Delegated
Currency rules	Allow participants to select bidding currency	Yes	Delegated
	Show currency exchange rates to participants	No	Delegated (and only visible if **Allow participants to select bidding currency** is **Yes**)

Table 5.5 Setting Event Rules

Type	Rule	Default Value	Visibility or Availability in Events
Market feedback	Show formulas to all participants	No	Delegated
Include bidder agreement	Choose whether to include the bidder agreement as a prerequisite	Yes	Hidden

Table 5.5 Setting Event Rules (Cont.)

Converting a Classic Sourcing Template to a Guided Sourcing Template

Existing classic sourcing templates can be converted to guided sourcing templates by creating a new version of the existing template and flipping the **Guided Sourcing Template** field to **Yes**. By default, this field is set to **No**.

5.6 Summary

Achieving full sourcing potential and defining a competitive source of supply for all commodities, whether direct or indirect, is key for driving sustainable savings and strategic partnerships within the supply chain to create better bottom lines and growth across an enterprise. SAP Ariba Sourcing and SAP Ariba Strategic Sourcing Suite, along with supporting tools such as SAP Ariba Discovery and the new Guided Sourcing event, can serve as a comprehensive platform for managing this key strategic area of procurement. Our goal in this chapter was to describe the features and benefits of SAP Ariba Sourcing and the product sourcing functionality (in SAP Ariba Strategic Sourcing Suite). Lastly, in this chapter, we recommended some tried-and-true best practices and described the SAP Activate methodology for quickly and efficiently implementing SAP's cloud solutions for sourcing so you can drive your organization toward achieving its full sourcing potential.

Chapter 6
Contract Management

SAP Ariba Contracts is SAP's cloud-based solution for managing enterprise-wide contracts from creation to renewals and amendments. In this chapter, we'll show you how to implement SAP Ariba Contracts to convert what has been a rather cumbersome and clunky process for managing contracts into a highly efficient process.

Contracts are vital documents that contain important (and, in most cases, legally binding) information on the relationship of an organization with its partners—its suppliers, customers, and employees. Contracts can also contain important cost-saving incentives, such as early payment discounts. Mismanagement of and noncompliance in contracts can be quite costly for organizations in terms of savings lost and/or regulatory penalties. In larger organizations, or as an organization grows and an increasing number of contracts are executed, maintaining and enforcing compliance in contracts can become a nightmare.

In today's "digital transformation" era, the process of managing these contracts, also known as *contract management*, is evolving into a sophisticated and scientific discipline. No longer an administrative task, contract management instead now plays a more dynamic role by orchestrating change based on market volatility and driving performance- and outcome-based agreements as it integrates processes across enterprises.

Most organizations have some kind of contract management tools in place, but, unfortunately, these tools are usually used more as a "software vault" to store contracts across the enterprise. Best-in-class companies, however, understand that the potential power of contract management to transform legal, procurement, finance, and sales operations can do the following:

- Build strategic relationships and drive contract collaboration.
- Improve supplier performance and negotiation efficiencies.
- Standardize contract processes and approvals.
- Lower administrative and legal costs.
- Automate all phases of contract management from contract request and contract authoring to contract execution (contract approvals, e-signatures, and contract amendments).
- Strengthen operational and contractual compliance and reduce risk.

In the following sections, we'll uncover what contract management means by discussing contract management strategies for indirect procurement. We'll then discuss SAP Ariba Contracts in detail, including how to implement and configure the solution.

6.1 What Is Contract Management?

Contract management is a process that enables parties to a contract to meet their respective obligations to deliver the objectives required from the contract.

In this section, we'll discuss the pain points in the contracting process for indirect procurement. We'll then discuss SAP's best-in-class contract management solution—SAP Ariba Contracts—and describe how this solution can help alleviate pain points. We'll also look at how contracts are created and handled within SAP Ariba Contracts.

6.1.1 Contract Management Strategies for Indirect Procurement

Indirect procurement relates to the procurement of goods and services that aren't directly used in the product or service sold by the organization. While indirect procurement may share common business and legal issues/risks, drafting and executing contracts for indirect procurement presents some unique and critical challenges when compared to direct procurement or other kinds of contracts. Some of the risks and issues observed in the contract management process for indirect procurement, as well as some of the contract strategies that SAP Ariba Contracts can enforce, include the following:

- Terms and conditions (T&Cs) management
- Contract planning and negotiation planning
- Post-award contract management and control

Let's take a deeper look at each of these considerations next.

Terms and Conditions Management

Terms and conditions (T&Cs) management has proven to be an area of weakness in a large number of indirect contracts. Most contracts lack consistency in terms of the usage of standard language. In addition, as the language used in each contract can be different, the language is usually not preapproved by the legal department, thus considerably increasing the risks associated with the contract. Implementing standard contract templates with approved mandatory, optional, and alternate T&Cs clauses for various types of indirect goods and services is a key step required to mitigate contract management risks.

SAP Ariba Contracts provides comprehensive contract authoring and clause library functionality, which enables the creation of standard required/optional clauses and contract boilerplate templates. Additionally, the feature allows for defining alternate clauses and provides an outline view of sections and clauses in the contract document. End users can review and replace standard optional clauses with alternate clauses found in the clause library.

Certain contract T&Cs will need to be negotiated based on the nature of the purchase and/or the supplier's relative power. SAP Ariba Contracts provides negotiation functionality with a comprehensive auditing and version control feature. SAP Ariba Contracts also provides standard prepackaged reports, while identifying standard clauses that contract creators constantly change. As a result, legal or contract template administrators can easily revise clauses or create approved alternate clauses.

Using SAP Ariba Contracts and enforcing the usage of standardized contract templates and approved/alternate clauses reduces contracting risk considerably and accelerates the contract creation and negotiation process.

Contract Planning and Negotiation Planning

Many contracts lack proper planning and consistent process for negotiation and execution of the contract. An efficient contract management process requires proper contract planning and negotiation planning processes that involve scoping requirements, contract negotiation planning, and planning for post-award contract administration. The negotiation plan should leverage the analysis performed during the strategic sourcing process.

SAP Ariba Contracts, with standard integration to SAP Ariba Sourcing, provides improved strategic sourcing and contracting process design. SAP Ariba Contracts helps improve the contract planning and negotiation process considerably by enabling faster, paperless contract creation, authoring, and back-and-forth collaboration/redlining of the contract with external contracting parties, such as the supplier. SAP Ariba Contracts also provides tasks for internal reviews and highly configurable workflows and can be customized for each contract template based on the customer's internal processes for handling contractual risks.

Post-Award Contract Management and Control

Contract compliance is a major area of improvement in many organizations. Most contract management solutions lack integration to operational procurement where transactions are executed based on contracts, thus ensuring contract compliance.

SAP Ariba Contracts provides seamless integration with SAP Ariba procure-to-pay (P2P) solutions, thus enabling contract compliance and reducing maverick spending by ensuring that negotiated terms are used during the execution and release of contracts. SAP Ariba Contracts also provides a set of prepackaged reports for monitoring contracts. Task-driven reminders and renewal reminders notify contract authors (owners) of expiring contracts well in advance, so that they can renegotiate those contracts and terms and thus realize further savings. In addition, SAP Ariba Contracts fully supports the contract amendment process, whether an administrative amendment, a renewal, or the termination of a contract. All changes to a contract are recorded for audit purposes. Standard amendment templates can also be maintained with standard clauses to standardize and accelerate the amendment process for all contracts maintained within SAP Ariba Contracts.

6.1.2 SAP Ariba Contracts

SAP Ariba Contracts is a leading software-as-a-service (SaaS) solution and is accessible anywhere, anytime. It's an integral part of the SAP Ariba procurement suite with tight integration to other SAP Ariba solutions such as SAP Ariba Sourcing, SAP Ariba Supplier Lifecycle and Performance, and SAP Ariba Buying and Invoicing. In addition, the SAP Ariba Contracts solution can be easily integrated to your SAP ERP backend or to your other third-party solutions, as shown in Figure 6.1.

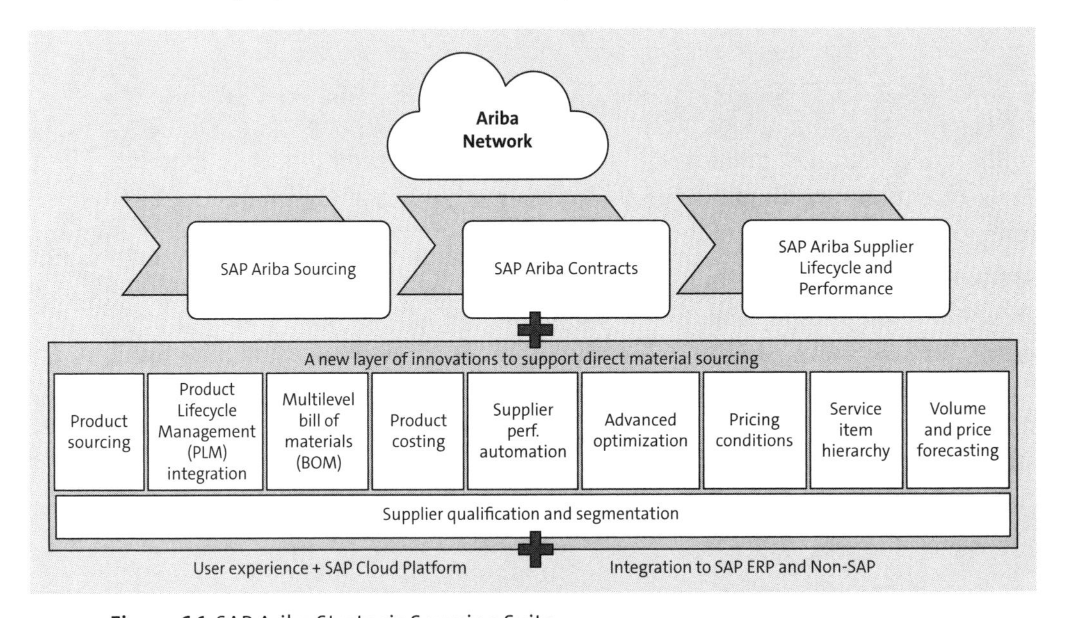

Figure 6.1 SAP Ariba Strategic Sourcing Suite

SAP Ariba Contracts can also be licensed as part of the SAP Ariba Strategic Sourcing Suite, which contains the following SAP Ariba solutions:

- SAP Ariba Sourcing
- SAP Ariba Contracts
- SAP Ariba Supplier Lifecycle and Performance
- SAP Ariba Sourcing, savings and pipeline tracking add-on
- Bonus/penalty and lookup table formulas

- SAP Ariba's product sourcing capability
- SAP Ariba Discovery
- SAP Ariba Spend Analysis (optional add-on)

SAP Ariba Contracts combines the contract creation, negotiation, and amendment process with the authoring process into a comprehensive contract management solution. With SAP Ariba Contracts, companies can develop best-value agreements by addressing the two major components of the contract lifecycle:

- **Contract management**
  This stage goes from contract request through contract authoring, negotiation, and approval using e-signatures and, finally, to contract execution.
- **Commitment management**
  This stage includes ongoing compliance and performance management through task-driven reminders and search and reporting capabilities, as well as contract renewal activities.

SAP Ariba Contracts reduces the cost of managing and updating contracts by eliminating the need for pen and paper for the creation, execution, and management of any type of contractual agreement. Furthermore, in today's digital economy, SAP Ariba Contracts provides a powerful platform to efficiently and effectively perform the following tasks:

- Manage procurement and sales contracts, license agreements, internal agreements and various other kinds of contracts.
- Automate and accelerate the entire contract lifecycle process, including the contract signature process by enabling digital signatures.
- Standardize the contract creation, authoring, and maintenance process.
- Collaborate with both internal and external stakeholders with a complete audit trail of changes made to the contract throughout the process.
- Strengthen operational, contractual, and regulatory compliance.

Some additional benefits of using SAP Ariba Contracts include the following:

- **Fast time-to-value**
  Quickly deploy SAP Ariba Contracts on the cloud with lower total cost of ownership (TCO). Automatic updates of the latest version of the software by SAP Ariba also reduce your IT costs.
- **Ease of use**
  Use this interactive tool to quickly draft contracts from the preapproved templates, make changes to the contract as needed, complete negotiations with stakeholders, and execute the contract with ease.

- **Controlled processes**
  Customize tasks and approval flows. Changes made to preapproved language in the contract will be displayed for approvers to review and approve the contract changes.
- **Efficient collaboration**
  Work quickly and effectively with all stakeholders and negotiate contracts with your suppliers on the Ariba Network.
- **Complete visibility**
  Stay informed throughout the contract lifecycle with automatic alerts, handy dashboards, and configurable reports.
- **Central repository**
  Never lose track of a contract with secure electronic storage and powerful search tools for access on demand.
- **On-time renewals**
  Receive notifications well in advance of key milestone dates and make the most of your business opportunities.
- **E-signature saving**
  Eliminate the time and expense of shipping and signing multiple contracts by adding e-signature capabilities.
- **Stronger compliance**
  Stay informed about any off-contract activity with controlled processes, automated tracking, and a full audit trail.
- **End-to-end commerce**
  Integrate your contract processes with additional SAP Ariba solutions, your ERP, or other third-party systems for unrivaled compliance and control.

In this chapter, we'll outline the SAP Ariba solution for understanding and managing contracts within SAP Ariba, with a focus on the SAP Ariba Contracts Professional package and its implementation.

6.1.3 Planning Your Implementation

The key to a successful implementation of SAP Ariba Contracts is planning, key stakeholder/executive-level ownership, and effective change management. The SAP Ariba Contracts team must be engaged to implement this solution. Implementation of SAP Ariba Contracts follows the SAP Ariba on-demand deployment methodology, which is based on the SAP Activate methodology. This method is described in Section 6.1.3.

Each implementation includes only one contract type, for example, Contract Workspace (Procurement). Implementing additional contract types requires a separate implementation. The standard scope for each SAP Ariba Contracts implementation is shown in Table 6.1.

Configuration	Deployment Description
Master data	■ Loading master data using the enablement workbook.
Custom header fields	■ Up to 5 custom header fields.
Custom report	■ 1 custom report.
Contracting process (one template configured by SAP Ariba during a typical implementation cycle)	■ Up to 30 tasks. ■ Up to 20 process configuration conditions. ■ Up to 20 related documents. ■ Standard review/approval rules on the tasks in template.
Form (one template configured by SAP Ariba during a typical implementation cycle)	■ Up to 50 fields/line items. ■ Up to 10 conditions.
Master agreement template and clause library setup	■ Create 1 main agreement document in the template (limited to 50 clauses per main agreement). ■ Create up to 10 conditions for clause usage. ■ Create up to 5 document properties that automatically populate content within the main agreement. ■ Configure and load up to 20 preferred and 20 alternate clauses. ■ Configure clause library approvals and notification settings.
E-signature configuration	■ Provide guidance for enabling e-signatures. ■ Enable tasks in contract process for e-signatures.
Legacy contract loads	■ Define standard out-of-the-box fields for contracts. ■ Define up to 10 custom fields. ■ Create 1 template for legacy loads. ■ Load up to 250 contracts (with a maximum of 500 documents) in the production environment. Use only the sample load in the test environment.

Table 6.1 Scope for Deployment

Tip

We recommend a "crawl, walk, run" methodology when implementing SAP Ariba Contracts. In the first phase, refrain from using the contract authoring functionality. Clause-related templates and contract authoring can be enabled as part of the second phase, after you are familiar with using SAP Ariba Contracts.

Loading legacy contracts into the solution can be time-consuming as contract extraction can take a long time, so plan accordingly and start working on legacy contracts early in the implementation.

6.2 Configuring SAP Ariba Contracts

In SAP Ariba Contracts, a contract initiation can start with a contract request and then the creation of a contract workspace, or by directly creating a contract workspace. In this chapter, we'll look at configuring contract requests and contract workspace templates.

A *contract workspace* is the contract "shell" within SAP Ariba Contracts that houses contract documents, such as master service agreements, amendments, addendums, statements of work (SOWs), and others.

Mature organizations have a shared services or center of excellence (CoE) team or a dedicated contract management team responsible for creating and executing all contracts within the organization. Legal, finance, supply chain, and other departments may be involved during the approval process for contracts. Such organizations with a more mature contract management process may require business users in the field to create contract requests. After a contract request is created and approved, someone with the contract agent role on the shared services, CoE, or contract management teams will create the contract workspace and execute the contract. In some organizations, the legal department may be actively involved with creation of master agreement and amendment templates using clause libraries within contract templates.

Figure 6.2 shows the simple workflow that can be configured within SAP Ariba Contracts.

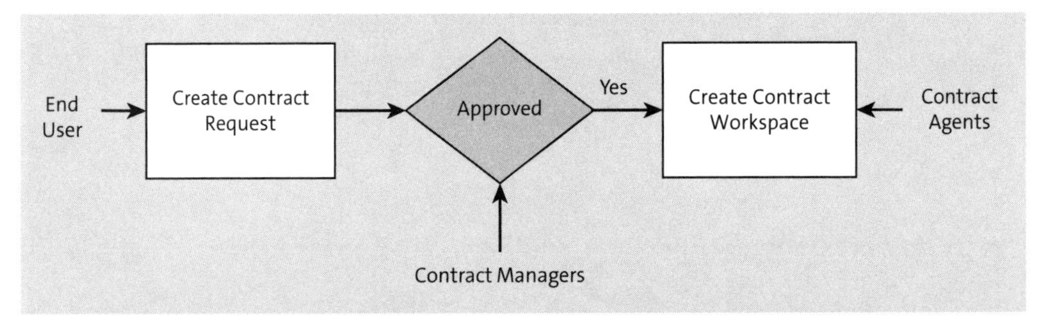

Figure 6.2 Contract Creation Process

In the following section, we'll look at some of the configuration steps needed in a typical implementation of the SAP Ariba Contracts application to create and manage a contract through its lifecycle:

- Creating project templates for contract requests and contract workspaces
- Creating contract authoring processes
- Configuring workflow approval tasks
- Configuring expiration reports
- Creating custom fields as required on contract requests and contract workspaces
- Loading the master data required to create contracts
- Enabling e-signatures (if in scope)

6.2.1 Creating Project Templates

Within the SAP Ariba Contracts and other SAP Ariba applications, objects such as contracts and sourcing events are called *projects*. To create a project, you'll first need to create a project template. For your end users to be able to create contract requests and contract workspace projects, templates must be first created by a user with the template creator role.

SAP Ariba has multiple dashboards based on the solutions configured and the user's role, as shown in Figure 6.3. The particular information displayed on the dashboards, as well as access to specific features and other dashboards, depends on that user's permissions and roles. A user with the customer administrator role can access the **Administration** section (**Manage • Administration**). Users with the template creator role can access the **Templates** section (**Manage • Templates**).

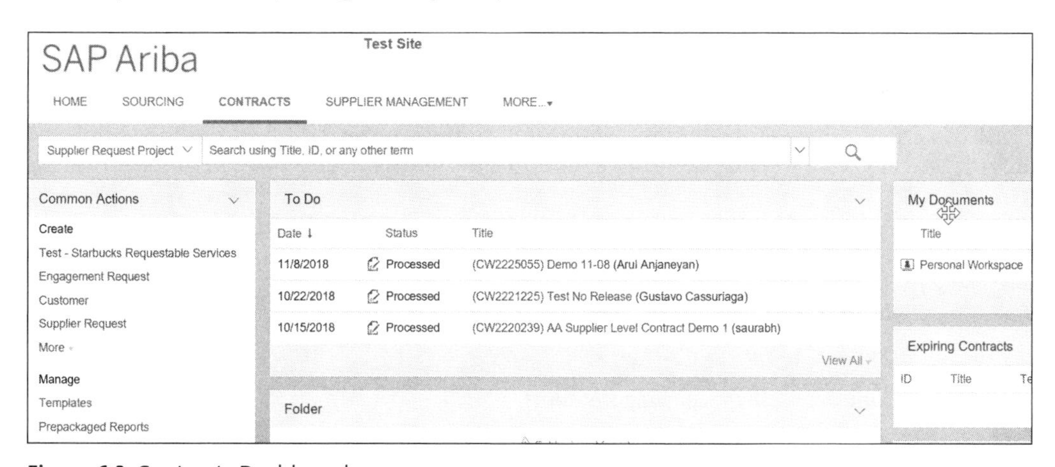

Figure 6.3 Contracts Dashboard

In the **Templates** section, click the **Documents** tab to access all the existing templates in your system.

You can create a new contract request/contract workspace templates from an existing template or make one from scratch.

Tip

When creating new templates, a best practice is to copy a standard template and then modify the copy as needed.

To copy an existing standard template, follow these steps:

1. Search for a standard contract request/contract workspace template, click the template name, and then select **Action • Copy**, as shown in Figure 6.4.
2. Rename the template, and proceed with the necessary changes.

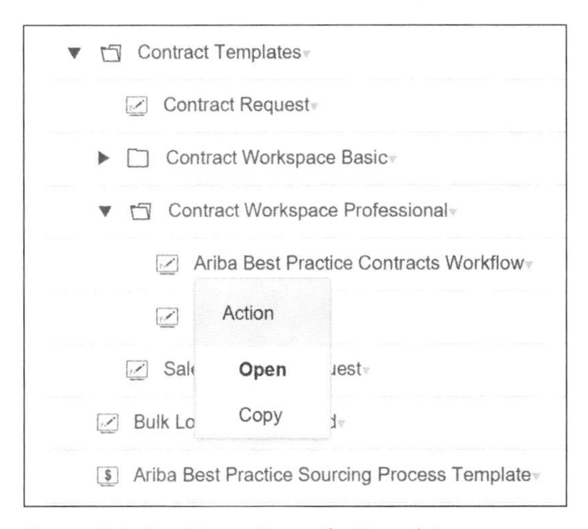

Figure 6.4 Creating a Copy of a Template

To create a new template from scratch, follow these steps:

1. Click **Actions • Create • Template**.
2. Select one of the following options for the **Project Type for Template** field:
   - **Contract request (procurement)**
   - **Contract request (sales)**
   - **Contract request (internal)**
   - **Contract workspace (procurement)**
   - **Contract workspace (sales)**
   - **Contract workspace (internal)**
3. Enter a name for the new template.

A new template is created. If only two tabs appear (**Overview** and **Conditions**), then click **Action • Display • Full View**. Now, all the tabs should be visible, including the **Overview**, **Documents**, **Tasks**, **Team**, **Conditions**, **Advanced Options**, and **History** tabs.

In this section, we'll look at configuring each tab within a project template in the contract request/contract workspace. The **History** tab is used for tracking changes made in the template. Because this tab isn't configurable, we won't cover it in this section.

Overview Tab

This tab provides an overview of the template's properties, such as **Name**, **Description**, **ID**, **Owner**, **Base Language**, **Access Control**, and **Conditions**. To change these properties, click **Actions • Edit Properties**, as shown in Figure 6.5.

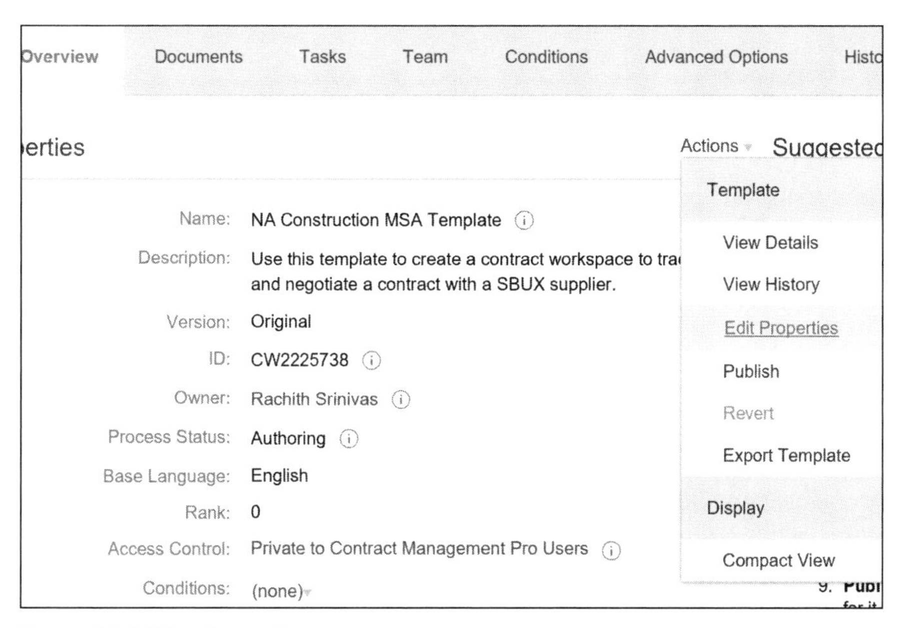

Figure 6.5 Editing Properties

As shown in Figure 6.6, the following two fields can be changed when you edit properties:

- **Owner**
  Person who owns the template. If other users need access to maintain the template, they can be added to the owner group under the **Team** tab.
- **Access Control**
  Allows the level of access provided to contracts created using this template.

All contracts created from this template will inherit these settings from the template.

> **Tip**
> A best practice is to restrict access to the template to team members. Select the **Private To Team Members** option from the **Set Access Control** dropdown menu.

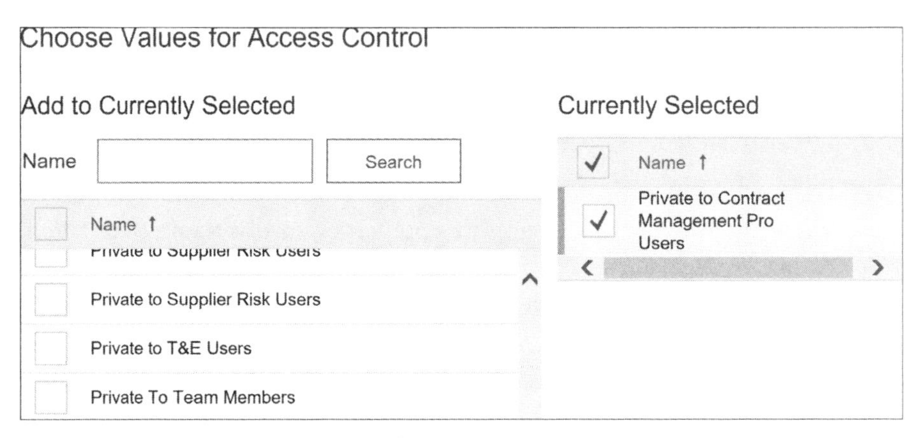

Figure 6.6 Choosing an Access Control

Conditions enable business rules to be defined that only allow end users to see or select this template when conditions are met. For instance, if a contract request template is meant for usage in **NA**, a condition on **Region = NA** can be set in the **Conditions** field under the **Overview** tab of the contract request template. This template will only be available when a user selects **Region = NA** while creating a new contract request.

Documents Tab

This tab provides a container to create different types of documents. You can also organize documents into folders. Documents can be created from scratch or can be copied from other objects within the SAP Ariba solution, such as a document in the Sourcing Library. A document can also be created as a master agreement or a contract addendum (contract authoring). Documents and folders created in a contract request template will be automatically loaded into all contract request projects created from this template.

In this section, we'll show you how to create a folder and how to create a new document by uploading an existing Microsoft Word document.

Creating a Folder

Folders can be created to organize the documents within a contract request project. To create a folder within a template, under the **Documents** tab, click **Action • Create • Folder**, as shown in Figure 6.7.

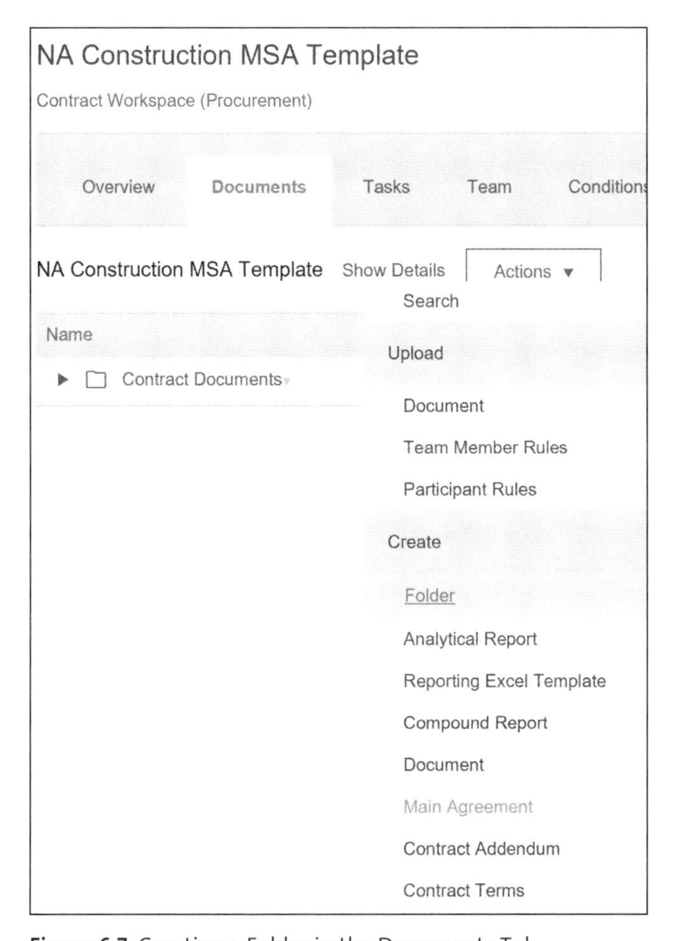

Figure 6.7 Creating a Folder in the Documents Tab

Creating a Document

The **Documents** tab in the contract request is where documents are housed. You can create a document in three ways:

- You can create a new document by uploading a document such as a legally approved template document, as shown in Figure 6.8. To create a new document, click **Action • Create • Document**.
- You can also create a new document by selecting a document housed in SAP Ariba, such as in Sourcing Library, as shown in Figure 6.9.
- You can create a new master agreement or contract addendum, as shown in Figure 6.10.

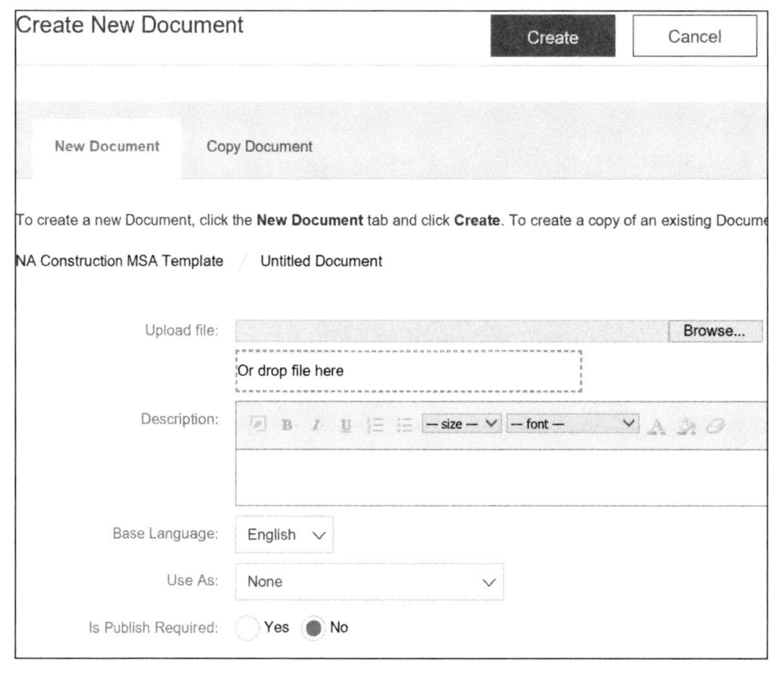

Figure 6.8 Creating a New Document from an Existing Word Document

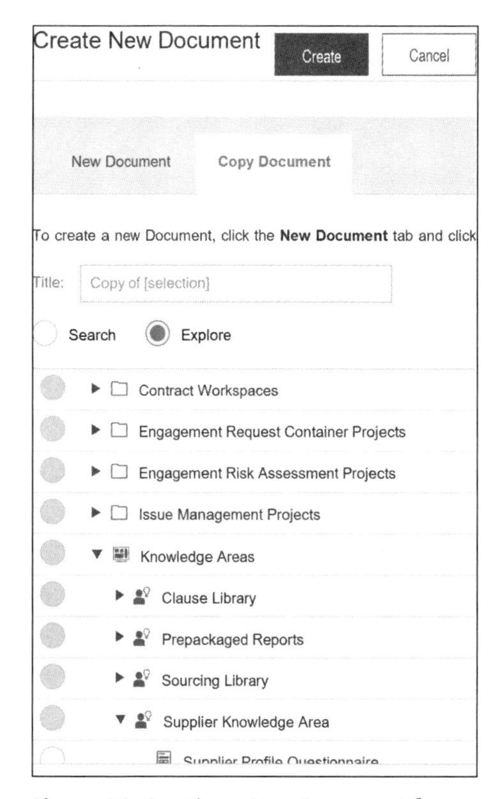

Figure 6.9 Creating a New Document from an Existing Document in the Sourcing Library

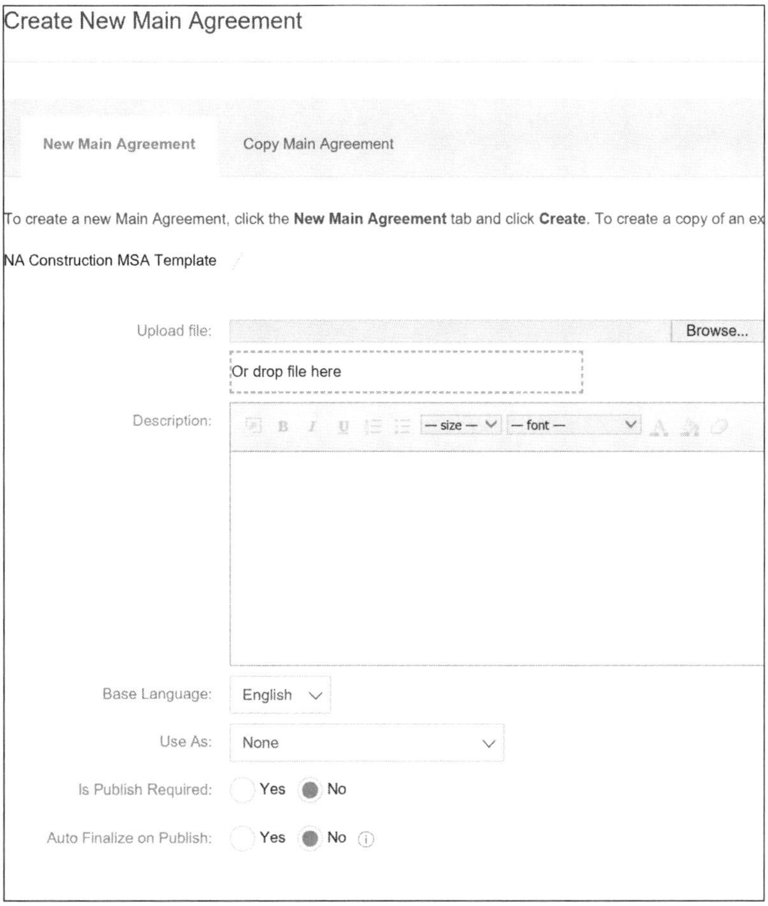

Figure 6.10 Creating a Master Agreement or a Contract Addendum

Creating contract addendum is similar to creating a master agreement, but we'll discuss the contract authoring process in more detail in Section 6.2.2.

Tasks Tab

This tab is used to define the phases as well as the tasks and milestones within each phase. Increased regulatory pressure to reduce costs and standardize the contracting process often motivates organizations to find ways to enforce standard processes for creating and managing contracts internally.

SAP Ariba Contracts allows contract administrators to define a standard contract process with phases, tasks, and milestones.

You can create the following types of tasks:

- **To-do tasks**
  Requires the task owner to perform some activity. The document folder can be linked to a to-do task.

- **Review and approval tasks**
  Routes documents associated to these tasks to the reviewer or approver. Reviewers can leave comments while reviewing the document. Approvers will need to approve/deny the document. The document folder can be linked to a review or approval task. All documents within the document folder will be part of the review or approval process. See Section 6.2.3 for more details on configuring approval tasks.
- **Negotiation tasks**
  Captures the negotiations between two or more parties concerning a contract or any other type of document. Comments from negotiating parties are captured as well as any changes made to the document.
- **Notification tasks**
  Sends email reminders to recipients. These tasks are used primarily after the contract's publication to remind users to take some sort of action, such as periodically checking contract activity or market pricing. These reminders can be set up to be sent only once or sent on a recurring schedule.

Tasks and milestones based on the standard process can be created at the contract template level. These tasks can be required or optional.

Contract workspaces created from a contract template will automatically inherit the template's tasks and milestones, as shown in Figure 6.11. Contract creators can add additional tasks to the contract process and/or delete optional tasks that have been pulled from the contract template. If this template is copied from the SAP Ariba Best Practices contract template, the following phases and tasks are copied by default:

- **Initiation – Authoring** phase (Figure 6.12)
- **Negotiation and Review** phase (Figure 6.13)
- **Approve and Finalize** phase (Figure 6.14)

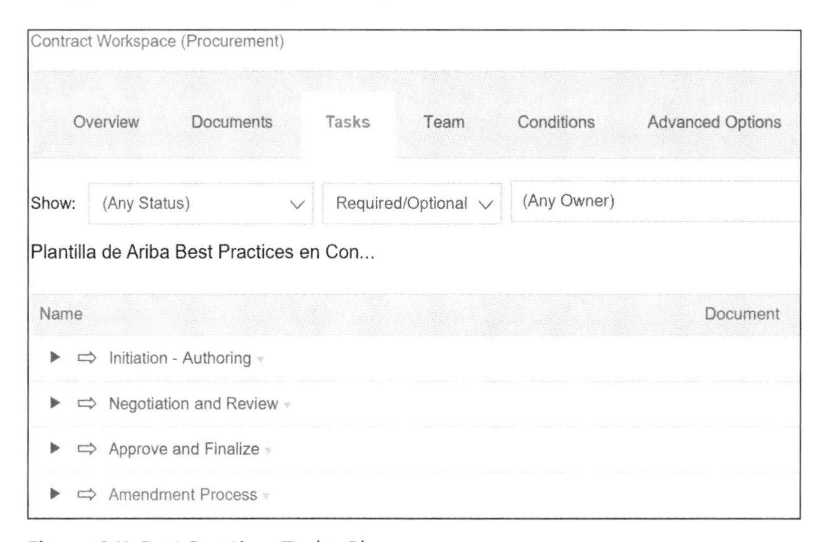

Figure 6.11 Best Practices Tasks: Phases

Name	Document	Conditions
▼ ⇨ Initiation - Authoring ▾		(none)
Update Team Members *		(none)
Upload Document - From Supplier or Other Paper *	Contract Documents	supplier_paper=true
Prepare Document for Negotiation *	Contract Documents	our_Main_Agreement=true

Figure 6.12 Tasks in Initiation - Authoring Phase

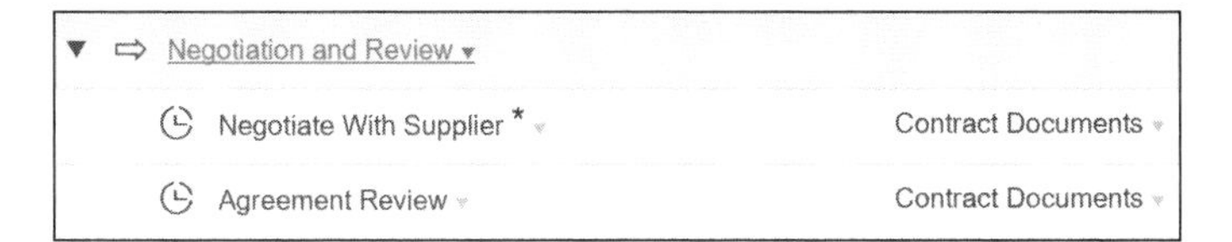

▼ ⇨ Negotiation and Review ▾	
Negotiate With Supplier *	Contract Documents
Agreement Review	Contract Documents

Figure 6.13 Tasks in Negotiation and Review Phase

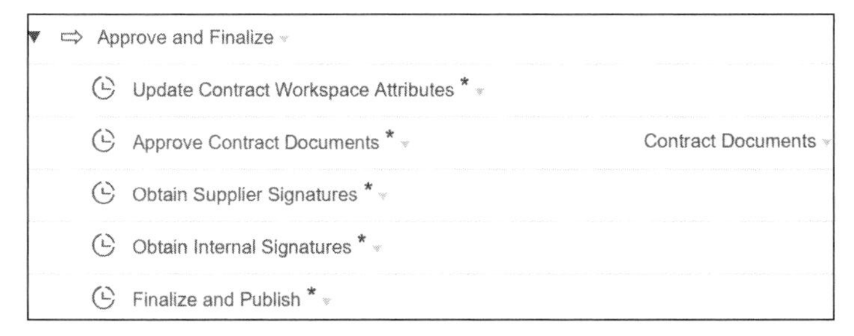

▼ ⇨ Approve and Finalize	
Update Contract Workspace Attributes *	
Approve Contract Documents *	Contract Documents
Obtain Supplier Signatures *	
Obtain Internal Signatures *	
Finalize and Publish *	

Figure 6.14 Tasks in Approve and Finalize Phase

Finally, the amendment process phase and tasks are available when the contract workspace is amended and the status changes to **Draft Amendment**. The **Approve Amendment** standard task is visible during the amendment process.

Team Tab

Collaboration throughout the contracting process is key for ensuring a faster and less error-prone execution of contracts.

SAP Ariba Contracts allows collaborators or team members to be added manually or automatically to the contract workspace under the **Team** tab. Project team groups, such as contract managers or contract observers, should be created first, before adding team members to these project team groups. Team members can be individual users or groups of users. Each project team group is defined a role, either **View Only** or **View/Edit**.

Similar to tasks, project team groups and team members can be set up by the contract administrator in the contract template level. Contract requests/workspaces created from the contract template will automatically inherit the project team groups and team members, ensuring certain team members will always be added to any contract workspace created.

When creating a new group under the **Team** tab, the following configuration steps allow template administrators to restrict end users from making changes to project groups added at the template level, as shown in Figure 6.15.

Project Group Details

Define this **Group** by entering a group **Title**. By changing which **Roles** are included in this group, you

Can owner edit this Project Group: Yes

Title: * Yes

No

Members:

Roles: (select a value) [select]

Project: Plantilla de Ariba Best Practices en Contract Management

Figure 6.15 Configuring a Project Group in the Team Tab

This limitation helps with enforcing business rules such as "all contract workspaces created must be accessible to all contract managers." The contract template administrator can also associate a system group in the **Roles** setting. This setting assigns the underlying roles in the system group to all users added as members to this group. Project team groups can be added to the review and approval tasks.

After the contract template is fully configured, the template must be published before it can be used for creating contract requests/workspaces. To publish a template, a template owner must select the **Overview** tab and then click **Actions • Publish**, as shown in Figure 6.16.

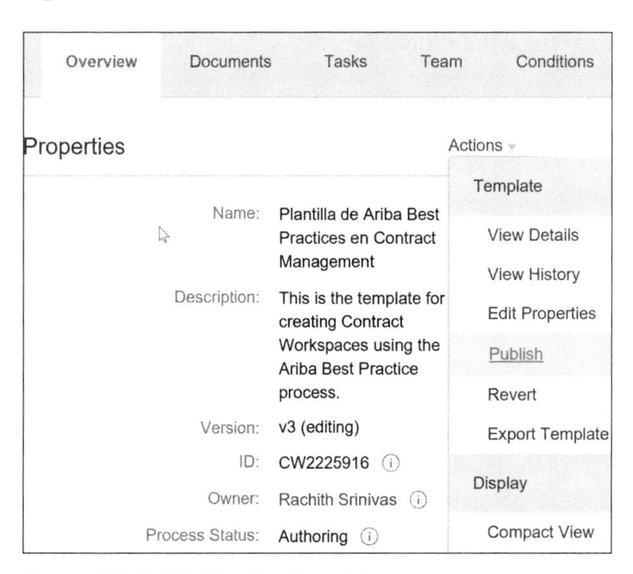

Figure 6.16 Publishing the Template

After the template is published, the status of the template changes from **Draft** to **Active**. The template is now available to users and will appear in the relevant list of templates (contract request or contract workspace) for users to choose from when creating a new contract request or contract workspace.

6.2.2 Contract Authoring Process

SAP Ariba Contracts provides the ability for contract template administrators and the legal department to create and maintain standardized enterprise-wide contract templates (or master agreement templates) that will be used as the starting point for contract creation by end users. A contract template comprises sections and clauses. Clauses are created, approved, and maintained in the clause library. Standardized clauses can be used in multiple contract templates.

A clause corresponds to one major point or piece of information, usually to one paragraph within the document. Similar clauses can be organized into sections. A common practice is to start with a currently used contract and create a master agreement template. Then, the clauses from this template are published to the clause library so that these clauses can be reused in other master agreement templates. Thus, clauses must be succinct and use standard language consistently.

The following are the benefits of the clause library:

- Provides a single source of truth for standardized contract content in the form of clauses that can be created and approved once and used in multiple contract templates, such as a nondisclosure agreement (NDA) clause can be used in different contract templates.
- Allows for defining fallback or alternate clauses for a contract clause, such as a tax clause that could vary based on the region for which the contract is being defined.
- Maintains clause versioning with automatic approvals when the clause is changed by an administrator with access to change clause content. Approvers can compare versions before approving the changes.
- Allows for conditionalized selection of clauses while a contract template is being generated. For instance, a specific clause can be automatically chosen based on the region or commodity condition.
- Provides visibility of the source of the clause in the outline view of the contract document such as inherited from template, replacement clause from clause library, edited, or added from external source.
- Allows for reporting on clause usage and how frequently a clause has been modified, as well as search for all contracts and templates where the clause is used.

A best practice is to reword similar clauses to create a standardized clause, thus reducing the number of clauses in the clause library.

Document Cleansing

SAP Ariba Contracts extensively uses Word's built-in features for contract authoring. Word features aren't always backward compatible. All Word documents contain embedded code that affects things such as formatting within the document. Because knowing what version was used to create the original document is impossible, to avoid potential problems within SAP Ariba Contracts, cleansing the document is imperative. Cleansing can be performed by simply copying and pasting the content from the original document into a text editor before copying the content into a new Word document.

The following considerations should be kept in mind while creating master agreement templates:

- Don't overcomplicate.
- Group clauses that should stay together.
- Put each section title in its own line.
- Make each clause its own paragraph when possible.
- Remove hard-coded numbering.

Style Mapping

Style mapping enables you to define styles in Word that the system will use to format sections and clauses when generating contract documents. Style mapping is specific to each master agreement template. Therefore, if a clause is used in multiple master agreements, the clause will be formatted differently based on the style mapping defined for that template.

Style mapping is configured in SAP Ariba Contracts, where you'll apply your styles to specific levels of your main agreement or contract addendum.

Bookmarking

In SAP Ariba Contracts, bookmarks are used to tell the system how to organize your master agreements or contract addendums into clauses and sections. Bookmarks can also be used to determine where a document begins and ends. Bookmarks aren't considered text and don't appear in printed versions of the document. When a Word document is uploaded into the template, the system uses bookmarks to identify sections and clauses. With bookmarks, you can specify exactly how you want the system to parse the document. If any content within the document isn't bookmarked, then the system will interpret each paragraph as a clause.

Bookmarking Format

The bookmarking format shown in Table 6.2 should be used.

Bookmark Type	Name in Microsoft Word
Section	`sectionAriba_<UniqueID>`
Clause	`clauseAriba_<UniqueID>`
Entire contract document	`sectionGlobalContract`

Table 6.2 Types of Bookmarks

Special Note on Partial Bookmarking in SAP Ariba Contracts

Some considerations on partial bookmarks and how SAP Ariba Contracts handles partial bookmarking include the following:

- All properly entered bookmarks are respected.
- If a particular paragraph isn't explicitly bookmarked, it's treated as a clause. SAP Ariba Contracts interprets all ad hoc and unmarked paragraphs as clauses.
- If a manual section bookmark is used, the first paragraph of the section is used as the title of the section. The rest of the content within the section is handled based on the bookmarking rules.
- Every Word document in SAP Ariba Contracts should only have one `sectionGlobalContract` bookmark.
- If the `sectionGlobalContract` bookmark hasn't been added, the system adds one spanning the entire document.

Uploading a Master Agreement

After formatting is complete, the document must be closed and uploaded to SAP Ariba Contracts as a master agreement or contract addendum document. The system will read everything within the `sectionGlobalContract` bookmark and ignore the rest. The system parses the documents and builds the sections and clauses. After the document is loaded, in the outline view, the sections and clauses can be reviewed.

> **Note**
>
> Document file synchronization must be enabled before uploading the master agreement in SAP Ariba Contracts.

Document Properties

Document properties can be used to define placeholder tokens within the clause content of the contract management templates that will be replaced by content, defined when the contract workspace was created, under the **Overview** tab of the contract workspace during the creation of the contract workspace.

Note

Document properties can be created in contract management templates (master agreement/contract addendum documents) or directly in the contract workspace, but these properties are usually maintained at the template level.

Document Property Types

The following are the different types of document properties that can be defined as placeholders within clauses:

- **Read-only document properties**
  These properties are only updated in the contract document when the corresponding field is changed in the contract workspace. Changing these in the contract document won't update the contract workspace. Read-only document properties may be used multiple times in a contract document.
- **Editable document properties**
  These properties are similar to read-only properties, except they also update the contract workspace if changed in the contract document. Editable document properties may only be used once in any given contract document and *must* appear in any document for which they are enabled.

Conditions on Clauses

Conditions are logical constructs that can be used within SAP Ariba Contracts to control the visibility of a contract management template or the documents, tasks, or team members within a template that appears in the contract workspace.

Conditions can also be used on sections and clauses within a master agreement template that define when a section or clause should be made visible when a master agreement document is created from the template. As a result, a single main agreement template can create workspaces with different main agreement content based on information entered by the original creator of the contract workspace. This capability can greatly reduce the number of main agreement templates required. The same is true of contract addendum templates.

6.2.3 Configuring Workflow Approval Tasks

SAP Ariba Contracts provides an approval workflow process task for enforcing strong compliance and oversight and for reducing contracting risks within an organization.

The workflow task, along with the e-signature task, provides a seamless process for internal and external signatories to quickly and efficiently review, approve, and sign contracts within SAP Ariba Contracts without the need for pen and paper, thus considerably speeding up the contracting process while reducing risks and potential noncompliance in contracts.

A *workflow task* is a task that can be defined in each contract template. This task is highly customizable and can be set as a required task. This task can be associated to a specific document or to a folder within the **Documents** tab, as shown in Figure 6.17. When assigned to a folder, all the documents in the folder will be included in the approval workflow process.

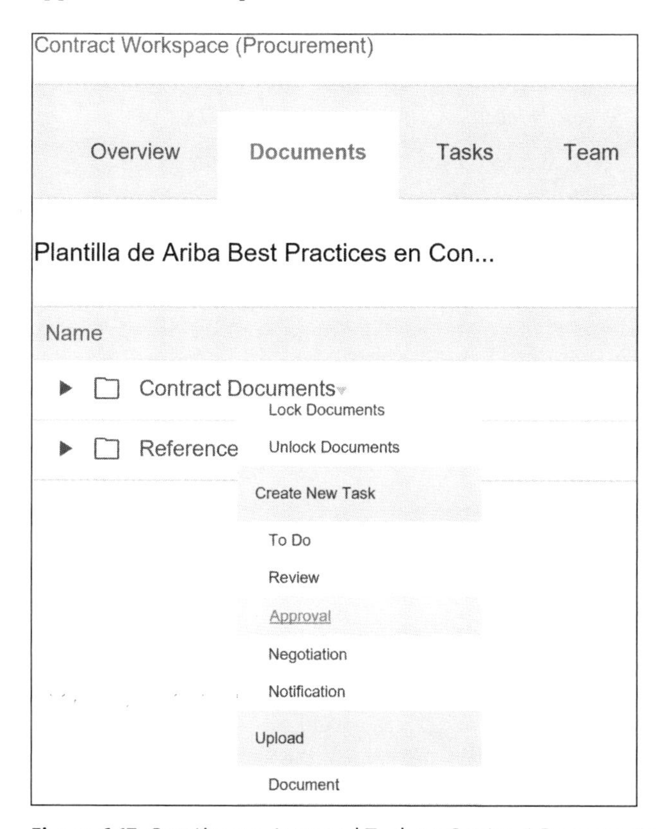

Figure 6.17 Creating an Approval Task on Contract Documents Folder

If an approval task is instead created at the folder level, on triggering the workflow task, all the documents in the folder will be routed for approval, and an email with all the documents attached will be sent out to approvers specified in the approval flow, as shown in Figure 6.18.

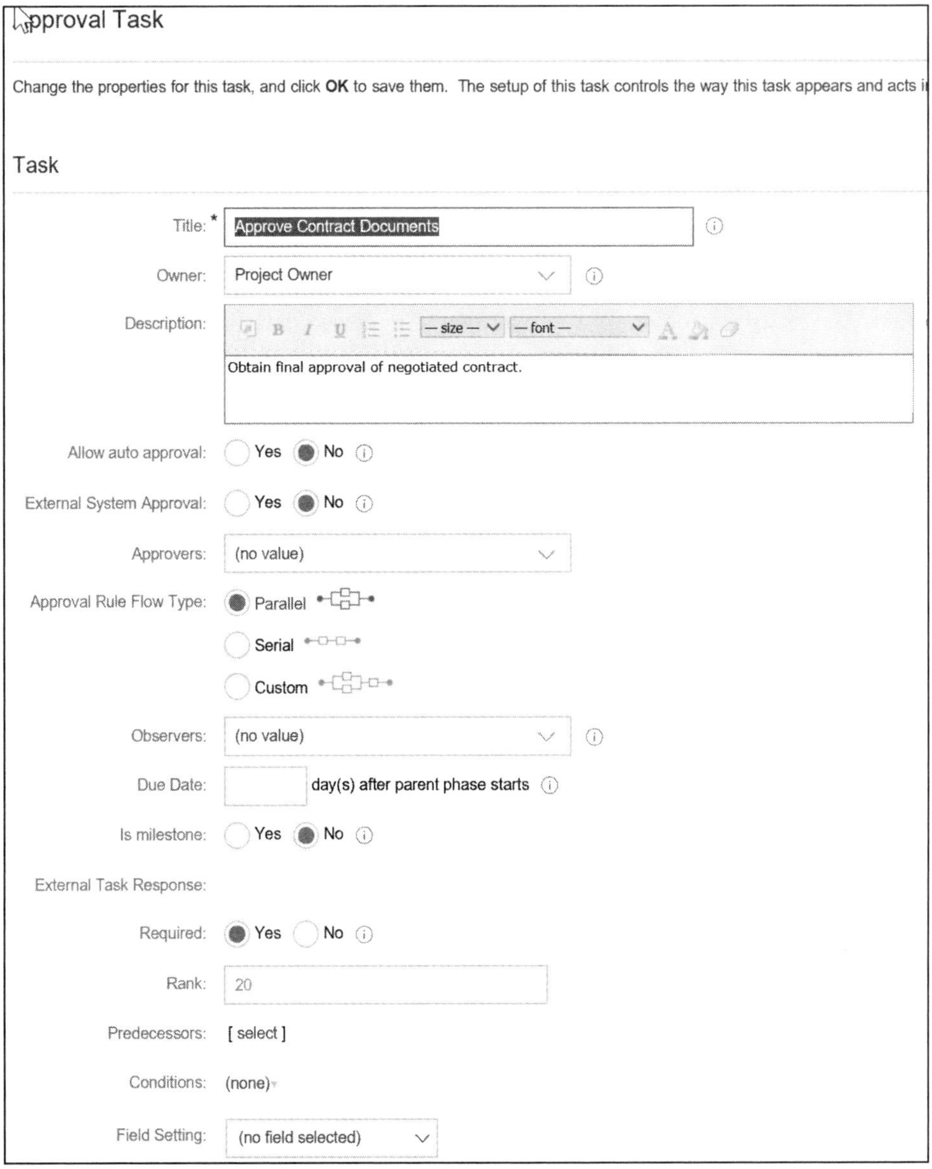

Figure 6.18 Approval Task Details

Approval Flow

SAP Ariba Contracts provides a flexible approval workflow task that can be customized based on the customer's approval process. The following approval flows are allowed:

- **Parallel**
  Multiple approvers/groups of approvers will receive the approval email simultaneously, allowing access for approvers to take action on the approval task simultaneously.

- **Serial**
  Each approver/group within the approval flow will receive an approval email in a specified series. An approver/group will receive an approval request email only after the previous approver/group has approved the document/s in the previous approval step.
- **Custom**
  Customers can define a more complex workflow with a combination of parallel and serial approvals. Customers with a more complex approval flow can use this feature to design their approval process.

Approvers can be added to the parallel and serial approval flows by clicking on the **Approvers** field and adding multiple users/groups within SAP Ariba Contracts.

Approvers will be required to review and approve the documents in the folder associated with the approval task. Observers can be added to the approval workflow as well. Observers aren't required to take any action throughout the process but will have access to review the status of the approval and review the documents within the approval task.

Adding Approvers to Custom Approval Flows

Creating custom approval flows is a flexible option that SAP Ariba Contracts provides to define customized and complex approval processes. If you select **Custom** as the flow, both parallel and serial approval steps can be added to the approval process.

SAP Ariba Contracts allows contract template administrators to define approval workflow tasks at the template level and make tasks required. This capability allows you to standardize the contracting process and thereby alleviate risk in the process.

When a workflow task is defined at the template level, the approvers in each step can be set to a project team member group. When a contract workspace is created, and a user/group is added as a project team member under the **Team** tab, the user/group automatically becomes an approver of the document.

Custom approval flows allow administrators to create complex approval flows based on specific requirements. The approval can also be conditionalized based on information on the contract. For instance, a senior VP approval step can be included for all contracts above a certain amount (a standard field), for example, greater than $1 million.

In addition, approvers for this step can be set up in several ways, which we'll look at next.

Adding Approvers and Groups

You can select approvers from a list of existing users and categorize them in the following ways:

- **Contract approvers**
  Users in the contract approver project team group.
- **System group**
  Users in any system/custom user group.
- **Project group**
  Users from one of the project team groups.
- **Supervisor**
  Supervisor of a particular user.

User Approver Lookup Table

An approver matrix can be created using the approval lookup table shown in Figure 6.19.

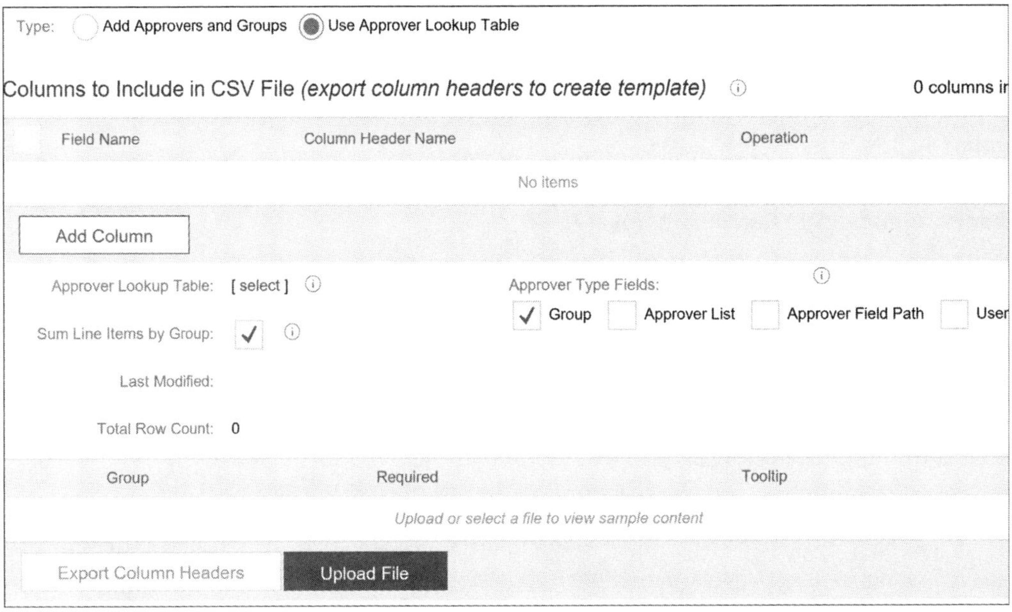

Figure 6.19 Approval Lookup Table

6.2.4 Create Custom Fields Required on Contract Requests and Contract Workspaces

SAP Ariba Contracts is flexible enough to maintain different information within different contracts. Contract administrators can define new custom fields. The new fields that can be created are shown in Table 6.3.

Field Type	Description
Boolean	Displays **Yes/No** radio buttons.
Date	Accepts a date. The date can be typed into the field or selected from a calendar. The date and time will be displayed to users in their local time zone.
Date (calendar)	Accepts a date. The date can be typed into the field or selected from a calendar. The **Calendar Date** field stores a date and time with no time zone information. The date and time will be displayed the same for all users, regardless of their local time zone.
Decimal number	Accepts integers (whole numbers) or decimal numbers (fractions), such as 2 or 0.008, respectively.
Integer	Accepts integers (whole numbers), such as 2 or 34567.
Money	Accepts a decimal number that the system then treats as a currency value in calculations and conversions. Note: The system can perform currency conversions on values entered in currency fields.
Percentage	Accepts integers (whole numbers) or decimal numbers (fractions) that are treated as percentages when used in calculations. For example, if users enter 25, the system uses 0.25 in calculations that use that value. Note: Percentage values can be greater than 100.
Free text	Accepts one line of free text up to 1,000 characters in length.
Multiline text	Accepts multiple lines of free text, up to 1,000 characters in length.
Text single select	Displays an enumeration picklist where only one value can be selected.
Text multiple select	Displays an enumeration picklist where multiple values can be selected. Note: Text multiple select fields don't support hierarchical data.
URL link	Accepts a URL address and displays a text hyperlink. The text displayed is the URL.
User multiple select	Any field that allows you to choose from the list of users that are loaded into your site.

Table 6.3 Types of Custom Fields

Field Type	Description
Flex master data (FMD) single/multiselect	FMD is typically more complex than data stored in normal fields. For example, you would use an enumeration and a text field to store the names of hospitals. However, you would use FMD to store the names of hospitals and attributes such as their addresses, phone numbers, and the name of the hospital administrator. When users select a hospital, they can click it to view the attribute data. The most common use for an FMD field is as an extended, dynamic picklist where the number of picks available exceeds what would be rational for a standard picklist (up to about 20 or 30). The reason for this is that the FMD field is searchable, user-maintained, and can literally accommodate tens of thousands of picks. An FMD field can act like a mini database. The selected value becomes a hyperlink that can open up a widow displaying additional data fields (nonreportable). For FMD single select fields, it is possible to make the additional data reportable by auto-populating separate, reportable custom fields that are bound to the FMD field. Note: When the FMD field acts as a mini database, each separate, reportable custom field counts against your allotted number of custom fields for that solution.

Table 6.3 Types of Custom Fields (Cont.)

As shown in Table 6.4, the number of custom fields that can be added to custom reports is limited, depending on the type of field.

Type of Field	Limit
Integer	26
Text/multiline text/URL link	41
Money	8
Date	10
Boolean	12
Percentage	10
Decimal number	
User	4
FMD (single select)	11
FMD (multiple select)	4
Text multiple select	3

Table 6.4 Reportable Fields Limit

6.2.5 Loading Master Data

Master data, such as suppliers, regions, departments, units of measure, currencies, and more, will be required to create contracts with line items. SAP Ariba Contracts allows customer administrators to load and manage master data with an import/export feature. Some master data, such as suppliers, purchasing organizations, and company codes, can be integrated with SAP ERP or other backend systems, with periodic updates. The steps require for integrating master data with SAP ERP is beyond scope of this book. In this section, we'll look at how customer administrators can load master data using the import/export feature.

As a customer administrator, you can access the **Data Import/Export** screen on the **Administration** section by clicking **Manage • Administration • Site Manager • Data Import/Export**.

> **Tip**
>
> Always export the master data object and save it to your desktop. Use that file to make changes to the master data, and then click **Import** to update that master data object.

Exporting Master Data

To export master data, click the **Export** tab on the **Data Import/Export** screen. Look for the master data object you want to export and click the **Export** button.

Importing Master Data

To import a master data, click the **Import** tab on the **Data Import/Export** screen. Four choices will be available:

- **Load**
  Creates and modifies objects in the database using values in the data file. If an object in the data file doesn't already exist in the database, it's created. If an object in the data file already exists in database, it's modified using the value in the data file.
- **Create**
  Creates new objects in the database using values in the data file. If an object in the data file already exists in the database, it isn't modified.
- **Update Only**
  Modifies existing objects only in the database using values in the data file. If an object in the data file doesn't already exist in the database, it isn't created. If you don't want to modify a particular object, don't include it in the data file. This operation can cause deactivated objects to be reactivated.
- **Deactivate**
  Deactivates objects in the database based on objects in the data file. If you don't want to deactivate a particular object, don't include it in the data file.

Select one of these choices, and click the **Browse** button. Choose the master data object file you updated and saved on your desktop, and click **Run** to execute the import process. If any error arises, a new tab will appear where you can correct the errors and reupload the data file.

> **Note**
> The enablement workbook can be used instead of individual master data object files to import some kinds of master data, for instance, regions, departments, enterprise users, and external organizations (suppliers).

6.2.6 Enabling Electronic Signatures

SAP Ariba provides e-signature capabilities in SAP Ariba Contracts through integration to third-party digital signature services, such as DocuSign and Adobe Sign. A signature task within a contract workspace routes the document electronically and securely via the Internet to the document signee, who can complete the signature task electronically using one of the two services enabled in the landscape.

> **Note**
> While detailed configuration of Adobe Sign and DocuSign services is beyond the scope of this book, for more information, refer to the SAP Ariba documentation, specifically **Help @ Ariba • Administration • Administration and Integration Documentation • Contract Administration**.

All SAP Ariba users who will be submitting Adobe Sign service/DocuSign tasks must have an email associated with their user account that matches the email of a valid and active Adobe Sign service/DocuSign account (or have the auto-account creation option turned on for DocuSign). Users who will be working with signature tasks must also belong to the **Document Signer Group**.

Signature provider services can be enabled by administrators. The paper signature option allows signature task owners to manually complete a signature task by uploading an image of the signed document. To complete a paper signature task, users can submit a file with an electronic image of a "wet" signature (i.e., an image of a signature originally made a paper document in ink). To enable paper signatures, simply navigate to the **Signature Providers** option under **Manage • Administration • Signature Providers**, and select the checkbox at the bottom.

Creating a Signature Task

Signature tasks can be created off of individual documents or folders by clicking on the relevant document/folder within a contract workspace or contract request and selecting

Signature under the **Create New Task** options. If you only have one signature solution enabled, the task will simply be created. If you have multiple signature solutions enabled, you'll select the desired solution. You can create multiple signature tasks within a single contract request or contract workspace, but only one signature task can be associated with a single document or folder. Note that you can create signature tasks within an individual workspace or at the process level. You can also assign any project group the document signer role, which will allow anyone with that role to sign the document. A combination of this project group/role setup and team member rules can be used to automate signers in line with your workflow (delegation of authority), similar to the approvals process.

Submitting an Adobe Sign/DocuSign Signature Task

To submit a signature task, you must edit the **Not Started** task created in the previous step. Both internal and external signers for the document can be added, and you can also create new contacts to identify for signing, like in a review task.

After the task is submitted, the user is directed to the Adobe Sign or DocuSign interface, where additional signature tags can be added. If a single document is associated with the task, then that document is routed for signatures. If a folder is associated with the task, all published documents within the folder are routed for signatures.

After submitting the task for e-signature, the task's status changes to **Signing**. Depending on the signatories' actions, the status will then change to **Signed** or **Denied**. The system checks for returned documents every two hours.

Receiving a Signature Request

Signers for both Adobe Sign and DocuSign tasks receive an email requesting signatures for a document or a set of documents if the task is associated with a folder. The title of the email will contain the SAP Ariba contract ID (or contract workspace ID) and the name of the document (the first document if a folder) to be signed.

The signer can open the email and follow the instructions to complete the signature task or forward the email to someone else who is able to complete the signature. The status of the task changes based on the action taken by the signer.

Completing the Adobe Sign/DocuSign Signature Task

After the signer returns the document, which changes the task's status to **Signed** or **Denied**, the task is officially complete. You can start a new round for the signature task or withdraw the signature task (in any status) by viewing the details of the task and selecting the **New Round** or **Withdraw** options.

If the document is signed, it will be returned automatically to the contract workspace with a new title of the format. This signed document will return to the workspace or a specific folder, based on your task setup.

6.2.7 Enabling Reports on the Contracts Dashboard

SAP Ariba Contracts comes with prepackaged reports that users with one of these roles can run: senior analyst, analyst, and customer administrator.

To run a prepackaged report, follow these steps:

1. Click on **Manage • Prepackaged Reports**.
2. Select a reporting area by clicking on the area and selecting **Open**. The prepackaged reports visible to the user depend on the solutions enabled on the SAP Ariba realm (system), as shown in Figure 6.20.

Figure 6.20 Prepackaged Reports

Adding Reports to Dashboards

Reports can be added as pivot tables or charts. These reports will run dynamically to reflect current information. For example, the Contract Expiration report can be added to the **Contracts** dashboard by a system administrator and set as a required content.

SAP Ariba Contracts Prepackaged Reports

Let's look at some of the useful prepackaged reports for SAP Ariba Contracts:

- **Contracts expiring in the next three months**
  Lists all contracts expiring within a certain time period. A contract manager might use this report to identify contracts that must be reviewed within the next month/quarter/year prior to their expiration.
- **Active contract workspaces by owner**
  Lists all active contract workspaces by user. Contract managers and administrators might use this report to see all the workspaces for which they have ownership.
- **Contract workspaces to start in the next three months**
  Lists all contracts with effective dates within the next three months. A contract manager might use this report to judge the upcoming workload or a manager may use it to determine resource loading.

> **Note**
> Several prepackaged reports involve using clauses only available in the clause library. These reports are only available in SAP Ariba Contracts Professional because the clause library isn't a feature in SAP Ariba Contracts Basic.

Users with the senior analyst or analyst role can create custom analytical reports. SAP Ariba Contracts maintains data in fact tables. Most of these fact tables are available for analysts to pull data into custom reports. Each custom report can be configured with one primary fact table joined to a secondary fact table. Columns from these tables can be set either as report filters or as display fields in the report.

> **Note**
> A best practice for creating custom reports is to search for an existing prepackaged report that comes close to meeting your needs, copying that report, and then editing the copy to make the necessary changes.

6.3 Creating Contracts

SAP Ariba Contracts helps resolve a number of challenges in the contract process, including standardizing and controlling contract creation, automating approvals, and enabling e-signatures. With SAP Ariba Contracts, you can also manage procurement and sales, internal agreements, and intellectual property/licenses. Its overall effect is to automate and accelerate the entire contract creation and execution process. In this section, we'll cover contract creation and execution.

6.3.1 Contract Creation

Spend management in SAP Ariba has several dashboards to cater to the various applications included in the SAP Ariba suite. Once logged in to SAP Ariba, the user will see the dashboard home page. The information displayed on the dashboards, as well as access to specific features and other dashboards, depends on that user's permissions and roles. Users with access to SAP Ariba Contracts will be able to click the **Contracts** dashboard to access information on contracts.

A contract (also known as a contract workspace) in SAP Ariba Contracts can be created in the following ways:

- **From a sourcing event award**
  A contract workspace can be created from a sourcing event award. In this scenario, after the sourcing event is awarded to one or more suppliers, a contract from the award can be created for each supplier by clicking on the **Create New Contract** button from the sourcing award. Information such as pricing terms for the goods or services that were negotiated with the awarded suppliers will be automatically pulled into the newly created contract workspace.
- **From a contract request**
  A contract workspace can be created without a sourcing event. Customers can configure a contract request process. End users not in the Contract Management department are usually required to request the creation on a contract. In this scenario, the end user first creates a contract request. Users with internal user permissions can create contract requests. After the contract request is approved by the appropriate contract managers, the contract workspace can be created from the contract request. Information such as documents, contract collaborators, supplier information, and so on can be carried over directly from the contract request to the newly created contract.
- **Standalone contract workspace**
  A standalone contract (or contract workspace) can be created directly without requiring a contract request or a sourcing event. Users with the role contract agents or contract manager can create contract workspaces.

Three types of contracts can be created in SAP Ariba Contracts: procurement, sales, and internal contracts.

Note

Each contract type (procurement, sales, and internal) requires a separate subscription. Depending on the contract types you subscribe to, different **Create** menu options will appear.

The **Create Contract Workspace** screen is displayed when a user clicks **Create • Contract Workspace** regardless of the type of contract workspace (procurement, sales, or internal) that the user chooses. For simplicity, this book shows screens for creating **Contract Workspace (Procurement)**, but the process and screens are similar for all type of contracts, except for some differences in standard fields, such as the following:

- Contract workspace (sales)
  - **Customer**
  - **Product**
- Contract workspace (procurement)
  - **Supplier**
  - **Affiliated Supplier**
  - **Business Segment**

The contract workspace creation screen is configurable, for the most part, based on your requirements. The screen is made up of standard and custom-defined fields. The fields for procurement, sales, and internal contract types vary slightly. Customer-defined and standard fields can be made optional or required. Required fields must be completed before creating the contract workspace. Entries in the standard fields and custom fields can control which contract templates will be visible for selection to create a contract workspace.

Similar to other projects in SAP Ariba, such as a sourcing project, sourcing request, and so on, contract templates must be created and approved before an end user can create a contract request or a contract workspace. For more information on creating a contract template, Section 6.2.1.

6.3.2 Contract Execution and Consumption

The contract template selected when creating the contract workspace drives the contracting process. In this section, we'll cover the various tabs you'll use during contract execution.

Overview Tab

As shown in Figure 6.21, you can view basic information about the contract, such as the **Contract Status**, **Version**, **Language**, **Owner**, and **Access Control**, similar to the contract workspace template. You'll also see other standard fields, such as **Description**, **Commodity**, and **Regions**. Some standard fields can be hidden through configurations. SAP Ariba automatically generates a contract ID when a contract is created. As shown in Figure 6.21, this contract ID will be visible in the top-right corner of the contract workspace, along with the number of outstanding tasks to be completed.

Figure 6.21 Contract Workspace: Overview Tab

Other sections under the **Overview** tab include the following:

- **Contract Attributes**
  Displays standard fields such as **Hierarchy Type** (**Master Agreement**/**Sub Agreement**/**Standalone**), **Contract Amount**, **Supplier** and **Sub** agreements, and all customer-defined fields.
- **Contract Terms Attributes**
  Displays standard fields such as **Term Type** (**Fixed**/**Auto Renew Type**/**Perpetual**), **Effective Date**, **Expiration Date**, and other fields related to renewal reminder notifications.
- **Process**
  Displays the contract process phases and any milestones defined in the contract process. The status for each process and milestone is displayed via images to denote phase status and milestone status; for instance, a green checkmark is displayed next to a completed milestone.
- **Quick Links**
  Displays any quick links created to important documents within the workspace.
- **Announcements**
  Displays any announcements created within the workspace for members to view.

Documents Tab

The **Documents** tab in the contract workspace is used to create the master agreement and contract addendum from standard approved clause templates or to upload a paper invoice from a supplier or nonstandard contract document. Contract template administrators can choose to include the standard contract document (with approved clauses) as a default document in the contract template. When a contract workspace is created from this contract template, the standard document will be automatically built and will appear under the **Documents** tab, as shown in Figure 6.22.

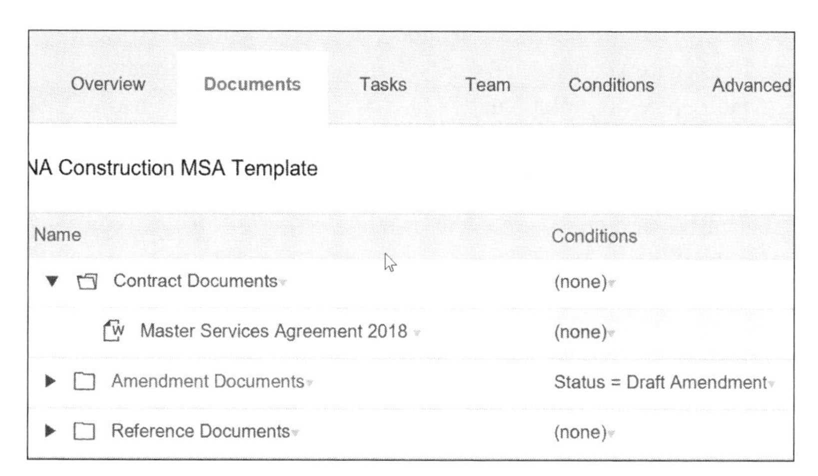

Figure 6.22 Sample Documents Tab in the Contract Workspace

Documents created or uploaded can be edited and modified from the **Documents** tab. All changes to the contract documents are saved as versions. Editing contract documents in SAP Ariba Contracts requires document file synchronization to be enabled by each user.

> **Note**
> Document file synchronization will only work with Internet Explorer.

As mentioned in Section 6.2.1, folders can be created in the **Documents** tab, as well. The complete folder or individual documents within the folder can be associated to to-do or workflow tasks. If a folder is associated to a workflow task, then upon the task submission for approval, all documents within the folder will be included in the workflow approval process.

Tasks Tab

This tab in the contract workspace is used to define the contract phases and the tasks and milestones within each phase. Increased regulatory pressure to reduce costs and standardize the contracting process often motivate organizations to find ways to enforce standard processes for creating and managing contracts internally.

SAP Ariba Contracts allows contract administrators to define a standard contract process with phases, tasks, and milestones. Tasks and milestones based on the standard process can be created at the contract template level. These tasks can be required or optional.

Contract workspaces created from the contract template will automatically inherit the template's tasks and milestones. Contract creators can choose to add additional tasks to the contract process, and/or delete optional tasks that have been pulled from the contract template. To execute a contract, all required tasks within the contract must be completed. Task dependencies and phase dependencies can be set, which ensures that a certain phase or task doesn't start before its predecessor phase or task is completed.

On completing all tasks, including the approval and signature tasks, the **Finalize and Publish** task can be started and completed. The owner can then publish the finished contract by going to the **Overview** tab of the contract workspace and clicking **Actions • Publish**.

Team Tab

Collaboration throughout the contracting process is key for ensuring a faster and less error-prone execution of contracts. As mentioned in Section 6.2.1, SAP Ariba Contracts allows collaborators or team members to be added manually or automatically to the contract workspace under the **Team** tab. Project team groups included at the template level will be automatically added to the contract. Team members can be individual users or groups of users. Each project team group is defined a role, either **View Only** or **View/Edit**. Project team groups can be added to the review and approval tasks.

Message Board Tab

This tab is another medium for communicating with the team members, as well as with stakeholders outside of SAP Ariba. Within a contract workspace, the **Message Board** tab allows contract authors or team members with edit access to create new message topics or to initiate a message via email. New topics created within the contract workspace's message board will be visible and accessible to all team members of that workspace.

New messages posted via email triggers the email application (e.g., Outlook). The mail generated will be automatically CC'd to the autogenerated contract workspace email address. Additional recipients can be added as required. Emails will be sent to each recipient. When **Reply All** is selected, responses are automatically forwarded and recorded under the **Message Board** tab within the contract workspace.

History Tab

The **History** tab captures in chronological order all the changes that have been made to the contract workspace. You can also use filters to search and report on changes made to the contract workspace.

6.4 Consuming Contracts

In a standalone implementation of SAP Ariba Contracts, the contract workspace is consumed within the solution itself. Standalone implementations are usually the case for organizations where SAP Ariba Contracts is used primarily as a contract repository.

In mature organizations, SAP Ariba Contracts is implemented along with SAP Ariba Buying and Invoicing or an SAP ERP backend. SAP Ariba Contracts is used to manage the complete lifecycle of the contract, where the contract is created, negotiated, and executed. The negotiated contract is then pushed to a transactional system such as SAP Ariba Buying and Invoicing or SAP ERP backend where the compliance of the negotiated contract is managed. Once pushed to these transactional systems, a contract is consumed based on the type of contract created and the system configuration. See Section 6.6 for more details.

After a contract has been consumed, if the contract owner decides to not extend the contract, the contract workspace can be closed. Closing a contract workspace in SAP Ariba Contracts will force the contract to be closed in the integrated transactional system, whether it's SAP Ariba Buying and Invoicing or SAP ERP.

6.5 Amending Contracts

SAP Ariba Contracts provides features for different types of contract amendment, although the most common contract amendment type is *renewals*. Contracts that have the status of **Published** or **Expired** can be renewed.

SAP Ariba Contracts provides timely reminders for contract renewals. By default, renewal reminders are sent to the contract author, but additional recipients can be added to receive the renewal reminders.

SAP Ariba solution has dashboards for each solution, including a **Home** dashboard. A tile for expiring contracts is added to the **Home** dashboard by default, as shown in Figure 6.23. This tile shows a count of contracts that have expired and that are about to expire in 7 days and 30 days, respectively. The **Contracts** dashboard has a tile for an **Expiring Contracts Report**, as shown in Figure 6.24. On logging in to SAP Ariba Contracts and navigating to the **Contracts** dashboard, you'll see all the contracts you own and that will expire within the next 90 days. You may alter this time period to fit your own business needs.

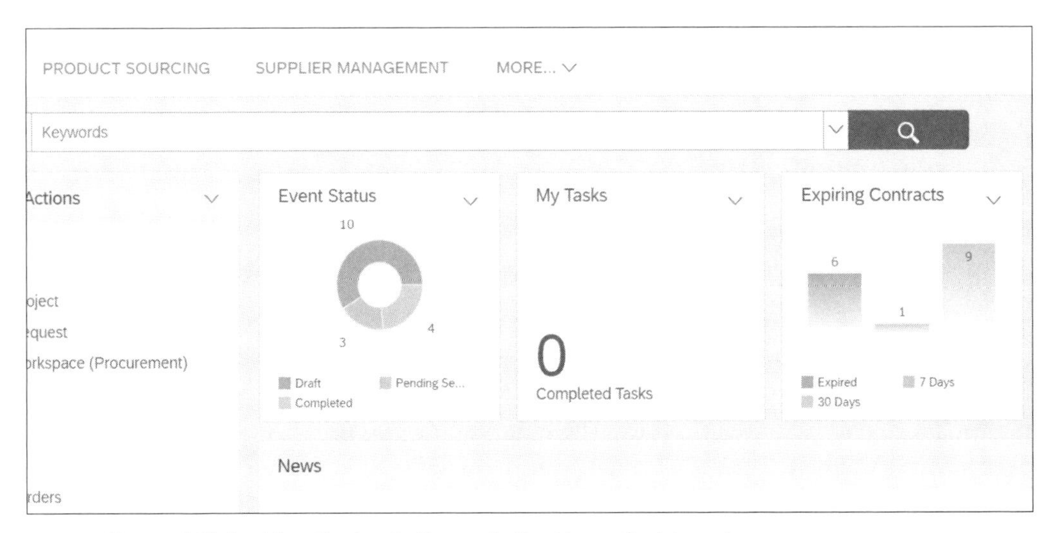

Figure 6.23 Expiring Contracts Shown in the Home Dashboard

Figure 6.24 Contract Dashboard: Expiring Contracts

Expired or published contracts can be renewed by amending the contract. A contract can be amended by following these steps:

1. Search for the expiring or published contract, and open the contract workspace.
2. From the **Overview** tab of the contract workspace, click **Actions • Amend** in the **Contract Attributes** section, as shown in Figure 6.25.
3. Choose **Renewal** as the amendment type.

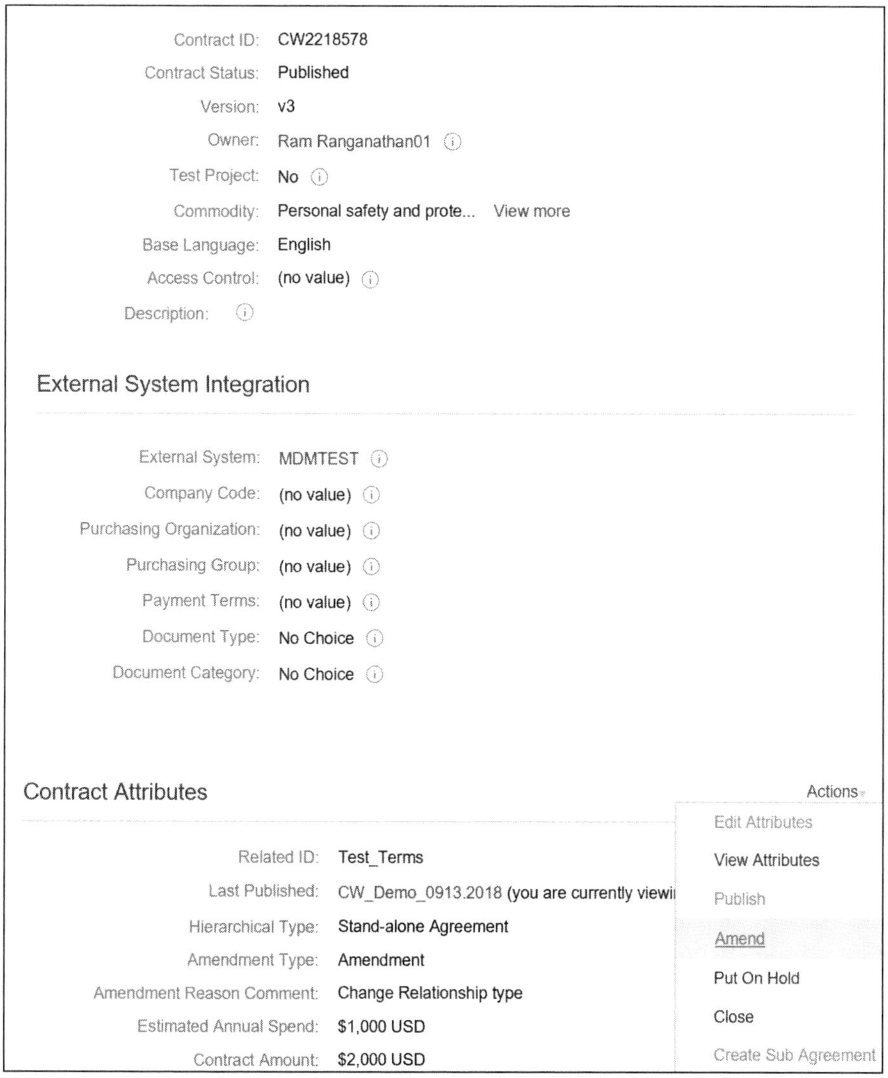

Contract ID: CW2218578
Contract Status: Published
Version: v3
Owner: Ram Ranganathan01
Test Project: No
Commodity: Personal safety and prote... View more
Base Language: English
Access Control: (no value)
Description:

External System Integration

External System: MDMTEST
Company Code: (no value)
Purchasing Organization: (no value)
Purchasing Group: (no value)
Payment Terms: (no value)
Document Type: No Choice
Document Category: No Choice

Contract Attributes

Actions
Edit Attributes
View Attributes
Publish
Amend
Put On Hold
Close
Create Sub Agreement

Related ID: Test_Terms
Last Published: CW_Demo_0913.2018 (you are currently viewi
Hierarchical Type: Stand-alone Agreement
Amendment Type: Amendment
Amendment Reason Comment: Change Relationship type
Estimated Annual Spend: $1,000 USD
Contract Amount: $2,000 USD

Figure 6.25 Amending an Executed Contract

The following changes can be made during renewals:

- **Contract term attributes**
  Effective and expiration dates can be changed.
- **Tasks**
  Certain tasks, including approval tasks, can be enabled during contract renewals. Required tasks must be completed before a contract can be published.
- **Documents**
  Documents in the **Documents** tab can be amended. Additional documents can be added as needed.

The contract is republished after all renewal changes, including any associated approval tasks, are completed.

As shown in Figure 6.26, other amendment types include the following:

- **Amendment (full amendment)**
  This amendment type is used when the complete contract workspace needs to be changed. The contract is republished after the changes are completed.
- **Administrative**
  This amendment type is used to change noncontract details, such as adding a team member or uploading a supplemental document. The status of the contract doesn't change when performing an administrative amendment. Republishing the contract isn't required after this amendment process is complete.
- **Termination**
  This amendment type is only available for published contracts. Closed or expired contracts can't be terminated. The expiration date and email notification settings can be changed during this amendment process. This process is used to close the contract before the expiration date due to adverse conditions or disagreements. The contract is closed after the changes are completed.

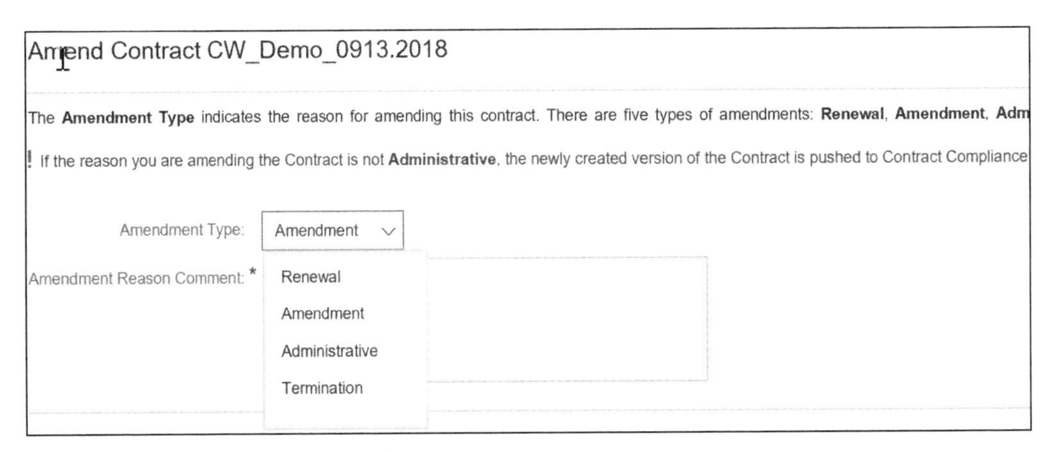

Figure 6.26 Contract Amendment Types

6.6 Using Contracts with Other Applications

As mentioned earlier, SAP Ariba Contracts is an integral part of the SAP Ariba procurement suite with tight integration to other SAP Ariba solutions such as SAP Ariba Sourcing, SAP Ariba Supplier Lifecycle and Performance, and SAP Ariba Buying and Invoicing. In addition, the SAP Ariba Contracts solution can be easily integrated with your SAP ERP backend or your other third-party solutions.

As shown in Figure 6.27, in a typical configuration, the initiation of a contract starts either during the supplier enablement process in SAP Ariba Supplier Lifecycle and Performance Management, after a sourcing event has been awarded in SAP Ariba Sourcing, or in the SAP Ariba Contracts directly in some cases. A contract can be started with a contract request or directly created as a contract workspace without a request process. When executing the contract in the contract workspace, multiple options exist for pushing the contract from SAP Ariba Contracts to a transactional system, such as SAP Ariba Buying and Invoicing or SAP ERP, to create a contract or catalog items.

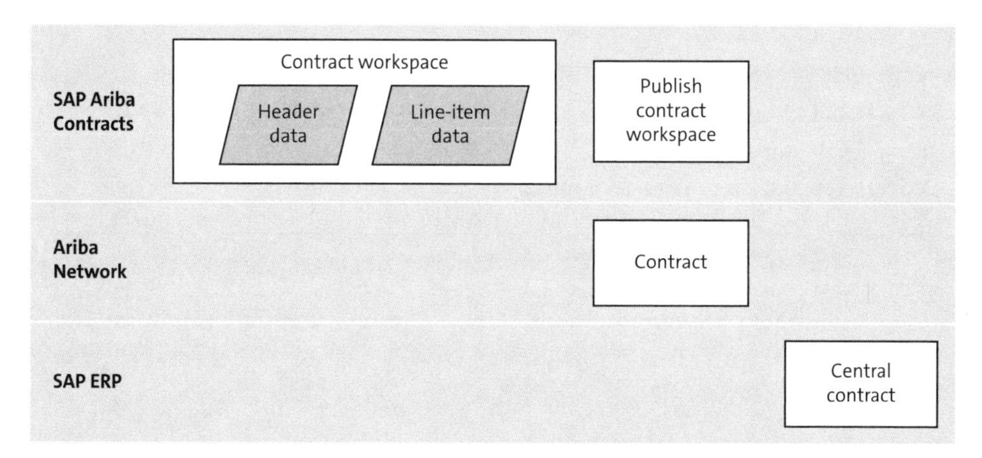

Figure 6.27 Contract Workspace Integration to SAP Ariba Buying and Invoicing and SAP ERP

6.6.1 Using Contracts in SAP Ariba Buying and Invoicing

Customers who have the complete source-to-order or source-to-pay solutions from SAP Ariba must implement suite integration. With the suite integration feature, you'll integrate your SAP Ariba upstream instance (consisting of SAP Ariba Supplier Lifecycle and Performance Management, SAP Ariba Sourcing, and SAP Ariba Contracts) with your SAP Ariba downstream instance (consisting of SAP Ariba Buying and Invoicing).

SAP Ariba Buying and Invoicing provides features and functionalities that enable modeling and managing contract pricing and terms to support both internal compliance as well as supplier invoice compliance. SAP Ariba Contracts provides visibility and control over the entire contract authoring and management process. In a suite-integrated environment, these two solutions can be integrated to allow seamless contract management from authoring to compliance enforcement. In addition, in a suite-integrated environment, a contract must be created first in SAP Ariba Contracts. A pricing terms document with items negotiated during the contracting process must be maintained. This upstream contract can be pushed downstream to create a contract term document, as shown in Figure 6.28. A *contract term document* is an intermediary document before the upstream contract is converted into a downstream operational document—the contract compliance document. In the contract terms, the owner of the contract can define additional information regarding the transactional contract.

From the contract workspace in SAP Ariba Contracts, four types of contracts can be created in SAP Ariba Buying and Invoicing:

- **Supplier Level Contract**
  Covers all products from a supplier.
- **Category Level Contract**
  Covers all products from a catalog.
- **Commodity Level Contract**
  Covers all products identified by specific commodity codes from a supplier.
- **Item Level Contract**
  Covers a specific item from a supplier.

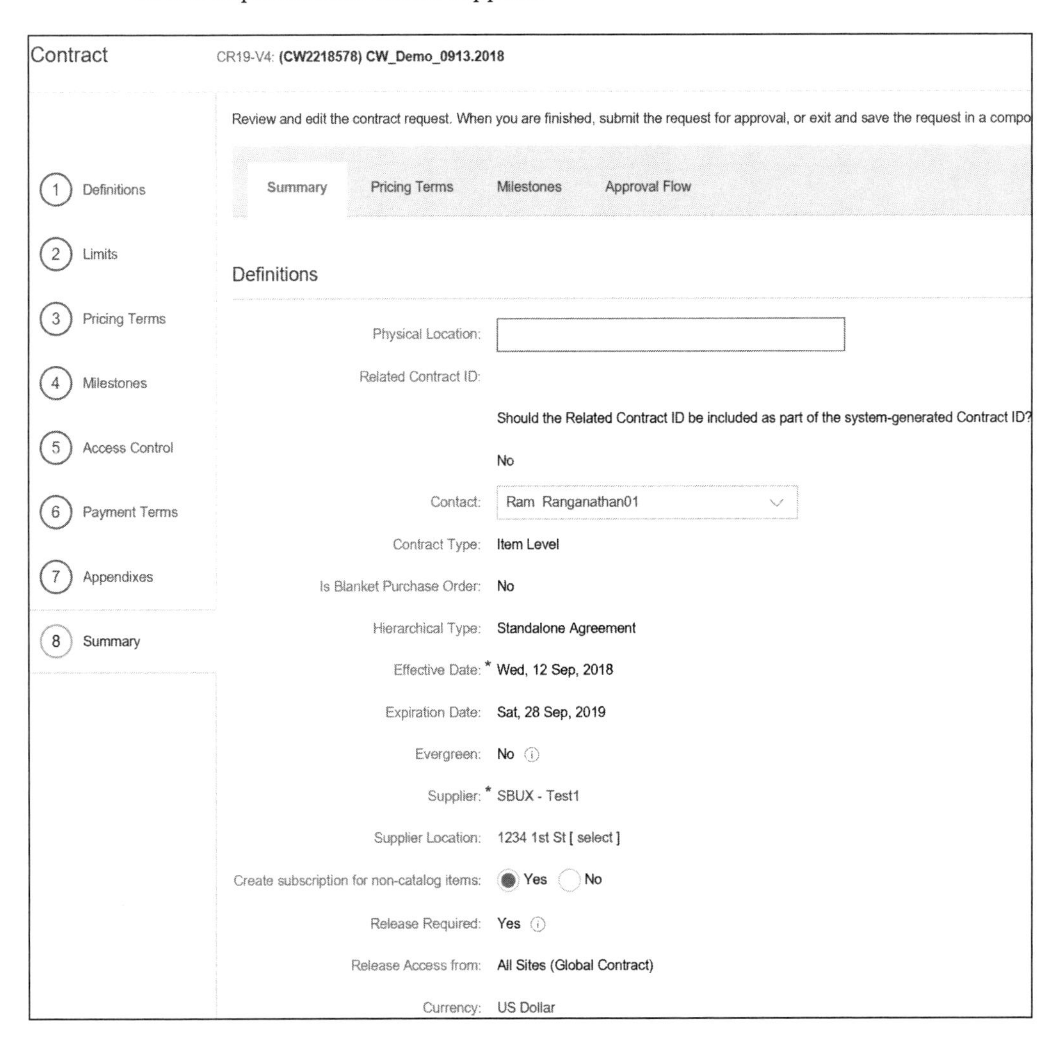

Figure 6.28 Contract Terms

In an integrated suite, with the right permissions, a user can see both dashboards related to SAP Ariba upstream (SAP Ariba Sourcing, SAP Ariba Supplier Lifecycle and Performance, SAP Contract Management) as well as SAP Ariba downstream (SAP Ariba Buying and Invoicing).

Depending on the type of contract and site configuration, the terms of the contract are applied in the following instances:

- Purchasing users or agents order items are associated with a contract. The contract terms are automatically applied when these items are added to their orders.
- Purchasing users or agents create a purchase order (or release order) against the contract.
- Supplier or buyer users create invoices against the contract.

6.6.2 Using Contract Compliance in SAP Ariba Buying and Invoicing with SAP Ariba Contracts

In a suite-integrated environment, contract creation must start from SAP Ariba Contracts. For existing contract compliance contracts, if SAP Ariba Contracts is deployed at a later time, a contract user in SAP Ariba Buying and Invoicing can create a contract workspace in SAP Ariba Contracts. At this point, contract compliance must be managed from SAP Ariba Contracts going forward.

For compliance contracts in SAP Ariba Buying and Invoicing with an associated contract workspace, the contract number will start with "CW." When a contract workspace in SAP Ariba Contracts is closed, the underlying contract compliance contract in SAP Ariba Buying and Invoicing is also closed. When the contract workspace in SAP Ariba Contracts is amended, a new version of the compliance contract is created in SAP Ariba Buying and Invoicing.

6.6.3 Using Contracts in the SAP ERP Backend

Customers using the Ariba Network adapter for SAP NetWeaver and SAP Ariba Contracts can integrate contract information from SAP Ariba Contracts with SAP ERP.

As shown Figure 6.29, a contract or amendment created in SAP Ariba Contracts can be pushed to the Ariba Network. From the Ariba Network, the SAP Process Integration (SAP PI) communication channel takes the contract and sends it to SAP ERP to create an outline agreement based on the header and line-item data of the contract received from SAP Ariba Contracts.

When the outline agreement is created in SAP ERP, the SAP document number and the outline agreement line-item numbers are mapped and sent back to the contract in SAP Ariba Contracts. If this integration process fails, SAP ERP sends error messages to the contract in SAP Ariba Contracts. See Chapter 11 for more information.

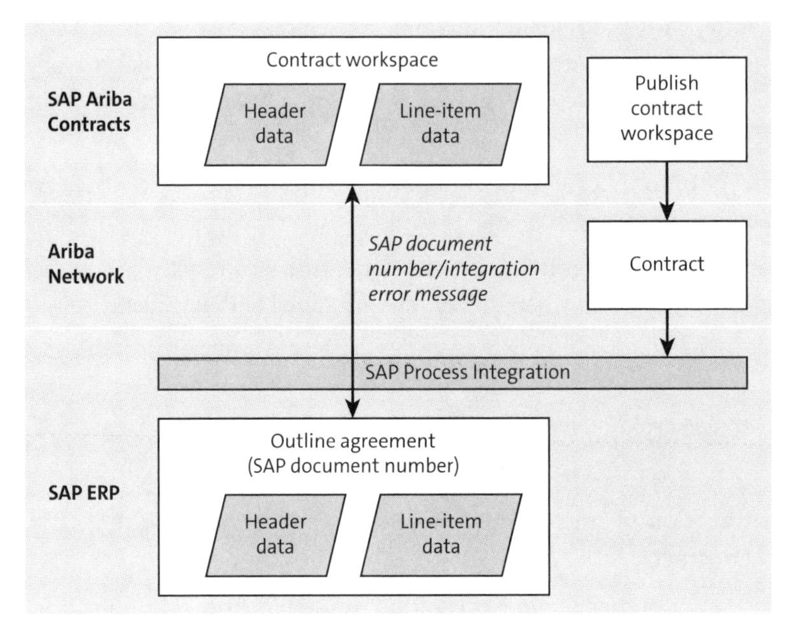

Figure 6.29 SAP Ariba Contracts Integration with SAP ERP

6.6.4 Using SAP Ariba Contracts in SAP S/4HANA Cloud

This new integration feature allows SAP S/4HANA Cloud customers using operational procurement to configure an end-to-end source-to-pay solution by integrating the SAP Ariba Strategic Sourcing Suite.

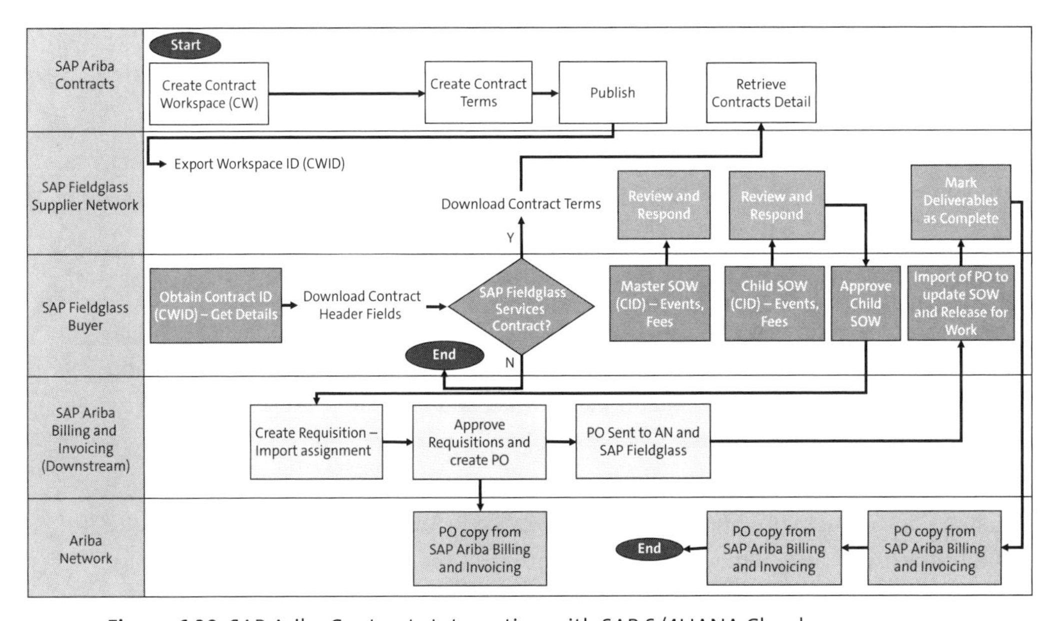

Figure 6.30 SAP Ariba Contracts Integration with SAP S/4HANA Cloud

This allows customers to create strategic contracts from sourcing events or standalone contracts with line items, then negotiate prices on the Ariba Network, and finally push the negotiated terms to SAP S/4HANA Cloud, to create an outline agreement or scheduling agreement (type LP and LPA documents), as shown in Figure 6.30. See Chapter 11 for more information.

6.6.5 Using SAP Ariba Contracts in SAP Ariba Fieldglass

A new feature was released where customers can now integrate their SAP Ariba Contracts with SAP Fieldglass. Contract workspace and contract terms created, negotiated, and executed in SAP Ariba Contracts can now be pushed to SAP Fieldglass to create a SOW contract in SAP Fieldglass.

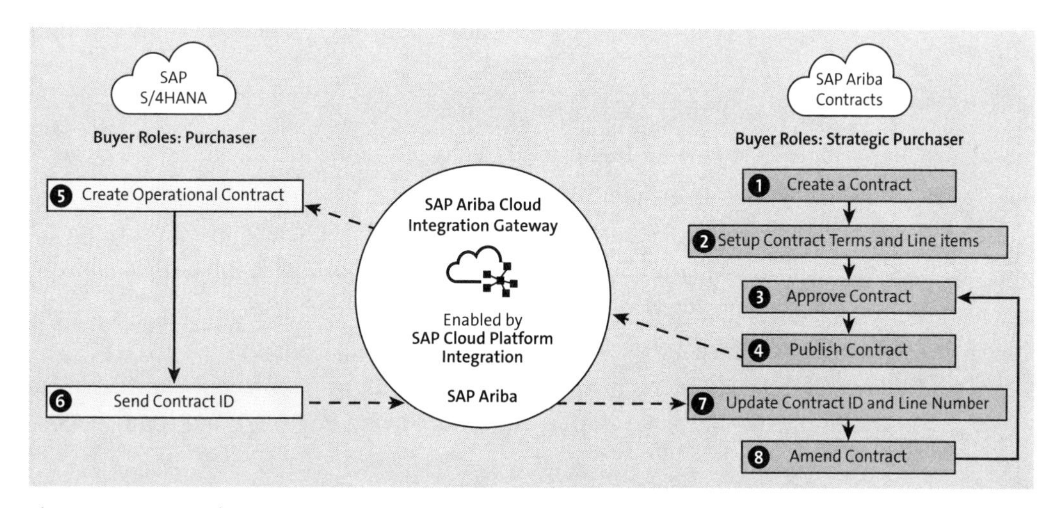

Figure 6.31 SAP Ariba Contracts Integration with SAP Fieldglass

Let's take a look at this process flow:

1. The contract user creates a contract workspace and maintains any line items information, which is pushed down to the contract terms document created from the workspace in SAP Ariba Contracts.
2. The contract user completes the contract process, including any approvals and signatures necessary.
3. The contract user then publishes the contract workspace and contract terms.
4. Upon publishing, the contract information such as contract ID, metadata, and contract terms information from SAP Ariba Contracts is pushed to SAP Fieldglass to create a master SOW. Line items from SAP Ariba Contracts are received as catalog rate definitions within the SAP Fieldglass master SOW. Child SOWs created from the master SOW in SAP Fieldglass will inherit the rates, budgets, and other attributes from the original SAP Ariba Contract workspace.

The full integration flow between SAP Ariba Contracts, SAP Fieldglass, SAP Ariba Buying and Invoicing, and the Ariba Network is shown in Figure 6.31. This integration uses Web Services, but the technical details are out of scope for this book. See Chapter 11, Section 11.6, for more information.

6.6.6 Contract Application Programming Interface

SAP Ariba provides application programming interfaces (APIs) to allow customers to extend the existing SAP Ariba solutions. The following APIs are specific to SAP Ariba Contracts:

- **Create Update and Delete Contract Workspace API**
  This API provides functionality to create a contract workspace using an existing template, update a contract workspace's header details, and delete a contract workspace.

  This API comes with the following endpoints:
  - `/contractworkspaces`: Endpoint used to create a new contract workspace or get a contract workspace to update the metadata.
  - `/contractworkspaces/$contractid`: Endpoint used to retrieve the contract workspace that matches the contract ID. The returned workspace will be deleted from the contract repository.
- **Contract Compliance API**
  This API allows customers to search for and get information about contract questions and contracts created in the SAP Ariba Buying and Invoicing solution (SAP Ariba downstream).
- **Contract Terms API**
  This API provides functionality to create contract terms documents in a contract workspace and retrieve contract terms and contract compliance details. This API is only applicable in a suite-integrated realm, that is, when the SAP Ariba Sourcing Suite is integrated to SAP Buying and Invoicing.

6.7 Summary

Our goal in this chapter was to describe the benefits of implementing SAP Ariba Contracts. The features defined in this chapter, such as contract authoring and integration with e-signature providers, can be easily enabled to create a complete end-to-end, paperless contracting solution that considerably improves the contracting process, resulting in greater efficiency, lower costs, and improved relationships. This chapter also recommended tried-and-true best practices and the SAP Activate methodology for quickly and efficiently implementing SAP's cloud-based solution for contract management.

Chapter 7
Guided Buying

The guided buying capability is one of SAP Ariba's most important features, deserving of its own chapter (although it might also be considered part of operational procurement). In this chapter, we'll focus exclusively on the guided buying capability, including how the application works, how to plan guided buying capability projects, and how to implement and use the solution for operational procurement with step-by-step instructions.

Guided buying is, for many companies, where you can start simplifying and extending your procurement activities (from source to pay), accompanied with analytics and a consistent user experience (UX) following the SAP Fiori guidelines and look and feel. Working in conjunction with SAP Ariba Buying and Invoicing and SAP Ariba Spot Buy Catalog, the guided buying capability in SAP Ariba now includes a private help community.

Activities involving the guided buying capability in SAP Ariba typically end with a supplier purchase order (PO) or with a sourcing, contract, or supplier lifecycle activity that will be completely supported by your procurement organization, if you're using the SAP Ariba solutions supporting operational procurement.

In this chapter, we'll outline the guided buying capability in SAP Ariba and how it interacts with SAP Ariba Buying, SAP Ariba Strategic Sourcing, and SAP Ariba Procurement Help Desk. We'll describe the solution's functionalities, including how to implement and configure them.

7.1 What Is Guiding Buying?

The core idea behind the guided buying capability in SAP Ariba is to provide a superior, simpler, and faster procurement experience to all kinds of casual users and nonprocurement professionals in your organization as they search for goods and services—with little to no involvement from your procurement department.

The guided buying capability delivers to your organization a simple, persona-based application- and category-tailored buying experience when used to procure goods and services. Through its guided buying capability, SAP Ariba can walk you and any other user in the organization through your procurement processes in a simple, outcome-based manner. Through the application, you'll be guided to select your company's preferred suppliers and catalogs or be guided through the Spot Buy capability. The users in your organization will be empowered to apply procurement policies in a transparent and implicit way, collaborating with and assisting the procurement department. All procurement policies are triggered in a distinguishable manner while the application is in use, so users will know right away when a policy is being violated. Corrective action will be recommended in the same screen, rather than finding out after you've finished and are ready to submit a request, resulting in a simple and satisfactory experience.

With the guided buying capability, users outside the professional procurement group can go to the guided buying capability in SAP Ariba, a one-stop shop to search for goods and services, where purchases can be made with little to no involvement from the procurement department.

In this chapter, we'll outline the implementation planning, configuration, administration, and usage of the guided buying capability in SAP Ariba.

7.1.1 Buying Strategies

Different buying strategies can be implemented in the guided buying capability by making use of its simple user interface (UI) and its category-driven behavior, supporting your local policies to guide users to the desired outcomes. Buying strategies are directly correlated to the potential benefits for your casual users, functional buyers, and procurement professionals.

With category behavior, the application will present category-specific content to help end users determine the appropriate suppliers, contracts, and procurement tools, as well as provide a simple way to compose and submit purchase requisitions or sourcing requests.

A *category strategy* considers the spend and buying channels, and a *user strategy* considers the type of users or personas whose purchasing processes require using SAP Ariba Buying instead of the guided buying capability in SAP Ariba.

Category Strategy

The UI for guided buying is simple and more category-driven than SAP Ariba Buying or SAP Ariba Buying and Invoicing. Because your users are guided through processes via the UI, you should define the buying channels or category strategies to deploy for the guided buying capability in SAP Ariba. The three levels of category strategy are the following:

- **No-touch strategy**
  A self-service or no touch category strategy involves catalogs with fixed rate cards for services from preferred suppliers and clearly defined goods and services with known prices resulting on a PO. A good candidate for a no-touch category strategy

would be any employee whose procurement activity goes no further than ordering goods with no concern for accounting information, receipting, invoicing, or complex procurement activities.

- **Low-touch strategy**
  A low-touch category strategy requires some collaboration with preferred suppliers in the small- to medium-spend area. This category strategy can require a simple quote and results in a PO. A good example would be any user procurement activity that goes no further than ordering goods with no concern for invoicing or complex procurement activities, users purchasing for their departments, or users with limited sourcing needs (three bids and a buy).
- **High-touch strategy**
  A high-touch category strategy requires a professional buyer for complex procurement without a prefer supplier or unknown supplier and can be high risk in nature.

Considering your company's category taxonomy, the guided buying capability in SAP Ariba can be the entry point for your users to procurement, sourcing, and travel expense modules, as well as other external applications. The guided buying capability doesn't replace SAP Ariba Buying and Invoicing or SAP Ariba Sourcing but offers a simpler guided functionality for a specific subset of users who don't require all the additional advanced procurement function capabilities these modules offer.

User Strategy

Based on what department each user belongs to, you'll need to decide which users need access to make purchases through the guided buying capability in SAP Ariba and which users will make purchases through SAP Ariba Buying; this distinction will be a vital component during configuration. This capability will avoid confusion in casual and functional users who are often unfortunately presented with multiple solutions with different access and UIs presenting different supporting functionalities. Thus, the guided buying capability is most appropriate for the following users:

- **Casual users**
  A casual user is an employee who makes infrequent purchases. For casual users, the guided buying capability is easy to use and efficient, providing a consumer-like purchasing experience in a one-stop shop for goods and services.
- **Functional users**
  A functional buyer is a user who makes purchases on behalf of a department, for example, marketing or facilities. Functional users don't belong in your procurement department; they could be administrators or employees in other departments that spend a good amount of time buying goods and services for their department. Guided buying empowers functional users by helping them follow all the necessary procurement policies without reading procurement manuals or involving the procurement department in every purchase.

- **Procurement professionals**
  The procurement department is composed of procurement professionals, the employees who oversee the purchasing function in your organization. Procurement professionals don't use the guided buying capability to make any purchasing decisions, but they can benefit from it. With proper configuration, the procurement department won't need to be involved in every low-value transaction or activity. They'll have the freedom to focus on strategic purchases while also improving policy compliance.

The guided buying capability leverages information from various SAP Ariba solutions to provide a rich UX. The objective is to provide casual requisitioners with the required information to procure goods and services within the organization while increasing compliance and simplifying the buying experience.

7.1.2 Guided Buying Capability

The guided buying capability is a persona-based application targeted to facilitate procurement needs for the casual users in your organization, as shown in Figure 7.1.

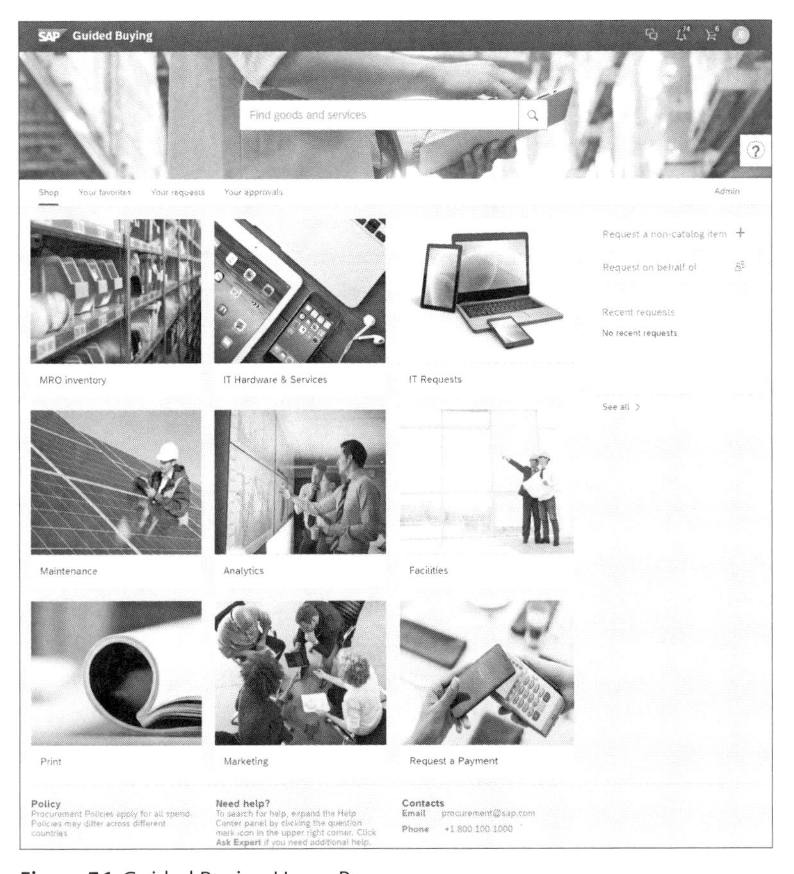

Figure 7.1 Guided Buying Home Page

The guided buying capability in SAP Ariba requires SAP Ariba Buying and can't be used without it. Basically, the capability depends on SAP Ariba Buying for your organization's master data, including users, suppliers, commodity codes, catalogs, and accounting information, which we'll describe in more detail in Section 7.2. The SAP Ariba Buying solution shares this master data with the guided buying capability in SAP Ariba. In SAP Ariba Sourcing, you'll configure sourcing templates to define the standard business processes that your organization uses to buy goods and services. These sourcing templates will be transparently integrated with your forms and policies in the guided buying capability.

In SAP Ariba Supplier Management solutions, you'll define qualified and preferred suppliers for ad hoc purchases and quote requests, so that users can make more informed purchasing decisions with the guided buying capability. When you're ready to configure the guided buying capability, you'll define and implement your organization's purchasing policies, create landing pages, and design forms. All these components will be covered in Section 7.1.3 and Section 7.2.

7.1.3 Planning Your Implementation

Planning is key, especially if you already have SAP Ariba. Compared to other SAP Ariba solutions, implementing the guided buying capability in SAP Ariba is simpler and is based on your targeted users. Your focus will be exclusively on the UX you want to provide—a simple and targeted no-touch or low-touch procurement experience.

The high-level steps in the implementation project are as follows:

- Define the program vision.
- Develop a business case and outcomes.
- Confirm your objectives, scope, and rollout plan.
- Deploy/kick off the project.

For your business case, confirm and capture the project goals and objectives like efficiency, visibility, fiscal control, and compliance.

The typical deployment timeline is between 8 and 12 weeks, broken up in the following phases:

- **Prepare phase (analysis/design)**
  - Gather requirements for buying channels.
  - Identify casual buyer use cases, procurement policies, spend categories, and process expectations for casual and functional users.
  - Determine fit-gap against key requirements.
- **Configure phase**
  - Configure and set up the guided buying capability in SAP Ariba.
  - Review and refine content (e.g., policy and process information).

  - Develop and gather category content.
  - Plan and execute user acceptance testing.
- **Deployment phase**
  - Develop communication and end-user deployment plans.
  - Establish and implement content admin procedures.
  - Cutover to production for use, and complete deployment activities.

Before you start configuring the guided buying capability in SAP Ariba, you'll need to plan, and you'll also need configured data in your SAP Ariba Buying, SAP Ariba Strategic Sourcing, and SAP Ariba Supplier Management solutions. Note that all the master data administration must be managed in SAP Ariba Buying and Invoicing. Above all, you must always consider the process flow used in guided buying, as displayed in Figure 7.2.

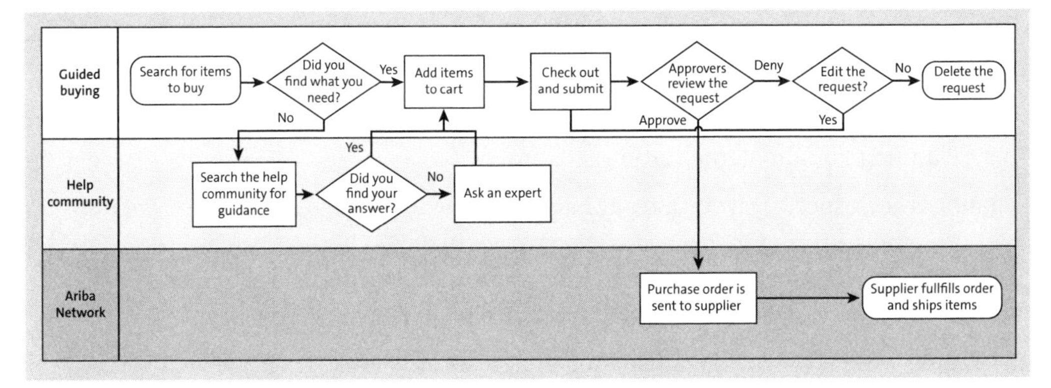

Figure 7.2 Guided Buying: No Request for Quotation Request Flow

You should continuously fine-tune your understanding of the audience for this application—your casual and functional users, that is, users who don't need the functionality offered in SAP Ariba Buying and Invoicing or SAP Ariba Sourcing. Even though the guided buying capability can serve as a gateway and entry point to full procurement, sourcing, supplier lifecycle, and contract modules, as well as other external applications, most of those processes start with casual and functional users. As summarized in Table 7.1, an effective way to start understanding your audience is thinking about yourself, the type of user you are, and whether you consider yourself a casual or functional user, excluding categories where advanced procurement functions are needed.

Key Activities	Description
Identify users	■ Identify the users within the company, whose profile (role) fits the guided buying capability in SAP Ariba. ■ Divide these users by location and/or role. ■ Divide these users by category.

Table 7.1 Key Planning Activities

Key Activities	Description
Identify categories and actions	■ Identify policies associated with categories. ■ Identify categories requiring tactical sourcing (bids and buy) and design forms. ■ Identify categories that require end-user input (parametric/partial). ■ Divide categories between goods and services ■ Study categories identified for services and determine additional needs (e.g., forms). ■ Identify preferred suppliers for each category and region. (Supplier rationalization may be needed.)
Define home page and landing pages	■ Define the navigation from the home page, based on category hierarchy versus category priority. ■ Identify and load images to be used for landing pages, branding, and tiles. ■ If the guided buying capability is to be used as a centralized entry point for procurement, identify the external system that must be accessible from within the capability.
Define community setup	■ Assign moderators and experts for each category. ■ Determine content per landing page (future).

Table 7.1 Key Planning Activities (Cont.)

7.2 Configuring the Guided Buying Capability

If you're deploying the guided buying capability for the first time, you'll need SAP Ariba support to enable the capability. SAP Ariba Buying and Invoicing will need to be configured. Additionally, your realms need to be suite integrated; you'll need to plan accordingly for the suite integration considerations and understand the existing architecture.

If your organization already uses SAP Ariba Buying or SAP Ariba Buying and Invoicing, your SAP Ariba technical lead will request that operational sourcing be enabled in your realm for low-touch sourcing templates and activities and then enable the guided buying capability for your environment. The solution configuration is self-service.

To configure and maintain the capability, you'll need administrator rights, which will enable the **Admin** link on the menu, as shown in Figure 7.3, to access the admin page.

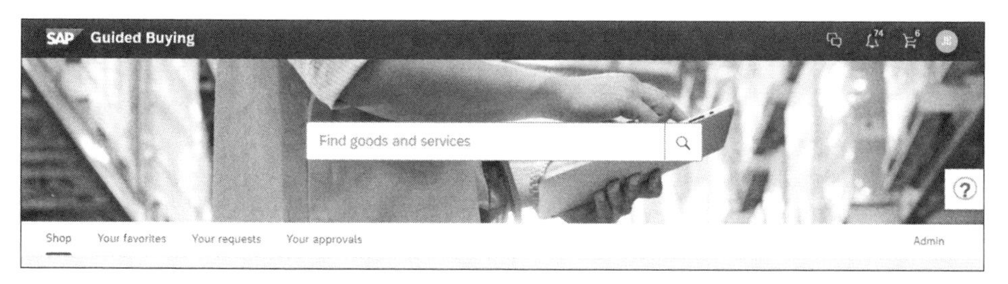

Figure 7.3 Admin Menu Access

On the **Admin** page, you can maintain policies, forms, landing pages, functional documents, translations, tactical sourcing, parameters, checkout page configuration, messaging, ad hoc classifiers for commodity codes, recommendations and reports, and useful links, as shown in Figure 7.4.

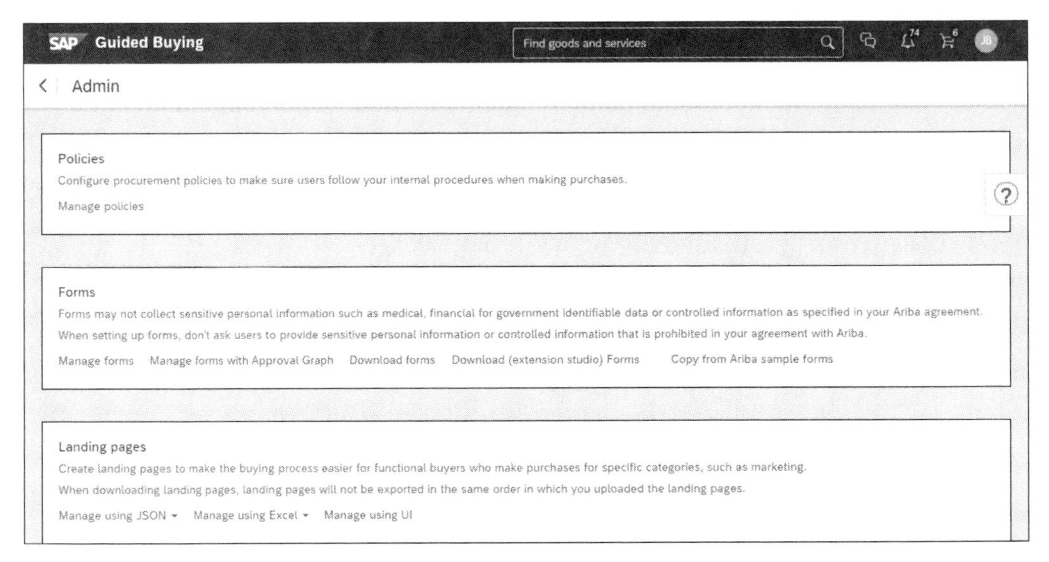

Figure 7.4 Admin Main Page

In the following sections, we'll describe the key components for your configuration.

7.2.1 Approval Flows

The approval flows are the result of business rules and conditions that determine the number of users and groups that are required for the request. Approval flows are optional, and, if desired, you'll need to configure them in SAP Ariba Buying and Invoicing. When approval flows are already configured in SAP Ariba Buying, they're automatically applicable to any purchase requisitions or receipt submitted in the guided buying capability in SAP Ariba, as shown in Figure 7.5.

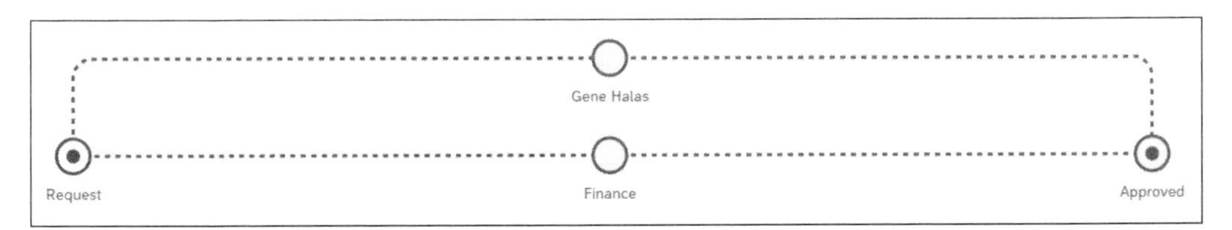

Figure 7.5 Catalog Requisition Approval Flow

You and any user who can view the purchase requisition can add additional approvers and watchers without editing requests before or after the active approver.

> **Note**
>
> Approvers can view *service sheets* that are waiting for their approval from their list of items to approve in the guided buying capability in SAP Ariba. Approvers are automatically redirected to SAP Ariba Buying solutions to review service sheet details.

7.2.2 Spot Buy

The Spot Buy capability is generally used to let casual users procure items that aren't under contract from vendors, don't already exist in your master data, and thus have no specific payment method or invoice process defined.

All users authorized to search and use the Spot Buy capability will need to belong to the Spot Buy user group. However, a Spot Buy supplier should be loaded in SAP Ariba Buying and Invoicing as a common supplier, and the relationship should first be established on the Ariba Network.

7.2.3 Images

You can customize the look and feel of the guided buying experience by uploading custom images via the **Upload images** option, as shown in Figure 7.6, and using Table 7.2 as reference. Images are used through the guided buying capability to provide visually appealing yet intuitive layouts and are critical components to designing forms, landing pages, tiles, and more. You'll need to consider graphic aspects such as resolution and size, and you'll need to have some layout design skills.

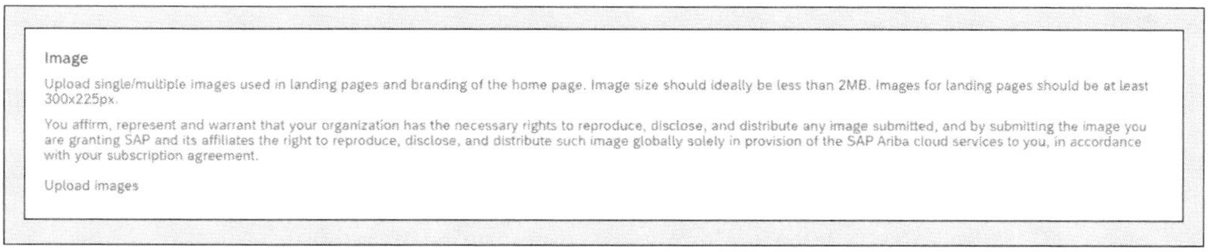

Image
Upload single/multiple images used in landing pages and branding of the home page. Image size should ideally be less than 2MB. Images for landing pages should be at least 300x225px.
You affirm, represent and warrant that your organization has the necessary rights to reproduce, disclose, and distribute any image submitted, and by submitting the image you are granting SAP and its affiliates the right to reproduce, disclose, and distribute such image globally solely in provision of the SAP Ariba cloud services to you, in accordance with your subscription agreement.
Upload images

Figure 7.6 Upload Images Option

Hierarchical considerations also come into play when developing landing pages and subpages, as summarized in Table 7.2.

Name	Type	Home Page	Description
Your company logo	Logo	Yes	▪ Recommended dimensions are 125 px wide × 25 px high. ▪ One image is used per realm.

Table 7.2 Image Types

Name	Type	Home Page	Description
			■ File name is based on your organization's name within your site URL (<customer>.jpg) and is case sensitive. ■ Recommended size is 125 px × 25 px.
Search banner	Banner	Yes	■ Recommended dimensions are 1440 px wide × 290 px high. ■ One image is used per realm. ■ File name is based on your organization's name within your site URL (<customer>_search.jpg). ■ Recommended size is 1920 px × 300 px.
Entry per category	Tile	Yes/no	
Entry per form	Tile	Yes/no	
Landing page image	Category landing page	No	■ Recommended dimensions are 300 px wide × 225 px high. ■ Landing page images should be imported before the associated landing pages are loaded. ■ An individual image can be used across multiple landing pages.

Table 7.2 Image Types (Cont.)

Note

To update an image, upload a revised image with the same name. Even though images that have been loaded into the guided buying capability can't be deleted, your users will only see images linked to a landing page, logo, or search banner.

To make this process easier, you can use the **Manage using UI** option, shown in Figure 7.7, to make changes that will be automatically saved and immediately viewable, as shown in Figure 7.8.

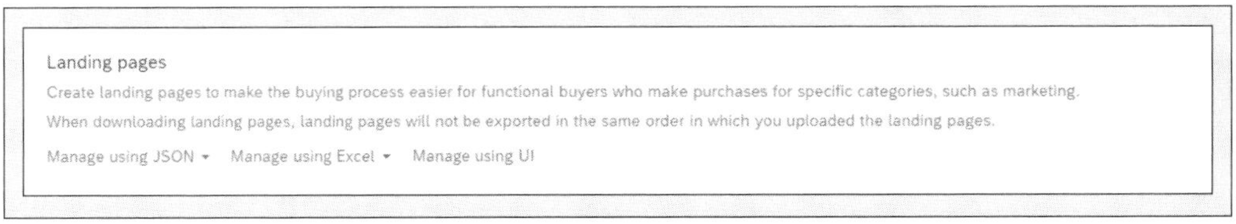

Landing pages

Create landing pages to make the buying process easier for functional buyers who make purchases for specific categories, such as marketing.

When downloading landing pages, landing pages will not be exported in the same order in which you uploaded the landing pages.

Manage using JSON ▾ Manage using Excel ▾ Manage using UI

Figure 7.7 Manage Using UI Option

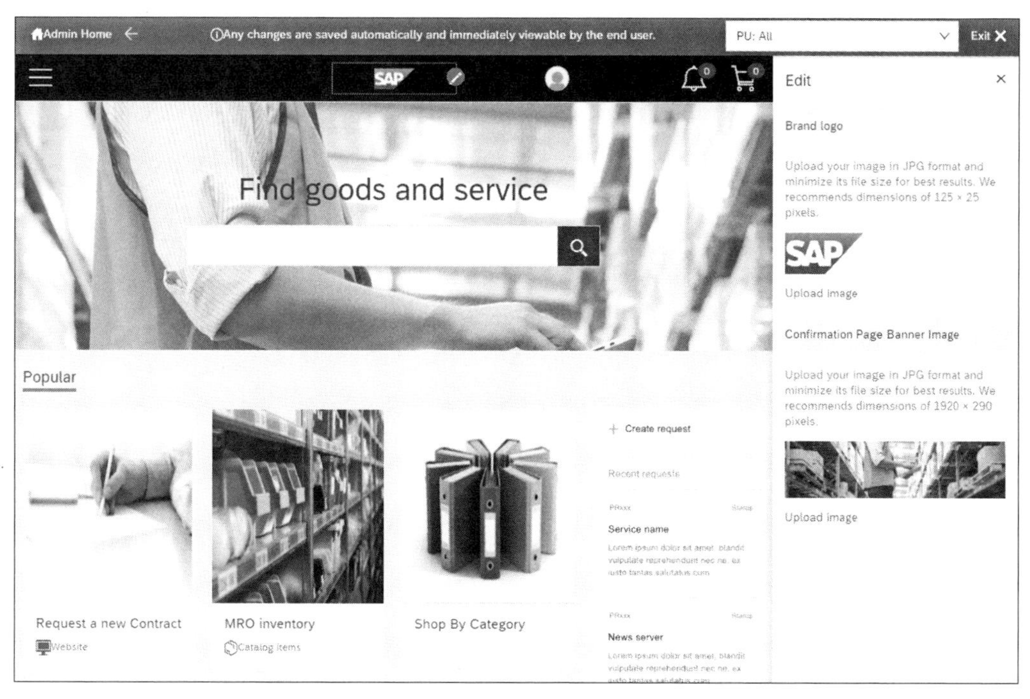

Figure 7.8 End-User UI Changes

You can make changes by moving tiles, modifying landing pages, and uploading images, request forms, preferred suppliers, and so on, as shown in Figure 7.9.

You can make any UI configurations using the JavaScript Object Notation (JSON) format or selecting the **Manage using UI** option.

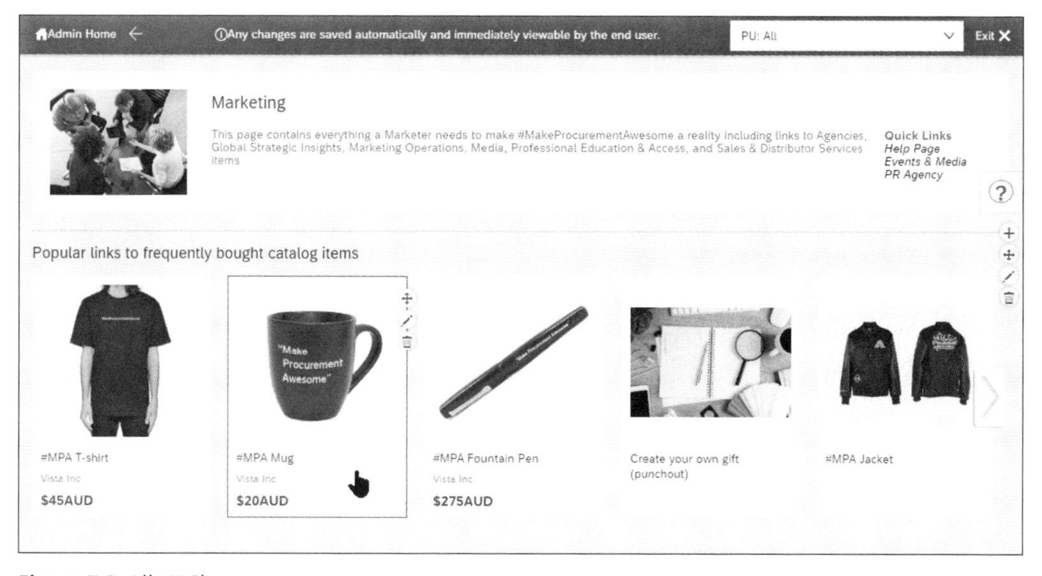

Figure 7.9 All UI Changes

7.2.4 Tiles

You can customize the type of tiles on each of the guided buying pages and the order in which they are presented. All landing pages need to have unique names and must be associated to one location.

The valid types apply to tiles include the following:

- **Landing page**
  Used for tiles that bring users to another landing page.
- **External site**
  Used for tiles that redirect users to an external web address.
- **Ad hoc item**
  Used for tiles that bring users to a request's ad hoc item or service page. Ad hoc items can be optionally combined with commodity code attributes to prepopulate the **Category** field in the **Request Ad Hoc Item** screen.
- **Catalog search**
  Used to bring the user to a specific search in the catalog.
- **Carousel**
  Used to add a row of form tiles to the page.
- **Form**
  Used for tiles that take users to a form.
- **Catalog item**
  Used for a tile that shows a single catalog item.
- **Sourcing form**
  Used for adding a custom form that will create a request for quote.
- **Requisition form**
  Used for adding a custom form that will add a new line to the requisition.
- **Extension studio form**
  Used for adding a custom form that will directly go to approval.
- **Preferred suppliers**
  Used to add a Preferred Suppliers row to the page.
- **Supplier carousel**
  Used to add a Preferred Suppliers row to the landing page. Your preferred suppliers (Section 7.2.11) for the specified commodity code are pulled into the row.

7.2.5 Landing Pages

Landing pages are required and refer to pages that contain various clickable tiles to help users navigate through a catalog. Users will see an initial landing page (home page) when logging in to the guided buying capability in SAP Ariba. Thereafter, a landing page is required for each category and should include pictures and clear descriptions, and the tiles needed to guide users through the procurement process for the category. Landing pages must be designed based on the process the user is expected to use on this page.

Users must belong to either the customer administrator or customer catalog manager group to configure landing pages. To configure landing pages, you can download a Microsoft Excel file template from the **Admin** page. The **Instructions** worksheet will have the instructions on how to structure the contents of the file and will include all the current landing pages in your guided buying application. You can download this file at any point, make changes, add any new content, and upload it. You can also use JSON or the **Manage using UI** option where you can edit landing pages and add/edit custom forms built in Extension Studio using the landing page designer tools. These tools include the UI designer and the Excel upload/download feature. You don't need knowledge of JSON to add or edit custom forms on a landing page.

Only forms need to be created and published in Extension Studio. Extension Studio forms can be added directly to the UI via the landing page designer, as shown in Figure 7.10 and Figure 7.11.

Users can locate any category tile or landing page by searching for one or more keywords, which you can configure.

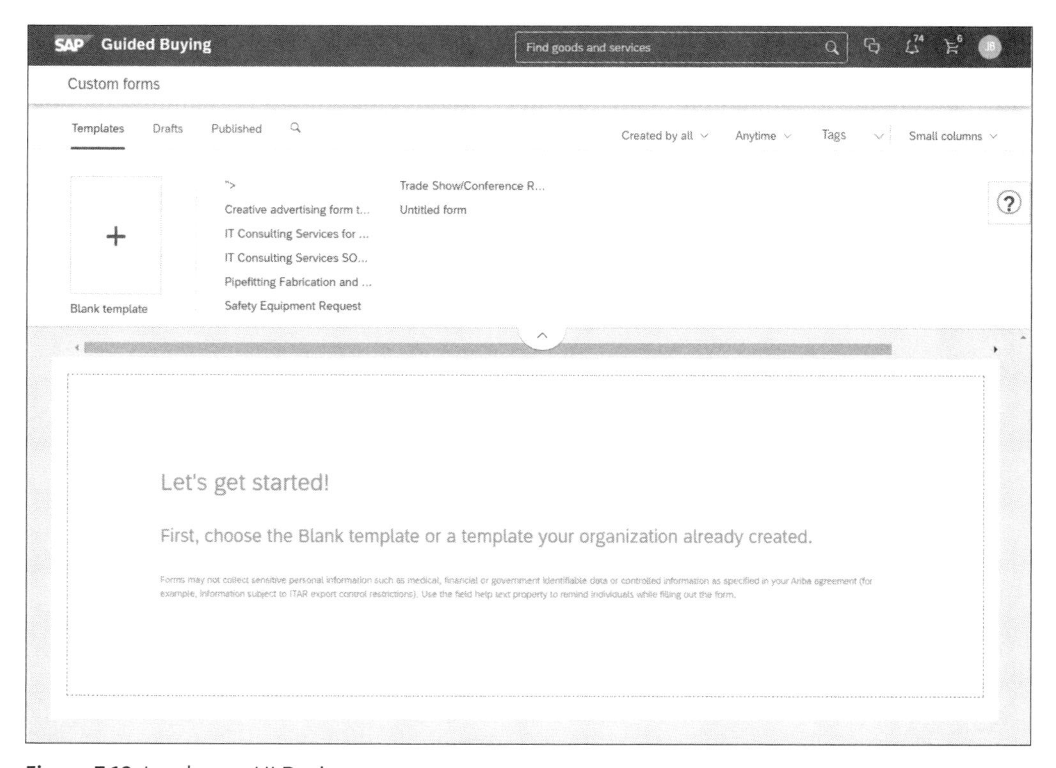

Figure 7.10 Landscape UI Designer

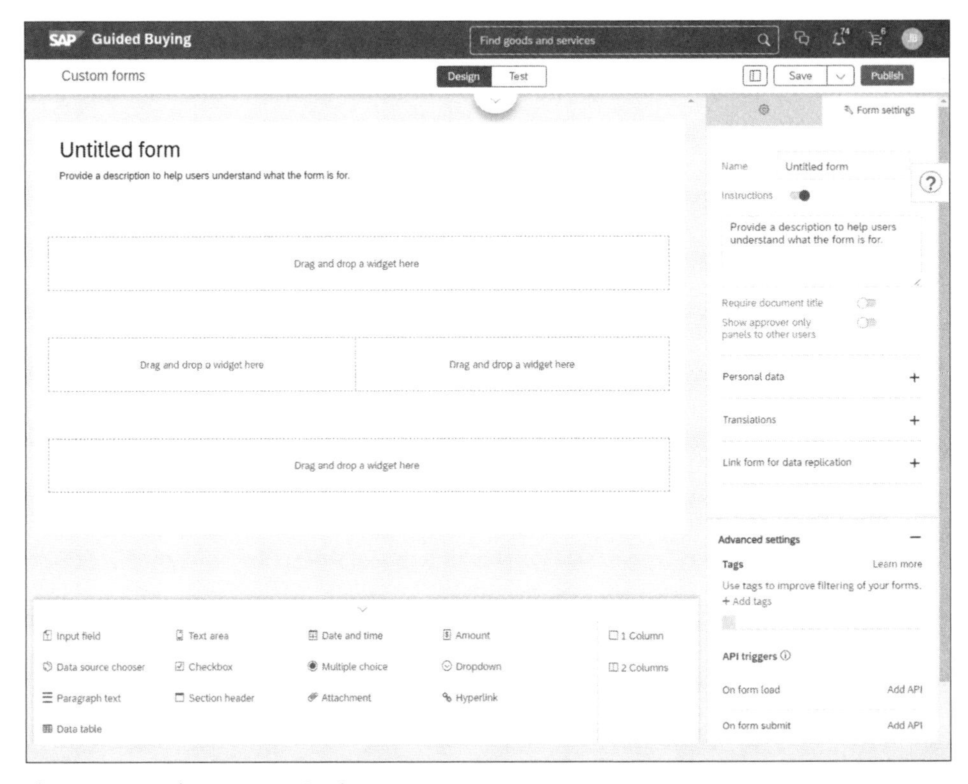

Figure 7.11 Landscape Form Designer

You can create a landing page URL to be used by your organization's internal portal. The URLs can be pointing to the main page or any other category-specific landing page in guided buying. This is useful if your organization has internal portals for different categories.

> **Note**
>
> To ensure usability, landing pages can't exceed four hierarchy levels deep. For optimum usability, SAP Ariba recommends showing 9 to 12 tiles on your landing pages, including the guided buying home page. Additionally, each landing page can be configured to be displayed to all guided buying users or only to a single purchasing unit.
>
> Be sure to load images for your landing page tiles and create any forms before configuring your landing pages.

7.2.6 Home Page

The home page is arranged in a three-column format, displaying the images displayed in the order determined by the JSON file (first in, first displayed). The sizes of the tiles can't be customized.

The home page footer contains three sections—**Policy**, **Guidance**, and **Contacts**—that include the corresponding relevant information, as shown in Figure 7.12.

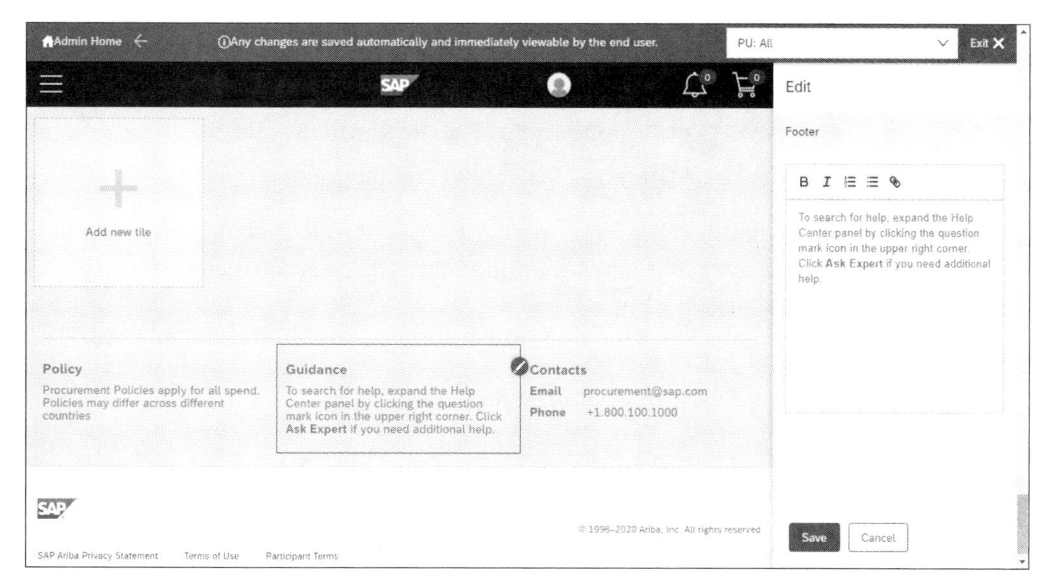

Figure 7.12 Policy, Guidance, and Contacts

7.2.7 Users and Groups

Users are assigned to the guided buying capability in SAP Ariba to execute different capabilities and functionalities. By default, the list of users assigned to the capability will be creating, approving, or managing purchase requisitions. All these users must be assigned to a purchasing user group to access the guided buying capability. These users will be able to create ad hoc noncatalog items, unless a user is assigned to the no ad hoc item group, in which case, the **Request Ad Hoc Item** option will be hidden. Users with administrator privileges must be assigned to the customer administrator or customer catalog manager group.

You can use the `RedirectGroups` or `NoRedirectGroups` parameters to specify which users are automatically redirected to the guided buying capability based on their group assignment, as follows:

- The `Application.GuidedBuy.RedirectGroups` parameter specifies the groups of users who should be redirected to the guided buying capability. Users who belong to at least one redirect group are redirected to the capability when doing any of the following:
  - Clicking an action button from a requisition email notification
  - Accessing the direct URL for the guided buying capability
  - Accessing the direct URL for SAP Ariba Buying solutions

- The `Application.GuidedBuy.NoRedirectGroups` parameter specifies groups of users who shouldn't be redirected to the guided buying capability in SAP Ariba. Users who don't belong to a redirect group are redirected to the capability when doing any of the following:
  - Clicking an action button from a requisition email notification
  - Accessing the direct URL for the guided buying capability
  - Accessing the direct URL for SAP Ariba Buying solutions

The guided buying capability assumes that the user profile has all accounting data loaded, but if you have additional accounting fields, you'll need to review your user and master data loads.

7.2.8 Filtering for Purchasing Units and Job Functions

Guided buying policies use purchasing units as organizational entities to help you group your users by location, by business unit, or by any other logical organization structure useful for you. The use of purchasing units helps separate spend activities by entities, business units, departments, or purchasing organization. You can make the purchasing units specific to the SAP Ariba solutions or map them to your ERP solution.

The purchasing units have a hierarchical organization where the "All" purchasing unit is the parent for purchasing units. This will happen even if you don't specify this in your configuration.

You can use the purchasing units to map a specific landing page, the forms a user can create, the applicable validation policies, and which supplier and touch policy apply to a request for quote.

7.2.9 Categories

Some customers refer to their spend taxonomy, materials groups, commodities, inventory catalogs, or spend categories. In the guided buying capability, you'll use categories to guide your users through the buying process. You can use the category template shown in Table 7.3.

Identifying relevant categories is important to make the navigation and usage of the application logical, simple, and always centered on your users' procurement needs.

Category Name
Subcategories
<List of categories applicable to this category>

Table 7.3 Guided Buying Category Template

Category Name
End-User Groups/Category Experts/Category Managers
<List of persona/roles applicable to the category by location and subcategory (if applicable)>
<Who are the key groups of people that buy?>
Suppliers
<List of suppliers by status (preferred/nonpreferred/minority owned), location, and subcategory>
Importance
<How critical is this category? What is the volume and spend (annual and quarterly numbers)?>
Policies
<List of applicable business rules/policies required for the category (can be linked to the policy template)>
<Provide further details for each policy in the policy template>
Process Flows
<List of to-be process flows specific for the applicable categories>
<Do you want to self-service this category?>
<Do you want to get three bids and a buy?>
Templates
<Attach templates such as statements of work or master agreements if applicable>

Table 7.3 Guided Buying Category Template (Cont.)

7.2.10 Catalogs

All catalogs must be loaded into SAP Ariba Buying and Invoicing as internal catalogs, Ariba Network catalogs, and punchout catalogs. The following types of catalogs are supported by the guided buying capability:

- Catalog interchange format (CIF)
- Commerce eXtensible Markup Language (cXML)
- BMEcat (an extension of XML)

You can have partial items catalogs to allow users to complete additional fields while requesting a service. This parametric type of partial service item is unique for guided buying and has to be manually defined in the CIF file under the **Ariba Item Type** value.

7.2.11 Suppliers, Preferred Suppliers, and Supplier Management

The use of preferred suppliers is optional. However, all supplier data must be loaded into SAP Ariba Buying and Invoicing using the ERP integrations so that their catalogs can be used and ad hoc purchase requisitions created. Suppliers will also be displayed on the ad hoc page to direct your spending to preferred suppliers, rather than nonpreferred suppliers. You'll also need to map your suppliers to your purchasing organizations.

Preferred supplier information must be separately loaded using the **Supplier Management Admin** section of your SAP Ariba Sourcing solution, identifying suppliers per category (commodity codes). Suppliers should be broken down by location (user ship-to country) and preference (i.e., numbers between 1 and 5 or text labels, e.g., high, medium, and low). You'll need to be in the supplier management ERP admin, supplier management operations admin, supplier risk manager, or customer administrator groups to import supplier data.

The main configuration activities are as follows:

- Identifying suppliers
- Importing supplier master data (to SAP Ariba Buying and Invoicing)
- Defining preferred supplier levels
- Defining and configuring preferred suppliers

Not all suppliers should appear in the guided buying capability, and you'll need to load your qualified suppliers using the *SupplierQualification.csv* file, which will include categories and locations. The file will be loaded using the supplier management administrator section of your SAP Ariba Sourcing solution.

> **Note**
>
> The use of preferred suppliers is recommended to support ad hoc requisitioning, for the use of forms for complex procurement, or for browsing items on a landing page.

7.2.12 Forms

Guided buying allows you to use two different type of forms:

- **Request forms**
  These forms are used to make a request for quotation (RFQ), for example, for large purchases, high risk purchases, or purchases that can result in a new contract.
- **Customized ad hoc request forms**
  These forms are used to capture additional information on an ad hoc request, for example, for small to medium purchase amounts.

The most common form is the RFQ within the guided buying capability, which is integrated with SAP Ariba Sourcing. In this form, the guided buying capability lets you define amount and category thresholds to determine whether the procurement department will be involved in a purchase, as shown in Figure 7.13.

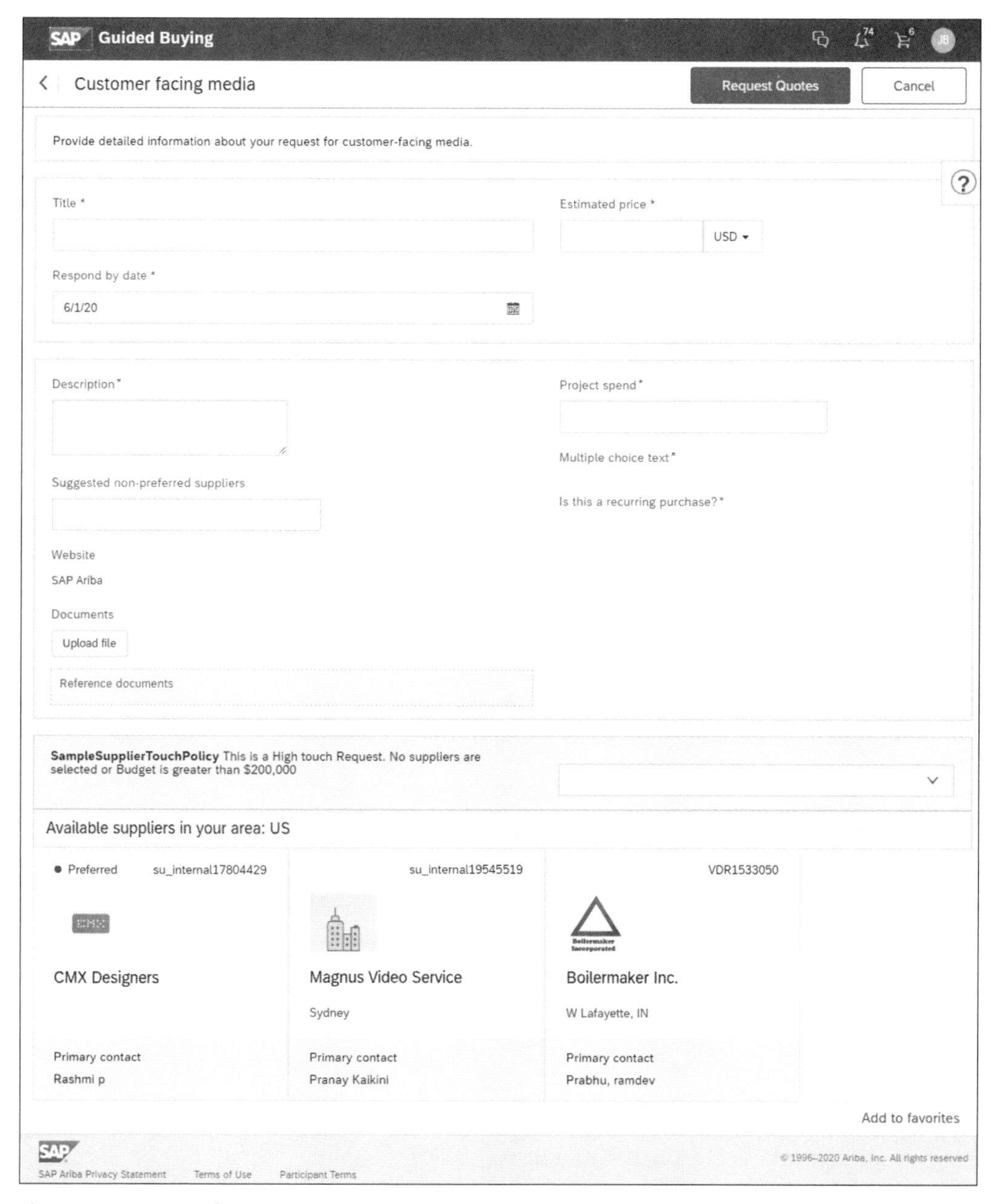

Figure 7.13 Request for Quote Form

The **Request Quotes** option, shown in Figure 7.14, will enable you to directly collaborate with a supplier to provide all the information needed to request a quote. If your company requires a minimum of three bids, based on the threshold, the system will guide you through the process, highlighting available preferred suppliers to help you decide on a supplier. When the response comes in, the multiple responses will be available side by side for you to compare them.

From the supplier standpoint, the supplier will go to the Ariba Network and click on **Intent to Participate**. Then, the supplier can submit a proposal by following the standard supplier sourcing UI with information populated by default from the customer's request and from the fields in the guided buying capability. In addition to the data located in the capability, additional questions will be asked based on the sourcing category configuration done in the sourcing application. After responses are submitted, the buyer will be able to view the quote.

In the guided buying capability, administrators can build forms using the form builder without IT assistance or the help of SAP Ariba. You'll have to belong to the customer administrator group or customer catalog manager group to use the forms builder.

The forms builder provides a fast and flexible way to extend procurement processes by using a drag-and-drop interface and configuring conditions for widgets to design forms without the need to write code.

With the forms builder, shown in Figure 7.14, you can configure a form through the following tasks:

- Adding existing fields to the form
- Determining field conditions
- Searching and selecting existing master data

The forms builder also includes widgets (fields and other controls) that you can drag and drop inside form layouts with one or two columns. You can select a widget on the form and modify its properties through the **Property** tab. The **Templates** tab is where you'll find form templates, draft forms, and published forms.

Forms are optional but are designed to guide users through complex purchases. Based on the list of your targeted categories, you'll need to determine which categories are complex enough to require a specialized form to capture a purchase requisition and then define the requirements for those forms.

To link a form to a landing page, you need to navigate to the **Admin** section, download the forms, and locate the form name. Then you have to copy the associated form ID and use it as the target URL in the landing page file.

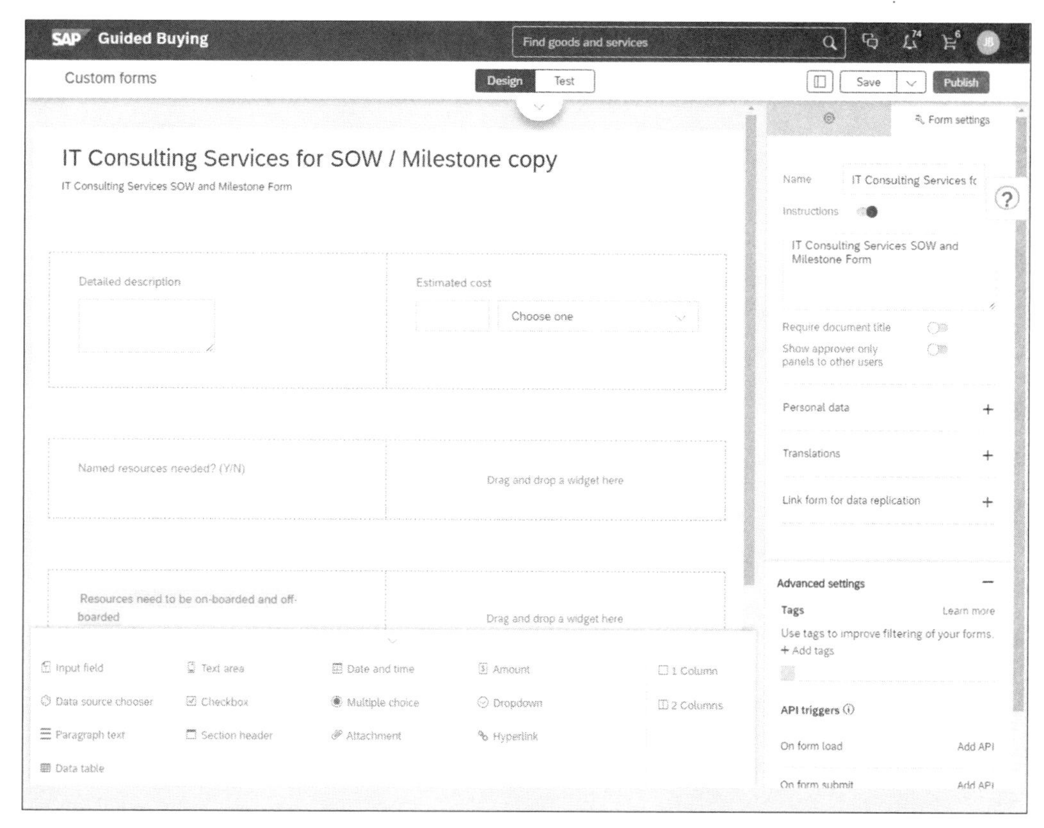

Figure 7.14 Custom Forms

7.2.13 Purchasing Policies

Purchasing policies ensure users will follow your internal procedures when purchasing. These built-in policies decrease the need for users to read procurement manuals. Policies in the guided buying capability can ensure consistent compliance for all purchases and help users understand the right processes to follow for any purchase. The validation policies also apply to forms with the ability to check on master data, detailed line item data, and supplier data from SAP Ariba Supplier lifecycle, such as the supplier qualification status and the preferred level. The policies get triggered after the user clicks on **Add to Cart** in a request form, **Check out** in the shopping cart, or **Send Request** in the checkout page.

To maintain the application policies, you must belong to the customer administrator group or customer catalog manager group. To see and modify your policies, select **Admin** from the options menu. Select the first section, **Manage Policy**. You'll see all the validation policies loaded by default in the system. As an administrator, you can also add or delete validation policies, as shown in Figure 7.15.

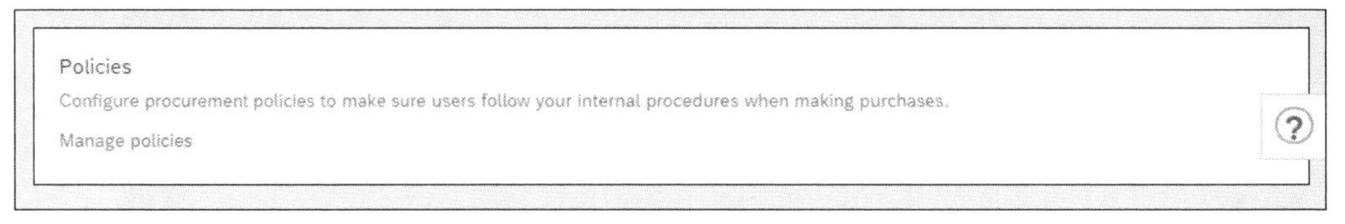

Figure 7.15 Manage Policies

On the **Policy Management** page, click **Export Policy** in the **Validation Policy** section to download and save the Excel template. Use the template to create your policy, and upload your revised policy file, as shown in Figure 7.16.

Validation Policy [Download Sample Policy] [Import Policy] [Export Policy]

Name	Description			
Gift Policy	Policy to determine whether the item could be sent as gift or not	Test	View Translation	
ContractPolicy	Test Contract Hold up	Test	View Translation	
Form Commodity Code Data Source Chooser Policy - Info	Form Commodity Code Data Source Chooser Policy Info Description	Test	View Translation	
IT Asset Time Frame Policy	policy to determine if a user is allowed to buy a particular item	Test	View Translation	
Form User Data Source Chooser Policy - Info	Form User Data Source Chooser Policy Info Description	Test	View Translation	
Product quantity limitation	Policy to prevent purchase of multiple quantity on a single line item	Test	View Translation	

Figure 7.16 Validation Policy

To view a policy, you can download the policy by using the **Export Policy** option and then open the file with your preferred editor. The policy can be opened and modified with a simple Excel file editor. As an administrator, you can also use the screen to drag and drop fields on the screen to define user policies.

In the Excel file, as shown in Figure 7.17, a policy is defined through the **Policy Name**, **Policy Description**, and **Policy Application Type** fields. Policies can be set to apply only to certain procurement units or certain regions using the **Policy Name**, **Lookup Key**, and **Lookup Key Value** fields using operators to evaluate expressions for form field master data, supplier data, or line-item data.

	A	B	C	D	E	F	G	H
1	Policy Name	Lookup Key	Lookup Key Value					
2	IT Asset Time Frame Policy	ProcurementUnit	ALL					
3								
4								
5								

Description | Key Definition | Expression | Message

Figure 7.17 Excel Export Policy

Policy administration is simple: You can click on the three dots icon for any field on the requisition page and affect how a policy is implemented. You can test your policy on admin pages before they are published to end users.

A good example might be a "gift" policy: When users create a gift request with a value greater than 50, a warning message could be issued, for example, "The cost exceeds the gift limit of $50. Please enter a justification." Users can enter a justification and submit the request as is, or they can choose a gift for a smaller amount.

The justification is an optional field. Figure 7.18 shows an example of how policy exceptions will appear on the requisition screen.

Figure 7.18 Requisition Policy Validation

Note

When you set up a policy, drive every behavior to help the user.

A policy can call an external system using application programming infterfaces (APIs). However, we recommend keeping policy data calls to within the system to avoid poor performance for the end user.

With warnings, you can give users the opportunity to provide a justification.

7.2.14 Help Community

The community is a private self-help community management feature targeting content to users so they can collaborate more efficiently in the purchasing process. You'll need to identify the users who will be your moderators, content managers, and experts. Ideally, category-specific documents, videos, and any other resources are added to the community. Approval flows for document management, users, and permissions will need to be configured.

To use private communities, you'll make a request to SAP Ariba to create private user communities to support the guided buying capability and to work with private communities as part of deployment. After the private community has been created, you can generate users and assign administrators the `Community Content Manager` and/or `Community Moderator` permissions. If your company doesn't want to use the moderation (Q&A) functionality, you can request that this functionality be disabled.

Although the goal of making the procurement process easy and more efficient for casual users has been paramount, users may still need assistance filling out the necessary information or need help deciding what vendor to use.

When you log in as a casual user who needs to finish the procurement process or just wants to buy certain goods or services, you'll see a question mark on the right side where you can see all the content that is available, as shown in Figure 7.19.

Through user questions and answers, you can enjoy a straightforward experience in finding the information you need. For example, you can see whether a preferred supplier exists and view questions already asked and answered. You'll also see other questions already asked by other users, and you can reach out to these users as well, as shown in Figure 7.20.

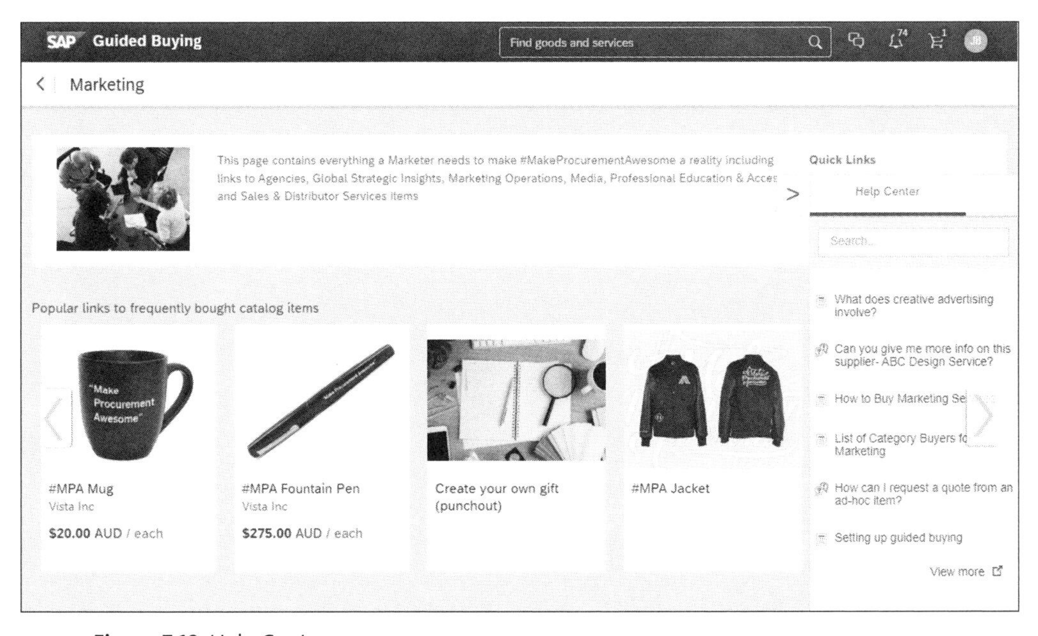

Figure 7.19 Help Center

If you don't immediately see for what you're looking for, you can search the help content. If you still can't find an answer, you can post a question, which will be routed to experts for a response.

Figure 7.20 Help Center Answer

> **Note**
> The information displayed on the help is directly related to the page you're currently on. If you navigate to a different page, the help information will change as well.

The information will be relevant to the page a user is working on. This content can be configured to change by country and/or by category. Using the help, gaps in knowledge are bridged.

Log in as a procurement expert to set up the help content and make it available to your users on the right side of the screen, and then provide answers to the questions users have asked, as shown in Figure 7.21.

> **Note**
> You can set up priority levels to display certain content higher up in the right screen list.

You can still log in to the guided buying capability and go to the help desk, but for more information, you have the option to click on **View More**, which will activate the **Moderation** and **Add content** options.

Adding Content

When you first start a community, the community will be completely empty because at that point no content has been uploaded. Click on **Add content** to start creating the community (see Figure 7.21).

SAP Ariba won't provide sample content because this area is specific to the buyer and to the community, and you own all the content here. Normally, one person or a team of people will be responsible for maintaining support questions as part of the procurement help desk.

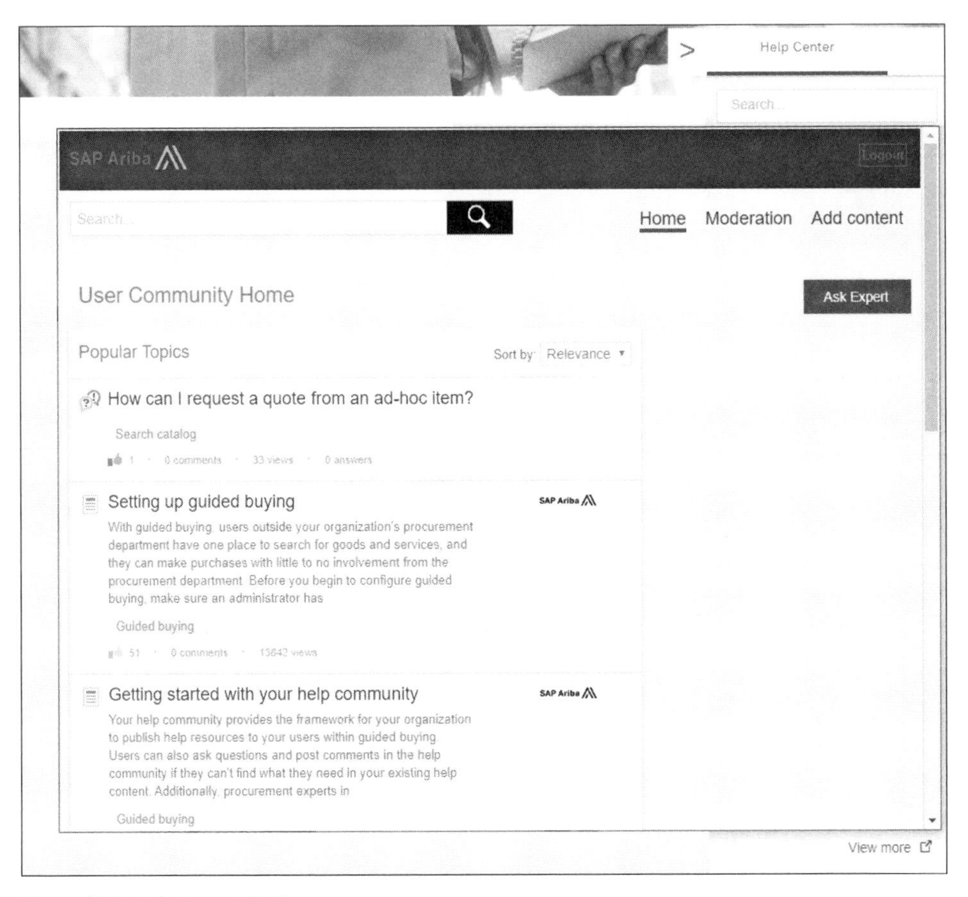

Figure 7.21 Ask Expert Option

Configuration

You'll have to designate a user with the moderator role to manage content and ensure its quality. The intent is to have the content in the guided buying capability enhance and produce a more rapid learning experience while using the application. The content should not be about the product, but your policies and your buying procedures. The content isn't about replicating a user guide but about finding procedural answers. The main goal is to push most communications inside the community to create a two-way stream to replace the usual static application help.

A procurement expert can add content by using the **Create Article** where they can add a **Title** and a **Body**, upload multiple attachments, and then **Submit** the content, thus making the content available to users. This article will be available to users in about half an hour.

This expert can also use **Upload video** to upload EMP4 video files, including specifying a **Title**, to be available to your users.

Making only certain content appear on certain pages isn't an implementation configuration task, but instead a knowledge management task. An expert should go back and choose **Help • View More**, click on any article, and select **Edit Tags** to determine under which conditions the article should appear. You can also link this article in multiple pages and make the article location specific.

Moderation

Moderation looks at any content posted by users on the private help community and highlights pending changes/additions, as shown in Figure 7.22. As in any other social media experience, the ability to answer users' questions is an important part of deployment, and functionality is in place to assist you in monitoring the moderation and responses. When a user asks a question, the question can be routed to the customer's help desk through configuration. The appropriate moderator will see the question under the **Moderation** tab, but the application can be also configured to email the moderator directly.

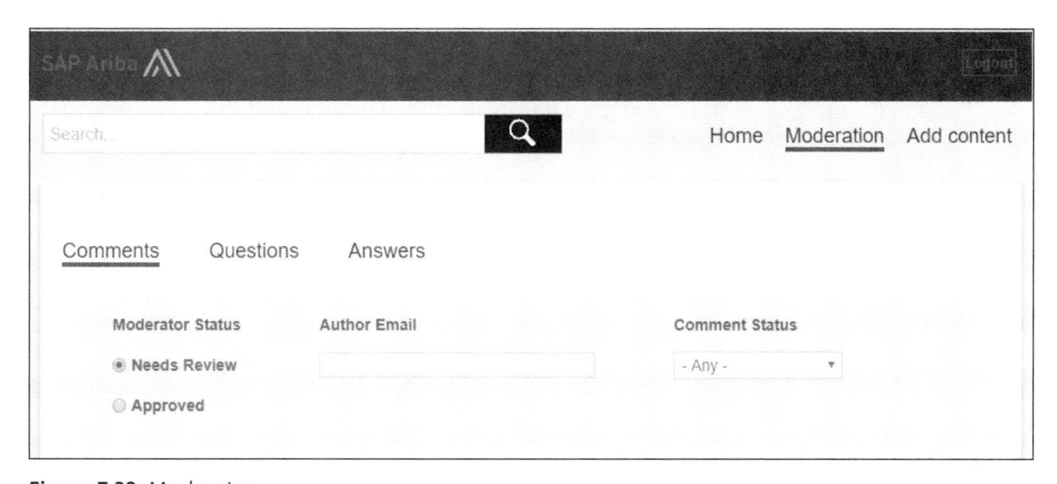

Figure 7.22 Moderator

> **Note**
> Access to the **Moderation** tab is based on configured roles.

7.2.15 Tactical Sourcing

To configure tactical sourcing, you'll need to determine those categories that will require a sourcing specialist. With your use cases and requirements, you can easily identify the type of sourcing event processes that can result on a PO, something that can't be accomplished by a regular purchase requisition. After this analysis, the SAP

Ariba team can update your sourcing templates and configurations to support self-service RFQs.

Enabling tactical sourcing must be performed by SAP Ariba. You'll need permissions to use SAP Ariba Buying (for pay-to-order processes) or SAP Ariba Buying and Invoicing (for procure-to-pay processes). In addition, you'll need to activate the sourcing pro, collaborative sourcing, or operational sourcing capabilities. Suite integration is required and is done by configuring the guided buying endpoint, where the endpoint needs to match your Ariba Network buyer account. This part of the configuration is done by the SAP Ariba Support organization.

If your company doesn't have SAP Ariba Sourcing, you can use the guided buying quote template only, which includes a request for proposal (RFP) tactical template. But if your company already has SAP Ariba Sourcing, in addition to the guided buying specific template, you can use different types of sourcing project templates and sourcing request templates (templates include documents, tasks, teams, and conditions).

If you already have SAP Ariba Sourcing, SAP Ariba Contracts, or SAP Ariba Supplier Lifecycle and Performance, these applications will need to be integrated into your SAP Ariba Buying or SAP Ariba Buying and Invoicing instances. Then, after a feature is enabled, you can use sourcing templates from guided buying, and the forms and RFQs will seamlessly flow to and from the capability.

To enable and configure tactical sourcing to work with guided buying, you have to log in into your SAP Ariba Sourcing application and find the sourcing template document that you want to use in guided buying. Go to **Manage Templates**, and load the templates in the **Guided Buying Templates** folder, as shown in Figure 7.23. By default, you should find the **Guided Buying Quote** template.

Templates / Guided Buying Templates — Show Details — Actions ▾

Name	Owner ↑	Status
Guided Buying Sourcing Request Quote ∨	aribasystem	Active
Guided Buying Quote ∨	aribasystem	Active

Figure 7.23 Guided Buying Templates

Open the **Guided Buying Quote** template as shown in Figure 7.24, and look for the **ID** field. The **ID** field is what you'll use to direct the guided buyer user to use the template. The default out-of-the-box ID for the guided buying template is SYS0240. If you want to update your quote template to add an expiration date and reusable field, you must copy the template and made modifications in the **Documents** area of the **Overview** page. Go to **Documents • Sourcing Request Items • Definitions**. Here you can add new selectable terms for **Quote Validity Date** or allow the quote to be reused or not. In this way, your supplier will be able to specify the date on which the quote is no longer valid.

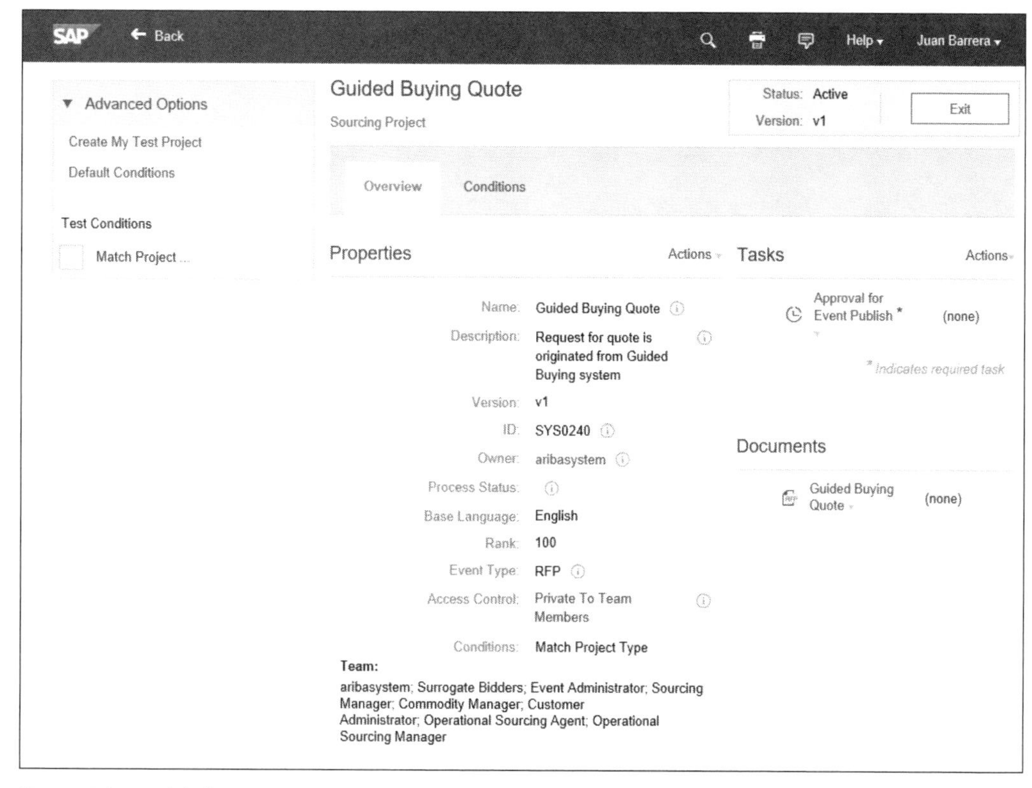

Figure 7.24 Guided Buying Quote

Go back to the **Guided Buying** screen, and access the **Admin** option. Navigate to the **Policies,** and click on the **Manage Policies** section. Navigate to the **Supplier and Touch Policy** area, and export the policies by selecting **Export Policy**, as shown in Figure 7.25.

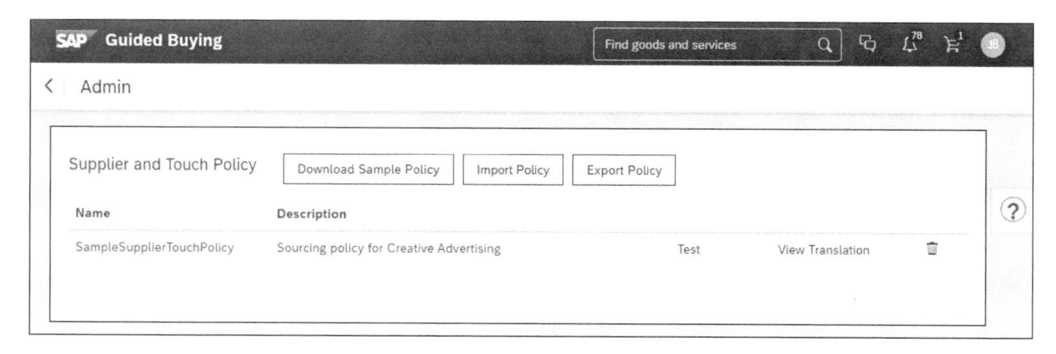

Figure 7.25 Supplier and Touch Policy

The *supplierTouchPolicy_Policies.xlsx* file will be downloaded to your computer; now you can open and update the file by going to the **Template Configuration** worksheet and changing/replacing the IDs of the workspaces in the **Sourcing Template Name** column with new values, as displayed in Figure 7.26. In this case, we're using the default ID

SYS0240. If you want to add and use different sourcing templates or change conditions, you can add additional lines to all the worksheets on this file.

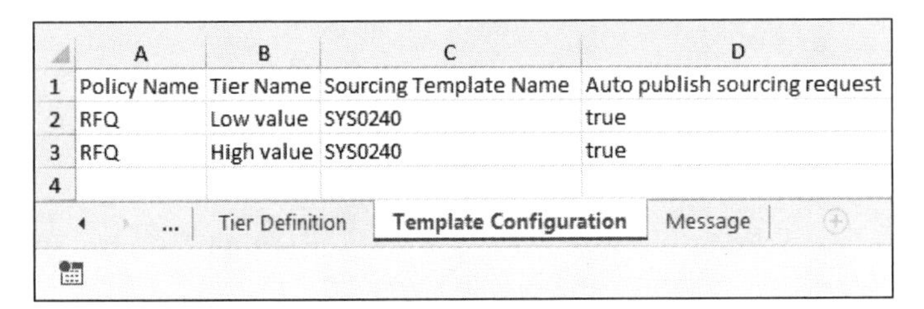

	A	B	C	D
1	Policy Name	Tier Name	Sourcing Template Name	Auto publish sourcing request
2	RFQ	Low value	SYS0240	true
3	RFQ	High value	SYS0240	true
4				

... Tier Definition | Template Configuration | Message

Figure 7.26 Supplier Touch Policy Configuration File

7.2.16 Mobile Solution

SAP Ariba has two downloadable mobile applications: SAP Ariba Mobile and SAP Ariba Procurement. With SAP Ariba Mobile you can track, review, and approve all your procurement operations; however, a new SAP Ariba Procurement application, displayed in Figure 7.27, was recently released with new and additional features. With the SAP Ariba Procurement application, you can access your company's catalog to search, order, and track your request anytime, anywhere.

> **Note**
>
> The recommendation is to use the new SAP Ariba Procurement application.

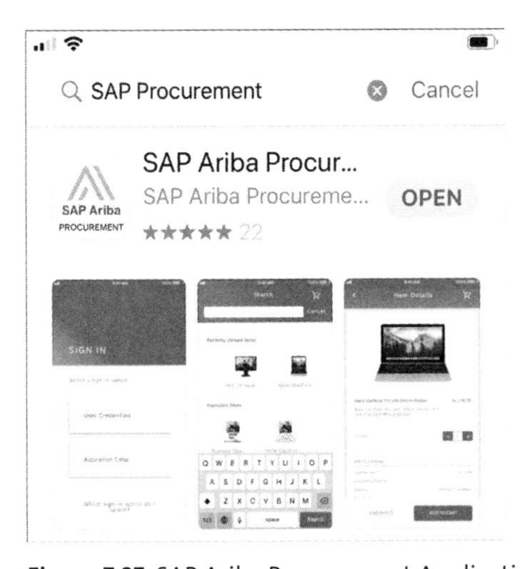

Figure 7.27 SAP Ariba Procurement Application

To use this application, you must request access from your administrator; after your access is granted, you'll receive an email with the information required to configure your application (see Figure 7.28).

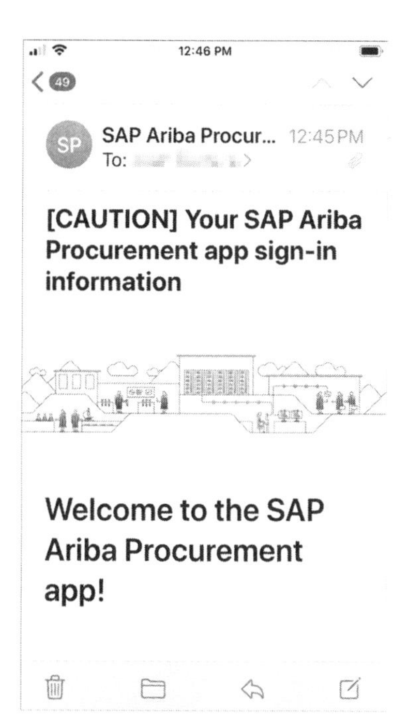

Figure 7.28 SAP Ariba Procurement Application Sign-In Information

After opening the application, you have to enter the information provided in your email and use your guided buying login and password to start buying from your organization's catalogs, track orders, and approve request directly from your mobile device.

7.3 Using the Guided Buying Capability

The guided buying capability in SAP Ariba is a simple, elegant, and smart multidimensional system where decentralization and collaboration are critical aspects to support your organizational procurement needs. It replaces the buying channel guides often stored where end users lack access or in corporate portals with different documents explaining procurement processes. The guided buying capability in SAP Ariba can serve as a one-stop shop for buying goods and services. Even if SAP Ariba doesn't cover all your procurement categories, such as SAP Concur travel and expenses or SAP Fieldglass for contingent labor, SAP Ariba can be a great place to start guiding your users to the right systems.

The goal of the guided buying capability is to serve as an easy-to-use, smart technology that will tell a user what to do when buying goods and services, even if not a professional buyer. Once configured, SAP Ariba can serve as a built-in buying channel logic for different category processes that will guide users in buying the right items with knowledge about preferred suppliers, the right process to follow, and the people needed for approvals.

For example, organizations often have lists of preferred suppliers in Excel files, especially for categories that might not be core to the procurement system. In these scenarios, you may not have preferred suppliers, and when you source, you book a contract. Then, you'll need to make those contracts operational, which can be a time-consuming process that doesn't always have great results. The guided buying capability makes this process transparent by making the list of preferred suppliers clearly available as part of the process. Again, the goal is simple: the technology should tell you what to do, especially if you're not part of the procurement department but still need to buy something.

Additionally, natural collaboration is a powerful concept in the guided buying capability that will enhance the buying experience. Collaboration can be a total process where the right end user who knows exactly what he needs can collaborate directly with the right supplier in the system and within the guided buying capability.

Communities enable users to interact with others with similar requirements and in a similar context to ask questions in order to determine if a product or service worked on their location. The community concept is important because companies often have FAQs and other documents that become obsolete and are crafted outside the system, perhaps codified in emails responding to the same questions over and over. The guided buying capability brings all that communication into one system in which people can continue to ask questions and expect category managers to answer, and which has the intelligent ability to collect all that knowledge to support the category.

The guided buying capability is also a solution for decentralization. With policy management capabilities, SAP Ariba can actually configure processes by company code, by region, by location, and by category. Every company can have procurement experts available at different locations, and policy management is how you can enable the decentralized management of your procurement processes. For example, if you have the same marketing services vendor in Germany and the United Kingdom, the selection processes and the procurement policies might differ country by country.

7.3.1 The Guided Buying User Experience

Guided buying has some simple and intuitive elements in the main screen. The top of the screen has a header section with the notification bell, the shopping cart, and the user icon, as shown in Figure 7.29.

The image in the banner is smaller, allowing more space for the main screen. The navigation controls in the screen are organized in a way that will help you navigate to five different sections: **Shop**, **Your favorites**, **Your requests**, **Your approvals**, and the **Admin** section, as shown in Figure 7.30. Here you can see:

❶ Smaller size banner image, leaving more space for the tiles.

❷ Clickable tabs on the homepage for each functional area.

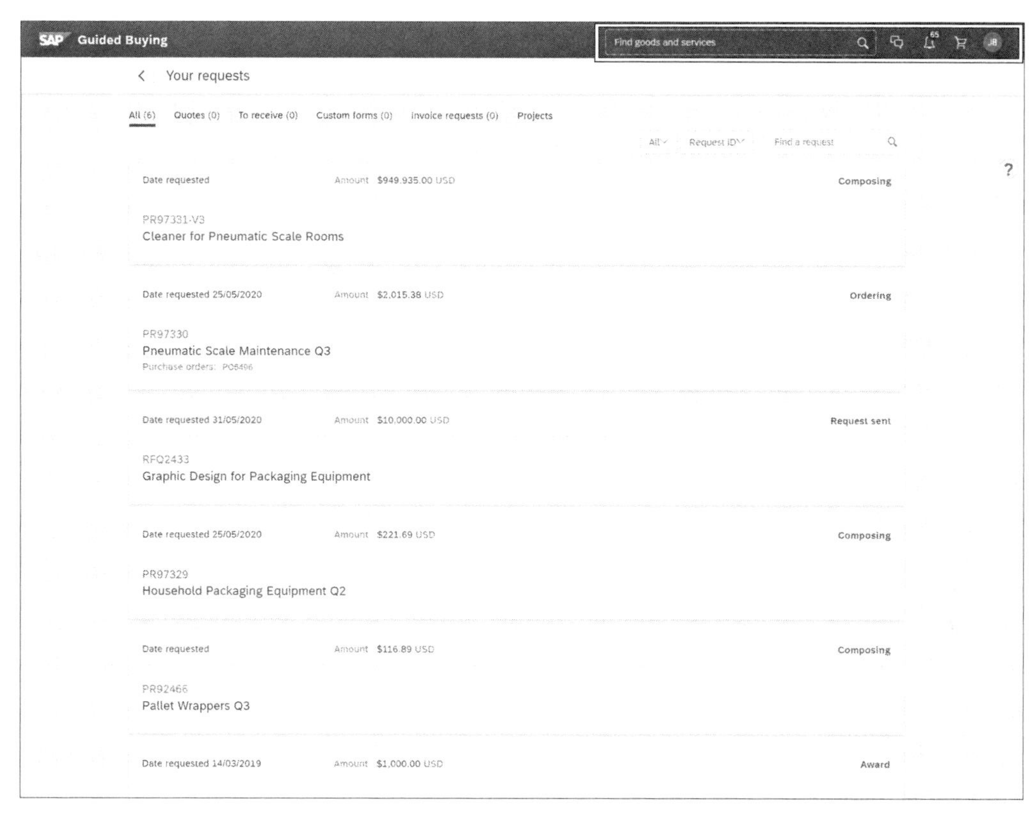

Figure 7.29 Simple and Straightforward Screen

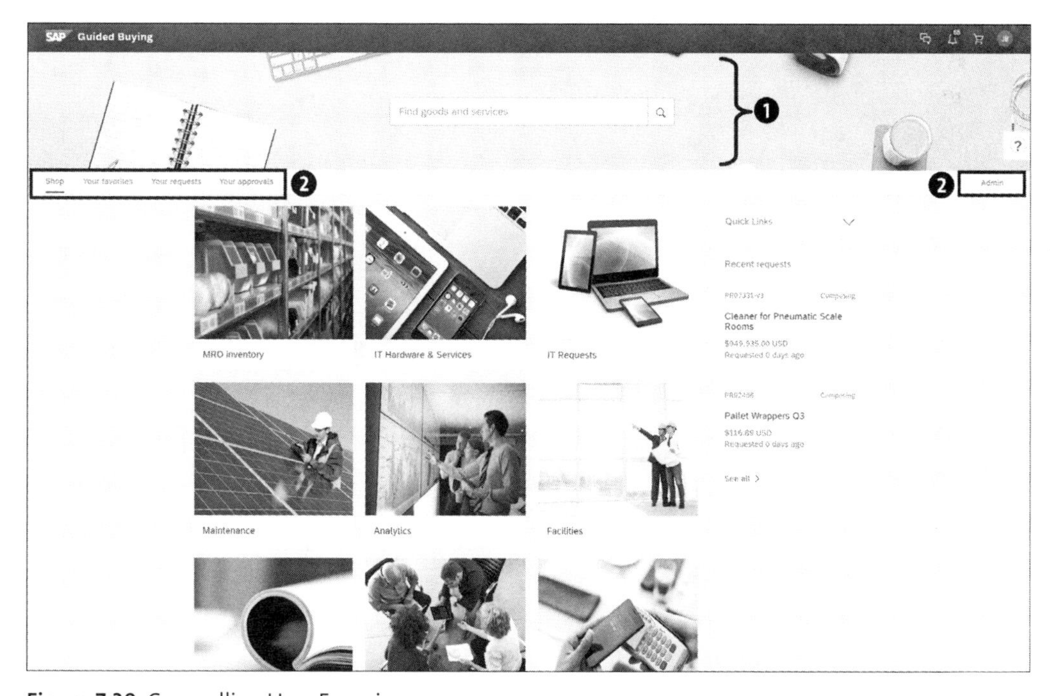

Figure 7.30 Compelling User Experience

The screens are intuitive. The text **Find goods and services** appear in the search box, and as users scroll down, the search box is always available at the top of the screen. The Action buttons appear at the top of the page. If the page or fields contain more than two options, the less frequently used actions will appear under the more options icon (...), as displayed in Figure 7.31 and Figure 7.32. The interface is simple and requires few clicks to navigate. Here you can see:

❶ Action buttons on the banner at the top of the page. In case of more than two main buttons, other buttons will be available from the ellipsis link (...).

❷ No more action buttons at the bottom of the page.

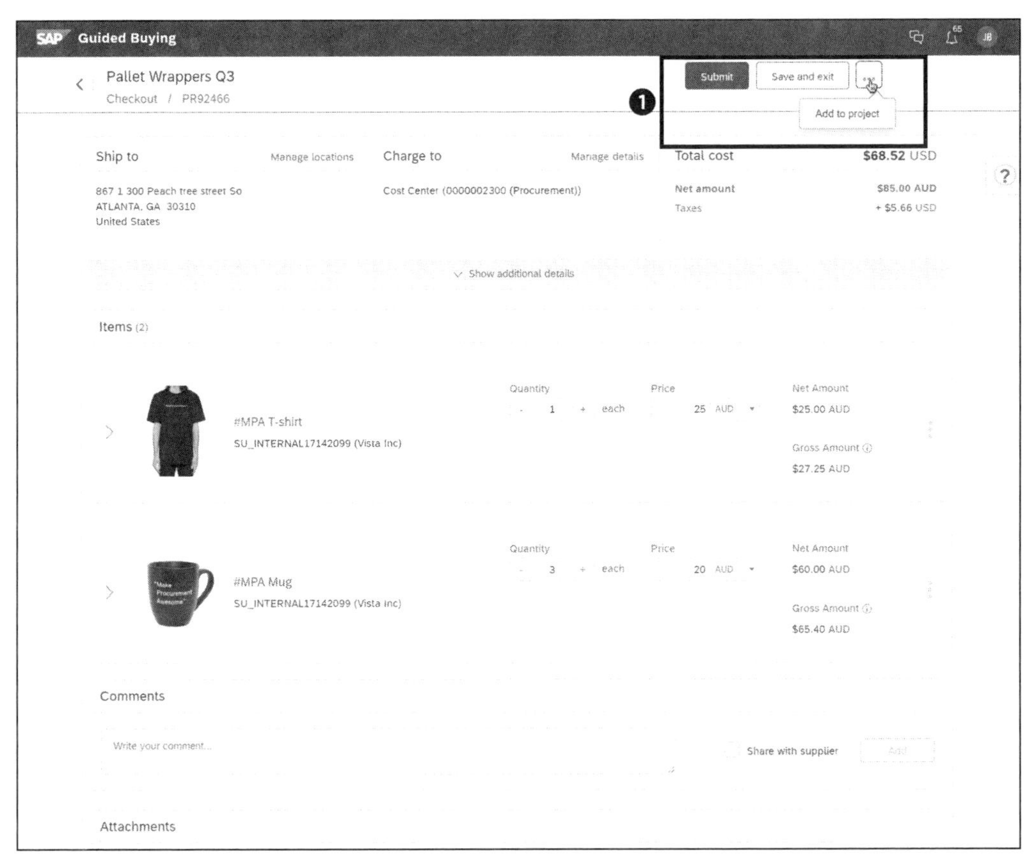

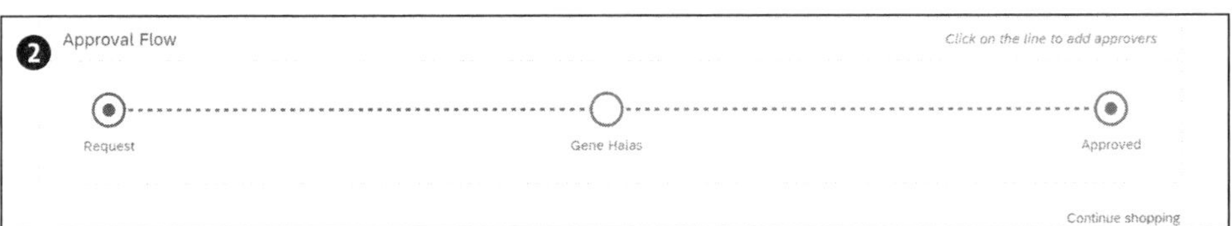

Figure 7.31 Action Buttons Location

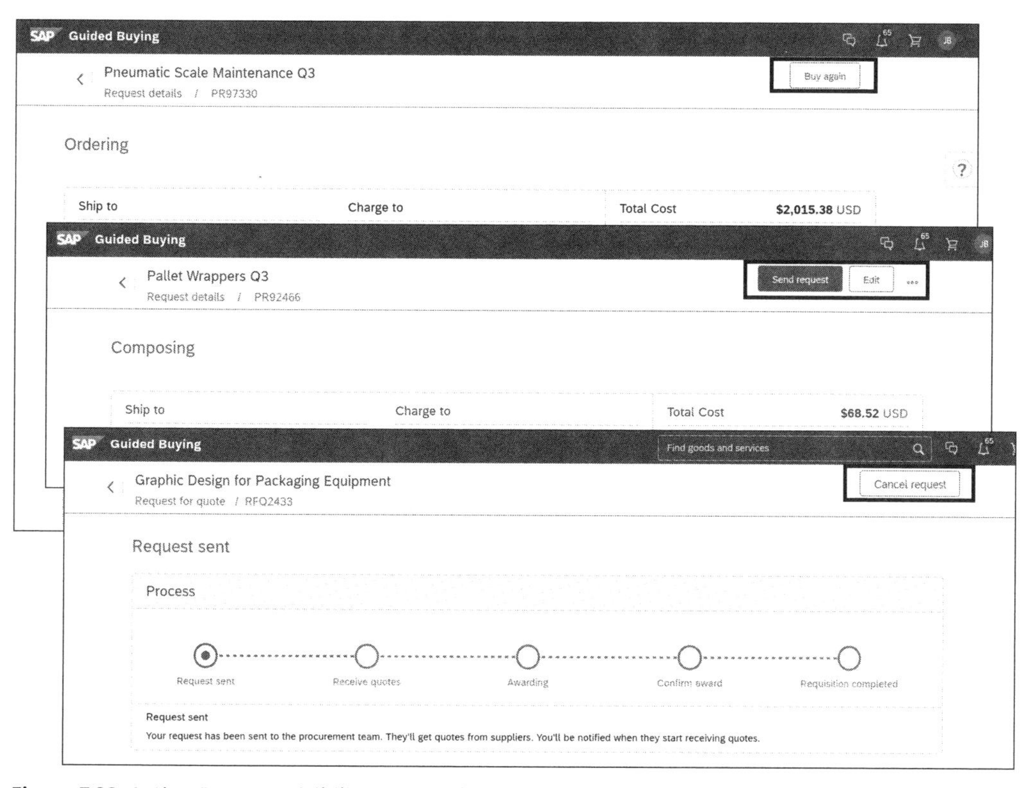

Figure 7.32 Action Buttons Visibility Approach

7.3.2 Creating Requests

Guided buying presents a straightforward landing page, as shown in Figure 7.33, where you can search for the goods or services you need.

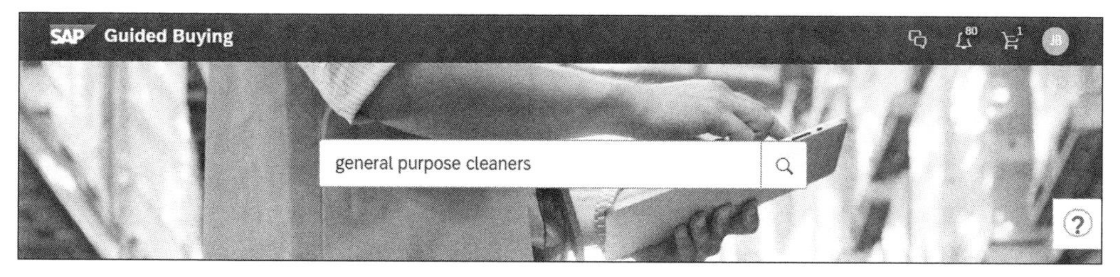

Figure 7.33 Search Screen

After you search and find, or browse the tiles looking for the item you need, you'll use the **Add to cart** button to add it to your shopping cart (see Figure 7.34).

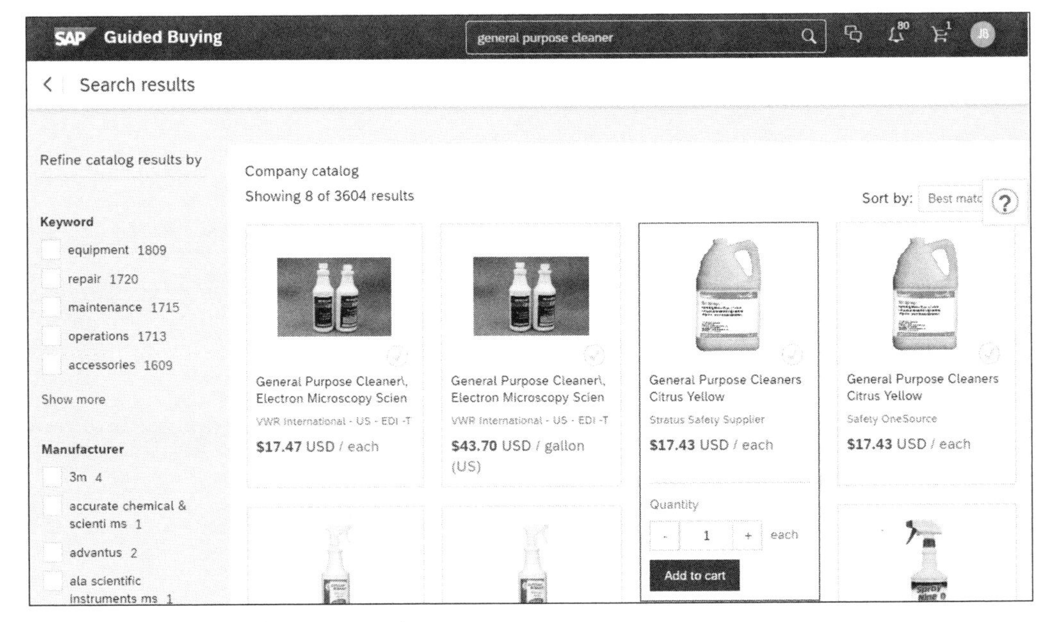

Figure 7.34 Adding Items to the Cart

The items in the shopping cart are always easily accessible and available by just clicking on the shopping cart icon on the top-right corner of the screen, which will bring up the **Check out** button (see Figure 7.35). When you select the **Check out** button, the items in your shopping cart will be shown (see Figure 7.36) where you can review or add additional information, such as the **Ship to** address and the **Deliver by** date, before you submit your request for approval.

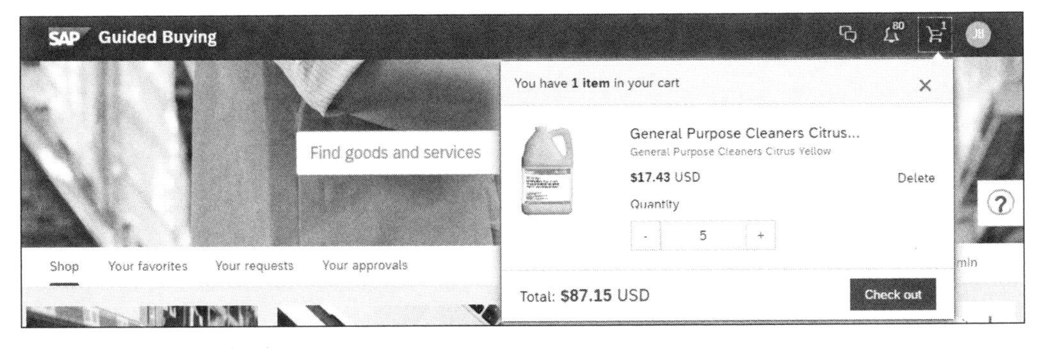

Figure 7.35 Check Out Items

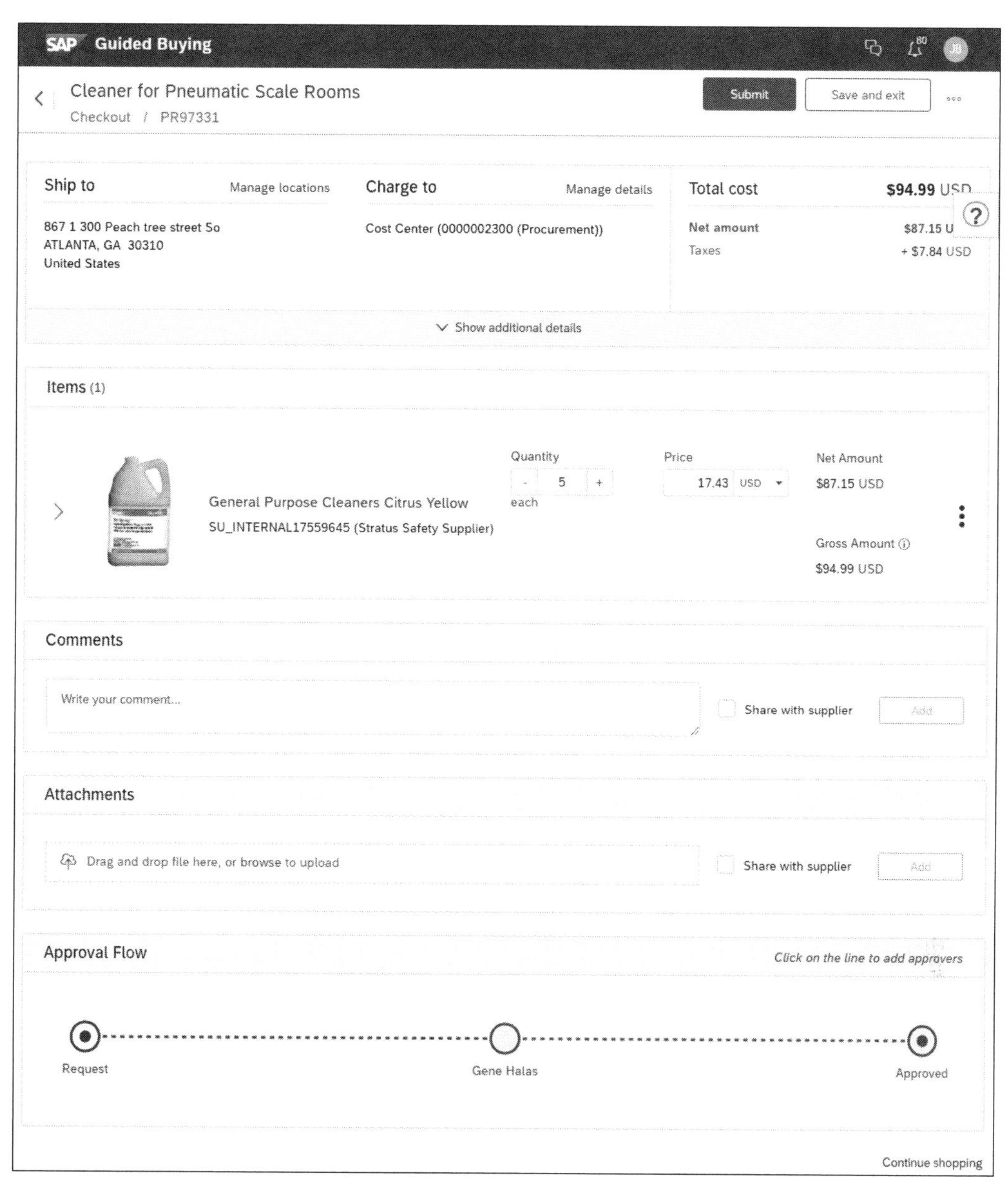

Figure 7.36 Submit Requisition

By selecting **Submit** or **Save and exit**, the purchase requisition will be created. You can access it by choosing the **Your requests** option from the main home page (Figure 7.37).

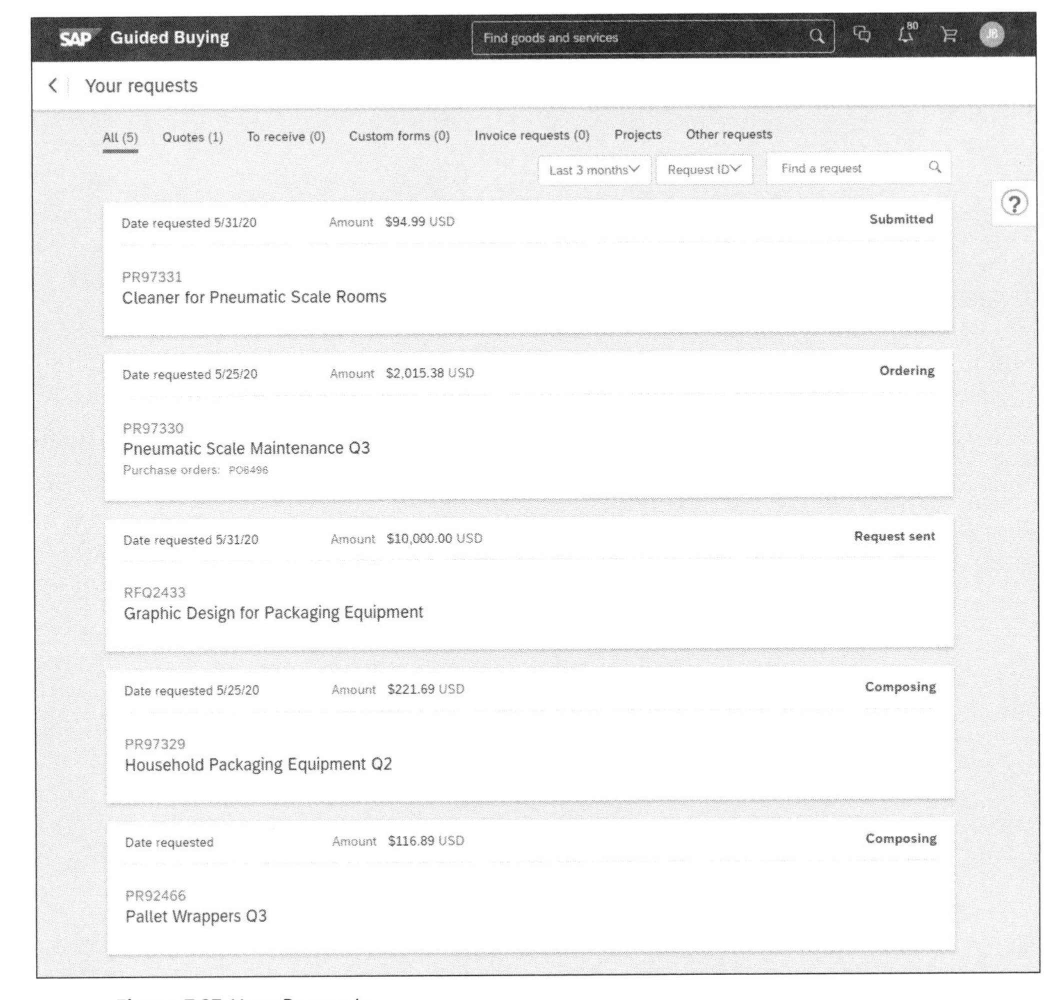

Figure 7.37 Your Requests

When guided buying is connected to SAP S/4HANA, any attachments added to the line item will be transmitted and available in SAP S/4HANA, and the requisition will stay in **Submitting** status until the associated attachments are sent to SAP S/4HANA and then will change to **Submitted**. If the transmission fails, you'll receive an email notification with the requisition ID. In this case, the preparer can delete the failed attachments, add them again, and resubmit the requisition. If you're an active approver, you can add approvers or watchers before or after. You can't change the list of approvers or watchers after the request reaches **Approved** status.

7.3.3 Purchase Orders

When a purchase requisition is fully approved, a PO gets created and sent to the supplier. If enabled in SAP Ariba Buying or SAP Ariba Buying and Invoicing, users can change and cancel POs by opening the PO under the **Your requests** page, as displayed in Figure 7.38. Before you can cancel your PO in guided buying, the PO must be in **Ordered** status.

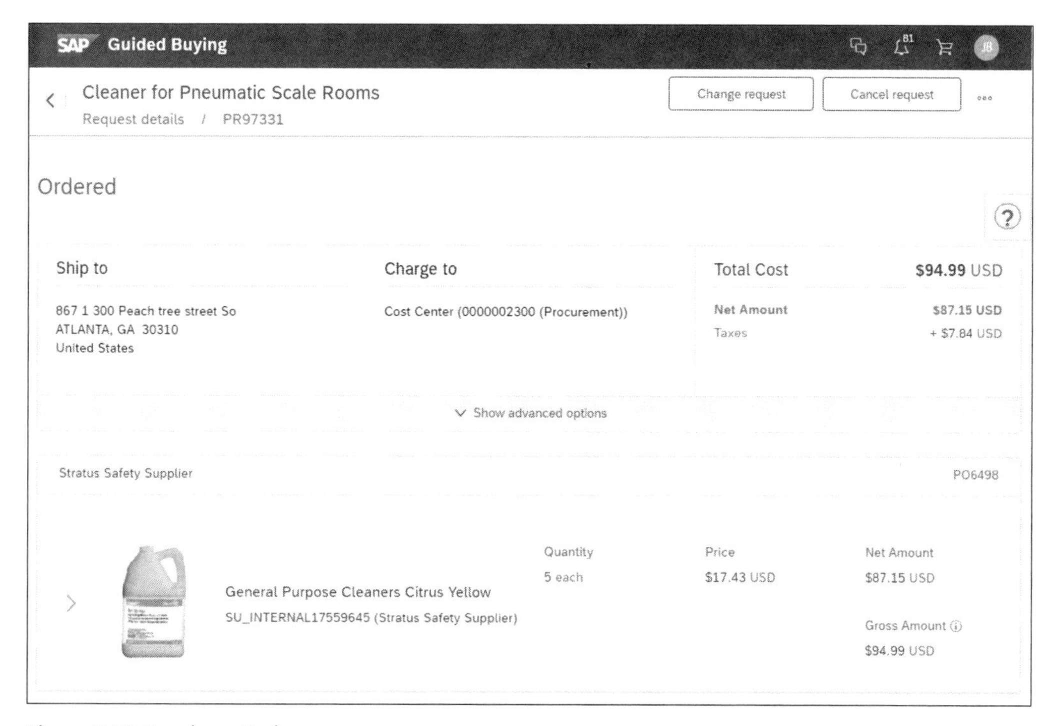

Figure 7.38 Purchase Order

> **Note**
>
> When guided buying is connected to SAP S/4HANA, in addition to the PO, the creator, requester, recipient information, ship-to address, and any attachments added to the line item will be transmitted and available in SAP S/4HANA.

7.3.4 Receiving

By default, all users who work and create purchase requisitions in guided buying can receive their own requisitions in the application itself or in SAP Ariba Buying by finding the purchase requisition and selecting the associated PO, as shown in Figure 7.39, and navigating to the **Receipts** tab shown in Figure 7.40.

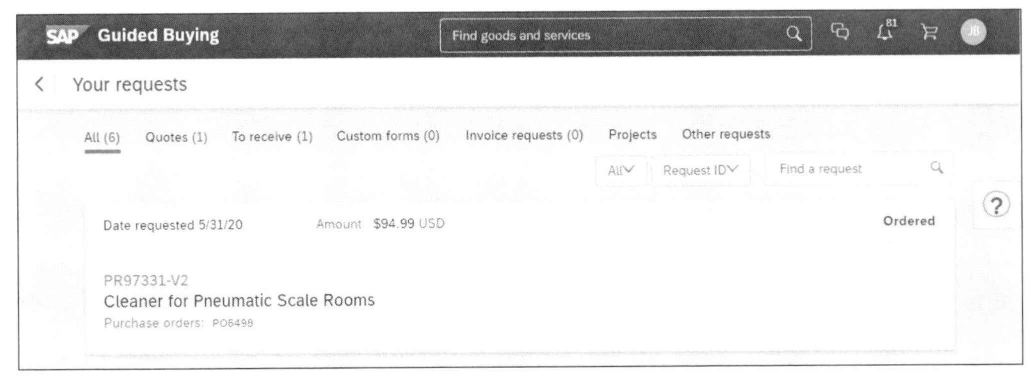

Figure 7.39 Purchase Order on the Your Requests Page

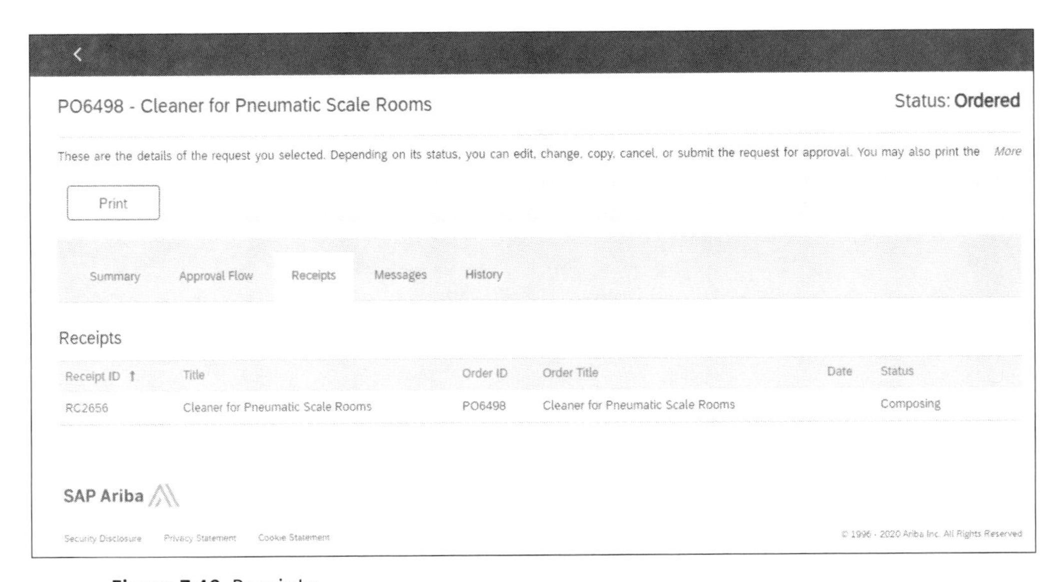

Figure 7.40 Receipts

In the **Receipts** tab, on click on the **Receipt ID**. Click on **Edit** in the screen shown in Figure 7.41, and enter a quantity in the **Accepted** field. To accept the entire quantity, you'll select **Accept All** and then click the **Submit** button, as shown in Figure 7.42.

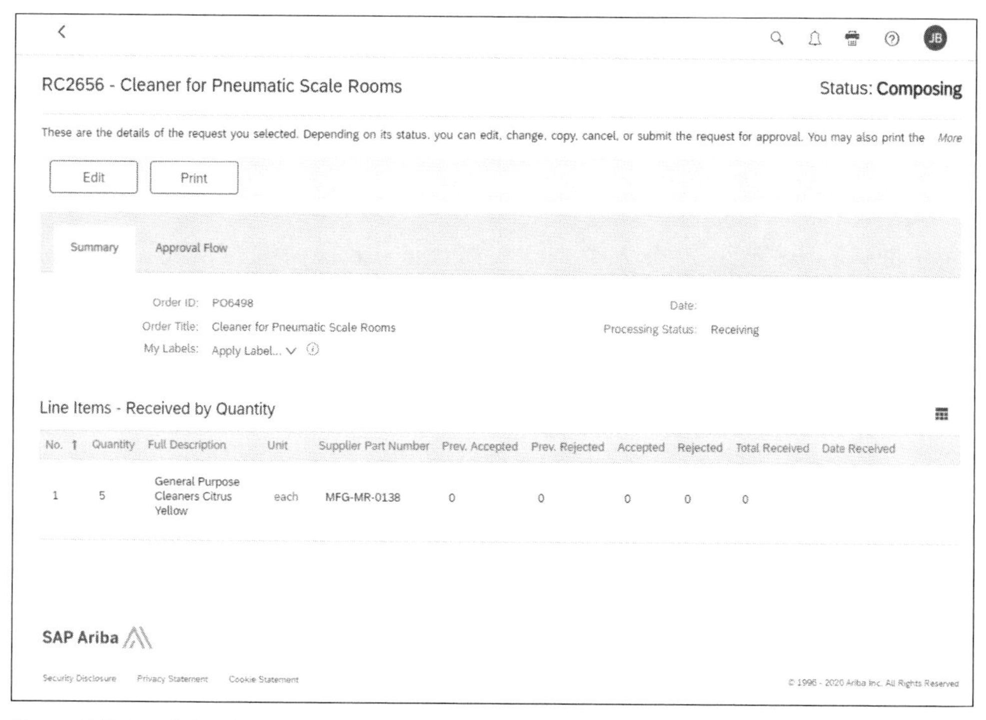

Figure 7.41 Receiving Screen

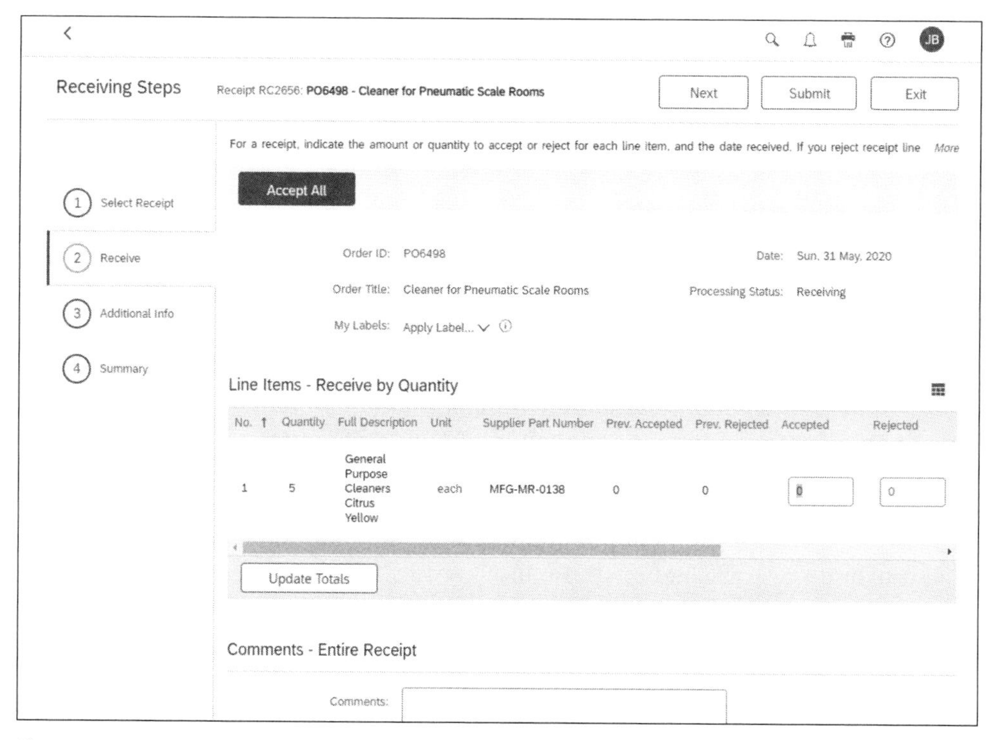

Figure 7.42 Quantity to Accept or Reject

An alternate configuration (advanced receiving) is available in the guided buying capability, in which all users can be pushed to SAP Ariba Buying for receiving, and no receiving would take place in the guided buying capability. In this scenario, when a user clicks on a PO number or clicks the **Submit** button in the capability, the system will navigate to SAP Ariba Buying, where the user must complete and submit the receipt in SAP Ariba Buying.

Finally, you can navigate to the PO and see all the associated documents and options, as shown in Figure 7.43.

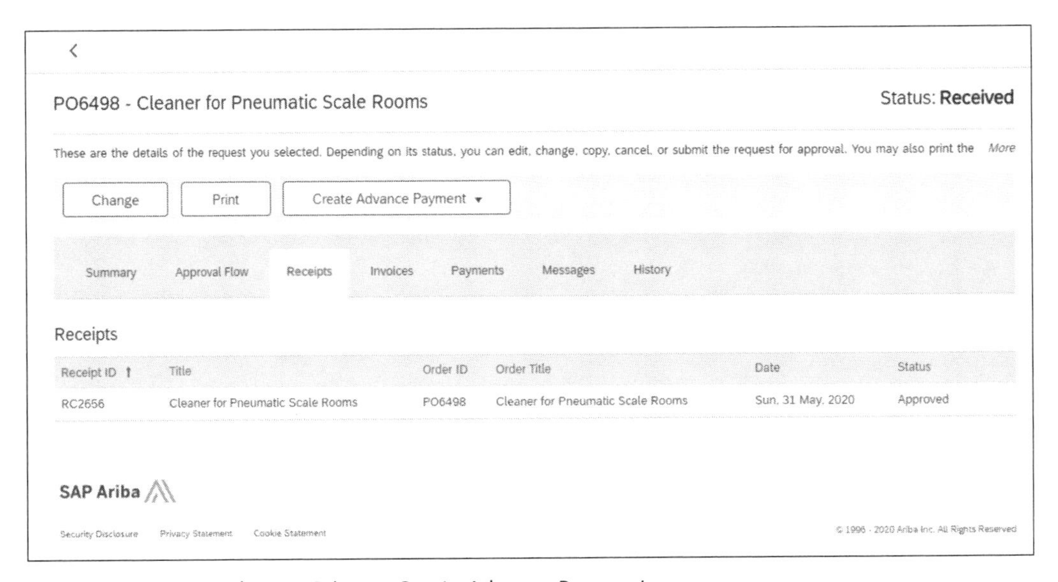

Figure 7.43 Change, Print, or Create Advance Payment

7.3.5 Simple Non-PO Invoices

You can use guided buying to create non-PO invoices without going to SAP Ariba Buying and Invoicing. A non-PO invoice is a request for a one-off payment for a single item or service and their corresponding taxes to a supplier. In this case, a PO doesn't exist. Examples of non-PO invoices are conference fees, professional dues, charitable donations, one-time services, or items purchased outside guided buying.

By default, a non-PO invoice form can be enabled by adding an **Invoice Request** tile and assigning it a commodity code. Otherwise, you can search for "payment request", as shown in Figure 7.44.

The invoice screen will be displayed as in Figure 7.45, where you can enter all the appropriate information, and the guided buying validation policies will be evaluated after you submit the invoice. The same validation rules configured in SAP Ariba Buying and Invoicing will be configured in guided buying. Auto-generation can be enabled for non-PO invoices. The invoice form supports multiple tax types, split accounting, comments, and attachments.

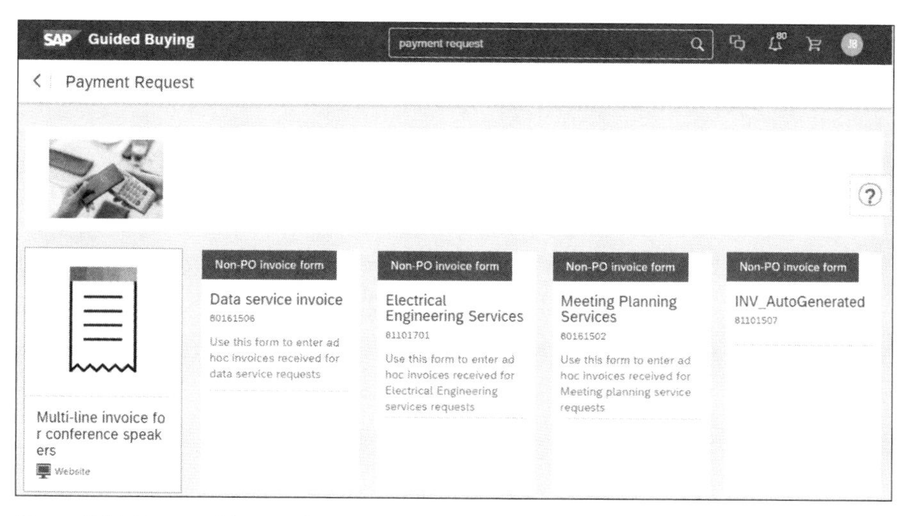

Figure 7.44 Payment Request

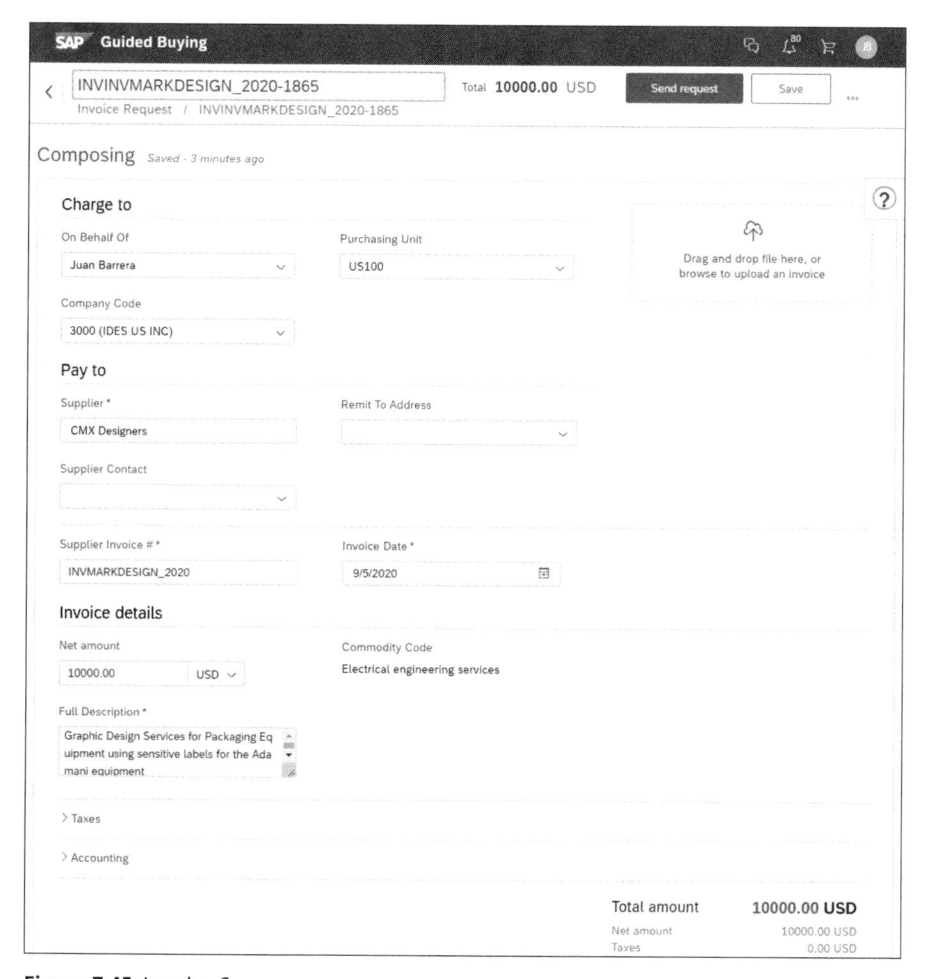

Figure 7.45 Invoice Screen

Finally, when the invoice is submitted, the invoice will offer a side-by-side PDF Invoice. At any point you can navigate back to the **Your requests** page and search your **Invoice requests**, as displayed in Figure 7.46.

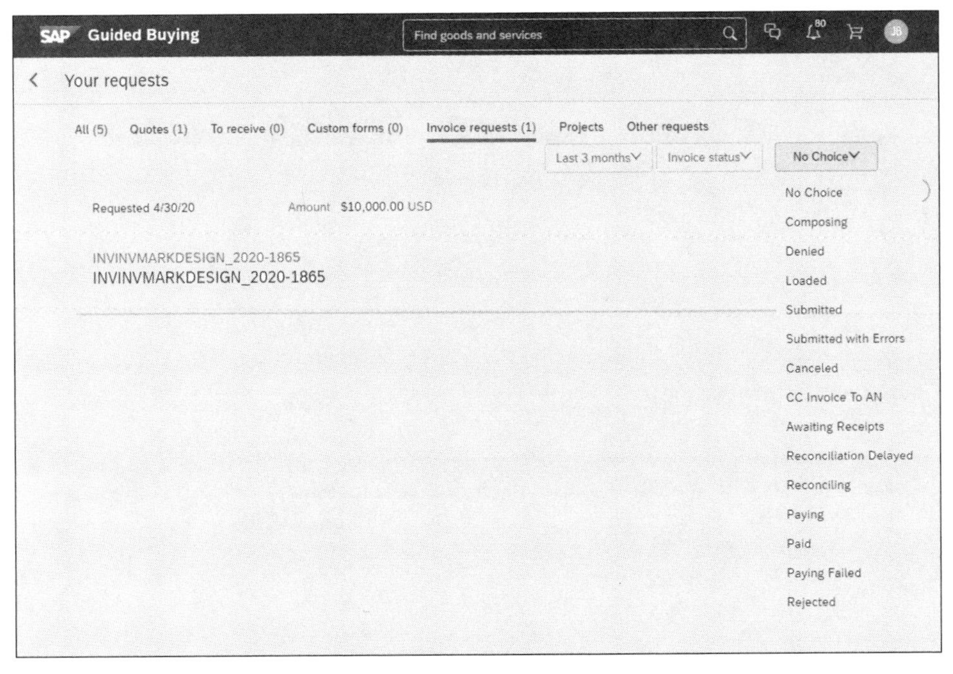

Figure 7.46 Invoice Request Status

7.3.6 Accessing Forms

Using forms is a flexible way to create a procurement request for something that isn't necessarily an item in a catalog. For example, you might need to create a check request, request a consultant, or create an educational reimbursement request, an event expense authorization, or even a maintenance service. As shown in Figure 7.47, you can use the search functionality in guided buying to start your request.

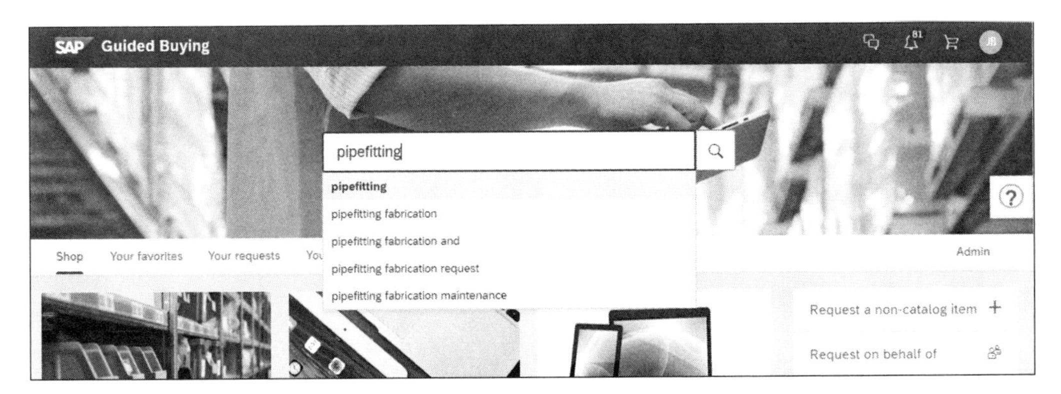

Figure 7.47 Form Request

A form is used to help you make your complex request simpler. The form has been preconfigured by your procurement organization, and the number of preconfigured forms depends on the number of unique cases your procurement organization has identified for you to submit the request you need. The forms are intended to help you submit your request and apply the company's procurement policies in guided buying, and they can be simple or complex, as displayed in Figure 7.48.

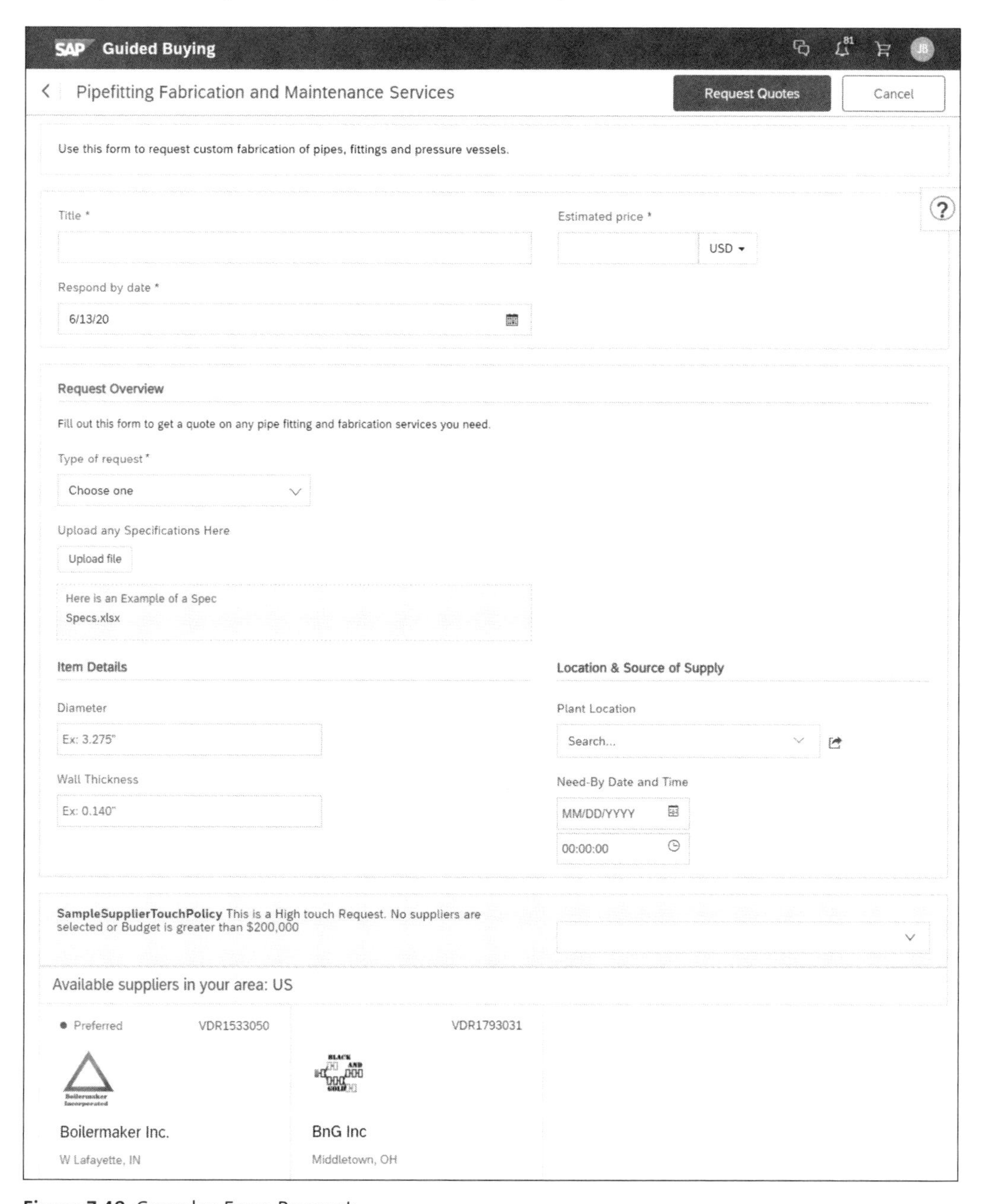

Figure 7.48 Complex Form Request

7.3.7 Initiating a Tactical Sourcing Request

You can request quotes from suppliers using the RFQ process in guided buying; normally, tactical sourcing is used when you need to react and solicit a quick quote from your vendors, and you're not necessarily looking to improve your company's strategic sourcing processes. To use tactical sourcing, using a form makes the process easy and simple. A tactical source usually appears on a tile, as shown in Figure 7.49.

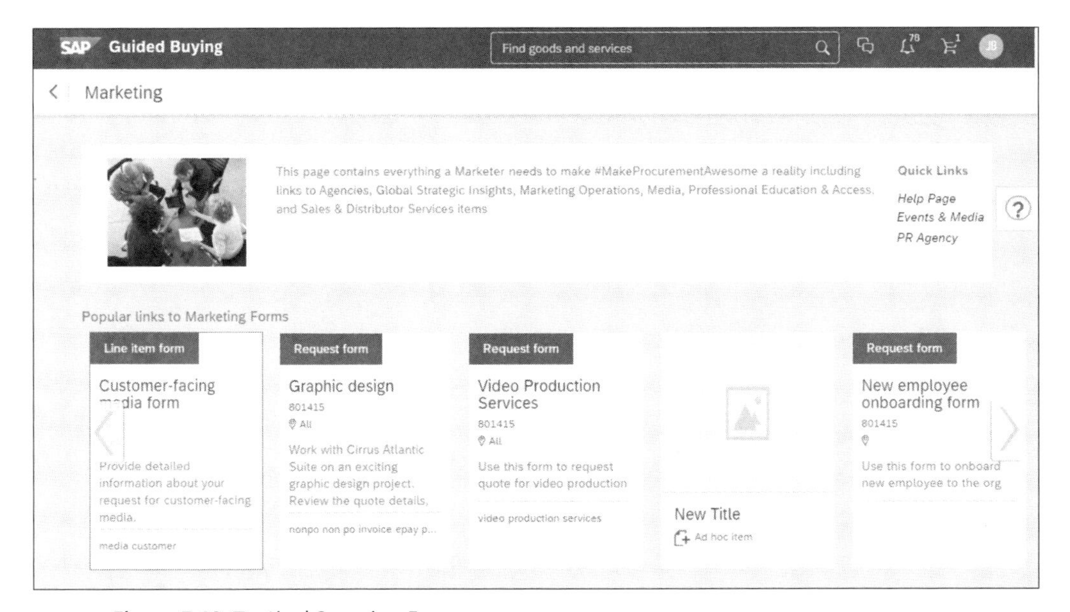

Figure 7.49 Tactical Sourcing Form

When you select the form, you won't need to know the commodity code or the list of preferred suppliers because all that information is already defined and predetermined in the form. Based on the information criteria you selected, the tactical sourcing will have four different flows to select from: self-service, low-touch, high-touch, or sourcing request. Figure 7.50 shows a self-service flow, where you fill out the form and submit it using the **Request Quotes** option. An RFQ event gets automatically created and sent to all the suppliers you selected to participate.

Thereafter, you'll be able to see the supplier responses via email or looking at your guided buying dashboard. When you accept the quote form the supplier, a purchase request gets automatically created with the information you provided (see Figure 7.51).

In contrast with the self-service flow, a low-touch, high-touch, or sourcing request will require some level of involvement from your procurement department to verify the information you provided and submit the form, monitor the sourcing execution at all steps of the process, and maintain the communication with the supplier. At the end of the process, it comes back to you to award the business to the supplier, and then the purchase requisition is automatically created in guided buying. The high-touch scenario requires more involvement from your procurement department. This scenario is used for large purchases or when you don't know which suppliers to use. Finally, sourcing requests are ideal for very large purchases. The guided buying administration guide has more information about the different tactical sourcing workflows.

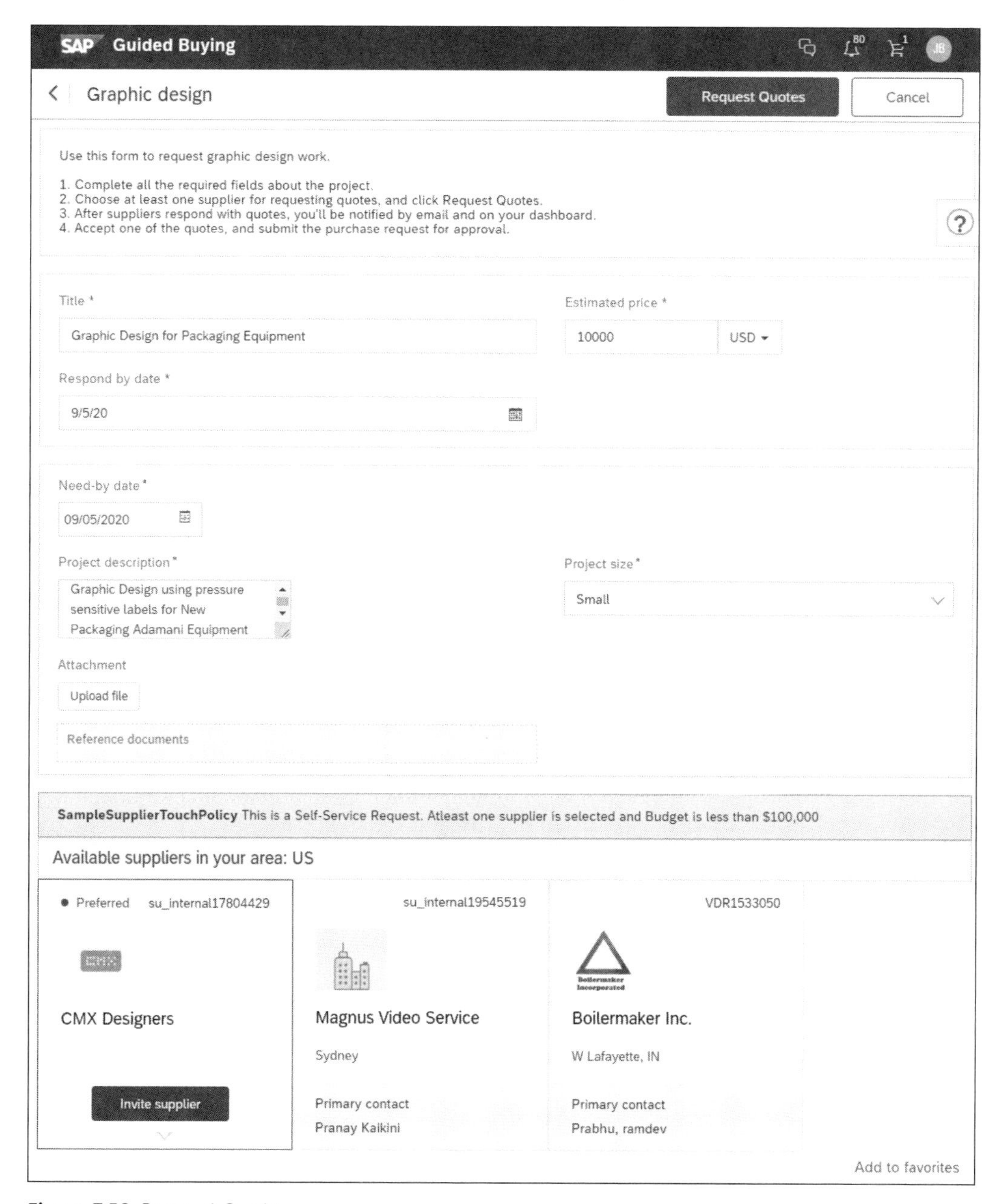

Figure 7.50 Request Quote

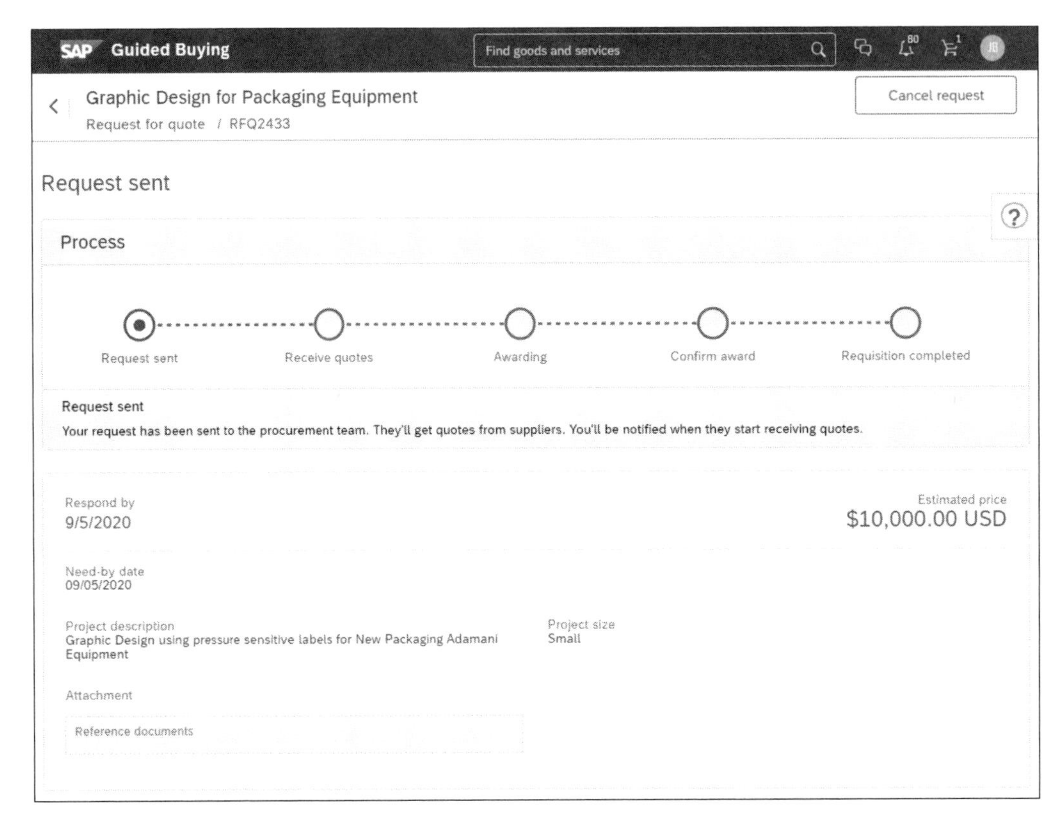

Figure 7.51 Sourcing Request Process

7.4 Summary

When an organization needs a one-stop shop for buying, the guided buying capability in SAP Ariba should be considered as a simple way to buy smarter by promoting preferred channel compliance. The guided buying capability will give your casual and functional users the ability to browse catalogs, create purchase requisitions and POs, submit forms, make service requests and noncatalog requests, create tactical sourcing events, review bids and buys, view and approve purchase requisitions, and receive items. With the private help communities and the interaction with SAP Ariba or SAP S/4HANA solutions, you'll increase the adoption to your company policies because all users, from the occasional to the frequent buyer, will find a satisfactory buying experience.

Chapter 8
Operational Procurement

In this chapter, we'll focus on operational procurement, a set of processes that SAP Ariba broadly calls "downstream." Operational procurement includes SAP Ariba Buying and its variants, and SAP Fieldglass, a leading vendor management system for contingent labor procurement.

After you've completed sourcing and setting up contracts, procurement in the traditional sense can begin. Several SAP Ariba Procurement solutions, specifically SAP Ariba Catalog, SAP Ariba Buying, and SAP Ariba Buying and Invoicing, support operational procurement activities. The SAP Fieldglass vendor management system (VMS) manages all categories of external labor, including contingent workers, independent contractors, and specialized talent pools. In other words, for contract workers, the SAP Fieldglass VMS addresses the intersection between human resources and procurement and can help your organization keep an eye on and analyze your contingent workforces.

In this chapter, we'll outline how to use SAP Ariba Procurement solutions. We'll connect the dots between the solutions' functionalities and show you the different configuration and implementation options for the SAP Ariba Procurement solution.

8.1 SAP Ariba Buying and Invoicing, SAP Ariba Buying, and SAP Ariba Catalog

SAP Ariba Buying and Invoicing supports end-to-end procurement processes through payment in SAP Ariba. In contrast, SAP Ariba Buying stops short at ordering and then integrates with an SAP S/4HANA system for goods/services receipts and invoicing/payment.

Viewed another way, SAP Ariba Buying and Invoicing covers the processes of requesting, approving, and procuring an item/service, as shown in Figure 8.1, as well as invoice processing and cash management. SAP Ariba Buying only covers the first part of this process. Figure 8.2 shows the SAP Ariba home page, which leverages a portal approach to serve as a gateway for users to all the SAP Ariba solutions they can access.

Supporting both SAP Ariba Buying and Invoicing and SAP Ariba Buying is SAP Ariba Catalog, which provides users with a catalog platform in the Ariba Network for accessing content from suppliers via their uploaded catalogs, directly to a catalog hosted by a supplier (punchout), or a customer-loaded catalog in SAP Ariba Catalog. Choosing an item from a catalog ensures pricing and item consistency, as well as contract consumption/adherence in the procurement process, and it's usually considered a best practice.

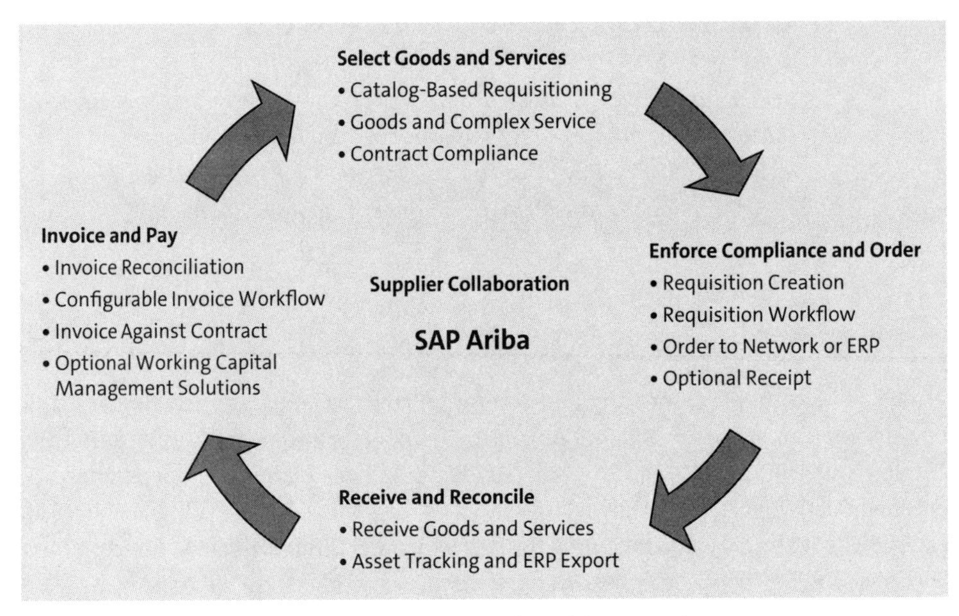

Figure 8.1 SAP Ariba Buying/SAP Ariba Buying and Invoicing Overview

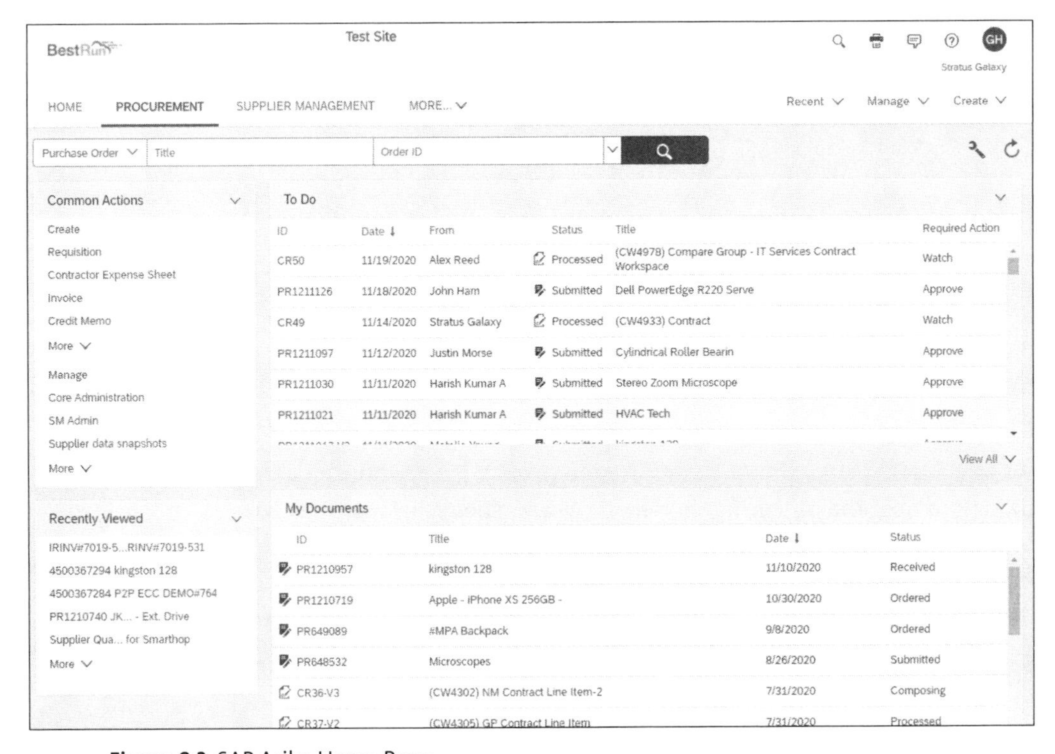

Figure 8.2 SAP Ariba Home Page

In the following sections, we'll cover when and where to deploy SAP Ariba Procurement solutions and the configuration options for the two variations of SAP Ariba Procurement: SAP Ariba Buying and SAP Ariba Buying and Invoicing. For contingent workforce management, we'll review SAP Fieldglass solutions and implementations.

8.1.1 SAP Ariba Buying and SAP Ariba Invoicing and Buying

In this section, we'll outline when and where to leverage SAP Ariba Buying and Invoicing and SAP Ariba Buying, the overall processes covered by these solutions, and how they work with other solutions, including SAP Ariba Discovery and SAP Ariba Catalog.

SAP Ariba Buying

Chapter 4 and Chapter 5 of this book, as well as the steps shown in Figure 8.3, cover sourcing activities.

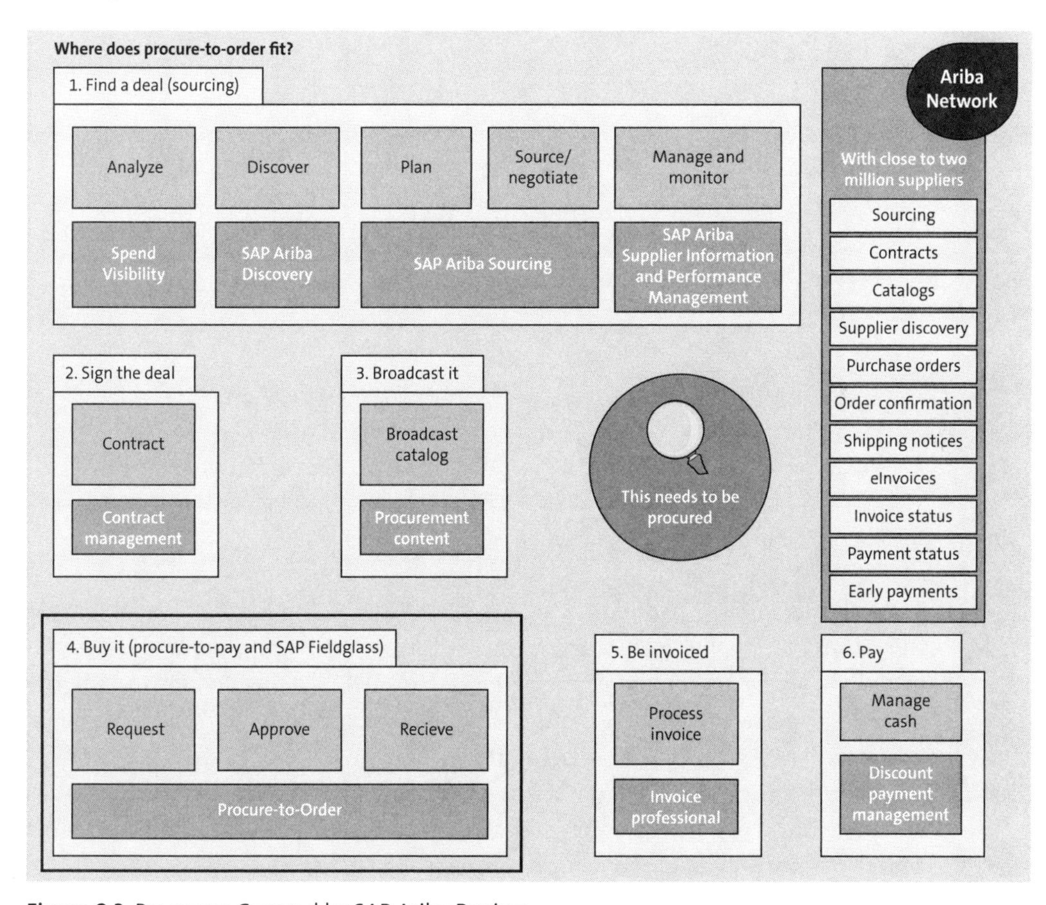

Figure 8.3 Processes Covered by SAP Ariba Buying

In these steps, a demand is registered or forecast, and a source of supply is identified using the request for (RFx) or by assigning one of the category suppliers directly as the source of supply for the purchase. New suppliers are typically vetted and then added to your supplier master. After the supplier relationship is established during this sourcing process, a user can log in to SAP Ariba Buying or SAP Ariba Buying and Invoicing and request an item, navigate the approvals process established internally by the company, order the item, and receive the item in SAP Ariba Buying or SAP Ariba Buying and Invoicing. A user can order both catalog items and non-catalog items, as shown in Figure 8.4.

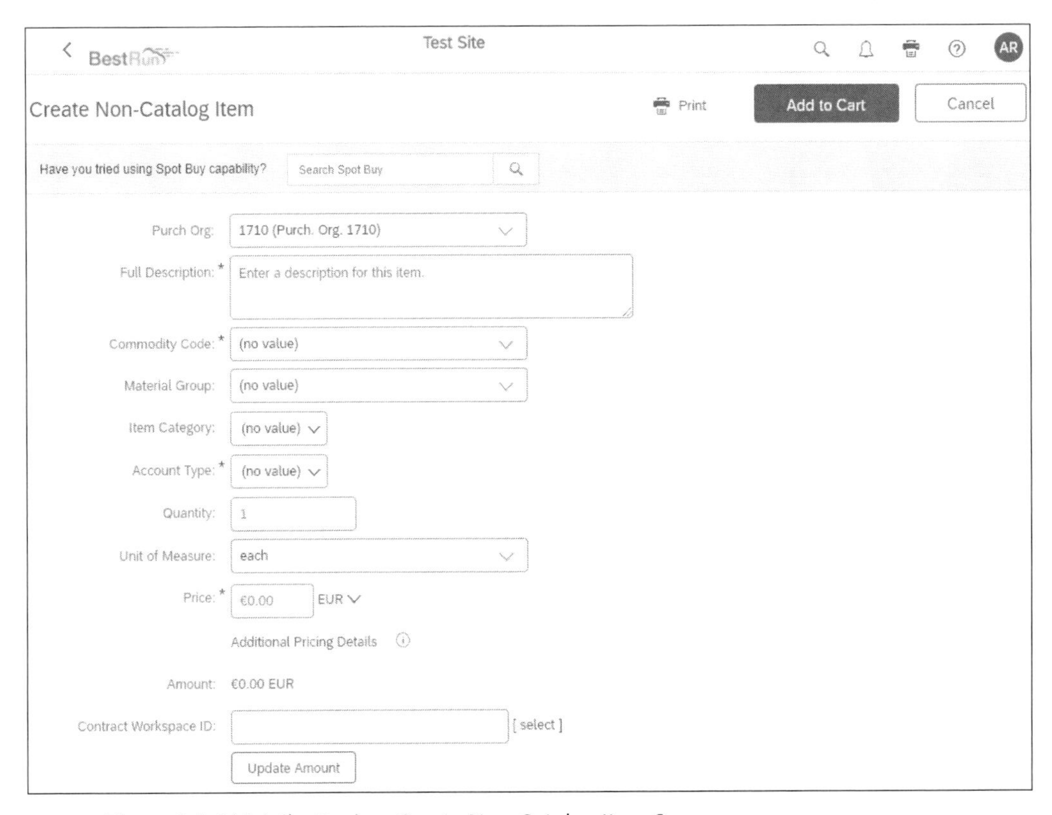

Figure 8.4 SAP Ariba Buying: Create Non-Catalog Item Screen

In SAP Ariba Buying, the invoice then arrives in the SAP S/4HANA environment for processing from the supplier. SAP Ariba Buying and SAP Ariba Buying and Invoicing both have the capability to update connected SAP S/4HANA environments so that the accounts payable department can make a three-way match based on the documents generated in SAP Ariba. For lightweight integrations where only payment to the supplier is required (e.g., SAP Ariba Buying and Invoicing), an OK2Pay message is sent to the SAP S/4HANA/payment system after all the transaction steps from request item/service to invoice processing have been completed in SAP Ariba.

SAP Ariba Buying's process also covers the requisition to order in SAP Ariba, with follow-on invoicing occurring in another system, typically an SAP S/4HANA–based process. As such, SAP Ariba Buying covers the requisition process, receiving, catalogs, and contract compliance, as well as interactions/collaborations on purchase orders (POs) with suppliers in the Ariba Network (including supplier determination, order confirmation, and advance shipping notifications [ASNs]) and PO delivery to suppliers via the Ariba Network or another supported transmission methods.

SAP Ariba Buying users and catalog managers can access SAP Ariba Catalog. Catalogs can be loaded from the business network using the Commerce eXtensible Markup Language (cXML) and catalog interchange format (CIF) file formats. Level 1 (hosted on the Ariba Network) and level 2 (hosted by the supplier) punchout catalogs can also be accessed via SAP Ariba Catalog. In both instances, the business network handles the authentication step. Keyword search, dynamic filters for categories, suppliers, price ranges, manufacturers, and favorites are all supported in SAP Ariba Buying connections with SAP Ariba Catalog. In addition, both SAP Ariba Buying and SAP Ariba Buying and Invoicing approaches in SAP Ariba Catalog support kits, discussed later in this section, and filtered views.

As shown in Figure 8.5, budget checks in SAP Ariba Buying can leverage budget data from SAP S/4HANA by performing a real-time budget check at the time items/services are added to a requisition, and updates to budget consumption occur in SAP S/4HANA occur as purchases are approved. Once approved, you can send the resulting PO via a number of delivery formats, including fax, email, cXML, electronic data interchange (EDI), and/or manual delivery. Purchasing cards (p-cards) are also supported, along with p-card reconciliation.

While suppliers can receive the POs in various formats from SAP Ariba Buying, suppliers will transmit ASNs and order confirmations only via the Ariba Network. Suppliers must have an active account to upload these types of documents on the network. Upon receipt, SAP Ariba Buying users can leverage the following features:

- Desktop receiving and central receiving
- Receiving against contracts for non-PO spend
- Automated, date-specified receiving for imported requisitions
- Asset information capture and reporting
- Returns management
- Receipt import

You can also configure procurement workspaces in SAP Ariba Buying and Invoicing and SAP Ariba Buying using the **Create** menu on the dashboard, as well as workspace requests, purchase requisitions, and templates, in order to manage procurement as a project.

These workspaces/templates, shown in Figure 8.6, can include tasks, approvals, and designated fields. Requisitions, POs, receipts, and invoices, as well as subprojects, such as sourcing and supplier performance management projects, can be linked to the workspace.

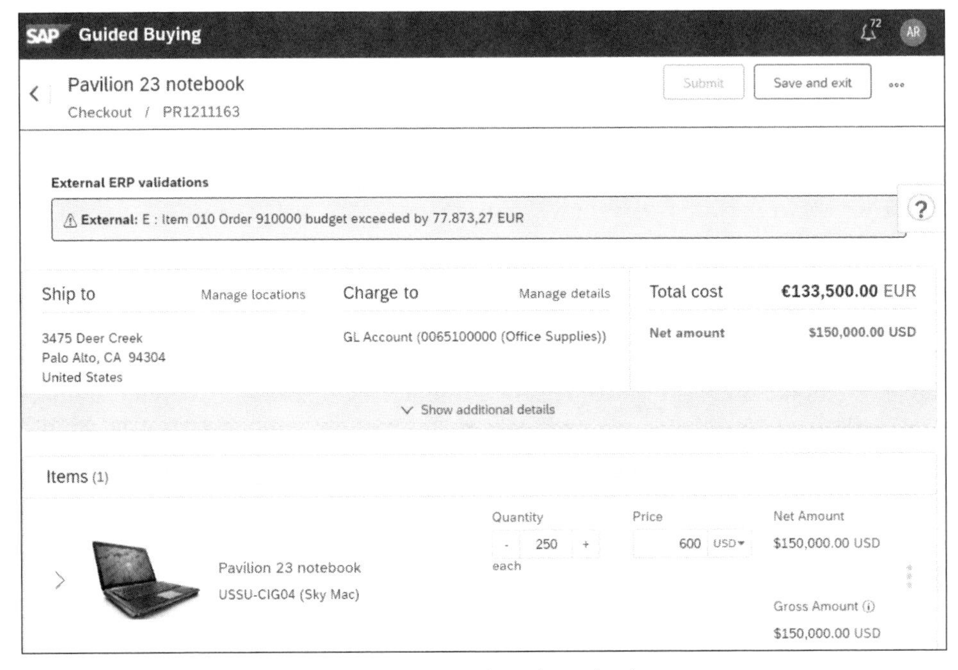

Figure 8.5 SAP Ariba Buying Shopping Cart with Budget Check

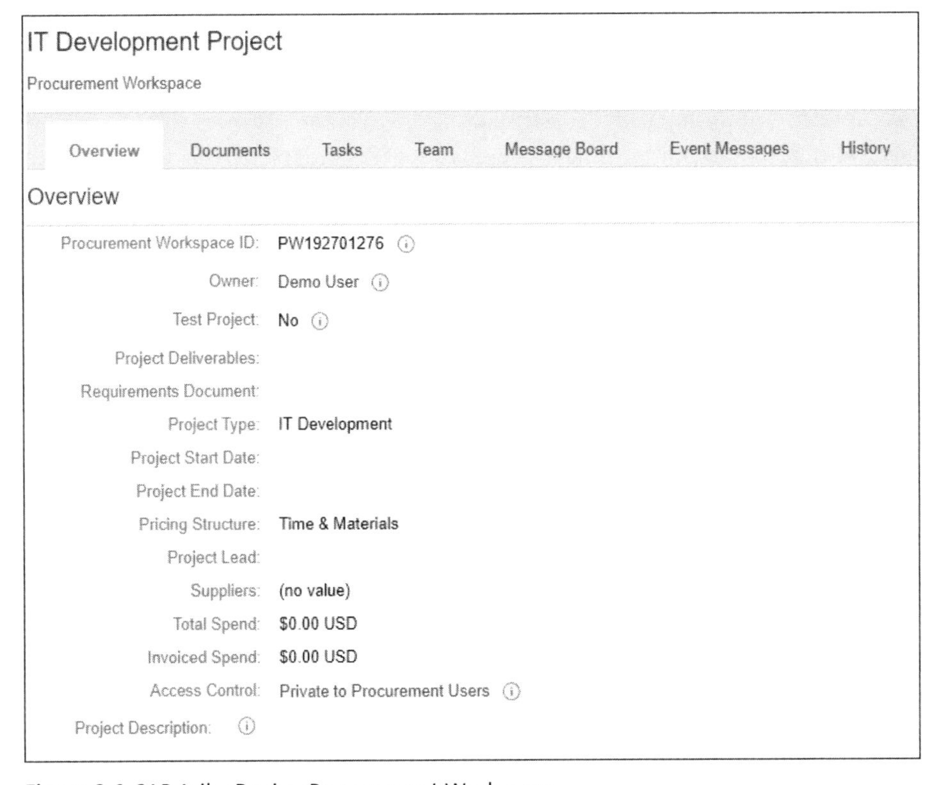

Figure 8.6 SAP Ariba Buying Procurement Workspace

SAP Ariba Buying and Invoicing

As shown in Figure 8.7, SAP Ariba Buying and Invoicing provides the same process coverage as SAP Ariba Buying, while extending into the invoice-to-payment area. As a result, suppliers submit invoices on the Ariba Network where the invoices are electronically routed back to the SAP Ariba Buying and Invoicing realm. After invoices are received, they are approved and reconciled in SAP Ariba Buying and Invoicing. Internal invoices (created by the buyer), contract-based invoices (supplier is allowed to punch in to their SAP Ariba Buying and Invoicing realm and access contract details and create invoices), remittance, and status advice are also supported as part of the process, in conjunction with the Ariba Network. As with SAP Ariba Buying, updates and process support can be integrated with the associated SAP S/4HANA environment.

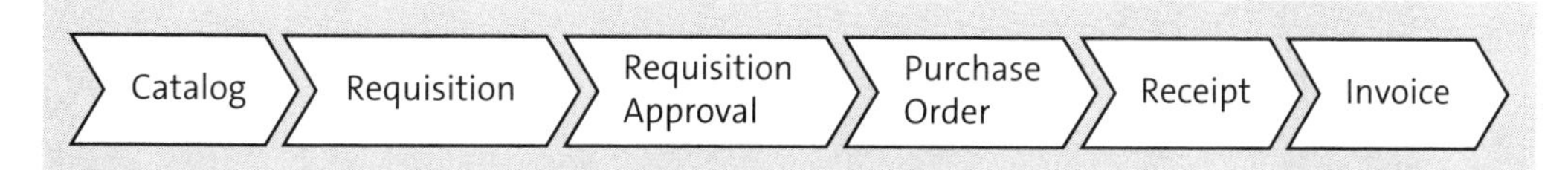

Figure 8.7 Process Covered by SAP Ariba Buying and Invoicing

Tactical Sourcing

You can configure items in SAP Ariba Buying and SAP Ariba Buying and Invoicing to require preorder collaboration with a supplier where you define the specifications, scope, and service levels associated with the item before a requisition is submitted for approval. Such an item will send a bid request to suppliers via the Ariba Network, and suppliers can respond with bids or proposals. You can require that your buyers always obtain bids from multiple suppliers in SAP Ariba Buying for high-value purchases.

You can also publish requests for quotations (RFQs) to SAP Ariba Discovery, which we discussed at length in Chapter 5, Section 5.3. SAP Ariba Discovery also plays a role in tactical sourcing, which is a shared scenario with SAP Ariba Sourcing. Tactical sourcing addresses an area of spending that is often neglected: low-cost items with tight timelines that aren't available in a company catalog. Often one-time or rush order purchases, these items are undermanaged and underserved categories that nonetheless require quick turnaround. These requests typically start as a request in the guided buying capability by a nonprocurement buyer who creates demand after discovering that the item/service isn't available. Buyers generally lack efficient or effective methods to source these categories. Tactical sourcing seeks to improve this situation by providing a process that allows the buyer to go out and source the item on a one-time basis, leveraging the power of the Ariba Network to find new and optimal sources of supply.

SAP Ariba Spot Quote is provided for easier supplier identification and bid execution, as follows:

- A RFQ in guided buying generates an automated RFQ and publishes it in SAP Ariba Sourcing.

- Suppliers provide responses through SAP Ariba Sourcing.
- Responses are sent back automatically to the source system.

As shown in Figure 8.8, a demand is generated in guided buying, but this demand could also come from an SAP Ariba Buying and Invoicing–based process. This case arises when a user doesn't find the item needed in one of the company catalogs. After the buyer determines they can't find the item in an existing catalog and/or contract, and that the item is acquired in such low quantities/frequency that a long-term contract or procurement cycle isn't necessary, the buyer can use SAP Ariba Sourcing to send a RFQ to suppliers on the Ariba Network. Suppliers can participate and send the resulting bids back through SAP Ariba Sourcing.

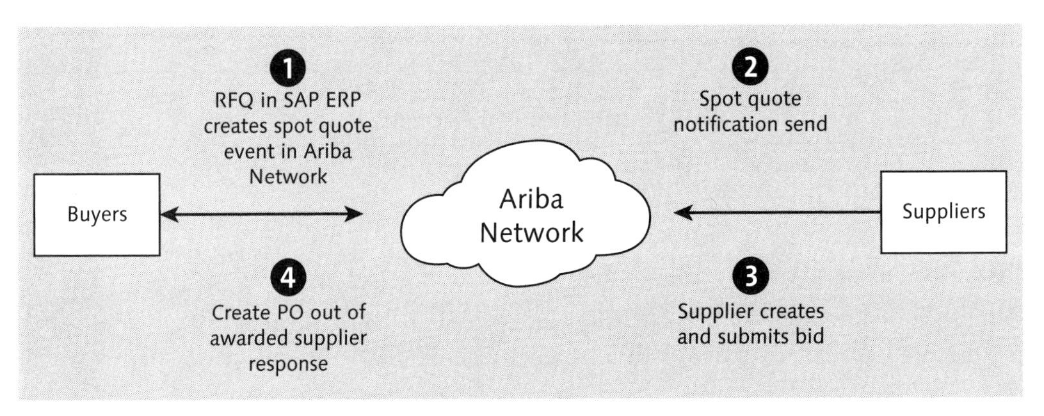

Figure 8.8 SAP Ariba RFQ Process Flow (SAP ERP)

SAP Ariba Catalog

The most efficient type of order is usually one sourced with a preexisting agreement from a defined item maintained in a catalog. This "procurement content," housed in a catalog, prevents typos from manually entered orders, while helping users quickly find what they need with the help of the search functionality, descriptions, pricing, part numbers, terms and conditions, availability, and other core items for supporting a successful ordering process.

SAP Ariba Buying leverages SAP Ariba Catalog to provide this functionality and content to end users, as well as provide a streamlined maintenance process to suppliers. SAP Ariba Catalog supports fuzzy searches, parametric refinement, and side-by-side comparisons of selected items.

SAP Ariba Catalog can also support non-SAP Ariba Procurement processes, such as SAP S/4HANA procurement processes, as shown in Figure 8.9. In this case, the buyer punches out to SAP Ariba Catalog to pull item/service content and then brings this information back into the SAP S/4HANA materials management (MM) or SAP Supplier Relationship Management (SAP SRM) procurement environment.

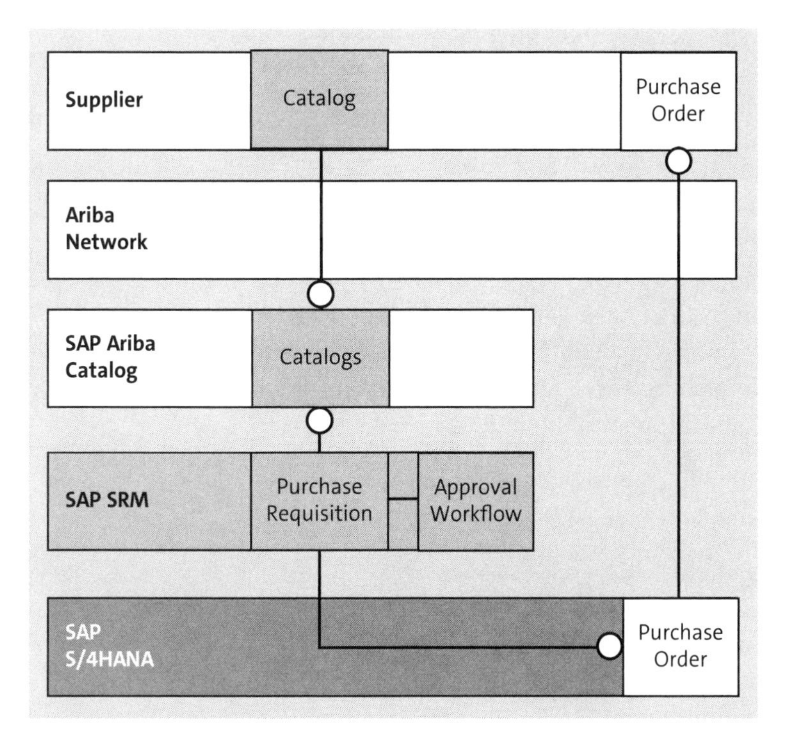

Figure 8.9 SAP Ariba Catalog Process Flow

To enable a supplier catalog with SAP Ariba, SAP Ariba provides the following setup services:

- Uploading and activating test and production catalogs
- Configuring catalog filters (views)
- Editing the catalog hierarchy
- Creating and configuring catalog kits
- Indexing catalog data
- Performing catalog refreshes
- Conducting supplier catalog education sessions

Additional functionality can be enabled by SAP Ariba upon request, such as kits and filtered views. *Kits* are packages of items pulled from a catalog into a shopping cart/requisition in one click, bringing multiple items as part of the selection. These bundles are useful for onboarding new employees (i.e., an onboarding kit with a laptop, badge, manuals, etc.); repetitive purchasing activities requiring the same bundle of goods (e.g., new store setup kits); and larger items that are assembled out of multiple defined smaller items (i.e., bills of materials [BOMs] in direct procurement scenarios). A kit appears as a single line item in SAP Ariba Catalog, but after it's added to a requisition or shopping cart, the kit unpacks into the underlying line items that can belong to multiple suppliers. *Filtered views* allow for different items or subsets of selections to be displayed to different users and organizations.

Another key area for catalogs is having a robust search functionality. In this area, SAP Ariba Catalog offers keyword search and dynamic filters by category, suppliers, price ranges, and manufacturers, as well as relevance ranking, side-by-side comparison, favorites, and catalog hierarchies (United Nations Standard Products and Services Code [UNSPSC]).

You can set up contracts in SAP Ariba Buying via the SAP Ariba Catalog user interface or by importing contracts directly via SAP Ariba Catalog, while controlling access to those contracts. SAP Ariba Catalog contracts can be based on suppliers, material categories, or items, and they include the following conditions:

- Releases required
- Contract hierarchies
- Effective expiration dates
- Limits
- Simple and advanced pricing terms
- Blanket POs

Contracts in SAP Ariba Catalog enforce term compliance on the shopping carts to which they are assigned. Contracts can be manually assigned to POs and track usage of contracts in purchasing activities with the designated supplier.

Most catalog scenarios involve tightly defined items that your company orders frequently to standardize as much of the order as possible. However, not all purchasing activities conform to this scenario. Some orders require further discussion and definition with the supplier prior to ordering. Still, parts of the process, such as initial selection and definition of a general item remain largely the same across orders. For this type of order, SAP Ariba Catalog offers collaboration options.

You can configure items in SAP Ariba Catalog to require preorder collaboration with a supplier to facilitate the definition of specifications, scopes, and service levels associated with the item before a requisition is routed for approval. Then, when the item is selected in SAP Ariba Catalog, a collaboration request is sent to one or more suppliers via the Ariba Network, and suppliers can respond with bids or proposals. Buyers can then continue the ordering process by reviewing, negotiating further, and accepting proposals. Buyers can enforce obtaining bids from multiple suppliers for high-value purchases to extract better discounts. Buyers can also invite suppliers who are currently not enabled on Ariba Network to register in order to build relationships with new suppliers and explore cost-saving opportunities.

You'll administer SAP Ariba Catalog through a separate login page, used solely for administration. The search functionalities in SAP Ariba Catalog support general, wildcard, refined, and parametric searches, as well as favorites.

Another variation on the catalog is the *punchout catalog*, which SAP Ariba Catalog also moderates. The punchout process involves the buyer, the supplier, and the Ariba Network. Punchout messages are routed through the Ariba Network for validation and authentication. This process follows these steps:

1. First, the buyer selects a punchout item in SAP Ariba Catalog. This selection sends a request to the Ariba Network to establish a connection with the remote catalog.
2. The Ariba Network authenticates the buying organization and forwards the request to the supplier's punchout site.
3. The supplier sends back a URL to a web page on the supplier's punchout site designed specifically for the buyer. The procurement system redirects the user to this URL. The remote shopping site appears in the user's window, and the user begins shopping.
4. After shopping, the buyer clicks the site's checkout button, and the items are sent back to SAP Ariba Catalog where further requisitioning activities can take place.

For your part, as a buyer, you'll need to define the in-scope catalogs and suppliers as well as define the types of catalogs required from each supplier. Suppliers will need to onboard onto the Ariba Network and create, upload, and maintain catalogs in this environment or use the punchout approach we described earlier. SAP Ariba Catalog specialists are available to assist suppliers in this process.

For importing catalogs into SAP Ariba Catalog, suppliers and catalog managers can leverage multiple approaches:

- **Network catalog or network subscription**
  Suppliers can load catalogs on the Ariba Network.
- **Local catalog or local subscription**
  You (or your catalog manager) can load this catalog into SAP Ariba Catalog directly.
- **Generated subscription**
  You can also load content from a contract through a generated subscription.

Each method has a different process with different advantages and use cases. Most organizations leverage network subscriptions to automate their catalog process as much as possible and reduce the workload for catalog managers.

8.1.2 Services Procurement with SAP Ariba Buying and SAP Ariba Buying and Invoicing

SAP Ariba Procurement solutions offer a number of ways to procure basic services using a standard purchasing process. A request for a service is made by a buyer by adding service line items to their requisition, and from that requisition, a service PO is generated and sent to the supplier on the Ariba Network. The supplier confirms the service order, performs the service, submits a service entry sheet, and submits it for approval by the buying organization. After the service has been validated and the service entry sheet approved, the supplier submits a service invoice for payment. Note that from a product roadmap standpoint, SAP Fieldglass is the preferred solution for complex service procurement scenarios. SAP Fieldglass is a cloud-based, open VMS that helps organizations find, engage, and manage their external workforce and services procurement resources—from temporary staff to consultants based on the statement of work (SOW) to freelance workers and contractors.

Service procurement differs from traditional material procurement in several ways. As shown in Table 8.1, pricing, quantity, and requirements are much more dynamic when procuring services than with procuring items.

	Material Procurement	Service Procurement
Prices	Prices are fixed.	Prices are negotiable and agreeable to using various approaches (fixed fee, time and materials, etc.).
Quantity	Item is fixed, apart from configurable items. If a contract or catalog item, the item is further defined.	Item can be project or bucket of hours—more dynamic than a material item regarding what is required.
Requirements	Material is shipped, usually in a consumable unit, and often the entire order is shipped together.	Service is delivered, usually over a defined time period, with the customer paying by unit of time or upon project completion.

Table 8.1 Services versus Material Procurement

Procuring services often require further steps that aren't required for materials. For example, you may need to select from a pool of potential providers and interview the candidate tasked with delivering the services or with joining your department for a set duration as a third-party provider. In many cases, the details of the services needed may not be known at the time the request is created.

The boxes shown in Figure 8.10 depict the standard procure-to-pay (P2P) process for a material. Note that a procurement of service often requires more steps and iterations during the order process to arrive at the PO and may have a more fragmented billing process with multiple service entry sheets submitted over time for invoicing.

To address these differences and challenges, services procurement offers the following options when creating service requests:

- Requester can create either a planned or unplanned service request
- Set the transparency (openness) of the collaboration process by selectively exposing information on a supplier's proposal to the supplier's competitors

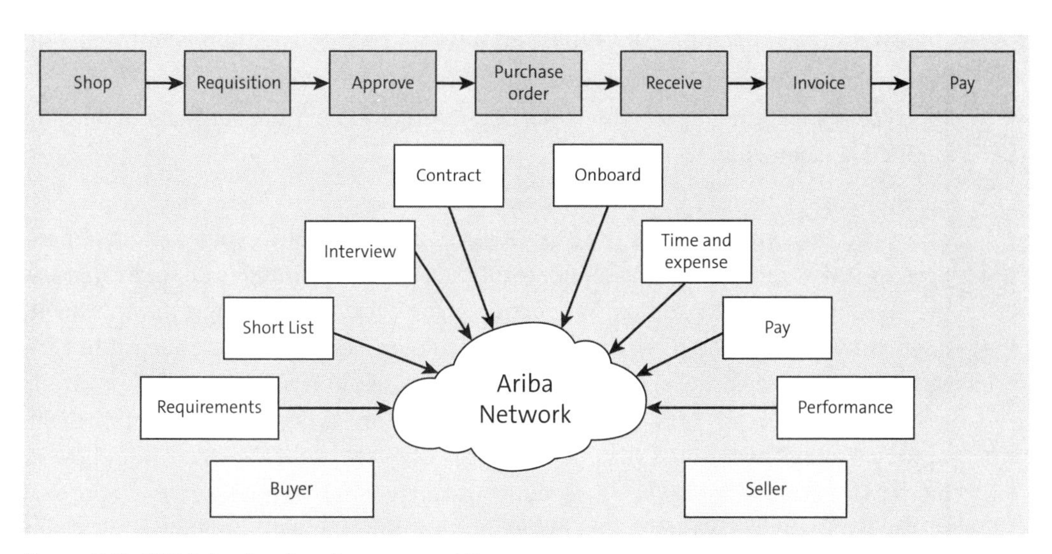

Figure 8.10 SAP Ariba Services Procurement Process

Typically, service requests are created as non-catalog items that require the requester to add details about the service before the PO is sent to the supplier. For catalog-based procurement, these details include permissions, default values, policies, and other constraints. After selecting a service item from the catalog, the buyer will fill in further information according to the rules specified for the item. For service procurement, much more emphasis is placed on iterative planning and collaboration during the procurement process that may not be required for material procurement.

The Value of SAP Ariba Buying and SAP Ariba Buying and Invoicing

In most organizations, indirect procurement is viewed as a cost center and thus an area to be managed and controlled for cost minimization and increased efficiency. However, indirect procurement, though a significant portion of overall company spending activities, is rarely prioritized by management and is left therefore to grow organically, using whatever processes and methods are available or cobbled together. As these processes and company customs around indirect procurement become increasingly convoluted and expensive, a need arises for comprehensive, system-based solutions in many areas of the procurement process from the sourcing and contracts areas, through procurement processes, and on down to invoice-to-pay scenarios.

In a P2P scenario, most companies initially concern themselves with the approval aspects of indirect procurement, making sure the requisitioning employee's manager signs off on the spending. In lite or systemless environments, approval can occur in the form of an email or as a literal signature on a PO.

This process of signing off on whatever the employee manages to source in an indirect procurement activity doesn't usually scale well and overlooks discounts and spend categorization areas, not to mention the system-based procurement structures that inform a supplier regarding the exact product requested, terms expected, and delivery points. Finance eventually grows weary of finding out about spending after the fact, as this type of spending impacts their cash flow management efforts and overall working capital management.

Enter SAP Ariba Buying and Invoicing and SAP Ariba Buying. With both of these solutions, you can bring order to the unruliness of an organically evolved procurement process. As a strategy and principle, we recommend focusing initially on the areas causing the most difficulty and costs in your organization. For indirect procurement, approval adherence will often drive initial discussions. Companies are subject to various reporting and compliance regulations, which are often not followed if approvals rely on the signature of a manager for any category type or spending threshold. Approval workflows should thus be a key area of focus, and SAP Ariba's workflow in buying scenarios provides clarity, sequencing, and record keeping for various approval scenarios. So-called "maverick" spending can be minimized with a defined, in-system approval structure, combined with record keeping for each individual order's approval workflow and approvers.

Costs resulting from one-off orders placed without contract or terms, or with poorly described items, are another fertile area of focus for identifying and addressing the key performance indicators (KPIs) of an SAP Ariba Buying implementation. The ideal order will be a PO created using a contract or catalog item from suppliers, with defined terms and relationships with your company. This ideal order is fulfilled with minimal manual intervention, which can be costly, to correct orders with the supplier. The resulting order leverages pricing that has been negotiated with the supplier.

After the order is placed, a supplier will submit a request for payment, typically in the form of an invoice. Invoices can also be another source of costs and inefficiencies if not managed well because invoices need to be received, scanned, and processed/reconciled first with original orders and then with receiving documents. If these documents are strewn about the company in various file cabinets or receipts, the process can become quite laborious. In addition, suppliers can assess penalties if the terms of their order aren't followed on the payment side. Late payments can create additional costs for the company, and, equally important, you may miss out on early payment discount opportunities provided by suppliers if negotiated during the payment cycle. SAP Ariba Buying and Invoicing covers the basic invoicing and payment areas in SAP Ariba, SAP Ariba Commerce Automation capability , and SAP Pay. These areas and additional SAP Ariba solutions for invoice management, dynamic discounting, and payment management are discussed in Chapter 9.

Negotiating for better pricing is often based on volumes and overall terms. To understand volume, your company first must understand the categories of items it buys and how their suppliers perform according to their relationships with the company. Doing so requires analytics and reporting. In an SAP Ariba Buying project, these insights can be gained by using SAP Ariba Spend Analysis and SAP Ariba Supplier

Lifecycle and Performance, which are outlined in Chapters 10 and 3, respectively. However, putting the data into an organized format and process for analysis is very much the focus area and value-add for an SAP Ariba Buying project. In SAP Ariba Buying, the tactical aspects of procurement are defined and structured so that analysis can be performed and valuable insights on volumes and performance can be brought to bear with suppliers in further negotiations and contract cycles.

8.2 Configuring SAP Ariba Buying and SAP Ariba Buying and Invoicing

In this section, we'll look at the different configuration options for SAP Ariba Buying and SAP Ariba Buying and Invoicing, from master data setup to enabling functionality related to specific buying, receiving, and invoicing topics. What follows isn't an exhaustive list of functionalities, but an overview of the common features most customers configure to get the most out of their SAP Ariba Buying or SAP Ariba Buying and Invoicing solution.

8.2.1 Procurement Master Data

Master data is the underpinning of all the transactions created in SAP Ariba Buying and SAP Ariba Buying and Invoicing. It ensures consistency between backend systems and integrated systems, as well as the filtering of purchasing data. In this section, we'll cover the various master data elements used in SAP Ariba Buying and SAP Ariba Buying and Invoicing, including the organizational hierarchy and accounting elements.

Organizational Hierarchy

SAP Ariba Buying provides multiple ways to segregate data, filter content, and create different divisions within the business. This is closely tied to the SAP S/4HANA divisional hierarchy in that the same objects that are fed to SAP Ariba Buying in the master data and used in transactional documents such as requisitions, receipts, and invoices.

Company Code

In the standard SAP S/4HANA configuration, the different accounting data and addresses available are filtered by company code. SAP Ariba Buying and SAP Ariba Buying and Invoicing are designed to match this configuration. The master data imports that are sent from SAP S/4HANA contain a mapping between the company code and the various accounting elements.

The following accounting elements are filtered by company code in SAP Ariba Buying:

- General ledger
- Cost center

- Work breakdown structure (WBS) element
- Internal order
- Asset
- Shipping/billing addresses
- Purchasing organization

Purchasing organizations are set up in SAP S/4HANA to reflect divisions within your business that have different purchasing processes. The purchasing organization concept is extended to SAP Ariba Buying to allow for further filtering of certain purchasing data and is set against the user profile, as shown in Figure 8.11. This includes the filtering of catalog data by suppliers. Only vendors mapped to the purchasing organization that is used on the requisition will be available for purchasing, and this includes the catalog items that are available.

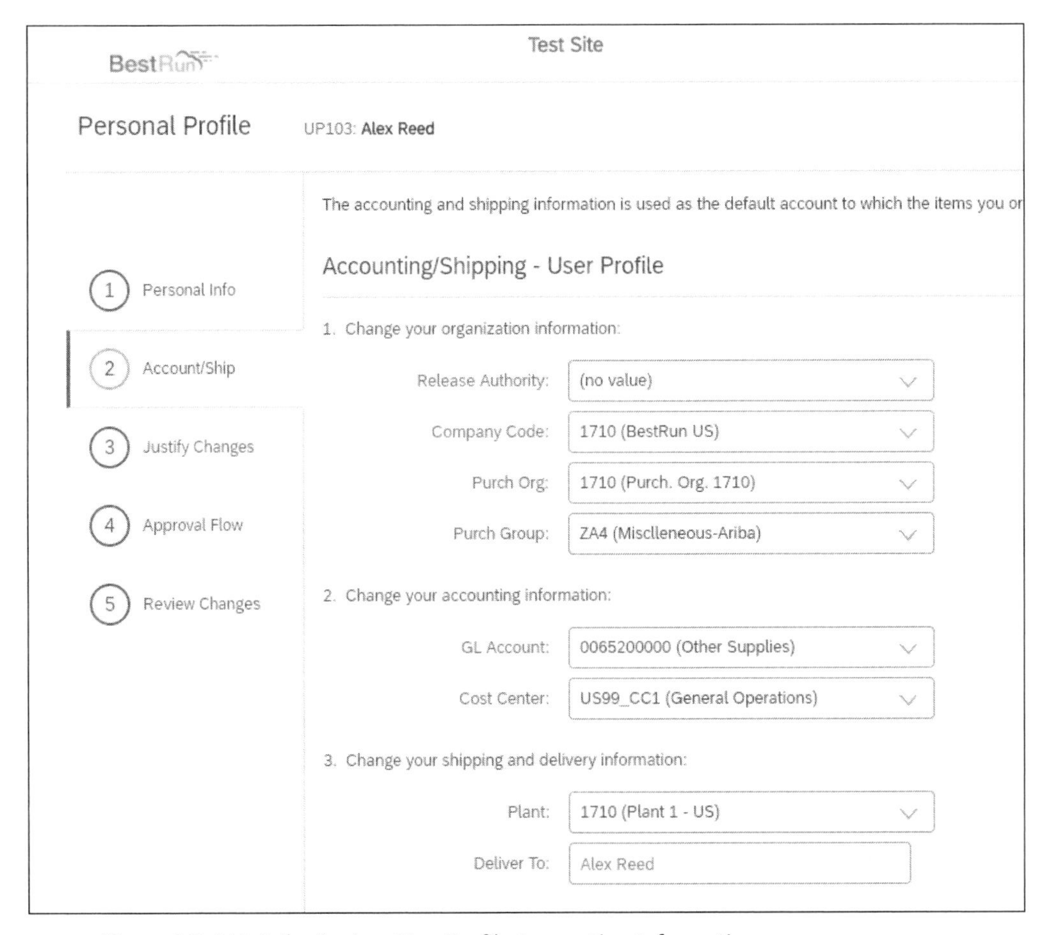

Figure 8.11 SAP Ariba Buying: User Profile Accounting Information

A vendor can be linked up to multiple purchasing organizations in SAP Ariba.

Purchasing Groups

The purchasing group is another SAP concept that identifies the different groups that create POs in SAP ERP. This is sent to SAP Ariba in the master data so that this can also be set on the requisition. The **Purchasing Group** should be defaulted on the user record if available so that this data can be sent to SAP ERP.

Accounting Elements

When a transaction is created in SAP Ariba Buying, it must be assigned to a costing element. This allows SAP S/4HANA to post the transaction to the correct financial objects and keep the accounting information consistent between systems.

Account Category

The account category in SAP Ariba Buying is used to define which of the accounting fields are exposed when a line item is added to a requisition. This list is based on a filtered list of account categories in SAP S/4HANA, and the behavior is driven by a string of characters that impact which fields are visible and required for the types of transactions in Table 8.2.

Account Assignment	Account Assignment Name	Mandatory Accounting Fields
K	Cost Center	**Cost Center**
P	Project	**WBS Element**
F	Internal Order	**Cost Center Internal Order**
A	Asset	**Asset**

Table 8.2 SAP Ariba Buying: Account Assignment Categories and Required Fields

Item Category

The item category is used to define what category of spend the line item applies to. This also flows to SAP ERP to assist with deriving whether the PO type is for a material or a service.

The **Item Category** field is defaulted based on which commodity code it's been mapped to in the system. If you want to change the definition of which commodity codes are materials and which are services, the commodity code mapping must be updated and corrected.

Account Type

The **Account Type** field will drive what type of account is being used when the requisition is created. The available values in this field can be changed to include an identifier for the purposes of categorization of the accounting types. For example, you may have accounting rules that dictate anything over a certain amount should be treated as a capital purchase and coded against a capital general ledger account.

Like the **Item Category**, the values in **Account Type** field are defaulted based on the commodity code mapping.

General Ledger Accounts

Every line item created in SAP Ariba Buying will also be posted to a general ledger code when sent to SAP S/4HANA. SAP Ariba Buying can derive the general ledger code based on the commodity being ordered so that the preparer of the requisition doesn't need to select the coding. This allows for more consistent coding to general ledger codes as there are rules that can be defined to set these depending on the inputs.

Regardless of the account category that is chosen, a general ledger must always be specified in SAP Ariba Buying. This is important so that the spend on the PO is assigned to the correct account when sent to SAP S/4HANA.

SAP Ariba receives all valid general ledger master data from SAP S/4HANA. The **General Ledger** field is mandatory but should always be defaulted and set to the correct coding value.

Cost Centers

Cost centers are used to charge expenditures to the accounting element where the funds are coming from. Valid cost centers are integrated to SAP Ariba Buying, as SAP S/4HANA master data, for selection on the documents. The **Cost Center** field will always default from the user profile if the data is available.

WBS Elements

WBS elements are used to charge expenditures to project-related accounting elements when documents are created in SAP Ariba Buying. WBS elements are integrated to SAP Ariba as part of the SAP S/4HANA master data, and all valid WBS elements which aren't locked and can be used should be sent to SAP Ariba Buying for use in the application.

The **WBS Element** field won't default based on any data on the requisition and must be manually selected when it's mandatory. This is driven by the account assignment, which, when set to **P**, will make the **WBS Element** field mandatory and required.

Asset

SAP Ariba Buying has a standard field for the asset accounting object in SAP S/4HANA. The **Asset** field is only visible and required when the account assignment is set to **A**. This account assignment is generally used to track spend done for maintenance against specific assets, if asset data is interfaced from SAP S/4HANA.

Asset data won't default based on any data on the requisition and must be manually selected when it's mandatory. This is driven by the account assignment, which when set to **A**, will make the **Asset** field mandatory and required.

Internal Order

SAP Ariba Buying has a standard field for the internal order accounting object in SAP S/4HANA. This field is only visible and required when the account assignment is set to **F**. Internal order elements won't default based on any data on the requisition and must be manually selected when it's mandatory. This is driven by the account assignment, which, when set to **F**, will make the **Internal Order** field mandatory and required.

Split Accounting

SAP Ariba Buying allows users to split the cost of the line item of the requisition among various accounting combinations. The **Split Accounting** field will be enabled in the SAP Ariba application so that users can charge one line to multiple different costing elements based on the following:

- Quantity
- Percentage
- Amount

This data will be integrated to SAP S/4HANA in the transactional interfaces to communicate the various splits that may apply.

8.2.2 Catalog and Non-Catalog Requisitioning

A catalog-based requisition is created for an item that is selected from a supplier catalog. A supplier catalog item will contain a description of the item, price (with contracted pricing), unit of measure, and other details about the item. There are a variety of catalog options available to suppliers and buyers, including punchouts with varying degrees of sophistication, static catalogs maintained by the supplier, and static catalogs created and maintained by the buyer. As catalogs are a major driver in realizing the compliance and efficiency benefits of SAP Ariba Buying, continuous catalog content refinement and expansion is a key area of solution adoption.

A non-catalog requisition is created by entering data manually into the requisition, which is then reviewed and potentially refined by your organization's procurement department. This is a less efficient process but does at least ensure compliance to buying with a PO and provide an option for uncommon purchasing needs where a user may need assistance from procurement. A primary objective of implementing SAP Ariba Buying and Invoicing is typically to drive compliance to a list of approved goods and services, as well as to enforce compliance with contracts where applicable.

8.2.3 Spot Buy

For nonsourced spend, the Spot Buy feature provides an option to buy using preenabled content from commercially vetted suppliers. This buy channel provides your procurement department with process efficiency and visibility for one-off purchases and nonstrategic categories where sourcing efforts and supplier management aren't required. Your buying organization can opt into as many of the marketplaces and sellers available for a country as you choose. Rather than establishing individual vendors, these items may be purchased directly via the Spot Buy capabilities, which has a variety of settlement options that vary by country. To a casual user, the experience is essentially just another option for selecting items during the requisitioning process; items are just presented in the Spot Buy catalog, as shown in Figure 8.12.

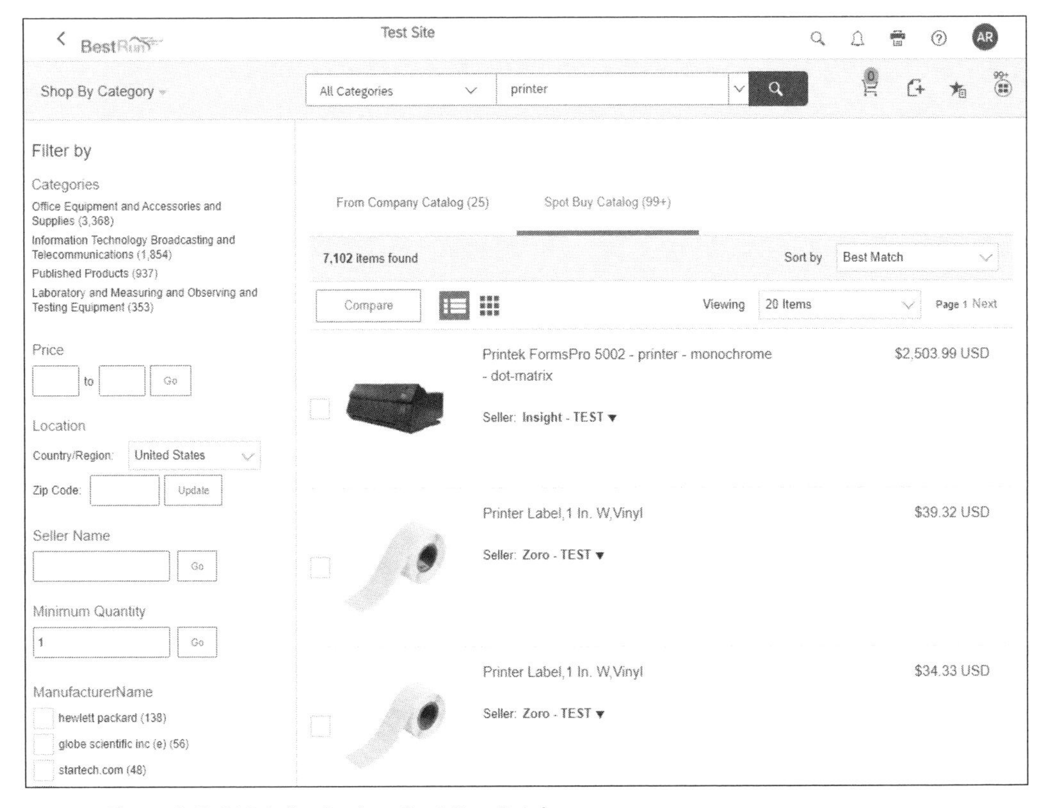

Figure 8.12 SAP Ariba Buying: Spot Buy Catalog

Typically, a Spot Buy approval node is included on the requisition for item purchases to ensure the request is within your company's purchasing policy. Spot Buy can set item restrictions by commodity code and amount. If restrictions are included, then the Spot Buy approval node becomes optional.

The Spot Buy feature isn't enabled by default and must be done by SAP Ariba Support. Once enabled, there are additional setup steps your organization's Spot Buy administrator will need to configure:

1. Create a supplier account for Spot Buy in SAP Ariba Buying.
2. Add remittance location address for the Spot Buy supplier account.

3. Create a relationship between the supplier account and your Spot Buy account on the Ariba Network.
4. Set up the Spot Buy content source (eBay, Mercateo Unite, Mercado Libre, Ariba Network suppliers).

After the initial setup steps have been completed, next you'll want to configure the commodity code and amount restrictions:

1. From the SAP Ariba Buying dashboard, click on **Manage** and then **Spot Buy Administration**.
2. Click on **Marketplace** and review the **Search Filters** options (Figure 8.13). From here, determine which items are displayed to your users. These options include limiting items to new items only or restricted suppliers.
3. Click **Setup • Restrictions** (Figure 8.14). Select a commodity code from the list of categories, and set allow or disallow and the price limit per category.

Figure 8.13 Spot Buy Admin: Search Filters

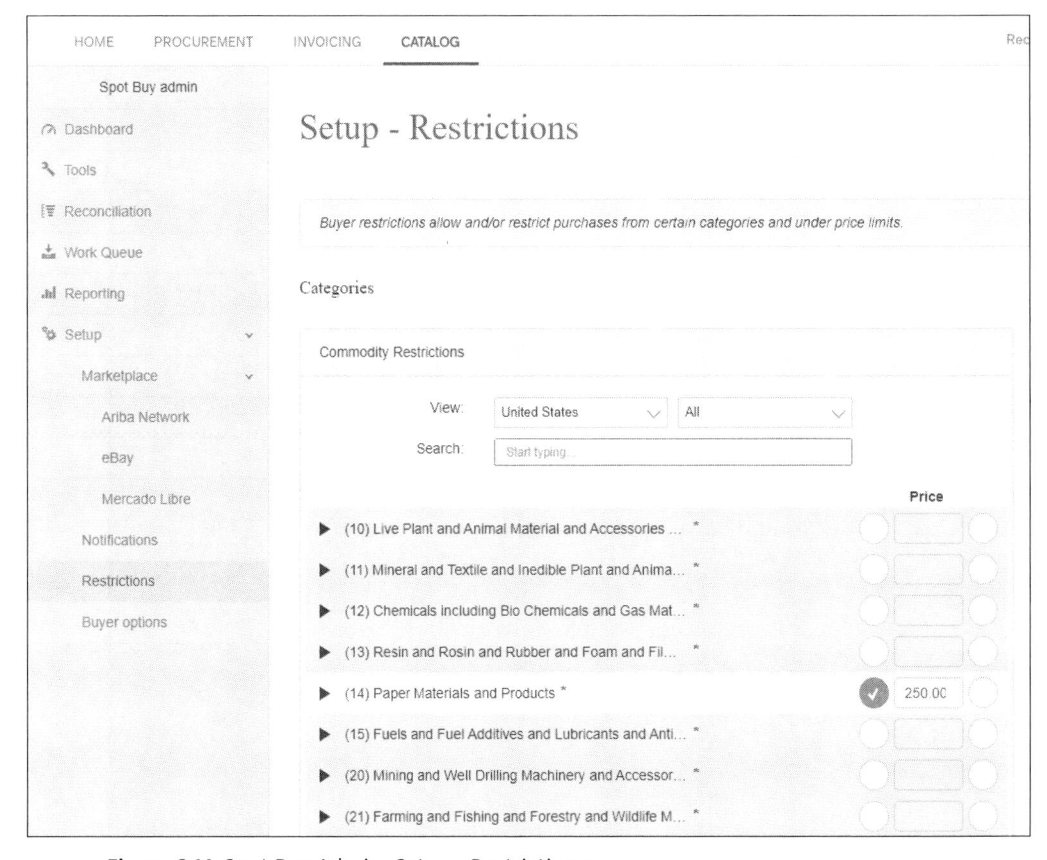

Figure 8.14 Spot Buy Admin: Setup - Restrictions

8.2.4 Total Landed Cost: Taxes, Allowances, and Charges

The *total landed cost* functionality enables your buying organization to create requisitions that have taxes, charges, and discounts on requisition line items, and which are then copied to POs. This gives you better budget compliance and adherence to approval limits and reduces variations between invoices and POs. You can configure SAP Ariba Buying to automatically apply tax information on purchase requisitions. This can be done via a call to an external tax provider or by using master data tax files in SAP Ariba.

Total Landed Cost Workflow

The following steps describe the workflow starting with managing taxes, charges, and discounts in requisitions to managing the same during invoice reconciliation:

1. The buyer adds line items to the requisition.
2. SAP Ariba Buying applies tax codes to the line items based on the tax code lookup or direct call to a third-party tax provider and then displays the tax amounts in the line items section on the requisition summary page.

Note

Tax code isn't applied to Spot Buy and punchout items. The tax amount on a Spot Buy or punchout item is automatically updated on the line items.

3. The buyer enters a discount amount for non-catalog items.
4. The buyer changes the tax codes applied on the requisition line items if the buyer is a member of the tax editor group, as shown in Figure 8.15.

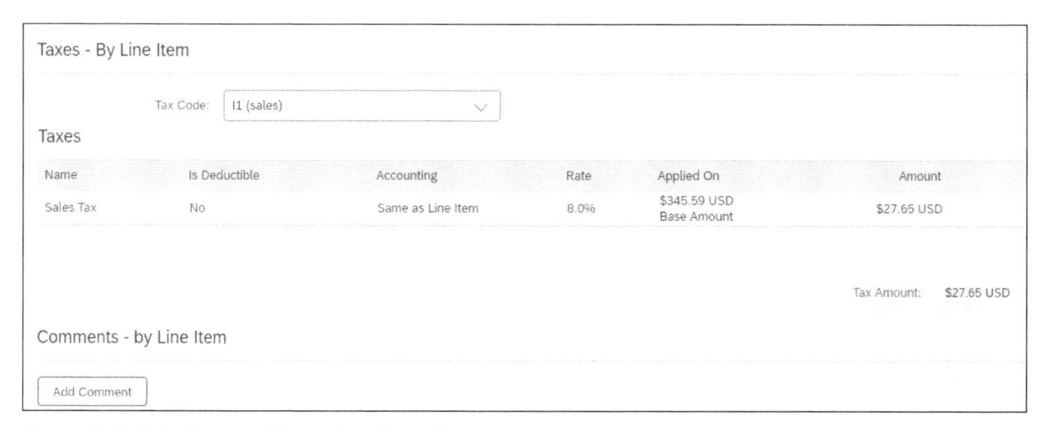

Figure 8.15 Total Landed Cost: Line-Item Tax

5. The buyer submits the requisition.
6. The system adds the tax editor group to the approval flow if configured.
7. The tax editor verifies the tax codes applied on the requisition line items and edits if needed.
8. The approvers approve the requisition.
9. A PO is generated from the requisition. The taxes, charges, and discounts applied on the requisition line items are copied to the PO.
10. The PO is sent to the Ariba Network.
11. The supplier creates an invoice for the PO on the Ariba Network and sends it to SAP Ariba Buying and Invoicing. The charges and discount on the PO are copied to the invoice, as shown in Figure 8.16. In case of paper-based invoices, an accounts payable clerk in SAP Ariba Buying and Invoicing creates an invoice for the PO. In this case, the taxes, charges, or discounts aren't copied to the PO-based invoice.
12. The approvers approve the invoice.
13. The system creates an invoice reconciliation document for the invoice.
14. The system calculates taxes and withholding taxes on the invoice reconciliation document at the header or line level based on the tax code lookup.

15. The system compares the calculated taxes in the invoice reconciliation document with the taxes applied in the invoice and displays invoice exceptions, if any.

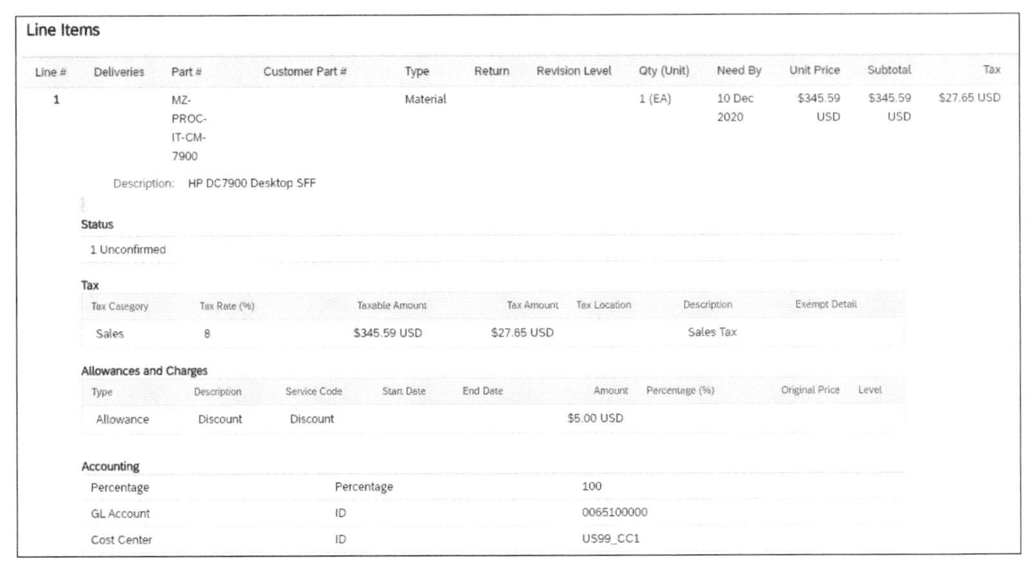

Figure 8.16 Ariba Network: Supplier View of the PO with Estimated Tax

16. The system compares the line-level and header-level tax and charge amounts for the invoice and PO and displays exceptions, if any.

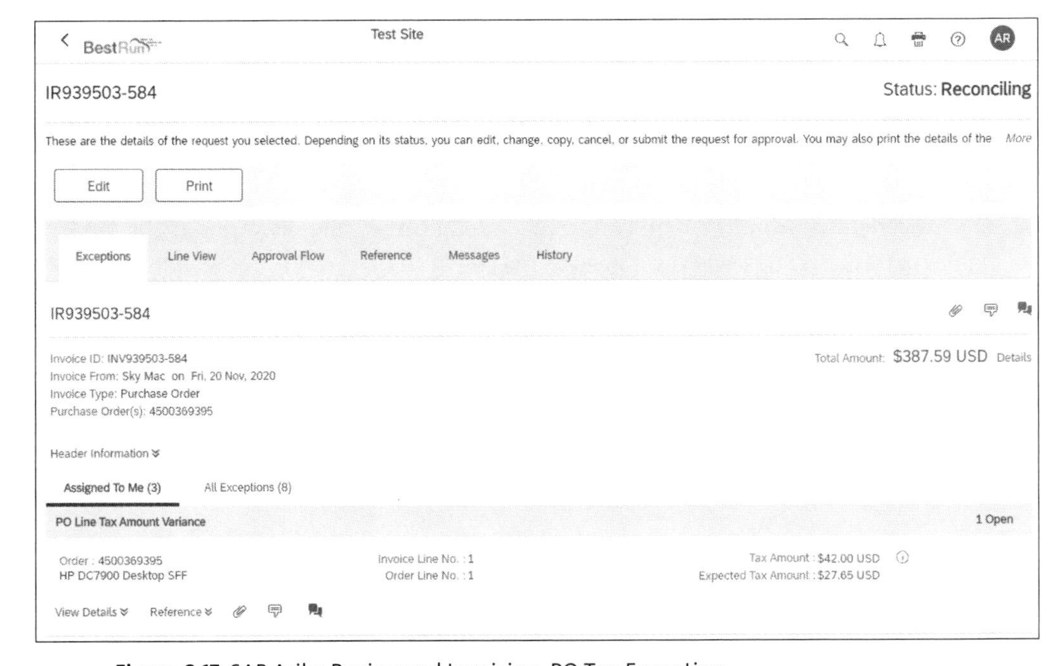

Figure 8.17 SAP Ariba Buying and Invoicing: PO Tax Exception

17. Users with the required permissions resolve the exceptions, and the payment manager reconciles the invoice. In the example in Figure 8.17, a PO tax exception has been raised where the tax the supplier provided on the invoice does not match the tax amount on the PO.
18. Approvers approve the invoice reconciliation document.
19. A payment request is generated for the reconciled invoice and sent to the external system. This payment request has details of the taxes, charges, and withholding taxes applied on the reconciled invoice.

The total landed cost functionality isn't enabled by default. Your buying organization's designated site contact (DSC) will need to request SAP Ariba Support to enable the total landed cost-related parameters. In addition, keep in mind the other total landed cost prerequisites, namely, loading and enabling tax codes and establishing a connection to a third-party tax provider.

8.2.5 Requisition Approval Process

SAP Ariba Buying drives compliance to existing policies by enforcing a preapproval process when the PO is initially requested as a purchase requisition. An important driver for the implementation of policies that drive compliance and control in the procurement process is to ensure that all purchases are in line with your organization's approval framework. When there are no strong procurement policies in place or no controls in place at the time the request is submitted, approval is done only at the point of invoice, well after the commitment has been made to the vendor.

Before the goods are ordered, and anything is sent to the vendor, the requisition must complete a preconfigured approval process, as shown in Figure 8.18. After all required parties have approved the requisition, a PO will be created and sent to the vendor.

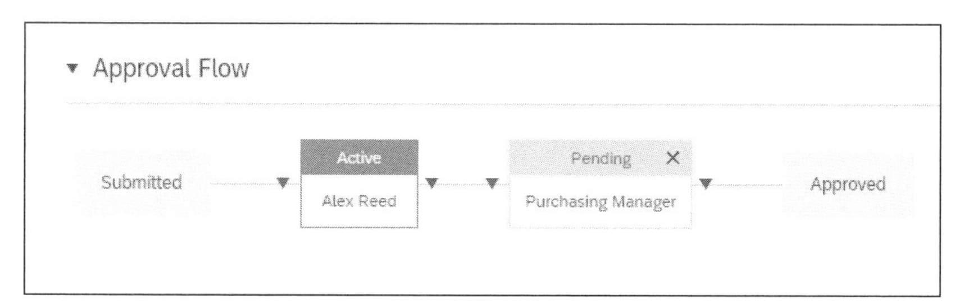

Figure 8.18 SAP Ariba Buying: Submitted Requisition Approval Workflow

Requisition Approval Flow Configuration Steps

Any user with the following roles can configure the requisition approval process:

- Customer administrator
- Purchasing administrator

To configure the requisition approval process, follow these steps:

1. Click the **Approval Processes** link under **Manage** from the SAP Ariba Buying dashboard.
2. Select **Requisition** from the **Approvable Type** dropdown list, and click on the title of the approval process.
3. Click **Edit** to add a new rule or modify an existing rule in the **Approval Process Diagram** section.
4. After changes are made, click **Activate** to make the requisition approval process active.

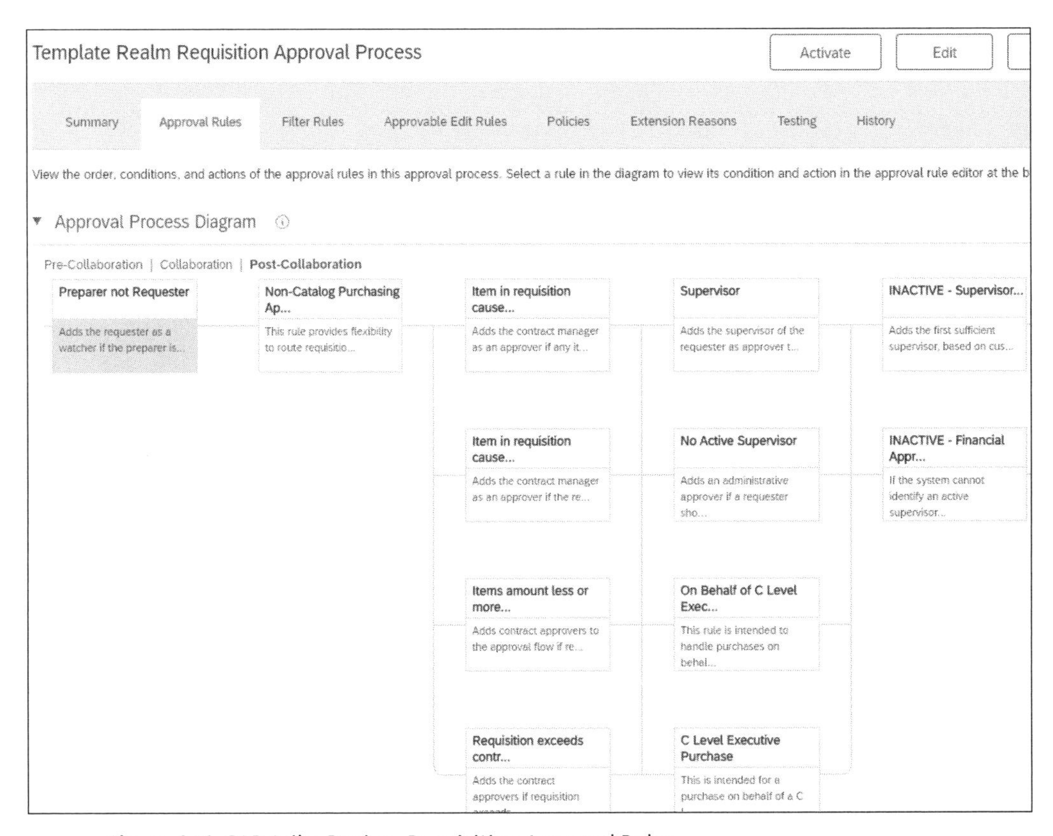

Figure 8.19 SAP Ariba Buying: Requisition Approval Rules

With requisition approval rules in SAP Ariba Buying, you can use the out-of-the-box approval rules as a baseline for your configuration. SAP Ariba Buying is preconfigured with a set of leading practice requisition approval rules you can use out of the box, as shown in Figure 8.19, or you can modify them to fit your requirements. Optionally, you can create new approval rules that match specific requisition approval requirements that may not be covered by one of the out-of-the-box requisition approval rules. As you can see in Table 8.3, the standard approval rules cover a number of different requisition approval scenarios.

Approval Rule	Description
On Behalf of Notification	The requester is added as a *watcher* if the preparer isn't the same as the requester (on-behalf-of user).
Non-Catalog Purchasing Approval	Operational buyers are added for non-catalog items as approvers. The criteria to add approvers, which can be combined in any combination includes the following: ▪ Ad hoc item (non-catalog) ▪ Minimum amount ▪ Commodity ▪ Supplier ID
Contract Release Related Rules	These out-of-the-box rules are intended to catch scenarios where contract release requisitions exceed contractual limits.
Supervisor	The supervisor of the requester is added for verifying the business validity of the request. If the requester doesn't have an active supervisor in the system, it's routed to the exception manager group for review. For a requisition on behalf of an executive in the "no supervisor" group, the executive will need to approve if within their limit and above the limit of the preparer that entered on their behalf. If a requisition on behalf of an executive in the "no supervisor" group exceeds the approval limit of the executive, it's routed to the exception manager group for review.
Financial Approval – Accounting Based	Cost center owner approval applies for all account assignments **K** (Cost Center) or **A** (Asset). Cost center financial approval applies for cost center–based approvals beyond the approval limit of the cost center owner (after owner). WBS element owner approval applies. Internal order owner approval applies.
Financial Approval – Supervisor Chain Based	The approval limit is evaluated of the requester's supervisor, that person's supervisor, and so on until a supervisor has a sufficient approval limit for the requisition total. The supervisor is always added. If the supervisor's limit is exceeded, this rule will also add the first sufficient supervisor in the requester's chain. It doesn't add supervisors between the supervisor and the first in the hierarchy with a sufficient approval limit. This approach balances process efficiency and order expediency with some degree (requester's supervisor responsibility) of business validation prior to reaching the final financial approver.

Table 8.3 SAP Ariba Buying: Standard Requisition Approval Rules

Approval Rule	Description
	If the system can't identify an active supervisor in the requester's chain, it's routed to the approval exception manager group to determine next steps. The WBS element owner approval probably still applies. The internal order owner approval probably still applies.
Functional Commodity Approval	Approval for commodities is provided where obtaining approval outside of purchasing is needed. This rule should be reserved for unique scenarios requiring additional approval (e.g., risk mitigation for hazardous materials).
Supplier-Specific Approval	Approval is given for specific suppliers. It can also be used to notify people by adding them as watchers.
Asset Management	Asset managers validate and edit as required the asset information for lines with the asset account category.

Table 8.3 SAP Ariba Buying: Standard Requisition Approval Rules (Cont.)

8.2.6 Receiving

In the receiving process, users acknowledge when goods/services have been received by entering that information into a receipt document in the system. Receiving also facilitates a three-way match for invoice reconciliation. Depending on the ship-to location, commodity, supplier, monetary threshold, and potentially other factors, receiving responsibilities can vary significantly. If a requester wants to return a product that was previously received and tracked as accepted in the system, the user can create a negative receipt. This allows the requester to take corrective action against an order that was received (accepted or rejected) with errors.

8.2.7 Purchase Order-Based Invoicing

SAP Ariba supports the creation and reconciliation of multiples invoice types:

- PO invoices associated with POs
- Contract invoices associated with contracts that have pricing terms
- Service sheet invoices associated with service entry sheets
- Non-PO invoices that aren't associated with a PO, a contract, or a service entry sheet
- Credit memos that represent an amount owed to the buyer by the supplier

An invoice document can be created using any of the following processes:

- By the supplier on the Ariba Network and automatically transmitted to SAP Ariba Buying and Invoicing. Suppliers can create these invoices online or integrate their own ordering and billing systems with the Ariba Network.

- By suppliers by punching into SAP Ariba Buying and Invoicing from their Ariba Network account to create contract-based invoices
- By SAP Ariba invoice conversion services who receive paper invoices from suppliers and convert them into electronic invoices onto the Ariba Network
- Manually by the accounts payable department in SAP Ariba Buying and Invoicing based on a paper invoice received from the supplier
- Auto-generated by SAP Ariba Buying and Invoicing for evaluated receipt settlement (ERS)-enabled suppliers upon receipt approval
- Auto-generated by SAP Ariba Buying and Invoicing for fixed and recurring fee items on no-release contracts
- Universal charge document, an invoice-like document that is auto-generated by Spot Buy in the Ariba Network
- Purchasing card (p-card), an invoice-like document that is auto-generated by SAP Ariba Buying and Invoicing when the p-card charge statement is loaded

Although the invoice is technically an approvable document, it's intended to be preserved exactly as submitted by the supplier. Apart from validating scanned content when SAP Ariba invoice conversion services are in scope, the only reason for an invoice document to require approval is if the requester (business owner) can't be identified. If the requester isn't identified, the invoice is routed to accounts payable to select an appropriate requester.

When the invoice requires no further approval, an invoice reconciliation document is automatically created. An invoice reconciliation is an approvable document that invoice exception handlers use to resolve discrepancies among the invoice, associated orders or contracts, and receipts. It can also be used to validate and modify accounting and other information. Essentially, it's a working copy of the document that allows the invoices to be edited before being posted to the ERP system.

The automatic reconciliation processor tries to match the invoice reconciliation document to existing POs, contracts, and receipts to facilitate a three-way match process. It then analyzes invoices using comparative logic called *invoice exception types* that evaluate the invoice and identify discrete discrepancies. If no discrepancies are identified, or if the discrepancies are within configured tolerances, the invoice reconciliation document can be sent directly to the responsible group or person for approval to pay the supplier, based on configured invoice reconciliation approval rules.

If discrepancies outside of tolerances are identified, they are listed as exceptions on the invoice reconciliation document and are routed to the appropriate group or person to reconcile the exceptions before further approvals.

After the exceptions are reconciled and the invoice reconciliation is fully approved, the status of the invoice changes to **Reconciled**. The invoice reconciliation moves to the

Paying status, and a payment request is exported to the customer's ERP system, where final payment and settlement occur.

8.2.8 Non-PO Invoicing

Non-PO invoices don't reference a PO, contract, or service entry sheet. Non-PO invoices allow users to pay suppliers for one-time purchases when there is no associated PO. Non-PO invoicing is noncompliant with a regulated spend management process, and its use should be minimized, but most organizations have a need for non-PO invoicing for various reasons. Some examples of non-PO invoicing categories include the following:

- Conference fees
- Professional dues
- Charitable donations
- One-off services

If suppliers send paper invoices, accounts payable can create electronic non-PO invoices in the SAP Ariba invoicing solution. If business rules allow, suppliers can also create non-PO invoices in the Ariba Network. Suppliers can create non-PO invoices in the following cases:

- If there is no corresponding PO
- If a PO wasn't routed through the Ariba Network
- If a PO was routed through the Ariba Network, but later expired and was deleted from the Ariba Network

Suppliers create non-PO invoices using the **Create Non-PO Invoice** task in their Ariba Network accounts to submit an invoice through the Ariba Network. However, their SAP Ariba invoicing solution treats these invoices as non-PO invoices or PO-based invoices, depending on the values the supplier enters in the following fields:

- **Sales Order # only**
  Treated as a non-PO invoice,
- **Customer Order #**
  - Treated as a non-PO invoice if the value in the **Customer Order #** field doesn't match an existing PO on the Ariba Network.
  - Treated as a PO-based invoice if the value in the **Customer Order #** field matches an existing PO on the Ariba Network.
- **Contract #** (valid or invalid)
  Treated as a contract-based invoice. If the contract ID is invalid, the invoice triggers an **Invoice Unmatched** exception.

8.2.9 Contract Invoicing

With the contract compliance feature within SAP Ariba Buying and Invoicing, it's possible to create contracts that define the anticipated invoicing structure for spend that doesn't require a PO. This channel isn't available for SAP Ariba Buying.

This buy channel is most commonly used for simple services and may have either fixed or variable invoices over time. Invoices can be created directly from the contract and are validated against the contract terms.

Note

For topics related to invoice exceptions, see Chapter 9, Section 9.2.6.

Note

For topics related to the Ariba Network transaction rules, see Chapter 9, Section 9.2.4.

8.2.10 Services Procurement

SAP Ariba Buying and Invoicing supports requisitions for service items as well as material items. As shown in Figure 8.20, you indicate that a requisition line item is a service item by selecting **Service** from the **Item Category** dropdown, which will require additional service-specific fields to be entered (**Service Start Date**, **Service End Date**, **Max Amount**, and **Expected Amount**).

Service items are converted into a service PO after the request is approved, and a service entry sheet will be required to confirm the delivery of those services.

Several concepts need to be understood regarding service POs.

- **Service requisition**
  A service purchase requisition or service requisition is the approvable document created when you submit a request to procure service items. Each service purchase requisition is assigned a unique ID to identify and track it as it moves through the requisition process.
- **Service PO**
  Service POs or service orders provide a basic process to procure service items that can be of a planned or unplanned nature of work.
- **Service hierarchy**
  An outline structure represents items in a service order, order confirmation, order inquiry, service sheet, or invoice. If these documents have parent line items (outline items) with child items, there is a hierarchy, and a parent item sits at the top level of the hierarchy. Each top-level parent item represents a different hierarchical structure that can have a nested level of multiple child items.

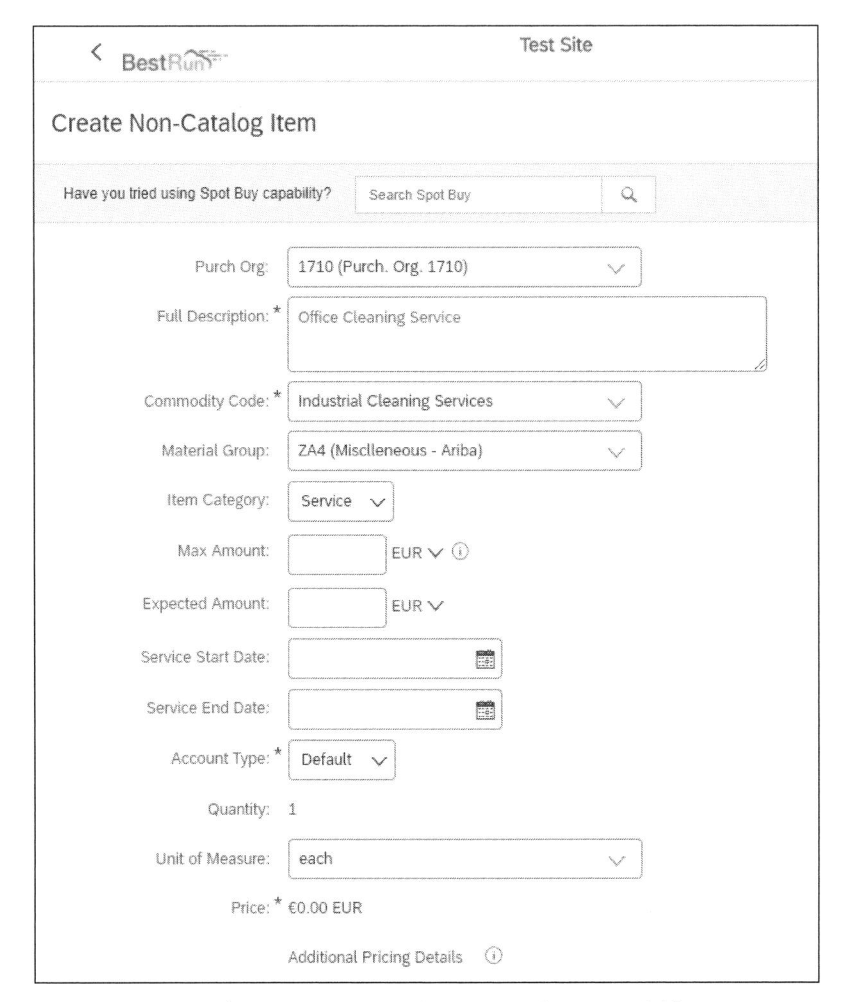

Figure 8.20 SAP Ariba Buying: Requisition Service Line-Item Fields

- **Parent or outline item**
  This is a container item in a service hierarchy. An outline or a parent item can contain a nested level consisting of both service items and material items. When you confirm or reject a parent item, you also confirm or reject all its child items. Service requisitions with only parent line items are also called unplanned services. In an unplanned services scenario, details of the service aren't known by the buyer. The parent line item has a general description of the service to be performed and the expected and maximum amount.
- **Child item**
  This is the item nested below a parent item and also called a subitem or planned service line. The child line has more detail about the specific service being requested.
- **Service item**
  This line item in a service order represents a service to be performed.

- **Material item**
  This line item in a service order represents material goods.
- **Service sheet**
  A service sheet is a document that supplier users performing a service create to describe that service. Service sheets can contain both service and material goods lines. Suppliers can create a service sheet for one or more service lines in an associated service order.

  This document replaces the receipt document that is normally required for goods requisitions.
- **Maximum amount**
  This is the maximum amount allowed for the service item. Suppliers can't submit a service sheet against a service item if the service sheet amount exceeds the maximum amount specified for the service item. Additionally, the expected amount or the total amount for a service line item can't be greater than the maximum amount. Note: Suppliers can't see maximum amounts in service orders.
- **Expected amount**
  This specifies the amount estimated to procure a service item.
- **Field engineer**
  The field engineer is a person in the buyer organization who oversees the supplier's work and knows details about the work completed. This role corresponds to the field engineer user group.
- **Approver**
  The approver is a user in the buyer organization who approves service sheets for invoicing.

Service PO High-Level Process Workflow

In a high-level service PO process workflow, the following occurs:

1. A buyer completes the following tasks:
   - Creates a service requisition for service items
   - Adds catalog or non-catalog items as child items for parent line items
   - Submits the requisition for approval
2. All approvers identified in the approval flow approve the requisition.
3. SAP Ariba Buying creates a service PO. For planned services, the order includes the parent or outline item at the top level with material goods and service lines appearing below it as child items.
4. The supplier completes the following tasks:
   - Logs in to the Ariba Network and opens the service PO.
   - Creates an order confirmation and confirms the top-level items. All children of the parent items are also automatically confirmed when the parent items are confirmed.

  - Starts the activities required to fulfill the service.
  - Submits service sheets for each task completed to fulfill the service using one of the following methods:
    - Using the service sheet wizard to create service sheets against service orders on the Ariba Network
    - Uploading service sheet comma-separated values (CSV) files on the Ariba Network

> **Note**
>
> Buyers can create service entry sheets on behalf of suppliers that aren't enabled for the Ariba Network; to do so, users have to be assigned to the service sheet entry user group.

5. SAP Ariba Buying and Invoicing validates the service sheet. After successful validation, the service sheet enters the approval flow.
6. A member of the field engineer group approves the service sheet.
7. A member of the service sheet editor group identified in the approval flow of the service order edits the accounting information for the service sheet lines.
8. Other approvers in the approval flow for the service sheet complete their approvals. Recommended nodes are as follows:
   - **Service sheet requester**
     Adds the requester specified in the service sheet as an approver in the approval flow if the service sheet has a valid requester.
   - **Order requester**
     Adds the order requester as an approver in the approval flow if the service sheet doesn't have a valid requester.
   - **Contract requester**
     Adds the contract requester as an approver in the approval flow if neither the service sheet nor the order specifies a valid requester and the service sheet items are associated with a contract.
   - **Field engineer**
     Adds members of the field engineer group as approvers in the approval flow if there is no valid requester on the service sheet, order, or contract.
   - **Service sheet editor**
     Adds members of the service sheet editor group as approvers in the approval flow because the service sheet contains invalid accounting information.
9. On the Ariba Network, the supplier invoices one or more approved service sheets.
10. SAP Ariba Buying and Invoicing creates invoice reconciliation documents for the invoices. The appropriate users reconcile any exceptions, edit accounting information, and approve the invoice reconciliation documents.

11. After payments for all the invoices are complete, SAP Ariba and Invoicing changes the status of the service order to **Invoiced**.

8.3 Using Advanced Buying and Invoicing Functions

Catalog and non-catalog requests created by casual users usually don't require immediate intervention by procurement departments. Creating these types of requests can be done easily and intuitively using the guided buying capability by searching company catalogs or filling out a short form for an ad hoc request. There are, however, advanced buying functions that require more expertise by power users to use SAP Ariba Buying more effectively. What follows is a brief overview of these advanced buying topics.

8.3.1 Uploading Requisitions

SAP Ariba Buying power users sometimes have a need to create requisitions with many line items. These requisitions could have complex accounting, as well as different ship-to locations and suppliers per line item. It can be a time-consuming effort to create these types of purchase requisitions. SAP Ariba Buying offers a solution to increase the efficiency of creating complex, multiline requisitions.

This feature allows powers users to define the structure of their requisitions in an offline mode and then import the Microsoft Excel file in SAP Ariba Buying to create, update, or cancel requisitions.

Authorized users can also perform bulk import operations to create, update, or cancel multiple sets of requisitions put together in a ZIP file format. In addition, users can export an existing requisition to their systems in Excel format and use it as a template to make changes to existing requisitions or create new requisitions by importing the Excel file.

In Figure 8.21, you can see the upload file is prepared using an Excel requisition template that has three tabs:

- **Requisition Header**
- **Requisition Detail**
- **Requisition Split Accounting**

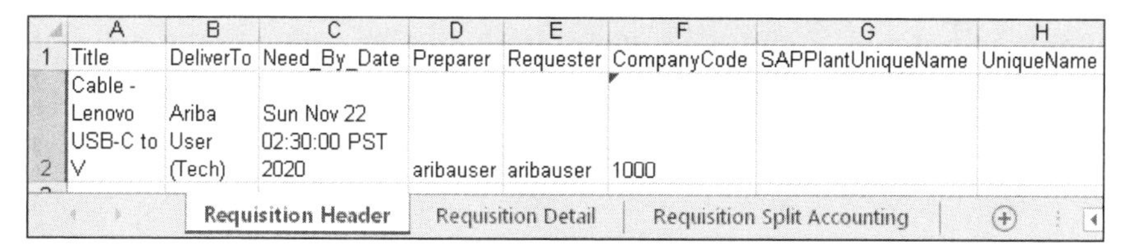

	A	B	C	D	E	F	G	H
1	Title	DeliverTo	Need_By_Date	Preparer	Requester	CompanyCode	SAPPlantUniqueName	UniqueName
2	Cable - Lenovo USB-C to V	Ariba User (Tech)	Sun Nov 22 02:30:00 PST 2020	aribauser	aribauser	1000		

Figure 8.21 SAP Ariba Buying: Excel Requisition Upload File

After the details of the requisition are added to the file, the next step is to upload into SAP Ariba Buying, as shown in Figure 8.22.

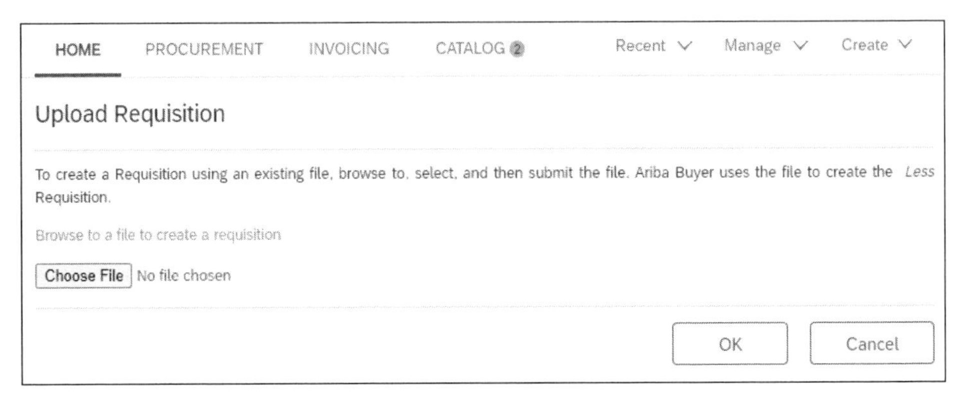

Figure 8.22 SAP Ariba Buying: Upload Requisition File Selection

A batch job runs in the background, and the system provides a job ID to monitor the progress. After the job has completed, the system will display the **Requisitions** number and **Status** of the requisition, as you can see in Figure 8.23.

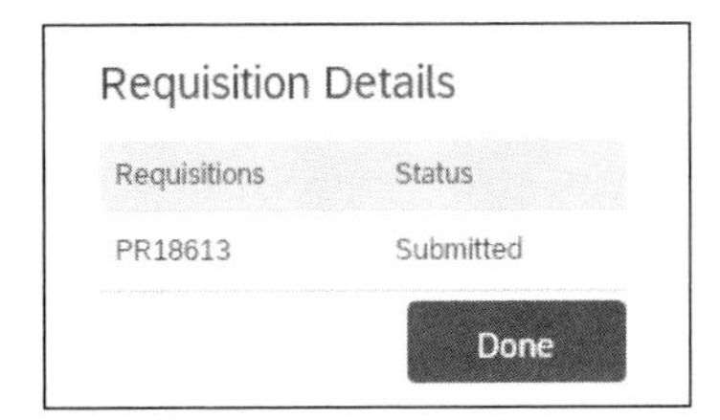

Figure 8.23 SAP Ariba Buying: Requisition Upload Details

Keep in mind that this is an advanced feature and not intended for casual users. Creating the upload file in the correct format can take time; however, for the experienced user, it can make ordering complex, multiline requisitions easier.

8.3.2 Creating Planned Service Orders

Two types of standard services orders can be created in SAP Ariba Buying and Invoicing: planned services and unplanned services. Planned orders are created when you know the exact details of the service to be requested and what will be performed or provided by the supplier. The second type, the unplanned service, is only an estimate of what services might be performed. With this type of service order, any requester who has basic knowledge of the requested service can create this type of service order. Planned services use a service hierarchy concept to construct the service request.

Note

Casual users can create unplanned service requests in guided buying or in SAP Ariba Buying and Invoicing. Planned service requests can only be created in SAP Ariba Buying and Invoicing.

Creating a planned service in SAP Ariba Buying and Invoicing is done with the following steps:

1. From the SAP Ariba Buying and Invoicing dashboard, click on **Create Requisition**. This will take you to the **Catalog** home page. From here, click **Add non-catalog item**.
2. Enter the required fields on the non-catalog form as shown in Figure 8.24, and select **Service** from the **Item Category** dropdown. Additional service-related fields are displayed: enter the **Max Amount** and the **Service Start Date** and **Service End Date**.
3. From the **Summary** screen, add child lines to the parent line you just added. The child lines will be the details of the service you're requesting. As you can see in Figure 8.25, the child line is nested under the parent line item.

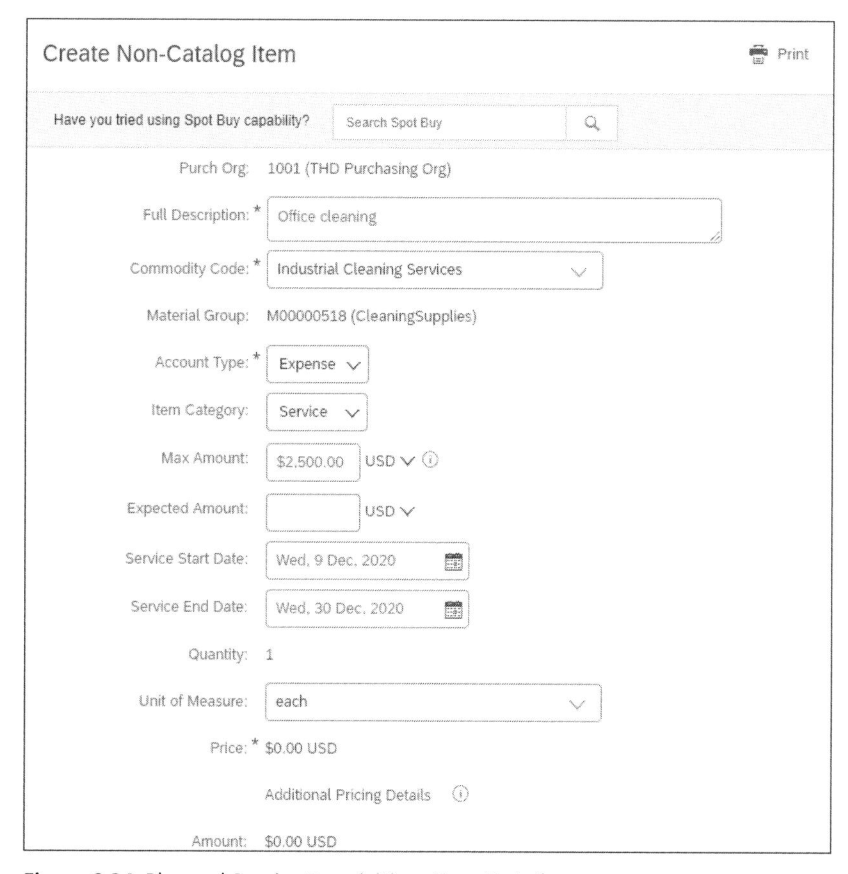

Figure 8.24 Planned Service Requisition: Item Details

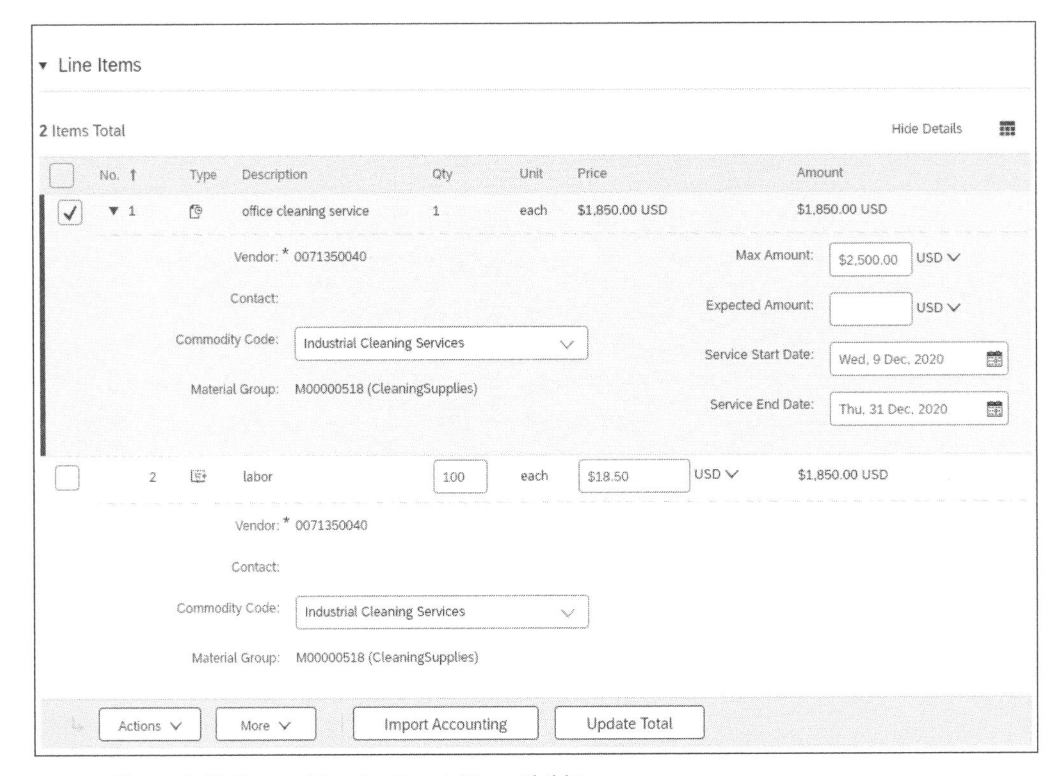

Figure 8.25 Planned Service Requisition: Child Item

4. After the service requisition parent and child lines are added, the request is submitted and goes through the requisition approval process. Once approved, the service PO is created and sent to the supplier on the Ariba Network.

8.3.3 Creating Service Entry Sheets

In SAP Ariba Buying and Invoicing, service sheets are a key component to the services procurement process. They serve as documented proof that the service requested was performed by the supplier. It's the leading document for invoicing and payment to the supplier. That is, after a service sheet has been submitted by the supplier on the Ariba Network and then approved by an authorized person in SAP Ariba Buying and Invoicing, there are no further approvals needed to move forward with invoicing and payment.

In the typical services procurement process, service sheets should always be provided by the supplier as it's their proof that they performed the work. However, there are some cases where buyers, your purchasing department, or even end users will need to submit a service sheet on behalf of suppliers by creating it directly in SAP Ariba Buying and Invoicing. Follow these steps to create a service sheet in SAP Ariba Buying and Invoicing:

1. From the SAP Ariba Buying and Invoicing dashboard, use the search function to find your service PO.

> **Note**
>
> Service sheets can't be created as standalone documents; they must reference the service PO. Therefore, the only way to create a service sheet is by going to the PO first.

2. Click **Create Service Sheet** from the PO. If the service sheet is for a planned service order, the child lines from the order are automatically copied to the service sheet, and quantities or amounts can be adjusted within tolerance. If the service is unplanned, proceed to add all line items to the service sheet by clicking **Add Items**.
3. After items have been added and quantities and amounts have been adjusted, you can review the accounting to ensure proper cost allocation for each line. Standard configuration derives the accounting for the PO line item, but if it needs to be changed, that can be done before submitting the service sheet.
4. After submitting the service sheet, additional approvals may be needed. Once fully approved, a copy of the service sheet is sent to the supplier on the Ariba Network. The invoice document can then be created by the supplier and submitted based on the service sheet copy.

8.3.4 Evaluated Receipt Settlement

ERS is a business process between the buying organization and the supplier that allows companies to conduct commerce without supplier-created invoices. In an ERS transaction, the supplier ships the goods ordered, and the purchaser confirms the existence of a corresponding PO or no-release contract with receiving and then verifies the identity and quantity of the goods via the receiving process. When the receipt is approved, the invoice is automatically generated and processed based on the quantity received and the price and accounting details from the PO or no-release contract line item.

The **ERS Allowed** field is added to the line item on requisitions and directly invoiceable contracts to enable an item as an ERS item. The defaulted value of this field is determined by supplier data. When receipts are created against **ERS Allowed** items, the system will automatically create an invoice by considering the quantity from the receipt and price from the ordered item. A new invoice will be generated and sent to the supplier through the Ariba Network for both enterprise and standard account-enabled suppliers. A new attribute **ERS** will be added in the supplier master, and this value will be considered during the requisition creation at the item level. The value of this field will default the **ERS Allowed** value to **Yes** or **No**.

Supplier data determines if suppliers are set to **ERS Allowed**. Data import tasks include the ERS flag to enable supplier locations for ERS transactions. Setting the ERS flag as Yes on the supplier master data defaults the **Yes** value on all requisition line items for the **ERS Allowed** field. If the flag is set to No, the **ERS Allowed** value on the line item will be defaulted to **No** but can be still set by users in the purchasing manager group on the requisition, as shown in Figure 8.26.

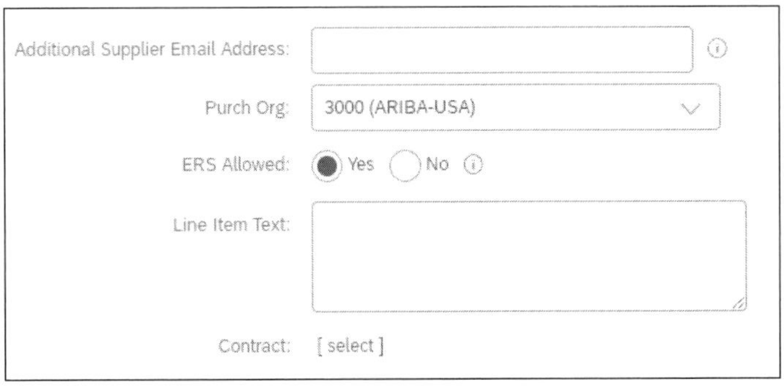

Additional Supplier Email Address:
Purch Org: 3000 (ARIBA-USA)
ERS Allowed: Yes No
Line Item Text:
Contract: [select]

Figure 8.26 ERS Allowed Field on the Requisition Line Item

When the PO is sent to the supplier in the Ariba Network, the cXML message includes an extrinsicfield to prevent the supplier from creating an invoice, as shown in Figure 8.27.

```
- <BillTo>
   - <Address isoCountryCode="US" addressID="3000">
       <Name xml:lang="en"> New York </Name>
     - <PostalAddress name="default">
         <Street>691 Broadway</Street>
         <City>NEW YORK</City>
         <State>NY</State>
         <PostalCode>10001</PostalCode>
         <Country isoCountryCode="US">United States</Country>
       </PostalAddress>
         </Address>
      </BillTo>
      <PaymentTerm payInNumberOfDays="0"> </PaymentTerm>
      <Extrinsic name="Ariba.availableAmount">2222</Extrinsic>
      <Extrinsic name="Ariba.invoicingAllowed">No</Extrinsic>
      <Extrinsic name="Ariba.collaborationAllowed">No</Extrinsic>
      <Extrinsic name="Ariba.contractType">Item</Extrinsic>
      <Extrinsic name="List of authorized users"/>
      <Extrinsic name="Hierarchical Type">StandAlone</Extrinsic>
      <Extrinsic name="Evergreen">No</Extrinsic>
     </OrderRequestHeader>
     -<ItemOut lineNumber="1" quantity="1">
```

Figure 8.27 PO cXML with ERS Flag

8.4 Implementing SAP Ariba Buying and Invoicing and SAP Ariba Buying

In the next few sections, we'll focus on implementing the two main versions of SAP Ariba procurement solutions: SAP Ariba Buying and Invoicing and SAP Ariba Buying.

8.4.1 SAP Ariba Buying and Invoicing and SAP Ariba Buying Projects

SAP Ariba Buying and Invoicing and SAP Ariba Buying, as well as variations of the two, generally involve the following implementation steps:

1. **Supplier enablement**
   When implementing SAP Ariba Buying, a buyer typically looks to onboard suppliers not yet transacting in the Ariba Network. Suppliers can join at a number of levels, but, at minimum, they need to be set up to receive POs.
2. **Flight planning**
   Flight planning is more of a project management topic, as it relates to supplier enablement. SAP Ariba projects are typically rolled out in phases, and coordination with suppliers during the different phases, as well as coordination of resources and timelines, is usually required. Flight plans help automate and standardize this process of onboarding suppliers into the new system in their designated roles and in an optimal manner.
3. **Catalog enablement**
   As discussed earlier in Section 8.1, catalogs of items you require can be managed in the Ariba Network via SAP Ariba Catalog. Suppliers can log in to the business network and upload catalogs. Suppliers can also host punchout catalogs and allow for SAP Ariba Procurement users to "punch out" to their hosted catalogs. Lastly, suppliers can send data to you for review and can upload the data into the catalogs you maintain in SAP Ariba Catalog.
4. **Change management**
   Often overlooked, change management is a crucially important element of SAP Ariba projects. Unlike specialized applications or processes, which affect a limited number of employees, the procurement processes established in SAP Ariba Buying typically impact a large number of employees and suppliers in day-to-day activities. Inadequate training and communication with the different community members, primarily administrators, buyers, requisitioners, approvers, and suppliers, can lead to the failure of an otherwise flawless implementation and system. Change management is typically lead by your in-house resources, with guidance from SAP Ariba or third-party consultants.
5. **Technical workstreams**
   This area includes integrating/interfacing procurement documents and processes with the customer's ERP environment and the Ariba Network. This area is sometimes included in the SAP Ariba Buying workstream, with SAP Ariba Buying supporting various initial requisitioning and PO steps in the process, before transferring the requirements down to an ERP system such as SAP S/4HANA. Topics such as integration, master data conversions, and data optimizations all fall under the technical workstream of an SAP Ariba Buying implementation.

In this section, we'll provide a high-level overview of several integration options, while Chapter 10 provides detailed approaches and configuration steps for the main options for implementing SAP Ariba Buying with SAP S/4HANA.

Supplier Enablement

Supplier enablement requires planning and execution steps for an SAP Ariba Buying project. In traditional on-premise software environments, suppliers still require notifications and support during the transition to a new system on your side. New master data requirements, order formats, and transmission methods all can affect suppliers and how they conduct business with a customer. Due to SAP Ariba Buying's interaction with the Ariba Network, supporting your supplier's transition to transacting with you via SAP Ariba Buying is paramount, as the supplier not only needs to understand the new process but may also need to join the Ariba Network if not already enrolled. Both supplier enablement and flight planning in SAP Ariba Buying pertain to enabling and migrating suppliers to the new SAP Ariba Buying environment. The supplier enablement process is covered at length in Chapter 2.

Flight Planning

Procurement is a business area that necessarily interacts with the outside world to obtain the goods and services a company requires as inputs for production (direct procurement) and as supporting goods and services underpinning operations (indirect procurement). Without cooperation from suppliers, no procurement solution will be entirely successful. A flight plan addresses this need by providing a guide for the enablement of suppliers and categories of spend to ensure successful adoption of SAP Ariba solutions.

On an SAP Ariba Buying project, flight plans are created using a combination of automated tools, which include best practices, as well as fine-tuning the plan to meet customer-specific needs and goals. Flight plans are created and refined throughout SAP Ariba implementation, including during the sales phases of the project and continuing on after the project enters production.

You'll manage the flight plan creation process throughout the various phases. Depending on the phase, the creation and build-out could be managed by different stakeholders and process owners. The main phases can be broken out into three distinct steps: data preparation, flight plan generation, and customer-specific updates, as shown in Table 8.4. Often, the data preparation phase occurs at a high level during the sales phase. You'll provide SAP Ariba with supplier data to verify which suppliers are already active on the Ariba Network. As you begin transitioning into a customer, the flight plan is generated and refined during the project. You'll then review the plan and provide additional feedback to fine-tune this flight plan to your needs and constraints, as shown in Table 8.4.

Preparation	▪ Request data from prospect/customer (the most recent 12 months of spend). ▪ Enrich customer spend data through a match and classification process (analytics). ▪ Identify highly addressable spend by in-scope solution. ▪ Identify spend that would not be considered addressable by in-scope solution. ▪ Review with customer the business case and drivers.
Flight Plan Generation	▪ An automated tool segments suppliers into waves and prioritizes based on SAP Ariba best practices. ▪ Prioritization is based on supplier attributes such as spend, volume, Ariba Network match, Ariba Network relationships, buying process group, and discount propensity. ▪ Depending on the solutions in scope, category enablement considerations are balanced with supplier enablement considerations.
Customer-Specific Updates	▪ Review the flight plan with the customer, and apply customer-specific updates to meet requirements outside of standard SAP Ariba best practices. ▪ Customer-specific considerations may include regional restrictions, user adoption, backend systems, and resource availability.

Table 8.4 Flight Plan Creation

Suppliers are selected for the various onboarding waves according to a variety of criteria. The automated analysis looks at percentage of spend, buying process, transaction volumes, geography, discount propensity, and existing Ariba Network scores, and then provides an overall score based on this assessment.

You can then overlay this analysis with internal considerations, such as the business units most focused on the project (if a business unit is driving this project, then its suppliers will likely receive special attention during the flight plan process). Your project will depend on the availability of resources, as well as required integration to backend systems. (Does a supplier need to receive documents directly into its system, or will the supplier log in to the Ariba Network to process documents?) While integrated suppliers are typically higher volume and spend, the integration may take longer than suppliers who can process documents in the Ariba Network, influencing the time frame and phase in which certain suppliers can onboard. Lastly, additional SAP Ariba Buying design items are required to support particular transactions with the supplier or to support general considerations on the customer side, and these factors can also influence the sequencing of supplier onboarding.

Ultimately, the shared goal of all parties involved is to realize the most savings, revenues, and efficiency as quickly as possible. A typical flight plan output, shown in Figure

8.28, will try to capture as much of the addressable spend of the project as quickly as possible, broken down in phases.

SAP Ariba Procurement Flight Plan Summary										
SAP Ariba Procurement Wave (Final)	SAP Ariba Procurement Scope	Supplier Count	Invoice Spend	Invoice Count	PO Spend	PO Count	AN Match Suppliers	AN Match Spend %	AN Match Invoice %
1	Network Enablement	101	106,949,593	53,943	0	16,967	63	63%	37%
2	Network Enablement	231	542,299,742	70,777	0	48,657	150	82%	14%
3	Network Enablement	59	247,528,164	40,009	0	19,661	33	37%	21%
4	Network Enablement	3	450,171	7,070	0	0	2	84%	91%
5	Network Enablement	16	7,552,818	20,392	0	0	11	82%	75%
6	Network Enablement	5	1,044,078	2,828	0	0	2	23%	1%
7	Network Enablement	1	553,817	11	0	0	1	100%	100%
Out of Scope-Low Doc Count/LE Ineligible	Out of Scope-Low Doc Count/LE Ineligible	234	19,662,929	469	0	34	139	56%	58%
Out of Scope-Low Inv Spend	Out of Scope-Low Inv Spend	237	2,026,178	27,494	0	13,621	108	46%	22%
Out of Scope-Non P2X Process Group	Out of Scope-Non P2X Process Group	3,727	118,687,955	124,991	0	7,298	369	42%	36%
Grand Total		**4,614**	**1,046,755,446**	**347,984**	**0**	**106,238**	**878**	**64%**	**32%**

Figure 8.28 Flight Plan Output

In the flight plan shown in Figure 8.28, the majority of the addressable spend, POs, and invoices are covered in the initial wave. A best practice and goal when creating flight plans is to address the "low-hanging fruit" first and then focus on the remaining, often more difficult or infrequently used, suppliers and their transactions. The flight plan provides both summary-level and supplier-level analyses for your review.

Creating a flight plan involves three main stages:

- **Planning phase**
  In this phase, you'll meet with stakeholders, define the spend and supplier selection, and analyze your current processes.
- **Architecting phase**
  In this phase, you'll define business and functional requirements, develop the design specifications, and validate design concepts with both the project stakeholders and the suppliers directly.
- **Deployment phase**
  In this phase, you'll configure the solution and enable suppliers, contracts, catalogs, and invoicing. Finally, you'll initiate training for the categories deployed in the system and communicate the launch.

Finally, the plan is tracked according to its actual execution over the different waves.

Catalog Enablement

In addition to the catalog functionality, the core features of SAP Ariba Catalog are as follows:

- **Contract compliance**
  SAP Ariba Catalog allows for items to have a contract number maintained, which references an existing contract in SAP Ariba Contract Management and allows for the enforcement of contract terms and their inheritance by contracts, as well as consumption and reporting during and after the transaction.
- **Collaborative shopping cart**
  This in-catalog shopping cart option can fine-tune catalog item orders further before going through approvals.
- **Approval rules**
  Approval rules allow catalog managers to manage content as well as manage additional optional workflows in the catalog to augment approval flows for user profile changes, shopping carts, and contract requests.
- **Reporting**
  Catalog managers have access to standard reports to view content usage and other key performance indicators (KPIs).

Three types of catalogs can be loaded or accessed in SAP Ariba Catalog: supplier provided/loaded catalogs, customer loaded catalogs, and punchout catalogs hosted directly by the supplier.

The punchout process involves the buyer, the supplier, and the Ariba Network. Punchout messages are routed through the Ariba Network for validation and authentication. This process, shown in Figure 8.29, involves the following steps:

1. The buyer selects a punchout item from a supplier catalog in the SAP Ariba Catalog solution. This selection sends a request to the Ariba Network to establish a connection with the remote catalog.
2. The Ariba Network authenticates the buying organization and forwards the request to the supplier's punchout site.
3. The supplier sends back a URL of a web page on the supplier's punchout site designed specifically for the buyer. The procurement system redirects the user to this URL. The remote shopping site appears in the user's window, and the user begins shopping.
4. After shopping, the user clicks the site's **Check Out** button, which removes the items from the shopping cart and triggers the rest of the sales process.

Two main loading options for catalogs exist. The first and more efficient option is to allow suppliers to provide catalogs via the Ariba Network, which can be done via a business network, a local network, or generated subscription, as discussed earlier in Section 8.1.1.

Parameters can be found in the **Catalog Manager** section of the Core Administration tool to control automatic downloading of catalogs.

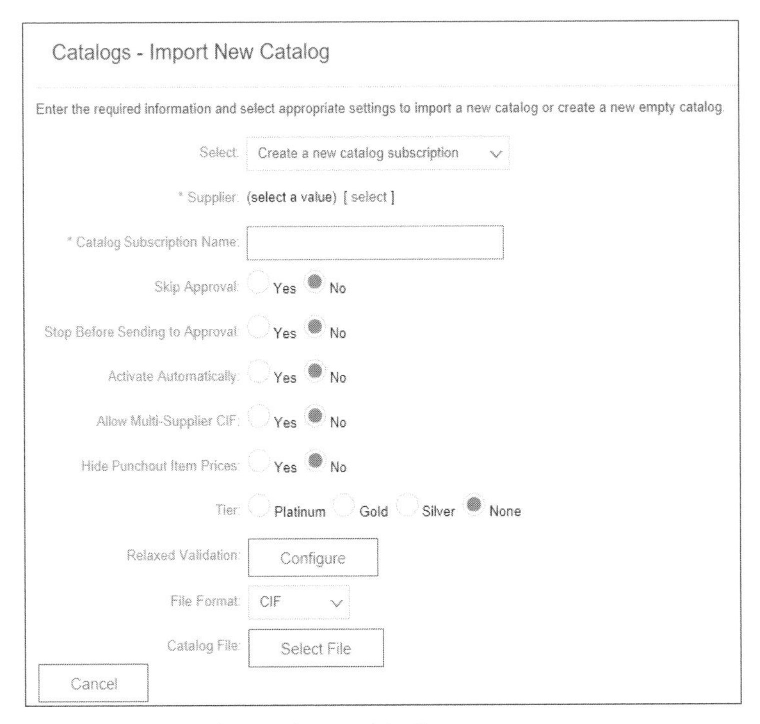

Figure 8.29 SAP Ariba Catalog Load Options

The **Subscription** section in SAP Ariba Catalog displays current catalog subscriptions, their versions, and statuses. In this section, you can perform the following activities:

- Activate or deactivate a subscription.
- Delete versions or entire subscriptions.
- Compare versions of the same subscriptions to understand the differences among them. The resulting report can be emailed or printed.
- View supplier details.
- View version validation information or content.
- Fix validation errors online by editing erroneous catalog items. Click on a specific version marked **Validation Error**.

The second option is to load customer catalogs via catalog interchange format (CIF) catalogs. Not all catalogs may be supplier provided via the Ariba Network. Some catalogs are internal to your company, containing items such as material masters from SAP S/4HANA or from suppliers who don't offer catalogs via the Ariba Network. These catalogs can be uploaded to SAP Ariba Catalog using the core administration console in SAP Ariba Catalog. The typical format for these catalogs is CIF. The core administration console is accessed by administrators from the SAP Ariba Catalog home page.

CIF files are different from Excel spreadsheets and should not be edited in Excel. CIF files are more similar to CSV files and should be edited only in a text editor. SAP Ariba Catalog leverages the LOADMODE parameter setting for loading CIF catalogs. The LOADMODE parameter in the CIF header determines whether SAP Ariba Catalog will perform a full, destructive load of the items or an incremental update. CIF catalogs contain a header, body, and body trailer sections. The CIF also contains the commodity classification code for each item so that these codes can be properly mapped into your commodity code structure.

The SAP Ariba Catalog timestamp on a CIF file determines whether the file is an update on an existing file, which prevents the same file from being uploaded more than once. As a result, your catalog managers should update existing catalogs using the same CIF file name as previous versions, leveraging the timestamp as the trigger for updates, along with setting the updates to incremental. Each customer typically creates their own CIF template for catalog managers to maintain and update customer-loaded catalogs. Catalog managers populate their catalogs using the template and then upload the template directly into SAP Ariba Catalog or into a loading area, where periodically SAP Ariba Catalog pulls the new materials up into the system.

A typical content management project or project phase includes outlining the different types and number of catalogs in scope, update frequency required, loading processes, and optional configuration items, such as additional workflows to be configured internally in SAP Ariba Catalog. Catalog managers and SAP Ariba Buying subject matter experts (SMEs) drive this project and delivery as well as ongoing maintenance of SAP Ariba Catalog, typically. More than other solution areas, catalogs are continually updated, added, or removed in procurement. As such, defining ongoing maintenance and subscription protocols are of utmost importance. Catalog managers also need to be included at the onset to learn how to manage their new SAP Ariba Catalog environment on an ongoing basis.

Internal Adoption

Within any system implementation project affecting end users, the internal adoption of the new system is required. With SAP Ariba Buying implementations, the need for organized, well-executed change management is particularly acute, given the large and casual user population typically requisitioned at a company. These users require communication and training, as do the buyers, suppliers, and approvers of the new system. As shown in Figure 8.30, internal adoption entails the following:

- Executing activities related to human performance impacts
- Managing the "people aspects" of new technology and processes
- Providing stakeholders with the knowledge to be successful
- Ensuring users have what they need to embrace a new way of doing the job
- Mitigating the risks associated with the project

A big risk for any project involving technology and people is the possibility that end users of the new system don't adopt the new systems and processes, thereby undermining the project's objectives. In a worst-case scenario, an organization can experience negative productivity impacts and losses as users devise workarounds and paper-based or manual processes to avoid working in and with the new system. Change management is thus of utmost importance and should be a central theme for any SAP Ariba project.

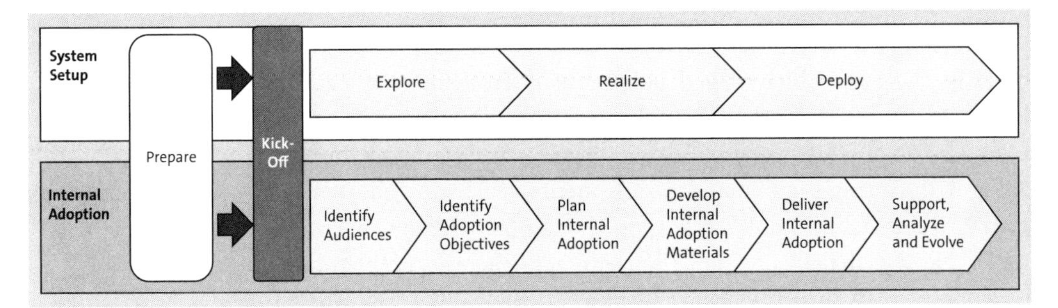

Figure 8.30 SAP Ariba Buying Internal Adoption Steps

The steps in an SAP Ariba internal adoption project comprise six distinct phases, as shown in Figure 8.30. In each step, the following actions are taken:

1. **Identify audiences.**
   For an SAP Ariba Buying project, the stakeholders for change management are primarily the suppliers and end users doing the requisitioning, as well as the buyers and approvers. In addition, accounts payable processes and people will need to be considered, as well as manufacturing, forecasting, materials management, and sales and distribution groups that may have specific touch points with the system and processes.
2. **Identify adoption objectives.**
   In this phase, you'll conduct a review of the strategic objectives of the program, audience, and timeline.
3. **Plan internal adoption.**
   To facilitate the change process within the company and supplier base, a plan around timing, required resources, and core phases will be finalized.
4. **Develop internal adoption materials**
   Any education, communication, or training materials are then created to support the program rollout.
5. **Deliver internal adoption**
   In this phase, all the strategy, objectives definition, and supporting materials will converge in the execution phase.

6. **Support, analyze, and evolve.**
   During the run phase of the project, further fine-tuning and analysis will be required to expand the footprint and success of the project for both buyers and suppliers.

Technical Workstreams

Technical-related workstreams are supported primarily by the SAP Shared Services group and your corresponding on-site customer, consultant, and partner resources. These workstreams focus on the following:

- **Master data**
  Conversion of master data, such as supplier information, accounting data, units of measure, categories, and so on, for eventual upload into SAP Ariba Buying.
- **SAP Ariba Cloud Integration Gateway for buyers**
  Integration platform designed for SAP ERP and SAP S/4HANA systems. This platform allows for the direct, or mediated, transfer of master data and transactional data from and to SAP Ariba Buying to SAP S/4HANA.
- **SAP Ariba Integration Toolkit (CSV)**
  Integration topics for SAP Ariba Buying linking a non-SAP ERP environment with SAP Ariba Buying activities (discussed in more detail in Chapter 10).
- **Web services**
  Calls from SAP Ariba Buying to the source system to check validity/availability of items and cost assignments, for example.
- **Ok2Pay (SAP Ariba Buying and Invoicing)**
  Message sent from SAP Ariba to an ERP backend, endorsing payment of an invoice.
- **Other technical services:**
  - Application management: General support and management of applications.
  - Adapter integration: Integration with ERP for document exchange via adapter or direct connection.

For many SAP Ariba Buying deployments, buyers choose to have invoices sent from suppliers directly to their ERP system. As a result, POs and invoices are sometimes exchanged between the Ariba Network and another procurement application, such as SAP S/4HANA.

Such process flows may necessitate integrations between SAP Ariba Buying and SAP S/4HANA, typically for exporting POs and receipts. Additional integrations from SAP S/4HANA and the Ariba Network may also be required. These integrations often leverage SAP S/4HANA add-ons such as SAP Ariba Cloud Integration Gateway for buyer integration or SAP Process Orchestration with the Ariba Network Adapter, as part of the architecture. Adapters enable document conversion from the original format to cXML to communicate with the Ariba Network.

Systems integrations leads are SAP Ariba resources familiar with adapter-specific implementations. These resources explain the project plan and can work with your team to define document mappings. For SAP integrations, further functional and technical specialists may be needed on the SAP S/4HANA side to cover integration nuances in the SAP S/4HANA environment.

8.4.2 SAP Ariba Buying Deployment

An SAP Ariba Buying deployment project doesn't follow the traditional ASAP methodology and goes beyond the high-level framework of SAP Activate. An SAP Ariba Buying project is more of an amalgam of a packaged service implementation using a start-deploy-run approach and the phases traditionally covered in an SAP Activate project: prepare, explore, realize, and deploy. In addition, there are multiple workstreams, or a project within a project, in the form of the supplier enablement and change management projects, as we've described earlier in Section 8.4.1.

An SAP Ariba Buying deployment divides into two main sections: the architect and enable phase and the solution setup. Unlike traditional on-premise projects, where kickoffs and project preparation often comprise a single week, more focus and investment are made in this area up front, so you can determine organizational readiness and the resources needed to deliver this project.

Figure 8.31 shows the various streams that flow in tandem after kickoff. Not only are the supplier enablement and internal adoption programs running in tandem to the configuration and custom requirements realization, the planning and design phase of the project is considerably condensed, and the solution setup steps overlap. (Design can be conducted while the build and prep govern for other project elements.)

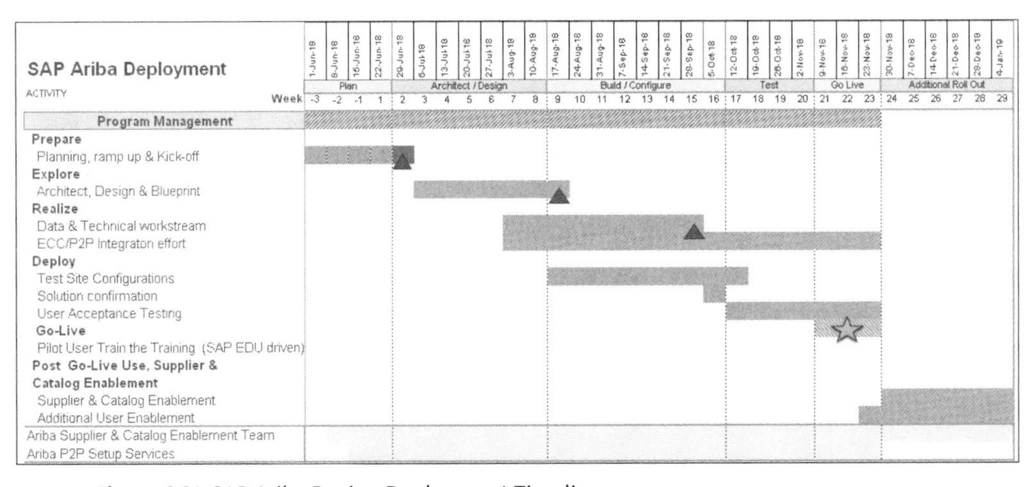

Figure 8.31 SAP Ariba Buying Deployment Timeline

In addition, given the nature of supplier enablement and internal adoption, the overall rollout of SAP Ariba Buying can comprise multiple go-lives, as each wave of suppliers and internal organizational units join the system in subsequent stages. This path can be useful for onboarding large supplier bases, as a "big bang" approach can quickly overwhelm your internal supplier resources and procurement departments. The focus of the initial project is typically to plan for the overall rollout/go-live approach and then target the first go-live.

Often known as a *pilot approach model*, an SAP Ariba Buying project designs the solution in the system setup track, as shown in Figure 8.31, with subsequent rollouts involving slight adjustments to the initial templates from the solution setup track. The main work is done in the solution adoption tracks and internal adoption track, which includes rolling the solution out to a new part of the organization and to the supply base. Figure 8.32 shows these events as SAP Ariba milestones—each subsequent rollout phase is a milestone, with the initial rollout as the initial milestone. After the first go-live/milestone has been completed, SAP Ariba project resource involvement can transition further to customer-led resources for the further rollouts.

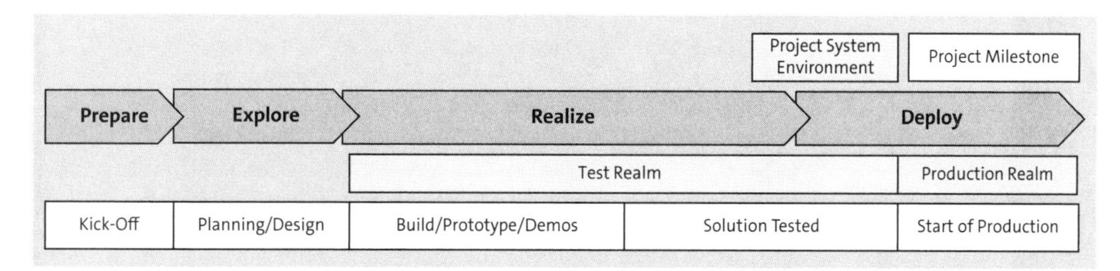

Figure 8.32 SAP Ariba Project Milestones

In addition to a multiple-wave approach for implementation, the SAP Ariba milestones methodology, shown in Figure 8.33, provides the following:

- **Better project planning and preparation**
  Using milestones acknowledges the organizational limitations of rolling out everything in a big bang approach, while also acknowledging the full scope of the rollout, via the milestones. This approach ensures that regions and supplier groups are accounted for in the overall roadmap and can help focus design discussions and the project in general.
- **Alignment of all delivery services work in a common methodology**
  One methodology can cover the entire scope of work, including the go-live and the preparation to adopt, all managed from a single project plan and built around clearly defined milestones.
- **Separates requirements from design**
  Milestones for further supplier groups and rollouts don't necessarily get folded into the initial design phase, allowing you to focus initially on reaching the first milestone and pilot organization.

The first phase, **Success Planning: Develop Business Case**, is driven largely by the SAP Ariba account team, value engineering, and your requisitioning team responsible for the purchase of the system/subscription. Although this phase occurs before the others, objectives and business cases defined during this phase underpin much of the delivery, as do the project sponsors deriving from this process. As such, most successful projects start with a solid handoff from this phase, with business goals and high-level solution objectives reiterated to the delivery team and future project team members. The first item of business for the delivery acceptance team on the customer side is to confirm business goals and define solution objectives and alignment for the handover from the value engineering team and their colleagues on the purchasing side of the decision to implement SAP Ariba.

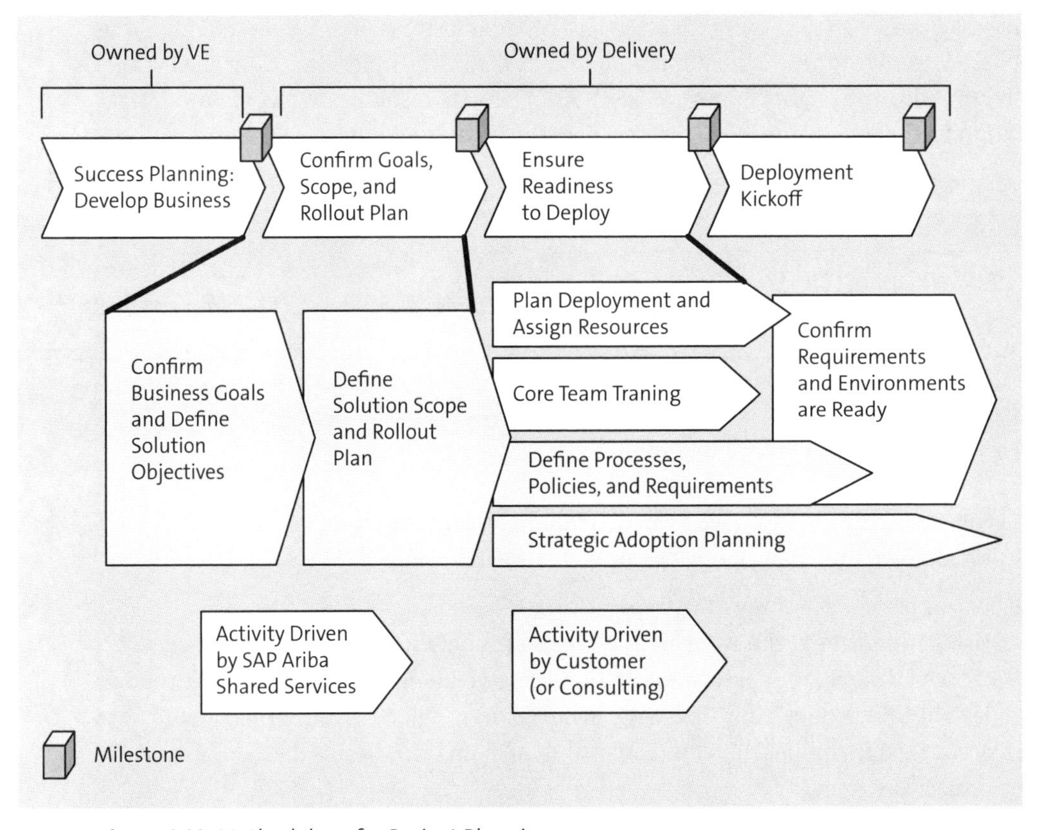

Figure 8.33 Methodology for Project Planning

On-site SAP Ariba and/or SAP Partner consultants play a supporting role in the **Confirm Goals, Scope, and Rollout Plan** phase, but a key role on large projects by helping guide customer resources in setting up a successful implementation. After these areas have been defined, the SAP Shared Services team is engaged to begin aligning the plan, assigning resources, and delivering core team training. As you implement SAP Ariba, you'll start parsing out the supporting streams of change management and supplier enablement. After SAP Shared Services confirms the requirements for setting up the instances, called *realms* in SAP Ariba, you've essentially finished the project planning stage and are ready for kickoff.

Architect and Enable Phase

Let's go into a little more detail, shown in Figure 8.34. For each phase, we'll give an overview of what happens, discuss the major milestones, and describe the importance of that phase within the bigger picture.

Beginning with the first phase after handoff, **Confirm Goals, Scope, and Rollout Plan**, the main goals/objectives of this phase are to confirm business goals and define solution objectives that will achieve those goals, and define the solution scope to realize the goals. Table 8.5 outlines the steps and responsibilities for each project participant type.

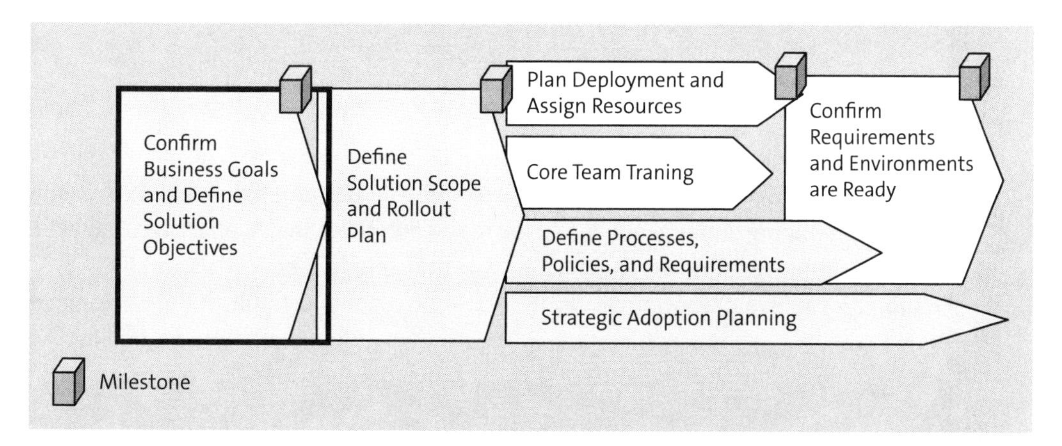

Figure 8.34 Confirm Business Goals and Define Solution Objectives Phase

Step	SAP Shared Services	Customer	SAP Ariba/ Partner Consulting
Define business goals and solution objectives.	Walk the customer through an exercise to confirm and document their business goals and solution objectives.	■ Define/confirm three to five top priority business goals and associated solution objectives. ■ Perform any due diligence needed to make these decisions.	Do the heavy lifting analysis to identify top priority goals and objectives—either via success planning or extra consulting.

Table 8.5 Confirm Goals, Scope, and Rollout Plan

Step	SAP Shared Services	Customer	SAP Ariba/ Partner Consulting
Confirm scope through go-live.	Walk the customer through an exercise to confirm the solution scope through one go-live. Cover these topics: ▪ Processes ▪ Categories ▪ Suppliers ▪ Departments ▪ Systems ▪ Content	▪ Make decisions to clearly define and confirm the solution scope. ▪ Develop and document any rollout planning beyond the first go-live.	▪ Support the customer through defining the solution scope. ▪ Develop expected return on investment (ROI) based on spend volume included in the scope and savings/ value identified in the business case across all rollout waves.

Table 8.5 Confirm Goals, Scope, and Rollout Plan (Cont.)

In this phase circled in Figure 8.34, you'll identify the business case, business/savings goals, and objectives to attain these goals. These considerations only need cover the upcoming/pending deployment wave. The major milestone in this phase is to have the goals and objectives defined. You must understand your business goals and objectives before you can evaluate whether your solution (scope) will fulfill them, as listed in Table 8.6.

Step	SAP Shared Services	Customer	SAP Ariba/ Partner Consulting
Analyze	▪ Ask customer to provide 12 months of spend data. ▪ Analyze the data and identify savings opportunities based on industry benchmarks (if not already done).	Provide the last 12 months of spend data in SAP Ariba's supplier spend/ transaction data file format.	N/A

Table 8.6 Confirm Business Goals and Objectives

Step	SAP Shared Services	Customer	SAP Ariba/ Partner Consulting
Review analysis	■ Review, understand, and confirm customer's business goals/business case and expected savings and benefits. ■ Document the customer's business goals in the solution goals and objectives guide.	■ Provide customer's business goals/ business case, showing the savings/outcomes customer expects to obtain from the SAP Ariba solution. ■ Perform any analysis necessary to confirm the business case.	Develop a high-level program vision and strategy, resulting in a detailed business case and high-level roadmap to drive benefit realization.
Perform due diligence	■ Lead meeting to help the customer identify how the solution will deliver the desired business results, and what major capabilities will be needed to drive the expected savings or operational outcomes. ■ Document these critical capabilities as solution objectives in the solution goals and objectives guide.	■ Participate in the review to identify how the solution can deliver the expected savings/ outcomes. ■ Perform any due diligence necessary to confirm how major solution capabilities will drive savings through better price compliance, operational efficiencies, negotiating leverage, and so on.	Perform any due diligence necessary to confirm how major solution capabilities will drive savings through better price compliance, operational efficiencies, negotiating leverage, and so on.

Table 8.6 Confirm Business Goals and Objectives (Cont.)

In the second phase, **Define Solution Scope and Rollout Plan**, as shown in Figure 8.35, the scope of the current/pending rollout wave is confirmed, documented, and finalized. Scope definition must include all six dimensions of scope: business processes, categories, suppliers, systems, regions/departments/users, and content. The major milestone is having the solution scope defined. You must understand the scope of the solution before you can plan and confirm whether you'll be able to deliver it with the resources available.

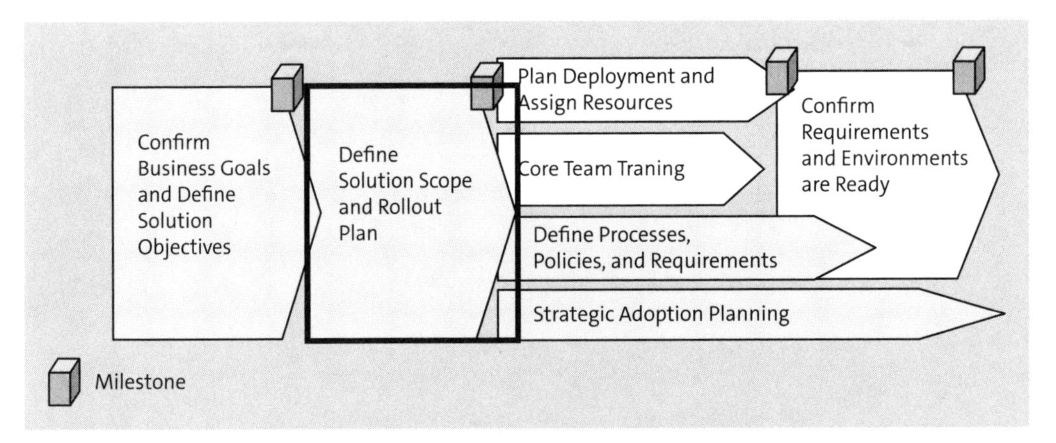

Figure 8.35 Define Solution Scope and Rollout Plans

Step	SAP Shared Services	Customer	SAP Ariba/ Partner Consulting
Analyze/ decide	Lead the customer through SAP Ariba's scope confirmation exercise to align the in-scope business processes, spend categories, and suppliers with the systems, departments, and content needed to support their transactions.	Decide which spend categories, business processes, suppliers, systems, departments, and content to include in the scope of the SAP Ariba solution and with what rollout timing.	Lead the customer through analyzing business goals, value levers, and organizational limitations to identify the solution scope and rollout plan that will maximize ROI and ease of rollout.
Document scope	Document scope through a single go-live, covering the following scope dimensions in the solution scope alignment document: Spend categories Purchasing processes Customer's suppliers who provide these categories: ■ Systems/ technology ■ Users/departments/ regions ■ Content (catalogs, contracts, etc.)	■ Review and sign off on the solution scope alignment document. ■ Develop and document any rollout planning needed beyond the initial go-live and any wave-by-wave cost-benefit or ROI plan needed to drive program support.	Develop a category rollout plan that documents the scope of each wave (processes, categories, suppliers, systems, departments, and content) along with the cost, effort, and expected savings based on transaction volume and operational improvements identified in the business case, across all planned rollout waves.

Table 8.7 Define Solution Scope and Rollout

To arrive at a meaningful definition of scope according to the steps listed in Table 8.7, the following areas should be covered in the scope document:

- Categories, which is how you'll tie items to volume, savings, and business impact
- Business processes, which will drive what systems, features, and functionality are needed
- Suppliers
- Systems
- Departments/regions/users
- Content (catalogs, contracts, reports, etc.)

The next phase, as shown in Figure 8.36 and Table 8.8, **Plan Deployment and Assign Resources**, is when a project plan covering all activities from kickoff through go-live and SAP Ariba rollout is completed and signed off on. All project resources understand their responsibilities and have committed to completing their assignments on time. The major milestone is when the deployment project plan is complete and resources have been assigned. A project plan is only valid if it's realistic and if the team buys into it and commits to delivering on time.

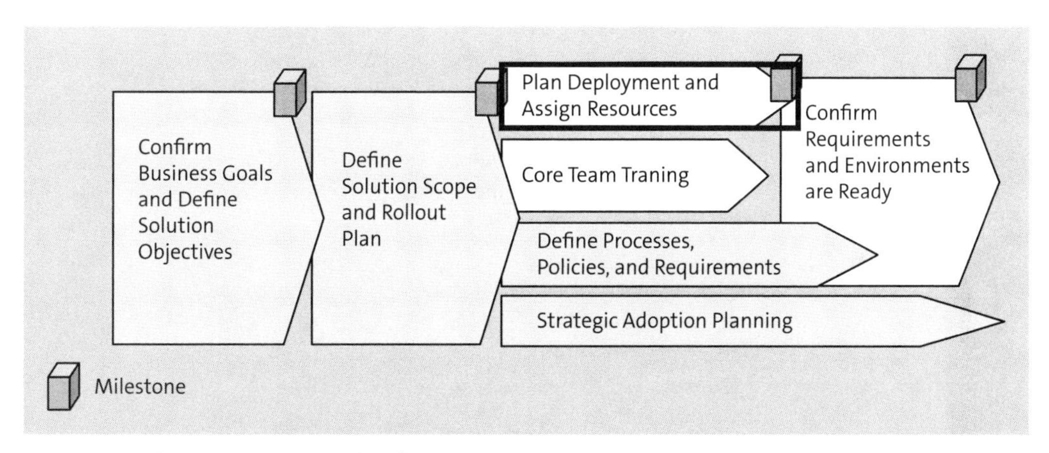

Figure 8.36 Plan Deployment and Define Resources

Step	SAP Shared Services	Customer	SAP Ariba/ Partner Consulting
Review deployment	Walk the customer through the deployment methodology to help the customer understand the nature, effort, and skills needed for the major project tasks.	Review methodology, tasks, and estimated time and effort needed to perform customer-owned tasks.	Assist the customer in estimating time and effort to complete project tasks for the specific scope and organization.

Table 8.8 Plan Deployment and Define Resources

Step	SAP Shared Services	Customer	SAP Ariba/ Partner Consulting
Assign resources	Assign SAP Ariba resources to SAP Ariba project roles.	Assign customer resources to customer project roles.	Assist the customer in defining project and post-project responsibilities and evaluating candidates.
Timeline definition	Lead the customer through developing a detailed project plan that delineates tasks, ownership, and timing of all deployment activities needed to reach go-live.	Contribute customer tasks and timing to the project plan, and sign off on the completed project plan.	▪ Work with the customer to identify internal events and initiatives that might impact or be impacted by the SAP Ariba program, and build them into the project plan. ▪ Suggest timing and risk mitigation steps to address these dependencies.
Project management framework definition	Propose the project management framework, including recurring meetings, status reporting, and risk and issue management, and partner with the customer to define the management framework for this project.	Participate in defining the project management framework.	Work with the customer to develop the project management framework appropriate for the scope and customer culture.
Kickoff presentation	Draft the deployment kickoff presentation and plan the deployment kickoff to align the core project team on project goals, scope, methodology, activities, roles, responsibilities, timing, communication, and administration.	Contribute to, review, and approve the deployment kickoff presentation.	Work with the customer to engage project stakeholders and tailor kickoff material to specific customer audiences.

Table 8.8 Plan Deployment and Define Resources (Cont.)

Step	SAP Shared Services	Customer	SAP Ariba/ Partner Consulting
Assessment	N/A	Identify and assess project impact on other departments, and proactively communicate with stakeholders and leaders of impacted departments to obtain their support and cooperation.	After the solution scope is defined, lead the customer through assessment of project impact on other departments and through proactively communicating with stakeholders and leaders of impacted departments to obtain their support and cooperation.

Table 8.8 Plan Deployment and Define Resources (Cont.)

The next two phases, **Core Team Training** and **Define Processes, Policies, And Requirements**, as shown in Figure 8.37 and Table 8.9, are important to ensure the success of the overall project. First, core team training is necessary to provide team members with a baseline understanding of the solution they are designing and implementing. Defining processes, policies, and requirements is essential to guiding the project's focus and ultimately understanding whether the project is successful or not by these definitions after delivery.

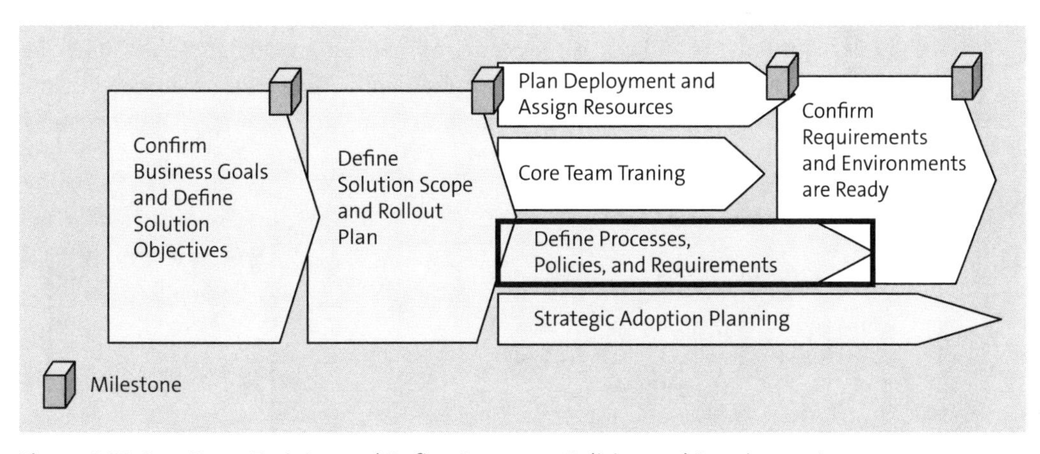

Figure 8.37 Core Team Training and Define Processes, Policies, and Requirements

Step	SAP Shared Services	Customer	SAP Ariba/ Partner Consulting
Training	Provide SAP Ariba features and functions training.	Send customer team members to SAP Ariba features and functions training as necessary.	Provide SAP Ariba features and functions training to additional customer team members as scoped.
Operations analysis and future state definition	N/A	This is determined by customer.	Lead the customer through analyzing its operations and define the future state of business processes, policies, and requirements that the solution will need to handle in-scope categories down to the level of detail needed to support solution design.

Table 8.9 Training and Define Processes, Policies, and Requirements

The second-to-last phase, as shown in Figure 8.38 and Table 8.10, **Confirm Requirements and Environments Are Ready**, is where, in addition to confirming that resources have been assigned and the project has been planned, you must confirm that you understand the requirements and that all test environments are ready and accessible. The project will hit delays if either customer requirements aren't well understood or the teams doesn't have access to both SAP Ariba and customer systems. The major milestone for this phase is being ready for kickoff.

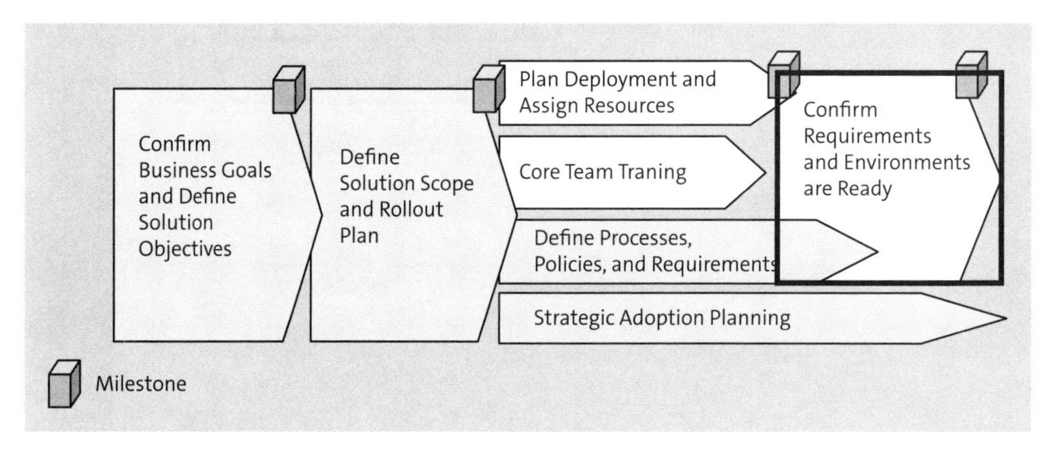

Figure 8.38 Confirm Requirements and Environments Are Ready

Step	SAP Shared Services	Customer	SAP Ariba/ Partner Consulting
Complete the requirements assessment questionnaire	Using the requirements assessment questionnaire, ask the customer basic questions about its business requirements to assess whether the requirements cover all in-scope functional topics and are well enough understood to proceed with detailed requirements confirmation and design.	▪ Answer all questions in the SAP Ariba requirements assessment questionnaire. ▪ Research and identify decision makers as necessary.	If SAP Ariba has done a detailed process, policy and requirements definition for the customer, you shouldn't have to assess the state of requirements unless much time has passed since they were defined and signed off on.
Review the requirements assessment questionnaire.	Review the requirements assessment questionnaire and all other business process and requirements documentation provided by the customer to ascertain readiness.	Provide the most up-to-date business process flows and requirements documentation.	Conduct a detailed review of the defined future state of process, policy, and requirements with the customer, and assist the customer in validating that it accurately represents the desired future state.
Confirm system availability.	Confirm that the customer's instances of SAP Ariba on-demand applications have been created and are available to use.	Confirm that the project team will have access to all customer system environments and tools that will be needed to conduct the project.	Help the customer identify and document who will need access to which environments and tools, and verify that access has been granted.

Table 8.10 Confirm That Requirements and Environments Are Ready

In the final phase, **Strategic Adoption Planning,** as shown in Figure 8.39 and Table 8.11, the system is assessed for its potential to integrate at an organizational level into everyday processes, essentially becoming your company's platform for procurement.

No milestones are associated with strategic adoption planning, other than that this phase needs to be completed along with the requirements/environments workstream.

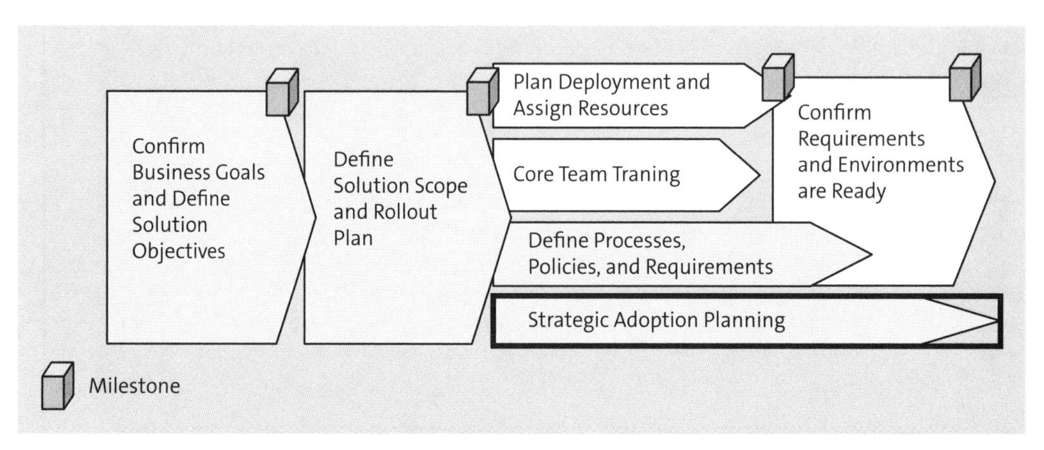

Figure 8.39 Strategic Adoption Planning

Step	SAP Shared Services	Customer	SAP Ariba/ Partner Consulting
Solution assessment	Answer customer questions on the typical impact of the SAP Ariba solution and SAP Ariba best practices for managing the impact on suppliers and stakeholders.	▪ Assess the solution's impact on suppliers. ▪ Assess customer relationships with impacted suppliers, and ascertain whether there are other active or pending initiatives that affect the suppliers. ▪ Identify the appropriate level of communication to each critical supplier. ▪ Develop a supplier campaign plan to recommend activities and approaches to get the customer's suppliers to transact through the SAP Ariba solution in the way that maximizes value for both the customer and the suppliers.	▪ Assess the solution's impact on suppliers. ▪ Assess customer relationships with impacted suppliers, and ascertain whether there are other active or pending initiatives that affect the suppliers. ▪ Identify the appropriate level of communication to each critical supplier. ▪ Develop a supplier campaign plan to recommend activities and approaches to get the customer's suppliers to transact through the SAP Ariba solution in the way that maximizes value for both the customer and the suppliers.

Table 8.11 Strategic Adoption Planning

Step	SAP Shared Services	Customer	SAP Ariba/ Partner Consulting
		▪ Assess relationships between the department sponsoring the project and other departments, and ascertain whether there are other pending changes that could complicate those departments adoption of the solution. ▪ Identify the appropriate level of communication to affected departments and the stakeholders to engage.	

Table 8.11 Strategic Adoption Planning (Cont.)

Solution Setup

After the architect phase has been completed for the core planning and workstreams, the actual workstreams begin. As shown in Figure 8.40, these workstreams consist of the following:

- Solution setup track
- Supplier adoption track
- Internal adoption track

In this section, we'll discuss the solution setup track, which begins when requirements receive final confirmation.

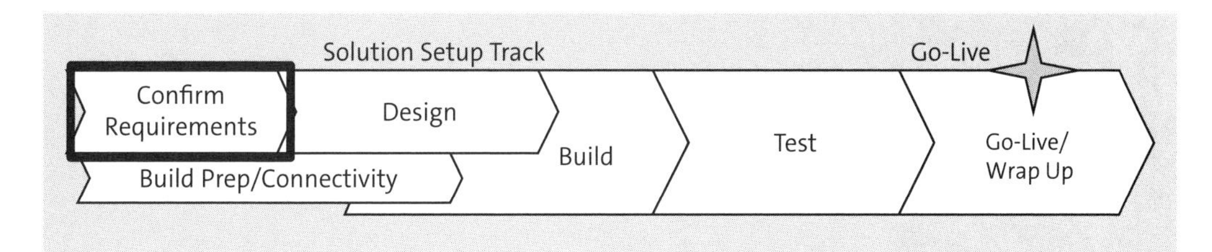

Figure 8.40 Confirm Requirements

For SAP Ariba Buying projects, the phase shown in Figure 8.40 covers a wide array of areas. Workflows and processes, material/service category definition, and connections to and integration with other SAP Ariba solutions in the landscape (i.e., SAP Ariba Sourcing, SAP Ariba Invoice Management, or SAP Ariba Supplier Lifecycle and Performance on the analytics side), all need to be defined in the context of SAP Ariba Buying, as shown in Table 8.12.

Step	SAP Shared Services	Customer	SAP Ariba/ Partner Consulting
Demo	Hold solution demos.	Attend solution demos. Learn what the SAP Ariba solution can do.	Attend solution demos.
Requirements review	Walk the customer through the business requirements workbook (BRW). Help the customer understand requirements questions.	Participate in the BRW review. Answer questions to provide requirements.	Participate in the BRW review. Add context and advice relevant to the customer's situation.
Best practice discussion	Advise on general best practices by responding to customer questions.	Perform research and analysis necessary to define requirements.	Help the customer define its requirements by analyzing the customer's organization and operations to make customer-specific recommendations.
Review and sign off	Review requirements decisions with customer to verify that requirements are complete, consistent, and clear.	Sign off on the completed requirements.	Proactively drive and facilitate requirements decisions.

Table 8.12 Sourcing Setup Track: Confirm Requirements

Much of the requirements gathering is done using SAP Ariba workbooks, or BRWs. The data and requirements defined in these workbooks underpin the SAP Ariba Buying system and its configuration.

After a requirement area has been confirmed, the **Design** phase follows, as shown in Figure 8.41 and Table 8.13, either for that area or for the entire set of requirements.

Requirements without overlapping and dependent areas that need definition to proceed can begin their design phase while remaining requirements are still being defined. This flexibility is somewhat a departure from a traditional on-premise project, where requirements are typically all laid out prior to commencing design. While beginning the design only after all the requirements have been defined is often easier and less risky, beginning the design for areas prior to their definition allows for time efficiencies on the project and may actually help make the overall set of requirements more realistic, if the design steps uncover dependencies that influence other requirements. For example, if a certain material classification is defined in a requirement in the beginning, and the design phase uncovers additional field dependencies for the material code extensions, further requirement definition around these fields in invoices and other related documents may need to follow suit.

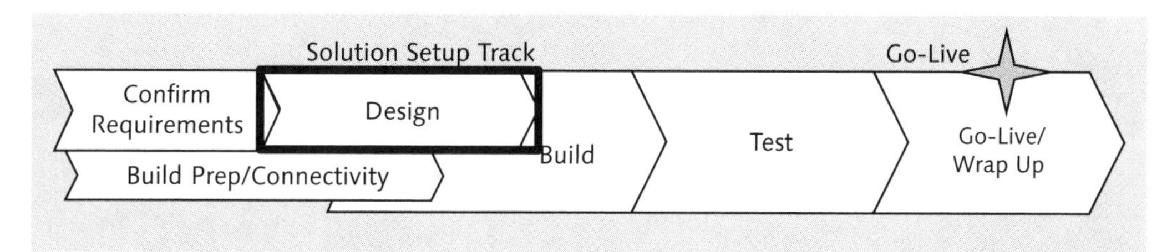

Figure 8.41 Solution Setup Track Design

Step	SAP Shared Services	Customer	SAP Ariba/ Partner Consulting
Configuration design	Identify and document configurations to SAP Ariba applications that would best handle the defined requirements.	Design and document the holistic solution. Review suggested system configurations and assess whether they are right for the customer.	Lead the design and documentation of the holistic solution: business processes, user actions, workarounds, and admin and support functions in addition to system functionality.
Design review and sign off	Explain system configurations and functionality to the customer.	Understand, provide feedback, and sign off on the holistic design and system configurations.	Facilitate the review of system configurations and holistic solution design. Drive signoff on both.

Table 8.13 Solution Setup Track: Design

During the **Design** and **Confirm Requirements** phases, an additional phase is running in parallel, as shown in Figure 8.42 and Table 8.14.

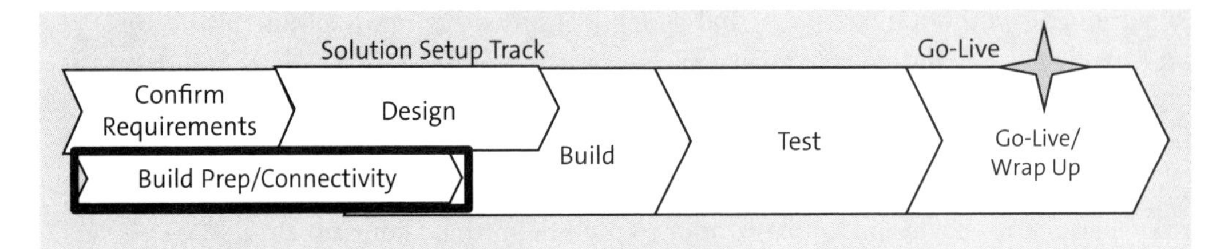

Figure 8.42 Solution Setup Track: Build Prep/Connectivity

During this phase, the connectivity and infrastructure necessary to support the solution, the requirements, and resulting design is established. Although on-demand SAP Ariba solutions are hosted in the cloud, some setup and connection may be required within the customer's realm to enable various solutions to communicate with one another, as well as external integration with third-party systems and data sources, as shown in Table 8.14.

Step	SAP Shared Services	Customer	SAP Ariba/ Partner Consulting
Technical knowledge transfer—integration	Teach customer the technical capabilities of the SAP Ariba tools.	Build out all necessary infrastructure and connectivity between the customer and SAP Ariba systems.	Work with the customer to configure SAP Ariba tools and customer systems to talk to one another.

Table 8.14 Solution Set Up Track: Build Prep/Connectivity

After the basic connectivity, configuration requirements, and design have been completed, the next step is to begin the **Build** phase, as shown in Figure 8.43 and Table 8.15.

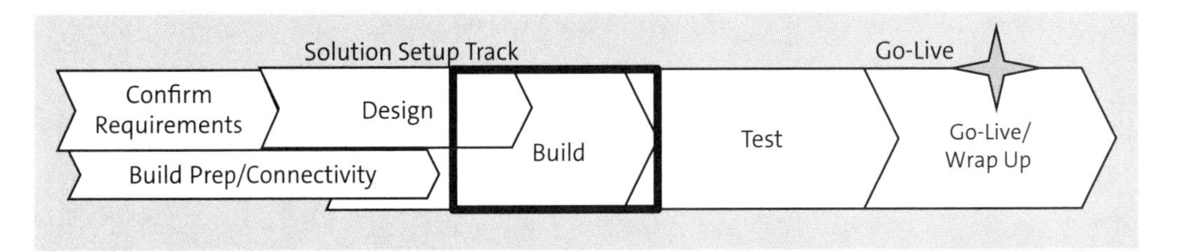

Figure 8.43 Solution Setup Track: Build

In this phase, the workbooks completed during the **Confirm Requirements** and **Design** phases are handed over to the SAP Shared Services team to begin building out the solution in the customer's realm. SAP Shared Services performs the majority of this build activity, as this group has access to the configuration points in the multitenant cloud environment to action these types of changes. Having SAP Shared Services as the central point of configuration for **Build** helps control access to the system and validate the configuration changes in line with the other configuration and interdependencies.

The last two items in Table 8.15 relate to the test phase. Testing is typically a customer-led effort, given you'll ultimately need to validate the solution to ensure it fulfills your requirements.

Step	SAP Shared Services	Customer	SAP Ariba/ Partner Consulting
Build	Build/configure elements of the SAP Ariba applications to which the customer doesn't have access.	Configure all customer-facing elements of the SAP Ariba applications (Ariba Network rules, approval rules, configuration data, etc.) and build all interfaces between customer applications and SAP Ariba interfaces.	Work with the customer to configure customer-facing aspects of the SAP Ariba applications, including master data and transactional interfaces.
Test plan definition	Provide a sample test plan.	Plan the entire testing program.	Plan the entire testing program.
Test script definition	Provide a sample test use cases and scripts.	Develop customer-specific test use cases and scripts.	Develop customer-specific test use cases and scripts.

Table 8.15 Solution Set Up Track: Build

As with an on-premise solution implementation, the **Test** phase shown in Figure 8.44 exists to validate that the key functions and business processes are supported in the new solution and that this solution will effectively support the business and the scenarios in scope at go-live. During this phase, you'll ensure that all these areas work as they have been designed and built, and address any issues that arise during testing. As shown in Table 8.16, break/fix work is typically handled by SAP Ariba or the consulting partner, whereas testing is led by you.

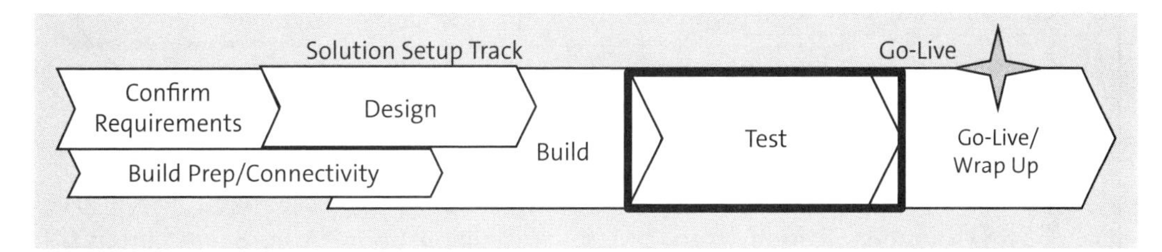

Figure 8.44 Solution Set Up Track: Test

Step	SAP Shared Services	Customer	SAP Ariba/ Partner Consulting
Execute test	Provide advice on general best practices for testing.	Manage and execute testing.	Manage testing and assist in executing the system test.
Defect resolution	Debug and resolve test defects with SAP Ariba applications.	Identify test defects. Debug and resolve test defects with customer systems.	Work side by side with the customer to research and debug test issues.

Table 8.16 Solution Setup Track: Test

After testing is complete, the project can go live and wrap up. However, first, the project needs to effectively pass testing, as with a quality gate in the SAP Activate methodology, and project leaders and sponsors need to validate whether testing has been successfully completed. If items arose during testing that haven't been completely addressed by go-live, an additional decision about whether the project can accept this outstanding issue at go-live must be made. Some minor issues can be addressed post-go-live if necessary, but some showstopper issues will require analysis on shifting the go-live date to address and determine the potential impact.

As shown in Figure 8.45, after the project management and steering committee have agreed on the go-live, the project is ready to cut over into production. SAP Ariba on-demand projects typically have two, instead of three, environments and limited transport paths. In contrast, in a traditional on-premise SAP software project, you'll move most configuration items from development to quality assurance to production. SAP Ariba instead involves a quality assurance realm and a production realm, and the configuration is essentially replicated from quality into production at the time of go-live. Conversions and interfaces need to be established in tandem as part of the go-live, as shown in Table 8.17.

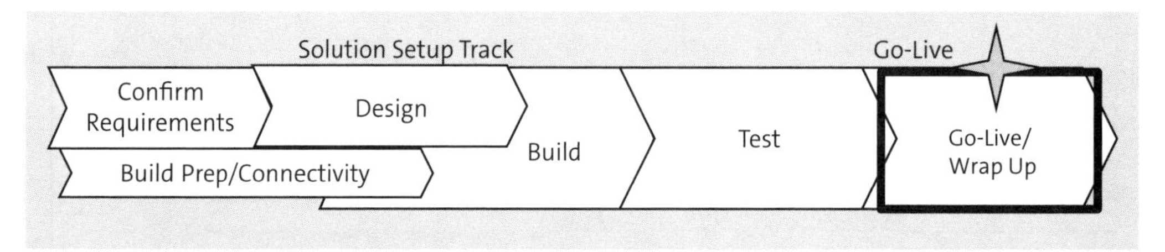

Figure 8.45 Solution Set Up Track: Go-Live/Wrap-Up

Note that multiple systems may require adjustments during the go-live of an SAP Ariba Buying solution, as certain legacy systems are replaced by SAP Ariba Buying and as SAP Ariba Buying is connected to other systems for periodic batch updating or transactions within the customer's landscape.

Step	SAP Shared Services	Customer	SAP Ariba/ Partner Consulting
Cutover planning	Work with the customer to plan SAP Ariba system production cutover.	Plan the holistic cutover of all impacted systems and data.	Work with the customer to plan the holistic cutover of all impacted systems and data.
Cutover execution	Execute noncustomer-facing production cutover tasks for SAP Ariba applications.	Execute customer-facing production cutover tasks of SAP Ariba applications as well as cutover tasks for customer systems.	Work with the customer to execute customer-facing production cutover tasks of SAP Ariba applications as well as cutover tasks for customer systems.

Table 8.17 Solution Setup Track: Go-Live/Wrap Up

In the Confirm Readiness phase, which is the change management part of the go-live/wrap-up phase shown in Table 8.18, you'll complete the following steps:

1. Draft the change management project plan.
2. Assign resources.
3. Confirm the commitment and finalize the plan.
4. Define/confirm the requirements.
5. Train the resources.
6. Assess readiness.
7. Perform strategic adoption planning.

Step	SAP Shared Services	Customer	SAP Ariba/ Partner Consulting
Change management project plan	Draft a project plan covering SAP Ariba deployment through one go-live, and walk the customer through assigning names and dates to plan tasks.	Assign names and dates to the SAP Ariba deployment project plan. Raise and include any customer-specific events or activities and sign off on the project plan.	Work with the customer to align the SAP Ariba deployment project plan to any related customer projects or events.

Table 8.18 Go-Live and Wrap-Up: Change Management Planning

Step	SAP Shared Services	Customer	SAP Ariba/ Partner Consulting
Assessment	Assess the customer's readiness to deploy, including resources, systems, business requirements, and sponsorship.	Get ready to deploy.	Lead the customer through analyzing and defining business requirements and gaining sponsor and stakeholder commitment.
Communication and impact assessment	Answer questions on typical organizational impact and adoption planning for SAP Ariba solutions.	Assess the impact on suppliers and end users. Develop a strategic adoption plan for obtaining support and participation from the customer organization and suppliers.	Lead the customer through assessing the solution impact and developing a strategic adoption plan.

Table 8.18 Go-Live and Wrap-Up: Change Management Planning (Cont.)

8.4.3 Standalone Implementations

Standalone implementations aren't entirely independent of any system but focus on supporting as much of the process in SAP Ariba as possible. As a result, SAP Ariba Buying implementations, where invoicing and reconciliation of invoices will occur outside of the system, are less likely candidates for standalone scenarios, whereas SAP Ariba Buying and Invoicing implementations can fit this category up to the point when payment is required, when a payment request is sent to your bank, or OK2Pay is transmitted to SAP S/4HANA for posting.

Figure 8.46 shows the landscape and process. A requisition originates in SAP Ariba Buying and Invoicing, with the PO moving out to the Ariba Network. From the Ariba Network, the supplier pulls the PO and can create order confirmations and ASNs. The item is received in SAP Ariba Buying and Invoicing, and the invoice submitted by the supplier via the Ariba Network is received in SAP Ariba Buying and Invoicing. Finally, an invoice remittance advice is issued to the supplier from SAP Ariba Buying and Invoicing via the Ariba Network, advising the supplier of invoice payment.

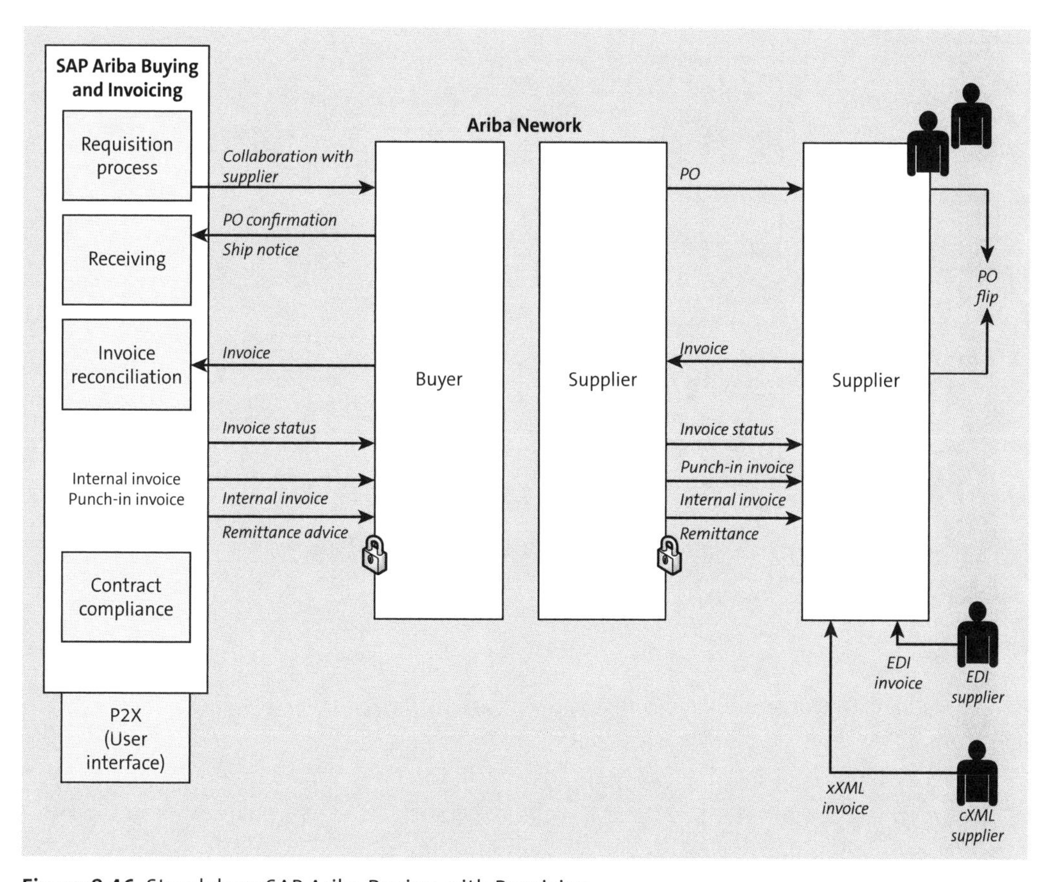

Figure 8.46 Standalone SAP Ariba Buying with Receiving

8.4.4 Integrated Implementations

A number of integration options for SAP Ariba Buying and Invoicing, SAP Ariba Buying, and SAP Ariba Catalog with other systems in your P2P process are available. The main integration approach involves an SAP S/4HANA environment for some or all documents while enabling the back-and-forth transfer of documents originating in SAP S/4HANA, including direct procurement requisitions from material requirements planning (MRP) runs in SAP S/4HANA, and invoice and remittance advice in the SAP Ariba Buying and Invoicing scenario.

Figure 8.47 shows a lighter integration with SAP S/4HANA, with exports of requisitions from SAP S/4HANA and imports of POs created in SAP Ariba Buying and Invoicing. In addition, an OK2Pay message is delivered from SAP Ariba Buying and Invoicing to trigger the actual payment of an invoice. These integration points can be built out to a level where the process documents are both reflected in SAP S/4HANA and SAP Ariba Buying and Invoicing, with leading documents being created directly in SAP S/4HANA as well.

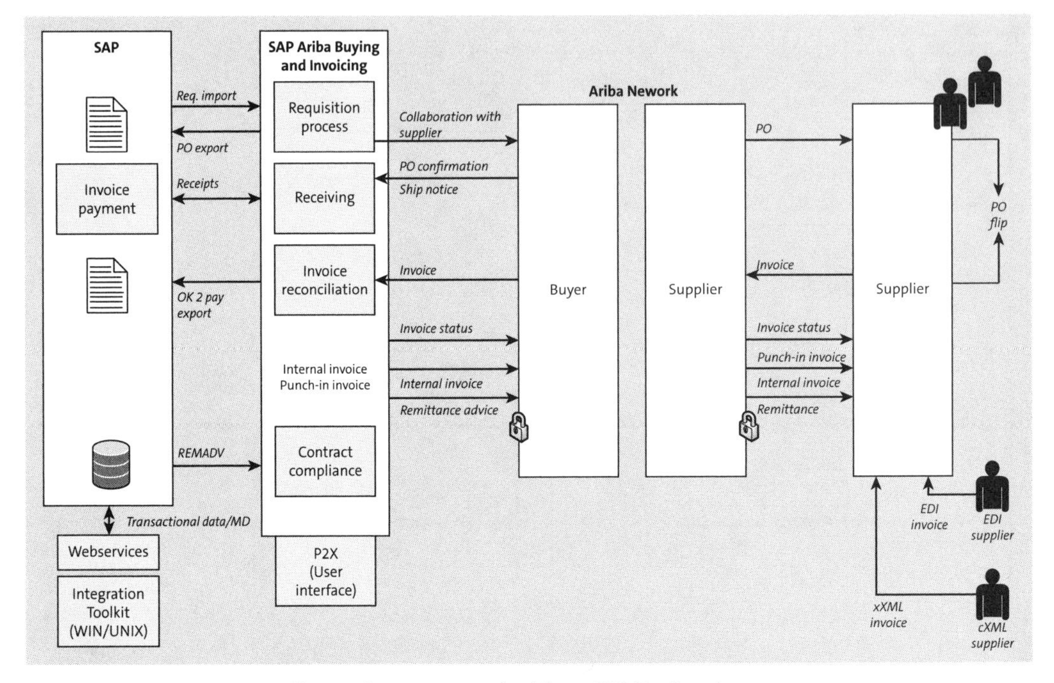

Figure 8.47 SAP Ariba Buying Integrated with an ERP Backend

As a catalog solution that can be accessed using an SAP-defined catalog protocol called Open Catalog Interface (OCI), SAP Ariba Catalog offers numerous ways you can integrate with SAP on-premise procurement systems, such as materials management in SAP S/4HANA, purchasing in SAP S/4HANA, and SAP SRM. In this scenario, a web service call is made from SAP S/4HANA/SAP SRM out to SAP Ariba Catalog, as if the system were accessing a third-party catalog hosted by a supplier. After an item has been selected, the item is returned using the same call and protocol types. As long as SAP Ariba Catalog communicates back to the originating system with the fields in a format that can be used on the purchasing document, which can either be set up in SAP Ariba Catalog directly with SAP Ariba or via a user exit in SAP S/4HANA or SAP SRM, your order then populates in the procurement system, and the order is approved/sent to the supplier.

An alternative approach is to use the entire shopping cart process along with the approvals in SAP Ariba Catalog, sending back the order information to SAP S/4HANA to be issued. Figure 8.48 shows a process whereby the user creates a purchase requisition in SAP S/4HANA, then punches out to SAP Ariba Catalog and conducts the requisitioning and ordering process (leveraging processes and content from SAP Ariba Catalog), returning to SAP S/4HANA for PO creation and follow-on processes.

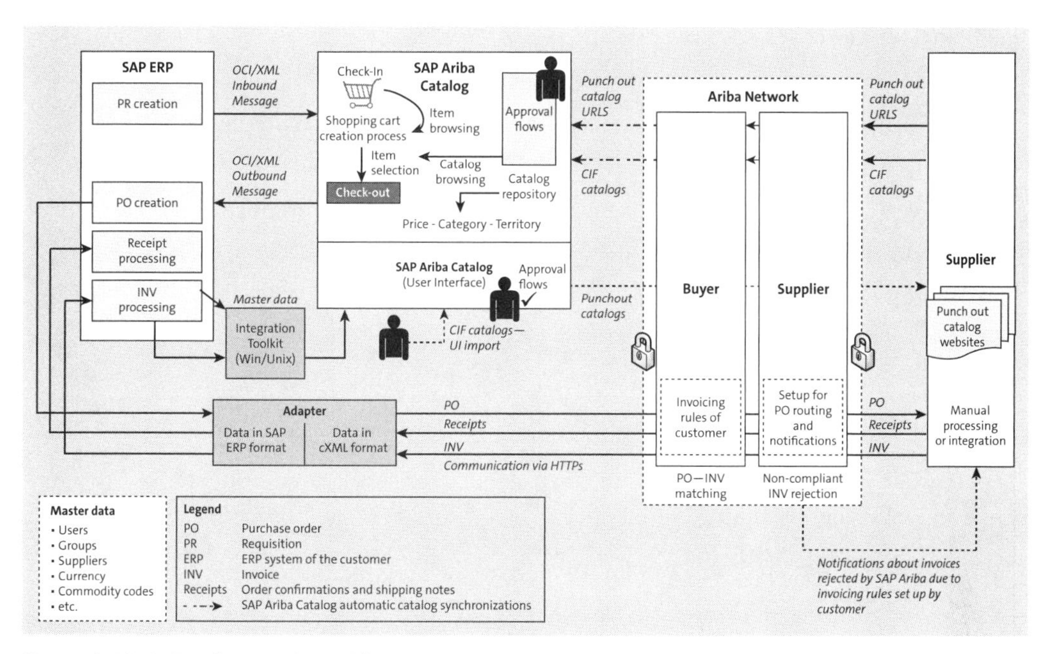

Figure 8.48 SAP Ariba Catalog with Procurement Automation

Suppliers upload CIF catalogs via the Ariba Network, which are made available to SAP Ariba Catalog customers—these catalogs and supplier-hosted punchout catalogs are called via SAP Ariba Catalog during the requisitioning process. After the catalog content is leveraged to create a complete order, the PO is generated in SAP S/4HANA. Leveraging the SAP Business Suite Add-On or an adapter, such as the Ariba Network Adapter, the PO is then sent to the supplier, and the invoice is submitted via the Ariba Network.

Figure 8.49 shows the main approaches for further integrations with SAP S/4HANA. From SAP S/4HANA, SAP Ariba Catalog is called via a web service, and the user then fills a shopping cart in SAP Ariba Catalog. Once sourced, this shopping cart is sent back to SAP S/4HANA and creates a PO in SAP S/4HANA. In SAP S/4HANA, the PO is finalized and then transmitted back to the Ariba Network. The supplier then acknowledges the PO and fulfills it on the Ariba Network. Upon receipt, the customer logs a goods receipt in its SAP S/4HANA environment, and again this goods receipt is transmitted out to the Ariba Network, and the supplier can now invoice. The invoice document is replicated back to your SAP S/4HANA environment. Intermediate communications/documents, such as ASNs and order confirmations can also be transmitted from the supplier to your SAP S/4HANA instance.

The content scope covers integration between the buyer's SAP Business Suite and the Ariba Network for the following:

- PO and invoice automation
- Discount management

- Technical connectivity options:
  - SAP Business Suite Add-On/web services direct
  - IDoc mapped and mediated via SAP Process Integration on premise
  - Add-on via SAP Cloud Platform Integration as middleware in the cloud
  - Add-on via SAP Process Integration as middleware on premise
- Documents:
  - PO
  - PO response
  - Goods receipt
  - Service entry sheet
  - Supplier invoice
  - Payment proposal
  - Remittance advice
  - ASN

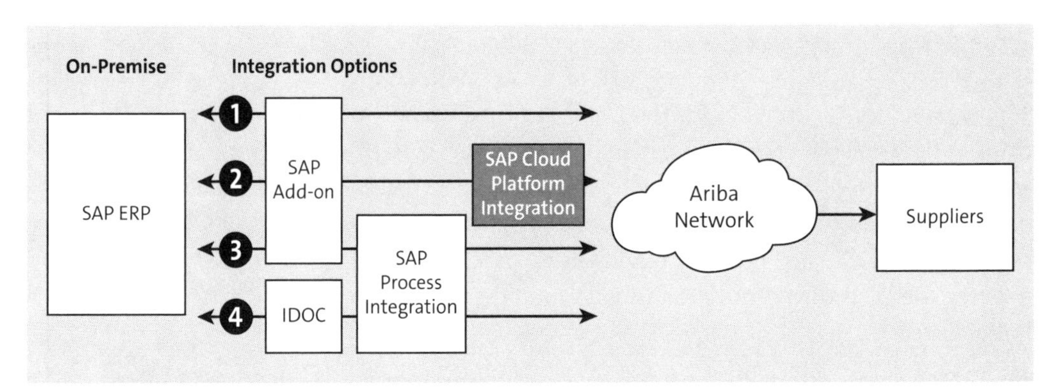

Figure 8.49 Ariba Network Integration for SAP: Full Buyer Enablement

Figure 8.49 shows the main integration options for SAP Business Suite with the Ariba Network. Several integration options exist, which can be confusing when analyzing which route to take for integration. The options are essentially driven by three factors. First, the SAP Business Suite Add-On replaces the SAP Ariba Adapter built on SAP Process Integration prior to the acquisition of Ariba by SAP. The second factor is the need for some customers to "mediate" the connection via SAP Process Integration for centralization purposes (customer seeks to use a middleware platform such as SAP Process Integration for all interactions with third-party systems), or security, as some security protocols don't allow for direct access to SAP ERP from the Internet. Lastly, while the Ariba Network and most SAP Ariba solutions "speak" cXML when communicating with third-party systems, SAP ERP needs translation for these messages. Messages can be translated into IDocs, which is the old method for exchanging data intersystem with SAP ERP, or via web services–based communication to communicate back and forth.

Essentially, as we'll discuss in more detail in Chapter 9, the main integration options for connecting SAP ERP to the Ariba Network are to leverage the SAP Business Suite Add-On to directly connect SAP ERP with the Ariba Network, to connect via SAP's middleware platform (SAP Process Orchestration) either in a mediated fashion between the SAP Business Suite Add-On and the Ariba Network, or via IDocs and an SAP Process Integration–based adapter. As a fourth option, you can leverage SAP Cloud Platform Integration.

The document types covered by these integration approaches are as follows:

- POs
- PO responses
- ASNs
- Goods receipts/service entry sheets
- Supplier invoices
- Payment proposals
- Remittance advice

In addition, SAP Ariba Discount Management can be implemented for SAP S/4HANA, leveraging these integration approaches. With SAP Ariba Discount Management, you can negotiate discounts at thc timc of invoice in exchange for early payment.

Viewed another way (as shown in Figure 8.50), two options for connecting directly are available:

- SAP Business Suite Add-On installed directly on SAP ERP
- SAP Cloud Platform Integration

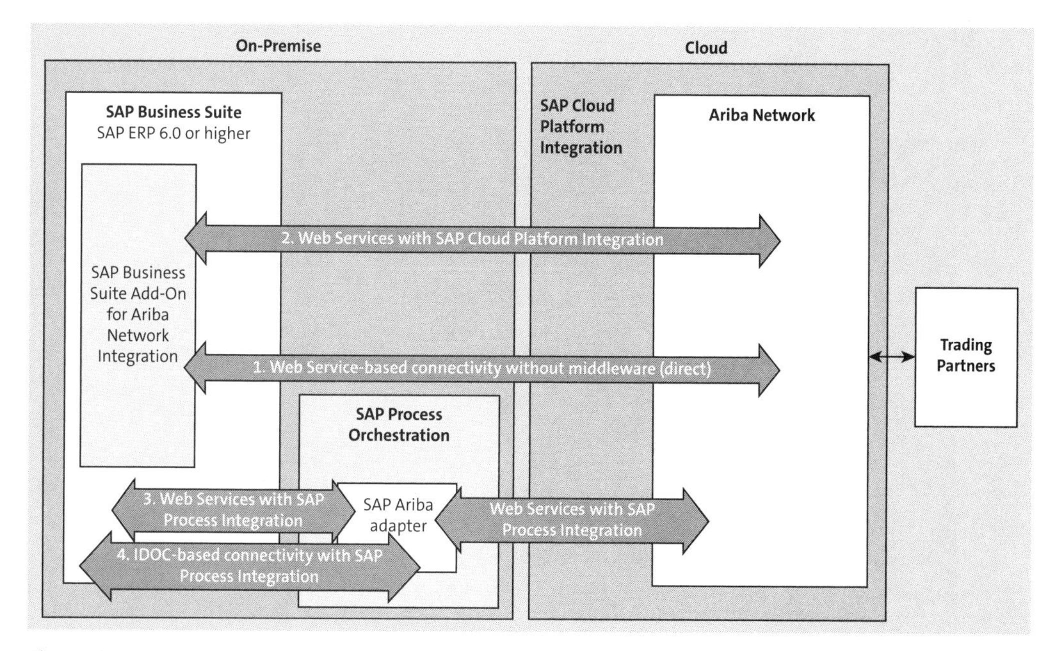

Figure 8.50 System Landscape and Technical Connectivity Options for the Ariba Network

Two SAP Process Integration–mediated options are also available:

- **SAP Business Suite Add-On mediated via SAP Process Integration**
  This approach is often taken for security requirements or if SAP Process Integration is the main integration platform for the ERP environment.
- **Ariba Network Adapter**
  Based on SAP Process Integration, the Ariba Network Adapter was the original approach for integrating SAP ERP to SAP Ariba, prior to the acquisition of Ariba by SAP.

8.4.5 Defining Project Resources and Timelines

Having the right resources and colleagues on an SAP Ariba Buying project can be the difference between a successful or mediocre execution. Table 8.19 describes the key SAP Ariba and consulting partner roles on an SAP Ariba Buying project.

SAP Ariba Role	Description
SAP Ariba customer engagement executive	▪ Provide customer ownership. ▪ Establish clear lines of accountability within SAP Ariba. ▪ Reduce confusion for the customer when dealing with SAP Ariba. ▪ Establish high-quality relationships with your customers with a goal of becoming a trusted advisor. ▪ Establish a point of contact to manage contract extension. ▪ Contact to discuss events or changes within customer organization that would impact the use of SAP Ariba products/services. ▪ Focus on customer success and collaborative commerce adoption. ▪ Create a faster return on customer investment. ▪ Control more spend, and manage cost effectively. ▪ Perform rapid deployment to improve the ability to capture achievable potential.
SAP Ariba quality management	▪ Support with coordination of activities between customer/partner and SAP Ariba stream resources. ▪ Ensure that the project plan and methodology are accurately followed. ▪ Review and validate the functional design document provided by the customer/partner. ▪ Review and validate the master data rationalization provided by the customer/partner. ▪ Review and validate the test cases provided by the customer/partner. ▪ Ensure that key deliverables are provided and signed off at each project phase closure.

Table 8.19 SAP Ariba and Consulting Partner: SAP Ariba Buying Project Roles and Responsibilities

SAP Ariba Role	Description
SAP Ariba delivery executive	■ Oversee multiple SAP Ariba solutions and often multiple SAP Ariba project managers. ■ Serve strategic role as member of the leadership team. ■ Create a single point of escalation. ■ Own and track progress on executing the vision/value proposition, and align, manage, and monitor all SAP Ariba workstreams and project deliverables to achieve the business goals. ■ Serve as first point of contact to either address queries directly or to further channel to the various SAP Ariba levels of ownership. ■ Monitor and drive compliance with agreed-upon change control, communication, and governance policy to facilitate on-time delivery. ■ Provide direct linkage for issue management for SAP; partners; customers; and all SAP Ariba teams such as field consulting, remote shared services, and P&I (engineering). ■ Take responsible for overall SAP Ariba quality management, and ensure that quality assurance is enabled across all workstreams.
SAP Ariba project advisor	■ Serve in tactical role as a member of the SAP delivery team. ■ Audit project deliverables and ensure compliance with project management office methodology. ■ Coordinate with the project manager and the customer engagement executive to handle escalations and project issues. ■ Ensure that project team roles and resourcing are aligned with the SOW and the project plan/schedule.
SAP Ariba project manager	■ Confirm the customer goals and project scope. ■ Support customer/partner in planning the stream activities and timelines. ■ Support the establishment of the stream management framework (meetings, status, issue management, etc.). ■ Coordinate activities and input from SAP Ariba stream resources. ■ Ensure timely project communication and status updates. ■ Identify and escalate stream issues and risks as appropriate. ■ Serve as point of contact for deployment.

Table 8.19 SAP Ariba and Consulting Partner: SAP Ariba Buying Project Roles and Responsibilities (Cont.)

SAP Ariba Role	Description
SAP Ariba functional lead	■ Propose solution configurations to handle customer/partner requirements. ■ Review customization functional requirement and propose alternative options. ■ Document configurations to application. ■ Configure system parameters. ■ Support the testing phase and SAP Ariba-related issues.
SAP Ariba technical lead	■ Support customers/partners on SAP Ariba master data and transactional data integration capabilities and formats. ■ Support customers/partners on configuring integration parameters in the SAP Ariba on-demand platform. ■ Build customer/partner customizations into SAP Ariba Buying. ■ Perform unit tests on SAP Ariba customizations built. ■ Support the test phase and resolve issues for SAP Ariba–developed configurations/customizations. ■ Execute cutover activities for go-live.
SAP Ariba Catalog enablement lead	■ Educate customer and suppliers on catalog capabilities. ■ Support catalog content creation. ■ Support suppliers enabling catalogs on the Ariba Network. ■ Assist with testing and resolve catalog-related issues.
SAP Ariba supplier enablement lead	■ Manage the enablement services workstream. ■ Advise the customer on how to perform supplier enablement activities. ■ Lead supplier data collection efforts. ■ Develop supplier communication and education materials. ■ Monitor supplier registration progress and address supplier enablement-related issues. ■ Manage resources to facilitate education and testing of integrated suppliers.

Table 8.19 SAP Ariba and Consulting Partner: SAP Ariba Buying Project Roles and Responsibilities (Cont.)

Customer roles are equally as important for a successful SAP Ariba Buying project. Table 8.20 lists customer roles along with descriptions of their tasks. Please note that project commitment time indicates what percentage of an individual's working hours would be occupied by the project. For roles where the project commitment time varies widely project to project, we haven't indicated a percentage.

Customer SAP Ariba Buying Role	Description
Project sponsor	■ Establish and communicate overall project vision. ■ Provide senior leadership communication in support of the project. ■ Mandate appropriate change management across leadership of all affected departments. ■ Monitor status reports and timelines. ■ Resolve escalated issues, including those involving customer/partner resources, lack of participation, or supplier compliance messaging.
Project manager	■ Confirm customer/partner goals and project scope. ■ Plan the detailed project activities and timeline. ■ Ensure adequate resources are assigned. ■ Manage participation of all required resources. ■ Coordinate stakeholders as needed (accounts payable, purchasing, receiving, finance, etc.). ■ Coordinate signoff on all SAP Ariba deliverables. ■ Manage project timeline and adherence to schedule. ■ Manage project requirement and adherence to initial scope. ■ Drive project status meetings. ■ Identify and manage risks as appropriate. ■ Develop go-live cutover plan and manage cutover execution. ■ Serve as the single point of contact for overall deployment.
Functional team	■ Drive configuration workshops. ■ Define and document business requirements pertaining to all in-scope processes, transactions, and SAP Ariba solutions. ■ Gather business input from all involved functional resources and departments: procurement, accounts payable, accounting, finance, and so on. ■ Define business process and requirements and make decisions on solution configurations to handle those requirements. ■ Write customer/partner functional design document. ■ Develop test cases and test scripts for testing. ■ Plan, manage, and conduct system testing and user acceptance testing (UAT) of the entire solution. ■ Validate that master data and application configuration data functions as desired. ■ Identify and resolve project issues. ■ Help develop the go-live cutover plan and manage cutover execution. ■ Become the expert on system functionality, documentation, and customer-specific system use.

Table 8.20 Customer Roles and Responsibilities

Customer SAP Ariba Buying Role	Description
	■ Train customers on SAP Ariba solutions and customer-specific processes implemented. ■ Support customer change management activities to take full benefit of SAP Ariba solutions. ■ Participate in configuration workshops, and contribute to business requirements and solution design. ■ Learn the detailed SAP Ariba functionality all the way through the P2P process, and learn how the system configurations, customizations, and data drive system functionality. ■ Answer questions on how the system will support the business and technical requirements raised by all involved customer departments. ■ Provide sign off on all SAP Ariba deliverables. ■ Participate in test case/script development. ■ Participate in system test and UAT. ■ Identify and train power users within each department to act as experts by providing assistance to peers and input to overall process. ■ Help define and execute internal change management program within their respective department.
Middleware and platform technical lead	■ Project commitment time: 80%–100% ■ Serve as ERP technical resource. ■ Participate in configuration workshops. ■ Contribute to ERP-related aspects of solution design. ■ Develop and test ERP interface/integration functionality (for both master data and transactional data). ■ Extract and format master data from ERP systems and load it into SAP Ariba. ■ Assist with testing, and resolve ERP-related issues. ■ Conduct cutover activities. ■ Serve as middleware technical resource. ■ Install, configure, and test SAP Ariba Integration Toolkit or web services interfaces. ■ Develop and test middleware transmission and mapping functionality. ■ Configure integration error handling and notification functionality. ■ Define scheduling scripts for automatic data integrations and email notifications. ■ Configure corporate network to support end-user authentication (if needed). ■ Configure corporate network to enable inbound and outbound transmissions with appropriate security controls. ■ Assist with testing and resolve integration-related issues.

Table 8.20 Customer Roles and Responsibilities (Cont.)

Customer SAP Ariba Buying Role	Description
Change management/training lead	■ Project commitment time: 10%–20% ■ Assess the impact of the project's business process, policy, organizational changes, and cultural changes on end users and other affected employees. ■ Develop and execute change management plan to address impacts from all changes related to the SAP Ariba program. ■ Manage communication plan development and project-wide communications to key stakeholders. ■ If end-user training is needed: – Develop a training approach. – Develop training documentation. – Plan and manage training delivery. – Ensure training needs are addressed/issues escalated appropriately.
System administrator	■ Serve as customer technical expert on SAP Ariba Buying and Invoicing. ■ Administer configurable aspects of SAP Ariba Buying and Invoicing such as group permissions, user group assignments, approval rules, invoice exception types, and configuration data. ■ Raise any technical issues with SAP Ariba Buying and Invoicing.
Process experts/pilot users/power users	■ Participate in configuration workshops, and contribute to business requirements and solution design. ■ Participate in test case/script development. ■ Participate in system test and UAT. ■ Identify power users within each department to act as experts by providing assistance to peers and input to overall process. ■ Help roll out supplier compliance mandate within their respective departments. ■ Help define and execute internal change management program within their respective department. ■ Act as the knowledge expert on the Ariba Network and ongoing SAP Ariba solutions.
Designated support contact	■ Establish a production support model. ■ Serve as the main contact to SAP Ariba support after deployment. ■ Grant designated support contact access to *http://connect.ariba.com*. ■ Serve as on-demand liaison to SAP Ariba for end-user issues and requested system changes. ■ Authorize administrative requests, including site customizations and enhancement requests.

Table 8.20 Customer Roles and Responsibilities (Cont.)

Customer SAP Ariba Buying Role	Description
	■ Track status of issues and requests on *http://connect.ariba.com*. ■ Work with SAP Ariba to manage volume of issues. ■ Manage and communicate impact of the SAP Ariba upgrades and downtime on the organization. ■ Designated support contact automatically receives the following types of notifications: ■ Notifications of scheduled/unscheduled downtime and security information/bulletins ■ Notifications of new product releases, features, and service pack availability
Supplier enablement lead	■ Serve as primary liaison among the customer, suppliers, and SAP Ariba enablement team. ■ Manage supplier compliance and escalation process for noncompliance. ■ Work with SAP Ariba enablement supplier enablement lead to manage the project and drive project milestones. ■ Serve as country/regional lead for multicountry programs. ■ Enforce program goals within the organization. ■ Approve lists of suppliers targeted for Ariba Network enablement. ■ Manage the gathering of supplier contact data and validate results from SAP Ariba data collection effort. ■ Plan and support supplier communications. ■ Approve supplier communications and education materials to be shared with suppliers. ■ Participate in system test and UAT. ■ Reinforce supplier enablement program compliance with identified suppliers and internal stakeholders. ■ Manage supplier relationships, and monitor and enforce supplier compliance. ■ Facilitate supplier training sessions (if any). ■ Participate in regular status meetings. ■ Promote the initiative internally with category managers and business relationship owners, with the accounts payable group, and externally with suppliers.
Catalog lead	■ Communicate and educate suppliers on catalog requirements and catalogs they must provide. ■ Test catalogs with suppliers to ensure successful publication. ■ Ensure that suppliers provide correct catalog content in correct format.

Table 8.20 Customer Roles and Responsibilities (Cont.)

Customer SAP Ariba Buying Role	Description
Accounts payable, IT, finance, and procurement	■ Select suppliers. ■ Collect data. ■ Maintain vendor master. ■ Develop new supplier processes. ■ Perform training (internal). ■ Prepare for go-liv. – Ensure supplier compliance.
Ariba Network account admin	■ Be responsible for administration tasks related to Ariba Network account (company profile). ■ Perform user administration (create and delete users, roles, and permissions). ■ Set account configuration (transaction business rules). ■ Perform application management. ■ Perform data management.

Table 8.20 Customer Roles and Responsibilities (Cont.)

As with other modules, the actual timelines for SAP Ariba Buying and Invoicing and SAP Ariba Buying projects, as well as for SAP Ariba Buying, multi-ERP edition projects (for more information, Section 8.4.6), depend on the complexity and scope of the implementation. Many accounts payable implementations are done in tandem or embedded in an overall SAP Ariba Buying and Invoicing project, representing the "invoice-to-pay" process in the procedure.

Figure 8.51 shows 22 weeks in 6 core phases. This timeline includes both the supplier adoption and internal adoption tracks running during the second half of the project timeline.

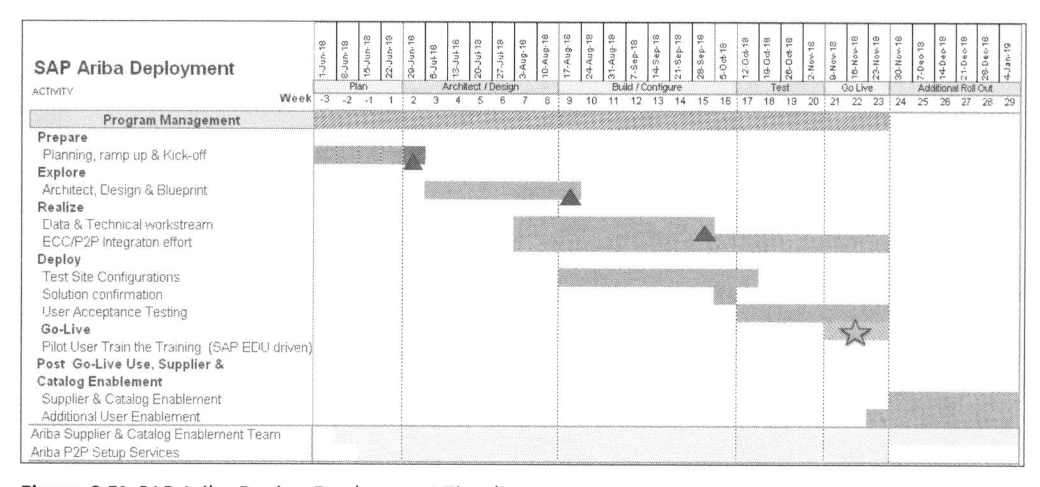

Figure 8.51 SAP Ariba Buying Deployment Timeline

For projects leveraging multi-ERP landscapes, additional time is required, as shown in Figure 8.52, primarily in the architect/design and test phases of the implementation.

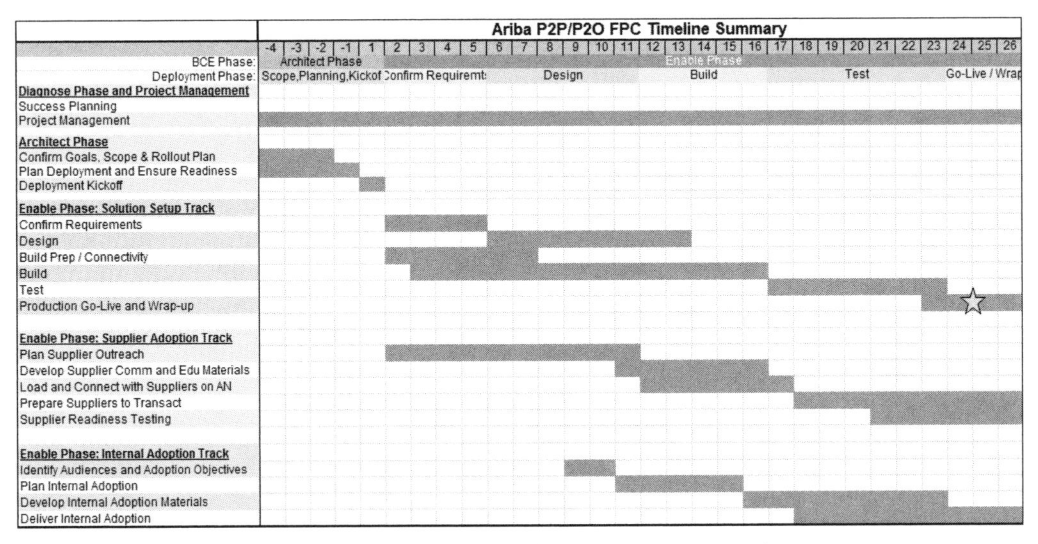

Figure 8.52 SAP Ariba Buying and Invoicing/SAP Ariba Buying Timeline Summary

8.4.6 SAP Ariba Buying, Multi-ERP Edition

If you have multiple SAP S/4HANA backends or business units in various parts of your systems landscape and need to isolate these systems on a one-to-one basis with the SAP Ariba frontend solutions, SAP Ariba offers multi-ERP editions for SAP Ariba Buying and Invoicing and SAP Ariba Buying. The parent site is a control center that defines enterprise-wide procurement policies (approval rules, system parameters, and customizations), manages catalog content, and consolidates the master data that is shared across the enterprise.

The parent site publishes these policy, content, and data components to the child sites, either through subscription or inheritance. The child site (or sites) is where all the actual procurement transactions (requisitions, orders, invoicing, and receiving) occur. Each child site is configured to support one ERP system, which is connected to the child site. Each child site can be integrated with only one SAP S/4HANA system. In most cases, SAP S/4HANA-specific master data is loaded directly into the child site. Note that modifications in child sites can override customizations inherited from their parents.

The multi-ERP child sites can be configured in three different variants:

- **Single variant**
  Data is highly shared between parent and child sites. All data objects are typically maintained in the parent site. This option is only available if the parent and child ERP variants are the same.

- **Multivariant**
  Common data is shared between parent and child sites, but ERP-specific data is loaded directly into the child site.
- **Disconnected/basic**
  Data isn't shared between parent and child sites.

Depending on the child site configuration, a variety of features are available, as shown in Table 8.21.

Feature	Basic	Single Variant	Multi-ERP Variant
Cross-site reporting	X	X	X
Cross-site global contracts		X	X
Shared content	X	X	X
Invoice reassignment	X	X	X
Site switcher	X	X	X
Services procurement (categories imported in children only)	X	X	X
Common data server integration		Parent	Parent
Suite integration		X	X

Table 8.21 Multi-ERP Feature Availability by Child Site Configuration Type

Configuration capabilities vary by child configuration type, as described in Table 8.22.

Configuration Capability	Basic	Single Variant	Multi-ERP Variant
Default language	Configured in child	Child can override parent configuration	Child can override parent configuration
Default currency	Configured in child	Child can override parent configuration	Child can override parent configuration
Catalogs	Configured in child	Configured in parent; child can't override	Configured in parent; child can't override
Catalog hierarchy	Configured in child	Configured in parent; child can't override	Configured in parent; child can't override
Catalog views	Configured in child	Configured in child	Configured in child

Table 8.22 Configuration Capabilities by Child Site Configuration Type

Configuration Capability	Basic	Single Variant	Multi-ERP Variant
Customizations	Configured in child	Child can add to parent configuration	Child can add to parent configuration
Approval rules	Configured in child	Child can override parent configuration	Child can override parent configuration
Rule CSV files	Configured in child	Child can override parent configuration	Child can override parent configuration
Enumerations	Configured in child	Child can override parent configuration	Child can override parent configuration
Flex master data templates	Configured in child	Child can add to parent configuration	Child can add to parent configuration
Parameters	Configured in child	Child can override parent configuration	Child can override parent configuration
String resources	Configured in child	Child can override parent configuration	Child can override parent configuration

Table 8.22 Configuration Capabilities by Child Site Configuration Type (Cont.)

For suite integration, both single- and multi-ERP variants support master data synchronization, SAP Ariba Contract Management and compliance workflows, Single Sign-On (SSO), suite-integrated dashboards, and flexible suite-integrated fields.

A data rationalization process must be completed before beginning an SAP Ariba Buying, multi-ERP edition, deployment. Data rationalization is the process of analyzing an enterprise's master data, business policies, and configuration requirements, as well as determining what can be shared across the enterprise versus what must remain unique to specific business units.

Examples of data rationalization include the following:

- User data, with power users common across realms
- Supplier data, with common suppliers versus site-specific suppliers
- Accounting data, with a chart of accounts typically being ERP-specific and loaded into child sites
- Site configuration data, including exception types, expense policies, and so forth

Before beginning an SAP Ariba Buying, multi-ERP edition, deployment, you must conduct a data rationalization exercise to determine the following:

- Number of child sites you need
- Parent and child site variant types

- Child site configuration types
- Common or shared data, policies, and customizations
- SAP S/4HANA–specific data

General setup is required to roll out SAP Ariba Buying, multi-ERP edition, in a parent-child configuration (as shown in Figure 8.53). Note that all the "children" need to be either SAP Ariba Buying and Invoicing or SAP Ariba Buying systems, as multi-ERP doesn't support mixing different versions of SAP Ariba Buying and Invoicing. Migration isn't available for SAP Ariba Invoice Management sites, nor for sites with different Ariba Network IDs. You can't maintain two different reporting sites in SAP Ariba Buying, multi-ERP edition.

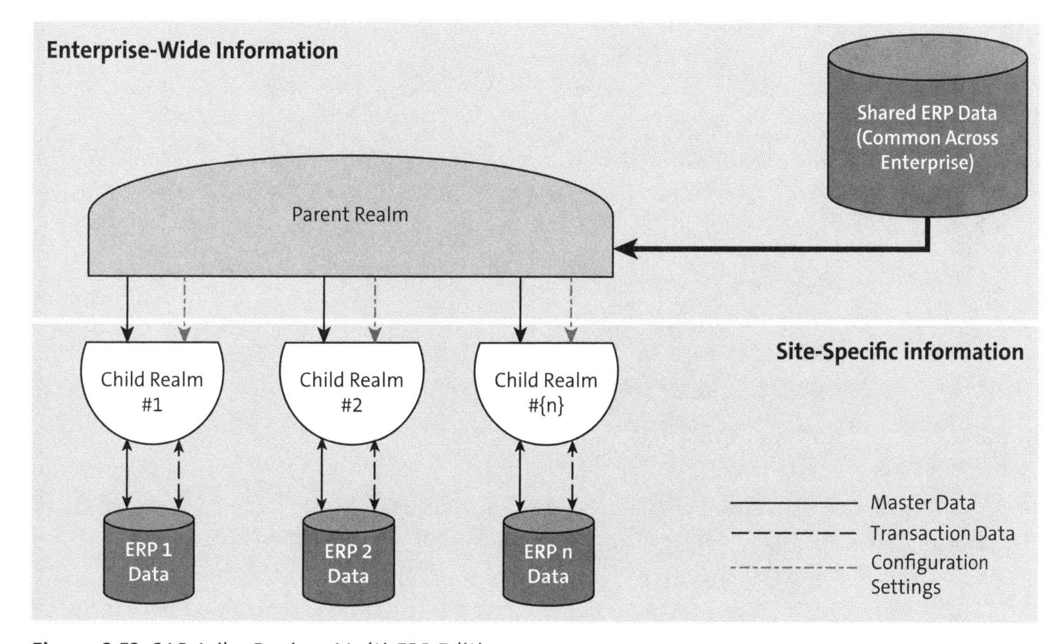

Figure 8.53 SAP Ariba Buying, Multi-ERP Edition

8.5 SAP Fieldglass Vendor Management System

The cloud-based SAP Fieldglass VMS manages all categories of external labor, including contingent workers, SOW projects and services, independent contractors, and specialized talent pools. The SAP Fieldglass VMS addresses the intersection between human resources and procurement, which can help your organization gain visibility into your contingent workforces.

As shown in Figure 8.54, SAP Fieldglass manages all categories of nonpayroll labor spend, including the following:

- Contingent labor in numerous categories
- SOW projects and services

- Independent contractors
- Specialized talent pools, such as retirees and alumni

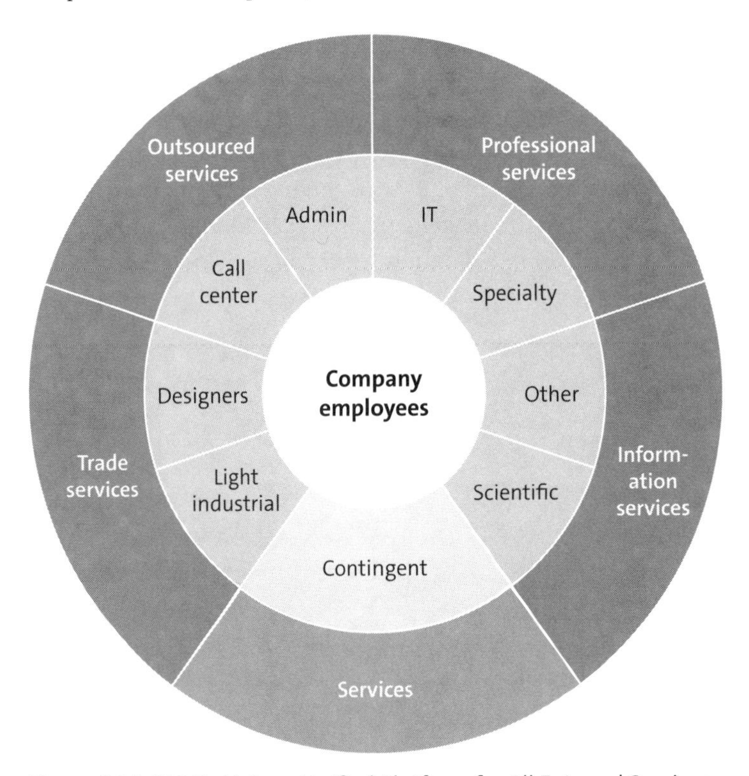

Figure 8.54 SAP Fieldglass: Unified Platform for All External Services

SAP Fieldglass VMS manages the entire lifecycle of procuring goods and services from the initial requisition to extensive reporting. The solution also goes beyond these with core functionalities in five key areas, which address issues with contingent labor procurement, potentially knotty issues for procurement systems.

These key areas are as follows:

- **Visibility**
  Visibility has always been a difficult issue for services, especially regarding contingent labor. Many countries and states have laws regulating the use of contingent labor, so understanding what types of services are being purchased is paramount. In SAP S/4HANA and other systems, workarounds involving misclassification of services to materials are sometimes used to make the procurement process conform to a single master data set or process, which can prevent accurate reporting on services, if these misclassifications have been put in place to streamline the SAP S/4HANA environment. SAP Fieldglass provides a platform to bring this information back to the forefront with accurate classifications and measurements to enable precise management of this spend category.

- **Efficiency**
  Groups operating in SAP S/4HANA often devise their own methods for procuring services, leading to process fragmentation. Some groups may go through a strict review and then resume approvals; others may cut blanket POs and bring in contract workers the next day. Having a common platform and process via SAP Fieldglass brings both scenarios into a more uniform environment, which increases efficiency.
- **Program and worker quality**
  Many procurement systems lack detailed rating options and performance tracking for contract workers, as these types of functions are more commonly associated with human resources-centric applications. A procurement or logistics system might rate damaged goods and supplier on-time performance, but registering and archiving issues with a particular individual contractor in the system is usually difficult. As such, companies are often doomed to repeat the same mistakes down the line, as they lack institutional awareness of past issues. SAP Fieldglass allows this information to be memorialized in the system and provides a register for future interactions with service suppliers and individual contractors, thus increasing quality.
- **Cost savings**
  After you've started accurately measuring your spend in operational procurement using SAP Fieldglass, your additional knowledge can be leveraged to drive savings during negotiations and consolidate spend for further reductions in costs.
- **Risk mitigation and compliance**
  As discussed in the first bullet point, numerous laws that vary by state and even local entity may regulate your service procurement activities. Compliance is often impossible to achieve in a fragmented procurement system, where fields and processes aren't already available in the system. In many cases, an offline, cumbersome process is devised to try to stay in compliance and mitigate risk. With SAP Fieldglass, all these details are collected into one environment and platform, which in turn is integrated with contracts and other compliance drivers on the commercial side of the equation.

Having one focused area for VMS activities, SAP Fieldglass provides an open integration platform that supports all major enterprise systems. With extensive integration with both SAP Ariba and SAP applications, SAP Fieldglass can seamlessly fit into an existing SAP ecosystem and yield many process efficiencies through integrations with accounting or accounts payable systems for invoicing or with SAP Ariba for PO attachment and compliance to total-spend management policies. We'll provide more detail on integration approaches for SAP Fieldglass, SAP Ariba, and SAP S/4HANA in Chapter 10.

Services are a challenging area for most SAP Ariba Buying systems, as the entire document stream is more dynamic and variable than materials procurement (as shown in Figure 8.55). Many engagements have to be negotiated via an RFx process, as the work is often one-off and not repeatable in nature. This type of negotiation produces a tailored SOW and PO. After the work is performed, the actual duration of a service item often doesn't amount to the original amount quoted, and time entries supplant goods receipts as the main matching criteria for accounts payable to pay an invoice.

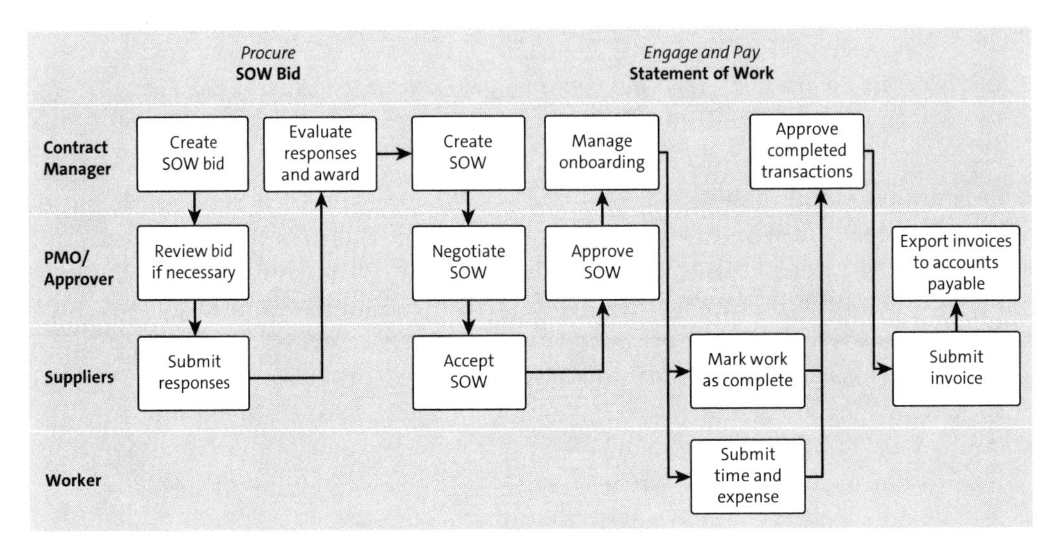

Figure 8.55 Services Procurement Process Flow

Now, add contingent labor management challenges, where vetting a service provider's employees, onboarding/offboarding these employees, and tracking their time and performance becomes even more pronounced, and you have the business case for adopting an overarching VMS.

For contingent labor, vetting the individual who is being engaged to perform the work is often as important as defining the work with the supplier (as shown in Figure 8.56). Interviewing and HR management make this area of services procurement more complex. Not only is the procurement systems area put upon to act dynamically with the changes of services procurement, but a tie-in with human resources systems and processes must be established to manage workers who don't technically belong to your company.

Common issues organizations face in managing their external workforce include the following:

- **No qualifications standards**
  Large numbers of suppliers may have different qualifications to perform various services, or inefficient processes make it difficult for managers and suppliers to comply with policies and existing agreements. Further, a limited view into how much others pay for a service can result in managers paying too much.
- **Cost savings**
  Whether managers are solely sourcing engagements to friends and family, suppliers aren't adhering to the terms and conditions of the agreement that exist with them, or the scope of projects is creeping out of control, SAP Fieldglass can help drive cost savings of 8%–12% in these areas.

- **Lack of efficiency**
  The process of engaging workers and services is complex. For example, a single request can take 20+ phone calls and emails. A great deal of time is spent onboarding and offboarding workers and on reconciling invoices. SAP Fieldglass has simplified the process with a two-click requisition procedure and prepopulated SOW templates.
- **Quality and analysis**
  Without a VMS, organizations commonly approach measuring quality through surveying to managers on a quarterly or semiannual basis. If managers respond at all to the survey, they typically zero in on the most recent or the worst activity from the suppliers they have worked with. With SAP Fieldglass, feedback can be provided at a line-item level throughout the lifecycle of the assignment. Measuring quality is made easier and provides specific feedback about what can be done to continuously improve on the overall service being provided.
- **Compliance**
  With the emphasis on security and safety today, ensuring compliance to your policies on these topics is paramount. How is safety training completed? After workers are terminated, who ensures that they no longer have access to your systems? SAP Fieldglass helps drive compliance to internal policies, such as "don't hire" rules, as well as compliance with broader legislative rules and regulations. SAP Fieldglass provides auditable proof to ensure that policies are adhered to and tracked in a centralized manner.

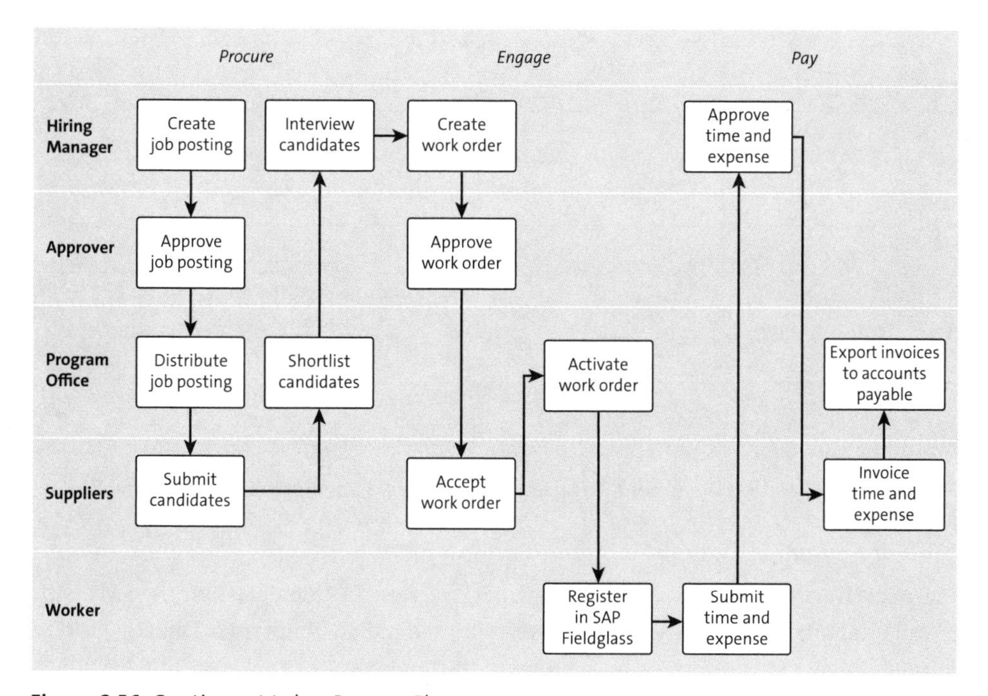

Figure 8.56 Contingent Labor Process Flow

SAP Fieldglass is a technology platform that sits between buyers and suppliers. A common concern is that SAP Fieldglass will complicate your relationships with your suppliers, but the reality is that SAP Fieldglass facilitates that relationship. For buyers, the requisition is simplified; suppliers can provide services in an expedited fashion, as they no longer held up by complex or unfamiliar onboarding/offboarding or approval policies. These highly manual processes are streamlined with SAP Fieldglass. Buyers can create requisitions in two clicks, competitively bid for services, benchmark their needs against the broader market, onboard/offboard workers, capture time and expenses, automate invoicing, and receive detailed reporting and analytics.

As shown in Figure 8.57, a requisition for contingent labor is created in SAP Fieldglass, approved, and then passed to SAP Ariba Buying and Invoicing, where the requisition is processed further and turned into a PO. The work order is moved back up through SAP Fieldglass to the supplier. After the supplier resource submits a timesheet, the requisition is again passed through SAP Fieldglass and approved, and the resulting invoice is passed to the business network for reconciliation in SAP Ariba. After SAP Ariba Buying and Invoicing matches an invoice with its PO and the approvals are complete, the invoice is paid via SAP Ariba's OK2Pay message to an accounts payable system, such as SAP S/4HANA.

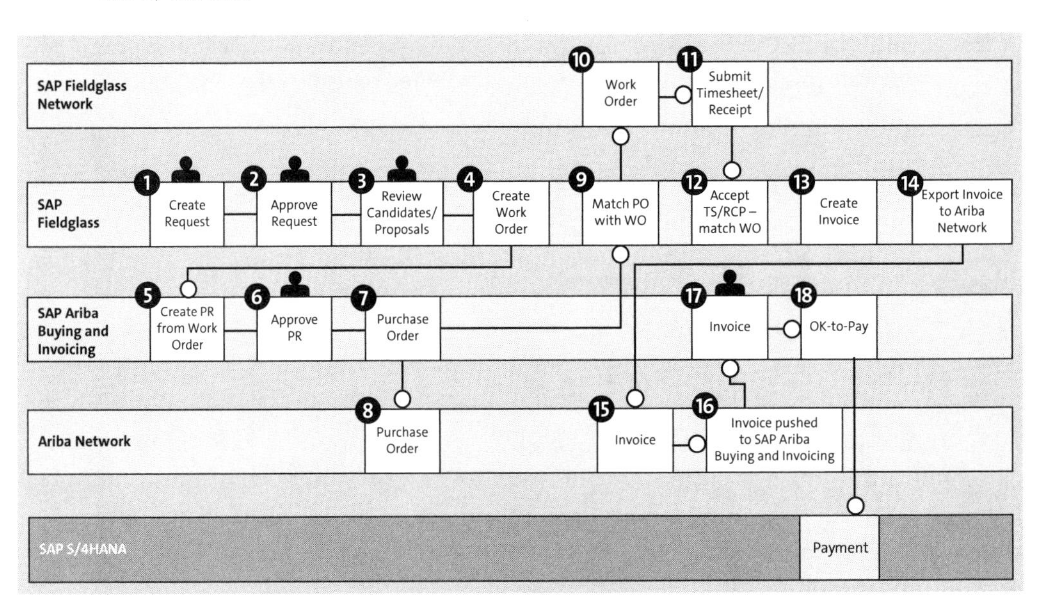

Figure 8.57 Integration Document Flow between SAP Ariba Buying and Invoicing and SAP Fieldglass

The Ariba Network can now intelligently integrate with SAP Fieldglass via the SAP Ariba Cloud Integration Gateway which we'll describe in detail in Chapter 10. This integration provides business benefits for both buyers and suppliers.

As a buyer, you can do the following:

- Use your existing integration with the Ariba Network to route SAP Fieldglass invoices to your SAP S/4HANA system.
- Give your suppliers visibility into the status of SAP Fieldglass invoices on the Ariba Network.
- Realize lower IT maintenance costs by avoiding two integrations.
- Pay your suppliers faster.

As a supplier, you can do the following:

- Get exposure to more business through the Ariba Network (if an SAP Fieldglass-only supplier).
- Check the status of your SAP Fieldglass invoices on the Ariba Network.
- Receive payment faster.

Combined with the collaborative, network-based procurement capabilities of SAP Ariba and the human resources expertise of SAP SuccessFactors, SAP delivers a platform for businesses to manage their entire workforce—both temporary and permanent staff—from initial recruiting and onboarding to ongoing development, performance management, retention, and retirement.

8.6 Implementing the SAP Fieldglass Vendor Management System

SAP Fieldglass has a subscription model typical of the vendor management space. In this model, customers receiving invoices don't pay subscription fees or basic setup services to join the SAP Fieldglass network. The SAP Fieldglass network is similar to the Ariba Network, but the differences are noticeable in terms of scale and focus—more than $20 billion in invoices are submitted over the SAP Fieldglass network annually, whereas the Ariba Network has a $750 billion run rate and supports a much wider variety of transaction types and scenarios.

Suppliers submitting invoices on the SAP Fieldglass network pay modest fees; this revenue model drives a tight focus on enabling customers on the network as quickly as possible. SAP Fieldglass's highly focused delivery organization engages as soon as sales activities have been completed with the customer.

In the following sections, we'll cover the preparatory steps you should take before beginning an SAP Fieldglass project. We'll discuss the resources, phases, and timelines associated with such a project.

8.6.1 Planning Your Implementation

Before beginning an SAP Fieldglass project, you'll need to cover five key steps, as follows:

1. **Define objectives, metrics, and KPIs.**
   Similar to SAP Ariba projects, one of the first things you should do when kicking off a VMS implementation is develop the scope of the project and determine the goals it should accomplish. Is the priority to control costs through rate cards or retiree reuse, or would you prefer to focus on optimizing cost savings by analyzing the use of fixed bids, project-based SOWs, or time and materials? Perhaps your priority is to mitigate risk by better managing compliance with tenure regulations and onboarding/offboarding processes. After these goals are defined, you then align the entire project to those goals.

   Similar to SAP Ariba Supplier Lifecycle and Performance projects, some companies are more concerned with improving the quality of the workforce by using a scoring system and supplier/candidate evaluations. Other common objectives include ensuring adoption of VMS automation, so that more spend is managed, thus decreasing the time to hire and shortening onboarding times. Comparing these goals with the capabilities of your suppliers helps further refine the scope. If your suppliers don't specialize in or support certain goals you initially defined, these goals or the in-scope supplier group need to be adjusted.

   After program goals are established, you'll want to determine how to measure and track results in a systematic and meaningful way. Tying specific metrics and KPIs to each objective is crucial for understanding the performance of the program. With SAP Fieldglass VMS in place, reports should be available for tracking these important metrics. If you have a third-party managed service provider (MSP) or vendor on-premise partner, you can also link the KPIs directly to the service-level agreements (SLAs) of these suppliers in areas such as response times, resume delivery, rates, and invoice accuracy.

2. **Build the team and develop a clear communication strategy.**
   As with most projects, you must ensure the correct people are involved from the beginning of an SAP Fieldglass implementation project to ensure that the process, scope, and timeline are understood by all parties. Your internal team should include a program sponsor, a project manager, and someone with technical aptitude. Ensure that your resources are dedicated to the project with enough bandwidth and have the appropriate skill level and knowledge. This team should provide updates to your senior management periodically concerning the status of the project. A project sponsor helps maneuver through political or organizational roadblocks. Buy-in at senior levels of management will greatly aid in adoption of the VMS. After executives understand the enormous cost savings and efficiencies offered by the SAP Fieldglass VMS platform, they will drive the project's success and the adoption of the system.

The project manager and team should also offer regular updates on the project status as well as communicate the value to specific segments of the organization, such as time savings and higher-quality workers, which is critical to drive further internal adoption and usage of the SAP Fieldglass platform. A VMS deployment requires change management similar to SAP Ariba Buying and Invoicing implementations. User adoption and the changing of buying habits and processes are keys to a successful rollout. Therefore, you must address this important aspect of change management and organizational impacts throughout the implementation by determining how to deal effectively with change in your end-user community. Communication, both crisp and continuous, is a lynchpin in a VMS project rollout. Project leaders and executive sponsors must continually communicate the importance of the project throughout the organization.

One challenge often not immediately apparent for SAP Fieldglass projects is obtaining commitment from your internal IT organization for integration and user management topics. Because SAP Fieldglass has a very light IT footprint, both from a cloud standpoint and in terms of costs to your organization, project leadership may overlook or downplay the important role IT still needs to play in some SAP Fieldglass projects. For example, SAP Fieldglass may need to integrate with existing on-premise systems or other cloud-based systems. Your internal IT department may be the only group with specialized knowledge to assist in the design and eventual access points to these systems. Having the business sign off on the case for SAP Fieldglass is a great first step, but obtaining an IT signoff on identified integration, process, and user management areas before starting the implementation is often just as important.

3. **Establish business processes, assign approval workflows, and understand data requirements.**
Documenting your business rules and system design is the best way to set expectations and avoid misunderstandings throughout the course of the project. In addition, the written materials can act as a reference during the system configuration process and can help curb changes to the scope and design that cause delays. Using process flows will help you make important decisions about approvals. Collecting data such as users, workers, and jobs can be one of the more tedious aspects of the implementation. We recommend determining up front what will be required, including file formats, to avoid complications when it comes time to load and test the data elements in the tool. Make sure you take the time to identify custom field requirements and functionality. As with all cloud implementations, customization options are more limited than with on-premise implementation, but a design phase still may be able to outline achievable enhancements and customizations.
4. **Determine integration requirements early.**
Many SAP Fieldglass VMS implementations include system integrations to existing applications that touch various phases of the worker lifecycle, such as SAP ERP Human Capital Management (SAP HCM), SAP SuccessFactors, SAP S/4HANA,

accounts payable, and SSO. Often, the sharing of data needs to be bidirectional. Integrations can be scheduled to occur in batch mode on a periodic basis or configured to occur only on demand. On-demand calls to an SAP S/4HANA system from an external system are sometimes not possible due to internal IT department rules, operating procedures, and firewall constraints. In these instances, the batch receipt of data from SAP S/4HANA can be scheduled at regular intervals.

Following are a few integration factors to consider:

- Identify required/desired interfaces and required resources.
- Prioritize interfaces realistically and clearly. Phased approaches work provided the interfaces being phased in either have no overlapping data or have been tested thoroughly prior to the phased rollout.
- Account for the corporate integration policy and process requirements in the project plan.
- Integration designs and realization are sometimes an iterative process, where the initial design and functionality require tuning after they are in place.
- Ensure that proper testing by applicable resources and stakeholders is conducted prior to any rollouts.

5. **Enable your supplier community.**
   As with SAP Ariba projects, supplier adoption is as critical to the VMS's success as your internal user adoption. Be sure to maintain an open dialogue between your program owners and suppliers to help ease resistance to new processes, potential fees, and the system itself. Some potential areas to address include the lack of clarity on how to handle workers staffed by noncompliant suppliers and suppliers that try to modify existing MSP or VMS agreements.

If executed properly, your suppliers will recognize their own benefits from the VMS implementation such as faster payments, reduced sales costs, and potential sales increases by receiving more requisitions through the system.

8.6.2 Defining Project Resources, Phases, and Timelines

An SAP Fieldglass implementation schedule typically lasts 8 to 12 weeks, depending on the complexity and integration requirements of the implementation. During that time, you'll work closely with its VMS project team (and that of its MSP, if the program isn't self-managed).

SAP Fieldglass implementations are more cross-organizational than most, leveraging at times SMEs from HR, IT, and procurement organizations. The first thing to obtain is strong project sponsorship that can span these different organizations. A C-level executive on the finance or operations side may need to embrace the cost savings and compliance benefits offered by SAP Fieldglass.

From HR, the group or individual managing contingent labor employees is a logical choice as an SME. From IT, for SAP Fieldglass integrations, we recommend having a lead familiar with both systems and the network requirements. For example, if SAP HCM needs to be integrated with SAP Fieldglass to provide contingent worker information from SAP HCM, we recommend having an SAP HCM SME who owns this information in SAP HCM on the project. If no integrations are planned, then employees will need Internet access and the training to leverage the system. In such cases, the project is more of an HR/procurement-driven initiative and change management exercise. Contingent workforce managers in HR and services procurement people in procurement need to be involved from the onset and through rollout to ensure the design is properly tailored to the company's requirements and that the rollout is accepted by both the internal users and the suppliers.

From a project phase perspective, SAP Fieldglass implementations can leverage the SAP Ariba methodology and roles similar to those in SAP Ariba projects (as shown in Table 8.23 and Table 8.24). Additional regions, suppliers, and company divisions can be expanded in SAP Fieldglass after the initial pilot implementation is complete by using the same milestone methodology.

SAP Fieldglass Role	Description
SAP Fieldglass project manager	■ Serve as point of contact for overall deployment. ■ Ensure that resources are available and properly assigned. ■ Manage overall project timeline to help ensure timely completion of project tasks. ■ Identify and address resource needs. ■ Provide project roles, responsibilities, and issue-escalation paths. ■ Participate in the steering committee. ■ Confirm customer goals and project scope. ■ Establish project management framework. ■ Serve as point of contact for issue escalation. ■ Identify and escalate, as appropriate, project issues. ■ Provide timely project communication, and facilitate regular status meetings. ■ Distribute project wrap-up collateral at the conclusion of the project.
SAP Fieldglass workstream lead	■ Handle responsibility for management of the SAP Fieldglass deployment workstream, workstream timeline, and workstream status on regular basis. ■ Coordinate and facilitate the SAP Fieldglass design and configuration phase. ■ Serve as point of contact for day-to-day configuration questions.

Table 8.23 SAP Fieldglass Roles

SAP Fieldglass Role	Description
Functional consulting: subscription services	■ Facilitate business process review and configuration sessions. ■ Configure the solution according to customer requirements. ■ Keep track of all issues and coordinate issue resolution. ■ Assist with resolution of issues during system test and UAT.
Functional consulting: incremental services	■ Assist customer with gathering all spend category information. ■ Collect high-level business requirements. ■ Assist customer with mapping business processes and procurement categories into SAP Ariba P2P processes. ■ Assist customer with making SAP Ariba P2P configuration decisions. – Assist customer with test script preparation.
Technical consulting: subscription services	■ Educate customer technical team on master data extract, mapping, and load into SAP Fieldglass system. ■ Review data for completeness, and assist in data issue resolution. ■ Assist with technical aspects of site configuration. ■ Perform customizations. ■ Educate customer on functionality and setup of SAP Fieldglass integration tools. ■ Educate customer site administrators on administrative tasks throughout deployment. ■ Assist with resolution of issues during system test and UAT.
Technical consulting: incremental services	■ Support customer technical team to design and build integration interfaces with their SAP ERP and backend systems. ■ Provide customer with best practices to develop the integration architecture.
Supplier enablement lead	■ Handle responsibility for management of the enablement services workstream, enablement workstream timeline, and enablement workstream status. ■ Provide input and obtain approval for suppliers' communications and education. ■ Manage buyer and supplier interactions. ■ Prepare customer and suppliers to transact using the SAP Fieldglass Network.

Table 8.23 SAP Fieldglass Roles (Cont.)

Customer Role	Description
Project sponsor	■ Project commitment time: 10% ■ Establish and communicate overall project vision. ■ Provide senior leadership communication in support of the project.

Table 8.24 Customer SAP Fieldglass Roles

Customer Role	Description
	■ Mandate appropriate change management across leadership of all affected departments. ■ Monitor status reports and timelines. ■ Resolve escalated issues, including those that involve customer resources, lack of participation, or supplier compliance messaging.
Project manager	■ Project commitment time: 100% ■ Confirm customer goals and project scope. ■ Help plan the project activities and timeline. ■ Ensure adequate resources are assigned. ■ Manage participation of all required resources. ■ Coordinate stakeholders as needed (accounts payable, purchasing, receiving, finance, etc.). ■ Coordinate sign-off on all SAP Fieldglass deliverables. ■ Manage project timeline and adherence to schedule. ■ Participate in project status meetings. ■ Identify, manage, and escalate project issues and risks as appropriate. ■ Develop go-live cutover plan and manage cutover execution. ■ Single point of contact for overall deployment.
Functional lead (ideally not the same person as the project manager)	■ Project commitment time: 100% ■ Participate in configuration workshops. ■ Define and document business requirements pertaining to all in-scope processes, transactions, and SAP Fieldglass solutions. ■ Gather business input from all involved functional resources and departments: procurement, accounts payable, accounting, finance, and so on. ■ Define business process and requirements and make decisions on solution configurations to handle those requirements.
Technical team	■ Project commitment time: 80%–100% if integration is required. ■ Serve as SAP ERP/HR technical resource. ■ Participate in configuration workshops. ■ Contribute to SAP ERP–related aspects of solution design. ■ Develop and test SAP ERP interface/integration functionality (for both master data and transactional data). ■ Extract and format master data from SAP ERP systems, and load it into SAP Fieldglass. ■ Assist with testing and resolve SAP ERP–related issues. ■ Conduct cutover activities. ■ Serve as middleware technical resource.

Table 8.24 Customer SAP Fieldglass Roles (Cont.)

Customer Role	Description
	■ Install, configure, and test SAP Fieldglass integration toolkit or web services interfaces. ■ Develop and test middleware transmission and mapping functionality. ■ Configure the integration error handling and notification functionality. ■ Define scheduling scripts for automatic data integrations and email notifications. ■ Configure corporate network to support end-user authentication (if needed). ■ Configure the corporate network to enable inbound and outbound transmissions with appropriate security controls. ■ Assist with testing, and resolve integration-related issues.
Change management/ training lead	■ Project commitment time: 10%–20% ■ Assess the impact of the project's business process, policy, and organizational and cultural changes on end users and other affected employees. ■ Develop and execute a change management plan to address impacts from all changes related to the SAP Fieldglass program. ■ Manage communication plan development and project-wide communications to key stakeholders. ■ If end-user training is needed: – Develop a training approach. – Develop training documentation. – Plan and manage training delivery. – Ensure training needs are addressed and issues are escalated appropriately.
Administrator	■ Project commitment time: 50% ■ Serve as customer technical expert on SAP Fieldglass on-demand site. ■ Administer configurable aspects of the SAP Fieldglass on-demand site such as group permissions, user group assignments, approval rules, invoice exception types, and configuration data. ■ Raise any technical issues with the SAP Ariba on-demand site.

Table 8.24 Customer SAP Fieldglass Roles (Cont.)

Customer Role	Description
Process experts/pilot users/power users	■ Participate in configuration workshops, and contribute to business requirements and solution design. ■ Participate in test case/script development. ■ Participate in system test and UAT. ■ Identify power users within each department to act as experts by providing assistance to peers and input to the overall process. ■ Help roll out the supplier compliance mandate within their respective departments. ■ Help define and execute internal change management program within their respective department. ■ Act as knowledge expert on the SAP Fieldglass network and ongoing SAP Fieldglass solutions.
Supplier enablement lead	■ Serve as primary liaison among customer, suppliers, and the SAP Fieldglass enablement team. ■ Manage supplier compliance and escalation processes for noncompliance. ■ Work with the SAP Fieldglass supplier enablement lead to manage the project and drive project milestones. ■ For multicountry programs, require a country/regional lead. ■ Enforce program goals within the organization. ■ Approve lists of suppliers targeted for Ariba Network enablement. ■ Manage the gathering of supplier contact data, and validate results from the SAP Ariba data collection effort. ■ Plan and support supplier communications. ■ Approve supplier communications and education materials to be shared with suppliers. ■ Participate in system tests and UAT. ■ Reinforce supplier enablement program compliance with identified suppliers and internal stakeholders. ■ Manage supplier relationships, and monitor and enforce supplier compliance. ■ Facilitate supplier training sessions (if any). ■ Participate in regular status meetings. ■ Promote the initiative internally with HR, category managers, and business relationship owners; with the accounts payable group; and externally with suppliers.

Table 8.24 Customer SAP Fieldglass Roles (Cont.)

8.7 Summary

In this chapter, we reviewed what SAP Ariba would call "downstream solutions" for both SAP Ariba and SAP Fieldglass. These areas cover both P2P processes as well as augmenting solutions in the areas of content and contingent labor procurement. Broadly, this area is called operational procurement, as these systems support different kinds of tactical procurement at an operational level.You should now have an idea what SAP Ariba Buying entails from a functional standpoint, and how some of the different configuration options can help you achieve efficiencies in your procurement processes. The topics covered here – creating requistitions, automated approval processes, purchase order generation, and services procurement, will help you understand how your organization can get the most value out of your SAP Ariba Buying solution.

Chapter 9
Invoice Management

In this chapter, we'll examine the SAP Ariba solutions for managing invoices and payment processes in accounts payable.

The final part of the procure-to-pay (P2P) process resides with accounts payable. Traditionally, suppliers submitted paper invoices in various formats, which created document management challenges and multiple inefficiencies. SAP Ariba approaches this area in a multitiered manner, leveraging the Ariba Network to demystify and streamline the process of invoice submission and management, while providing a host of payment tools and functionalities to support both buyer and supplier in the final step in the transaction.

9.1 What Is Invoice Management?

After a supplier fulfills a purchase order (PO), the supplier requests payment from the customer via an invoice. An invoice is thus essentially a bill. Invoice management focuses on processing invoices with the least number of steps and employee-driven adjustments. As with the PO, receiving and processing an invoice for a customer at the system level isn't necessarily routine or straightforward. To remain touchless while in the system, the invoice submitted by the supplier must adhere to a protocol that the customer's system can understand and include a reference that enables the customer to match the invoice to the originating purchasing process and documents. Otherwise, an invoice process can quickly necessitate more time-consuming and expensive manual review and corrective steps.

At its worst, a paper-based invoice-to-payment process entails the supplier writing up an invoice by hand and sending it via traditional mail. Then multiple cumbersome official communications take place, and a supplier eventually issues a new invoice or a credit using the same method. The customer receives the new invoice, and then sorts and requeues it for processing. By this time, the customer may be out of compliance with the terms of the original PO for payment and spending considerable amounts of time and money to finally initiate payment via a check run, which itself is another laborious task that involves issuing, printing, and mailing checks to suppliers. With all this overhead, some invoice processing operations can cost as much as $60 per invoice processed; processing costs sometimes eclipse the amount of the actual invoice to be paid, even while they may miss opportunities for discounts and working capital management.

In this chapter, we'll cover SAP Ariba solutions for accounts payable, which includes SAP Ariba Invoice Management, SAP Pay, SAP Ariba Discount Management, and the SAP Ariba supply chain finance capability. In this chapter, we'll outline the solutions' functionalities, implementation approaches, and usage.

9.1.1 Invoice Management Strategies

The overarching strategy for driving efficiency with invoicing is first to move off of paper-based, manual processing, and you can do this via SAP Ariba's invoice management and payment solutions. Using the rules and tools of SAP Ariba Invoice Management, an accounts payable team can quickly return invoices with missing or erroneous information to the supplier for correction.

With SAP Pay, you can further avoid late-payment fees and inefficiencies caused by paper checks, as well as finance your supply chain with the SAP Ariba supply chain finance capability. SAP Ariba Discount Management manages prorated, automatic, ad hoc, and dynamic discounting.

9.1.2 SAP Ariba Portfolio

The area of focus in a P2P scenario for these SAP Ariba solutions is at the end of the overall process, as shown in Figure 9.1, at the time of invoice submission and payment.

The payables and invoice management solutions in SAP Ariba are largely split between SAP Pay and SAP Ariba Invoice Management, as shown in Figure 9.2.

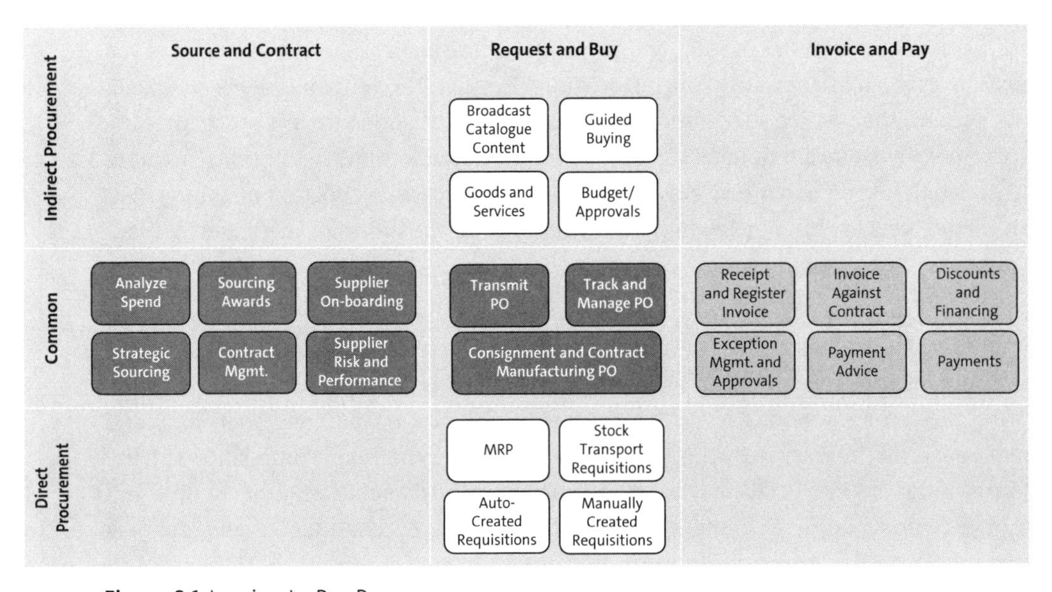

Figure 9.1 Invoice-to-Pay Process

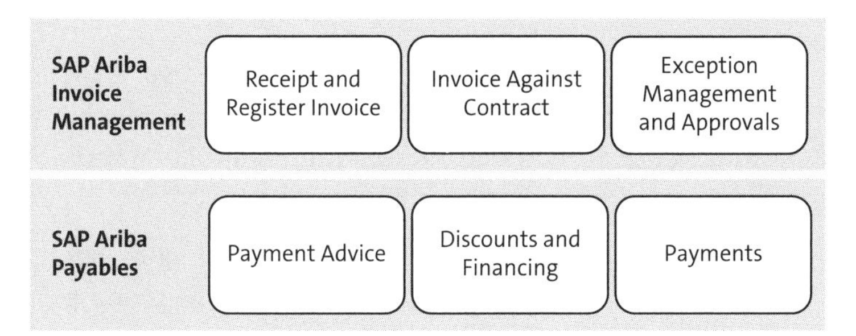

Figure 9.2 Invoice-to-Pay Process with SAP Ariba

In the following sections, we'll expand in more detail on these two main SAP Ariba solution areas for invoices and payments.

9.2 SAP Ariba Invoice Management

SAP Ariba electronic invoice (e-invoice) automation comprises five core solution areas, including cloud-based electronic invoicing, invoice management, contract and service invoicing, global invoicing and compliance, and supplier enablement services. Leveraging these solutions, you can optimize invoice processing and lower operating costs by significantly reducing paper and more effectively handling exceptions and supplier inquiry calls. These solutions support more than 70 currencies, digital signature authentication, value-added tax (VAT)/tax compliance, and data archiving.

The invoice automation process can be outlined at the system level, as shown in Figure 9.3.

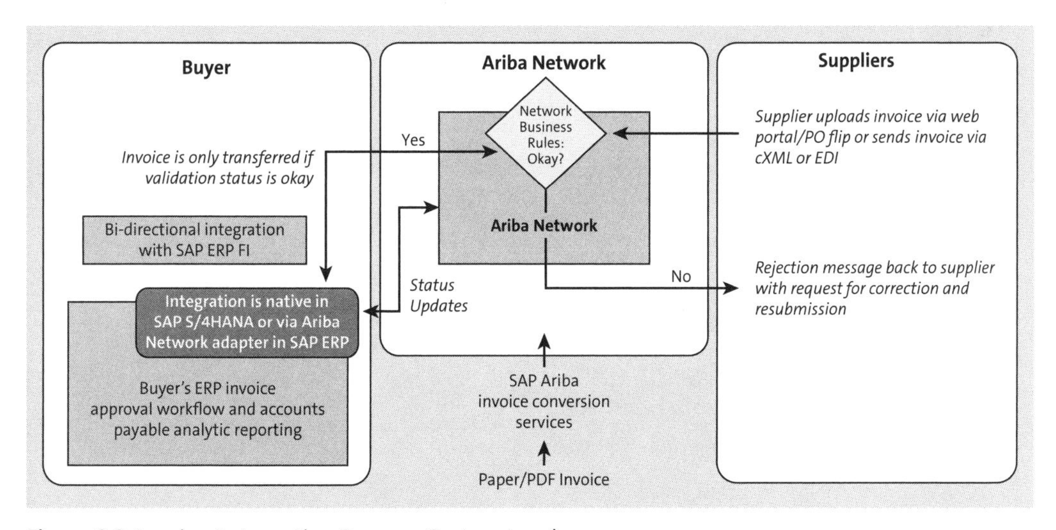

Figure 9.3 Invoice Automation Process: System Level

Supplier connectivity and invoicing options include the SAP Ariba supplier portal, which is accessible via a web browser and an Internet connection. For suppliers that insist on sending paper invoices or other scenarios/jurisdictions where paper invoices are required, the SAP Ariba invoice conversion services (SAP Ariba ICS) add-on is available to scan and convert those paper invoices electronically.

SAP Ariba e-invoice automation focuses on the general invoice processing functionality enabled in the Ariba Network between buyer and supplier. Buyers can configure rules to ensure that suppliers submit conforming invoices at the line item, country, and supplier group levels. SAP Ariba provides more than 30 templates for these invoices, as well as more than 80 rule settings to conform to various country-based and invoice-type requirements in order to prevent a supplier from submitting a faulty invoice into the system from the start, as shown in Figure 9.4.

Given the scope of the Ariba Network, with more than two million suppliers, many suppliers in your supply base are likely already transacting in this Ariba Network. During supplier enablement services supporting SAP Ariba invoice automation, as well as the other areas covered, the project onboards targeted suppliers into the Ariba Network as part of the project and follow-on phases. As with SAP Ariba Buying projects, customers typically onboard their suppliers in waves to adequately support the onboarding and to set up processes for the new invoicing approach. Throughout this enablement, SAP Ariba is with you at every step, as you'll see at the end of Section 9.2.1.

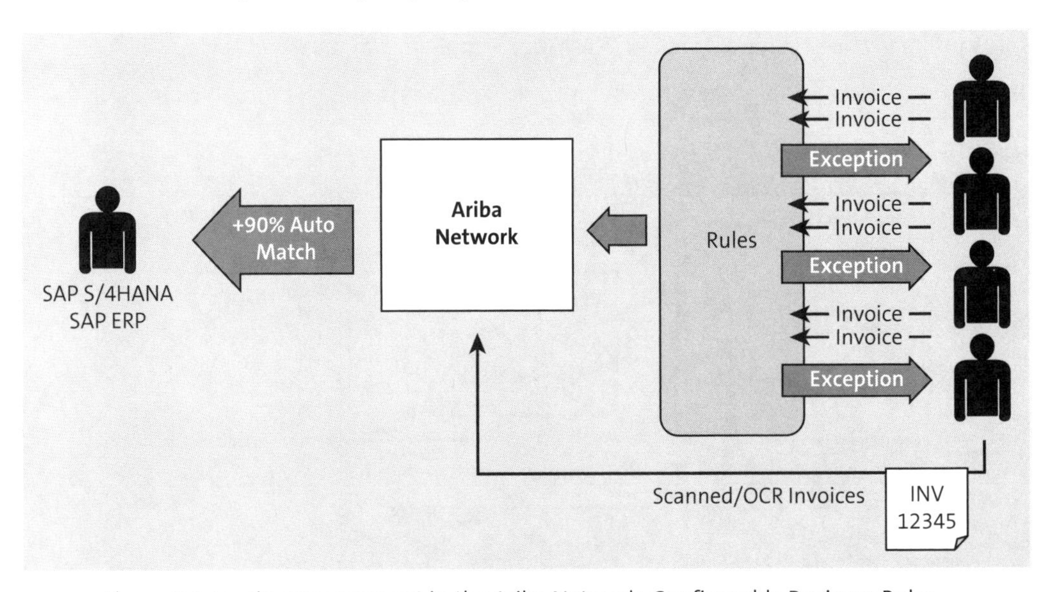

Figure 9.4 Invoice Management in the Ariba Network: Configurable Business Rules

9.2.1 Electronic Invoice Management at a Glance

For a typical SAP Ariba P2P workflow at the invoice end of the process, the steps are as follows:

1. The supplier creates an invoice. Invoices are submitted in one of the following ways:
   - The supplier submits an invoice via the Ariba Network.
   - The supplier uses the SAP Ariba ICS , submitting their invoice in a traditional format (e.g., paper), and the add-on converts the invoice into an electronic format.
   - The buyer can receive invoices from and submit invoices to the SAP Ariba ICS, or the supplier can submit invoices directly to SAP Ariba ICS.
   - The buyer submits the supplier's invoice by keying the supplier's paper invoice data into the SAP Ariba invoicing solution.
2. The invoice is submitted.
3. SAP Ariba receives the invoice data and generates an invoice reconciliation document.
4. The SAP Ariba automatic reconciliation process tries to match the invoice reconciliation document to existing POs, contracts, and receipts data, such as price, terms, quantity, and so on, and then validates them according to your configuration.
5. SAP Ariba lists details from the invoice, POs, contracts, and receipts on the invoice reconciliation document:
   - If no discrepancies are found, or if the discrepancies are within your company's preconfigured tolerances, the invoice reconciliation document is sent directly to the invoice manager for approval.
   - If unacceptable discrepancies are found, they are listed as exceptions on the invoice reconciliation document and must be manually reconciled.
6. After an invoice has been reconciled and approved by the invoice manager, a payment approvable is generated. A *payment approvable* takes the form of an OK2Pay file that is sent to your SAP ERP system. Validate that it's OK2Pay based on the supplier's payment method and terms (net 30, etc.).
7. You may configure SAP ERP systems to communicate payment status updates back to your SAP Ariba invoicing solutions and the Ariba Network. In SAP Ariba e-invoice automation scenarios, you can view payment status details by navigating to the **Payments** tab of a given invoice. Suppliers can view payment information for their Ariba Network invoice as well.

The five core solution areas of SAP Ariba's invoice automation capabilities cover and buttress this process comprehensively, as we'll outline next.

Cloud-Based Electronic Invoicing

Beginning with cloud-based electronic invoicing, you can configure workflows to address specific business requirements and enable mobile approvals for your employees. SAP Ariba also offers SAP Ariba ICS, as mentioned earlier, to outsource the investment and effort required to convert paper-based invoices into electronic ones via optical character recognition (OCR). You can also invite suppliers to join the Ariba Network and begin transacting electronically with your organization, reducing expensive, paper-based invoicing over time.

SAP Ariba Invoice Management

SAP Ariba Invoice Management allows you to process all your invoices digitally, handling e-invoices submitted via the Ariba Network, as well as paper invoices submitted by suppliers to an outsourced service or via internal accounts payable processing routes. Intuitive drag-and-drop workflow capabilities allow users to create workflows for various invoice processing scenarios, prioritize invoices for processing, and adjust for continuous improvement. With mobile approvals, financial reporting, and dashboard-driven monitoring, SAP Ariba Invoice Management further enhances the effectiveness of accounts payables and the user community satisfaction.

Invoicing in the Ariba Network includes rules that you can categorize into the following levels:

- Default transaction rules
- Supplier group rules
- Country-based invoice rules

Supplier group rules override default transaction rules, and country-based rules are evaluated against the default transaction rules or supplier group rules. Buyers can also override rules for suppliers where required. These rules prevent suppliers from submitting faulty invoices in the first place.

SAP Ariba Contract Invoicing

Non-PO invoicing often occurs during complex projects where a blanket order or agreement is created for the project, but the supplier is unable to reference a PO. SAP Ariba Contract Invoicing allows a supplier to access applicable rate sheet information and terms from a contract and create invoices directly referencing the contract instead of being tied to a purchasing document at all. Contract invoicing is available for contracts on the supplier, commodity, catalog, and item levels as well as for complex, service-oriented or service-driven project contracts.

Complex services and stocked maintenance, repair, and operations (MRO) items are often difficult to manage in the system, and companies revert to expensive, paper-based processes as a result. SAP Ariba Services Invoicing pulls these procurement areas back into the digital world by enabling suppliers to create service entry sheets via an online tool, as well as by uploading comma-separated values (CSV) files or commerce eXtensible Markup Language (cXML) files and third-party integration. These service entry sheets can then be "flipped" into an invoice for touchless processing. You can also leverage catalog and contract data to further enrich the invoice's details and support touchless processing (where little to no further review or adjustment is required by the accounts payable department staff).

A buyer can define service-specific rules, such as the following:

- Require suppliers to provide approver information on service sheets.
- Require suppliers to send attachments.
- Allow suppliers to add ad hoc line items.
- Allow suppliers to increase item quantities.
- Require a field contractor and/or field engineer as part of service.
- Create exception tolerances.
- Allow suppliers to add items from the contract/catalog.
- Mask the expected amount so suppliers can't see the allocated budget.

As a general rule, service invoices can be created only for approved service sheets.

9

Global Invoicing and Compliance

You can configure more than 40 business rules based on best practices and local (by country) tax requirements in SAP Ariba Invoicing. SAP Ariba also supports e-signatures in 36 countries, further digitizing the invoice process and driving compliance. For invoice document retention, you can either download your invoices for storage or have SAP Ariba automatically store invoices by leveraging a third-party provider.

Paperless and tax-audited invoicing is supported in the countries listed in Table 9.1.

▪ Australia	▪ Denmark	▪ Italy	▪ Romania
▪ Austria	▪ Estonia	▪ Japan	▪ Singapore
▪ Belgium	▪ Finland	▪ Latvia	▪ Slovakia
▪ Brazil	▪ France	▪ Luxembourg	▪ South Africa
▪ Bulgaria	▪ Germany	▪ Malaysia	▪ Spain
▪ Canada	▪ Greece	▪ Mexico	▪ Sweden
▪ Chile	▪ Hong Kong	▪ Netherlands	▪ Switzerland
▪ China	▪ Hungary	▪ New Zealand	▪ United Arab Emirates
▪ Cyprus	▪ India	▪ Norway	▪ United Kingdom
▪ Czech Republic	▪ Ireland	▪ Poland	▪ United States

Table 9.1 Paperless and Tax Audited Invoicing: Supported Countries

An additional 60 countries support pro forma invoicing, as shown in Table 9.2.

■ Argentina ■ Aruba ■ Bahamas ■ Bahrain ■ Barbados ■ Bermuda ■ Botswana ■ Brunei ■ Cambodia ■ Cayman Islands ■ China ■ Colombia ■ Costa Rica ■ Croatia ■ Dominican Republic	■ Ecuador ■ Egypt ■ El Salvador ■ Ghana ■ Guam ■ Guatemala ■ Gibraltar ■ Honduras ■ Iceland ■ Indonesia ■ Israel ■ Jamaica ■ Jordan ■ Kazakhstan ■ Kenya ■ South Korea	■ Kuwait ■ Lebanon ■ Liechtenstein ■ Lithuania ■ Macao (special administrative region of China) ■ Monaco ■ Malta ■ Mauritius ■ Morocco ■ Mozambique ■ Namibia ■ Nicaragua ■ Nigeria ■ Pakistan ■ Panama ■ Peru	■ Philippines ■ Portugal ■ Puerto Rico ■ Russia ■ Saudi Arabia ■ Slovenia ■ Sri Lanka ■ Taiwan ■ Thailand ■ UAE ■ Vietnam ■ Virgin Islands

Table 9.2 Pro Forma Invoicing: Supported Countries

Supplier Enablement Services

In the world of traditional software solutions, enabling suppliers is an often-overlooked topic. The supplier portal is often built under the *Field of Dreams* hope—"If you build it, they will come"—without taking a hard look at an organization's capacity to manage the onboarding of thousands of suppliers. The suppliers typically don't want to spend precious time trying to learn new systems voluntarily, and many point-to-point supplier portals on the customer side are often underused as a result.

A significant benefit to onboarding suppliers to your invoicing initiatives in the Ariba Network is that many of your suppliers may already be transacting with other customers in the Ariba Network and will readily join your invoicing initiatives as a result. For suppliers that aren't on the Ariba Network, SAP Ariba supplier enablement service offers an array of value-added services to drive supplier onboarding and adoption, including tailored enablement strategies, implementation of support and project management, and business process and geographic coverage, as well as supplier education, tracking, and support. With more than 700 full-time employees covering 35 languages in 21 countries, SAP Ariba supplier enablement greatly extends the reach of this area of an invoicing project.

In addition to the value-added services for supplier enablement, SAP Ariba offers different tiers for suppliers to onboard at the level they intend to transact on the Ariba Network, as shown in Figure 9.5. Of the three tiers, the first is the standard account, which includes the following:

- Email-driven access using a free standard account
- Interactive ordering-by-email capability and e-invoice status tracking capability
- Almost all Ariba Network documents are available using the online user interface (UI), including e-invoice profile setup:
  - POs, full and partial order confirmations, advanced shipping notifications (ASNs), service entry sheets
  - Non-PO invoices and credit notes
  - Invoice status notifications, payment proposals, remittance details
  - PO e-invoices for Mexico, Chile, Columbia, and Brazil
- Dynamic discounting
- Unlimited Ariba Network relationships
- Unified supplier experience

Regardless of the level of the supplier, the Ariba Network provides suppliers with a comprehensive platform for collaborating with customers, including the following:

- A single account for fulfillment, selling, and mobile
- Multiuser support
- Quick upgrade to enterprise accounts

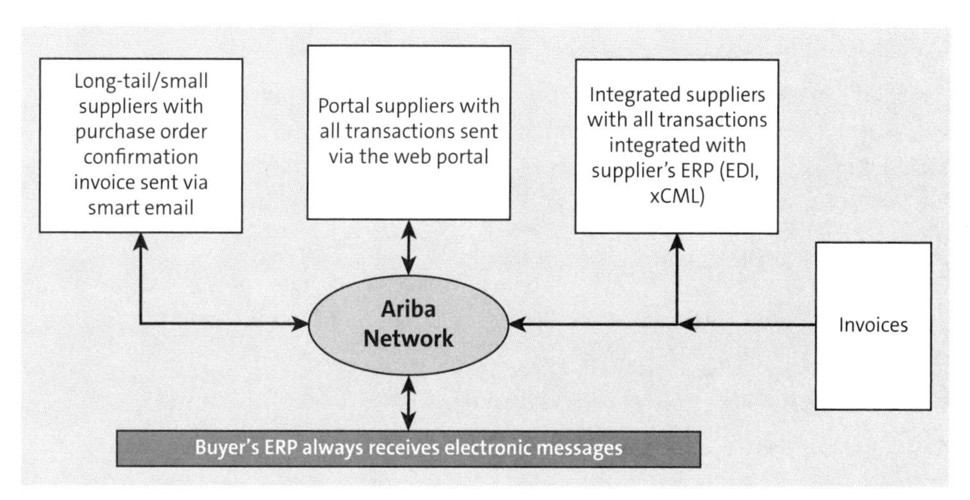

Figure 9.5 Supplier Enablement Tiers

The next tier is the enterprise account, which includes all web-based transactions, including *PO-Flip*, which is the SAP Ariba term for the automatic conversion of POs to invoices. The final tier is the integrated supplier tier, where a supplier can conduct machine-to-machine transactions via electronic data interchange (EDI) and cXML. Paper-based invoices can be converted via SAP Ariba ICS.

Via workflow rules, with SAP Ariba, your organization can achieve an almost 100% touchless processing of invoices. You can also extract tax table data from your SAP ERP system or a third-party tax application and load it into SAP Ariba Invoice Management on a recurring basis as well to validate taxes submitted on an invoice at either the header or line item level.

Self-service tools for suppliers include online dashboards, which centralize all documents and communications. In these dashboards, suppliers can view approvals, payments, rejected orders, and so on. By streamlining invoice processing with SAP Ariba e-invoice automation, organizations can maintain or extend days payable outstanding (DPO). With this efficiency in place, the payment process provides you with greater flexibility in payment scheduling, including obtaining discounts for earlier payment than negotiated terms (using SAP Ariba Discount Management, discussed later in Section 9.4.1). Working capital/terms and on-time payment analytics aligned with SAP Ariba Invoice Management enable in-depth analysis of the performance of these associated account payables topics.

Invoices with errors or omissions are rejected and returned to suppliers for correction and resubmission. With compressed invoice approval cycle times, driven by SAP Ariba Invoice Management and SAP Ariba invoice automation capability, organizations can leverage the speed and efficiency of their invoice processing to maximize early payment discounts.

Your suppliers can submit PO and non-PO invoices online through the Ariba Network UI or via EDI/cXML. Suppliers can also convert, or "flip," a PO received on the network to create a PO invoice.

Suppliers can use the Ariba Network to view status updates for invoices that are supplier-entered, buyer-entered, entered through SAP Ariba ICS, or copied from the buyer's SAP ERP system. This transparency reduces follow-up calls and confusion in your payment processes. For archiving purposes, both buyers and suppliers can request a ZIP file of invoices.

Optional features for SAP Ariba e-invoice automation include the services invoicing capability in the SAP Ariba Contract Invoicing solution, which includes the following capabilities:

- Creation of invoices from contracts through the UI
- Creation of service sheets for service POs through the UI
- Creation of service sheets by uploading them as CSV files or transmitting them electronically as cXML documents or EDI documents.
- Creation of contact invoices by uploading them as CSV files or transmitting them electronically as cXML or EDI documents
- Auto-generated service entry sheets from service invoices
- Search functionality for checking and reporting on service sheets and contract invoices
- Ability for buyers to set up contracts in the UI and validate service sheet submissions against these contracts with access control and master data approval

Finally, SAP Ariba provides optional services for SAP Ariba Invoice Management, including the following:

- Expanded site configuration
- Invoice CSV upload
- Acceleration or update to SAP Ariba country guides
- Tax invoicing for Mexico (CFDI – Digital Tax Receipt by Internet)
- Tax invoicing for Brazil
- SAP Ariba ICS
- SAP Ariba open invoice conversion services (SAP Ariba Open ICS)

9.2.2 Integrated Implementations

Many customers aren't in a position to decouple the procurement process from their existing SAP ERP (including SAP S/4HANA and SAP ERP) environments, especially finance and controlling (FI/CO). In this scenario, documents are sent down to SAP ERP, beginning with the PO. The PO copy updates SAP ERP's budget in finance, allowing the finance department to understand their outstanding liabilities in real time as liabilities are created in SAP Ariba. Next, the order is confirmed, and the supplier sends an optional inbound delivery notification in SAP Ariba. These documents also update in SAP ERP, which notifies the inventory management and production planning departments of a pending order.

Inventory management/receiving then creates a goods receipt in the SAP ERP environment. This goods receipt takes the item into inventory (in some cases) and can queue financial processes in the event that the goods receipt is valuated for the item. In some instances, finance can begin including these valuated items in capital depreciation runs, in the case of an asset, and inventory value calculations in SAP ERP, while the invoice is being reconciled and payment is being actioned in SAP Ariba Invoice Management.

9.2.3 Data Sources and Solution Landscape Inventory

In this chapter, we've covered several different data sources for the invoice-to-pay process to consider both during the project and at runtime. During the project's requirements gathering phase, master data elements for the new system need to be defined. Should these sources of data remain the system of record after SAP Ariba's solutions are in place? In other words, will SAP Ariba need updated tax tables from a tax system/file cabinet and need purchasing groups from SAP ERP, or will SAP Ariba become the system of record and these systems effectively become the children in a parent-child relationship, where SAP Ariba updates the various systems? Will the systems be maintained independently? Please note that maintaining systems independently is always a dangerous proposition if you expect any interaction between the systems going forward. These decisions you'll make during the project phases and at runtime have implications for the overall system performance and, ultimately, for your return on investment (ROI).

Understanding where the rules and tolerances are defined and kept in the current system and process is crucial. Rules, terms, and tolerances could be maintained in an existing system being replaced or in a file cabinet in a paper-based operation. Conversion will involve inventorying what needs to be converted from where, as well as interface activities for any required ongoing updates after the SAP Ariba solution is live. You must also consider and account for banking relationships in the design of your solution to ensure that communication is seamless and that payments are made as required by the process and systems.

Regarding the landscape of an overall invoice-to-pay solution in SAP Ariba, much of the interplay between systems will take place between systems of record with information and transaction capabilities relevant to the process. These systems include SAP ERP systems; invoice processing solutions for scanning documents, such as OpenText Vendor Invoice Management (VIM); and tax solutions that are either connected to SAP ERP already and updating SAP Ariba tables or are integrated at some level with SAP Ariba directly. Banking and payment solutions may also require integration at some level.

The process and the system landscape can be configured in many different ways to integrate SAP Ariba's invoice-to-payment solutions into your accounts payable operations. Invariably, these solutions relieve the administrative burden of manual, paper-based processes and drive greater flexibility in a diverse set of strategic areas of the business. Whether working capital, supplier satisfaction/alignment, or greater overall efficiencies for finance and accounts payable, SAP Ariba invoice-to-payment solutions offer a wide variety of potential uses and payoffs for the financial operations aspects of procurement.

In the following section, you'll see how to configure the different invoicing touchpoints from the Ariba Network transaction rules to the invoice approval and reconciliation process in SAP Ariba Invoice Management.

9.2.4 Configuring Ariba Network Transaction Rules for Invoicing

When orders are sent to Ariba Network–enabled suppliers, a number of different transaction rules determine which actions the suppliers can take. These transaction rules are typically configured based on a combination of your company's purchasing and invoicing policies and SAP Ariba's leading practices. For example, your company may have a "no PO, no invoice" policy where a supplier-submitted invoice must reference a corresponding PO, or you may have strict guidelines and tolerances around accepting invoices where item quantity or unit price must not exceed the quantity or unit price on the PO. The rules that enforce these polices are the Ariba Network default transaction rules. After default transaction rules are set, supplier groups and country-based invoicing rules can be maintained to address unique supplier invoicing situations.

Note

Default transaction rules are configured on the buyer side of the Ariba Network. Your buyer administrator can set the default network transaction rules by accessing the buyer account on the Ariba Network.

To configure default transaction rules, follow these steps:

1. Click the **Configuration** link under **Administration** from the Ariba Network buyer dashboard.
2. Click **Default Transaction Rules** from the list of company settings.
3. Review each section in the list and modify transaction rules as needed (Figure 9.6). Invoice-related rules are divided into the following sections:
   - **General Invoice Rules**
   - **PO Invoice Field Rules**
   - **PO and Non-PO Invoice Field Rules**
   - **Invoice Address Rules**
   - **Blanket Purchase Order Invoice Rules**
   - **Invoice Payment Rules**
   - **VAT Rules**
   - **Online Invoice Form Rules**
4. Click **Save** to exit and apply changes.

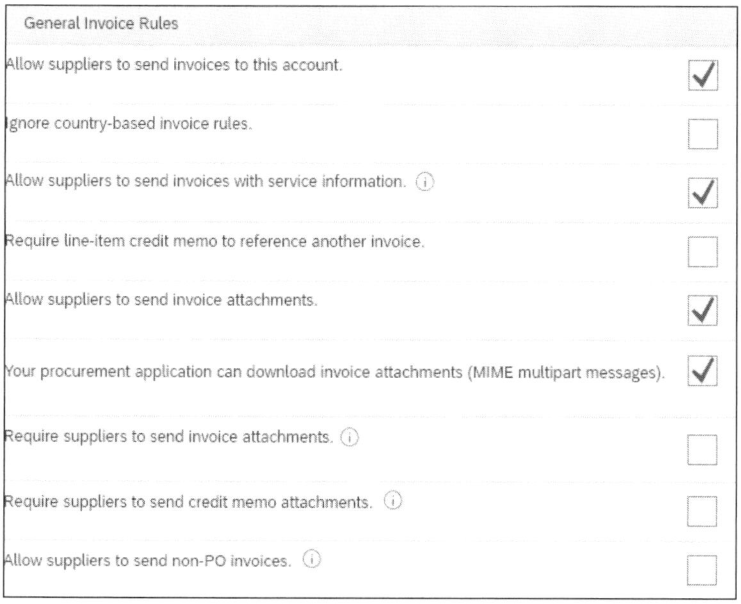

General Invoice Rules

Allow suppliers to send invoices to this account.

Ignore country-based invoice rules.

Allow suppliers to send invoices with service information.

Require line-item credit memo to reference another invoice.

Allow suppliers to send invoice attachments.

Your procurement application can download invoice attachments (MIME multipart messages).

Require suppliers to send invoice attachments.

Require suppliers to send credit memo attachments.

Allow suppliers to send non-PO invoices.

Figure 9.6 A Sample List of Buyer-Configured Ariba Network Default Transaction Rules

The default transaction rules set the baseline configuration on how suppliers will be allowed to enter invoices and will apply to the majority of your suppliers. Next, we'll look at setting up overrides to the default transaction rules using supplier groups.

Configuring Supplier Groups

A single set of transaction rules might not apply to all your suppliers. Depending on your relationship with the supplier or the types of goods and services you purchase, some network transaction rules might need to be different. For example, as a policy, your company may not require order confirmations with the exception of a handful of suppliers. To achieve this, the Ariba Network allows you to group your suppliers and apply a different set of network transaction rules specific to the group.

To configure supplier groups, follow these steps:

1. Click the **Supplier Groups** link under **Supplier Enablement** from the Ariba Network buyer dashboard (Figure 9.7).
2. Click **Create,** and enter a **Name** and **Description** for the supplier group.
3. Click **Supplier Group Transaction Rules,** and select **Enter supplier group specific transaction rules**. Review the list of rules, and modify them as needed to meet your country-specific requirements.
4. Click **OK** to save the changes.
5. Click on **Add Members**, and add suppliers to the group.
6. Click **OK**.

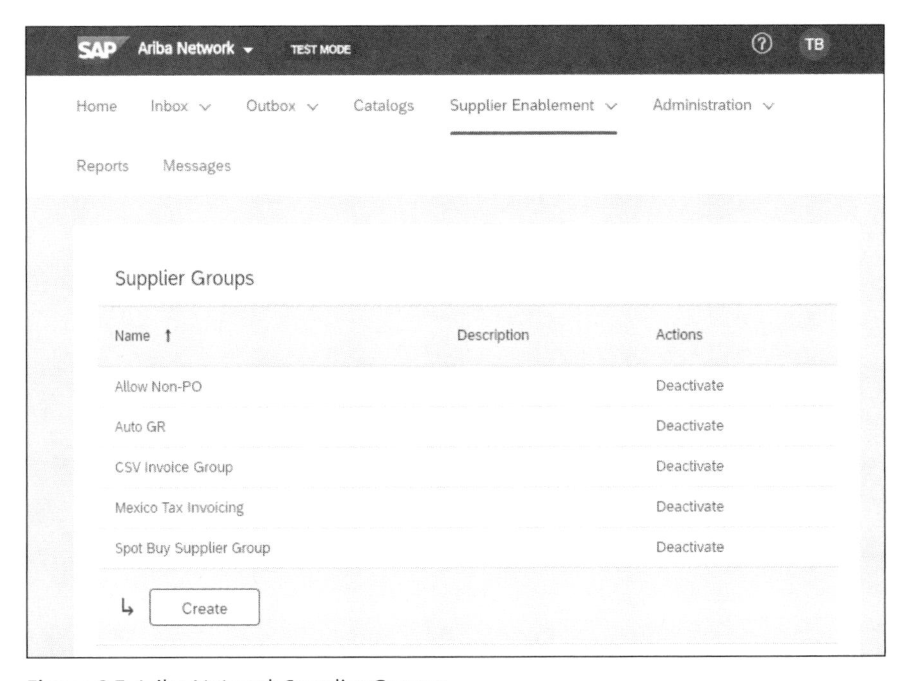

Figure 9.7 Ariba Network Supplier Groups

> **Note**
>
> While supplier groups are useful and needed in many cases, you'll want to avoid having too many. It's recommended to keep these to a limited number and only apply them to those unique supplier invoicing situations.

Configuring Country-Specific Invoicing Rules

In addition to supplier groups, you can also set up country-specific invoicing rules. Organizations have operations all over the globe and require invoicing to follow local government regulations, particularly as they relate to how taxes are handled. Country-based invoicing rules allow you to set up transaction rules based on the country in which your supplier is based.

Like the default transaction rules and supplier groups, country-based invoicing rules can be configured by your buyer administrator on the buyer Ariba Network account.

To configure country-based invoice rules, follow these steps:

1. Click the **Configuration** link under **Administration** from the Ariba Network buyer dashboard.
2. Select **Country-based Invoice Rules** from the list of configuration items.
3. Select and add a country to configure from the **Select Country** dropdown (Figure 9.8).
4. Select the country added, and click **Edit**. Review the list of rules and modify as needed to meet your country specific requirements.
5. Click **OK** to save the changes.

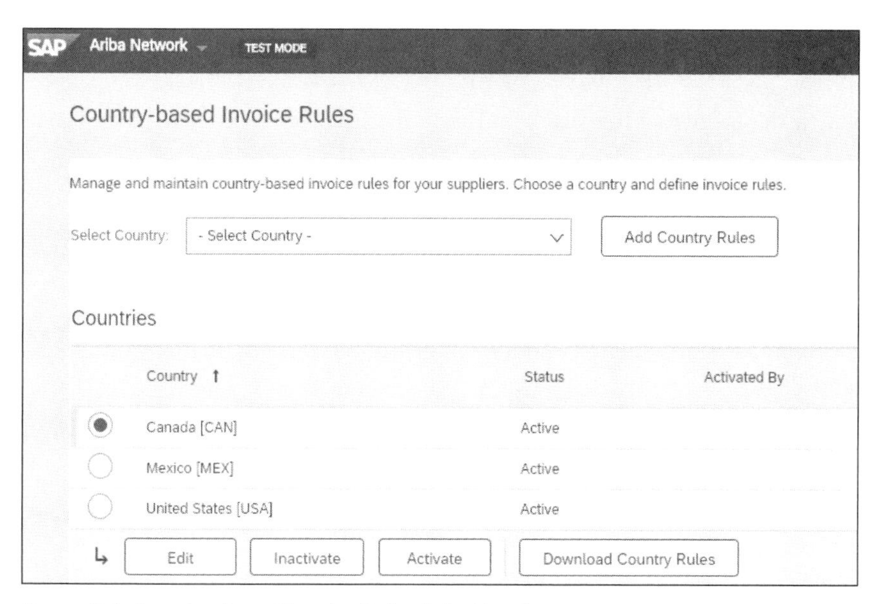

Figure 9.8 Country-Based Invoice Rules Selection Screen

After applying the country-based invoice rules, your new rule changes will take effect on all newly submitted invoices. The transaction rules configured on the Ariba Network shape how the supplier enters information and creates invoices. Next, we'll look at the different configuration options in SAP Ariba Invoice Management for invoice approvals and invoice reconciliation.

9.2.5 Configuring Invoice Approvals in SAP Ariba Invoice Management

After a supplier has submitted an invoice on the Ariba Network, or your organization's accounts payable department has manually entered a paper invoice directly in SAP Ariba Invoice Management, a series of configurable approval rules can be used to ensure visibility and that final financial approval has been met before payment is authorized.

PO-based invoices likely don't need additional approvals after the invoice is received in SAP Ariba Invoice Management. Typically, the spend on a PO-based invoice was approved when the PO was generated, so barring any exception, user intervention isn't needed. The invoice approval process in SAP Ariba Invoice Management is used to capture approvals on non-PO invoices (either supplier submitted or manually entered in SAP Ariba Invoice Management) or to provide one last review before payment.

Note

There are two types of invoice documents created when invoices are received in SAP Ariba Invoice Management. An Invoice document (INV) and an Invoice Reconciliation document (IR). An INV is an exact representation of the supplier-submitted invoice. The IR is a working copy of the invoice where the approval and reconciliation process occurs.

Any user with the following roles can configure the invoice reconciliation approval process:

- Customer administrator
- Invoice administrator

To configure the invoice reconciliation approval process, follow these steps:

1. Click the **Approval Processes** link under **Manage** from the SAP Ariba Invoice Management dashboard.
2. Select **Invoice Reconciliation** from the **Approvable Type** dropdown list, and click on the title of the approval process.
3. Click **Edit** to add a new rule or modify an existing rule in the **Approval Process Diagram** section (Figure 9.9).
4. Click **Activate** to make the approval process active.

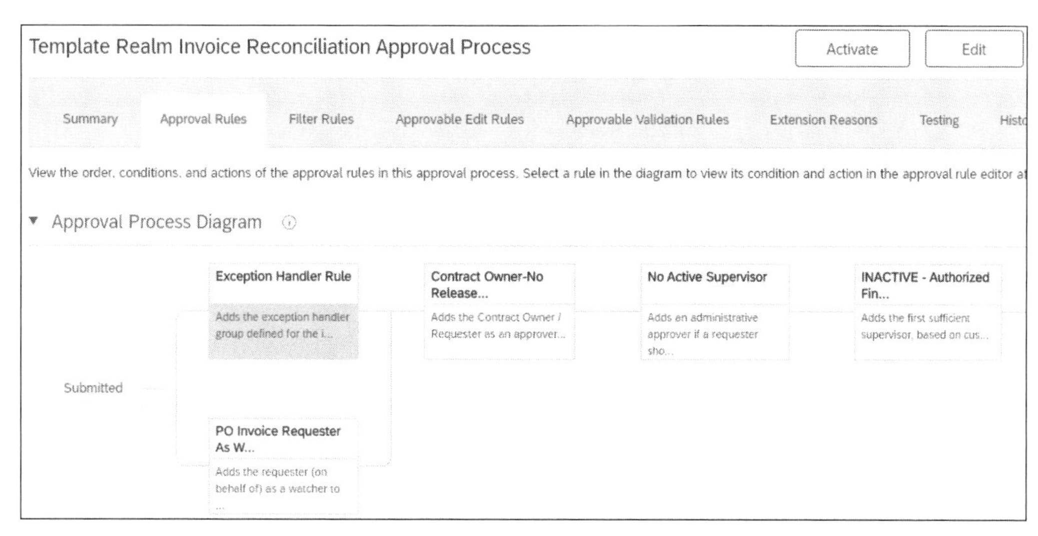

Figure 9.9 SAP Ariba Invoice Management: Invoice Approval Process Configuration

When you configure approval rules in SAP Ariba Invoice Management, you can use the out-of-the-box approval rules as a baseline for your configuration. SAP Ariba Invoice Management comes with preconfigured invoice approval rules you can use as is, or you can modify the workflow rules to fit your requirements. Optionally, you can create new approval rules that match specific invoice approval requirements that may not be covered by one of the standard invoice approval rules. As you can see in Table 9.3, the standard approval rules cover a number of different business process scenarios.

Approval Rule	Description
Exception Handler Rule	Adds the exception handler group defined for the invoice exceptions as approvers to the approval flow
PO Invoice Requester as Watcher Rule	Adds the requester (on behalf of) as a watcher to help resolve exceptions
Invalid Requester Rule	Adds the invoice agent group as an approver for a non-PO invoice receipt with an invalid requester
Advance Payment Exists	Adds an approver or group when the invoice has advance payments
Credit Memo Rule	Adds an invoice agent as an approver to the approval flow for credit memos and line-level credit memos
Requester Rule	Adds the requester (on behalf of) as an approver on a non-PO invoice receipt if the requester isn't the same as the preparer or the invoice has originated from the Ariba Network

Table 9.3 SAP Ariba Invoice Management Standard Invoice Approval Rules

Approval Rule	Description
Supervisor Rule	Adds the supervisor of the requester as an approver to the approval flow on a non-PO invoice receipt
Validation Rule	Adds the invoice agent group as an approver to the approval flow to fix validation errors
PO Invoice Cap Rule	Adds approver groups to the approval flow as specified in the approver lookup table on a PO-based invoice receipt
Non-PO Invoice Cap Rule	Adds approver groups to the approval flow as specified in the approver lookup table on a non-PO based invoice receipt

Table 9.3 SAP Ariba Invoice Management Standard Invoice Approval Rules (Cont.)

An extension of the approval workflow—invoice exception handling—is configured to address any discrepancies that may arise when the invoice is compared to the referenced contract, PO, or goods receipt. In the next section, we'll look at different configuration options for handling these exceptions.

9.2.6 Configuring Invoice Exceptions

There are a number of configuration options to automate the invoice approval and reconciliation process. SAP Ariba Invoice Management comes with a set of standard, out-of-the-box invoice exception types that can be modified to fit your invoice reconciliation process. Prior to payment, SAP Ariba Invoice Management uses comparative logic to automatically check the invoice against all the enabled exception types. It will match the invoice to the proceeding PO and goods receipt or service entry sheet. Depending on the tolerance levels you configure, SAP Ariba Invoice Management will either raise an exception and route to an appropriate exception handler or, if within tolerance, will allow the invoice to become reconciled without user intervention.

As with invoice approval rules, users with the following roles can configure the invoice exception types:

- Customer administrator
- Invoice administrator

To configure the invoice reconciliation approval process, follow these steps:

1. Click the **Core Administration** link under **Manage** from the SAP Ariba Invoice Management dashboard.
2. Under **Procure to Pay Manager**, click **Invoice Exception Types**. Click **List All**, and review the invoice exceptions (Figure 9.10).

3. To modify, select **Edit** from the **Actions** menu, and add relevant exception type information: tolerance settings, exception ranking, auto-reject/accept settings, and exception handling routing.
4. Click **Save** after changes are made.

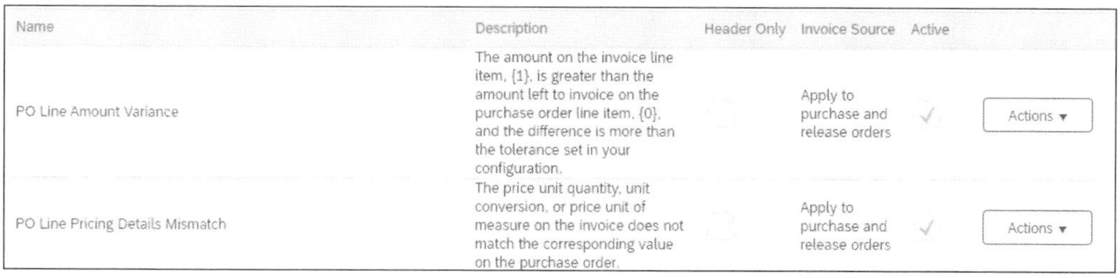

Name	Description	Header Only	Invoice Source	Active	
PO Line Amount Variance	The amount on the invoice line item, {1}, is greater than the amount left to invoice on the purchase order line item, {0}, and the difference is more than the tolerance set in your configuration.		Apply to purchase and release orders	✓	Actions ▾
PO Line Pricing Details Mismatch	The price unit quantity, unit conversion, or price unit of measure on the invoice does not match the corresponding value on the purchase order.		Apply to purchase and release orders	✓	Actions ▾

Figure 9.10 SAP Ariba Invoice Management Invoice Exception Types

Like approval rules, SAP Ariba Invoice Management comes with a library of preconfigured invoice exception types, as you can see in Table 9.4. You can modify these exceptions to meet your invoice exception handling requirements, or you can create new exception types that meet a specific business need. Exception overrides can also be configured to apply different tolerances to different types of invoices based on the commodity code, supplier, or a combination of both.

Invoice Exception Type	Description
Invoice Unmatched	The system is unable to match a PO to the invoice amount.
PO Line Pricing Details Mismatch	The invoice price unit quantity, unit conversion, or price unit of measure does not match the corresponding value on the purchase order.
PO Line Amount Variance	The total amount of the invoice is greater than the total amount of the PO.
PO Payment Terms Mismatched	The invoice's payment terms are different from the PO's payment terms.
Contract Price Variance	The invoice item's price is greater than the contracted price.
Contract Quantity Variance	The invoice item's quantity is greater than the remaining contract item's or item's context quantity limit.
Item Unmatched	The system is unable to find a line item on the PO that matches the invoice line item.

Table 9.4 SAP Ariba Invoice Management Invoice Exception Types

Invoice Exception Type	Description
Over-Tax Variance	The **Amount** field on the invoice tax line item is greater than the line item's calculated **Tax Amount**, and the difference isn't within the tolerance defined in your configuration.
PO Quantity Variance	The quantity on the invoice line item is different from the quantity on the PO line item, and the difference is more than the tolerance set in your configuration.
PO Received Quantity Variance	The invoice item's quantity is greater than the order item's received quantity.
Shipping Variance	The shipping amount on the invoice is greater than the shipping amount left to invoice on the PO, and the discrepancy is due to differences in shipping costs.

Table 9.4 SAP Ariba Invoice Management Invoice Exception Types (Cont.)

Invoice reconciliation and exception handling is the last step of the process in SAP Ariba Invoice Management prior to payment. Invoices received from the supplier on the Ariba Network have been approved and automatically reconciled against related purchasing documents, such as the contract, PO, and goods receipt. Data entry by the accounts payable group has been reduced, and the manual steps to match invoices have been drastically decreased.

In the next section, we'll look at SAP Ariba Discount Management: SAP Ariba's collaborative dynamic discounting and working capital management solution.

9.3 SAP Ariba Discount Management

With the efficiency and agility improvements provided by SAP Ariba Invoice Management, you can turn your accounts payable department into a profit center. Initially, you'll achieve this via improved management of days payable outstanding (DPO). In a high-interest rate environment, DPO management by itself provides you with better cash flow and additional profitability. When you're dealing with low interest rates, in which returns on deposits are much less than 1%, obtaining and managing supplier discounts for early payments can provide the lion's share of profitability gains.

9.3.1 SAP Ariba Discount Management at a Glance

Many suppliers would prefer to provide a discount for early payment, rather than resorting to factoring or other methods to buttress cash flow. With SAP Ariba Discount Management, you can fully automate early payment discount management from initial offer up to agreement, including transactions involving prorated or dynamic discounting. SAP Ariba Discount Management provides you with control over the amount of cash to apply and the suppliers to target for discount programs. To maximize savings via discounts, SAP Ariba's working capital management services team can provide expertise and payment term data, as part of a consulting engagement, to help you formulate an effective DPO improvement and dynamic discounting program.

As shown in Figure 9.11, SAP Ariba Discount Management automates early-payment discount management with a robust toolset from initial offer to agreement, as well as a services arm (SAP Ariba working capital management services) to support you in the formulation of an effective DPO-management program.

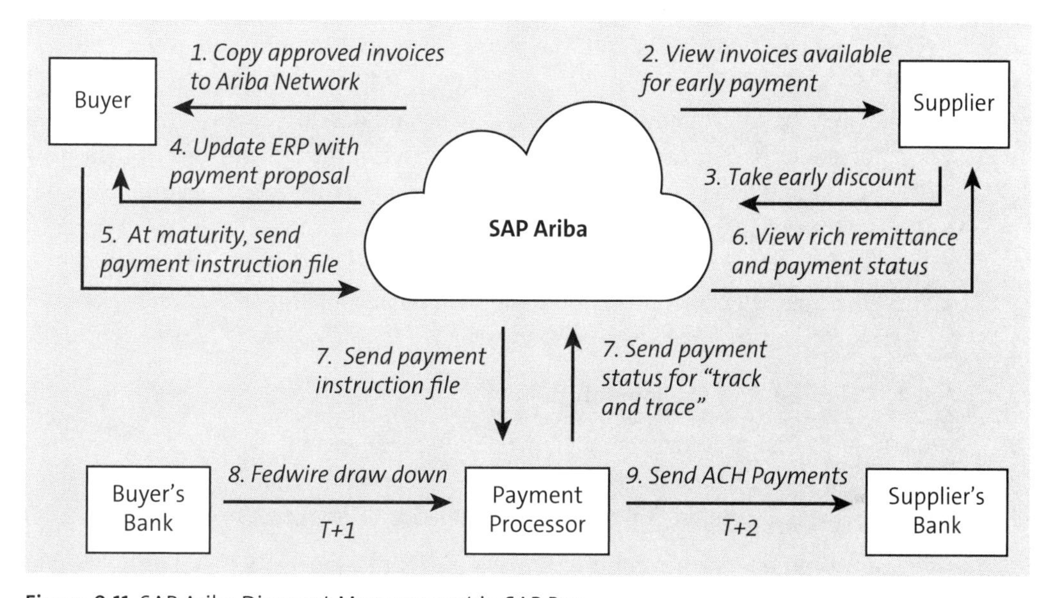

Figure 9.11 SAP Ariba Discount Management in SAP Pay

SAP Ariba Discount Management supports prorated, dynamic discounting; automatic and ad hoc dynamic discounts; discount groups; full treasury control; alerts/notifications; and reporting. For suppliers, in addition to better cash flow management and control, SAP Ariba Discount Management also provides a cash flow optimizer tool and invoice/payment visibility. Standing early payment terms apply discount terms automatically to every applicable approved invoice. Standing term offer parameters define discount term offers and include the percent discount off of face value, APR percent, net day, discount day, prorated discount scale (Y/N), and the supplier or group of suppliers to whom the offer applies.

Using SAP Ariba Discount Management, you can set up multiple offers from which a supplier or group of suppliers can choose, and then notify you automatically with the terms of the offer, date of change, and request to review the offer, when an early payment term is created or changed.

You can also create buyer-defined dynamic discounts, applying discount terms agreed to by a supplier on an ad hoc, invoice-by-invoice basis, using a wide array of parameters. Your suppliers have access to the notifications and to a supplier cash optimization tool, which allows them to identify the invoices required for early payment approval and the corresponding discounts to meet a supplier-defined cash flow need. Suppliers have the ability to initiate their own dynamic discount proposals based on this analysis or in general.

SAP Ariba Discount Management resides at the nexus of buyer and supplier on the Ariba Network. However, in many implementations, the integration points extend beyond the Ariba Network, back to the buyer's ERP environment, and even out to the supplier's SAP ERP environment. Integrated with SAP ERP, SAP Ariba Discount Management allows finance/accounts payable to expand and develop discount opportunities and agreements directly in the accounts payable environment in SAP ERP by using SAP Ariba Discount Management as the mediation point in the Ariba Network. As result, SAP ERP can send payment proposals up to the network, which allows the supplier to respond within the Ariba Network, while receiving notification of remittance advice status either in the Ariba Network or in the supplier's SAP ERP.

9.3.2 Planning Your Implementation

For SAP Ariba Discount Management, the design phase focuses on the one-time activities of design, setup, and configuration. The success of any joint customer-supplier project is just as dependent on supplier participation as it is on customer initiative.

To confirm the requirements for SAP Ariba Discount Management, defining thresholds, supplier/scenario targets, and savings strategies for the different types of discounts in SAP Ariba Discount Management (automatic, ad hoc user-driven, ad hoc supplier-driven) is a key part of the workbook deliverable.

To have a successful implementation for your SAP Ariba Discount Management deployment, the following key items will need to be prepared prior to starting requirements gathering:

- **Most current annual spend information populated in a standardized template**
  This template is provided by your SAP Ariba Discount Management consultant and includes common categories of information such as supplier names, addresses, supplier country, ERP spend amount, and currency.
- **Supplier-specific discounting contacts**
  This information includes contact names, email addresses, and phone numbers for the outreach to suppliers. The preferred contact will have the authority to agree to early payment terms (i.e., accounts receivable) and may not be the same individual

responsible for Ariba Network administrative functions such as accepting trading relationship requests (TRR) and completing assigned Ariba Network tasks.

- **Supplier-specific discounting information**
  This information includes payment term codes, paying company code, payment term description, standard term discount term, discount percentage, discounts earned, discounts lost, DPO, processing time of payments, and payment types.

You'll likely need to collaborate across a number of groups such as purchasing, tax, and treasury to understand your organization's policies and processes. A firm understanding will give the project team a better understanding of how to achieve the cost savings that can be realized from a successful SAP Ariba Discount Management deployment.

9.3.3 Integration Scenarios

The options available with SAP Ariba Discount Management can run almost entirely between SAP Ariba solutions. The benefits of running everything in SAP Ariba is the light integration requirements with SAP ERP and IT—all systems, except for a financial update after the invoice payment has been requested, are managed by SAP Ariba. Not only does the user get a single look and feel for the entire process in SAP Ariba, the tight linkages to the Ariba Network throughout the process enables collaboration on invoicing, helps avoid the complexities and fees of an automated clearing house (ACH), and optimizes document production.

SAP Ariba can run entire transactions in SAP Ariba Buying and the Ariba Network. For smaller organizations with less-intensive accounting requirements and IT infrastructures, SAP Ariba may provide the most efficient solution for procurement activities. However, for larger concerns and systems, which may need real-time updates throughout the transaction in SAP Ariba, then integration with SAP ERP may be the best option for satisfying this requirement.

9.3.4 Configuring Early Payment Requests

The configuration of SAP Ariba Discount Management is done almost entirely on your Ariba Network buyer account. Your Ariba Network account administrator sets up roles and responsibilities, sets the configuration options, and creates a test account for testing document flow with suppliers.

There are a number of discount options that you'll need to review before setting the discounting configuration. A well-defined discounting strategy is enhanced by a thorough review of your suppliers, the discount terms those suppliers have used, and your recent annual spend. This information will help you decide which discount terms should be offered, which suppliers should be in scope, and which payment terms should be consolidated.

After a review is done, and a baseline has been established, you can take the following steps to configure discounting:

1. Log in to your Ariba Network buyer account.
2. From the landing page, navigate to **Administration • Configuration • Ariba Discount Management**.
3. Next, on the **Discount Offers** tab, add a buyer-initiated dynamic discount by selecting **Buyer Initiated Discount Offer** from the **Add** dropdown menu, as shown in Figure 9.12.
4. On the **Configure a Buyer Initiated Dynamic Discount** page, enter the relevant discounting information: **Name**, **Description**, **Effective Start Date**, **Effective End Date**, **Range (low)**, **Range (high)**, **Discount Rate**, **Processing Time**, and to which suppliers the discount will be offered (Figure 9.13).
5. Click **OK**.

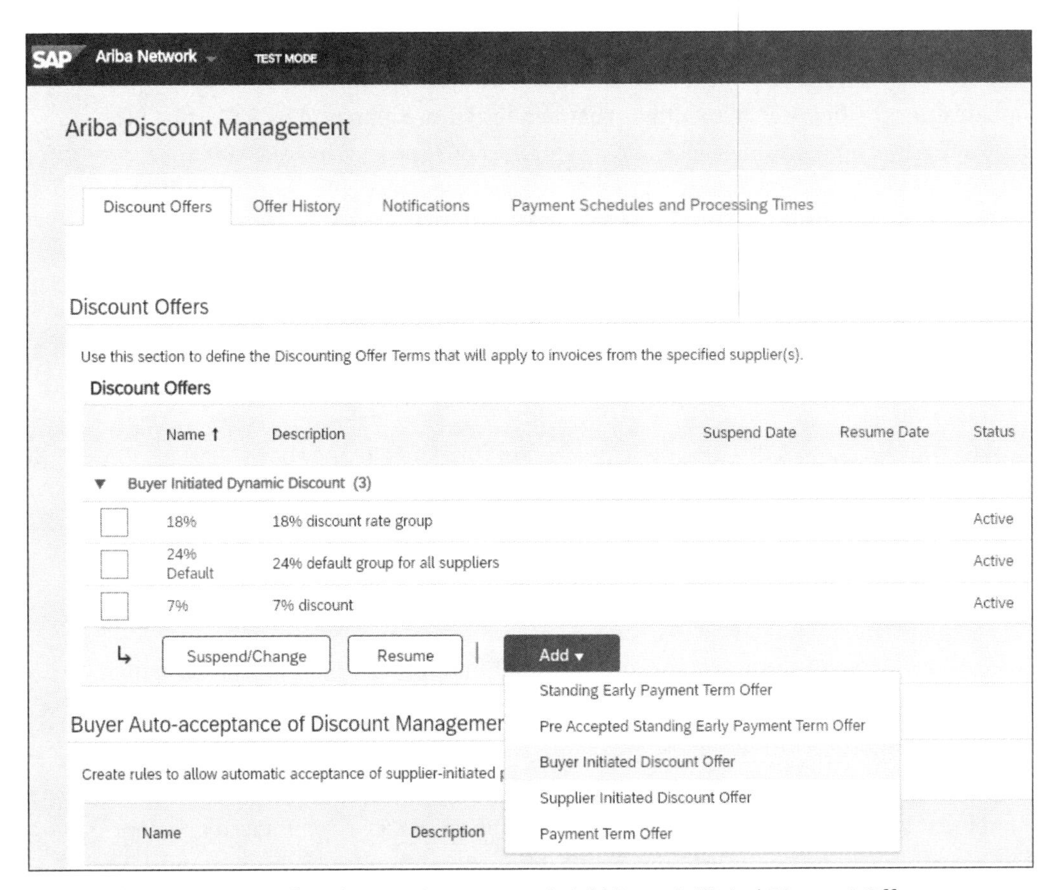

Figure 9.12 SAP Ariba Discount Management: Add Buyer Initiated Discount Offer

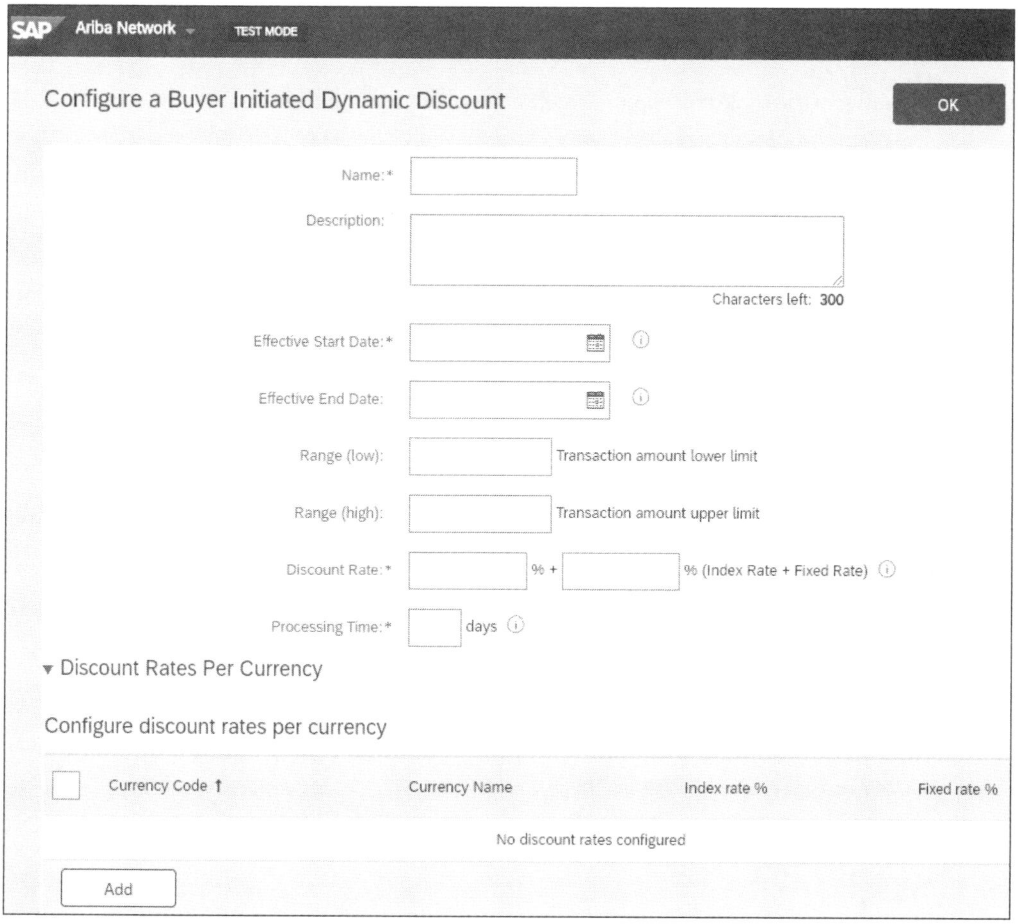

Figure 9.13 SAP Ariba Discount Management: Buyer Initiated Dynamic Discount Configuration Details

After the details have been added, the dynamic discounting will apply based on the **Effective Start Date**. A buyer-initiated dynamic discount offer is sent to some or all suppliers, and they can be accepted individually or on an invoice-by-invoice basis. These are also referred to as ad hoc, pay me now, or dynamic discounting.

9.3.5 Configuring Scheduled Payment Extraction and Vendor Master Setup

After you send a payment proposal, and the payment meets one of your SAP Ariba Discount Management rules, the Ariba Network sends an email notification to the supplier. The supplier follows the instructions in the notification to log in.

The **Scheduled Payments** tab in the supplier's account shows discount offers for all payments. The Ariba Network displays the **Buyer-Initiated Dynamic Discount**, but doesn't display the name you assigned to the term, such as APR 30% net 30, nor does it display the discount rate.

Each row in the table corresponds to a series of payment periods, as shown in Figure 9.14. The payment periods are listed from the closest day (at the top) to the last day (at the bottom), with the total payment corresponding for each day. Suppliers can see that the discount is greatest at the top and decreases as the days get closer to net at the bottom.

Suppliers choose a specific day for settlement by selecting a row. They weigh the benefit of early payment against the associated higher discount. After selecting a row, they click **Accept Early Payment Offer**.

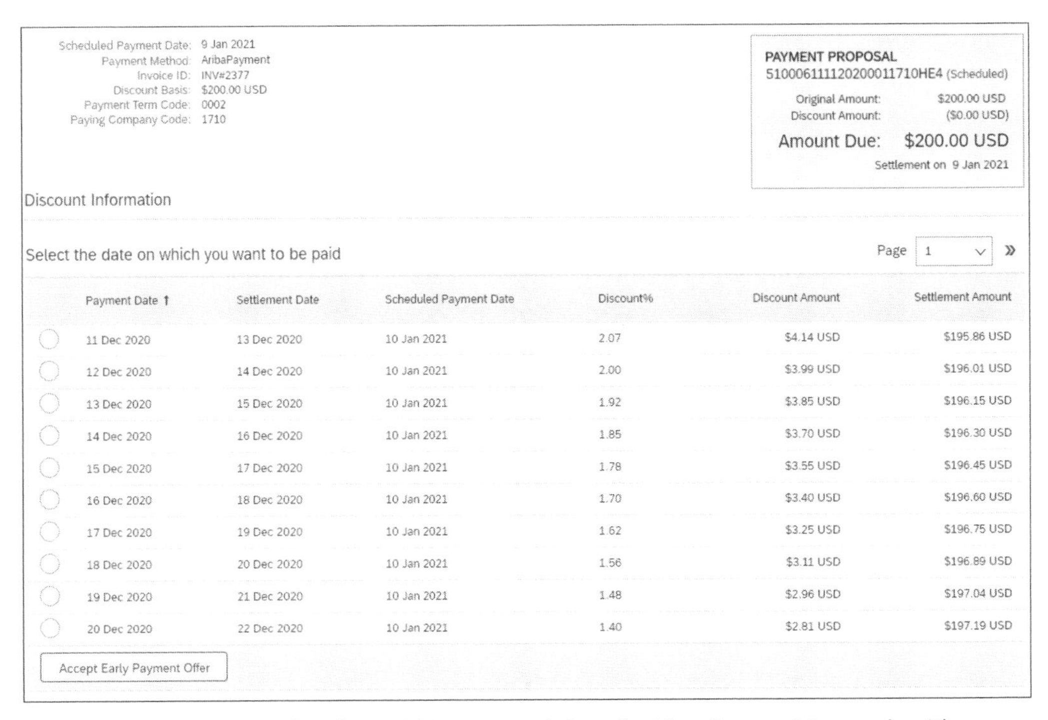

Figure 9.14 SAP Ariba Discount Management: Supplier View Payment Proposals with Discount Information

After suppliers select a specific discount, the Ariba Network sends a payment proposal to your payment system to schedule the payment. Next, we'll discuss SAP Pay.

9.4 SAP Pay

With SAP Pay, you can further avoid late-payment fees and the inefficiencies caused by paper checks. SAP Pay covers the bottom half of the invoice process shown in Figure 9.15.

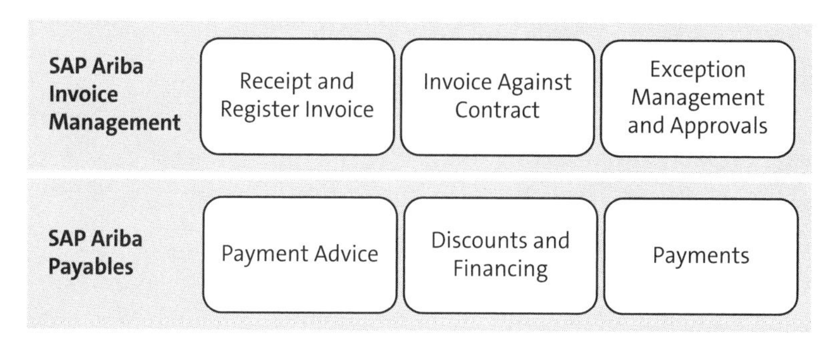

Figure 9.15 SAP Pay Process Areas

SAP Ariba Discount Management manages discounts: prorated, dynamic discounting, automatic and ad hoc dynamic discounts, and discount groups.

As shown in Figure 9.16, the core areas for SAP Pay are solutions focused on managing, financing, and discounting payments during the process.

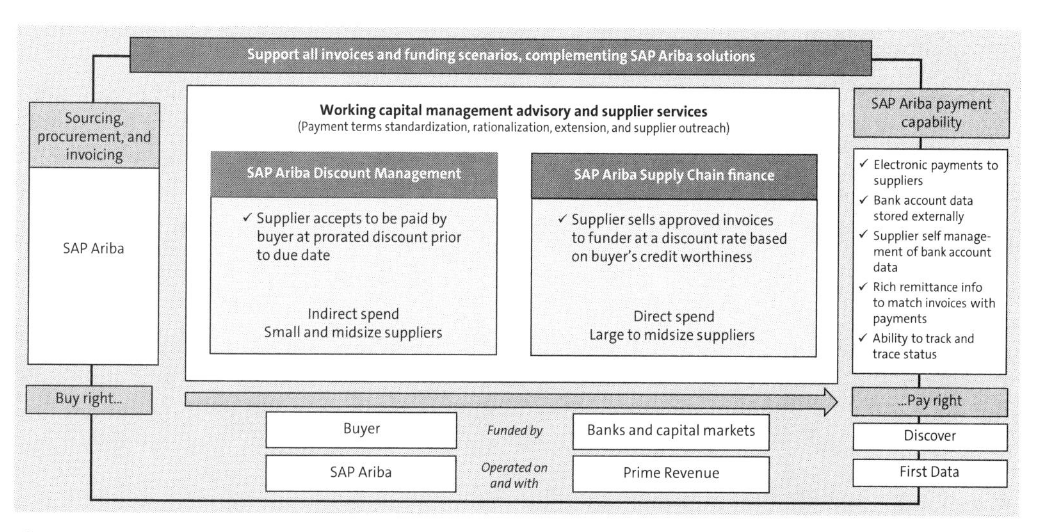

Figure 9.16 SAP Pay

9.4.1 SAP Pay at a Glance

After you process an invoice, the next step is payment. Traditionally, paying suppliers has been challenging to automate because suppliers can't check on pending payments or receive status updates about the payment process and thus begin calling for updates. SAP Pay leverages the power of the Ariba Network to extend back-office systems to transmit ACH payments to suppliers. For suppliers, having SAP Pay integrated with the Ariba Network provides updated information on the status of payments and allows them to update their account details directly.

SAP Ariba invoicing, payment, financing, and discount solutions support a general, overarching process for accounts payable. After an invoice is reconciled, payment is scheduled through a payment request.

The capabilities in SAP Pay can completely automate your payment processes, including electronic payments, detailed remittance statements, a supplier portal, and supplier bank-routing information verification. Using these tools, you can accelerate supplier participation rates (in-house ACH programs often struggle with onboarding suppliers), increase accounts payable productivity and capability, and reduce fraud cases and late-payment fees. Combining SAP Ariba Invoice Management with SAP Pay delivers a one-two punch to inefficiency and waste in accounts payable. Using the rules and tools of electronic invoicing, your accounts payable team can quickly return invoices with missing or erroneous information to the supplier for correction. With SAP Pay, you further avoid late-payment fees and inefficiencies caused by paper checks. A summary is provided in Table 9.5.

Task	Common Challenges	How SAP Ariba Helps	SAP Ariba Solutions	Value Drivers
Invoicing	▪ Manual processes ▪ Lack of visibility	▪ Automated invoicing ▪ Buyer and supplier visibility	▪ SAP Ariba invoice automation capability ▪ SAP Ariba Invoice Management ▪ SAP Ariba Contract Invoicing ▪ SAP Ariba Services Invoicing	▪ Reduce invoice processing cost by > 50% and cycle time to 1–5 days ▪ Better forecasting for buyers and suppliers
Working capital	▪ Inconsistent payment terms ▪ Rigid, buyer-initiated payment terms	▪ Strategically manage and extend payment terms ▪ Buyer-/supplier-initiated discounts	▪ SAP Ariba supply chain finance capability ▪ SAP Ariba dynamic discounting and SAP Ariba Discount Management	▪ $1M–$2M in early payment discounts per $1B targeted spend ▪ 1- to 4-day DPO extension

Table 9.5 Accounts Payables Challenges and SAP Ariba Solutions

Task	Common Challenges	How SAP Ariba Helps	SAP Ariba Solutions	Value Drivers
Payment	■ Secure storage of supplier bank account information ■ Converting checks to electronic format	■ Supplier bank account data removed from buyer servers ■ Check to electronic conversion service	■ SAP Pay, SAP Ariba payment automation capability	■ Enhanced controls related to supplier bank account data ■ $2–$3 savings per check converted to electronic

Table 9.5 Accounts Payables Challenges and SAP Ariba Solutions (Cont.)

In this section, we'll begin with an overview of how the available SAP Pay solutions support the invoicing process. Next, we'll delve into the specifics for the core functionality areas in SAP Pay, including payment, discounting, and supply chain finance.

SAP Ariba Payment Management

In this section, we'll focus on the payment aspect of the invoicing process and the applicable SAP Ariba solutions you can use after an invoice has been submitted. This functionality manages payment aspects at the end of a P2P process in SAP Ariba. Many SAP ERP customers revert to SAP ERP for accounts payable from their SAP Ariba solutions. SAP Pay offers the option of managing payments in SAP Ariba, further consolidating the processes in SAP Ariba and circumventing the often-cumbersome ACH process used in most payment scenarios, including SAP ERP.

SAP Ariba Payment Capability

After the initial invoice has been validated and reconciled, the payment portion of the process must be addressed. SAP Pay provides an array of payment capabilities via the Ariba Network–based payment solution, facilitating payments from buyers to suppliers. Buyers can create individual payment requests or batch payment requests in their individual SAP ERP systems, which are then transferred via the Ariba Network to a payment service provider, and then on to the originating and receiving financial institutions. During this process, SAP Ariba provides visibility into payments via a track-and-trace functionality, facilitates faster reconciliation, and confirms expected invoice and payment amounts.

As shown in Figure 9.17, SAP Ariba enables you to pay your suppliers in Canada, Germany, United Kingdom, and the United States over the Ariba Network, providing improved visibility, cash flow controls (when payment is to be made), dual user verification, simple UIs facilitating self-service, and, ultimately, reduced costs and greater efficiencies when compared to traditional ACH processes.

A similar process is also available for the *virtual card*, which is a customer credit card intended only for charges by a single supplier or group of suppliers and check payments, as shown in Figure 9.18 and Figure 9.19.

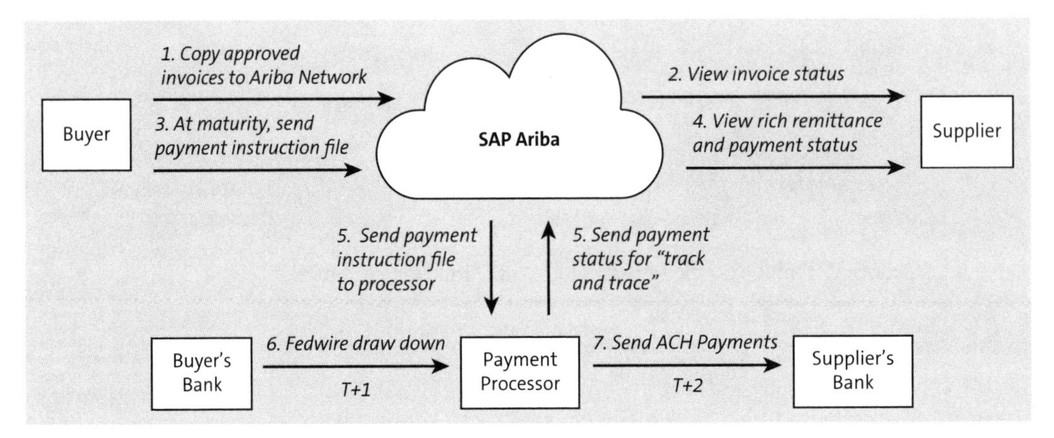

Figure 9.17 SAP Ariba Payment Process

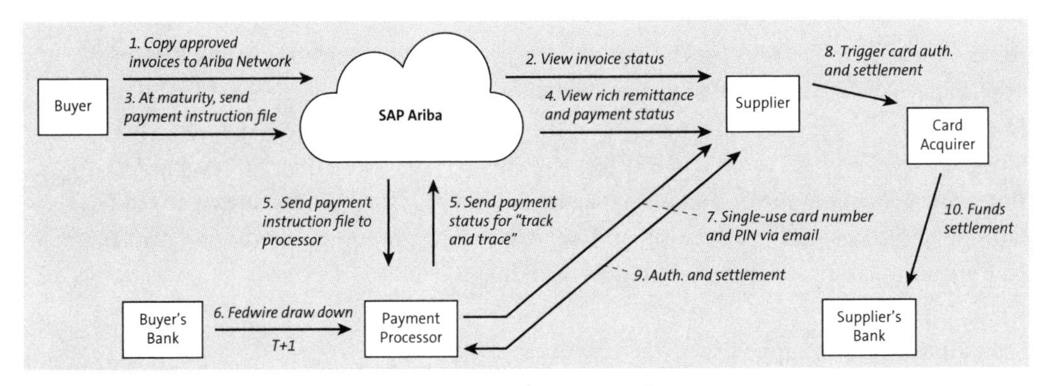

Figure 9.18 SAP Ariba Payment Capability (Virtual Card)

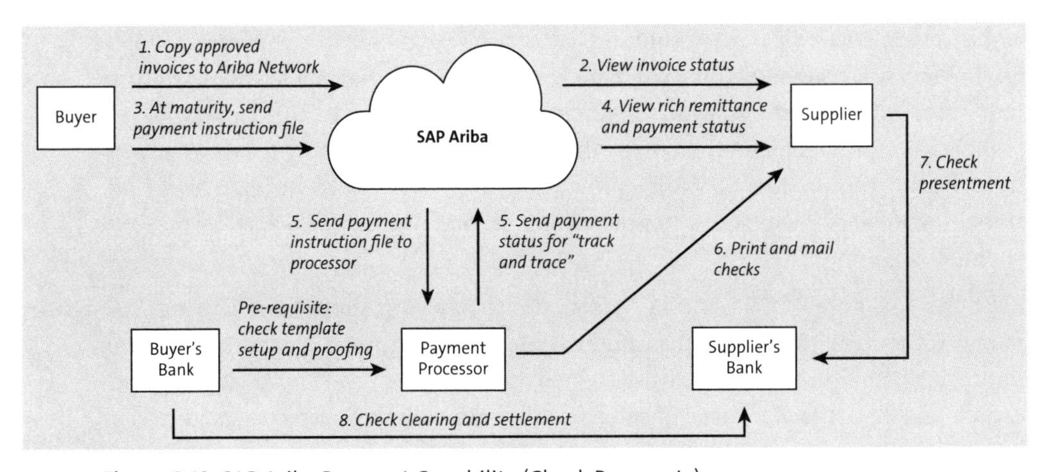

Figure 9.19 SAP Ariba Payment Capability (Check Payments)

In these scenarios, instead of an ACH payment, the payment is made directly with a virtual card or a check.

In these process flows, the process for all three types of payment shares the following steps:

1. The invoice is either entered directly in the Ariba Network by the supplier, or a copy is uploaded by the accounts payable user or directly from the ERP system on the customer side.
2. The supplier can then view the status of the invoice via the Ariba Network.
3. At maturity, a payment instruction file can be sent.
4. The payment status is further updated in Ariba Network for the supplier, which reduces calls to the customer's help desk and accounts payable department.
5. The actual payment is then processed via ACH, virtual card, or check payment.

For each transaction, traditional ACH processes must reestablish information that already is static and available in the Ariba Network in secure areas. With ACH, determining the status of a payment can be difficult, and surcharges on payments typically run 3% of the total payment amount.

With SAP Ariba, leveraging the Ariba Network allows for a more efficient payment process than ACH. The differences are outlined in Table 9.6.

Functional Area	ACH	SAP Ariba Payment Capability
Traditional ACH disbursement services	Yes	Yes
Multiple, supplier-friendly remittance formats	No	Yes
Secure storage of supplier bank account details	No	Yes
Supplier responsible for updating bank account details	No	Yes
Supplier self-service: online payment visibility	No	Yes
Supplier enablement, education, and support	No	Yes
SAP Ariba receives first-line invoice/payment inquiry phone calls	No	Yes
Collection and sharing of "hard to get" supplier information: W-9s, proof of banking, etc.	No	Yes

Table 9.6 ACH versus SAP Ariba

Combining the power of the Ariba Network for supplier information and transaction support with focused accounts payable solutions on the payment side makes SAP Pay an offering that extends well beyond traditional ACH approaches to processing payments.

SAP Ariba Payment Automation

The SAP Ariba payment automation capability requires SAP Ariba's invoice automation capability or SAP Ariba Buying and Invoicing. Customers can maintain existing payment (ACH) connections with their banks and use payment automation to deliver remittance information only to suppliers; provide status updates to buyers and suppliers; deliver remittances to suppliers (email, EDI, cXML, or directly via the Ariba Network); and support supplier self-service for maintaining remittance name, address, and bank account information.

You can choose to use the SAP Ariba payment functionality both for the delivery of remittance information to suppliers and the transmission of ACH payment instructions to your bank. Payment instructions are transmitted to your bank in accordance with the bank's transmission standards. Suppliers can view payment status updates through the Ariba Network. When exceptions occur, SAP Ariba processes ACH return information from the bank and updates payment status. You can configure SAP Pay to send ACH instructions in the following ways:

- Using supplier-maintained information in SAP Ariba that identifies suppliers' bank accounts
- Using bank account information from SAP ERP and passed along in the payment file
- Issuing ACH only if bank account information in the payment proposal and SAP Pay Professional match exceptions that have been flagged using business controls
- $0 pre-note validation of supplier bank accounts, which is a zero-dollar electronic funds transfer validation of the supplier's account

For any bank new to SAP Ariba, a bank integration project is required. SAP Pay delivers reports on supplier bank account changes for suppliers with trading relationships with a buyer.

SAP Ariba Supply Chain Finance Capability

The SAP Ariba supply chain finance capability is similar to other discount management capabilities in that you can extend cash flow management flexibility and terms to your suppliers. However, instead of paying early out of your own money, the supply chain finance capability allows you to extend early payment options via a third party to your suppliers for a small fee. Suppliers can get access to cash early without affecting your DPO, while improving the supplier's days sales outstanding (DSO).

Why is DSO important? Companies are valued in part on their cash flow, and the more positive the DSO, the higher a company's valuation. The supply chain finance capability applies to both untraded and traded invoices, as shown in Figure 9.20, and a supplier can review the invoices and request early payment on applicable invoices via the SAP Ariba supply chain finance capability. As shown in Figure 9.21, when a supplier elects to have a particular invoice paid early via the service, the supply chain funding bank pays the supplier at that point, while the buyer pays as scheduled. The buyer's payment, however, now goes to the funding bank instead of the supplier's bank. The supplier in this case has already received payment from the funding bank. Essentially, the funding bank in SAP Ariba's supply chain finance functionality acts as a provider of the factoring service, paying the supplier early and collecting the customer payment for a smaller fee than offered by traditional factoring services.

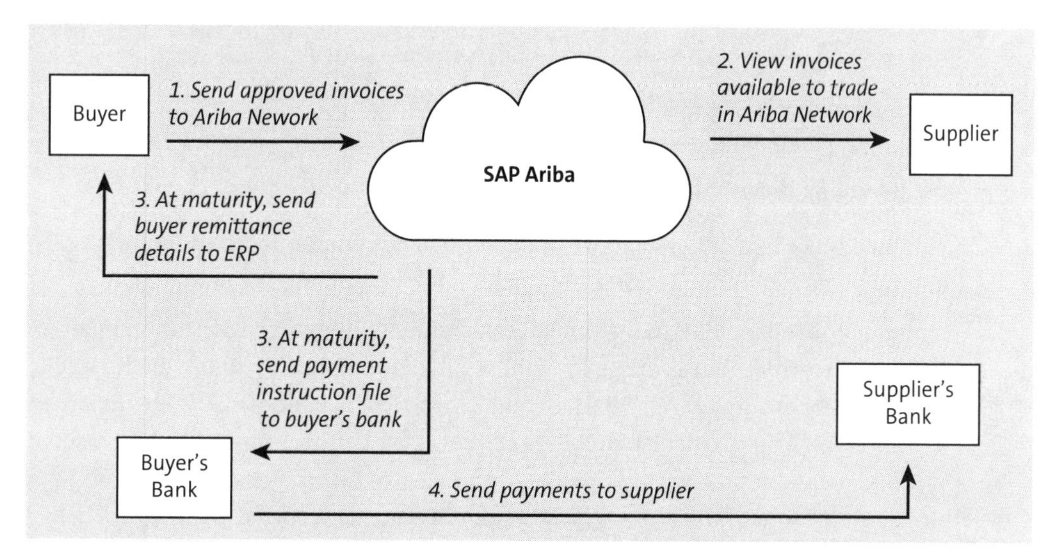

Figure 9.20 SAP Ariba Supply Chain Finance Capability: Untraded Invoices Process

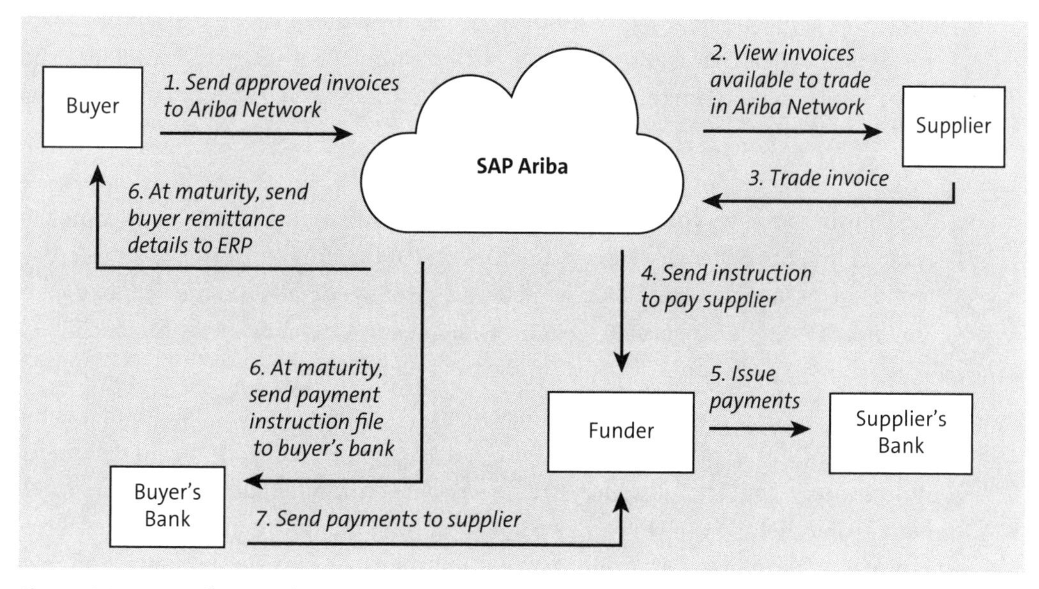

Figure 9.21 SAP Ariba Supply Chain Finance Capability: Traded Invoices Process

9.4.2 Planning Your Implementation

After you've decided on the desired invoicing and accounts payable solutions, you'll need to start planning the implementation. In this section, we'll cover various solutions and implementation approaches for SAP Pay, and we'll focus on how additional cash flow and discounts can be achieved during the invoicing and payment process.

Following are the first four phases in your implementation project:

- Planning for success by developing a business case
- Confirming your goals, scope, and rollout plan
- Planning your deployment and confirming readiness
- Performing the deployment/kickoff

Note that these four phases apply to all three delivery tracks: solution setup, supplier adoption, and internal adoption.

For invoice analysis, identifying the largest sources and savings, whether supplier- or process-driven (or both), is the key insight required for defining scope in all three solution areas. For example, SAP Ariba Discount Management functionality, if your largest returns come from particular suppliers or processes, this fact should drive the scope and understanding during this phase of the project and serve as a guide for later phases. At the end of the project, these goals and metrics are what you'll use to define success and justify the investment.

In the next phase, you'll plan the deployment and ensure everything is ready during the "pre-kickoff," the final step prior to kickoff. During this phase, the SAP Ariba team or a consulting partner will provide you with best practices related to readiness and help you build a program to support payment submissions or discount facilitation via SAP Pay.

Regarding customer key deliverables and activities, which you'll be responsible for, the first step is to assign the applicable internal team members that have a complete understanding of the current scope and timeline of the program. Prior to kickoff, you should also gather identified supplier data to deliver to SAP Ariba in the vendor collection template for the applicable aspects of supplier strategy development, specifically flight planning and supplier strategy/onboarding development.

At the supplier data level, you'll need to collect all required supplier data elements for outreach to the identified suppliers. This data includes the supplier ID, supplier name, supplier contact, full address, phone, email, remittance address, category manager at customer, DPO, PO versus non-PO, contract expiration data, contract term, invoice number, and invoice copy for top suppliers in the outreach program. You may need to include additional data elements, based on industry, spend category, and solution.

For individual solutions, you should initiate discussions with accounts payable for all three solution areas, while focusing on the invoice submission process/areas for SAP Ariba Invoice Management, on payments and bank interactions for SAP Pay, and on discounting and supply chain financing. These internal customer groups typically reside in accounts payable but could be distinctly different at a subgroup level. Payment methods, banking relationships, legacy systems, and company culture all need to be accounted for in the evaluation.

For SAP Pay projects, you'll need to begin discussions with treasury/finance to obtain the desired and required discounting rates targeted for suppliers. SAP Ariba delivery and consulting partner deliverables include sharing best practices related to readiness, building an early payment program, and defining a measurement strategy.

Next begins the kickoff phase. The most important deliverable of the kickoff phase is the actual kickoff meeting. This meeting serves a real purpose to align the different teams and stakeholders, create understanding and familiarity among the teams, and, perhaps most importantly, generate enthusiasm. These considerations apply to all three tracks on these types of projects, as users and suppliers of the system all benefit from communication around the project kickoff.

On the SAP Ariba/partner consulting side, now SAP Ariba shared services will begin to play a significant role in the project. For smaller implementations, shared services will often run the kickoff meeting with the customer, as this group handles most of the project deliverables going forward. On larger projects with more complex requirements and on-site teams, the SAP Ariba delivery teams or consulting partners will prepare and deliver the kickoff presentation.

Key customer deliverables and activities are equally important during kickoff, such as the following:

- Participation in the kickoff meeting to confirm scope of the aspects of the program, goals/objectives, and high-level timeline
- Delivery, by the customer program sponsor, of the program's message to internal stakeholders supporting particular aspects of that program

Joint key deliverables and activities for customer and SAP Ariba delivery/consulting partner include the following:

- Agreeing on aspects of the plan, timeline, and milestones
- Reviewing key success factors for the aspects of the program
- Initiating governance structure and approach for aspects of the program

Solution Setup

After the kickoff meeting is complete, the project/program is officially underway. In this section, we'll walk you through the remaining phases for the solution setup track.

Confirm Requirements Phase

The next step is to confirm the requirements, which should directly reflect the planning and scope defined in the previous phases. For an SAP Ariba project, these requirements are typically gathered in workbooks. The SAP Ariba delivery, shared services, and partner consultants will define future processes with you, showing the solution via demos where applicable. You would then send responses/data in the workbook format back to SAP Ariba. Ultimately, these workbook responses are leveraged to build out the solution in your "realms," which are the individual customer areas in a cloud-based SAP Ariba environment.

Defining Project Resources and Timelines

Timelines for SAP Pay projects depend on the complexity and scope of the implementation. Many accounts payable implementations are done in tandem or embedded in an overall SAP Ariba Buying project, thus representing an invoice-to-pay process in the procedure. For baseline implementations of these solutions, the following timelines can serve as examples:

- **SAP Pay**
  Generally, this is a 20-week implementation, with the customer providing the project manager, the functional lead, the technical lead, the training lead, and pilot users.
- **SAP Ariba payment automation functionality**
  The SAP Ariba payment automation capability, such as invoice automation, is typically implemented during an overall SAP Ariba Buying implementation project. The timeline for an SAP Ariba Buying project runs 26 weeks, with the customer providing the sponsor, the project manager, the technical lead, the functional lead, supplier enablement, change management, support, testers, the training lead, and business subject matter experts (SMEs).

Project roles and responsibilities on SAP Pay projects are summarized in Table 9.7, which discusses the roles and responsibilities for SAP Ariba/consulting partners.

SAP Ariba or Consulting Partner Role	Responsibilities
Project sponsor	■ Serve as the program sponsor and champion within SAP Ariba ■ Help define overall project vision ■ Provide senior leadership communication in support of the project ■ Drive program workstreams to achieve business case goals ■ Develop supplier and network strategy in support of business goals ■ Monitor status reports and timelines

Table 9.7 Consulting Partner Roles and Responsibilities

SAP Ariba or Consulting Partner Role	Responsibilities
	■ Provide guidance for high-level issue resolution ■ Serve as liaison to the commercial team for scope issues
Project manager	■ Confirm customer goals and project scope ■ Plan the project activities and timeline ■ Ensure adequate resources are assigned ■ Establish project management framework (meetings, status, issue management, etc.) ■ Coordinate activities and input from SAP Ariba project resources ■ Manage project timeline and adherence to schedule ■ Ensure timely project communication and status updates ■ Identify, manage, and escalate project issues and risks, as appropriate ■ Serve as a point of contact for overall deployment
SAP Ariba functional lead	■ Host requirement/design workshops to educate customers on SAP Ariba solution functionality and configurability and help customer understand their business requirements in the context of the SAP Ariba solution ■ Answer customer questions about the functionality of the SAP Ariba on-demand application and the Ariba Network ■ Host requirement/design workshops to educate customers on the SAP Ariba solution functionality and configurability and help customers understand their business requirements in the context of the SAP Ariba solution ■ Answer customer questions about functionality of the SAP Ariba on-demand application and the Ariba Network
SAP Ariba technical lead	■ Educate customers on SAP Ariba master data and transactional data integration capabilities, formats, and mapping ■ Educate customers on customer-facing aspects of site configuration ■ Build customer's configurations and customizations into the SAP Ariba on-demand system ■ Answer customer questions about SAP Ariba on-demand site functionality ■ Assist with testing and resolve SAP Ariba–related issues

Table 9.7 Consulting Partner Roles and Responsibilities (Cont.)

SAP Ariba or Consulting Partner Role	Responsibilities
Ariba Network adapter tech resources	■ Educate the customer on the Ariba Network adapter architecture, capabilities, and potential configurations ■ Provide SAP Ariba cXML documentation and list of data fields for each in-scope transaction ■ Support Ariba Network adapter installation ■ Support adapter-related design decisions ■ Support the customer with adapter-related data mappings and configurations ■ Support the customer in testing transactions between their SAP ERP and the Ariba Network ■ Support the customer with adapter test issue resolution ■ Develop customizations to the customer's Ariba Network account UI ■ Assist with testing and resolve Ariba Network–related issues
SAP Ariba supplier enablement lead	■ Manage the enablement services workstream ■ Advise the customer on how to perform supplier enablement activities ■ Lead supplier data collection efforts ■ Develop supplier communication and education materials ■ Manage the enablement services workstream ■ Advise the customer on how to perform supplier enablement activities ■ Lead supplier data collection efforts ■ Develop supplier communication and education materials
SAP Ariba supplier management team	■ Contact suppliers to instruct and encourage them to register on the Ariba Network

Table 9.7 Consulting Partner Roles and Responsibilities (Cont.)

Section 9.5 discusses the roles and responsibilities for customers.

Customer Roles	Responsibilities
Project sponsor	■ Establish and communicate overall project vision ■ Provide senior leadership communication in support of the project ■ Mandate appropriate change management across leadership of all affected departments ■ Monitor status reports and timelines ■ Resolve escalated issues, including involving customer resources, lack of participation, or supplier compliance messaging

Table 9.8 Customer Roles and Responsibilities

Customer Roles	Responsibilities
Project manager	■ Confirm customer goals and project scope ■ Help plan the project activities and timeline ■ Ensure adequate resources are assigned ■ Manage participation of all required resources ■ Coordinate stakeholders as needed (accounts payable, purchasing, receiving, finance, etc.) ■ Coordinate signoff on all SAP Ariba deliverables ■ Manage project timeline and adherence to schedule ■ Participate in project status meetings ■ Identify, manage, and escalate project issues and risks, as appropriate ■ Develop go-live cutover plan and manage cutover execution ■ Serve as the single point of contact for overall deployment
Functional lead	■ Participate in configuration workshops ■ Define and document business requirements pertaining to all in-scope processes, transactions, and SAP Ariba solutions ■ Gather business input from all involved functional resources and departments: procurement, accounts payable, accounting, finance, etc. ■ Define business process and requirements and make decisions on solution configurations to handle those requirements ■ Write customer functional design document ■ Develop test cases and test scripts for testing ■ Plan, manage, and conduct system testing and user acceptance testing (UAT) of the entire solution ■ Validate that master data and application configuration data functions as desired ■ Identify, escalate, and resolve project issues ■ Help develop go-live cutover plan and manage cutover execution ■ Become the expert on system functionality, documentation, and customer-specific system use
Customer functional systems analyst	■ Participate in configuration workshops ■ Experiment with system behavior and configuration to support customer decisions on system configurations ■ Learn detailed SAP Ariba functionality all the way through the P2P process, and learn how the system configurations, customizations, and data drive the system functionality ■ Answer customer questions on how the system will support the business and technical requirements raised by all involved customer departments

Table 9.8 Customer Roles and Responsibilities (Cont.)

Customer Roles	Responsibilities
Customer functional resources/testers/pilot users	■ Participate in configuration workshops and contribute to business requirements and solution design ■ Participate in test case/script development ■ Participate in system testing and UAT ■ Identify power users within each department to act as experts by providing assistance to peers and input to overall process ■ Participate in configuration workshops and contribute to business requirements and solution design ■ Participate in test case/script development ■ Participate in system testing and UAT ■ Identify power users within each department to act as experts by providing assistance to peers and input to the overall process

Table 9.8 Customer Roles and Responsibilities (Cont.)

Project Delivery Design Phase

The next phase, the design phrase, revolves around key deliverables for all three solution areas (SAP Ariba Invoice Management, SAP Ariba Discount Management, and SAP Pay). To start, SAP Ariba and consulting partners propose the configuration of any or all three SAP Ariba solutions based on the requirements gathered and follow-on design sessions, and they also provide configuration documents. The customer then defines any custom interfaces to third-party systems for master data, transactions, or both. Finally, SAP Ariba delivery and/or consulting partners work in tandem with the customer to define the configurations and document the holistic design.

9.4.3 Configuring SAP Pay

When setting up SAP Pay, you'll need to verify whether the supplier is set up for discounting by accessing the supplier master record in **Maintain Business Partner**. After you've entered a supplier number, select **Supplier (Fin.Accounting) (Maintained)** from the **Change in BP role** dropdown menu, as shown in Figure 9.22, and select the company code.

As shown in Figure 9.23, ensure that all outgoing payments are flagged.

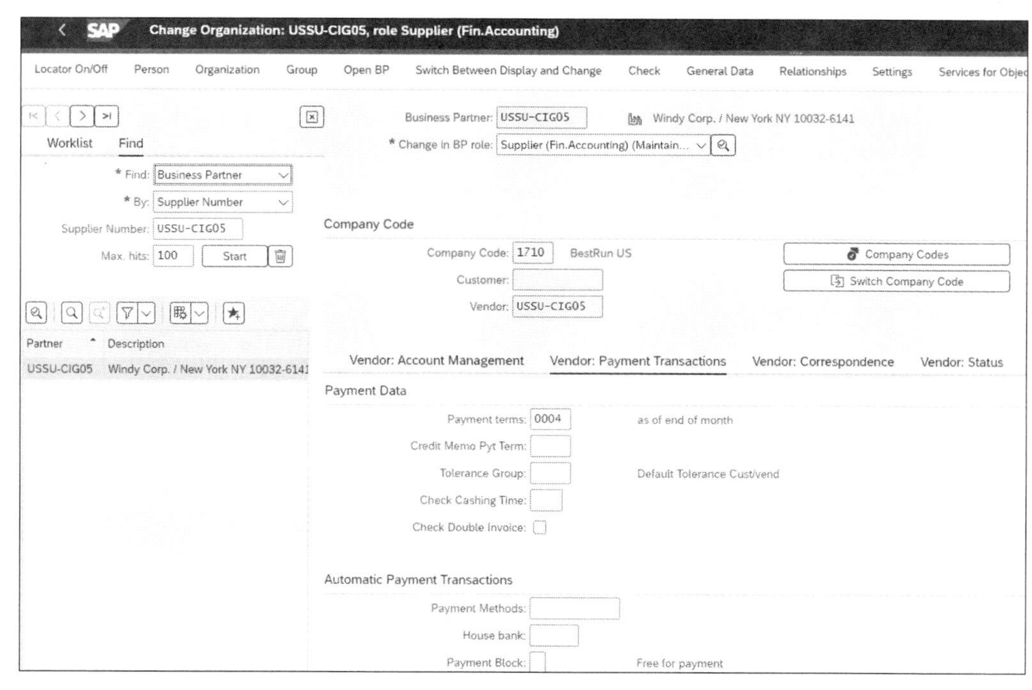

Figure 9.22 Defining Business Partner Financial Accounting

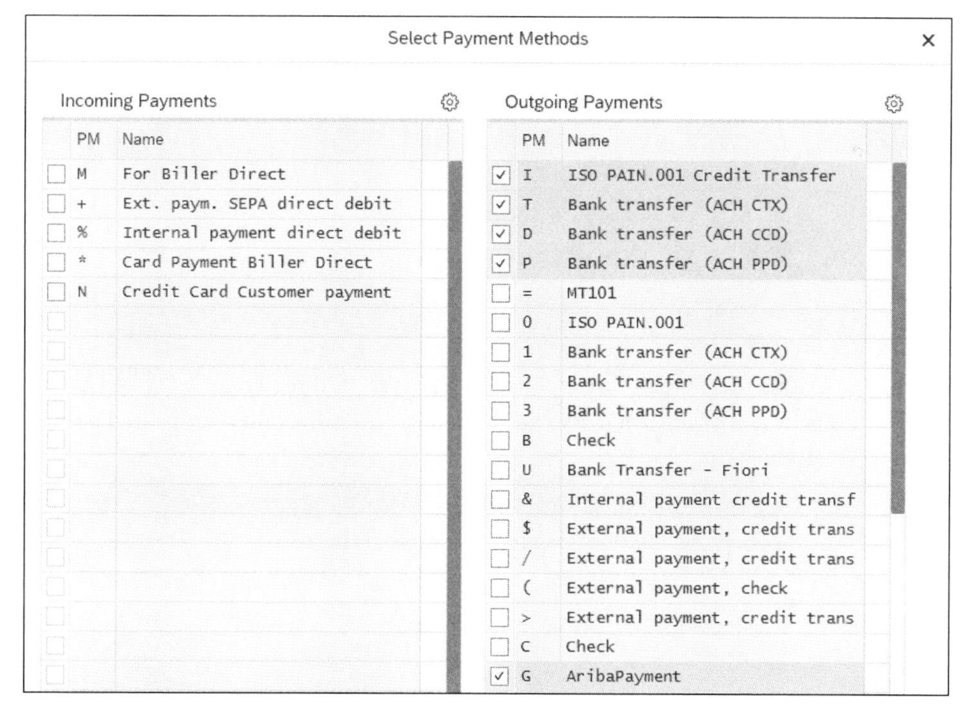

Figure 9.23 Outgoing Payments

9.4.4 Creating and Managing Payments

Another key feature of SAP Pay is the discounting functionality. Unlike the supply chain finance capability, in this scenario, the supplier offers the buyer a discount on the invoice in exchange for receiving payment earlier than the payment terms required on the invoice. The following scenario covers an invoicing transaction where a discount on the supplier's side is provided for early payment on the buyer's side.

Creating an Invoice

After the outgoing payment settings have been verified, the buyer is ready to create an invoice and leverage the discounting functionality using this supplier.

In accounts payable on the buyer's SAP S/4HANA system, an accounts payable manager can view their open invoices, as shown in Figure 9.24.

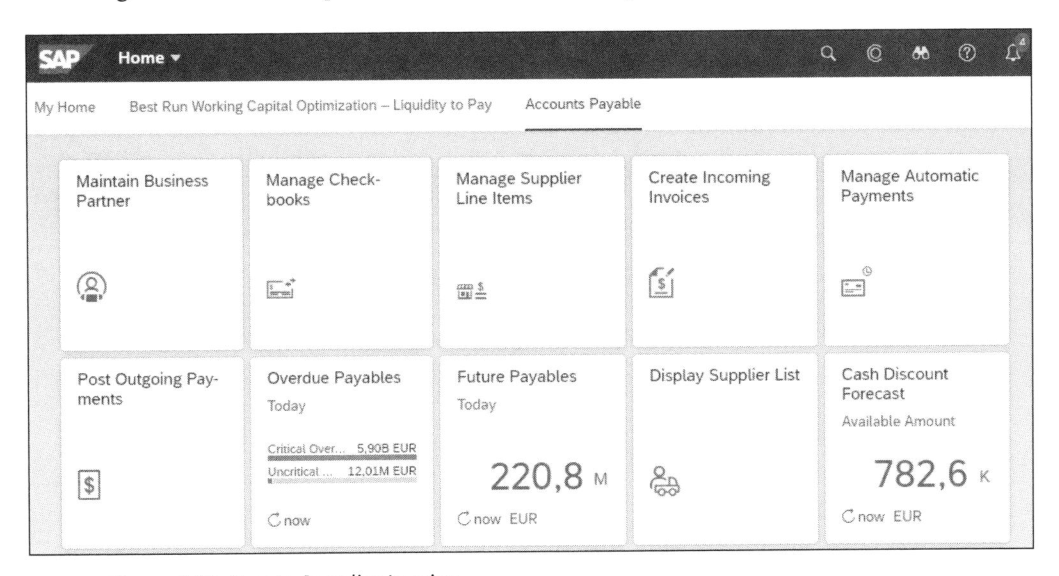

Figure 9.24 Create Supplier Invoice

Accounts payable then creates an invoice in the system directly, as shown in Figure 9.25. Note that this invoice could also be created in SAP Ariba and then posted back to the buyer's ERP system.

After the invoice has been created and saved, as shown in Figure 9.26, if the supplier is on the Ariba Network, the invoice will also be viewable there.

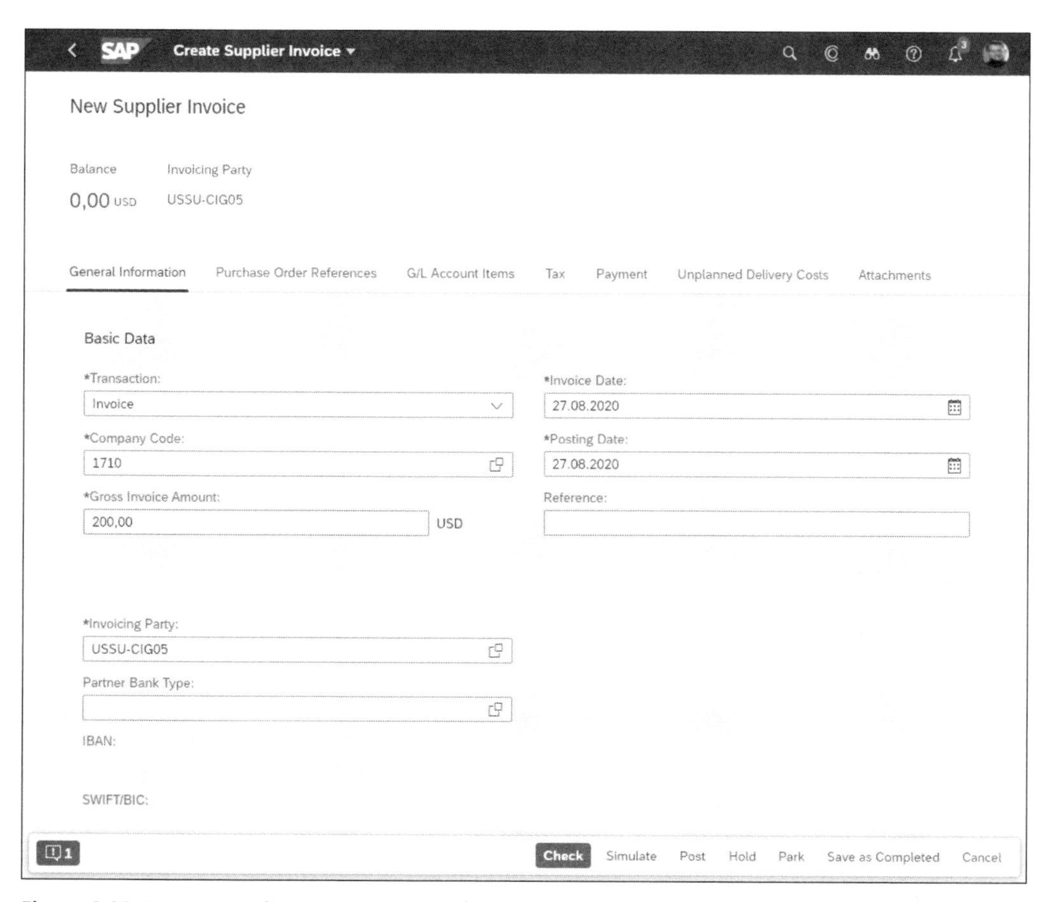

Figure 9.25 Create Supplier Invoice in SAP S/4HANA

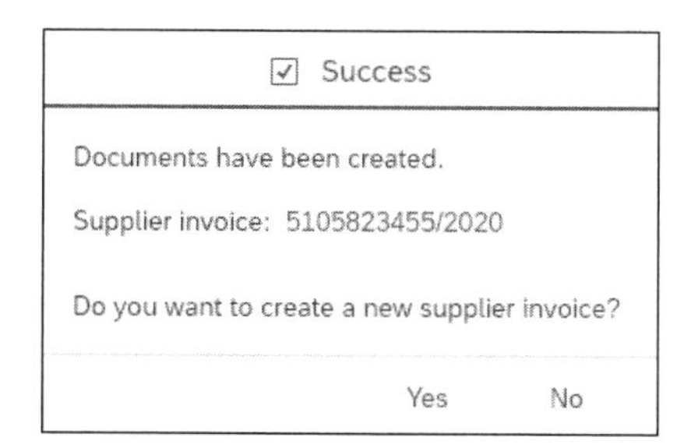

Figure 9.26 Invoice Created

Early Payment Discount

The buyer can flag this invoice to request an early payment discount. In this case, the supplier can then log in and offer an early discount to the buyer in exchange for early payment, as shown in Figure 9.27.

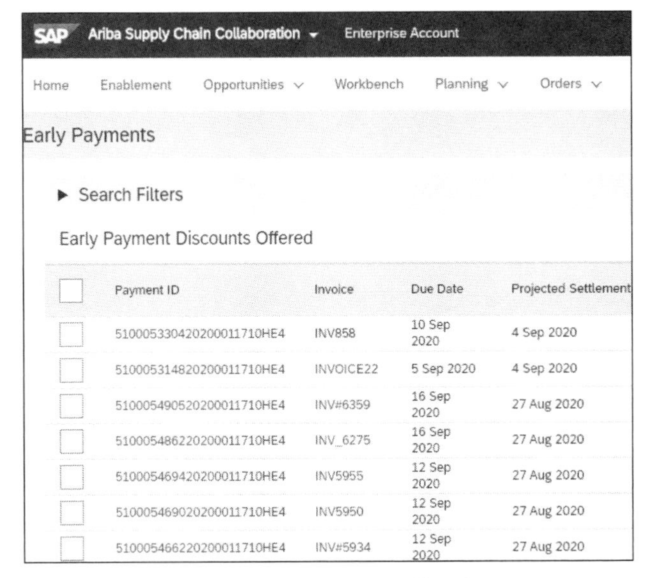

Figure 9.27 Early Payment Discounts Offered

In the payment options, the supplier can select the best option to cover cash flow needs while still preserving margins, as shown in Figure 9.28, Figure 9.29, and Figure 9.30.

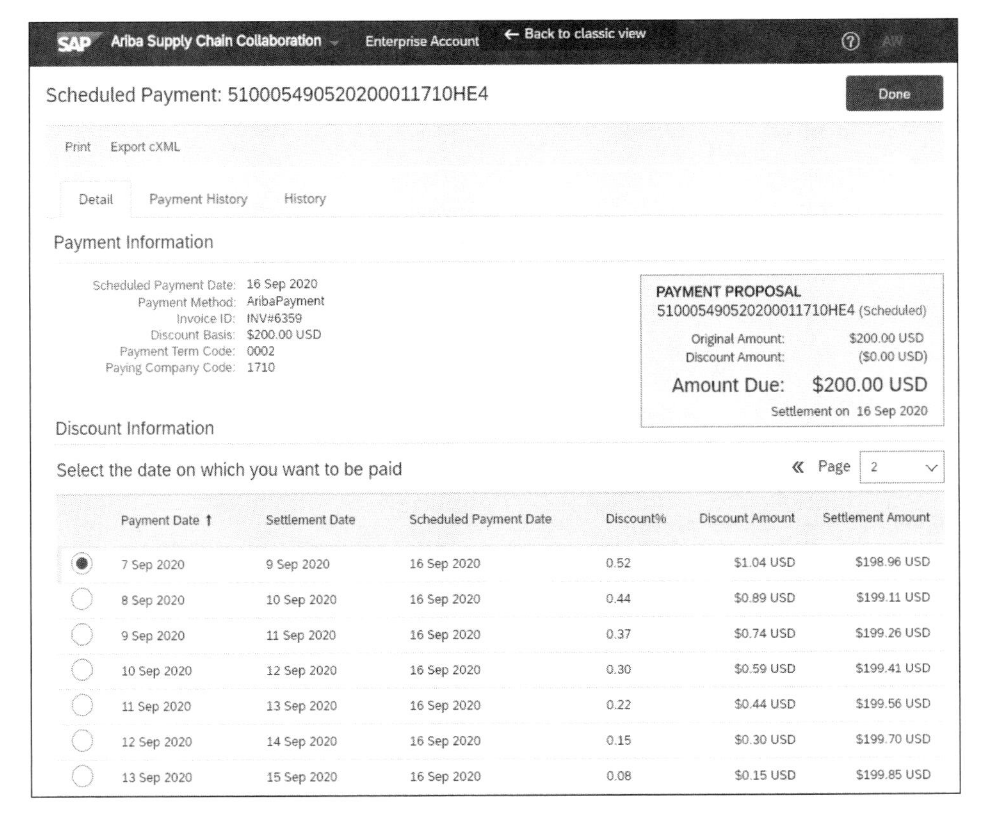

Figure 9.28 Early Payment Discount Options

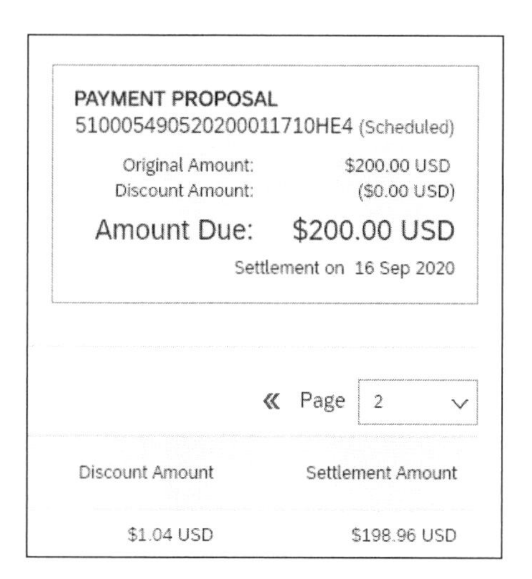

Figure 9.29 Early Payment Option Selected

Underneath the **Early Payment Option** list, the selected payment date is shown graphically, as a comparison with the original payment date, as shown in Figure 9.30.

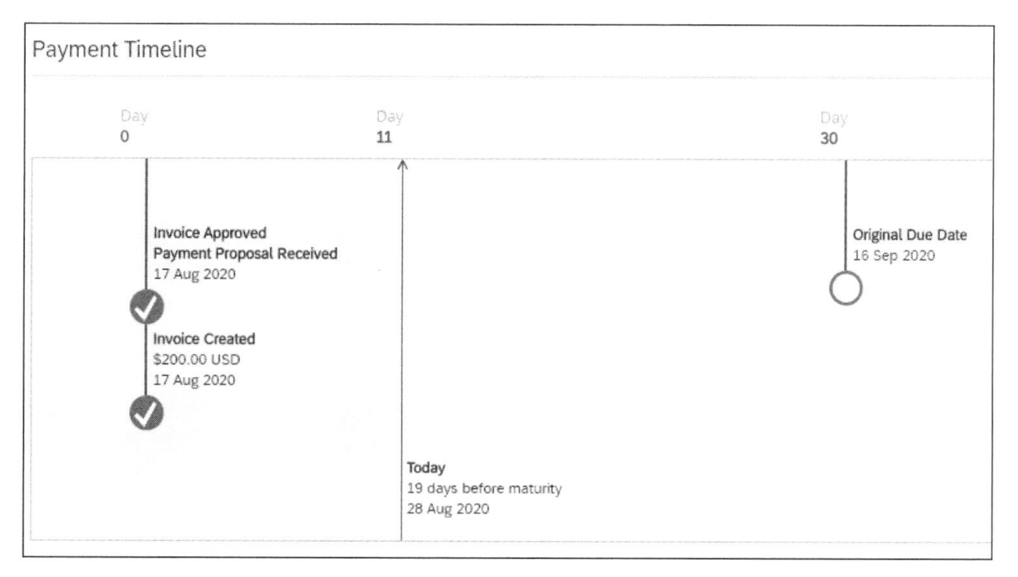

Figure 9.30 Comparison Graphic: Payment Proposal

After reviewing these choices, a supplier can select **Accept Early Payment Offer**, as shown in Figure 9.31, and the payment proposal will be sent to the buyer's accounts payable department.

The supplier will then see the payment timeline based on their selected payment date, as shown in Figure 9.32.

Payment Information

Scheduled Payment Date: 16 Sep 2020
Payment Method: AribaPayment
Invoice ID: INV#6359
Discount Basis: $200.00 USD
Payment Term Code: 0002
Paying Company Code: 1710

Discount Information

Select the date on which you want to be paid

	Payment Date ↑	Settlement Date	Scheduled Payment Date
◉	7 Sep 2020	9 Sep 2020	16 Sep 2020
○	8 Sep 2020	10 Sep 2020	16 Sep 2020
○	9 Sep 2020	11 Sep 2020	16 Sep 2020
○	10 Sep 2020	12 Sep 2020	16 Sep 2020
○	11 Sep 2020	13 Sep 2020	16 Sep 2020
○	12 Sep 2020	14 Sep 2020	16 Sep 2020
○	13 Sep 2020	15 Sep 2020	16 Sep 2020
○	14 Sep 2020	16 Sep 2020	16 Sep 2020

Accept Early Payment Offer

Figure 9.31 Early Payment Discount Confirmed

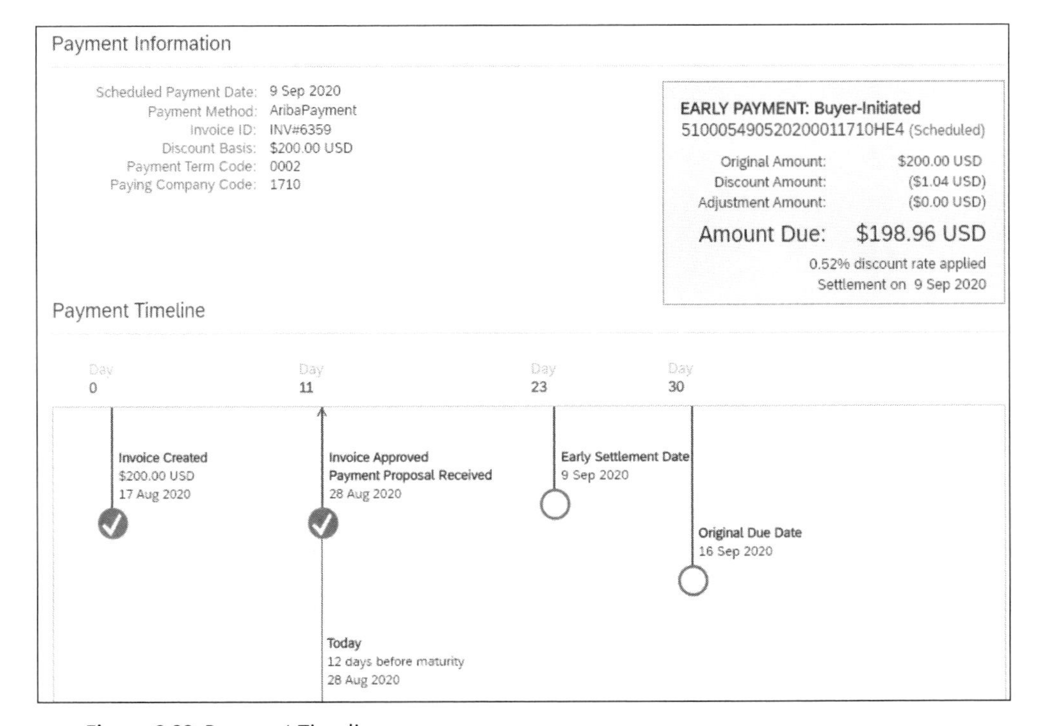

Figure 9.32 Payment Timeline

Note that the discount offered is modest, at only .52%. However, on large bills, this small discount could represent a large amount of money. Further, as with all procurement-driven savings, this money impacts bottom-line profitability directly.

Once accepted, the Ariba Network transmits pay me now instructions to the buyer's SAP S/4HANA accounts payable area, which can be accessed by clicking on the **Manage Supplier Line Items** tile, shown in Figure 9.33.

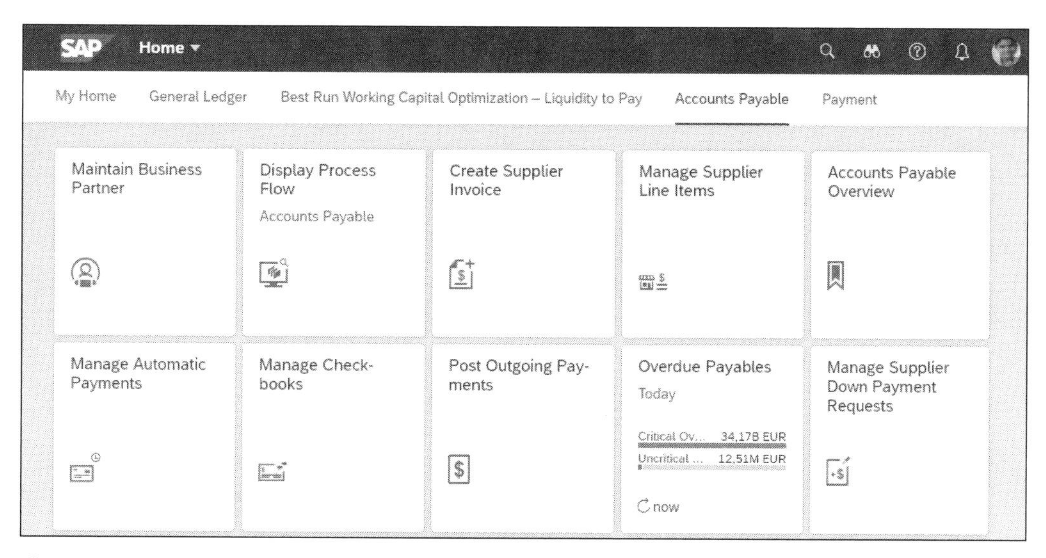

Figure 9.33 Manage Supplier Line Items Tile

Logging in then as the accounts payable person on the buyer side, you can query the invoice and view the offered discounts. Using the assisted search functions, you can view open line items by company code, supplier, and other criteria, as shown in Figure 9.34.

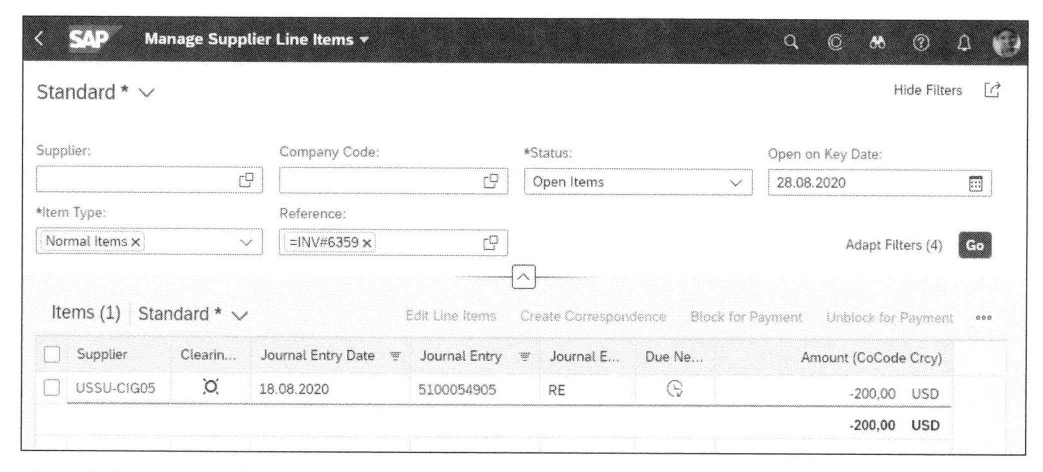

Figure 9.34 Supplier Line Items

Click on **Journal Entry** for the supplier line item, and then select the **Manage Journal Entry Item** option. Before a payment run, the journal entry will still show the original amount of the invoice debit, in this case, $200.00 USD, as shown in Figure 9.35.

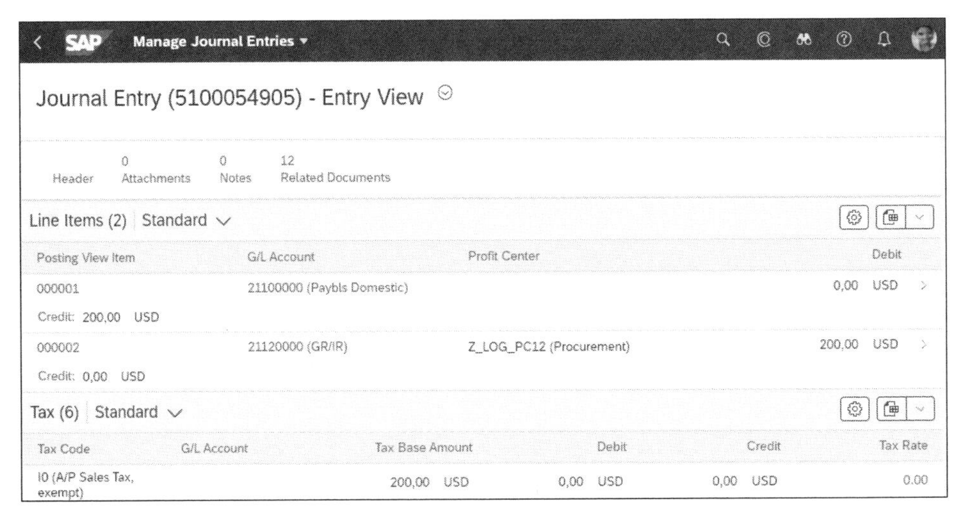

Figure 9.35 Journal Entry View

By clicking on the first **Posting View Item**, for example, accounts payable can review the terms and discount proposed, as shown in Figure 9.36.

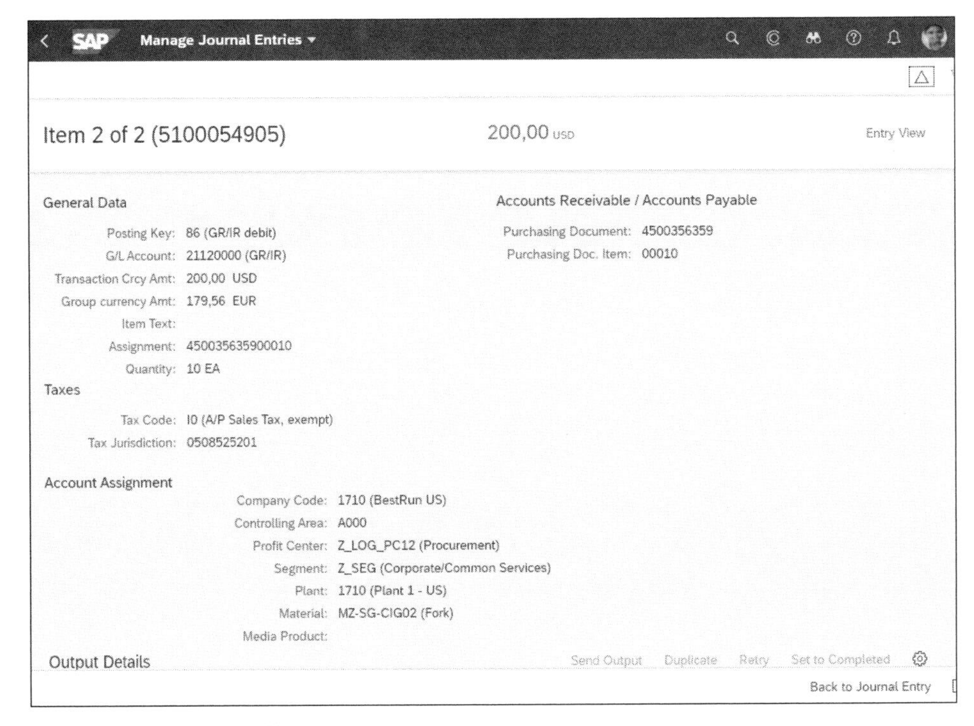

Figure 9.36 Discount Offer in Posting View Item

After accounts payable has accepted the proposed terms, accounts payable would select the **Manage Automatic Payments** tile, as shown in Figure 9.37.

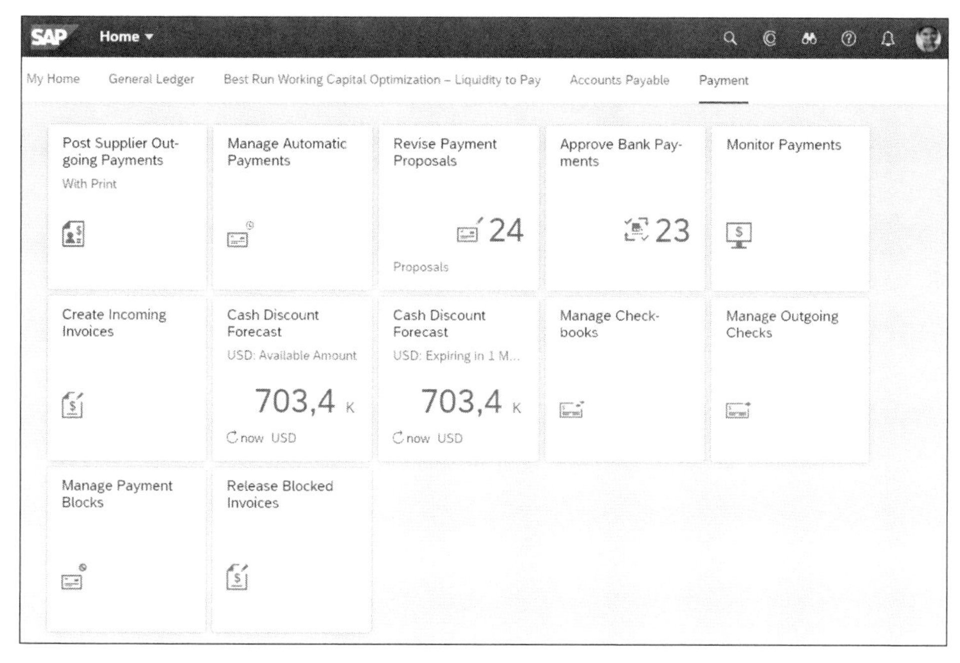

Figure 9.37 Manage Automatic Payments

As shown in Figure 9.38, a payment run is scheduled to cover the selected payment methods.

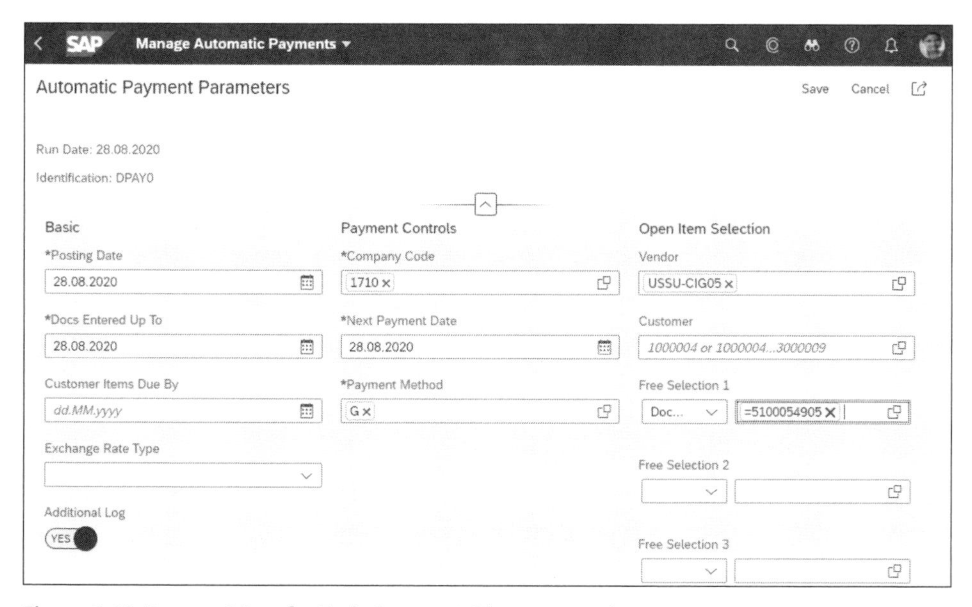

Figure 9.38 Payment Run for Early Payment Discount Invoice

After creating a new payment parameter, proposals are then generated, along with their relevant logs, as shown in Figure 9.39 and Figure 9.40.

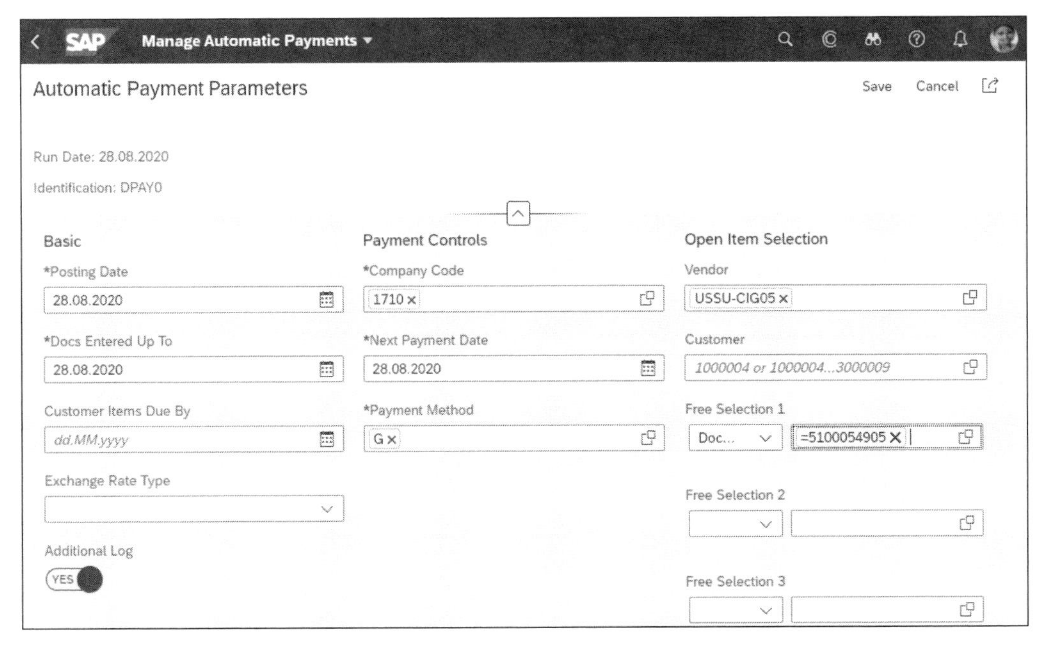

Figure 9.39 Manage Automatic Payments

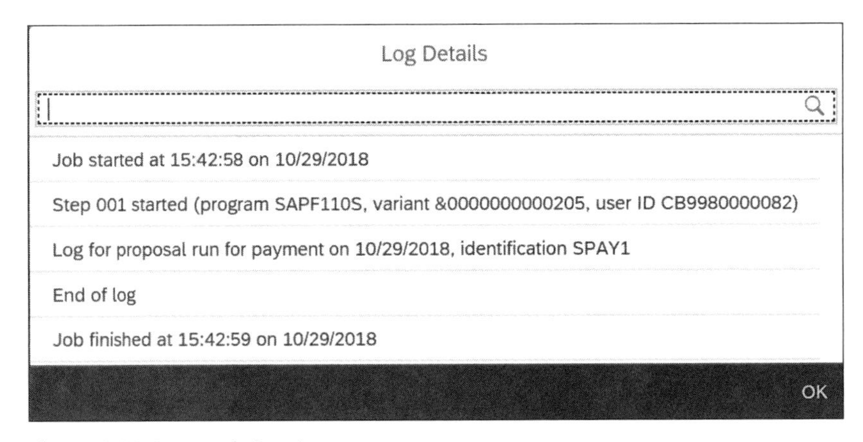

Figure 9.40 Proposals Run Log

After the payment proposals are run, the actual payment run still needs to be scheduled, as shown in Figure 9.41.

In the final step of this process, after the payments have run, the journal entry is updated with the resulting discount, as shown in Figure 9.42.

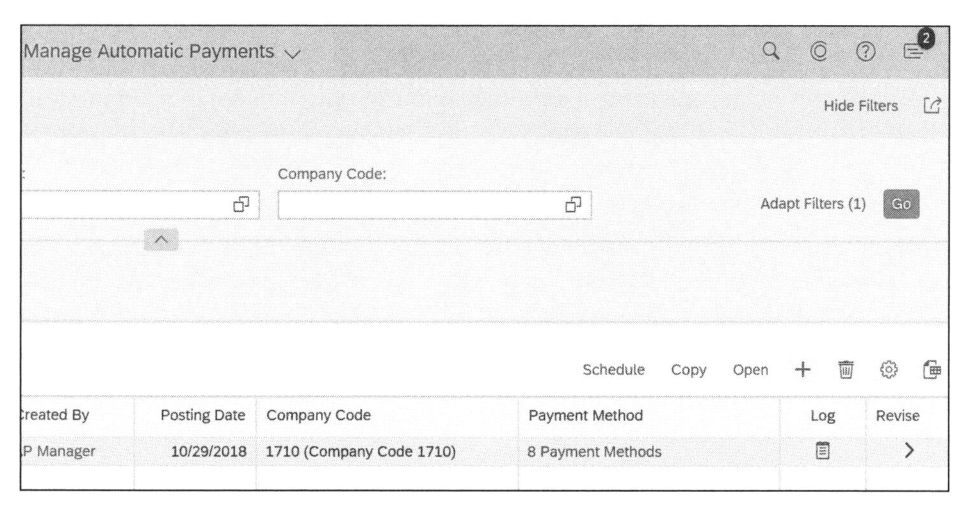

Figure 9.41 Schedule Proposals in SAP S/4HANA

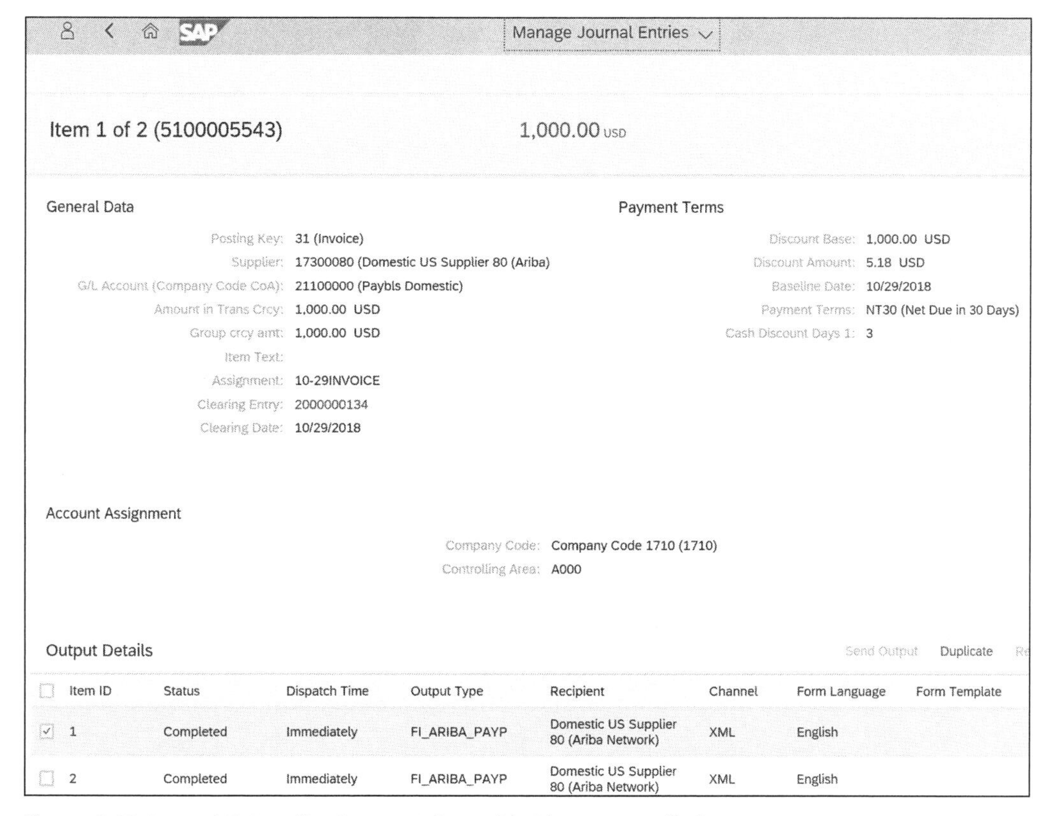

Figure 9.42 Journal Entry after Payment Run with Discount Applied

Payment has now been made and recorded in the customer's financial ledger.

9.5 Summary

Accounts payable has long focused on the reconciliation of supplier-submitted invoices with sourcing and purchasing documents, along with supporting workflow approvals and intermediate steps, to initiate payment to the supplier. This reconciliation effort can be difficult, tedious, and, ultimately, expensive in a paper-based transaction—even more so with a verbally initiated transaction. SAP Ariba's solutions in this area, with SAP Ariba Invoice Management and SAP Pay, leverage the processes you've built out on the transaction side in SAP Ariba to provide a much more cost-effective accounts payable experience. With payment speed and capabilities to reconcile in an efficient manner, SAP Ariba customers can further leverage this agility to provide financing to their supply chains and obtain early payment discounts negotiated as part of a supplier strategy or on the fly. This agility achieves cost savings at the end of the P2P process, at the critical juncture where money is changing hands. This juncture is often a key area for realizing savings and additional operational efficiencies.

Chapter 10
Spend Analysis

In this chapter, we'll showcase a major analysis area for procurement—spend analysis. SAP Ariba Spend Analysis combines supplier data with category- and invoice-based spending to build a comprehensive view of your procurement spending.

The most important strategic area of procurement is spend analysis. After your procurement systems and processes are in place, the resulting transactional data across these systems needs to be analyzed. An organization that constantly analyzes its procurement activities will grow smarter with each cycle, create more savings and competitive opportunities, and ultimately increase both bottom- and top-line revenue: Bottom-line revenue increases by saving the organization more money, and top-line revenue increases by identifying growth opportunities with a company's key suppliers. To paraphrase Socrates, an unexamined procurement operation isn't worth running.

In this chapter, we'll outline SAP Ariba's spend analysis tool—SAP Ariba Spend Analysis—and show you how to implement it to get a closer look at your procurement operations at large and learn where to make changes. Depending on which SAP Ariba solutions you've implemented, you may also run reports on other areas of procurement, such as sourcing events, contracts, purchase orders (POs), requisitions, and suppliers. In any case, SAP Ariba Spend Analysis is SAP Ariba's main reporting tool.

10.1 What Is Spend Analysis?

Spend analysis is the process of analyzing historical spend by collecting, categorizing, cleansing, and evaluating data on spend from all business units/departments across the organization. The goal of spend analysis is to increase profitability within an organization by identifying wasteful spending and reducing procurement cost, increasing operational efficiency, and identifying contracts for renegotiation.

Good spend analysis begins with a good spend visibility across the spend ecosystem, as shown in Figure 10.1.

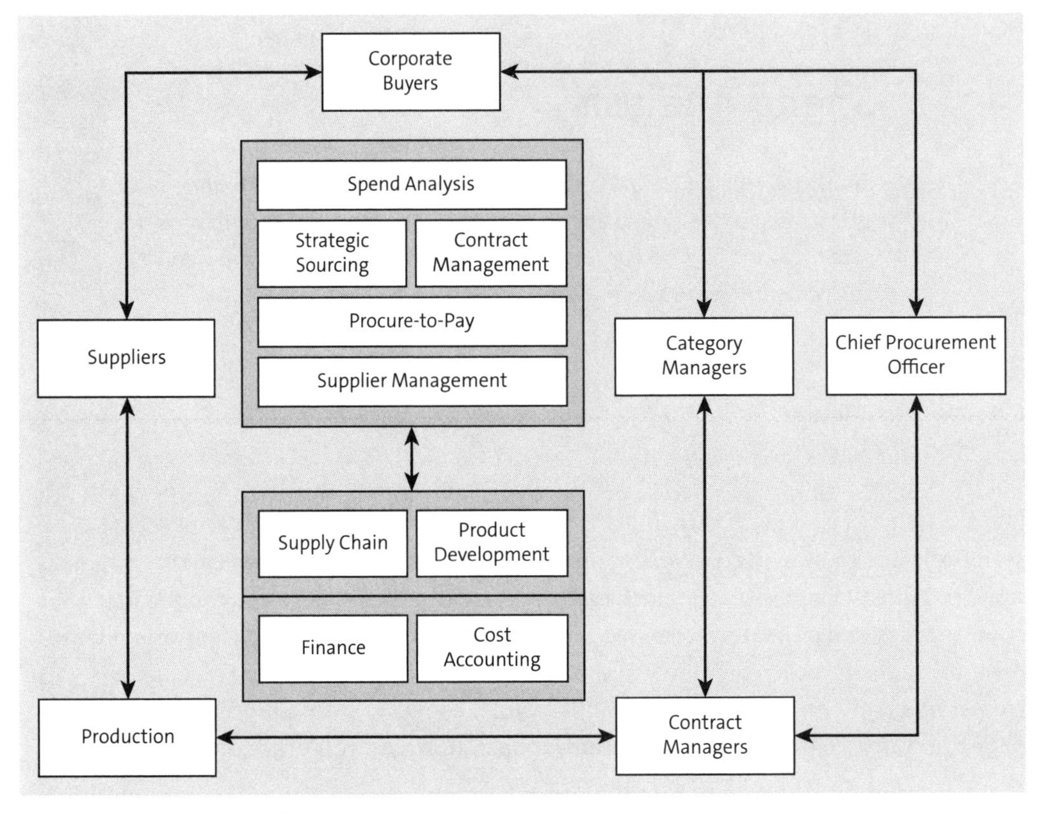

Figure 10.1 Spend Management Ecosystem

Factors that can drive the focus on spend analysis include the following:

- Need to identify and forecast savings opportunities
- Need to identify and prioritize top spend categories
- Need to improve negotiation leverage for supplier contracts
- Need to track off-contract spend
- Need to bolster bottom-line growth
- Need to reduce the supply base

Spend analysis is crucial for an organization to align procurement (purchasing) strategies with its overall strategic goals to maintain a competitive edge over peers in the industry.

Challenges in Gaining Spend Visibility

Some challenges that companies face in terms of gaining spend visibility across the organization are as follows:

- Inconsistent data quality
- Limited access to market information

- Poor visibility on suppliers
- Disparate data sources
- Manual effort-intensive processes

In the following sections, we'll look at some key strategies for spend analysis that your company can use and describe how SAP Ariba's Spend Analysis tool can help with this process.

10.1.1 Spend Analysis Strategies

In today's fast-paced global economy, to be competitive, companies should take a more strategic approach where their procurement organization plays a major role in driving toward the company's overall goals.

Following are some ways to understand your spend better and control costs through spend analytics:

- **Analyzing aggregate spend**
  With spend data and analytics, companies can use this data to understand what their company buys, who buys it, and the buy process and buy channels. This understanding of aggregate spend allows companies to streamline their processes and further reduce spend.
- **Managing supplier performance**
  With spend data and analytics, companies can measure their suppliers on metrics and key performance indicators (KPIs) that matters to the company. This allows companies to identify risks in its supply chain such as over dependency on a supplier, quality issues, and other risks. Based on this information, companies can take corrective actions with their supply chain to mitigate the company's risks.
- **Enforcing contract compliance**
  Contract compliance to ensure the right suppliers are awarded business and contracts are adhered to is key for your company's bottom line and performance. Spend analytics provides useful insights and opportunities for your company to drive for better contract compliance.
- **Forecast planning**
  With historical spend data and analytics, your company has better visibility into its spend trends. This allows you to make more informed decisions about managing supplier relationships so as to improve processes.
- **Internal benchmarking**
  The spend analysis tool can help companies create internal benchmarks so that you can monitor their performance and continuously improve to stay ahead of the curve.
- **Sharing spend behavior across departments**
  With spend data and analytics, departments and business units can understand

spending behaviors and review internal processes. By providing visibility to preferred suppliers and tools to better manage spend, departments and business units can modify the way they buy goods and services, and award contracts to a smaller set of key, high-performing suppliers. Departments can significantly increase efficiency using SAP Ariba Spend Analysis.

Through spend analytics, you can both identify spend and drill down to spend by supplier. This data allows you to understand who your key suppliers are and provides opportunities to build strategic relationships with your key suppliers that allow you to negotiate deeper discounts. Consolidating your spend to these suppliers provides savings opportunities for your company.

In the next section, we'll look at how SAP Ariba can provide you with a key component of spend analysis—spend visibility.

10.1.2 SAP Ariba Spend Analysis

SAP Ariba's answer to spend analysis is its cloud-based solution—SAP Ariba Spend Analysis. This solution allows you to extract, classify, and enrich spending data from your other SAP Ariba solutions such as SAP Ariba Buying and Invoicing and from your SAP ERP backend or other legacy spend systems. You can then analyze the spend data collected from various sources using dashboards, risk intelligence, compliance/spend reporting, and benchmarking, as shown in Figure 10.2.

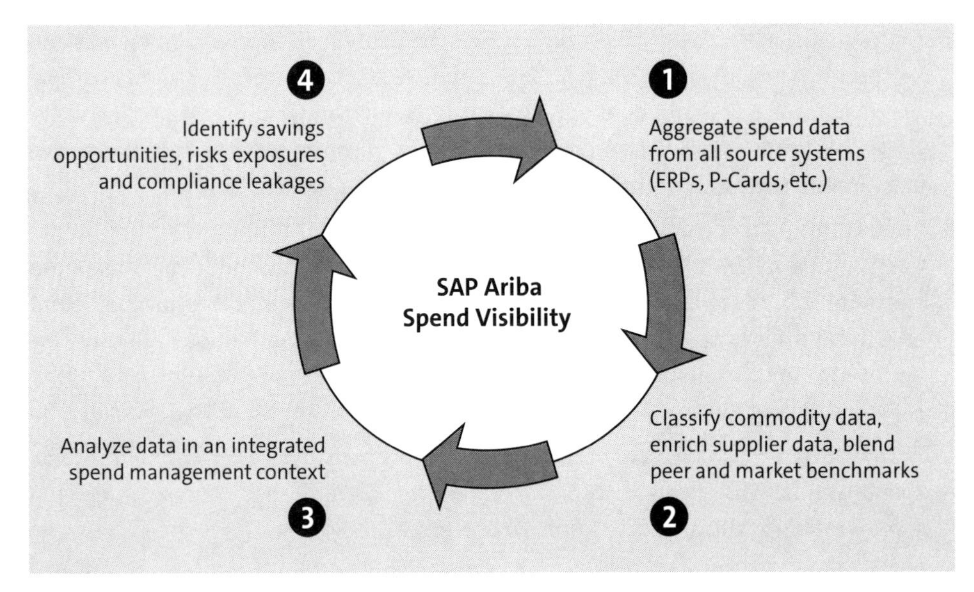

Figure 10.2 SAP Ariba Spend Analysis

SAP Ariba Spend Analysis is an effective spend management tool with the following features:

- Spend management dashboard with a 360-degree view of all spend activities
- Native integration to the most comprehensive supplier database—Dun & Bradstreet (D&B)
- SAP Ariba's next-generation enrichment services that use advanced neural networks and machine learning technologies for fast and accurate commodity classification of the spend data and supplier enrichment
- Blazing fast analysis on spend data with SAP HANA
- Best practice services with the tool to accelerate your return on investment (ROI)
- Cloud-based deployment to realize lower total cost of ownership (TCO) and rapid time to value

SAP Ariba Spend Analysis is available in several versions, depending on your data analysis needs and landscape:

- SAP Ariba Spend Analysis basic
  - SAP Ariba Spend Analysis basic for SAP Spend Performance Management (primarily for SAP environments)
  - SAP Ariba Spend Analysis basic (focused on non-SAP environments)
- SAP Ariba Spend Analysis professional
  - An augmented solution to the basic versions

In short, spend analysis is the general product area, and SAP Ariba Spend Analysis basic and professional are the reporting tools that can deliver complete analytics projects, including project management and data enrichment.

The core spend performance management functionality in SAP Ariba Spend Analysis basic includes multiple engines to distill the data into actionable form:

- **Data validation engine**
  Uploading SAP Ariba Spend Analysis data files includes automatic data validation, designed to identify formatting errors in the files, as well as reports detailing errors occurring during the data load.
- **Supplier matching engine**
  A supplier engine matches supplier records in your data against SAP Ariba's more than 200 million-record supplier database and enriches validated suppliers for supplier parentage and other information, according to your service-level agreement (SLA).
- **Rationalization engine**
  A rationalization engine uses learned models; text reading; and linguistic analysis of invoices, general ledgers, and supplier information to rationalize transactions.
- **Business rule engine**
  A business rule engine invokes business rules or mappings specific to your company to categorize transactions.

- **Inference engine**
  An inference engine uses weighted triangulation and multivector inference on potential outcomes of classifications to ensure the highest reliability of outcome.
- **Machine learning engine**
  A machine learning engine uses decision trees, classification by example, Bayesian algorithms, and joint field hybrid methods using historically categorized transactions and export refinement to predict classification outcomes.

If you see commodity classification errors in the data, you can submit those errors via a change request workflow for correction. Approved requests are exported and included in the next enrichment cycle. Depending on the service levels defined in the SAP Ariba Spend Analysis deployment description, SAP Ariba will run data enrichment refreshes at given intervals.

Optional features and services include the following:

- **Custom commodity taxonomies up to six levels**
  This option must be implemented by leveraging SAP Ariba services.
- **Supplier diversity and green reporting**
  This reporting option needs to be enabled by an SAP Ariba representative, and supplier diversity and green data are managed as an SAP Ariba service.
- **Supplier risk and financial data**
  Risk data requires a separate contract with D&B, a company information provider. The SAP Ariba representative can act as a liaison between your company and D&B to gather the supplier risk data. Financial data is enabled by SAP Ariba representatives directly in SAP Ariba.

SAP Ariba Spend Analysis professional includes all these capabilities plus a few additional features. One main difference between the basic and professional versions of SAP Ariba Spend Analysis is the dashboard, which is exclusively part of SAP Ariba Spend Analysis professional. This dashboard centralizes the views and reporting in one area and includes the following:

- A personal calendar for each user
- SAP Ariba data, such as to-do lists and document folders, which users can add to their dashboards
- Company news content, including information from RSS feeds, which can be configured to include news content from your sites
- The ability for users to create multiple dashboards to cover different areas

Other differences can be found in reporting. SAP Ariba Spend Analysis professional includes the following:

- Prepackaged invoice and PO reports
- Basic supplier financial data

- Reporting against common commodity benchmark data, such as the consumer price index and producer price index
- Custom analytical reporting, including reporting across multiple fact tables and compound reports
- United Nations Standard Products and Services Code (UNSPSC) commodity display in English and up to four different languages

Each SAP Ariba Spend Analysis subscription includes an SAP Ariba best practice center-managed project, which is, in essence, a coaching engagement that provides recommendations on the technical and functional use of the software.

Note

The dashboard, which, as mentioned, is only provided in SAP Ariba Spend Analysis professional, also includes augmented reporting services and support.

In the following section, we'll look at the how you can plan for a successful implementation of SAP Ariba Spend Analysis using either the basic or professional version.

10.2 Planning Your Implementation

As mentioned in previous chapters, the key to a successful implementation of SAP Ariba Spend Analysis is planning, key stakeholder/executive-level ownership, and effective change management. The SAP Ariba Spend Analysis team must be engaged to implement this solution. Implementing SAP Ariba Spend Analysis is different from implementing other SAP Ariba modules in that SAP Ariba Spend Analysis doesn't follow the typical SAP Ariba on-demand deployment methodology, which is the SAP Activate methodology.

Table 10.1 shows the implementation methodology for spend visibility.

Prepare and kickoff	▪ Kick off the project team ▪ Provide data schema training
Data collection and validation	▪ Build data extraction scripts ▪ Upload data to SAP Ariba Spend Analysis ▪ Validate data collected and provide feedback
Data enrichment	▪ Load data to SAP Ariba Data Enrichment Services tool ▪ SAP Ariba Data Enrichment Services experts analyze and cleanse data
Supplier enrichment	▪ SAP Ariba Data Enrichment Services provides a supplier-enrichment component, including supplier parentage for customer analysis

Table 10.1 Spend Visibility Implementation Methodology

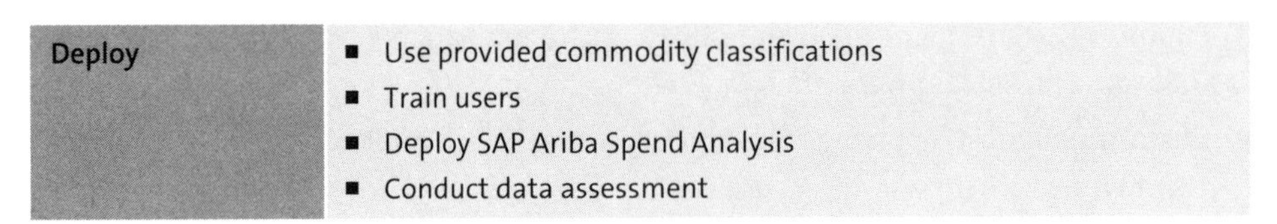

Deploy	■ Use provided commodity classifications ■ Train users ■ Deploy SAP Ariba Spend Analysis ■ Conduct data assessment

Table 10.1 Spend Visibility Implementation Methodology (Cont.)

To implement the SAP Ariba Spend Analysis solution, the following roles will be required on the SAP Ariba side:

- SAP Ariba data enrichment services lead
- SAP Ariba project manager
- SAP Ariba Spend Analysis shared services lead
- SAP Ariba Spend Analysis technical lead

The following roles will be required on the customer side:

- Project sponsor
- Project manager
- Functional data lead
- Functional subject matter experts (SMEs)
- Technical data lead (IT)

In the following section, we'll review the phases in the SAP Ariba Spend Analysis deployment process that you, the customer, must plan for.

10.2.1 Prepare and Kickoff Phases

The prepare phase is the first implementation phase, usually after you've completed the discover phase with the SAP Ariba sales organization. The initial project scope, timeline, and budget is agreed upon during the discover phase. The purpose of the prepare phase is to confirm value drivers, goals and objectives, project scope, and success metrics for a successful implementation. In this phase, the project team is identified. The project plan, the project governance framework, the issues and risk plan, and the project roles and responsibilities matrix are defined and finalized with the customer.

Analytics projects are somewhat more nuanced, however. Analytics can often provide the ROI for projects, especially procurement projects, where a single insight can sometimes pay for the entire implementation. As tantalizing as this prospect may be, analytics projects usually come later in the overall rollout process of procure-to-pay (P2P) solutions because analytics requires data, and data must be generated before you can report on it. For SAP Ariba Spend Analysis, defining the data set that will underpin these reports is a key task and ideally should be undertaken prior to project kickoff.

The following roles and responsibilities will be needed for the tasks we've just discussed:

- **SAP Ariba data enrichment services team**
  The SAP Ariba data enrichment services team is responsible for all aspects of customer data enrichment for on-demand engagements leveraging SAP Ariba Data Enrichment Services technology for supplier enrichment and commodity classifications.
- **SAP Ariba project manager**
  Both an in-house project manager and an SAP Ariba or consulting partner project manager are assigned to an SAP Ariba Spend Analysis project. SAP Ariba Spend Analysis project managers should be well versed in the various aspects of the SAP Ariba Spend Analysis engagement and will assist you in the following:
  - Understanding the overall SAP Ariba Spend Analysis process
  - Mapping your SAP ERP data to the analysis data schema
  - Providing feedback on uploaded source data extracts
  - Identifying the optimal combination of hint fields for enrichment
  - Various informational sessions on both the analysis and the enrichment processes
  - Assisting with reviewing enrichment results and refining classifications where appropriate
  - Deploying enriched data to users
  - Conducting a data assessment at the completion of the first pass of enrichment
- **Customer project manager**
  You should assign a project manager to keep all resources focused on the project goals; to make key decisions in a timely manner; and to report progress to internal stakeholders, the project sponsor, and the steering committee.
- **Customer technical data lead**
  Your IT data lead and his team are primarily responsible for the data collection at your end, as your source data must be extracted and transformed into the SAP Ariba Spend Analysis data schema format.
- **Customer functional data lead (procurement) and SMEs**
  Your functional data lead is the key resource that SAP Ariba team engages with during this project. This resource is responsible for identifying historical spend data and its sources where the data must be extracted and pushed to SAP Ariba Spend Analysis. This resource works closely with the SAP Ariba Spend Analysis shared services lead throughout this project to define and enable the data extraction process from your various spend systems and to validate any enrichments made to the extracted data. In addition, obtaining participation and support from SMEs on your

procurement team, who will eventually be the tool's primary users and beneficiaries, is critical to the success of the project. Throughout the project, these procurement SMEs will have access to increasing levels of data, starting with the raw customer data after it's initially loaded into SAP Ariba Analysis and SAP Ariba Spend Analysis. The first enrichment milestone is available at the midway point of the enrichment phase and provides supplier enrichment results, including supplier parentage and additional enrichment attributes (diversity, industry codes, and credit ratings). The final stage is the first-pass go-live, when commodity classifications are available, and your data can be leveraged to achieve numerous predefined project objectives, including the identification of sourcing savings opportunities, supplier rationalization activities, and compliance monitoring.

- **Customer project sponsors (and other key stakeholders)**
  Stakeholders are executive-level members of your organization who sponsor the project and promote buy-in and adoption to realize ROI. Your project manager will normally update stakeholders on progress, issues, and decision-making crucial to the completion of the project. Alternatively, stakeholder meetings can be held for such updates.

In the elapsed time between contracting SAP Ariba for an SAP Ariba Spend Analysis engagement and the project kickoff, you can begin to review the data schema document to frame questions for the kickoff. However, we recommend delaying data extraction work until after project kickoff, as key decisions and clarifications will often be made during kickoff meetings.

During this time, strategizing about how to accomplish the data extraction is crucial. If not properly planned and resourced, extraction poses the greatest risk for delaying the project timeline.

Note

SAP Ariba can offer assistance in your data extraction efforts, so we don't recommend delaying a visibility effort if IT resources aren't available to extract data. For more information, ask your SAP Ariba account executive.

Finally, several decisions must be made before you can request an SAP Ariba Analysis instance. You should brainstorm several key concepts and be prepared to have an answer for each during the project kickoff. As a result, the SAP Ariba project manager can request the development of the new SAP Ariba Analysis instance immediately following kickoff. Some considerations to keep in mind include the following:

- Customer name to appear embedded in the URL (20 characters maximum, lowercase only, no spaces)
- Your company's fiscal calendar

- The spend currencies that will be available for reporting in SAP Ariba Analysis (USD plus up to four additional currencies, as required)

Typical deliverables in this phase are project kickoff, SAP Ariba Spend Analysis project overview, and analysis data schema. Let's begin with the project kickoff.

Project Kickoff

The project kickoff routinely takes one or two days, depending on project size and scope. The kickoff consists of two parts: project overview and data schema training. The SAP Ariba project manager can be on-site or can be available by teleconference, at your discretion. These two parts break out as follows.

SAP Ariba Spend Analysis Project Overview

The project overview should be attended by all project participants if possible. The discussion will include the following:

- An overview of the SAP Ariba Spend Analysis solution and tools
- A scoping discussion, including project resources, data, challenges, and time estimates
- A high-level overview of the SAP Ariba Spend Analysis process (shown later in Figure 10.4)
- Data collection talking points
- An overview of the SAP Ariba data enrichment services process
- Next steps

Analysis Data Schema

The second portion of the kickoff is much more detail-oriented and consists of the SAP Ariba project, your functional and technical data leads, and your project manager walking through the analysis data schema in a detailed fashion. The objectives of this session are to make key decisions on what spend will be in scope (and out of scope), to map data fields from your SAP ERP systems to the analysis data schema, and to ensure that you understand all schema structures and formatting.

SAP Ariba Spend Analysis uses a fixed schema designed to enable best practice spend analysis. While not all fields are required, we recommend populating all the available fields. Although the process is flexible enough to accommodate other formats, and SAP Ariba has experience extracting and transforming data from multiple ERPs, you're far better off building scripts to automate data flows during refreshes and having a structured schema to ensure that all useful information is captured.

The SAP Ariba Analysis data schema is a series of associated flat files containing comma-separated value (CSV) tables in a star schema format. The center of the star is your invoice and PO data. The supporting tables that these tables refer to are the arms of the star.

> **Note**
>
> The standard scope for implementing SAP Ariba Spend Analysis doesn't prevent you from deviating from the standard SAP Ariba schema. However, if you want to deviate from the standard SAP Ariba schema, you must inform SAP Ariba during sales cycles to adjust the scope.

A typical SAP Ariba Spend Analysis project runs 17 weeks, as shown in Figure 10.3.

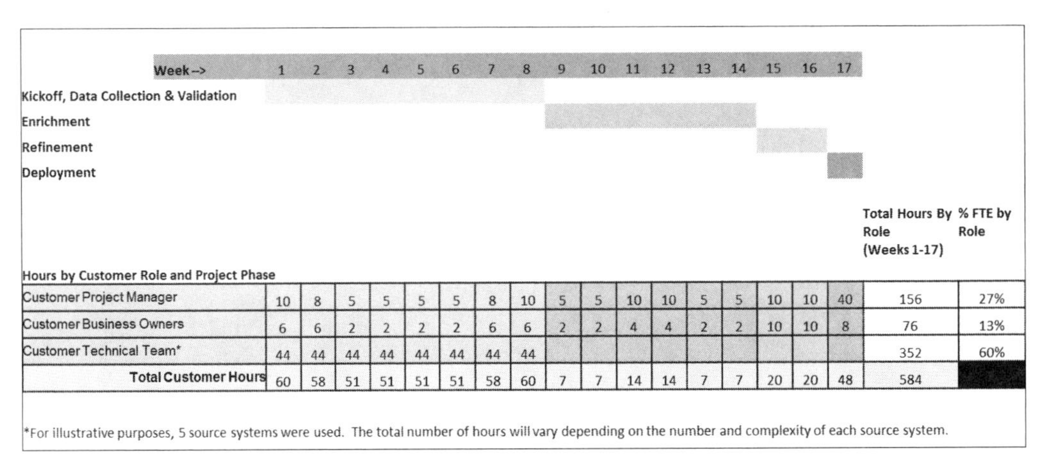

Hours by Customer Role and Project Phase

Week-->	1	2	3	4	5	6	7	8	9	10	11	12	13	14	15	16	17	Total Hours By Role (Weeks 1-17)	% FTE by Role
Customer Project Manager	10	8	5	5	5	5	8	10	5	5	10	10	5	5	10	10	40	156	27%
Customer Business Owners	6	6	2	2	2	2	6	6	2	2	4	4	2	2	10	10	8	76	13%
Customer Technical Team*	44	44	44	44	44	44	44	44										352	60%
Total Customer Hours	60	58	51	51	51	51	58	60	7	7	14	14	7	7	20	20	48	584	

*For illustrative purposes, 5 source systems were used. The total number of hours will vary depending on the number and complexity of each source system.

Figure 10.3 SAP Ariba Spend Analysis: High-Level Resource Plan and Timeline

For a list of typical activities by phase, consult Table 10.2.

Phase	Activities
Data Collection and Validation	■ Kickoff ■ Data schema training and mapping ■ Data extraction and validation ■ Spend approval
Enrichment	■ Invoice and supplier enrichment ■ Custom mapping ■ Enrichment training
Refinement	■ Results review and providing feedback ■ Feedback application
Deployment	■ Analysis training ■ Report templates, dashboards ■ User creation

Table 10.2 Typical Activities by Phase

10.2.2 Data Collection

After the data has been extracted in the SAP Ariba Spend Analysis data schema format, the designated individual at your site will upload the source data files directly into SAP Ariba Analysis. At this time, the SAP Ariba project manager will automatically receive a notification.

IT resources are typically required for data collection to perform the extraction and upload the data files into SAP Ariba Analysis. In the event that assistance is needed extracting data from SAP source systems, which are historically the most difficult, SAP Ariba offers an extract, transform, and load (ETL) tool that extracts data from customer source systems, transforms it into the SAP Ariba Analysis loading format, and helps load the data into SAP Ariba Analysis. The actual product is IBM WebSphere DataStage with SAP adapter. SAP Ariba engineering has built job templates (included in the product when purchased through SAP Ariba) that provide a good starting point for customers with SAP sources.

10.2.3 Data Validation

Upon uploading a data file in SAP Ariba Analysis, the user will receive validation messages in various formats depending upon the severity of the issue. The SAP Ariba project manager will assist by reviewing these validation messages and performing checks on the file formats and contents. The unenriched source data is loaded into SAP Ariba Analysis, where the data is made available for your review and remains available throughout the project. As a result, some valuable aspects of the data will be available to you in the analytic tool within days of project kickoff, depending on whether the raw data can be aggregated. Various spend reports should be run at this time to ensure that the data in SAP Ariba Analysis matches the data extracted and is consistent with your common knowledge of your spend base. Note that data collection and validation is an iterative cycle. As issues are discovered, data will have to be extracted, loaded, and validated again. After you're convinced that the data is acceptable to proceed, your project manager should send a spend approval notice to the designated SAP Ariba project manager. This email is a milestone that completes the data collection/validation phase and moves the project into the next phase of the project—data enrichment.

Again, this cycle time will vary greatly depending upon the scope of the project. However, SAP Ariba has found the average time is two to eight weeks for such an effort, keeping in mind that this time frame depends largely on your ability to complete data collection.

Typical deliverables and activities for the SAP Ariba team and for you, as the customer, in this phase include the following:

- SAP Ariba deliverables and activities:
  - Assist with customer data mapping, collection, and extraction.

  - Train the customer on online data validation messages to find data format issues, missing data, data inconsistencies, and anomalies.
  - Create spend validation reports detailing the amount of spend per source system, supplier, business unit, part, or other attributes defined by the customer.
  - Train users on basic navigation and report creation.
- Customer activities:
  - Review the online data validation messages.
  - Resolve all data issues and file formats to meet the SAP Ariba data acquisition schema requirements.

10.2.4 Data Enrichment

The first step for SAP Ariba data enrichment services is to perform a data assessment. This step provides a more accurate view into the magnitude of the work required based on the quantity and quality of your data. Because the actual quantity and quality of the data often varies significantly from the estimates provided during the sales cycle, project time estimates may change at this point, requiring more time or less time. In most cases, the estimate you receive at kickoff will hold true with minimal changes, that is, assuming no significant changes in the data occur. After the data assessment, SAP Ariba data enrichment services will run 100% of your spend data through SAP Ariba's enrichment engines, including engines that look for content, persistence, and supplier matches.

SAP Ariba data enrichment services will provide results and confidence levels for these results based on the data. These results are what the SAP Ariba data enrichment services team will process, update, and approve over the next several weeks for supplier enrichment and parentage, provided at the midway point of the enrichment phase, and for commodity classifications, which will be available at the first-pass go-live.

Data enrichment makes up a majority of the project timeline due to the volume of processing, review, and quality assurance required. In this phase, the greatest variability exists, depending on the scope of the project. Exceptions exist, but in most cases, data enrichment will require 3 to 10 weeks.

Typical deliverables and activities for the SAP Ariba team and for you, as the customer, in this phase include the following:

- Complete invoice and supplier enrichment.
- Support custom mapping activities.
- Create the data assessment report.
- Update the project timeline based on data enrichment requirements.

- Create a custom taxonomy mapping table and application on SAP Ariba analysis, if applicable.
- Set invoice classifications per SLA.
- Set supplier parentage and enriched information per SLA.

10.2.5 Deployment

The SAP Ariba project manager will work with your project manager to finalize the customizable dashboard, default reports, and user setups as the now-enriched data is deployed for widespread use in the identification of sourcing savings and analytics. This work is performed in the lead time up until deployment. Your project manager should allocate two to five days for this work.

One to two weeks after the first-pass deployment, the SAP Ariba project manager will conduct a spend assessment, which highlights possible opportunities based on the recently available enriched customer data. Note that the data assessment doesn't include customer input or interviews and is therefore more directional in nature; however, it can illustrate several examples of how the data can be analyzed for potential action.

Typical deliverables and activities for the SAP Ariba team and for you, as the customer, in this phase include the following:

- Train the customer on the feedback and refinement process.
- Create a base set of custom report and export templates and dashboards.
- Train the customer on SAP Ariba Analysis, including report templates, dashboards, and user management.
- Create data access controls.
- Publish enrichment results.

Note

Additional deployment services, such as enriching supplier data with diversity and environmental information, and data transformation services to transform data from your customer-specific format to the SAP Ariba format, can be purchased as optional services. Contact your SAP Ariba customer engagement executive for the SAP project for more information.

SAP Ariba Spend Analysis is now live in the customer's landscape and available for sourcing and for category managers and other users to run spend analysis.

In the next section, let's look at how to configure your SAP Ariba Spend Analysis solution.

10.3 Configuring SAP Ariba Spend Analysis

Let's now explore the configurations required to enable the SAP Ariba Spend Analysis solution in your company's landscape.

10.3.1 Configuring File Validation

File data validation is a critical step in spend analysis. In SAP Ariba, there are certain file validation configurations that must be completed. Let's explore these configurations in the following subsections.

Setting a File Validation Error Threshold

There are three levels of error: error, warning, and info. You can specify a threshold for each level of error reported. After SAP Ariba Spend Analysis encounters the limit for that level of error, file validation stops; thus, when the file is uploaded, the number of lines processed won't match the file's size. After SAP Ariba Spend Analysis encounters the limit for warnings and info messages, it stops recording messages but doesn't stop validating.

Following are the three parameters for setting file validation error thresholds:

- `System.Analysis.FileValidation.ErrorThreshold`
  Default = 100
- `System.Analysis.FileValidation.WarningThreshold`
  Default = 200
- `System.Analysis.FileValidation.InfoThreshold`
  Default = 200

Configuring Extended File Validation Reports

Extended file validation includes duplicate checks, dimension reference checks, and other field-specific checks. SAP Ariba Spend Analysis will export the results of the extended file validation to CSV validation reports, which include sample lines and groups of rows where errors occur, as well as detailed messages on the errors.

By default, extended file validation is enabled. You can enable or disable this feature using the following parameter:

- `System.Analysis.FileValidation.ExtendedValidationEnabled`
  The parameters for setting thresholds on the amount of information included in validation reports are as follows:
  - `System.Analysis.FileValidation.NumDuplicateCheckSampleLines`
    Default = 15
  - `System.Analysis.NumDuplicateCheckGroups`
    Default = 15

- `System.Analysis.FileValidation.NumDuplicateCheckMessage`
  Default = 15
- `System.Analysis.FileVali8dation.NumReferenceCheckSampleLines`
  Default = 150
- `System.Analysis.FileValidation.NumReferenceCheckGroups`
  Default = 150
- `System.Analysis.FileValidation.NumReferenceCheckMessages`
  Default = 150

10.3.2 Setting Opportunity Search Date Ranges

In SAP Ariba Spend Analysis, *opportunity search* allows users to perform targeted searches for savings and other organizational opportunities using data ranges. You must configure opportunity search date range settings before users can run opportunity searches. The settings you configure constrain the data set in which users can perform searches, and this setting applies globally to all searches. The default configuration is the *most recent year.*

Setting Up Opportunity Search Date Ranges

In this section, we'll look at how you can set up a date range for searching for opportunities:

1. Log in to SAP Ariba spend management with the analysis administrator role.
2. Click on **Manage • Administration**.
3. In the SAP Ariba administration page, click **Spend Visibility Manager • Opportunity Search**.
4. Under the **Edit Settings** tab, click **Edit**.
5. Specify a date range setting.
6. To set a date range that is automatically updated with new data whenever users view an opportunity search, click **Relative Date Range**, and choose a time period: **Year**, **Quarter**, or **Month**. Then, choose the number of most recent and future time periods you want to include, and select whether you want to include the current time period in the search.
7. To set a specific date range, click **Fixed Date Range** to enter dates or click the calendar icon. Select **Automatically** so that ranges are adjusted to include complete months, to exclude the current partial month, and to optimize search performance.
8. Click **OK**.

Because opportunity search measures are computed, the settings you specify are available in the data load database schema with the next scheduled SAP Ariba Spend Analysis data load that includes postprocessing. Users can then run opportunity searches on the data load schema. After the schemas are switched, users can run opportunity searches on the presentation schema. The opportunity search task's **Data Load Schema** and **Presentation Schema** tabs display the settings that are currently in effect in each database schema.

10.3.3 Configuring the Star Schema Export

You can configure the star schema export process in your system to limit the number of export ZIP files that are stored on the server at one time.

Use the following configuration parameters to set limits on the total number of stored ZIP files and the amount of time they are stored before being automatically deleted:

- `Application.Analysis.MaxStarSchemaExportsKeepTimeInDays`
- `Application.Analysis.MaxBgStarSchemaExportsKept`

10.3.4 Configuring Non-English UNSPSC Code Display

By default, SAP Ariba Spend Analysis displays the United Nations Standard Products and Services Code (UNSPSC) code in English even though the rest of the report is displayed in the user's locale. You can configure SAP Ariba Spend Analysis to display UNSPSC codes in other languages, and users with matching designated locales will see UNSPSC codes in those languages when running reports.

To display non-English UNSPSC codes in reports, you must load the UNSPSC code data for the languages you plan to use. The number of languages other than English in which you can display UNSPSC codes is specified by the following parameter:

- **`System.Analysis.Admin.AllocateMLSColumns`**
  Default = 4

Setting Languages for UNSPSC Code Display

The following parameter specifies the languages you can select for UNSPSC code display: `System.Analysis.ReportMSLLanguageList`.

After you've set other languages for UNSPSC code display, you must reload your data to associate the non-English UNSPSC codes with existing report data. Until you reload the data, reports will continue to display UNSPSC codes only in the previously specified languages.

To set languages for UNSPSC code display, follow these steps:

1. In the SAP Ariba administration page, choose **Reporting Manager • Customer Settings**.

2. On the **Edit Settings** tab, select the languages in which you want to display UNSPSC codes.
3. Click **OK**.

10.3.5 Integrating SAP Ariba Spend Analysis and SAP Ariba Data Enrichment Services

You can integrate SAP Ariba Spend Analysis and SAP Ariba Data Enrichment Services to implement server-to-server communication between the two applications. With server-to-server communication, SAP Ariba Spend Analysis users can export supplier and invoice data enrichment request files directly to a server, where the SAP Ariba Data Enrichment Services team takes over for enrichment and then exports the files back to the server. You then load the enriched data back into SAP Ariba Spend Analysis for use in reports. Email notifications let the appropriate users know when enrichment request files have been exported from SAP Ariba Spend Analysis to the server and when enrichment response files are ready to be loaded back into SAP Ariba Spend Analysis.

To integrate the two solutions, you'll need to perform the following steps:

1. When using your own installed instance of SAP Ariba Data Enrichment Services, see the SAP Ariba Data Enrichment Installation and Migration Guide.
2. When using SAP Ariba Data Enrichment Services, contact SAP Ariba to have your SAP Ariba Analysis integrated to the SAP Ariba Data Enrichment Services.
3. Configure SAP Ariba Spend Analysis to support enrichment activity notification emails and to allow SAP Ariba Data Enrichment Services to recognize and identify your instance of SAP Ariba Spend Analysis.

Integration-Specific Permissions

Two different SAP Ariba Spend Analysis permissions are associated with SAP Ariba Data Enrichment Services:

- `AnalysisDownloadfilesForEnrichment`
  Allows users to export enrichment request files from report pivot tables.
- `AnalysisMonitorEnrichmentFiles`
  Allows users to receive email notifications whenever SAP Ariba Spend Analysis generates an enrichment request file.

In addition to these integration-specific permissions, the `AnalysisMonitorDataFiles` permission allows users to receive include notifications whenever SAP Ariba Data Enrichment Services has pulled an enrichment request file from the SAP Ariba Spend Analysis server and whenever an enrichment response file is uploaded to SAP Ariba Spend Analysis, either through a server pull or a manual file upload.

Configuration Parameter Settings

You should also make sure the following general upstream platform parameters that support notification emails are set correctly:

- `System.Base.SMTPDomainName`
  Identifies the domain of the Simple Mail Transfer Protocol (SMTP) server used for sending notification emails.
- `System.Base.SMTPServerNameList`
  Identifies one or more comma-separated SMTP servers used for sending notification emails.
- `Application.Base.NotificationFromAddress`
  Identifies the name of the user account used as the "From" field in notification emails.

10.3.6 Enrichment Change Request Setting

SAP Ariba Spend Analysis enrichment feedback allows users with the appropriate permissions to submit requests from the report pivot table to change how SAP Ariba Data Enrichment Services enriches data. For example, a user running a report might notice that a supplier is associated with the wrong parent company and can submit a request to correct the mistake. After the request has been approved and sent to SAP Ariba Data Enrichment Services, the request is evaluated, and changes are made to enrichment results where appropriate. These corrections appear in reports when the corrected enrichment response data is loaded for reporting.

You can use enrichment change request settings to enable or disable enrichment feedback, to edit the filters that constrain the data subject to feedback, and to export and archive approved enrichment change requests.

You can modify enrichment change request settings in the **SAP Ariba Spend Analysis Manager** workspace on the administration page. To manage enrichment change request settings, you must have your user preferences set to view data from the data load schema.

Viewing Data from the Data Load Schema

In this section, we'll look at how you can set up preferences to enable viewing data from the data load schema:

1. In the SAP Ariba spend management dashboard, click **Home**.
2. On the command bar, click **Preferences**.
3. Click **Change Report Preferences**.
4. Select the **Use Data Loading Schema** checkbox.

5. Click **Ok** to apply these preferences.
6. Click **Done**.

Enrichment feedback is enabled by default. Users with the `AnalysisSubmitCCR` permission can submit requests to change the enrichment results for items in reports.

Enabling/Disabling Enrichment Feedback

In this section, we'll look at how you can enable/disable enrichment feedback:

1. In the SAP Ariba administration page, choose **Spend Visibility Manager • Enrichment Change Request Settings**.
2. To enable enrichment feedback, select the **Enrichment Change Request Enabled** checkbox. To disable enrichment feedback, deselect the checkbox.
3. Click **Save**.

You can edit enrichment change request filters to constrain the range of reporting data for which users can submit change requests. Users can run reports on data outside of the constraints you apply, but they can only request changes to enrichment results within the filter parameters you set. Enrichment change request filters apply for all users in all reports.

Editing Filters for Enrichment Change Requests

In this section, we'll look at how you can edit filters for enrichment change request:

1. In the SAP Ariba administration page, choose **Spend Visibility Manager • Enrichment Change Request Settings.**
2. Click **Edit Filters.**
3. Specify the date range for the data for which users can submit enrichment change requests:
   - Click **Relative Date Range** to let users submit enrichment change requests for data in a time period relative to the current date. Choose the time period (**Year**, **Quarter**, or **Month**), and then choose the number of most recent and future time periods you want to include. Select whether you want to include the current time period.
   - Click **Fixed Date Range** to let users submit enrichment change requests for data in a specific time period and enter dates or use the calendar icon. Select **Automatically** to adjust the range to include complete months in the fixed date range.
4. In the **Field Browser**, click the **Others** tab.
5. Add fields to the generic report from the **Others** tab, and filter the data as you would in any report. The **Applied Filters** area keeps track of the filters you apply to constrain the set of data for which users can request enrichment changes.

6. Click **Return to Enrichment Change Settings**.
7. Click **Save**.

You can export approved enrichment change requests to SAP Ariba Data Enrichment Services so that enrichment results will be corrected in subsequent data enrichment runs. After exporting the requests, you can archive these change requests so that they won't be included in any subsequent exports of approved requests.

The **Actions** area of the **Enrichment Change Request Settings** page displays the number of currently approved, unarchived enrichment change requests.

Users with the `AnalysisMonitorEnrichmentFiles` permission also receive notifications for the export of approved enrichment change requests. While these requests are being exported, you can move the operation to the background while you perform other tasks.

Exporting Approved Enrichment Change Requests

In this section, we'll look at how you can export approved enrichment change requests:

1. In the SAP Ariba administration page, choose **Spend Visibility Manager • Enrichment Change Request Settings**.
2. Click **Export Approved Requests**.
3. To download the exported request file, click **Download Enrichment File**.
4. Click **Return to Report** to return to the **Enrichment Change Requests** page.

The **Spend Visibility Enrichment Change Requests** folder contains all enrichment change requests. You can archive approved enrichment change requests to move them to the **Archived** subfolder in the **Spend Visibility Enrichment Change Requests** folder. Archived requests remain in the **Archived** folder until you delete them.

Archiving Approved Enrichment Change Requests

In this section, we'll look at how you can archive approved enrichment change requests:

1. In the SAP Ariba administration page, choose **Spend Visibility Manager • Enrichment Change Request Settings**.
2. Click **Archive Approved Requests**.

Viewing Archived Enrichment Change Requests

In this section, we'll look at how you can view enrichment change requests:

1. On the command bar, choose **Search • Knowledge Project**.
2. Click the vault link.

3. Click the **Spend Visibility Enrichment Change Requests** folder, and choose **Open**.
4. Click the **Archived** folder, and choose **Open**.

10.3.7 Managing Data Access Control

Access control determines which users are authorized to see specific SAP Ariba Spend Analysis data in reports. You should only use the access control discussed in this section to control access to SAP Ariba Spend Analysis data; to control access to other data, you should define access control for users in the application where the data originates.

To implement access control, you'll supply an access control class. SAP Ariba defines a Java application programming interface (API) for access control in reports, and the default SAP Ariba application implementation includes one access control implementation, `ariba.analytics.mdql.RoleBasedAccessControlManager`, which reads a set of access control rules from a configuration file. When you use this class, you're implementing data access restrictions by writing rules, which can be based on user name, group, role, or permission.

The upstream platform default configuration for reporting uses the `RoleBasedAccessControlManager` class. However, if you're not using a default configuration, you must change your configuration to use either the default `RoleBasedAccessControlManager` implementation or a custom implementation of your choosing.

Note

Data access control rules alter report queries and can affect report performance, especially if they interfere with materialized views (a database object that contains the result of a query). When you implement data access control rules, you should be aware that these rules may prevent the use of out-of-the-box materialized views and that you might need to implement custom materialized views that support the queries generated by your access control rules to minimize their impact on performance.

Role-Based Access Control Manager

The `RoleBasedAccessControlManager` reads data access control rules from an access control rule configuration file. To use the `RoleBasedAccessControlManager`, set the following parameter:

```
System.Analysis.AccessControlManager = ariba.analytics.mdql.
RoleBasedAccessControlManager
```

When you use the `RoleBasedAccessControlManager`, you can update and maintain your rules from the **Data Access Control** task in the SAP Ariba administration page **SAP Ariba**

Spend Analysis Manager workspace. Users with `AnalysisManageAccessControlRules` permission can access this task.

Administering Access Control Rules

In this section, we'll look at how you can administrator access control rules:

1. In the SAP Ariba administration page, choose **Spend Visibility Manager • Data Access Control**.
2. Click **Export Rules**.
3. Save the exported *AccessControlRules.table* file to the location of your choice.
4. Use a text editor to make any necessary changes to access control rules in the file.
5. Click **Import Rules**, navigate to the location where you saved the *AccessControlRules.table* file, and click **Open**.
6. Click **Upload** to import the modified rules configuration file.

The edits you make to the access control rules take effect immediately after you upload the rules file.

Rule Precedence

Each access control rule can either allow or block access for a user based on the user's user name, permissions, roles, or groups:

- `Allow` means that the user can see only the data specified by that rule, and no other data.
- `Block` means that the user can see all data except for the data specified by that rule.

The data a user can see in reports is defined by a cumulative view created by all the rules that apply to that user in the order they are listed in the access control rule configuration file.

The precedence of rules is as follows:

- If a user's information matches the user name, permission, role, or group in a rule, that rule applies to the user.
- If no rules apply to a given user, no restrictions limit that user's view of the data.

If two rules that apply to a user conflict with each other because they specify `allow` and `block` operations on the same fact/field-path/field-value combination, the last rule to be read in the configuration file is the rule that applies. You should structure rules to avoid this situation.

Note

Rules are evaluated against facts, not dimensions.

10.4 Integrating SAP Ariba Spend Analysis with SAP Analytics Cloud and On-Premise Data Sources

Transactions, especially payments, can be located outside of the SAP Ariba solutions you implement, which is perfectly okay, as long as you can get to the applicable data in these systems and extract it into SAP Ariba Spend Analysis. SAP Ariba offers a number of ETL tools, the main being the SAP Ariba Integration Toolkit, and other services to facilitate this process. In addition to adding data from third-party systems, SAP Ariba Spend Analysis offers a number of key report settings that can be leveraged to tailor reports for your organization. Following are some of the fundamental settings:

- Languages for UNSPSC display in addition to English can be set for up to four additional languages.
- Fiscal calendar settings are only available in sites with SAP Ariba Spend Analysis, and these settings determine how invoice and PO dates translate into fiscal years in reports, including possibly an offset configuration to allow fiscal years to begin at different times other than the calendar year.
- Custom field hierarchies allow for more detailed fiscal settings, such as fiscal months that begin several days into a calendar month.
- Custom fields for measure, date, dimension, and string can be enabled by creating custom fields for your site, loading the data to the data load schema, and then loading the data to the presentation load schema.
- Customizing field labels can be customized fields in SAP Ariba Spend Analysis reports by uploading CSV files with field label strings.

Users in the SAP Ariba Spend Analysis project manager group can customize field labels. Four files are available for string customization:

- The *aml.analysis.HostedSpendExt.csv* file contains flexible field labels.
- The *aml.analysis.InvoiceAnalysis.csv* file contains strings for field labels in the invoice fact table, such as invoice number and invoice spend.
- The *aml.analysis.PurchaseOrderAnalysis.csv* file contains strings for field labels in the PO fact table, such as the PO ID.
- The *spend.aml.analysis.SpendAnalysis.csv* file contains SAP Ariba Spend Analysis–related field labels, such as procurement system and source system.

10.5 Mining Procurement Operations for Data

In the following sections, we'll provide an overview of the types of data that with SAP Ariba Spend Analysis can analyze, the basic facets on which you should consider focusing your analysis, and the work areas available for drilling down into data and key reports. Next, we'll discuss some key reports, along with the areas they can impact.

Finally, we'll cover some key data sources and the options for importing this data into SAP Ariba Spend Analysis.

10.5.1 Data Types

SAP Ariba Spend Analysis provides tools that allow you to analyze the following types of data:

- SAP Ariba Spend Analysis invoice and PO data, which may be new or existing data enriched by SAP Ariba data enrichment with improved supplier and commodity classification.
- SAP Ariba Spend Analysis data also includes common data, such as custom units of measure, currency conversion rates, fiscal hierarchies, taxonomies for commodities and services, diversity certifications, and opportunity search ranges.
- Data from other SAP Ariba spend management solutions such as SAP Ariba Contracts, SAP Ariba Sourcing, and SAP Ariba Supplier Lifecycle and Performance.
- Custom fact data, which can come from external data sources such as SAP ERP systems or SAP Ariba spend management solutions.
- Data from other SAP Ariba spend management solutions, which is updated through regular, automated data pulls. SAP Ariba Spend Analysis and custom fact data is updated through regular data loads.

To make new SAP Ariba Spend Analysis and custom fact data available for reporting and analysis, you'll perform the following high-level steps:

1. **Acquire data.**
   Data may come from data sources such as SAP ERP or other third-party systems but must be consolidated in data files with the correct format.
2. **Load data.**
   After the data is consolidated into data files, designated users can add the files to data load operations and schedule those operations to load the data into the data load database schema.
3. **Switch schemas.**
   After the data load operation is complete, designated administrative users can verify the data and then switch from the data load database schema to the presentation database schema, at which point the data becomes available in reports to all users. Next, an automatic operation copies the data back from the presentation schema into the data load schema so that it's ready for the next data load.

10.5.2 SAP Ariba Spend Analysis: Areas for Analysis

After the data has been loaded and organized, you can proceed to analysis. The main areas for analysis and strategic review are categories, direct material order patterns, noncompliant spend (or "maverick" spend), supplier relationship management, and spending by brands. Using these insights can further reduce costs, focus and optimize your supplier base, and help you understand supplier vulnerabilities and criticalities, as shown in Figure 10.4.

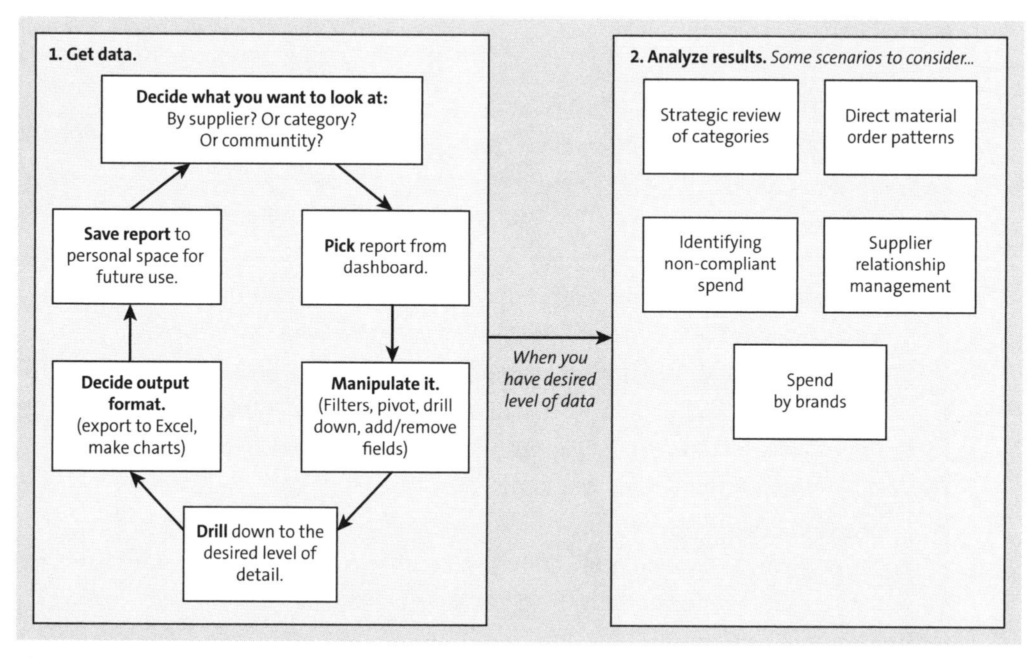

Figure 10.4 SAP Ariba Spend Analysis: Process

Ideally, these insights are funneled back into the SAP Ariba set of solutions (SAP Ariba Sourcing, SAP Ariba Buying and Invoicing), SAP Ariba Invoice Management professional, etc.) to optimize your procurement operations further in the system, cycle by cycle. For example, if you identify that most noncompliant spend is for items that haven't been categorized or aren't in the catalogs or contracts available, then adding these items to the system and communicating these changes to your users should significantly improve this area. If you rely solely on a single supplier for a key input or category, that supplier is in a position to raise prices or impact your supply chain. Thus, efforts should be made to invite, vet, and add similar suppliers to your environment. Similarly, if you have a growing number of suppliers delivering the same types of goods and services, standardization and optimization efforts will help focus your supply chain on key suppliers, drive up your bargaining power with volume increases, and ultimately drive down costs.

10.5.3 Work Areas

SAP Ariba Spend Analysis offers a variety of work areas for drilling down into your data and generating key reports, such as the following:

- **SAP Ariba Spend Analysis Manager workspace**
  You can use the SAP Ariba Spend Analysis Manager workspace to manage data related to SAP Ariba Spend Analysis reports. The following authorization groups have access to the manager workspace:
  - Customer administrator (everything except the opportunity search task)
  - SAP Ariba Spend Analysis project manager (everything except the opportunity search task)
  - SAP Ariba Spend Analysis category change request (CCR) reviewer (enrichment change request settings task)
  - SAP Ariba Spend Analysis opportunity analyst (opportunity search task)
- **Source systems**
  A source system represents a distinct data source for your reporting data. Each source system has a source type that defines the format of the data.

  Members of the SAP Ariba Spend Analysis project manager group have access to upload files and load data into all source systems, but members of the SAP Ariba Spend Analysis data file manager group only have access to upload files and load data into their assigned source systems.
- **Import/export star schema**
  You can import and export the database star schema to synchronize your database with external systems or with the database star schemas in your SAP Ariba solutions.

  Your system has a dedicated database star schema if it includes SAP Ariba Spend Analysis or custom reporting facts.
- **Enrichment data**
  Enrichment data improves commodity and supplier classifications, including a more detailed classification of commodities and services.

  Members of the SAP Ariba Spend Analysis project manager group will receive email notifications when users generate enrichment request files.
- **Data access control**
  Data access rules determine which users are authorized to see specific SAP Ariba Spend Analysis data. You can restrict access to SAP Ariba Spend Analysis data in reporting facts by writing access control rules based on user name or group.

  All changes to rules take effect immediately after you import the rules file.
- **Enrichment change request settings**
  The feedback loop with SAP Ariba for correcting and adjusting classifications in the data is the enrichment change process. To control access to feedback submission, you can use enrichment change request settings to enable or disable enrichment feedback, to manage when newly approved enrichment changes are loaded to the presentation schema, and to manage the current in your site. For example, a user

might submit an enrichment change request to correct a report that displays the wrong commodity classification for a particular commodity from a supplier.

This section also displays all the rules that have been generated by currently submitted enrichment change requests.

- **Manage benchmarking**
SAP Ariba customer support loads benchmarking data to use with your SAP Ariba Spend Analysis reports.
- **Opportunity search**
Opportunity searches are targeted searches, based on commodities, that highlight opportunities for savings, improved efficiency, supplier diversity, and other company goals in your spend data. You can run prepackaged opportunity searches or create your own opportunity search.
You use the opportunity search task to configure the accounting date range for opportunity search settings, as shown in Table 10.3.

Opportunity Search Type	Description
Price variation	Identify areas for savings through more effective choice of suppliers.
Supplier fragmentation	Identify areas for savings through supplier consolidation.
Order fragmentation	Identify inefficient purchasing in your company.
Opportunities for sourcing	Identify commodities for which a large sourcing event might achieve savings.

Table 10.3 Opportunity Search Types

10.5.4 Key Reports and Corresponding Impact Areas

In addition to understanding general spending, several reporting areas in SAP Ariba Spend Analysis can have immediate impact and ROI for procurement solutions. Areas such as contract awareness, rationalizing pricing across contracts, and understanding supplier ownership structures to consolidate further volume for contract negotiations, all provide relatively quick returns with little required change management or process changes. In essence, you're already working the processes, but "leakage" occurs due to a lack of understanding or visibility in these areas.

Contract Awareness Report

The Contract Awareness report, shown in Figure 10.5, is built in SAP Ariba Spend Analysis using purchasing price variance (PPV) data. You define the variance as where the same supplier billed two different prices for the same item. Another metric is different customer sites purchasing at different prices, which is called purchase price alignment

(PPA). After this data is defined, additional data points can be calculated, such as supplier optimization cost (SOC) or the savings associated from always buying from the supplier that offers you the most favorable pricing.

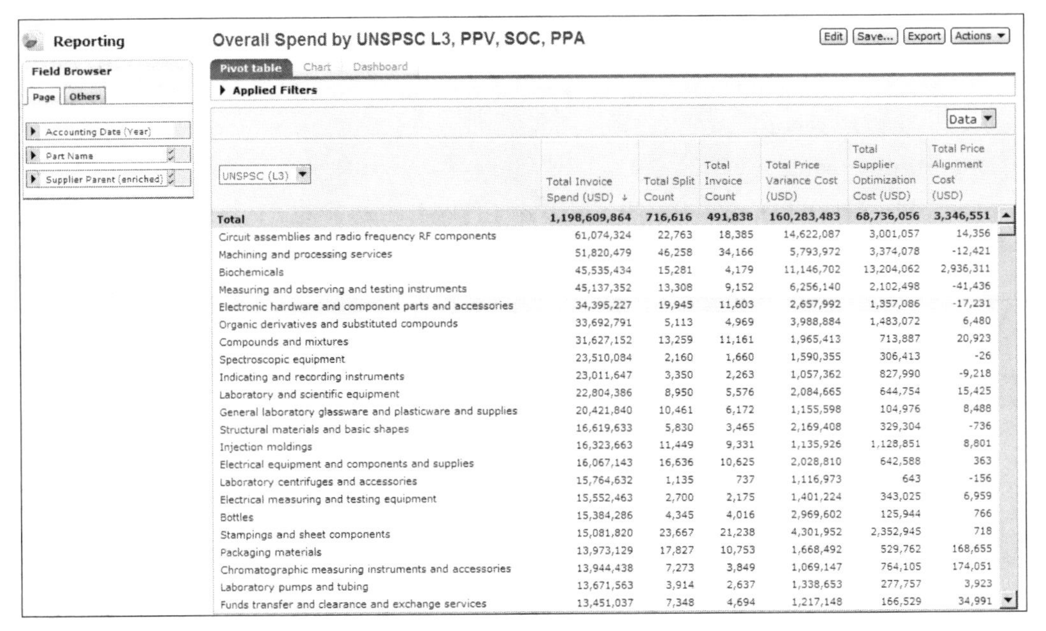

UNSPSC (L3)	Total Invoice Spend (USD) ↓	Total Split Count	Total Invoice Count	Total Price Variance Cost (USD)	Total Supplier Optimization Cost (USD)	Total Price Alignment Cost (USD)
Total	**1,198,609,864**	**716,616**	**491,838**	**160,283,483**	**68,736,056**	**3,346,551**
Circuit assemblies and radio frequency RF components	61,074,324	22,763	18,385	14,622,087	3,001,057	14,356
Machining and processing services	51,820,479	46,258	34,166	5,793,972	3,374,078	-12,421
Biochemicals	45,535,434	15,281	4,179	11,146,702	13,204,062	2,936,311
Measuring and observing and testing instruments	45,137,352	13,308	9,152	6,256,140	2,102,498	-41,436
Electronic hardware and component parts and accessories	34,395,227	19,945	11,603	2,657,992	1,357,086	-17,231
Organic derivatives and substituted compounds	33,692,791	5,113	4,969	3,988,884	1,483,072	6,480
Compounds and mixtures	31,627,152	13,259	11,161	1,965,413	713,887	20,923
Spectroscopic equipment	23,510,084	2,160	1,660	1,590,355	306,413	-26
Indicating and recording instruments	23,011,647	3,350	2,263	1,057,362	827,990	-9,218
Laboratory and scientific equipment	22,804,386	8,950	5,576	2,084,665	644,754	15,425
General laboratory glassware and plasticware and supplies	20,421,840	10,461	6,172	1,155,598	104,976	8,488
Structural materials and basic shapes	16,619,633	5,830	3,465	2,169,408	329,304	-736
Injection moldings	16,323,663	11,449	9,331	1,135,926	1,128,851	8,801
Electrical equipment and components and supplies	16,067,143	16,636	10,625	2,028,810	642,588	363
Laboratory centrifuges and accessories	15,764,632	1,135	737	1,116,973	643	-156
Electrical measuring and testing equipment	15,552,463	2,700	2,175	1,401,224	343,025	6,959
Bottles	15,384,286	4,345	4,016	2,969,602	125,944	766
Stampings and sheet components	15,081,820	23,667	21,238	4,301,952	2,352,945	718
Packaging materials	13,973,129	17,827	10,753	1,668,492	529,762	168,655
Chromatographic measuring instruments and accessories	13,944,438	7,273	3,849	1,069,147	764,105	174,051
Laboratory pumps and tubing	13,671,563	3,914	2,637	1,338,653	277,757	3,923
Funds transfer and clearance and exchange services	13,451,037	7,348	4,694	1,217,148	166,529	34,991

Figure 10.5 SAP Ariba Spend Analysis: Contract Awareness Report

As shown in Figure 10.5, the PPV, PPA, and SOC totals are available on the Contract Awareness report. Keep in mind that these amounts, if saved, could represent immediate additional profitability to your company but may require significantly more in sales/revenues to realize, depending on the company's net margin. So, if $10 million in savings is realized via these insights in procurement, and your company is working with a 10% net margin, you would have to sell 100 million more of products/services to realize this amount.

Supplier Parentage Report

Different types of procurement situations may include this type of opportunity. For example, if you're buying a product category that is supplied by an industry undergoing a lot of mergers and acquisitions activity, you may be unknowingly buying from subsidiaries of the same parent company at different prices and on different contracts. Or, you may be buying off of multiple contracts for the same product for the same supplier and would thus have an opportunity to consolidate that way. If the suppliers are conglomerates, or if the supplier base in this industry is heavily dominated by a few key players, fewer opportunities may arise for this type of report to uncover. In addition, if the owner of the supplier is an investment entity or conglomerate, this type of contract roll up/consolidation will be less fruitful.

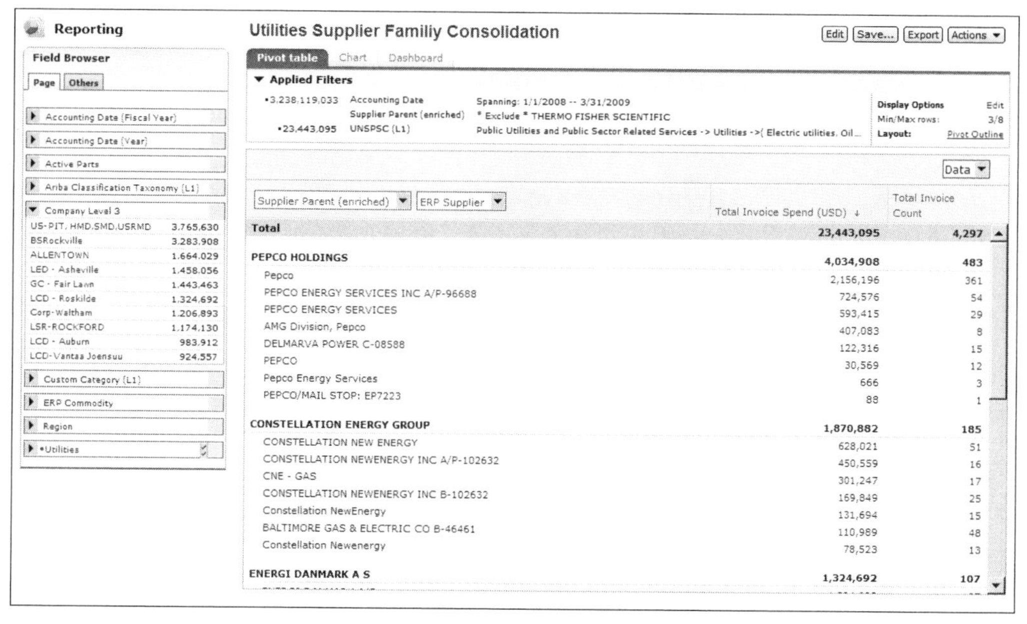

Figure 10.6 Supplier Parentage Report: Utilities

Supplier parentage can be brought to bear on negotiated utilities contracts, as shown in Figure 10.6, and on spend volume reports. The report shows which utilities share a common parent as well as your spend volumes and indicates opportunities to consolidate your contracts as parent-to-parent versus location to subsidiary.

The Supplier Parentage report is also where spend volumes and trends come into play—if you're spending higher amounts than before, include this fact in the next negotiation round as leverage and justification for obtaining better pricing and rates. Likewise, understanding which product categories are influenced by price changes in terms of how much your company eventually purchases (price/volume elasticity), and which categories are largely price insensitive, can determine whether you look for external price reductions from your supplier or for internal measures to curb demand for hat category. If your company's volume of buying in a category is largely driven by price, achieving a price reduction from the supplier will lead to an almost commensurate increase in purchasing, negating the savings effects. Having pricing and category information, as well as historical trend data in the form of volume over time by price, can assist you with understanding whether reducing volume or reducing price leads to the greatest savings. Follow-on reports are available in SAP Ariba Spend Analysis, such as the Spend Volume report and the Spend Variance Analysis Volume vs. Price Effects report.

Supplier Fragmentation Report

Fragmented markets often represent the best sourcing opportunities. From an economics standpoint, this means that with the correct analysis and understanding of the market, you stand to obtain pricing at or even below cost. A fragmented supply market typically encompasses many competitors vying for your business in that product or service category. Conversely, if the market for an item is split neatly between a few large suppliers, your negotiating leverage may be severely reduced. No matter how well laid out your arguments are, you may still end up paying what the supplier asked for initially because you don't have alternatives. Figure 10.7 shows an example supplier fragmentation report.

After you account for geographical or other justifications for a larger-than-normal number of suppliers providing a category, you'll next determine whether your sourcing and cost-saving strategies for a particular category have potential from a supplier availability and overall market standpoint.

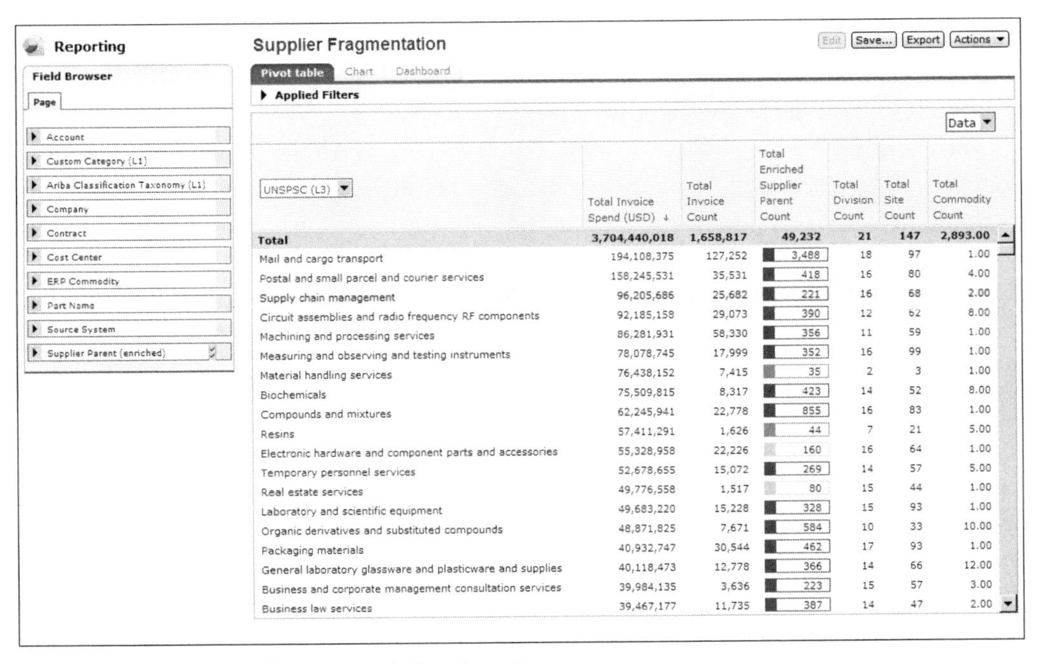

UNSPSC (L3)	Total Invoice Spend (USD) ↓	Total Invoice Count	Total Enriched Supplier Parent Count	Total Division Count	Total Site Count	Total Commodity Count
Total	**3,704,440,018**	**1,658,817**	**49,232**	**21**	**147**	**2,893.00**
Mail and cargo transport	194,108,375	127,252	3,488	18	97	1.00
Postal and small parcel and courier services	158,245,531	35,531	418	16	80	4.00
Supply chain management	96,205,686	25,682	221	16	68	2.00
Circuit assemblies and radio frequency RF components	92,185,158	29,073	390	12	62	8.00
Machining and processing services	86,281,931	58,330	356	11	59	1.00
Measuring and observing and testing instruments	78,078,745	17,999	352	16	99	1.00
Material handling services	76,438,152	7,415	35	2	3	1.00
Biochemicals	75,509,815	8,317	423	14	52	8.00
Compounds and mixtures	62,245,941	22,778	855	16	83	1.00
Resins	57,411,291	1,626	44	7	21	5.00
Electronic hardware and component parts and accessories	55,328,958	22,226	160	16	64	1.00
Temporary personnel services	52,678,655	15,072	269	14	57	5.00
Real estate services	49,776,558	1,517	80	15	44	1.00
Laboratory and scientific equipment	49,683,220	15,228	328	15	93	1.00
Organic derivatives and substituted compounds	48,871,825	7,671	584	10	33	10.00
Packaging materials	40,932,747	30,544	462	17	93	1.00
General laboratory glassware and plasticware and supplies	40,118,473	12,778	366	14	66	12.00
Business and corporate management consultation services	39,984,135	3,636	223	15	57	3.00
Business law services	39,467,177	11,735	387	14	47	2.00

Figure 10.7 Supplier Fragmentation Report

Additional Reports

The Supplier Footprint report provides several useful data points with invoicing activity, as well as commodity and site/division count on a supplier, with the goal of facilitating consolidation. Invoicing activity highlights opportunities to streamline ordering with top suppliers, including drilldown by category. Commodity counts uncover niche suppliers, whose spend could potentially be consolidated with a more diverse supplier. Site/division count shows how entrenched, or intertwined, a supplier is with your company; this is important information because a supplier with a large number of relationships with your company's sites will be more difficult to replace with another.

In the short term, moving away from a heavily entwined or favored supplier may not be feasible, but understanding from a long-term strategy approach may be the first step in eventually achieving dissolution or independence from the relationship. Likewise, the Spend Concentration report allows you to review categories of spend where you currently have too little competition and too much dependence on one supplier.

Finally, as with demand management, the insights gleaned from these reports should drive changes in procurement behaviors in your organization, rather than solely supplier-focused rationalizations. To help change procurement behaviors, SAP Ariba Spend Analysis provides prepackaged reports such as three PO vs. Non-PO Spend reports; an Off Contract Spend report by commodity/organization unit/supplier; and an Organizational Analysis report from source systems.

Understanding multiple aspects of a procurement scenario and finding is necessary to making an informed decision. You could simply act on the first report showing that you're overly dependent on a supplier who is underperforming. However, before using this report to justify a host of other actions, you'll need to understand (via other reports in SAP Ariba Spend Analysis) just where this supplier fits in, both internally at your company and externally in the market. For example, if you don't take into account the supplier's relationship level with your company from a site standpoint, or if you ignore the supplier market makeup for the product being supplied, a sudden "rip and replace" move by the supplier could have negative consequences that outweigh the sought-after savings. Likewise, if you chase a savings target on a commodity via negotiating a price reduction with the supplier without understanding your company's price elasticity with that commodity, achieving a reduction in price could simply lead to more wasteful usage of that commodity and no real savings, due to increased purchasing volumes.

As shown in Table 10.4, analysis and cross-functional insight are required when leveraging these data points and reports from SAP Ariba's analytics tools to take action. When pursuing a savings opportunity, be sure to assess not only the opportunity but also the business impact, supply risk, ease of implementation, and savings potential. To fully understand the savings potential, you'll need to consider additional historical factors in your spend volumes, supply base concentration, spending concentration with supplier, and savings history, as shown in Figure 10.8.

Business Impact	Supply Risk	Savings Potential	Ease of Implementation
▪ Total cost ▪ Customer value ▪ Product differentiation ▪ Performance ▪ Technology ▪ Safety, governmental, industrial regulations	▪ Market depth and intensity ▪ Pressure from substitutes ▪ Bargaining power ▪ Off-shore supply ▪ Entry barriers ▪ Constraints ▪ Corporate buying power ▪ Intellectual property ▪ Outsourcing services ▪ Logistics, inventory, and lead time	▪ Spending characteristics ▪ Specification accuracy ▪ Specification development ▪ Supply market analysis ▪ Competitive sourcing ▪ Saving history ▪ Labor as percentage of total cost ▪ Raw material pricing trends	▪ Contract length ▪ Switching costs ▪ On-site requirements ▪ Organizational sensitivity ▪ Technical complexity ▪ Availability of expertise ▪ Customer imposed constraints

Table 10.4 SAP Ariba Spend Analysis: Insight-Driven Action

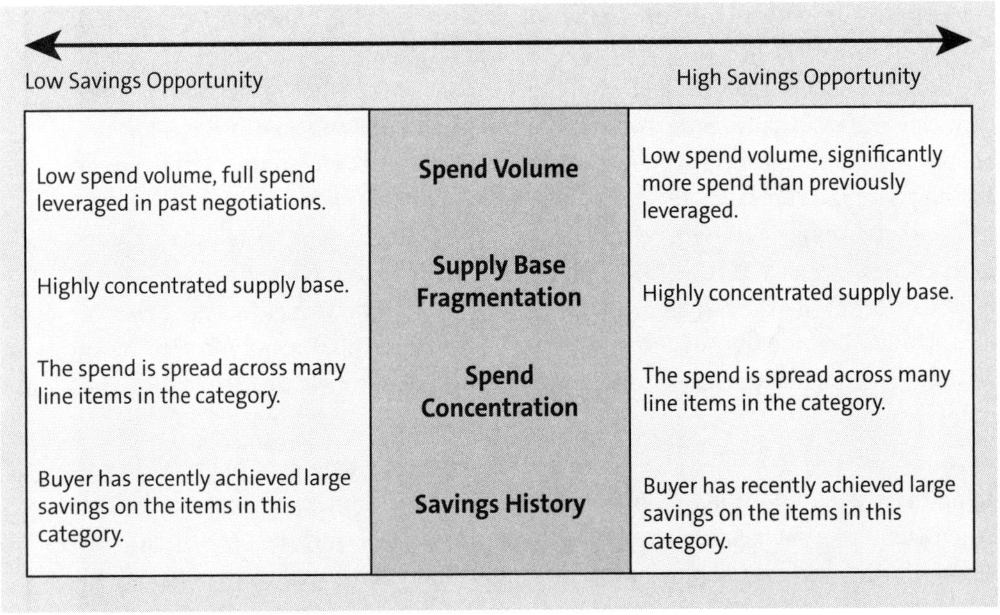

Low Savings Opportunity		High Savings Opportunity
Low spend volume, full spend leveraged in past negotiations.	**Spend Volume**	Low spend volume, significantly more spend than previously leveraged.
Highly concentrated supply base.	**Supply Base Fragmentation**	Highly concentrated supply base.
The spend is spread across many line items in the category.	**Spend Concentration**	The spend is spread across many line items in the category.
Buyer has recently achieved large savings on the items in this category.	**Savings History**	Buyer has recently achieved large savings on the items in this category.

Figure 10.8 Key SAP Ariba Spend Analysis: Factors for Sourcing Savings

Insights often seem obvious once stated but hidden up until that point. Similarly, if you aren't spending a lot of money with a supplier or type of item in the category, prioritizing this area for analysis or savings initiatives doesn't make sense. When only a few suppliers operate in a market, you're dealing with an oligopoly, and oligopolies don't have to negotiate as hard as perfectly competitive markets. Finally, if you've recently secured large savings in a category or with a supplier, the savings opportunities may very well be exhausted for this area at the moment.

Only through measured analysis of these vital areas will you likely achieve the desired results and realize the actual opportunity estimated from the data. Hasty decision-making based on one-dimensional reporting can resemble a game of whack-a-mole, with constant effort only surfacing other problems to replace the targeted one.

10.5.5 Key Data Sources and Options for Importing into SAP Ariba Spend Analysis

SAP Ariba spend management and SAP Ariba Spend Visibility use the concept of a source system,, which isn't necessarily a separate system but is more of a mechanism for segregating data. SAP Ariba spend management reporting uses source systems to represent distinct data sources. Each source system has a source type that defines the format of the data loaded into it. Users in the SAP Ariba Spend Analysis project manager group can add and delete source systems and can manage which source systems users in the SAP Ariba Spend Analysis data file manager group can access on the SAP Ariba Spend Analysis manager administration page.

Depending on the solutions implemented, your site will use some combination of the source systems shown in Table 10.5.

Source Type	Source System Name	Description
SAP Ariba spend management	SAP Ariba spend management	Transactional data and data on projects and tasks from SAP Ariba Contracts, SAP Ariba Sourcing, and SAP Ariba Supplier Lifecycle and Performance data. Automated data pulled from these solutions is loaded into this source system. In SAP Ariba spend management solutions that don't include SAP Ariba Spend Analysis, you always load custom fact data into the SAP Ariba spend management source system.
Global	Default	Global data, such as master data, from SAP Ariba Contracts, SAP Ariba Sourcing, SAP Ariba Spend Analysis, and SAP Ariba Supplier Lifecycle and Performance. Star schema ZIP files are loaded into this source system.

Table 10.5 Source System Types

Source Type	Source System Name	Description
SAP Ariba Spend Analysis	Defined upon creation of the source system	Users in the SAP Ariba Spend Analysis project manager group can create different source systems of type SAP Ariba Spend Analysis for loading SAP Ariba Spend Analysis data files. You can also load custom fact data into an SAP Ariba Spend Analysis source system.
Global	Self-service procurement: None	Global, unpartitioned data from SAP Ariba Invoice Management and SAP Ariba Procurement solutions. Automated data pulls from these solutions are loaded into this source system.
Buyer-generic	Self-service procurement: Generic	Generic format data for SAP Ariba Invoice Management and SAP Ariba Procurement solutions. Automated data pulls from these solutions are loaded into this source system.
Buyer: SAP	Self-service procurement: SAP	SAP format data for SAP Ariba Invoice Management and SAP Ariba Procurement solutions. Automated data pulls from these solutions are loaded into this source system.
Buyer: Psoft	Self-service procurement: Psoft	PeopleSoft format data for SAP Ariba Invoice Management and SAP Ariba Procurement solutions. Automated data pulls from these solutions are loaded into this source system.

Table 10.5 Source System Types (Cont.)

You can add or delete source systems of the SAP Ariba Spend Analysis source type, but you can't add or delete SAP Ariba spend management, global, or buyer source systems. You can add a single source system at a time, or you can add batches of source systems by uploading a source system via a CSV file. Source system names can't be longer than 20 alphanumeric characters or contain spaces.

Custom reporting facts are available in the following:

- SAP Ariba Contracts professional
- SAP Ariba Sourcing professional
- SAP Ariba Spend Analysis professional
- SAP Ariba Supplier Management
- SAP Ariba Supplier Lifecycle and Performance

SAP Ariba spend management solutions include a set of reporting facts to store data about the basic transactions that users are investigating when they run a report. These facts include invoice, PO, contract workspace (procurement), event item summary, and others. SAP Ariba spend management automatically loads data into these facts from your solution package at regular intervals. In SAP Ariba Spend Analysis solution packages, SAP Ariba Spend Analysis project managers can load external invoice and PO data into SAP Ariba Spend Analysis for analytical reporting.

Custom reporting facts aren't enabled by default. To enable custom reporting facts, you'll need to work with SAP Ariba services. You can use custom reporting facts to load other kinds of data from third-party systems into SAP Ariba spend management and then run analytical reports to show this data side-by-side with SAP Ariba spend management data. You can use custom reporting facts to load the following:

- Third-party supplier quality data, such as percentage of claims in total freight cost, for use in KPIs in SAP Ariba Supplier Lifecycle and Performance scorecards.
- Third-party supplier risk data for use in surveys.
- Third-party savings pipeline and tracking data for reporting alongside SAP Ariba Sourcing project data.
- Third-party external contract data for reporting alongside SAP Ariba Spend Analysis invoice and PO data.
- Third-party order fulfillment data for reporting alongside SAP Ariba Contracts management workspaces, SAP Ariba Sourcing projects, or SAP Ariba Supplier Lifecycle and Performance projects.
- Spend forecast data for reporting alongside SAP Ariba Spend Analysis invoice spend data.
- SAP Ariba supplier data in a separate supplier fact for drilling down and filtering supplier data by commodity category, region, minority-owned status, and so on. By default, SAP Ariba stores supplier data in a dimension, which doesn't allow for this kind of analytical reporting.

As with other aspects of database design, custom reporting facts require careful planning and analysis. After you create a custom fact, it can't be deleted in the system, and modifications to custom fact data are subject to limitations. For example, you can overwrite existing data with new rows of data that use the same lookup key values as existing data, and you also can add new data to the existing data set. However, you can't delete existing data.

Because of these limitations, custom facts are most suitable for data that is static or that changes slowly. For example, data from completed rounds of supplier performance evaluations is unlikely to change. You will, however, continue to add to it. Supplier data changes infrequently, and even if your company stops doing business with a supplier, you'll want to retain that supplier's record for archival purposes. On the other hand, you're likely to run up against these limitations with large-volume data sets that change rapidly, including procurement documents such as invoices, requisitions, and POs.

Custom facts have further limitations in that they don't include any default fields for common measures, such as spend and count, and they don't support currency conversions. You must create all the measure fields you want when you create the custom facts.

If you've enabled the custom fact feature or are implementing SAP Ariba Spend Analysis, you'll store data in a dedicated star schema. This star schema uses fact tables to store specific types of records, such as invoices, projects, or surveys, and it uses dimension tables to store records that are common to most or all facts, such as suppliers, commodities, and regions.

As shown in Figure 10.9, dimensions can contain different levels of data organized in top-down hierarchies that progress from general to specific. Report queries associate fact and dimension tables so that you can drill down, navigate hierarchy levels, add and remove hierarchy fields, and perform other analytical tasks in reports.

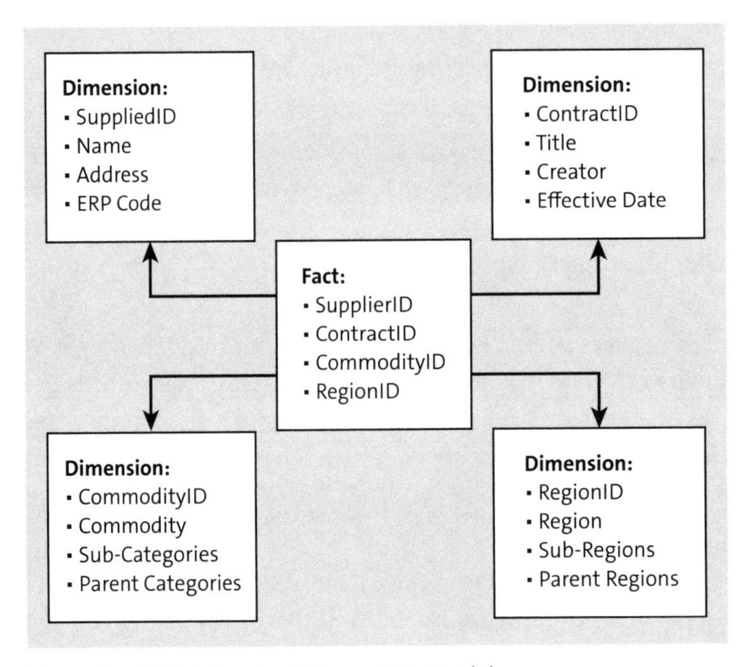

Figure 10.9 SAP Ariba Star Schema Data Model

10.5.6 Creating Sourcing Requests from a Spend Analysis Report

When a customer has both SAP Ariba Sourcing and SAP Ariba Spend Analysis enabled, this new feature enables users to create a sourcing request from a Spend Analysis report based on invoice or PO facts. After this feature is enabled by toggling the parameter `SourcingSpendVisibilityIntegration`, the **Create Sourcing Request** option appears in the **Actions** menu, as shown in Figure 10.10.

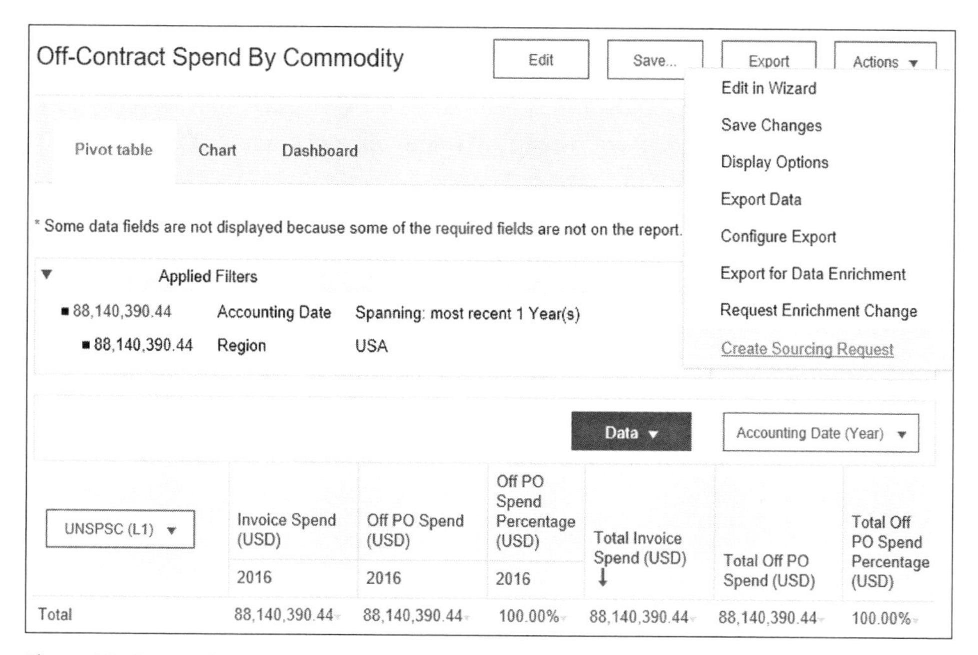

Figure 10.10 Sourcing Request from a Spend Analysis Report

The following information will be copied over to the sourcing request when available on the report:

- UNSPSC (lowest level will be copied to the **Commodity Code** field in sourcing request)
- Region
- Part description
- UOM (if aligns with a sourcing unit of measurement [UOM]; if not, defaults to **Each**)
- Quantity
- Historic price (computed price based on formula = Invoice spend/Invoice qty)

Sourcing request origin (set to **Spend Visibility**)

10.6 Summary

Like SAP Ariba Supplier Lifecycle and Performance, SAP Ariba Spend Analysis was developed to address more complex reporting requirements in procurement than the one-dimensional reports available in various solution areas. SAP Ariba Spend Analysis combines disparate data sources and purchasing documents with actual spend/payment data to provide managers and executives with a clear understanding of just where the money is going in procurement. Whether in procurement or in politics, following the money is always a good starting point. What is gleaned from SAP Ariba

Spend Analysis reporting can then be used to achieve greater savings through targeted initiatives. You should use further analysis with SAP Ariba Spend Analysis to focus on what is important and useful for your organization, using market analysis and reports in the supplier side of the system to identify where the most savings can be achieved with the least amount of effort. As with the other analysis tools in SAP Ariba Supplier Lifecycle and Performance, how you interpret and what you do with the insights is just as important as getting to them in the first place.

Chapter 11
SAP Ariba Integration

In this chapter, we'll explain some core topics in integrating SAP ERP and SAP S/4HANA with SAP Ariba solutions by providing overviews of the main integration approaches, scenarios, and process areas/integration points within SAP Ariba and SAP ERP.

Today, most large enterprises run hybrid models in which cloud solutions are integrated with ERP systems residing either in the cloud or on-premise, such as SAP S/4HANA for core master data updates and process integration. For many processes and solution areas in SAP Ariba, financial and master data, such as supplier data, reside in the connected ERP environments. Upon completing a transaction in SAP Ariba and issuing a purchase order (PO), fulfillment steps are supported and mediated in SAP Ariba with follow-on processes, such as payment and inventory management, occurring in the backend ERP. To make the handoffs and data updates seamless between the two environments, multiple integration points and approaches will need to be set up. In addition to indirect procurement areas, SAP Ariba also supports direct procurement via SAP Ariba Supply Chain Collaboration for Buyers.

In this chapter, we'll outline the various tools available for integrating SAP Ariba with SAP S/4HANA and SAP ERP systems, as well as describe some key procurement integration scenarios applicable to the SAP Ariba and SAP ERP environments.

By integrating the process in this manner with the Ariba Network, your organization can reduce transaction costs, increase the speed of procurement processes, reduce manual steps, increase productivity, and enforce compliance. For example, suppliers can't submit nonconforming invoices via the Ariba Network. The invoice must adhere to customer-defined rules to be entered at all.

11.1 SAP Ariba Integration Projects and Connectivity Options

Every integration project is different, depending on the customization, the customer's unique landscape, and requirements surrounding data and security. In addition, customers often have built-out procurement processes on the planned procurement side for their direct procurement in manufacturing and production operations running in the ERP system. With regard to integration, many standard integrations are now native between SAP S/4HANA and SAP Ariba using prebuilt web services, whereas other integrations still require build-outs using the core integration platforms and technologies available. The main integration technologies are the SAP Ariba Cloud Integration Gateway, SAP Process Integration and SAP Process Orchestration, and SAP Cloud Platform Integration. SAP Ariba Cloud Integration Gateway is the preferred platform for all integration approaches and investment at SAP Ariba, whereas SAP Cloud Platform Integration, SAP Process Integration, and SAP Process Orchestration continue to be used by many customers within the integration layer.

Note

Before embarking on an integration effort, please visit SAP's Best Practices area for the latest SAP Ariba integration documentation by scenario and SAP Ariba solution at *https://rapid.sap.com/bp/*.

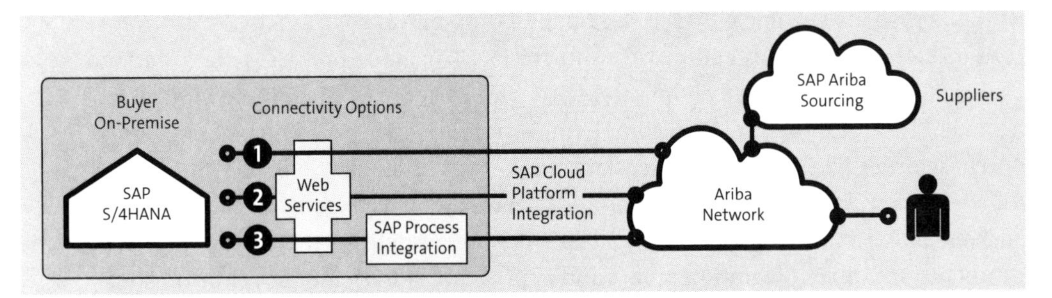

Figure 11.1 SAP S/4HANA Integration with SAP Ariba: Connectivity Options

As shown in Figure 11.1, the general connectivity options between SAP S/4HANA and SAP Ariba focus on the Ariba Network and SAP Ariba Sourcing. Web services–based connections exist for the following:

- SAP Ariba Sourcing
- Automated purchase-to-pay (P2P) with SAP Ariba Commerce Automation (PO/invoice automation)
- SAP Pay and SAP Ariba Discount Management collaboration

In addition to integrating directly via web services, integration can also support mediation for the web service via SAP Cloud Platform Integration or SAP Process Integration and SAP Process Orchestration (mediation possible for SAP S/4HANA). The supported transactions are as follows:

- Requests for quotation (RFQs)
- Purchase orders (POs)
- Order confirmations
- Advance shipping notifications (ASNs)
- Goods receipt notices

- Service entry sheets
- Supplier invoices
- Invoice updates
- Payment proposals
- Proposal acceptances
- Remittance advices and updates

The data flows between SAP S/4HANA and SAP Ariba are bidirectional, whether via SAP Cloud Platform Integration, SAP Process Integration, or directly via web services, as shown in Figure 11.2.

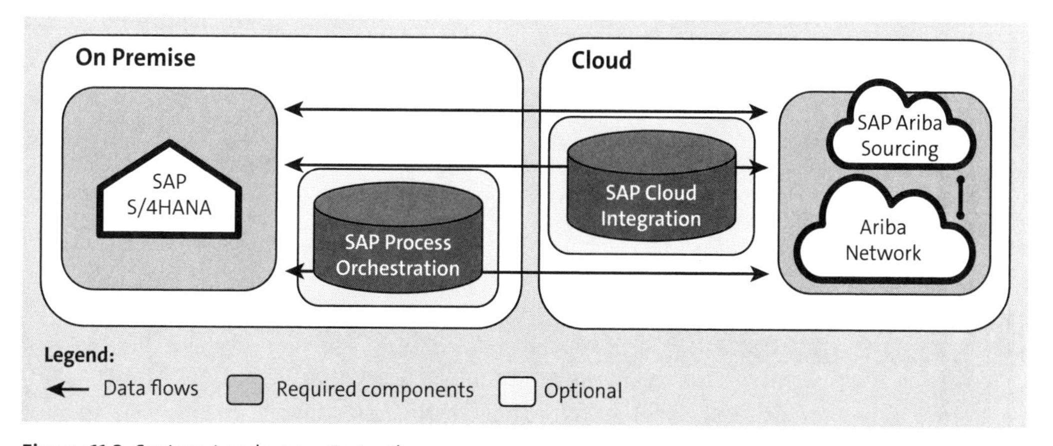

Figure 11.2 System Landscape: Data Flows

The same applies for SAP Ariba Cloud Integration Gateway, detailed later in Section 11.5.

The software required for enabling these integrations consists of the following:

- First, you must have an SAP Ariba buyer account, which is included in the PO and invoice automation, collaborative commerce, and collaborative finance subscriptions.
- For direct connectivity via prebuilt web services, no further software is required.
- For mediated web service connectivity via SAP Cloud Platform Integration, an SAP Cloud Platform Integration subscription is required.
- For mediated web service connectivity via SAP Process Integration and SAP Process Orchestration, SAP NetWeaver 7.5 (or higher) with Ariba Network Adapter for Ariba Network integration is required.

In addition to a cloud-based/on-premise integration, an organization may also need to integrate some cloud solutions with other cloud solutions, as shown in Figure 11.3. For integrating cloud to cloud, an organization may use SAP Cloud Platform Integration. In turn, SAP Cloud Platform Integration can then be used to integrate to an on-premise instance, thus negating the need for additional middleware such as SAP Process

Integration. In this scenario, the organization is better off choosing a single middleware platform if possible, which would sway the argument toward SAP Cloud Platform Integration over SAP Process Integration.

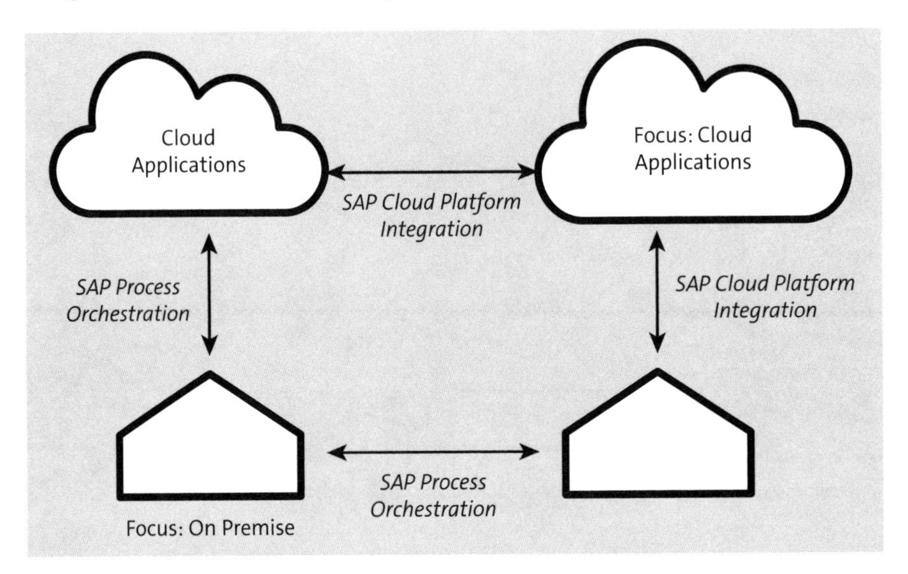

Figure 11.3 SAP Integration Paths

Thus, multiple integration scenarios applicable or tangential to the ERP-SAP Ariba integration topic exist. The deciding factors include the following:

- **Dominant system landscape**
  Is the ERP environment on premise or in the cloud? What are the plans for transforming the ERP area toward moving to the cloud?
- **Prepackaged content availability**
  What is the integration path requiring the least amount of custom work?
- **Total cost of ownership (TCO)**
  How much investment will be required to maintain the linkages and integrations going forward? Are additional upgrades in either SAP Ariba or SAP S/4HANA required that could impact the integration and require rework?
- **Go-live timelines**
  Which approach will yield the quickest integration?
- **Connectivity options**
  Are there firewall constraints that influence or impact the options available?

A thorough analysis of the processes/solution areas you want to integrate between SAP Ariba and SAP S/4HANA also should be conducted. Three core solution areas in particular should be evaluated. The first area is PO/invoice automation with SAP S/4HANA and the Ariba Network. This scenario is the most common and can yield immediate results and return on investment (ROI), as transactions are issued out of SAP S/4HANA, processed in the Ariba Network, and finally sent back to SAP S/4HANA.

The second integration area is SAP Ariba Sourcing connected to SAP S/4HANA. If something needs to be sourced during a transaction in SAP S/4HANA, you can leverage the rich sourcing functionality in SAP Ariba Sourcing to find the optimal suppliers, and then revert the item back to SAP S/4HANA for concluding the transaction (or take it forward in the SAP Ariba Procurement solution, depending).

The third integration scenario is linking SAP S/4HANA for payment processing and discount management in SAP Ariba. For direct procurement scenarios between SAP S/4HANA and the Ariba Network, you can use SAP Ariba Supply Chain Collaboration for Buyers, which has additional integration requirements.

Regarding the integration team itself, in order to underpin a successful implementation, a strong integration lead is required. The integration lead role requires a consultant or subject matter expert with a solid understanding of the master data and transactional integration points and methods, as well as the vernacular of both SAP ERP (e.g., ABAP, master data, transaction codes, tables) and SAP Ariba (e.g., SAP Ariba Cloud Integration Gateway, SAP Ariba Commerce Automation capability, SAP Ariba Sourcing, SAP Ariba Buying and Invoicing, SAP Ariba Catalog, SAP Ariba Supply Chain Collaboration for Buyers, and the Ariba Network).

The project should eventually reach a cutover stage, where the newly built integrations are moved to production. Here, a clear strategy for the impacted legacy systems, such as SAP Supplier Relationship Management (SAP SRM), is required to minimize business disruptions. A poorly planned or nonexistent cutover strategy altogether can cause significant issues on an SAP Ariba integration project, especially a project impacting financials. Some other common issues affecting integration projects include the following:

- **Understaffed project teams**
  Not budgeting adequate resources for both the project team and the customer's IT department (key participants: ABAP developers, SAP Process Integration experts, and basis team). Taking shortcuts and hoping SAP Ariba Support or SAP Ariba Expert Care will bail out the project if difficulties arise will likely cause delays and frustration in the project.
- **Missing the latest cloud integration or hotfix release**
  Not being on the latest hotfix release can cause delays when attempting to resolve an error with SAP Ariba Support. SAP Ariba Support expects your organization to be on the latest hotfix release.

Perhaps the most popular integration between SAP Ariba and an ERP environment is for purchasing document exchange between an ERP environment and the Ariba Network, as discussed in the next section.

11.2 Ariba Network Purchase Order and Invoice Automation with Purchasing Integration Options

Customers with built-out materials management processes in ERP or an SAP SRM classic implementation, where the leading PO document is created in ERP, can opt for integrating these documents with the Ariba Network. Integrating SAP S/4HANA with the Ariba Network allows you to access a number of efficiencies:

- Receiving order confirmations from your suppliers automatically
- Processing ASNs sent from your suppliers in the Ariba Network into goods receipts in SAP S/4HANA
- Sending goods receipt notices automatically to suppliers in the Ariba Network
- Receiving invoices from suppliers in the Ariba Network automatically
- Updating suppliers in the Ariba Network on invoice status updates

As shown in Figure 11.4, a PO from materials management in SAP is transmitted from SAP ERP to the Ariba Network, and follow-on documents through to the invoice are exchanged between the two systems and between supplier and customer.

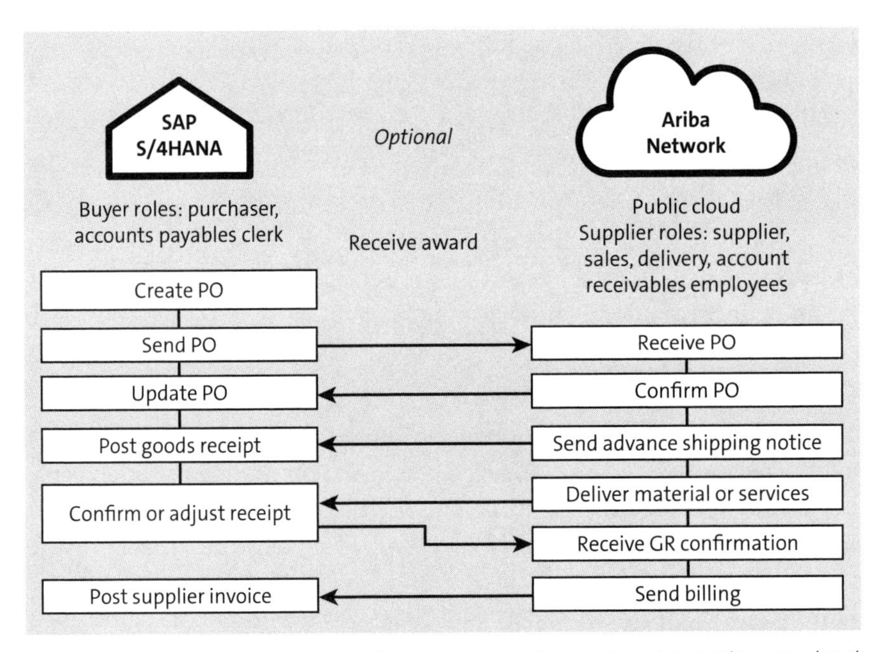

Figure 11.4 PO/Invoice Automation between Ariba Network and SAP S/4HANA (ERP)

Integration with procurement document processing on the Ariba Network options include the following:

- PO and invoice automation capability
- SAP Ariba Discount Management (optional)
- SAP Ariba Catalog (optional)

11.2.1 Conducting Transactions with Ariba Network Integrated with SAP S/4HANA

The steps reflected in the process flow include both steps for the buyer and for the supplier. A typical set of steps for conducting transactions is as follows:

1. The buyer (either a professional buyer or a casual user) creates a requisition in SAP S/4HANA, as shown in Figure 11.5.

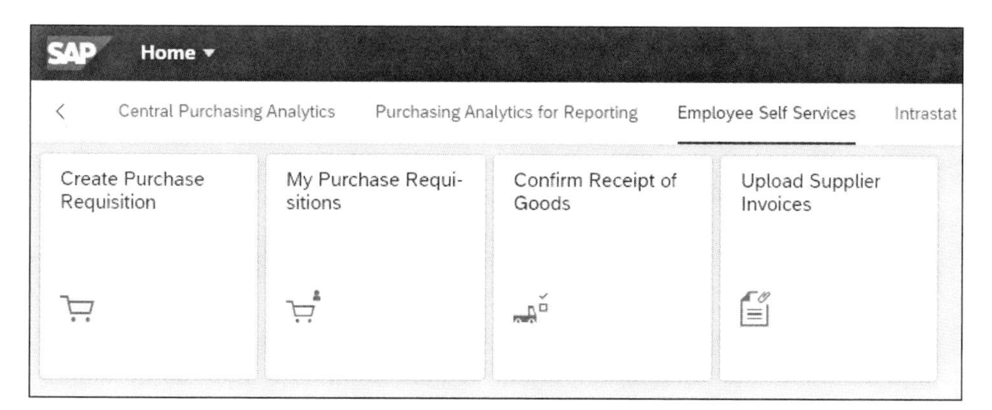

Figure 11.5 Creating a Requisition via the Self-Service Procurement Tile

2. The buyer describes the item, selects a material/service master, or browses a catalog. As shown in Figure 11.6, the buyer punches out (leaves the SAP ERP system to access SAP Ariba Catalog in the cloud) and selects an item from the catalog.

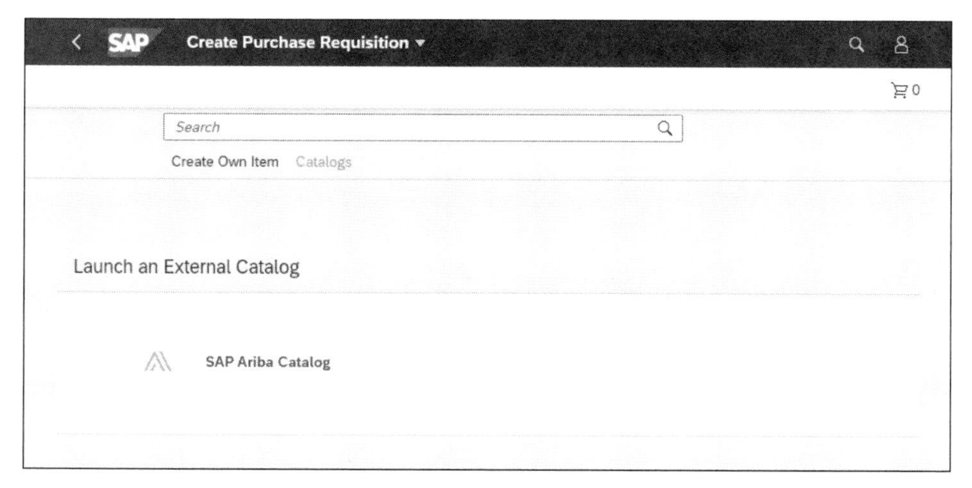

Figure 11.6 SAP Ariba Catalog Punchout in SAP S/4HANA

3. The buyer can search for and select the items as required, as shown in Figure 11.7 and Figure 11.8.

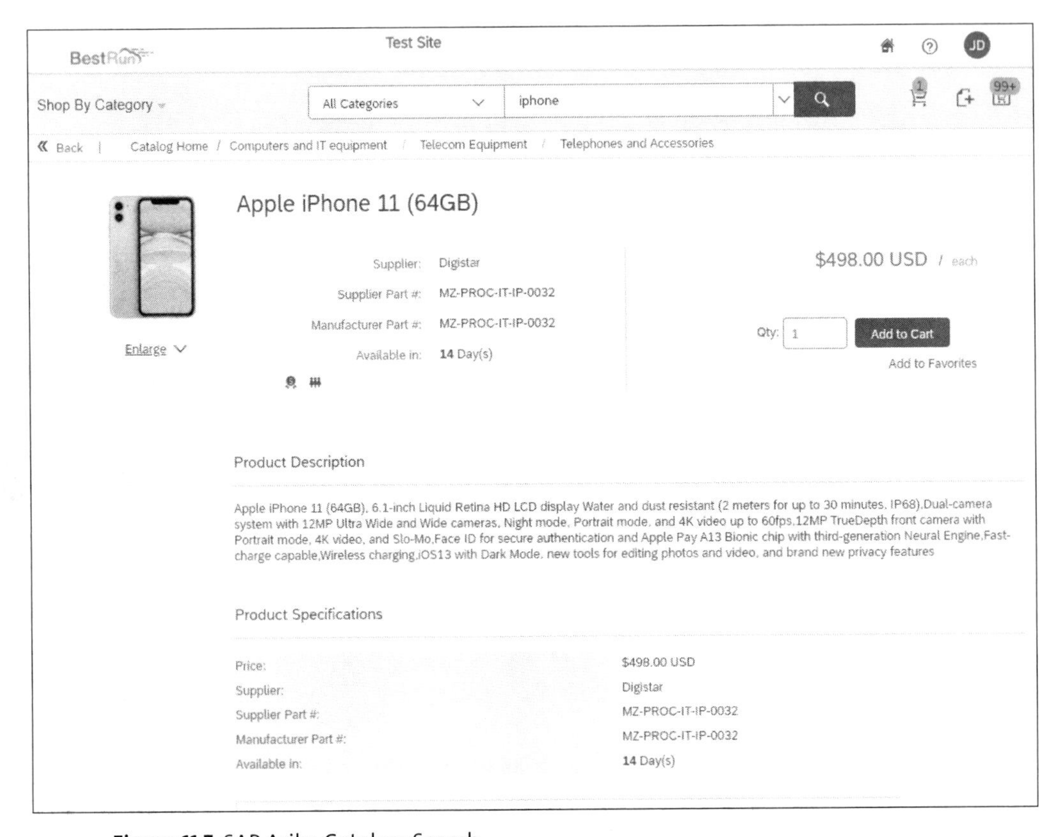

Figure 11.7 SAP Ariba Catalog: Search

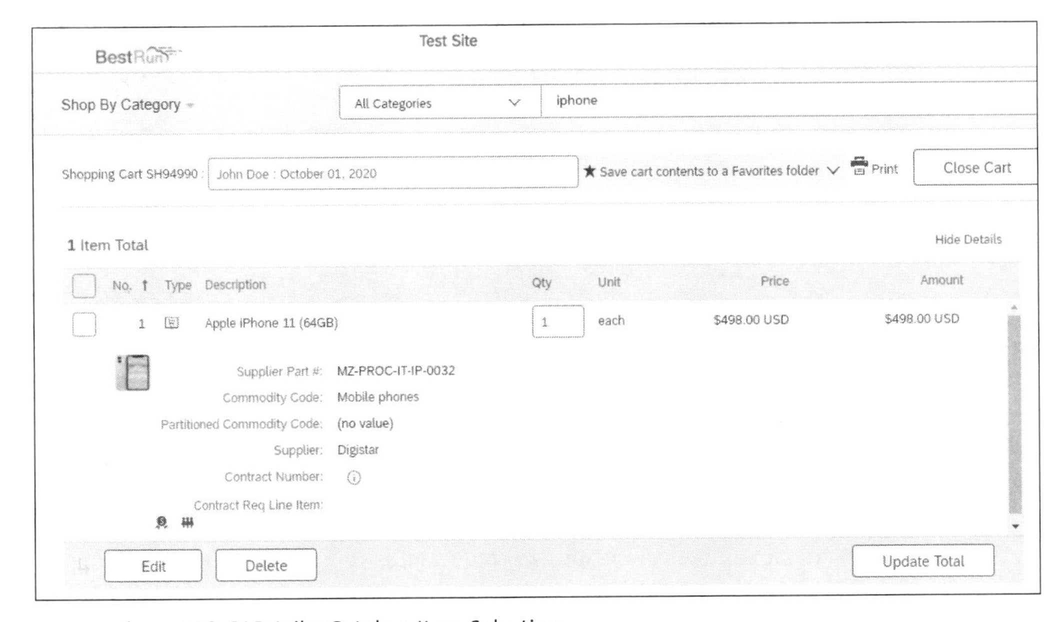

Figure 11.8 SAP Ariba Catalog: Item Selection

4. Next, the item is transferred from the SAP Ariba Catalog back to the purchase requisition in SAP S/4HANA, as shown in Figure 11.9 and Figure 11.10.

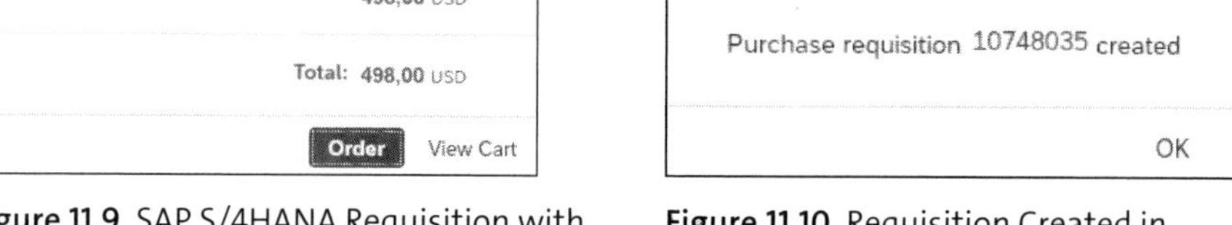

Figure 11.9 SAP S/4HANA Requisition with SAP Ariba Catalog Item

Figure 11.10 Requisition Created in SAP S/4HANA

SAP S/4HANA can be configured to automatically convert sourced requisitions into POs, or requisitions can be converted manually. The requisition now shows the corresponding/resulting PO number, as shown in Figure 11.11.

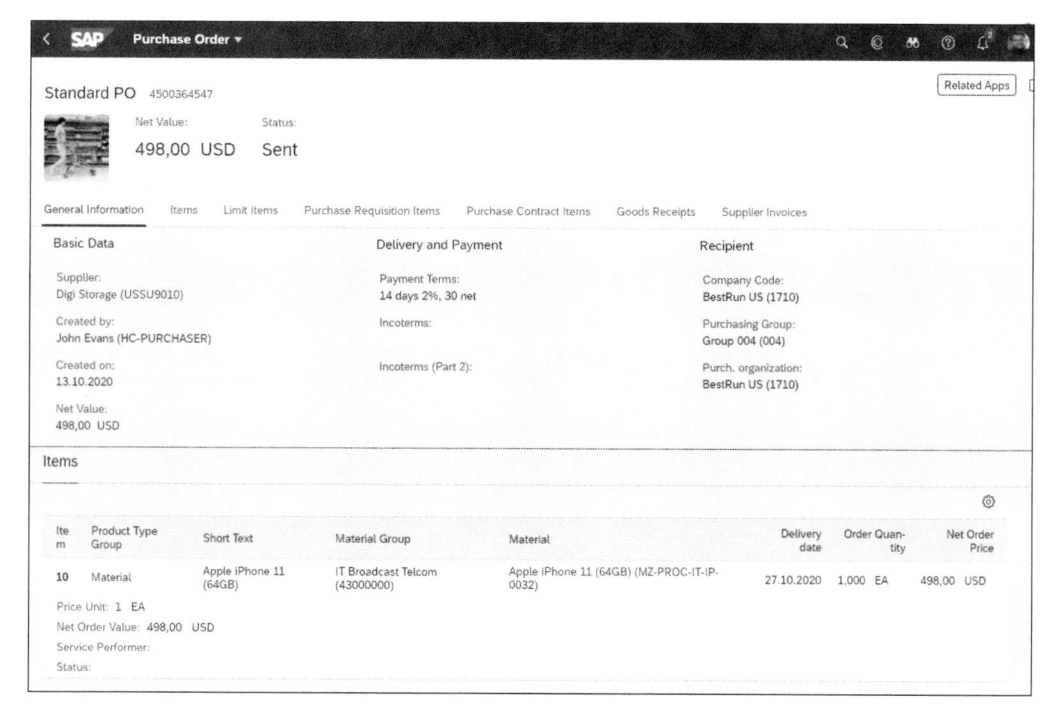

Figure 11.11 SAP S/4HANA: PO Generated from a Requisition

5. The PO is then transmitted to the Ariba Network for processing by the supplier.

In the next section, we'll cover the supplier's subsequent interaction with the buyer.

11.2.2 Ariba Network Customer and Supplier Interaction

Now that the document is up on the Ariba Network, the supplier can start collaborating with the customer by reviewing and processing the PO via the following steps:

1. The supplier logs in to the Ariba Network using the supplier login information, as shown in Figure 11.12.

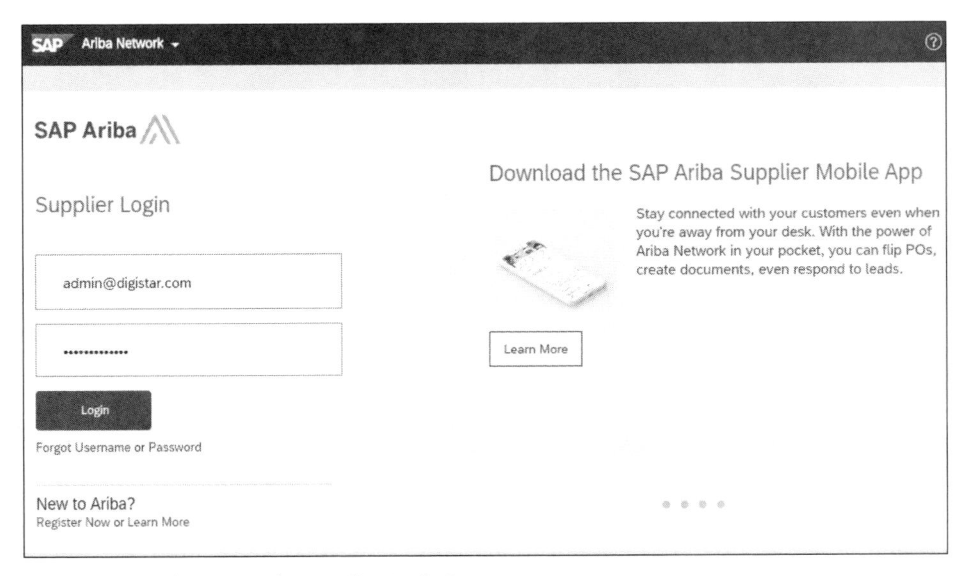

Figure 11.12 Ariba Network: Supplier Login Page

2. From the login page, the supplier can review new and existing POs, as shown in Figure 11.13.

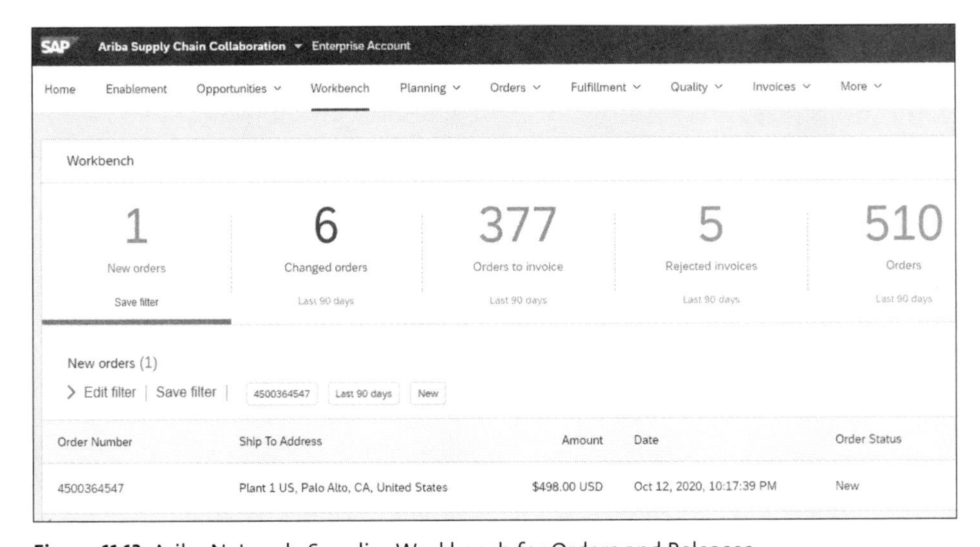

Figure 11.13 Ariba Network: Supplier Workbench for Orders and Releases

3. To find a PO, the supplier can also enter a partial number or other identifying characteristic, as shown in Figure 11.14.

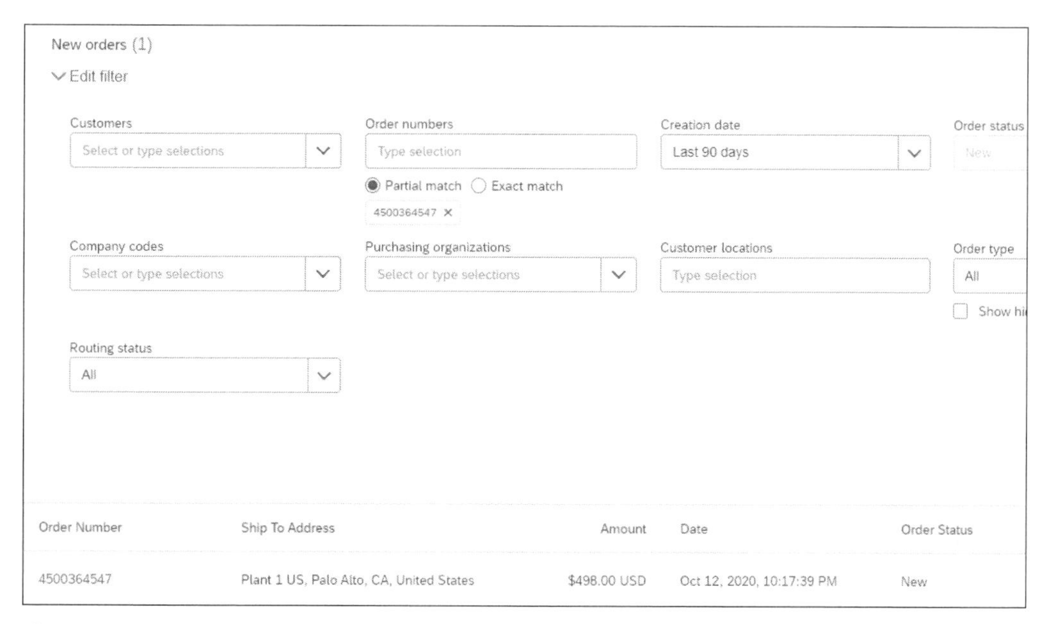

Figure 11.14 Ariba Network: PO Search

After the supplier has located the PO, the supplier can confirm that it plans to fulfill this order.

11.2.3 Confirming and Order Fulfillment

In this part of the process, the supplier confirms the order and fulfills it by following these steps:

1. The supplier selects the confirmation option for the PO and then confirms the order, as shown in Figure 11.15 and Figure 11.16.
2. After the supplier has confirmed the intention to fulfill the PO, an ASN can be created to provide the customer with an estimated time of arrival and confirmation that the items have shipped, as shown in Figure 11.17 and Figure 11.18.

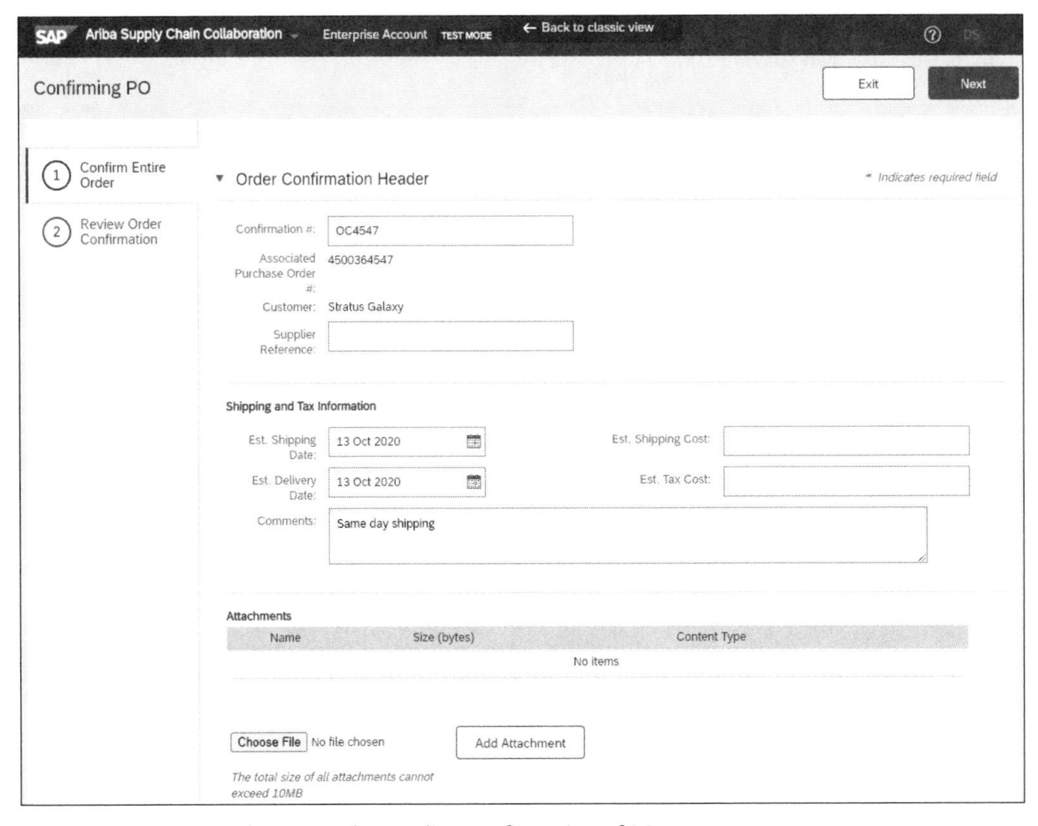

Figure 11.15 Ariba Network: Supplier Confirmation of PO

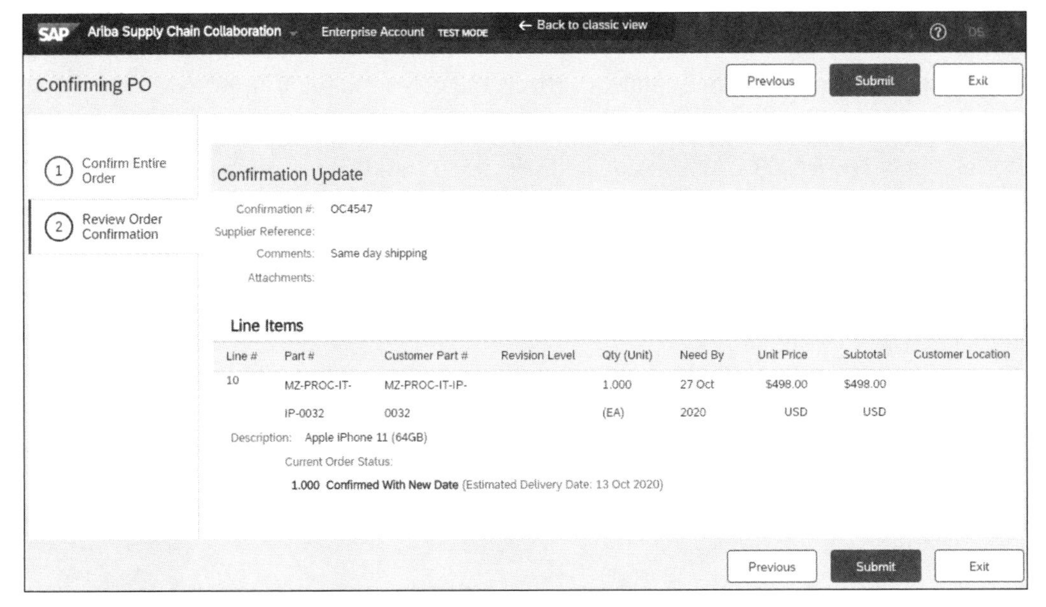

Figure 11.16 Ariba Network: Supplier Confirmation Release

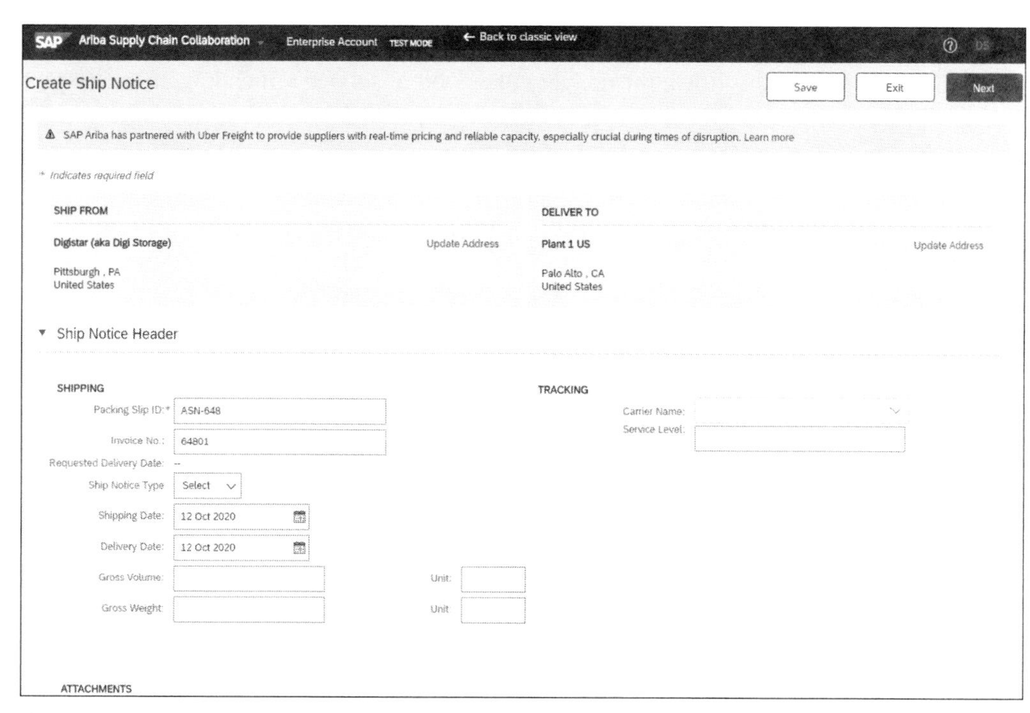

Figure 11.17 Ariba Network: Advanced Shipping Notification, Step 1

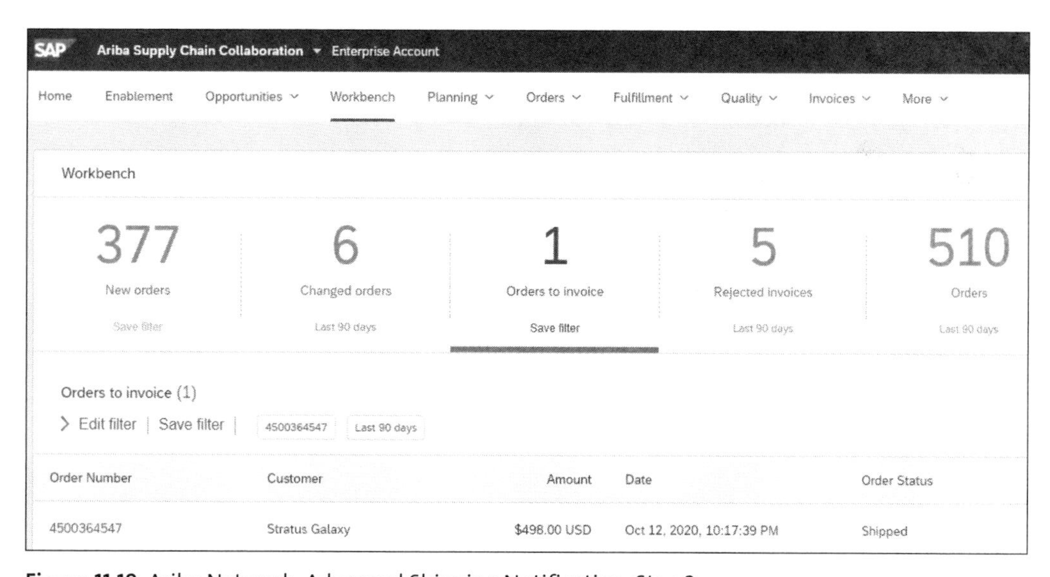

Figure 11.18 Ariba Network: Advanced Shipping Notification, Step 2

11

11.2.4 Confirming Goods Receipts

In this step, you'll confirm the delivery of a good or service, which in turn can trigger inventory management processes as well as provide the grounds for payment for the follow-on invoice document.

When a supplier has delivered the requested items, you can log in to SAP S/4HANA and complete the goods receipt, as shown in Figure 11.19 and Figure 11.20.

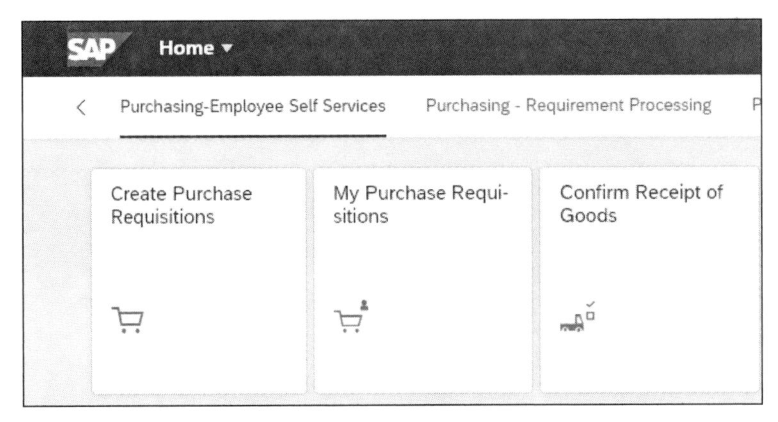

Figure 11.19 SAP S/4HANA: Confirming the Receipt of Goods

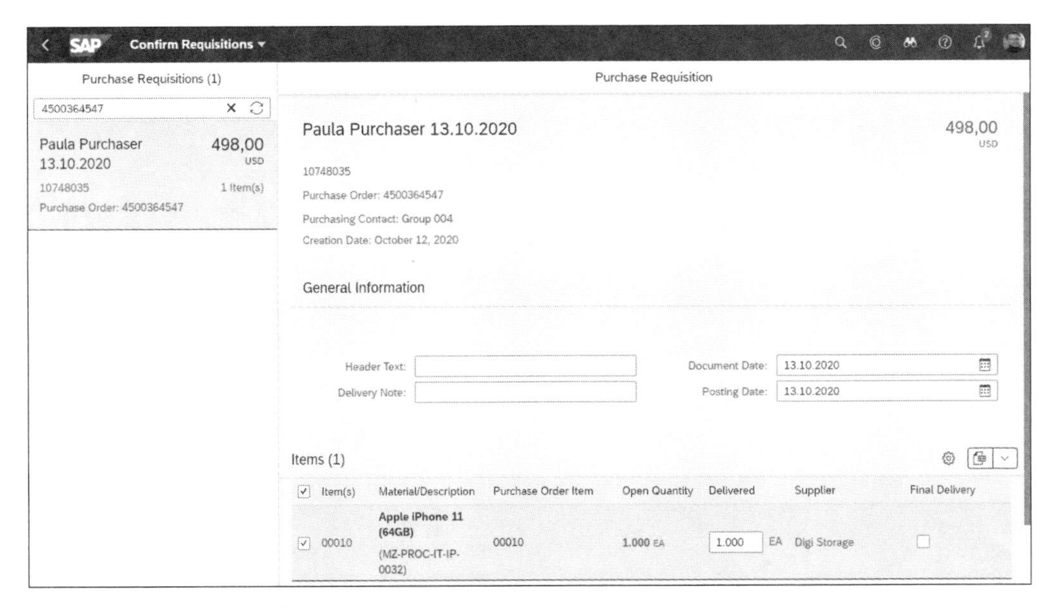

Figure 11.20 SAP S/4HANA: Goods Receipt

After these steps are complete, an invoice can be submitted and matched against not only the PO but also the goods receipt. This *three-way match* is more solid than relying on just matching the PO and the invoice, which is called a *two-way match*.

11.2.5 Invoice Creation

Invoices created in a supplier's system, or worse, written out on paper and faxed over to the customer, create all kinds of processing challenges for the customer. By using the Ariba Network to submit an invoice, the supplier and customer save valuable iterations in this process, making the submission much more accurate and making processing cycles faster.

At this stage, the seller creates an invoice by following these steps:

1. After the goods receipt has been entered, the supplier can log in to the Ariba Network portal and enter their invoice, as shown in Figure 11.21.

   The supplier user can select a PO and then begin creating the invoice using the various dropdown options shown in Figure 11.22.

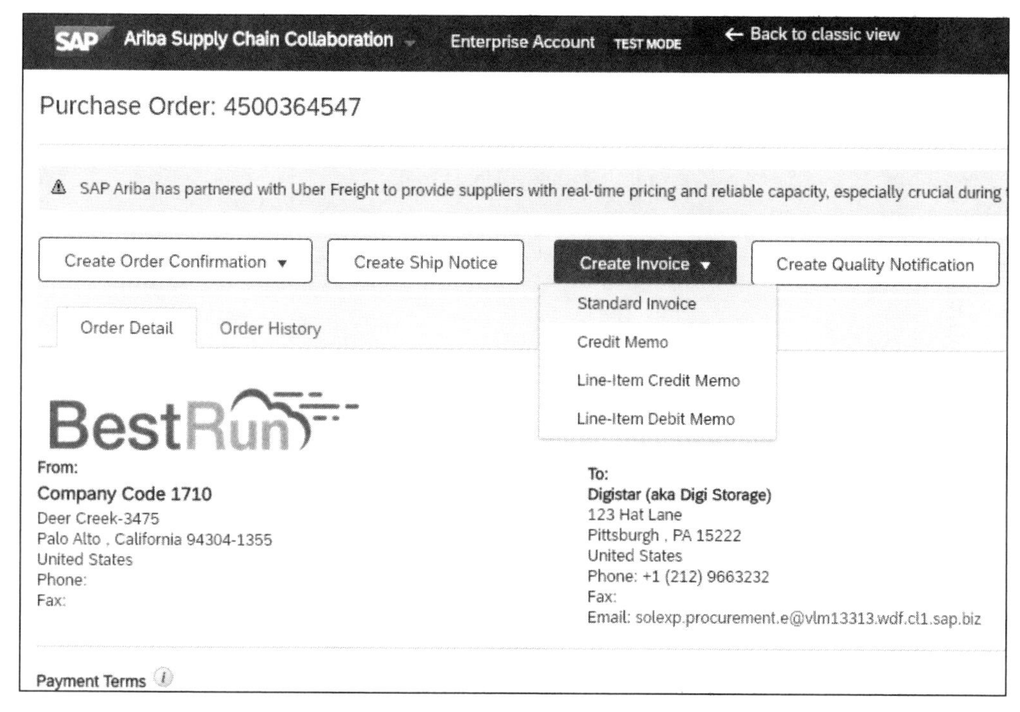

Figure 11.21 Ariba Network: Creating an Invoice from a PO

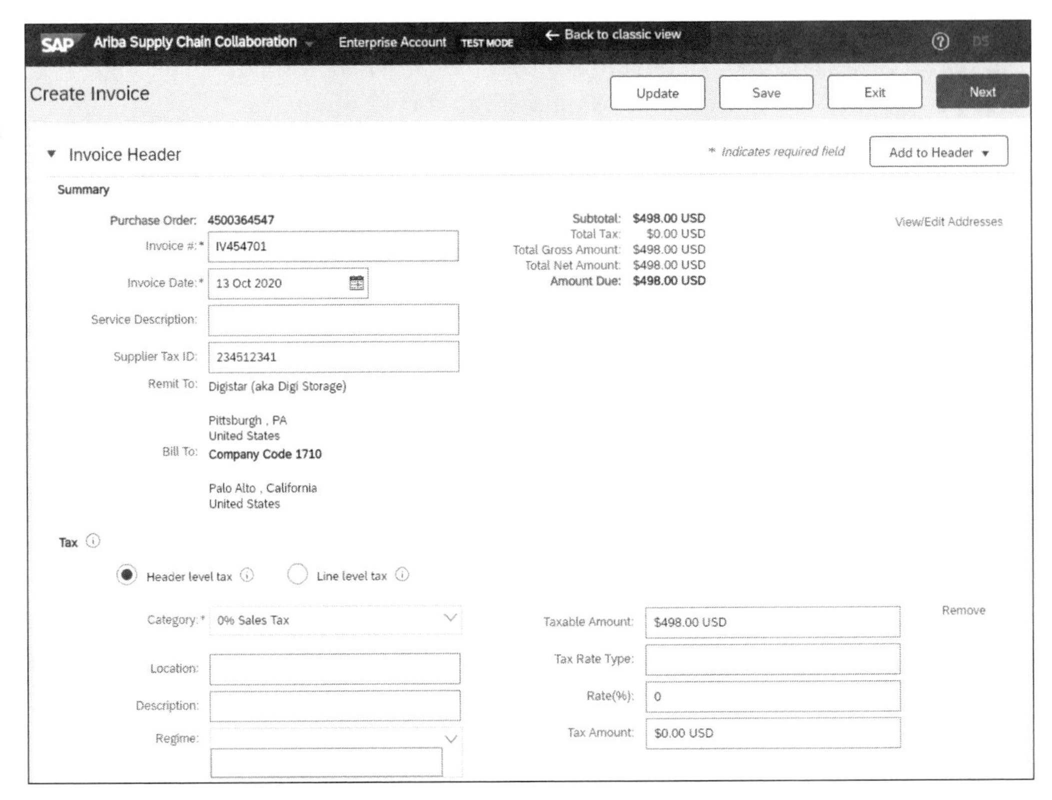

Figure 11.22 Ariba Network: Creating an Invoice, Step 1

2. Next, the supplier creates an invoice, as shown in Figure 11.23.

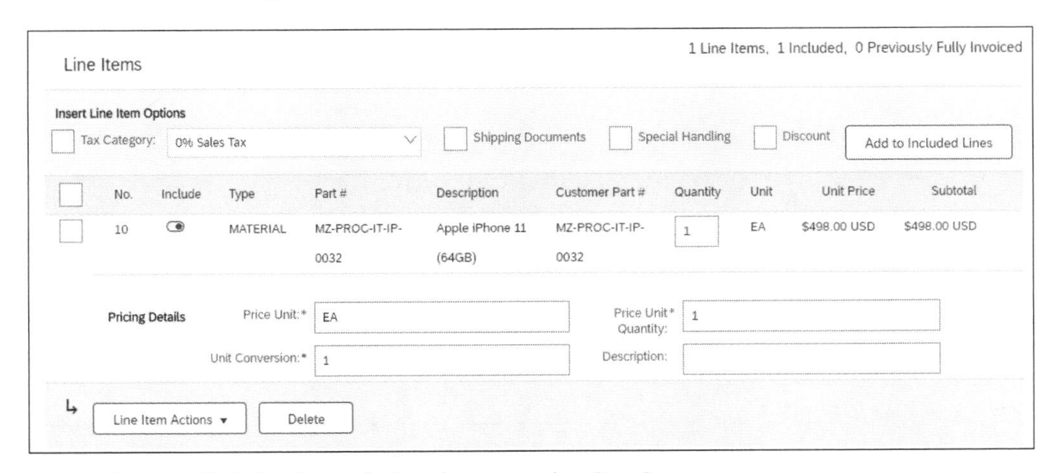

Figure 11.23 Ariba Network: Creating an Invoice, Step 2

For an overview of the invoice, the supplier can review the summary page as in Figure 11.24.

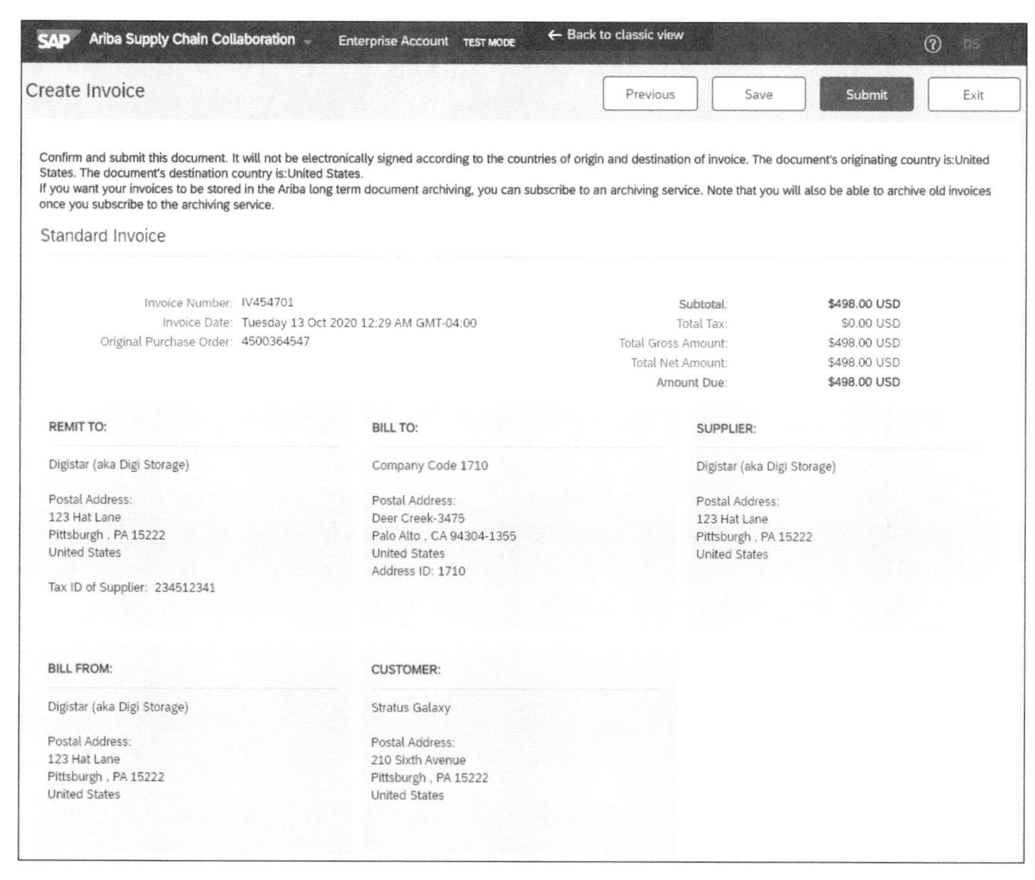

Figure 11.24 Ariba Network: Invoice Summary Page

3. After the invoice is submitted, the complete document flow is now available in both SAP S/4HANA within the **Supplier Invoices** tab of the PO, as shown in Figure 11.25, and in the Ariba Network.

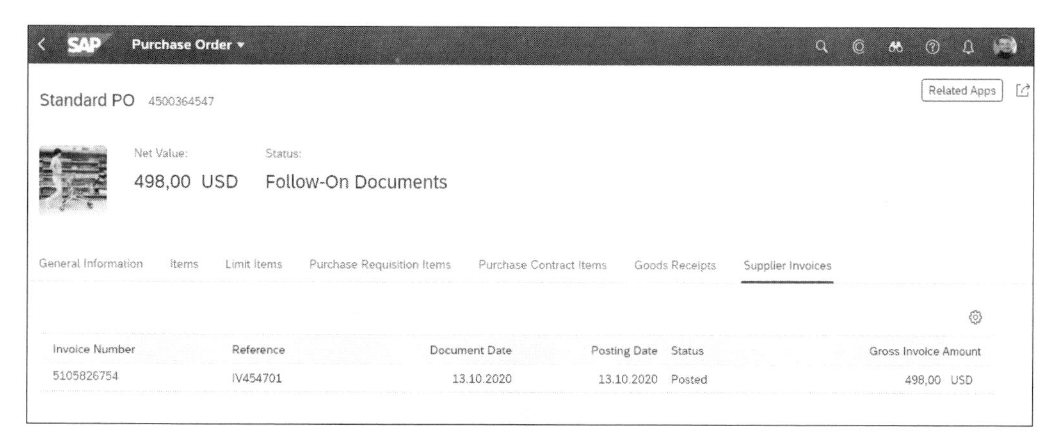

Figure 11.25 PO Generated from a Requisition in SAP S/4HANA

In the next section, we'll focus on SAP Ariba Sourcing integration, another native integration available in SAP S/4HANA. For nonnative integrations to SAP S/4HANA, such as SAP Ariba Buying, or for integrations involving previous SAP ERP releases, refer to Section 11.5.

11.3 SAP Ariba Sourcing Integration with SAP S/4HANA

Another integration option for customers running SAP S/4HANA or SAP ERP and looking to take advantage of SAP Ariba Cloud solutions to bolster their procurement operations is SAP Ariba Sourcing integration. In this scenario, the buyer user can generate a request for a quotation, information, purchase, or other solicitation type (called an *RFx*), and send this RFx for sourcing up to SAP Ariba Sourcing. SAP Ariba Sourcing enables customers and suppliers on the Ariba Network to seamlessly collaborate on bidding. Buyers can identify the best option for their needs and then allow the ordering and fulfillment of the item to occur within the ERP or SAP Ariba environments. SAP Ariba provides the crucial linkage to and mediation with a huge array of suppliers during this process, which an ERP-based sourcing event couldn't replicate easily alone because connecting to suppliers and setting up protocols are simply too laborious to scale.

11.3.1 Key Process Steps

Integration connects the digital core of SAP S/4HANA with SAP Ariba Sourcing, as shown in Figure 11.26, with the following key process steps:

1. The purchaser creates an RFQ for SAP S/4HANA purchase requisitions and automatically sends this RFQ to SAP Ariba Sourcing, triggering a sourcing request.

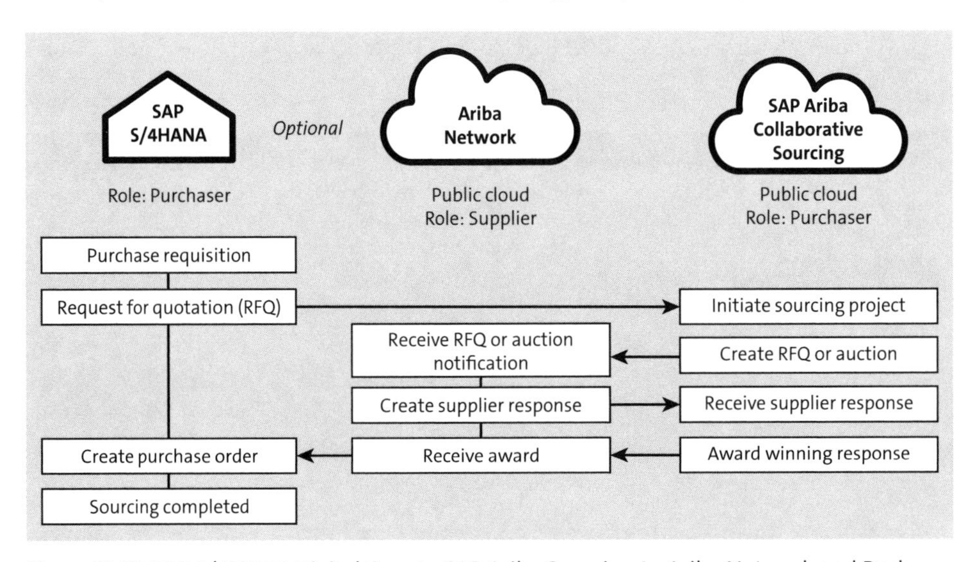

Figure 11.26 SAP S/4HANA Digital Core to SAP Ariba Sourcing to Ariba Network and Back

2. The sourcing manager in SAP Ariba Sourcing accesses his worklist and converts sourcing requests into sourcing projects, inviting suppliers to bid.
3. Suppliers process the RFQ.
4. The sourcing manager can then monitor sourcing projects and supplier responses, as well as support projects that result in (reverse) auctions or RFQs, adding suppliers to the Ariba Network as required.
5. Finally, after the sourcing process is complete, the sourcing manager can award one or more bids in SAP Ariba Sourcing. These awarded bids are sent automatically or directly by the sourcing manager to SAP S/4HANA, where POs and contracts are created as follow-on documents for the original RFQ.

In the next section, you'll see from a user perspective how this process works.

11.3.2 Conducting Transactions with SAP Ariba Sourcing Integrated with SAP S/4HANA

With the RFQ originating in SAP S/4HANA, seamless interaction in a closed loop must occur between the backend ERP, where the demand originates; the SAP Ariba Sourcing cockpit, where the bidders are assigned and managed for the request for proposal (RFP) process; and the Ariba Network, where bidders and suppliers respond and collaborate on the request. First, a buyer notes an unsourced requirement/requisition and creates an RFQ in SAP S/4HANA under the **Source of Supply Management** tab, as shown in Figure 11.27.

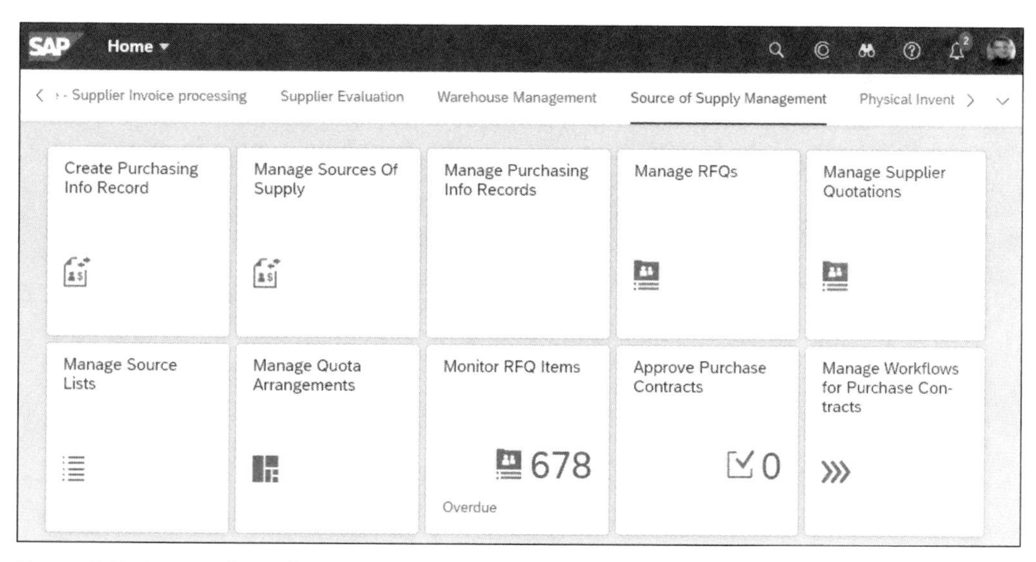

Figure 11.27 Source of Supply Management: Manage RFQs

11.3.3 Creating a New Request for Quotation in SAP S/4HANA

After a buyer identifies and confirms a new demand or an existing unsourced item requiring a source of supply, the buyer can leverage the following process for creating and managing a sourcing activity in system:

1. In the Manage RFQ app, the user can create a new RFQ by selecting the plus (+) icon, as shown in Figure 11.28.

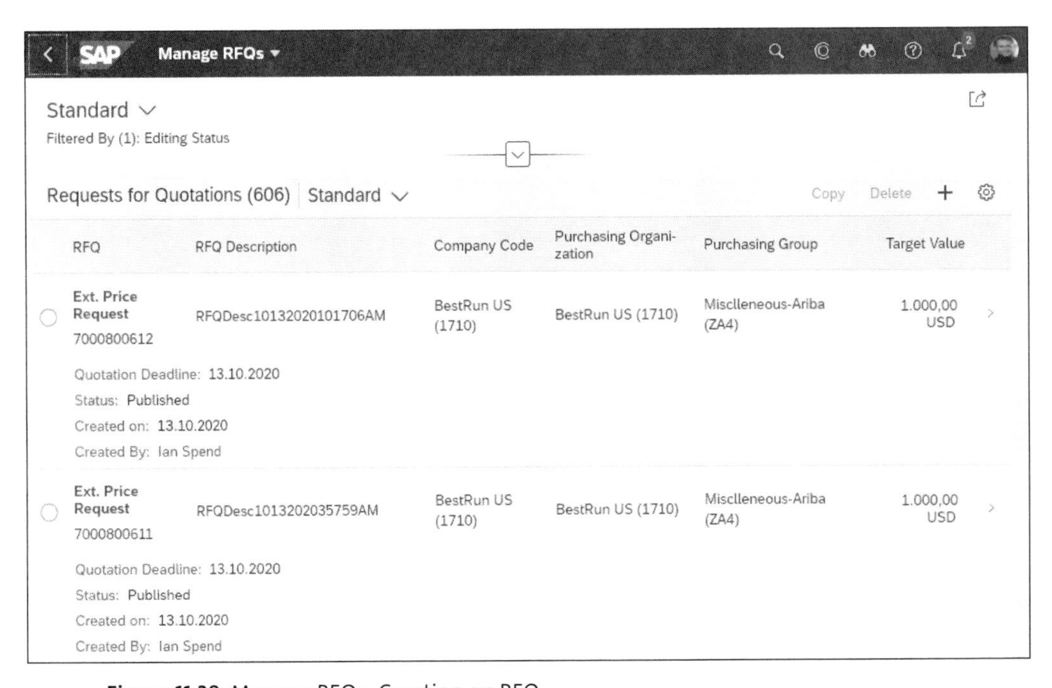

Figure 11.28 Manage RFQs: Creating an RFQ

2. As shown in Figure 11.29, the buyer then creates the shell for an RFQ. This RFQ is a "shell" because the bidders and other details don't need to be added at this point. These items will be added after the RFP moves to SAP Ariba Sourcing, where it can access the more than two million potential suppliers in the Ariba Network.

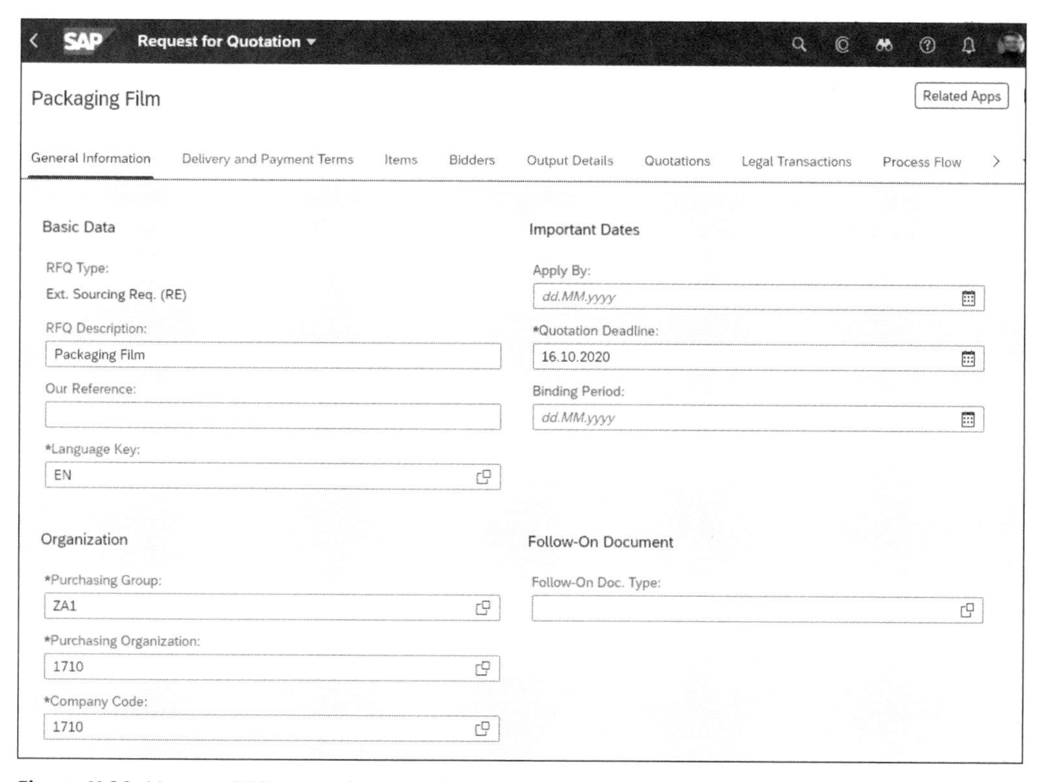

Figure 11.29 Manage RFQs: Creating an RFP

Under the **Delivery and Payment Terms** tab of the RFP, you'll maintain the currency as well as the target value, as shown in Figure 11.30.

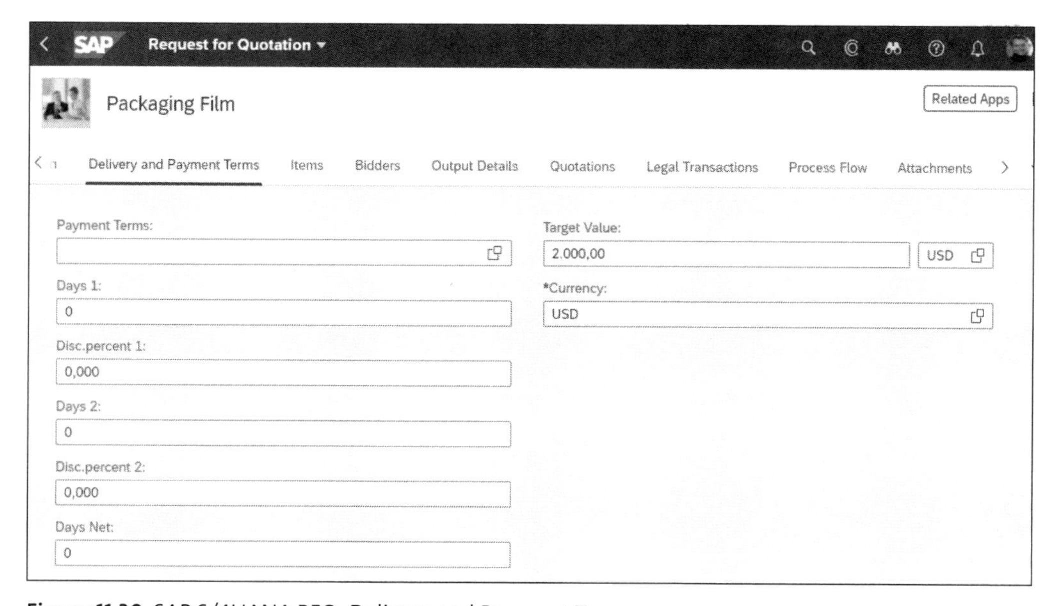

Figure 11.30 SAP S/4HANA RFQ: Delivery and Payment Terms

Inviting Bidders

While the SAP S/4HANA system may not contain all the potential bidders available in the Ariba Network, it typically serves as the leading system for material master data. This data can be included under the **Items** tab of the RFP, as shown in Figure 11.31.

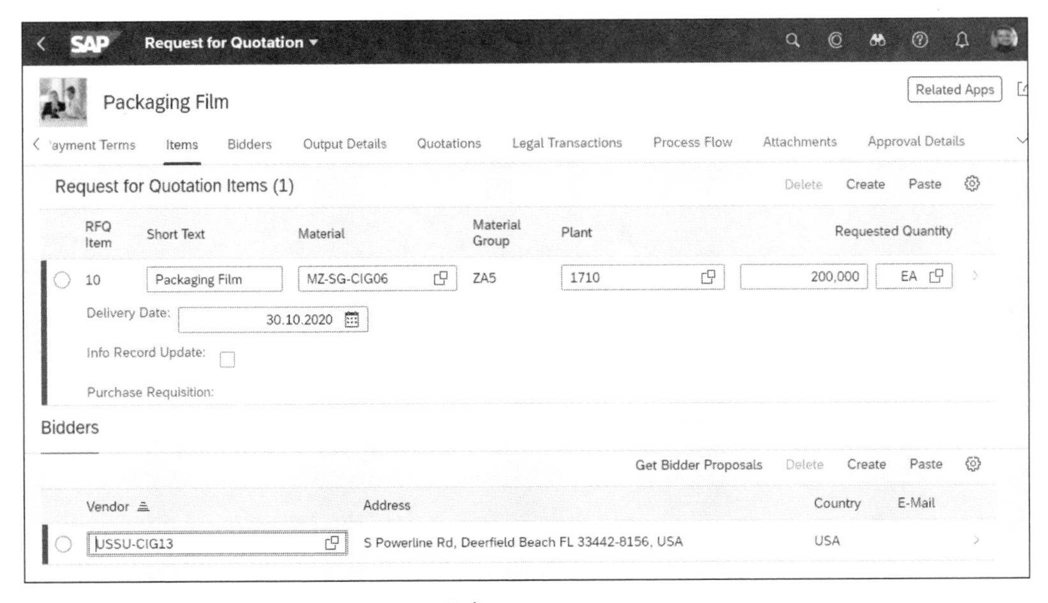

Figure 11.31 RFQ: Create RFP Items Tab

Before publishing the RFP, you can check the **Output Details** tab one last time, as shown in Figure 11.32. Verify the status, dispatch schedule, output type, and other facets of the RFP as well as preview the PDF document that will be sent to the supplier.

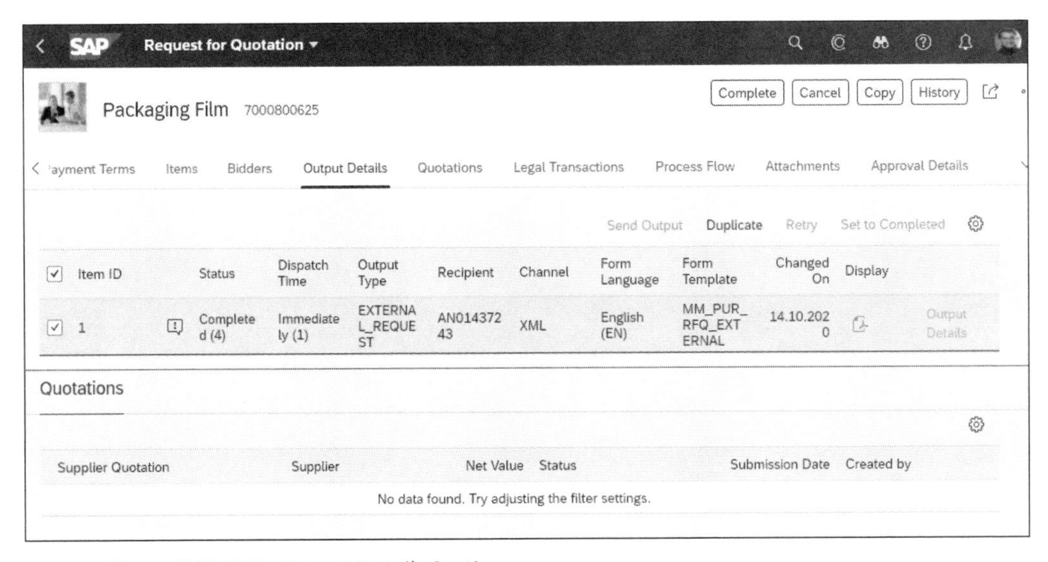

Figure 11.32 RFQ: Output Details Section

After you're satisfied that the RFP reflects the correct information in the SAP S/4HANA system, you'll then publish the RFP, which sends the RFP to SAP Ariba Sourcing.

11.3.4 R for Quotation in SAP Ariba Sourcing

When buyers log in to SAP Ariba Sourcing, the first area they'll see is their dashboard, as shown in Figure 11.33.

Figure 11.33 SAP Ariba Sourcing: Requests

From the dashboard, a buyer can then search for the RFP, as shown in Figure 11.34.

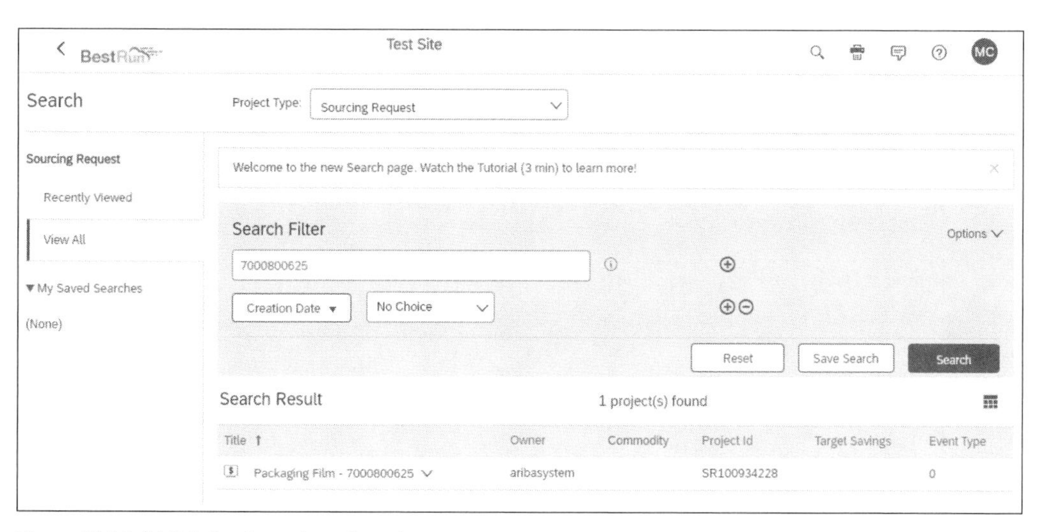

Figure 11.34 SAP Ariba Sourcing: Search

Two actions must be completed before issuing the RFP. The first is preparing the RFP for issuing, and the second is an approval step. You can access these steps under the **Tasks** tab, as shown in Figure 11.35.

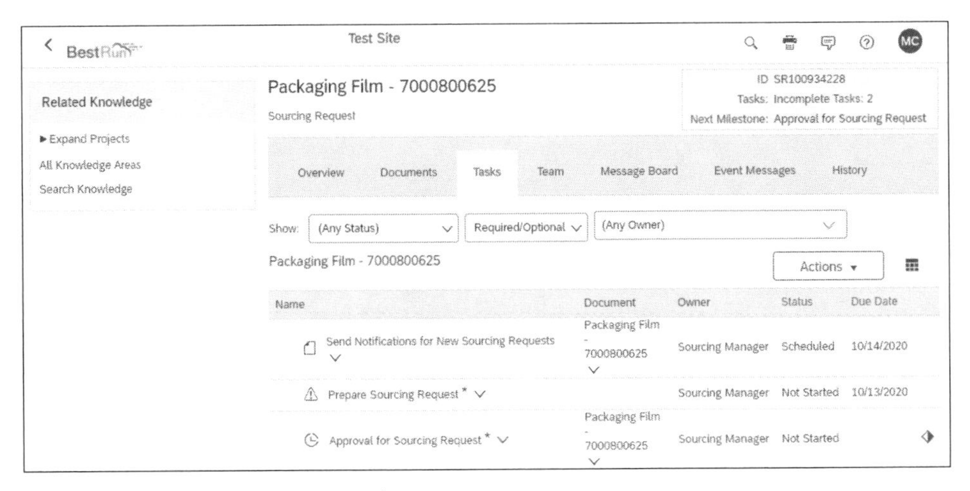

Figure 11.35 SAP Ariba Sourcing: Tasks

Select and review the first task, and mark this step as complete. You also have the option of marking this task as **Started** as an interim status, as shown in Figure 11.36.

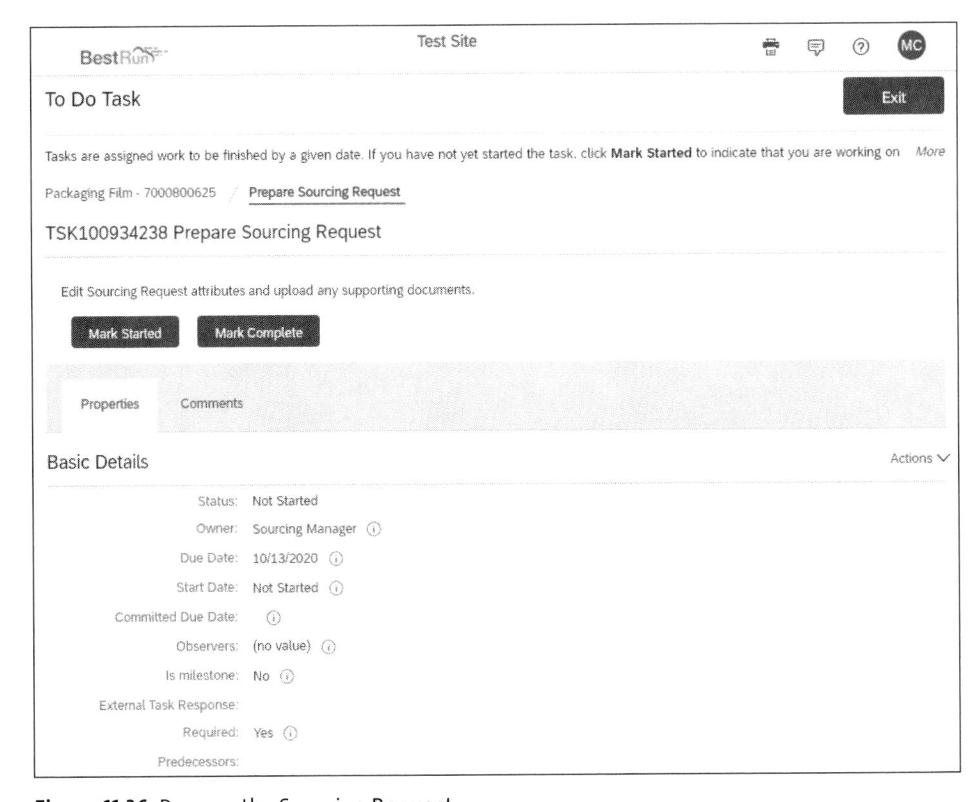

Figure 11.36 Prepare the Sourcing Request

After you've reviewed these two steps, and the approvals are completed, your RFP can now be converted to an SAP Ariba Sourcing RFP, as shown in Figure 11.37.

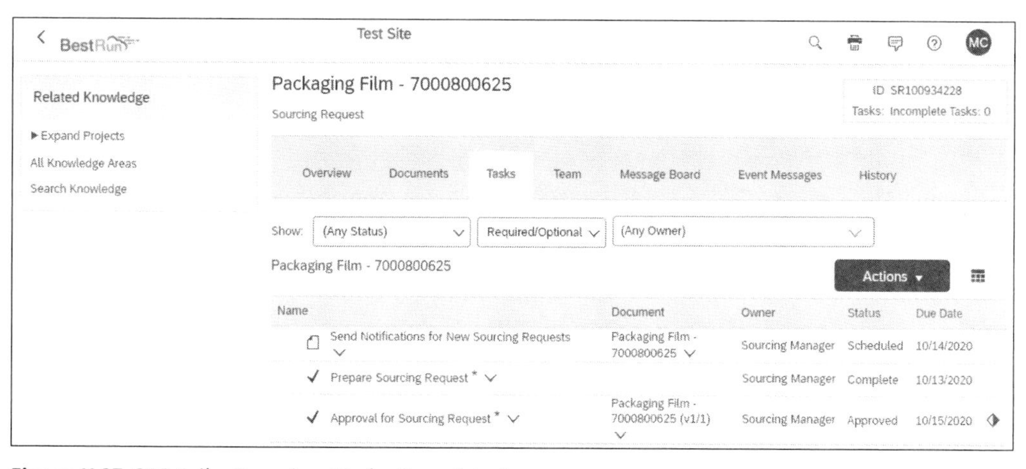

Figure 11.37 SAP Ariba Sourcing: Tasks Completed

With these tasks complete, select **Create** and convert the sourcing request to an SAP Ariba Sourcing RFP, as shown in Figure 11.38.

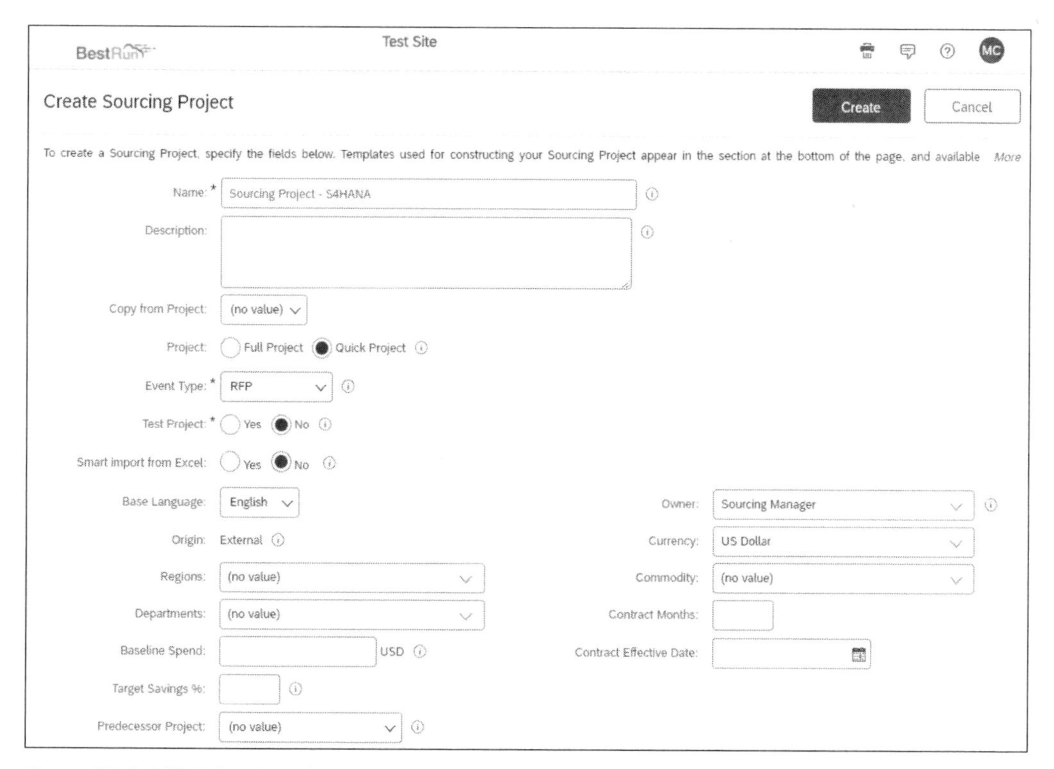

Figure 11.38 SAP Ariba Sourcing: Create Sourcing Project

The next step is to add bidders to the RFP event, as shown in Figure 11.39.

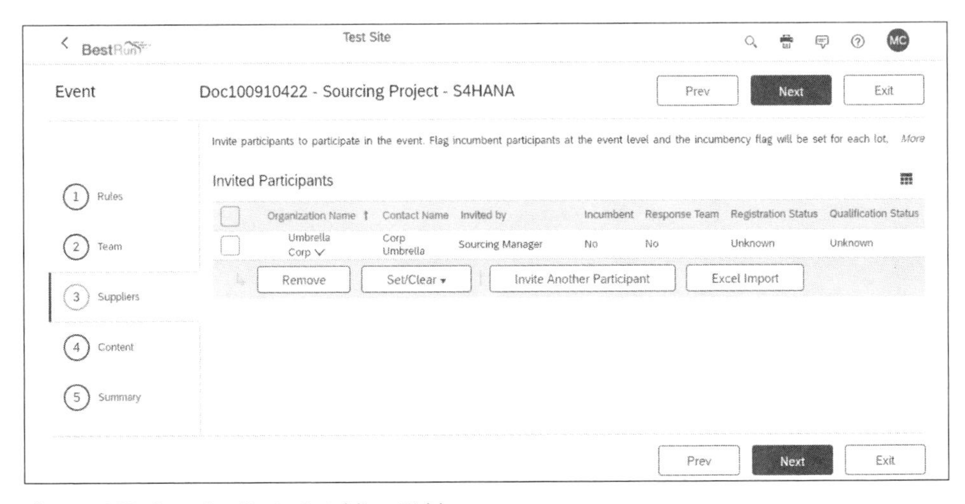

Figure 11.39 Sourcing Project: Adding Bidders

During these steps for creating an RFP, you also can define and refine the team and the content, finally arriving at the summary page, as shown in Figure 11.40.

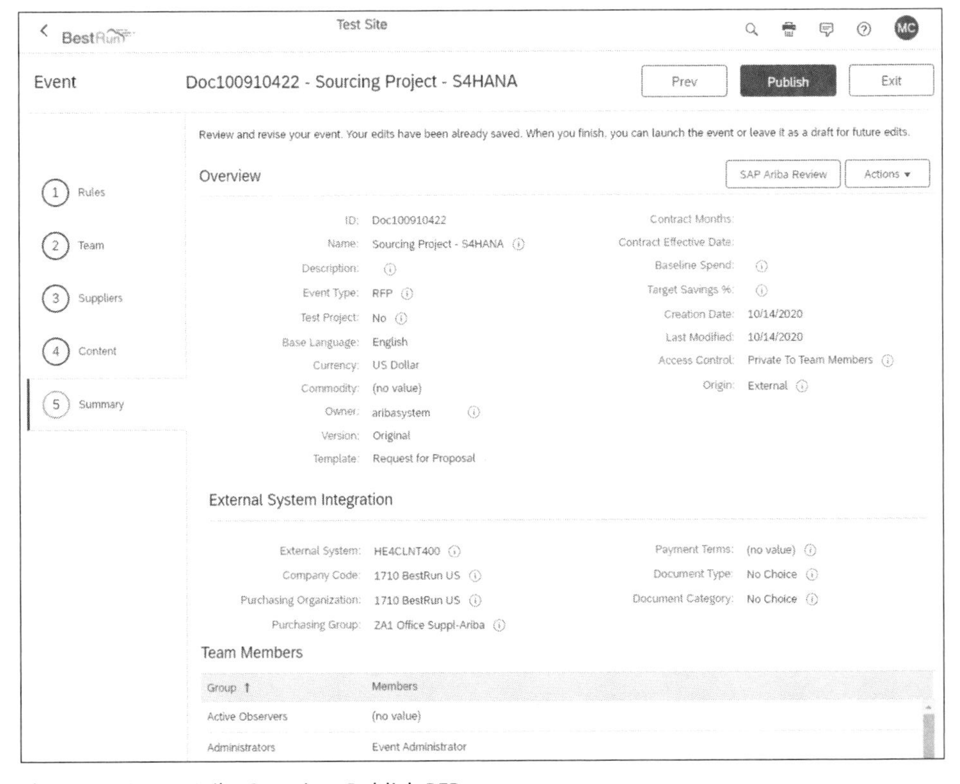

Figure 11.40 SAP Ariba Sourcing: Publish RFP

11.3.5 Supplier Processing of Request for Proposals

After you've published the RFP, suppliers can review the request in their Ariba Network area by following these steps:

1. Invited suppliers can review the RFP by logging in to the supplier portal on the Ariba Network at *https://supplier.ariba.com*. On their home page, a supplier can select **Proposals** from the dropdown menu at the top, as shown in Figure 11.41.

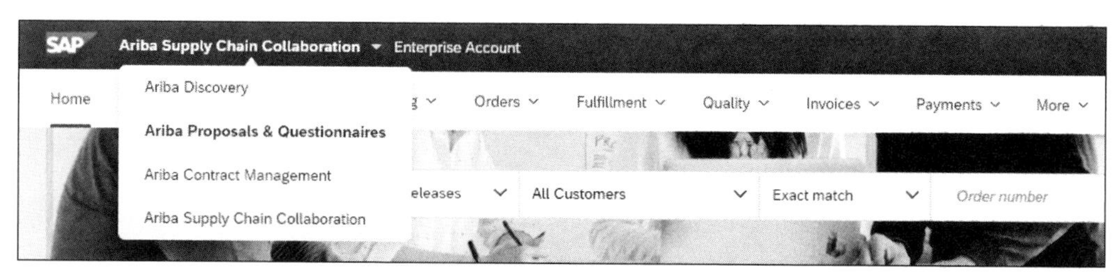

Figure 11.41 SAP Ariba Supplier: View Ariba Proposals

The **Proposals** section lists the bids in various statuses, such as **Completed**, **Pending Selection** (where the supplier has already put in a bid), and **Open** (new RFxs), as shown in Figure 11.42.

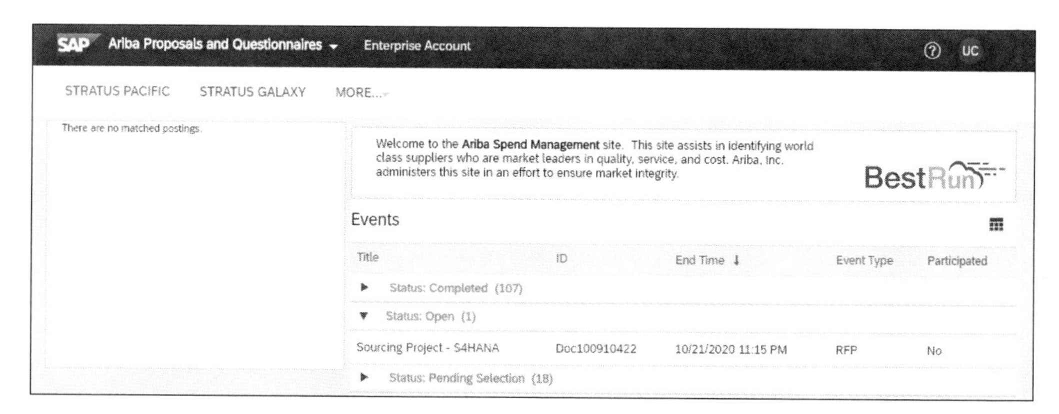

Figure 11.42 SAP Ariba Sourcing Projects: Events

2. After the supplier has selected and opened the invitation, the supplier can either decline to participate or review the prerequisites, as shown in Figure 11.43, or accept them, as shown in Figure 11.44.

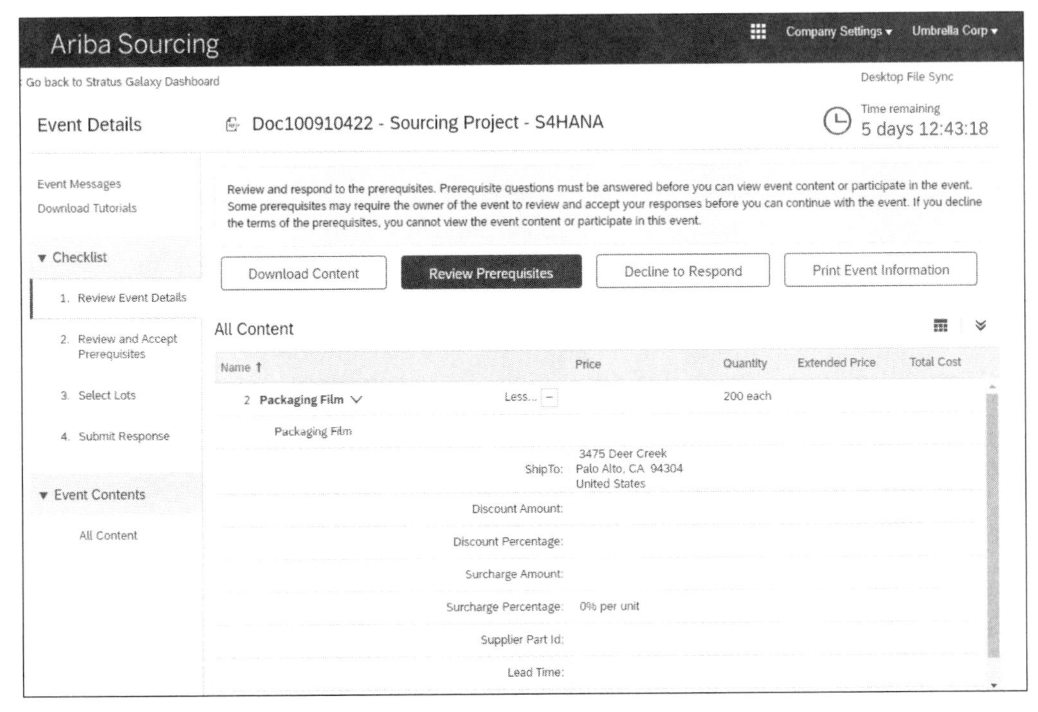

Figure 11.43 SAP Ariba Supplier: Review Prerequisites

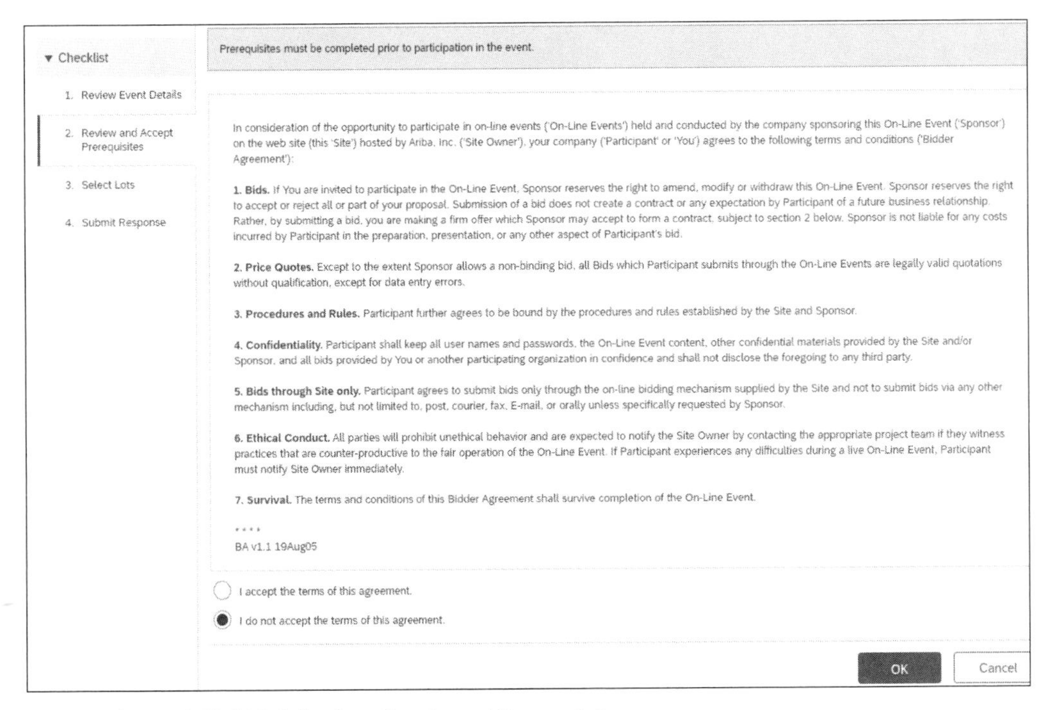

Figure 11.44 SAP Ariba Supplier: Accept Prerequisites

3. The next step is to place the bid, which is called **Select Lots** in SAP Ariba. In the step, the supplier selects some or all of the items requested in the RFx and responds, as shown in Figure 11.45. During the response, the supplier can also send a message to the customer.

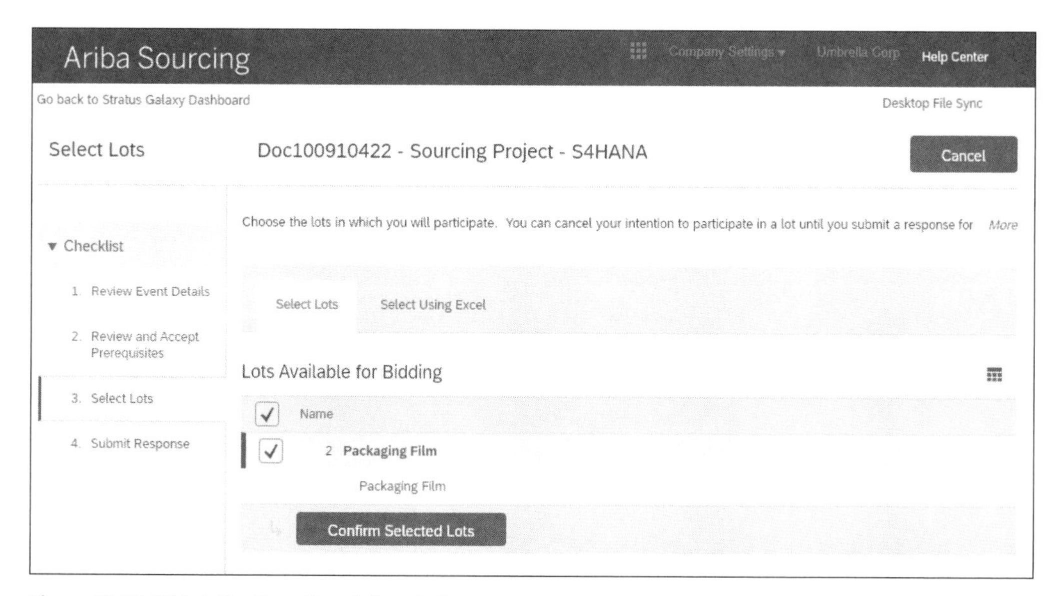

Figure 11.45 SAP Ariba Supplier: Select Lots

11.3.6 Bid Processing in SAP Ariba

In this step, the buyer consolidates and compares the bids, selecting, presumably, the best bid that meets their target goals for quality, timing, and price by following these steps:

1. After bids are submitted by suppliers, as shown in Figure 11.46, the bids are sent to the customer for review. As shown in Figure 11.47, you can compare bids side by side. You can completely or partially award an RFx to a supplier, depending on what is optimal/preferred. In other words, a supplier could receive a partial award of the tender, while another supplier could receive the rest of the order.
2. As shown in Figure 11.48, you'll confirm the awards for the RFx, prior to generating these awards in the backend SAP S/4HANA environment, as shown in Figure 11.49.

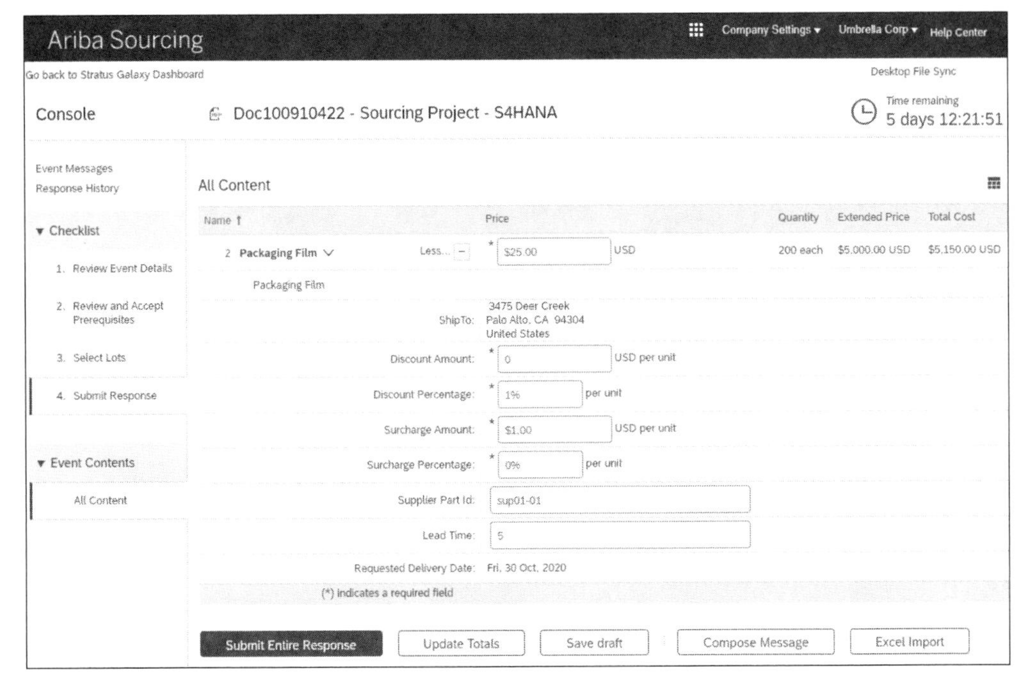

Figure 11.46 SAP Ariba Supplier: Maintain Bid

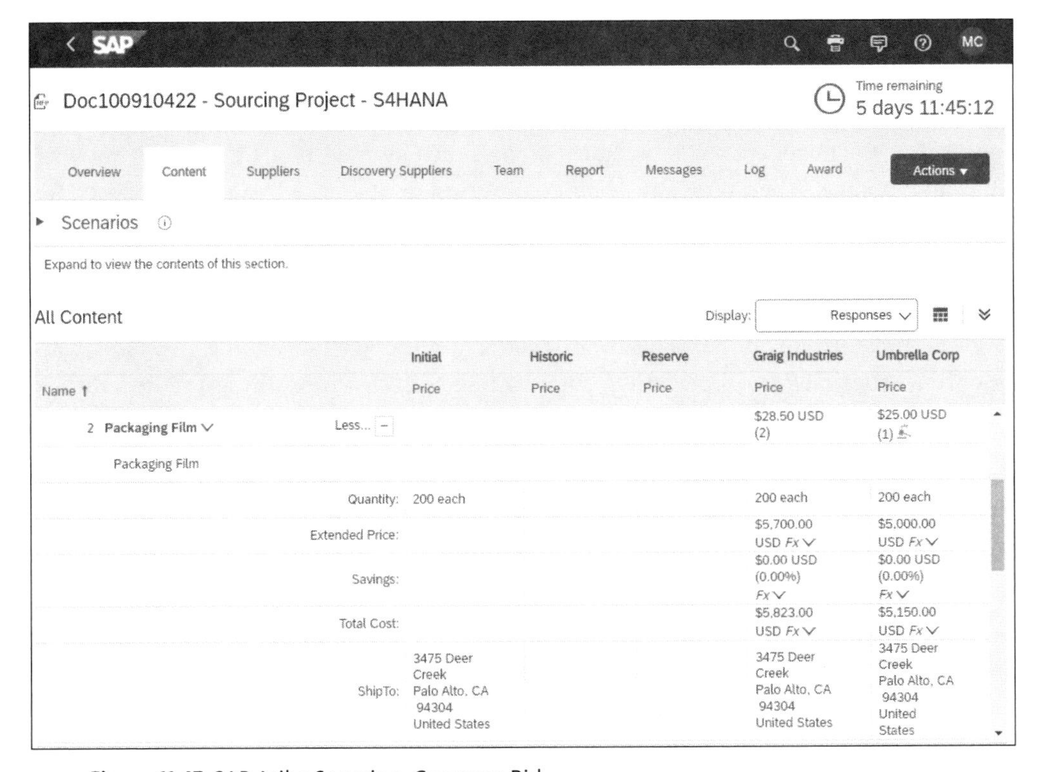

Figure 11.47 SAP Ariba Sourcing: Compare Bid

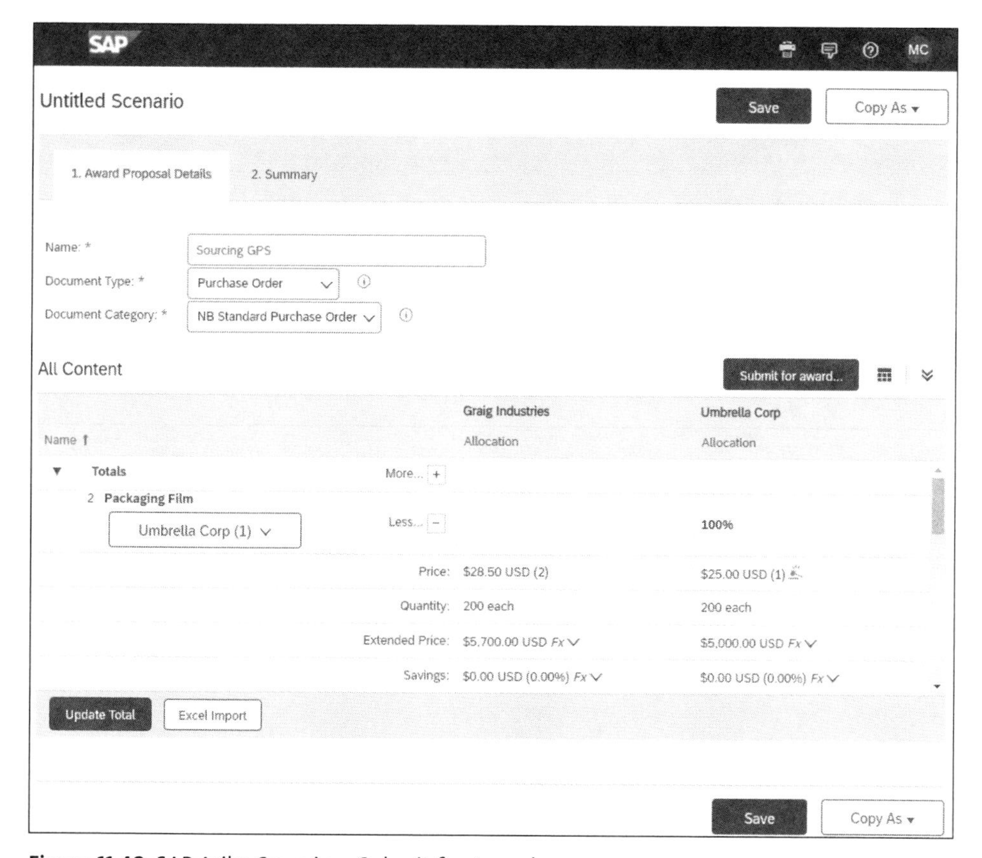

Figure 11.48 SAP Ariba Sourcing: Submit for Award

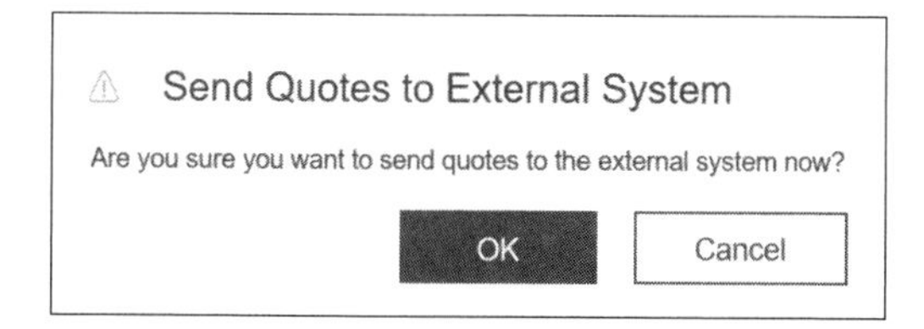

Figure 11.49 SAP Ariba Sourcing: Send Quotes to External System (SAP S/4HANA)

11.3.7 Creating Follow-On Documents for Requests for Proposals

After the awarded RFP is sent back to SAP S/4HANA, you can generate POs referencing the terms and pricing agreed to in the RFP and issue the POs from SAP S/4HANA.

Another integration scenario is to create the follow-on documents directly in SAP Ariba, leveraging the SAP S/4HANA system to generate the initial RFP, but then running the follow-on documents up to or including the invoice. As shown in Figure 11.50, you aren't limited to using POs as follow-on documents in either scenario. You can also convert the RFP into a contract in SAP Ariba.

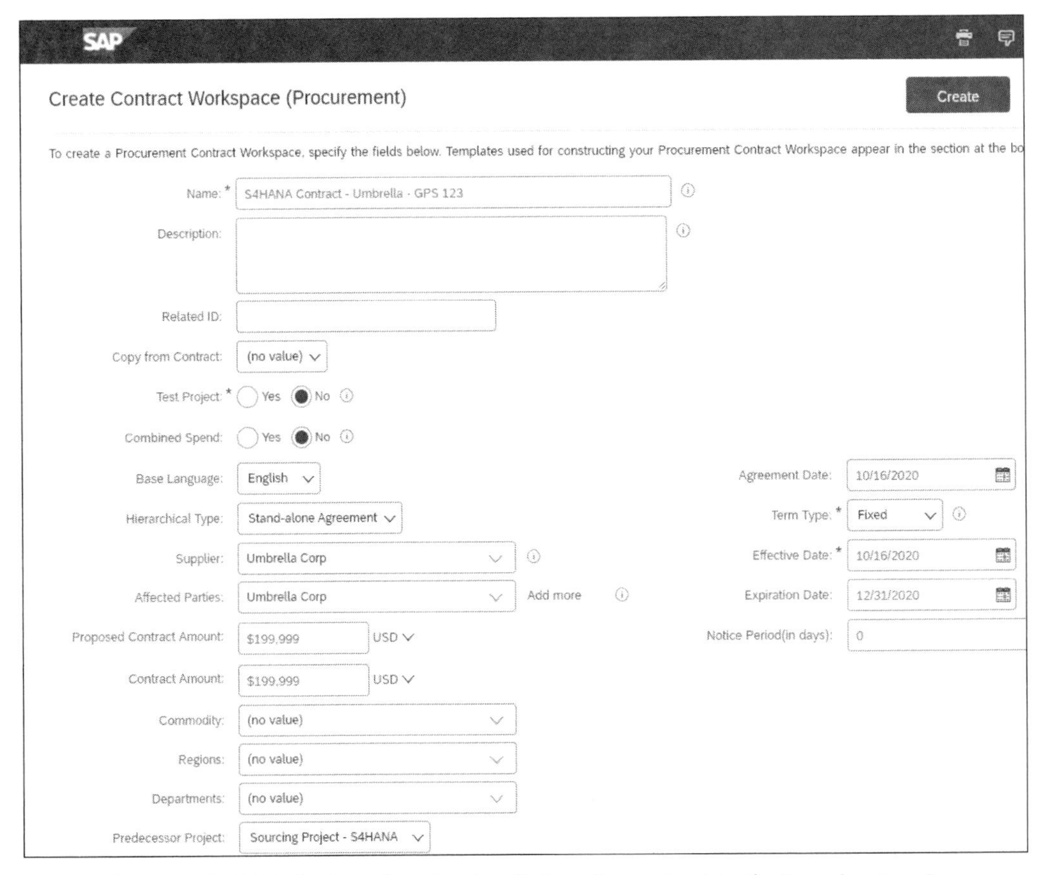

Figure 11.50 SAP Ariba Sourcing: Create a Follow-On Contract to the Sourcing Event

Depending on the scenario and your preference, integrating SAP S/4HANA and SAP Ariba Sourcing enables you to address sourcing requirements originating in SAP ERP with a much larger array of suppliers and functionality than is available in SAP S/4HANA by itself. Often, a single successful sourcing event can deliver substantial savings to the bottom line of an organization. Integrating SAP S/4HANA with SAP Ariba Sourcing greatly increases the odds and frequency for this type of sourcing success to occur. In the next section, we'll review your integration options for SAP Ariba Supply Chain Collaboration for Buyers, which extends the procurement functionality and capabilities of SAP Ariba into the world of direct procurement.

11.4 SAP Ariba Supply Chain Collaboration for Buyers

SAP Ariba Supply Chain Collaboration for Buyers brings enhanced visibility and efficiency for direct material/merchandise/supply chain collaboration in the Ariba Network. In the past, the SAP Ariba solution platform focused primarily on indirect procurement. With the release of Collaborative Supply Chain (CSC), SAP Ariba broadened this focus and the solution's capabilities to include direct procurement, which required expanding the functionality and integration with the Ariba Network. Direct procurement is driven out of multiple SAP ERP modules and is typically used to manage and control manufacturing, sales and distribution, production planning, and ever-present financial activities. The latest iteration of this solution is called SAP Ariba Supply Chain Collaboration for Buyers.

Part of the SAP Leonardo umbrella, as shown in Figure 11.51, SAP Ariba Supply Chain Collaboration for Buyers connects SAP ERP processes and scenarios in direct procurement and integrated business planning with the Ariba Network and its multitude of suppliers for collaboration. More recently, SAP Ariba Supply Chain Collaboration for Buyers is also supported by SAP Leonardo. SAP Leonardo enables customers to access intelligent technologies such as machine learning, blockchain, data intelligence, the Internet of Things (IoT), and analytics, via industry innovation kits, open innovation, and embedded intelligence.

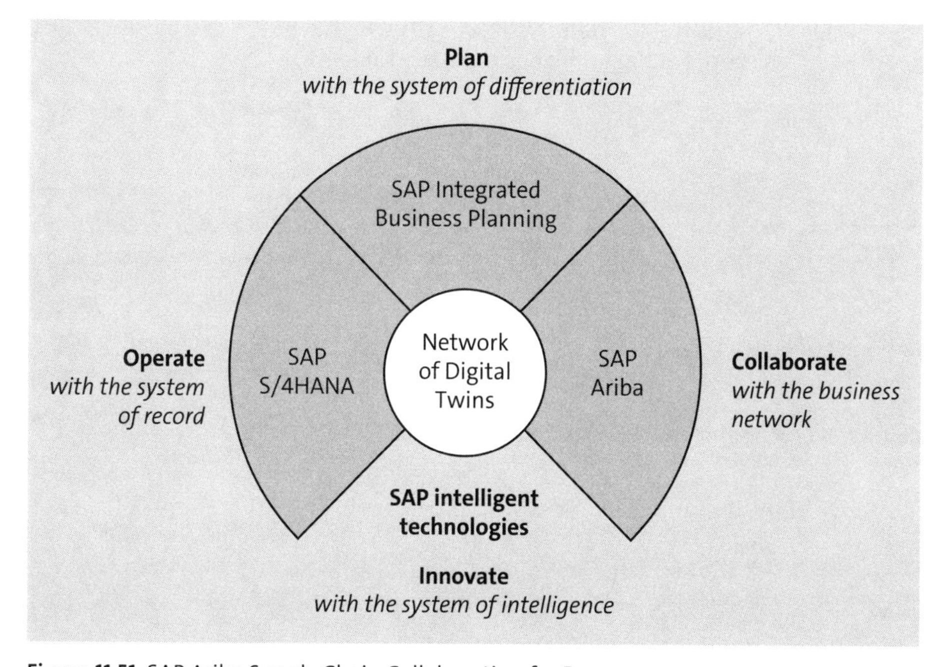

Figure 11.51 SAP Ariba Supply Chain Collaboration for Buyers

As shown in Figure 11.52, SAP Ariba Supply Chain Collaboration for Buyers comprises the direct processes of source-to-contract, plan and forecast, buy and deliver, and invoice and payment. In each instance, a combination of an SAP Ariba solution with its corresponding module in SAP S/4HANA work together to provide complete coverage of the process. In most instances, the demand originates and then is sourced or acquired in SAP Ariba, returning then to SAP S/4HANA for receipt and further processing/usage within the production process.

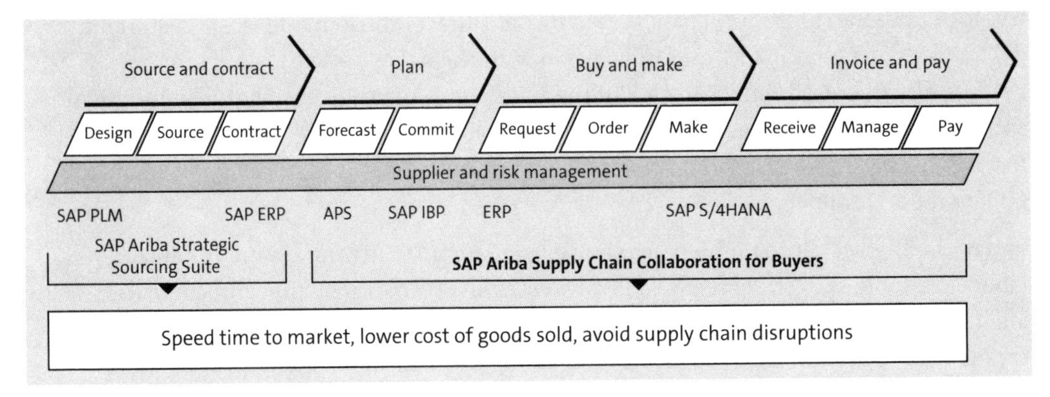

Figure 11.52 SAP Ariba Supply Chain Collaboration for Buyers

The processes covered between buyer and supplier under SAP Ariba Supply Chain Collaboration for Buyers fall into three categories: plan and forecast, buy and deliver, and invoice and payment processes. For coverage of the plan and forecast business processes in SAP Ariba Supply Chain Collaboration, see Table 11.1.

Plan and Forecast	Buyer	Supplier
Forecast collaboration	Share forecast	Forecast commit
Supplier-managed inventory	Share demand, inventory, and minimum/maximum levels	Provide and execute against shipment plan via scheduling agreement
External manufacturing visibility		Provide manufacturing and inventory status

Table 11.1 SAP Ariba Supply Chain Collaboration Planning

For coverage of the buy and deliver business processes in SAP Ariba Supply Chain Collaboration, see Table 11.2.

Buy and Deliver	Buyer	Supplier
PO collaboration (all PO types)	Place, inquire, and receive order	Confirm order and provide ASN
Vendor consignment	Report consumption of consigned stock	
Scheduling agreement release	Provide rolling delivery schedule with different firming levels	Execute against firmed orders (released)

Table 11.2 SAP Ariba Supply Chain Collaboration Buy and Deliver

Buy and Deliver	Buyer	Supplier
Subcontract/contract manufacturing with multitier orders	■ Place order with components ■ Provide notification of components' shipment ■ Provide component inventory and order visibility	■ Notify buyer of components' receipt ■ Provide component consumption
Quality notifications, quality inspections, quality reviews	■ Communicate and respond to quality notification ■ Request test results, certificate of conformance (COC), certificate of analysis (COA) ■ Collaborate and share quality documents	■ Request and respond to quality notifications ■ Submit test results, COC, COA

Table 11.2 SAP Ariba Supply Chain Collaboration Buy and Deliver (Cont.)

For coverage of the invoice and payment business processes in SAP Ariba Supply Chain Collaboration, see Table 11.3.

Invoice and Payment	Buyer	Supplier
Self-billing/evaluated receipt settlement (ERS) invoicing	■ Provide self-bill/ ERS invoice-based on goods receipt ■ Provide payment notification	Collaborate and share quality documents
Vendor invoice processing	Provide invoice processing status and payment notification	Provide invoice based on PO/ASN
Return order collaboration	Place return order and provide return ship notice	Provide credit memo

Table 11.3 SAP Ariba Supply Chain Collaboration Invoice and Payment

11.4.1 SAP Ariba Supply Chain Collaboration for Buyers Capabilities

Viewed another way, the SAP Ariba Supply Chain Collaboration for Buyers capabilities can be clustered into three groups: planning, execution, and quality. In planning, the solution provides forecast and inventory collaboration. In supply chain execution, suppliers and buyers use SAP Ariba Supply Chain Collaboration for Buyers to collaborate on direct items and services, multitier items and services, and scheduling agreements as well as subcontracting and consignment orders. In quality management, both suppliers and buyers can leverage SAP Ariba Supply Chain Collaboration for Buyers for notifications, inspection, reviews, and certificates.

Supply chains in manufacturing are often complex. Whether a high-tech manufacturer or a biomedical manufacturer, each industry has its own nuances and overlay processes, but collaboration between suppliers and buyers is always a core need. When the buyer is a manufacturer, the supplier can be a raw materials supplier, a third-party logistics provider, a subassembly supplier, a contract manufacturer, or a component supplier. SAP Ariba Supply Chain Collaboration for Buyers provides manufacturers and their supply chain stakeholders with numerous benefits, including the following:

- Simple collaboration within a single business network
- Embedded supplier onboarding services
- End-to-end process orchestration across all supplier channels to validate and enforce compliance
- Network intelligence, data, and insights to reduce supply chain risk

Table 11.4 outlines the functionality added with SAP Ariba Supply Chain Collaboration for Buyers that supports direct procurement processes through the Ariba Network and back.

Function	Description
Forecast	The forecast collaboration component helps avoid supply shortfalls and reduces/optimizes supply chain stock levels. This component compares and displays the buyer's firm or planned net demands and the supplier's firm or planned receipts. Buyer and supplier can have the same view of the planning situation and can quickly obtain a quick overview of critical situations. This functionality is integrated with SAP Integrated Business Planning (SAP IBP).
Work order	The work order collaboration component provides a consolidated view of the production demands and the current production progress. This component offers customers early insight into changes affecting shipping dates and final quantity.
Replenishment	The replenishment collaboration component offers suppliers a value-added service by performing the replenishment planning tasks for their business partners. By increasing visibility into actual consumer demand, as well as into customer inventory levels, suppliers can make better decisions on how to distribute goods across various customers, which in turn leads to increased customer service levels, lower transportation costs, reduced inventory, and lower sales costs.
Inventory	The inventory collaboration component ensures greater visibility of the inventory as well as enables supply-and-demand planning in a cross-tier environment.

Table 11.4 SAP Ariba Supply Chain Collaboration for Buyers: Functionality

Function	Description
Quality	Suppliers and customers can use the new quality collaboration component to notify each other of quality problems (or other issues with products or subcontracting components) during the manufacturing process or after delivery, so that the customer or supplier can quickly react to the complaint.

Table 11.4 SAP Ariba Supply Chain Collaboration for Buyers: Functionality (Cont.)

To keep documents consistent between the SAP Supply Network Collaboration and the Ariba Network, you must perform the business processes for POs and ASNs through the Ariba Network user interface (UI) exclusively, not through the SAP Supply Network Collaboration application.

Not all functions available in SAP supplier collaboration are supported by SAP Supply Network Collaboration integration. Table 11.5 compares the five functions supported and their related SAP Supply Network Collaboration components and starting screens.

SAP Supply Network Collaboration Integration Function	SAP Supply Network Collaboration Component	SAP Supply Network Collaboration Starting Screen
Forecast	Demand	Order forecast monitor
Replenishment	Replenishment	Replenishment vendor-managed inventory monitor
Work Order	Work order collaboration	Work order list
Inventory	Supply network inventory	Supply network inventory
Quality	Quality collaboration	Quality notification overview

Table 11.5 SAP Ariba Supply Chain Collaboration for Buyers and SAP Supply Network Collaboration Functions Compared

SAP Ariba Supply Chain Collaboration for Buyers includes several updated features and functions crucial to many direct procurement processes.

As of the predecessor product Collaborative Supply Chain 2.0, development for SAP Ariba Supply Chain Collaboration for Buyers functionality focused on scheduling agreement releases , ASNs, and order collaboration. In today's SAP Ariba Supply Chain Collaboration for Buyers, buyers and suppliers have access to many direct specific processes and fields that were previously unavailable in the SAP Ariba environment for ASNs, scheduling agreement releases, and order collaboration.

While setting up SAP Ariba Supply Chain Collaboration for Buyers, suppliers must be enabled via an adapter in the SAP Ariba Cloud Integration releases as a prerequisite to supporting the collaboration processes. This adapter is coordinated via SAP Ariba supplier enablement services. SAP Ariba began providing technical and installation support for adapters in releases as early as 2014 for both SAP ERP and Oracle Fusion.

The steps for deploying the SAP Ariba Supply Chain Collaboration for Buyers extension starts with the buyer, who must follow these steps:

1. The buyer enables a supplier for particular SAP Ariba Supply Chain Collaboration for Buyers features.
2. Data fields supporting enhanced functionalities, such as the ordering of direct materials, are passed from the buyer's SAP ERP system to the Ariba Network.
3. Through an enhanced Ariba Network UI, the enabled supplier can now use the additional fields and enhanced functionalities to collaborate with the buyer on documents such as POs, invoices, and ASNs.

11.4.2 Supplier Record in SAP ERP

To prepare a supplier record in SAP ERP for SAP Ariba Supply Chain Collaboration for Buyers activities, a relationship flag must be set. This functionality is standard and doesn't require additional installation or configuration. After the flag is set, industry-specific functionality can be configured. To complete this activity, your user role must have supplier enablement task management permissions. Buyers don't require further enablement for SAP Ariba Supply Chain Collaboration for Buyers.

As an administrator, you can disable SAP Ariba Supply Chain Collaboration for Buyers functionality for selected suppliers. To disable SAP Ariba Supply Chain Collaboration for Buyers or retail-industry functionality for a supplier, follow these steps:

1. Click the **Supplier Enablement** tab.
2. Click **Active Relationships**.
3. Select the suppliers for whom SAP Ariba Supply Chain Collaboration for Buyers or retail-industry functionality should be disabled.
4. Click the **Actions** menu at the bottom of the supplier table, and select **Enable/Disable Collaborative Supply Chain**. Next, perform the following steps:
   - To disable SAP Ariba Supply Chain Collaboration for Buyers and the retail-industry functionality for the selected suppliers, deselect the **Collaborative Supply Chain** checkbox.
   - To disable only retail-industry functionality for the selected suppliers, deselect **Retail Industry** fields. (Note that the **Retail Industry** setting enables supplier for the **Ship Notice Due List**.)
5. Click **OK**.

After suppliers have been appropriately flagged, they can be directed to the SAP Ariba Supply Chain Collaboration for Buyers supplier portal. The supplier portal contains all existing Ariba Network supplier functionality, plus SAP Ariba Supply Chain Collaboration for Buyers and related industry-specific functionality. When suppliers are enabled for SAP Ariba Supply Chain Collaboration for Buyers, they are directed to the supplier portal, where they can access various features.

11.4.3 Supplier and Buyer Collaboration in the Ariba Network

Suppliers and buyers often collaborate on the creation of POs, invoices, and ASNs for direct material orders. This collaboration requires the exchange of information specific to direct material orders. These features facilitate the collaboration process by allowing more detailed descriptions of the materials being procured. These features also allow the supplier to collaborate through the Ariba Network UI, which is particularly helpful for suppliers that don't have electronic data interchange (EDI), commerce eXtensible Markup Language (cXML), or other integrated business-to-business mechanisms in place.

Direct material fields for order and invoice collaboration facilitate the collaboration process for POs, invoices, and ASNs by allowing more detailed descriptions of the materials being procured. This Ariba Network feature is applicable to all supplier users. The buyer's industry type must be set as **Retail** to enable this functionality. For buyers using external ERP systems, such as SAP ERP, the adapter is required.

Ship notice business rules for order confirmation and ship notice rules in general support the direct material collaboration process. This Ariba Network feature is applicable to all supplier users.

Direct material includes all items, such as raw materials and parts, required to assemble or manufacture a complete product. A rule may be implemented based on customer and market requirements or may be driven by individual needs. A rule can be applied to many suppliers or to a predefined group of suppliers (even if the group only contains one member). Tooltips in the UI indicate when ship notice rules have been enabled.

The following ship notice rules particularly support collaboration between buyers and suppliers with regard to direct materials:

- **Require suppliers to provide a unique asset serial number for each PO line item on a ship notice.**
  This rule applies within the scope of the order line item. Using the same serial number on any of the ship notice line items for this order line item isn't allowed. If validation fails, the error message the supplier receives is **Asset serial numbers for item shipments must be unique**.
- **Require the total count of shipment serial numbers not to exceed the total quantity shipped for line items on ASNs.**
  Within a ship notice item, the number of serial numbers must be less than or equal to the quantity shipped. If validation fails, the error message you'll receive is **The total number of shipment serial numbers exceeds the quantity shipped for the line item**.

- **Require the packing slip ID to be unique on ASNs.**
  Duplicate shipment notice numbers (packing slip IDs) aren't allowed. The rule is checked against all ship notice numbers for the particular supplier without a time window restriction. If validation fails, the error message you'll receive is **This packing slip ID has already been used. Please enter a unique ID**.

Buyers often want to procure predefined quantities of materials in releases received on predetermined dates. For example, a buyer might order 100 items but want them delivered 10 at a time on a weekly basis. A buyer could therefore send a scheduling agreement release to a supplier.

11.4.4 Scheduling Agreements in SAP Ariba Supply Chain Collaboration for Buyers

A scheduling agreement release defines the quantities and dates for the shipments desired for a specific item. The scheduling agreement functionality touches both any supplier for whom you've enabled this functionality, as well as SAP Ariba PO automation.

A scheduling agreement release can have multiple schedule lines for any single line in a PO or other business document. A single schedule line contains fields for delivery date and time and for scheduled quantity and unit of measure (UOM). Buyers create schedule agreement releases with external ERP systems.

Prior to CSC 2.0, suppliers could view schedule lines but could not use them when creating ASNs through the Ariba Network UI. With the release of CSC 2.0, scheduling agreement releases are visible in the Ariba Network UI. This integration has been further built out in SAP Ariba Supply Chain Collaboration for Buyers. Buyers can periodically communicate releases to their suppliers, and suppliers can create ASNs for materials requested through scheduling agreement releases. Under the new **Items to Ship** tab, a supplier can create a single ship notice containing items from several POs that all are due by the same date.

New functionality allows buyers to configure whether their suppliers can confirm scheduling agreements and scheduling agreement releases for firm schedule lines. This gives you better visibility into expected delivery dates and the supplier's ability to meet requested demand. Note order confirmations for trade-off or forecasted schedule lines aren't supported. The processing of nonscheduled items isn't affected by scheduling agreement releases.

11.4.5 Scheduling Agreement Release Collaboration

Scheduling agreement release collaboration is disabled by default, but a buyer can enable it by following these steps:

1. Log in as a buyer.

2. Click **Manage Profile**.
3. On the **Configuration** page, click **Default Transaction Rules**.
4. Under **Scheduling Agreement Release Setup Rules**, select the **Allow Scheduling Agreement Release Collaboration** checkbox.

Suppliers can view scheduling agreement releases in the list of documents on the **Orders and Releases** tab and choose whether to search for only scheduling agreement releases. Suppliers can then create ASNs from scheduling agreement releases.

The cXML 1.2.028 Document Type Definition (DTD) introduced cXML elements and attributes that support scheduling agreement releases, as summarized in Table 11.6. For more information about cXML, see the DTD and cXML User's Guide available at *www.cxml.org*. In addition, see the cXML Solutions Guide on the Ariba Network.

11

cXML Element for Scheduling Release	Description
isSchedulingAgreementRelease	Indicates that the type of PO is a schedule agreement release. When isSchedulingAgreementRelease is set to yes, orderType must be set to release, for example: <OrderRequestHeader orderDate="2014-05-08T14:37:31-07:00" orderID="SAR301-1-FORECAST" orderVersion="1" orderType="release" agreementID="SA301" type="new">
ReleaseInfo	The OrderRequestHeader element has a new attribute, isSchedulingAgreementRelease, and indiRelease-Info has been added to ItemOut as an optional element. ReleaseInfo stores details about a release of items or materials.
ScheduleLineReleaseInfo	The element ScheduleLineReleaseInfo has been added to the ScheduleLine element. ScheduleLine-ReleaseInfo stores details about a specific release of items or materials for a schedule line. ScheduleLine-ReleaseInfo contains the following attributes.
AgreementType	The attribute agreementType has been added to the MasterAgreementIDInfo element and MasterAgree-mentReference element to indicate whether the refer-enced agreement is a scheduling agreement release. For example, MasterAgreementIDInfo: <MasterAgreementIDInfo agreementID="SA301" agreementType="scheduling_agreement"/>

Table 11.6 Elements for Scheduling Agreement Release

cXML Element for Scheduling Release	Description
ShipNoticePortion	ShipNoticePortion contains two optional elements, MasterAgreementReference and MasterAgreement-IDInfo, which can contain a reference or an ID, respectively, to the master agreement from which the release is derived. For example: <ShipNoticePortion> <OrderReference orderID="OD1" orderDate= "2014-01-28T11:15:32-08:00"> </OrderReference> <Contact role="buyerCorporate"> </Contact> <ShipNoticeItem quantity="1" line number="1" </ShipNoticeItem> </ShipNoticePortion>

Table 11.6 Elements for Scheduling Agreement Release (Cont.)

Table 11.7 and Table 11.8 summarize additional optional subelements and attributes for ReleaseInfo, ScheduleLineReleaseInfo, and ShipNoticePortion.

Type	Name	Description
Subelement	ShipNoticeReleaseInfo	References the previous ship notice created from a delivery schedule. This reference is against the previous shipment made for the schedule line in the schedule agreement release.
Subelement	UnitofMeasure	UOM for the quantity specified for the schedule line item.
Subelement	Extrinsic	Any additional information for the schedule line item.
Attribute	releaseType	A mandatory field. A string value to identify the type of delivery schedule against the schedule agreement release. Possible values include the following: ▪ Just in time (JIT) ▪ Forecast
Attribute	Forecastcumulative-ReceivedQuantity	A mandatory field. A number value to identify the cumulative quantity of all goods received against the scheduling agreement release over a period up to a certain date.

Table 11.7 Release Information

Type	Name	Description
Attribute	productionGoAheadEndDate	An optional field. Date denoting the end of the production go-ahead period (go-ahead for production).
Attribute	materialGoAheadEndDate	Date denoting the end of the material go-ahead period (go-ahead for purchase of input materials).

Table 11.7 Release Information (Cont.)

The following example shows the ReleaseInfo element in action:

```
<ReleaseInfo releaseType="forecast" cumulativeReceivedQuantity="0"
productionGoAheadEndDate="2014-06-13T14:37:31-07:00"
materialGoAheadEndDate="2014-11-14T14:37:31-07:00"
<UnitOfMeasure>EA</UnitOfMeasure>
</ReleaseInfo>
```


Type	Name	Description
Attribute	commitmentCode	A string value to identify the type of the delivery. The value can be any of the following: ■ Firm: Go-ahead for production. Vendor can ship against the schedule line. Customer is responsible for cost of production as well as cost of material procurement. ■ Trade-off: Go-ahead for material procurement. Vendor can ship against the schedule line if rule is enabled. Buyer is responsible for cost of material procurement. ■ Forecast: Informational. Customer can change the schedule line without incurring any liabilities with the vendor.
Attribute	cumulativeScheduledQuantity	Total quantity to be shipped for a particular line item up through the schedule line.

Table 11.8 Schedule Line Release

Table 11.8 demonstrates direct procurement's use of scheduling agreements. Scheduling agreements allow a supplier to deliver items based on a predetermined schedule, with additional orders released at the line level at predefined times. This scenario isn't typical for indirect procurement, but quite prevalent in direct procurement. The following example shows the ScheduleLineReleaseInfo element in action:

```
<ScheduleLine quantity="100" requestedDeliveryDate="2014-05-10T14:37:31-07:00">
<UnitOfMeasure>EA</UnitOfMeasure>
<ScheduleLineReleaseInfo commitmentCode="firm" cumulativeScheduledQuantity="100">
<UnitOfMeasure>EA</UnitOfMeasure>
<ScheduleLineReleaseInfo>
</ScheduleLine>
```

Master agreement references can be used to support releases done on master contracts or agreements. Capturing this field allows you to manage amounts outstanding at the master agreement level and allows for contract-based direct procurement. For these scenarios, an optional field is available in SAP Ariba Supply Chain Collaboration for Buyers, as shown in Table 11.9.

Type	Name	Description
Element	MasterAgreementReference	This optional field can contain a reference to the master agreement from which the release is derived.
Element	MasterAgreementIDInfo	This optional field can contain the ID of the master agreement from which the release is derived.

Table 11.9 Master Agreement

The following example shows the ShipNoticePortion element in action:

```
<ShipNoticePortion><OrderReference orderID="OD1" orderDate= "2014-01-28T11:15:32-08:00">
</OrderReference>
<Contact role="buyerCorporate">
</Contact>
<ShipNoticeItem quantity="1" linenumber="1"
</ShipNoticeItem>
</ShipNoticePortion>
```

11.4.6 Contract Manufacturing Collaboration

Large organizations can reduce their costs by transferring the manufacturing, assembly, or processing of parts to third-party contract manufacturers. Contract manufacturers are suppliers who build the parts for these buyers.

With the contract manufacturing collaboration feature, buyers and contract manufacturers can manage the supply of component materials and finished goods through the

Ariba Network. The contract manufacturing collaboration feature includes the following functionalities:

- Buyers can create subcontract POs, which can be transmitted to contract manufacturers via the Ariba Network.
- Contract manufacturers can view all the data of subcontract POs, including all the component information under the PO line item in the schedule line section.
- Contract manufacturers can confirm orders for finished goods purchased by buyers.
- Buyers can create ASNs for subcontracting components and can send these notices to contract manufacturers.
- Contract manufacturers can create receipts for subcontracting components and can send these receipts to buyers.
- Contract manufacturers can create ASNs for finished goods to be shipped to buyers.
- Contract manufacturers can send component consumption information to buyers for components used to manufacture the finished goods on the subcontract PO.

To support these functionalities, the following message types have been enhanced or added for direct materials buyers and contract manufacturers:

- **Subcontracting PO**
  A subcontracting PO is sent from a buyer to a contract manufacturer to request the production and delivery of finished goods.

  The subcontracting PO has been enhanced to carry not only item level and schedule line level information but also subcontracting component information. Subcontracting components are the raw materials that are used for manufacturing the finished goods specified at the item level.
- **Component ship notice**
  A component ship notice is a type of ship notice that informs the contract manufacturer of the shipment of subcontracting components.
- **Component receipt**
  A component receipt is a type of goods receipt that informs the buyer of the receipt of subcontracting components. The contract manufacturer can issue the component receipt against one or more component ASNs.
- **Component consumption (backflush) message**
  The ship notice message has been enhanced to include consumption details in a backflush component message. A backflush component consumption message is a type of ship notice request that informs the buyer of the completion of finished goods from subcontracting components. Unlike real-time component consumption reporting, backflush reporting is done only once, at the end of the production process.

- **Component consumption (real-time) message**
  A real-time component consumption message informs the buyer of the consumption of components at any phase of the production cycle.
- **Component inventory message**
  A component inventory message has been added to inform suppliers about quantities of components available for manufacturing.
- **SAP Ariba Supply Chain Collaboration for Buyers limitation**
  If a subcontracting PO has been uploaded or downloaded as a comma-separated values (CSV) file, a supplier won't see the component level on the downloaded subcontracting PO.

11.4.7 Order Collaboration Process for Direct Materials

In the order collaboration process for direct materials, suppliers typically strive to establish long-term, repeat business with their customers. From the supplier's perspective, a successful business model involves a continual flow of incoming POs from a number of buying customers. The supplier must efficiently process these incoming POs: confirming line items when confirmation is required, creating ASNs, and shipping goods. In addition, the supplier must process POs that have been changed. Further, the supplier must not confirm orders for quantities or delivery dates that aren't achievable. The process of confirming orders quickly and correctly, therefore, can be quite challenging because of the sheer volume of data and the level of detail required, which is where the items to confirm list comes in.

An *items to confirm list* enables a supplier to efficiently confirm incoming POs across multiple customers. The items to confirm list displays a list of POs, including information critical for a supplier to accurately confirm many order details simultaneously.

11.4.8 Purchase Order Line Item Views

As a buyer, you can enable or disable the visibility of PO line items in the items to confirm list. This feature is enabled by default and to configure it, you should perform the following actions:

1. Click **Manage Profile • Default Transaction Rules**.
2. Under the **Order Confirmation** heading and **Ship Notice Rules**, select or deselect the **Allow suppliers to send order confirmations on line item level** checkbox. If deselected, POs between the buyer and the corresponding supplier will remain visible but aren't actionable in that supplier's items to confirm list.
3. Select or deselect the **Allow suppliers to send order confirmations to this account** checkbox. If disabled, no POs from this buyer will be visible in the supplier's items to confirm lists.
4. Click **Save**.

11.4.9 Invoice Enhancements for Self-Billing in the Ariba Network

The Ariba Network also supports invoice enhancements for self-billing and scheduling agreement releases for direct material suppliers. For direct material suppliers and buyers, the Ariba Network can automatically transmit self-billing invoices based on data in POs, scheduling agreements, and goods receipts. The buyer can create a self-billing invoice in an external business system and send this invoice to a supplier over the Ariba Network. Suppliers and buyers can deploy self-billing invoices when they've agreed to not require the supplier to manually create invoices for order transactions, but instead agree to use the automated *evaluated receipt settlement (ERS)* routine mentioned earlier in Table 11.3. ERS determines the invoice amount for ordering transactions from the prices entered in the order, terms of payment, tax information, and delivery quantity entered in the goods receipt.

With the invoice enhancements for self-billing and scheduling agreement releases, suppliers and buyers can employ self-billing invoices for scheduling agreements as well as for POs.

11.4.10 Consignment Collaboration

With consignment collaboration, the supplier sends consignments to the buyer but still retains ownership of the consignments until the buyer pays for the consignments the buyer consumes through a prearranged PO method. Consignments stocked in the buyer's warehouses or production facilities aren't paid for until they have been used. The transfer of ownership and often of location to the buyer at the time of consumption is called a *consignment movement*. SAP Ariba Supply Chain Collaboration for Buyers enables suppliers to view a list of materials on consignment and create invoices for consignment movements, and it enables buyers to create self-billing invoices for consignment movements.

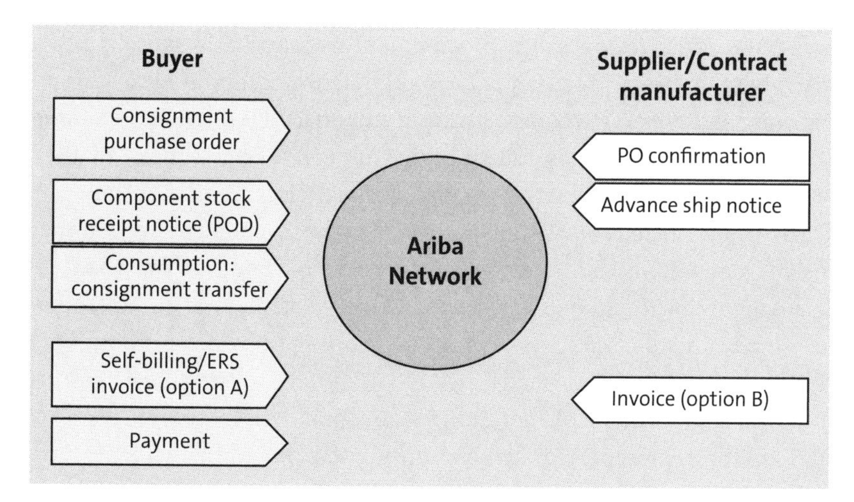

Figure 11.53 Consignment Process in SAP Ariba Supply Chain Collaboration for Buyers

As shown in Figure 11.53, the workflow for consignment collaboration is as follows:

1. The buyer creates or changes a consignment PO and sends the order to a supplier via the Ariba Network.
2. The supplier finds a PO for a consignment item, clearly differentiated as such in the **Item Category** field. The supplier creates a consignment fill-up order and sends a consignment PO confirmation to the buyer via the Ariba Network.
3. The supplier creates a ship notice for the consignment PO and sends the ship notice via the Ariba Network to the buyer, who receives notice of an inbound delivery.
4. The buyer reports the transfer of materials from consignment to the buyer's own stock to the supplier via the Ariba Network.
5. The supplier views consignment movements in Ariba Network and can execute either of the following:
   - The buyer executes a consignment settlement and sends a settlement invoice to the supplier via the Ariba Network.
   - The supplier creates an invoice against one or more consignment movements and sends the invoice to the buyer via the Ariba Network.

SAP Ariba Supply Chain Collaboration for Buyers takes a broad approach to integrating and supporting direct procurement processes originating in SAP ERP via the Ariba Network by supporting major direct processes from scheduling agreements to consignment to contract manufacturing. SAP Ariba Supply Chain Collaboration for Buyers has been deployed extensively for customers worldwide, extending the rich, collaborative functionality of the Ariba Network to SAP ERP direct procurement processes.

11.4.11 Conducting Transactions in SAP Ariba Supply Chain Collaboration for Buyers

Issuing a direct PO or conducting a direct procurement transaction is typically more automated than with indirect procurement, as production planning systems and inventory management systems interact during a material requirements planning (MRP) run to create orders. After these direct orders are created and processed, they typically are transmitted directly to a supplier via a transmission method, such as EDI, email, or fax. EDI, while efficient once established, is a point-to-point connection that can be expensive to maintain. To interact with suppliers, collaboration options typically must be augmented with an in-house portal. Other transmission and collaboration methods don't provide seamless communication options for running the transaction and basing production off lead times, ASNs, and so on. Using the Ariba Network with SAP Ariba Supply Chain Collaboration for Buyers allows you to connect these complex orders and processes, providing a platform-based, one-to-many connection option to direct suppliers. In essence, SAP Ariba Supply Chain Collaboration for Buyers leverages the Ariba Network to provide additional benefits to EDI and company-maintained supplier portals for direct procurement and complex indirect procurement at a fraction of the setup and maintenance costs. Beginning with planning, in the following subsections, we'll review how to conduct transactions in three core areas.

SAP Integrated Business Planning with SAP Ariba Supply Chain Collaboration for Buyers

SAP IBP is a cloud-based solution that integrates with the SAP Supply Chain Control Tower to provide forecasting and demand management, inventory optimization, sales and operations planning, and response and supply planning. SAP Ariba Supply Chain Collaboration for Buyers also integrates with SAP IBP in two different planning areas: forecast collaboration and supplier-managed inventory collaboration (also called inventory visibility).

Forecast Planning and Collaboration in SAP Ariba Supply Chain Collaboration for Buyers

The first type of planning activity supported in SAP Ariba Supply Chain Collaboration for Buyers is forecast collaboration and commitment between the buyer and the supplier, as shown in Figure 11.54.

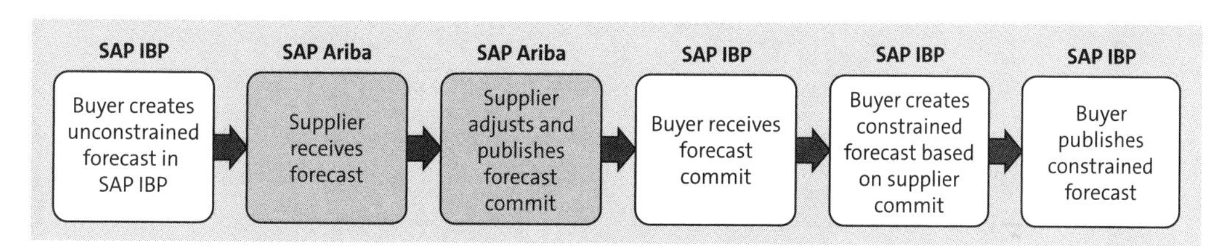

Figure 11.54 SAP IBP and SAP Ariba Supply Chain Collaboration for Buyers: Forecast Review and Commit

In this planning activity, the following steps are taken:

1. Log in to the SAP Fiori Manage Cases app in SAP S/4HANA to share your forecasts with suppliers in SAP Ariba Supply Chain Collaboration for Buyers. Note that the planner in SAP S/4HANA is working directly in Microsoft Excel to perform further analysis and can share the forecast with SAP Ariba Supply Chain Collaboration for Buyers directly from this workspace.
2. The supplier logs in to the Ariba Network and reviews the forecast.
3. The supplier then commits quantities to the SAP IBP forecast in SAP Ariba Supply Chain Collaboration for Buyers, as shown in Figure 11.55.

Key figures	1 Apr 2019	2 Apr 2019	3 Apr 2019	4 Apr 2019	5 Apr 2019	6 Apr 2019	7 Apr 2019	8 Apr 2019	9 Apr 2019	10 Apr 2019
Order forecast	1923							2313		
Cumulative forecast	1923	1923	1923	1923	1923	1923	1923	4236	4236	4236
Forecast commit	1600							1800		
									Cancel	Save
Cumulative forecast commit	0	0	0	0	0	0	0	0	0	0
Cumulative commit vs cumulative forecast	-1923	-1923	-1923	-1923	-1923	-1923	-1923	-4236	-4236	-4236
Previous forecast	1923							2313		
Cumulative previous forecast	1923	1923	1923	1923	1923	1923	1923	4236	4236	4236
Cumulative forecast vs Cumulative previous forecast	0	0	0	0	0	0	0	0	0	0
Forecast deviation	-1923	0	0	0	0	0	0	-2313	0	0
Forecast change	0	0	0	0	0	0	0	0	0	0
Previous forecast commit										
Upside forecast										

Figure 11.55 SAP Ariba Supply Chain Collaboration for Buyers: Supplier Confirming the Forecast

4. When the commitments revert to the SAP S/4HANA environment with SAP IBP, any shortages can be highlighted in the Custom Alerts app, as shown in Figure 11.56.
5. The planner can do further analysis of this shortage and general planning areas in the Dashboard app in SAP S/4HANA. As shown in Figure 11.57, the planner has found a drop in on-time deliveries with this particular supplier and can investigate further.

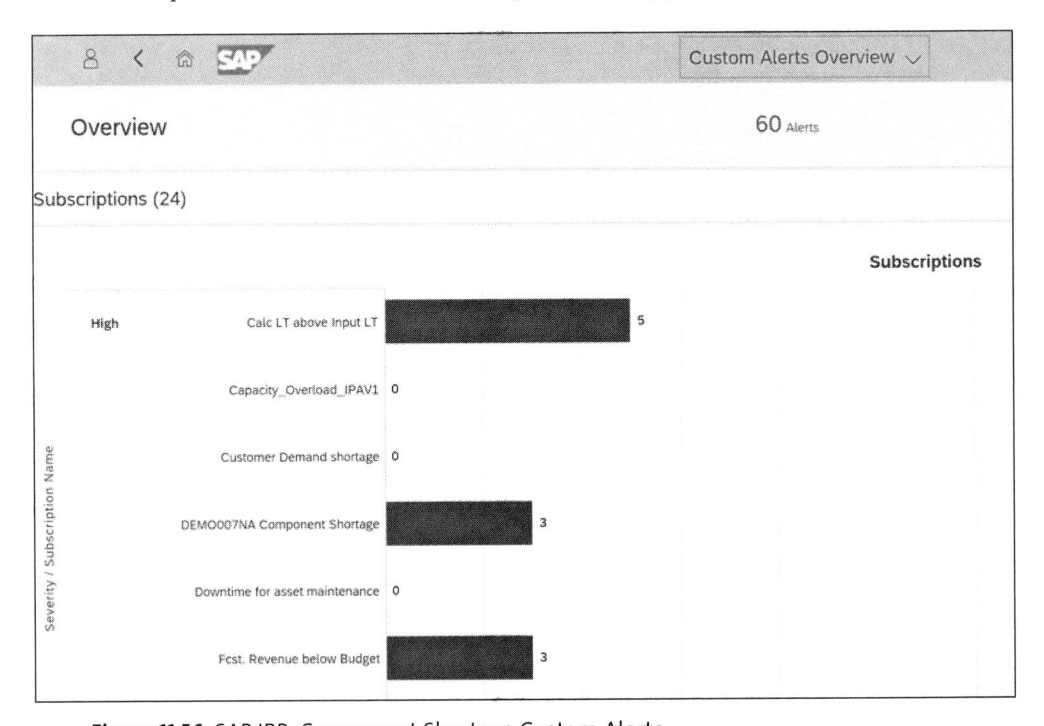

Figure 11.56 SAP IBP: Component Shortage Custom Alerts

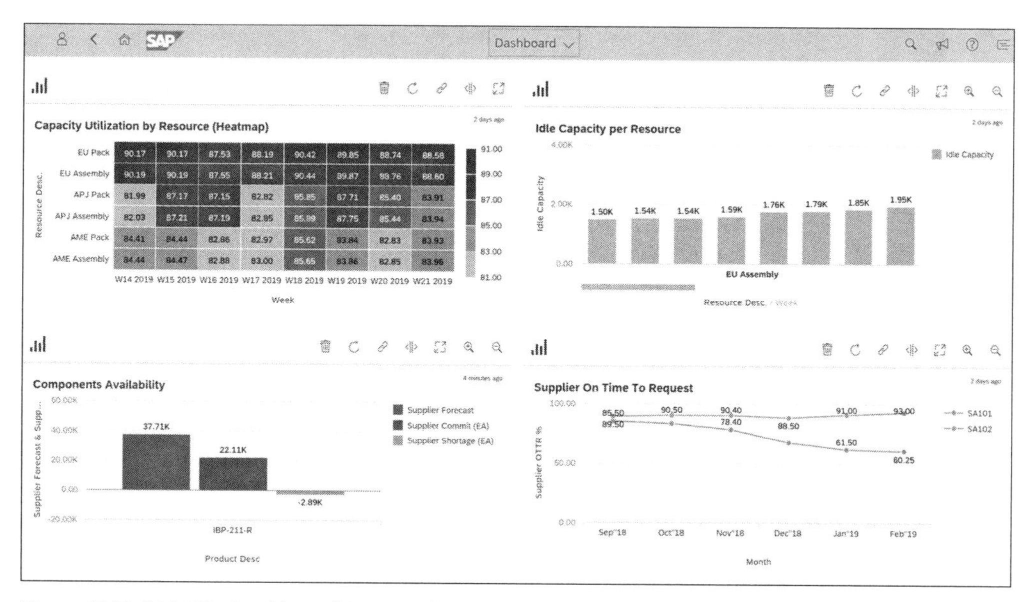

Figure 11.57 SAP IBP: Dashboard in SAP S/4HANA

6. Drilling into the components' availability reveals a supplier shortage, as shown in Figure 11.58.

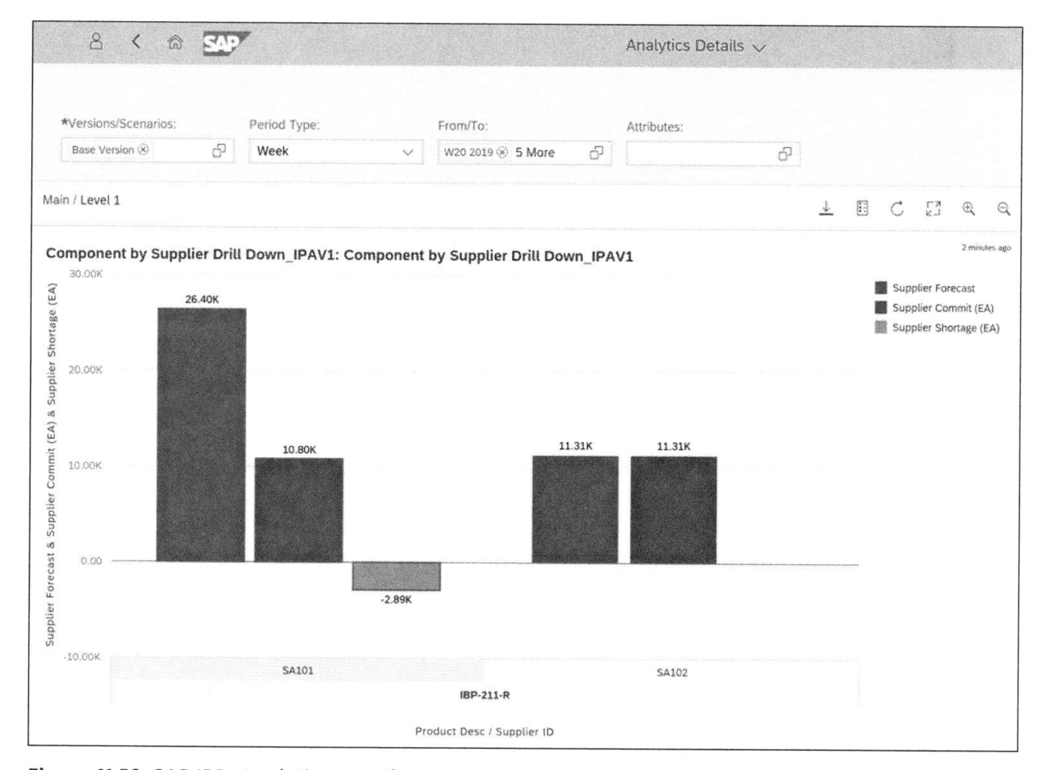

Figure 11.58 SAP IBP: Analytics Details

7. The planner can then use SAP IBP to plan with another supplier who can meet the requirements, even if the cost is marginally higher, as shown in Figure 11.59.

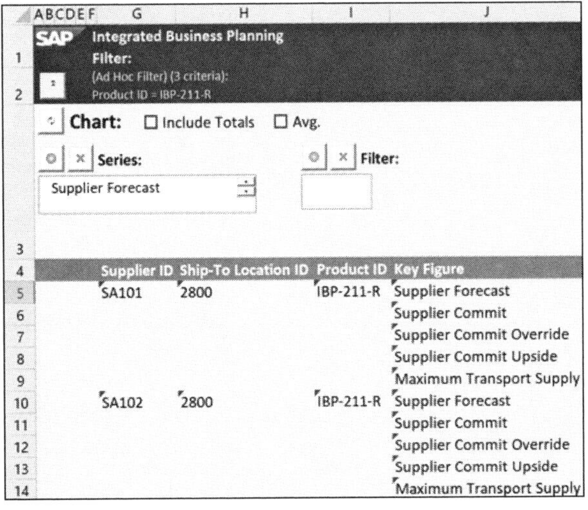

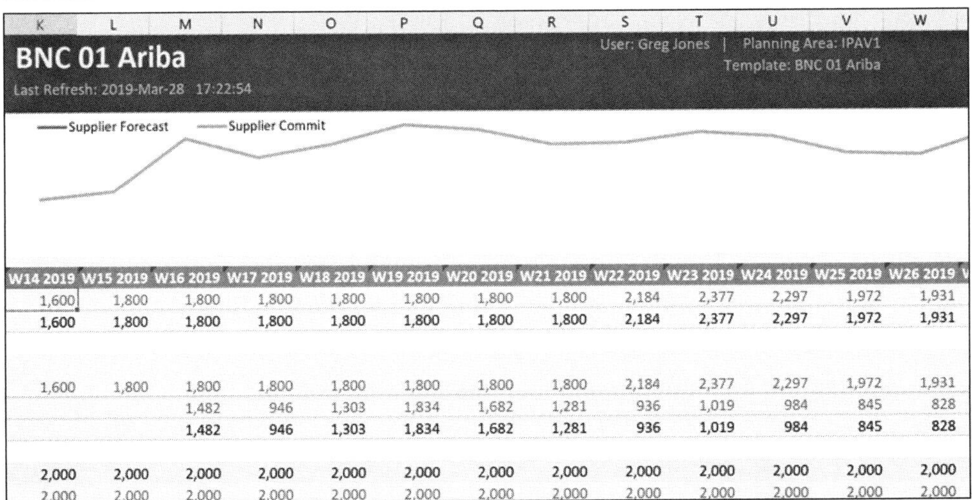

Figure 11.59 Optimization with Higher-Cost Supplier in the SAP IBP Excel Add-In

Inventory Visibility and Collaboration in SAP Ariba Supply Chain Collaboration for Buyers

The second type of planning activity supported in SAP Ariba Supply Chain Collaboration for Buyers is inventory visibility, where the supplier publishes its stock on hand to the buyer, and the buyer adjusts its planning accordingly, as shown in Figure 11.60.

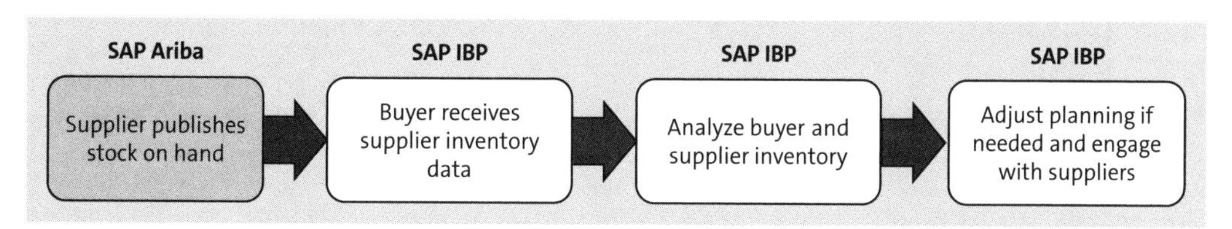

Figure 11.60 SAP Ariba Supply Chain Collaboration for Buyers: Inventory Visibility

In SAP Ariba Supply Chain Collaboration for Buyers, a supplier can provide stock on hand visibility by going into the **Planning Collaboration** area, as shown in Figure 11.61.

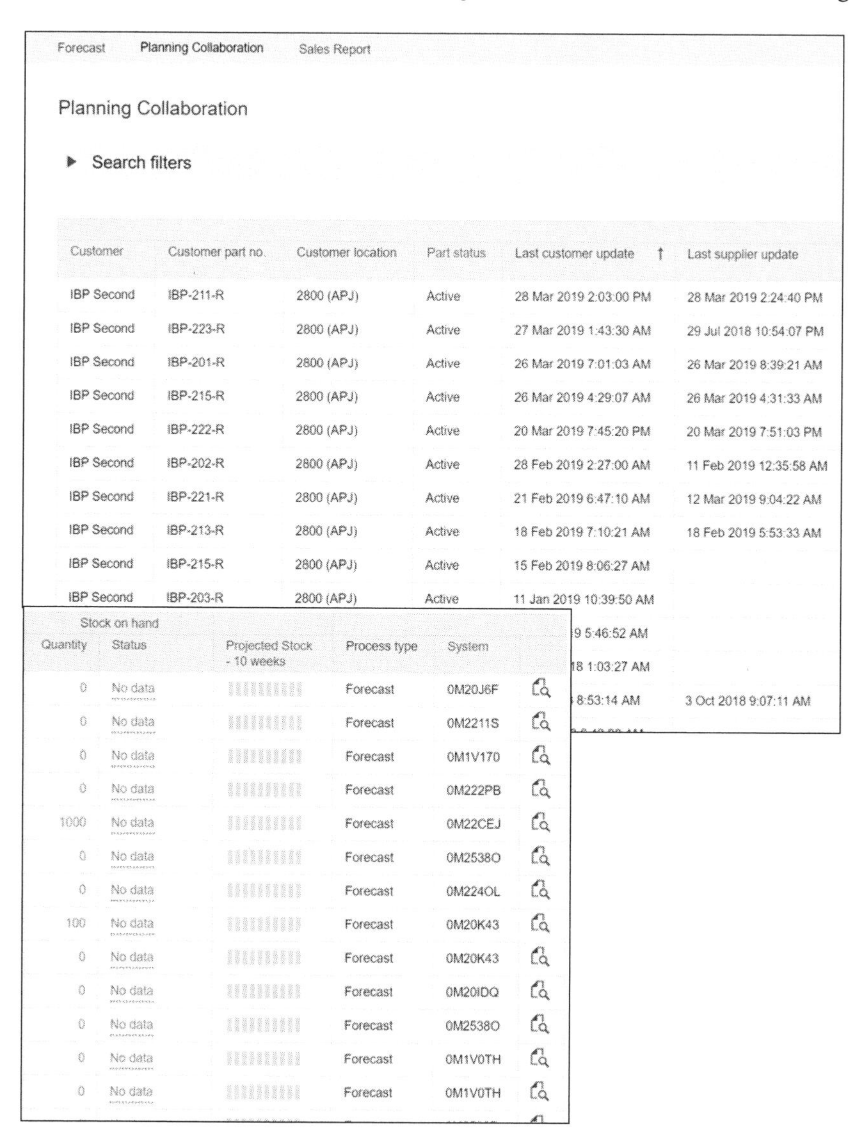

Figure 11.61 SAP Ariba Supply Chain Collaboration for Buyers: Supplier Inventory Planning/Collaboration

SAP S/4HANA Scheduling Agreements in SAP Ariba Supply Chain Collaboration for Buyers

Scheduling agreements can augment the classic direct procurement process, as shown in Figure 11.62.

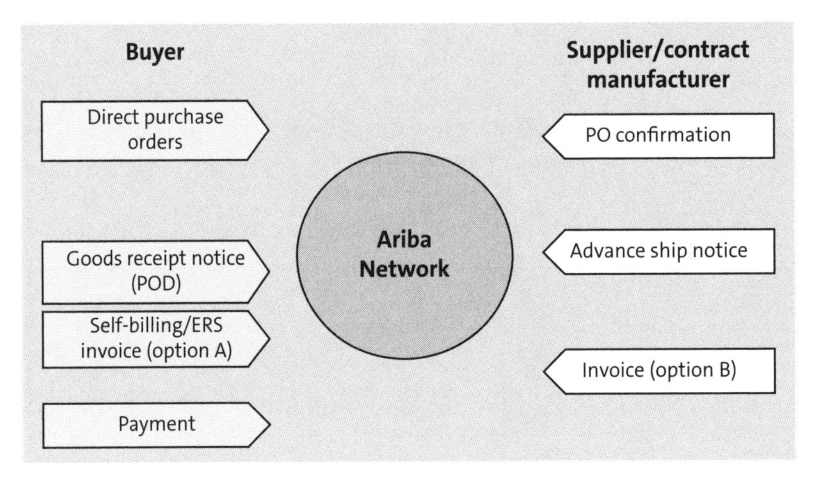

Figure 11.62 Direct Procurement Process in SAP Ariba Supply Chain Collaboration for Buyers

A scheduling agreement document is similar to a PO because the quantities are set up front, and it's like a contract because deliveries can be added or subtracted on the scheduling agreement throughout the time range. Scheduling agreements extend a rolling delivery schedule to the supplier. As a result, a scheduling agreement must be updated occasionally, and the update must be sent to the supplier. Delays in communication can result in poor order management, which trickles down into the manufacturing process. Scheduling agreements provide the supplier transparency into the buyer's longer-term horizons and also offer more flexible invoicing options, as shown in Figure 11.63.

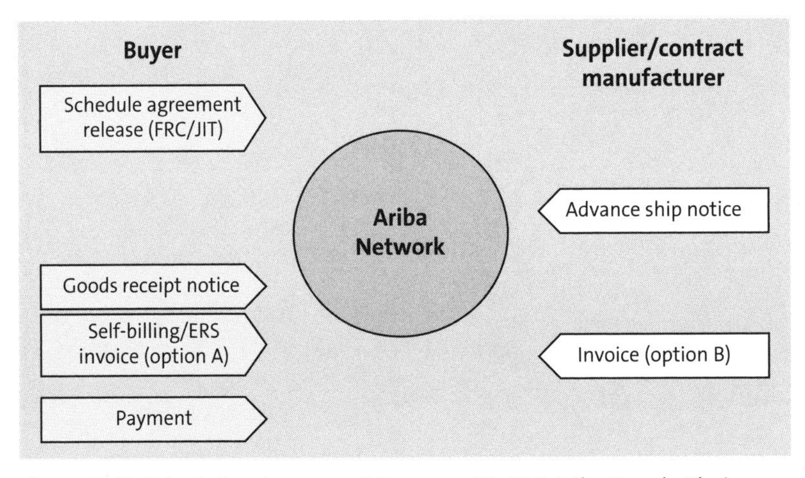

Figure 11.63 Scheduling Agreement Process with SAP Ariba Supply Chain Collaboration for Buyers

In this planning activity, the following steps are taken:

1. Beginning in SAP S/4HANA, a direct procurement buyer can create a scheduling agreement, as shown in Figure 11.64.
2. After filling out the initial page, the buyer then sets the delivery schedule and payment terms, as shown in Figure 11.65.
3. The buyer can further refine the delivery schedule, as shown in Figure 11.66.

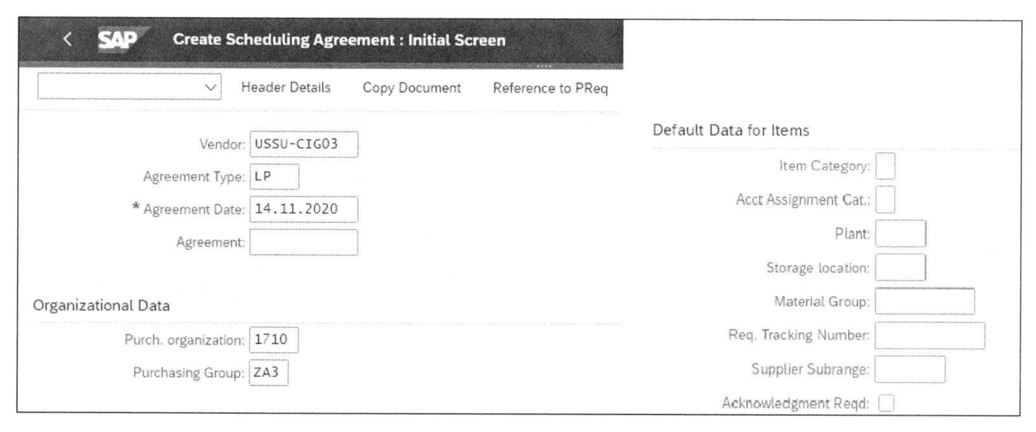

Figure 11.64 Create Scheduling Agreement App

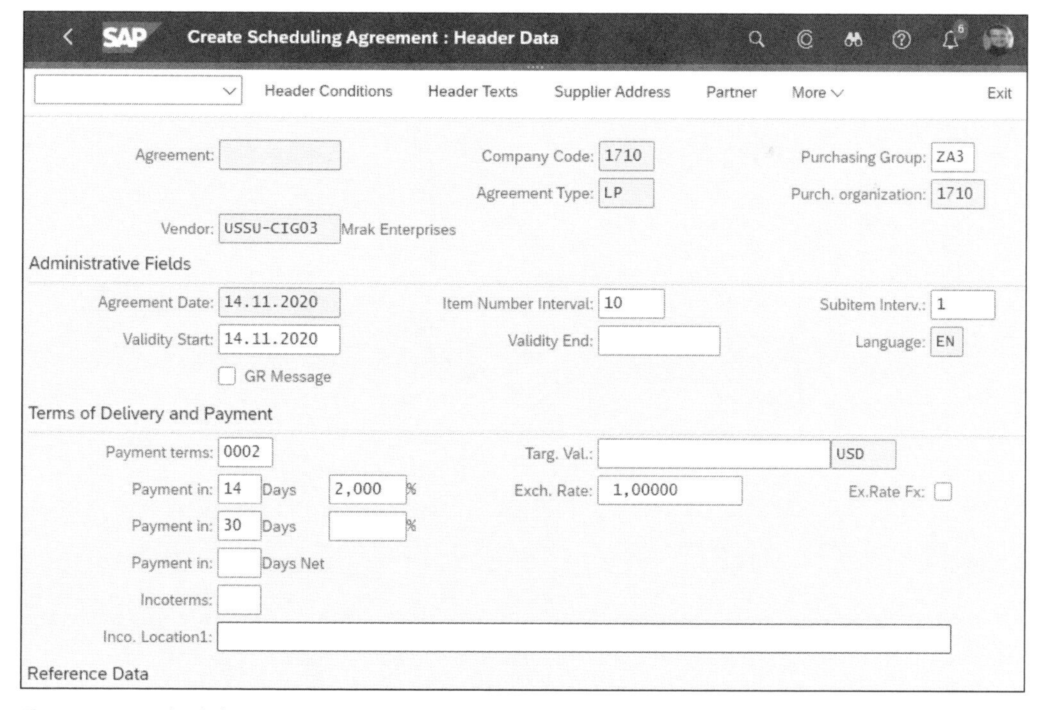

Figure 11.65 Scheduling Agreement Header

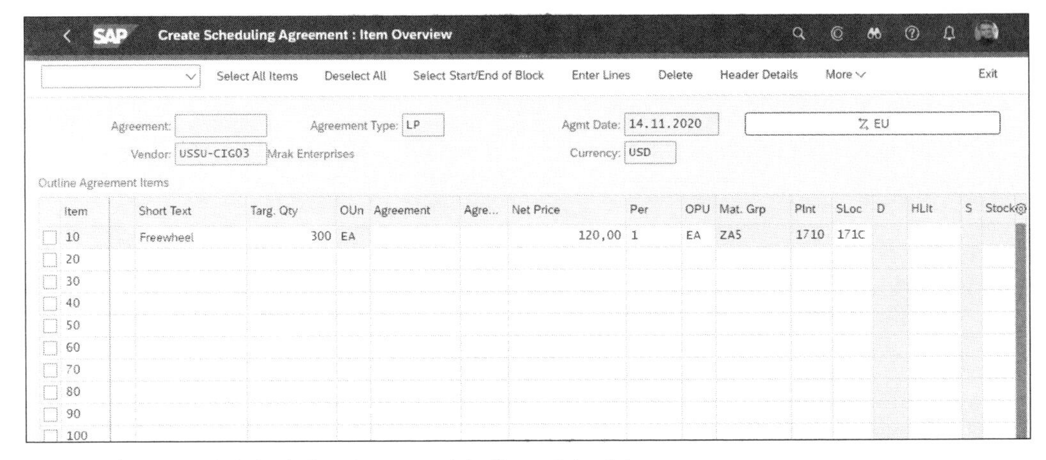

Figure 11.66 Scheduling Agreement Delivery Schedule

4. In the delivery schedule screen, shown in Figure 11.67, the buyer can maintain delivery times and quantities, which can be adjusted further as demand and requirements change at the manufacturer.

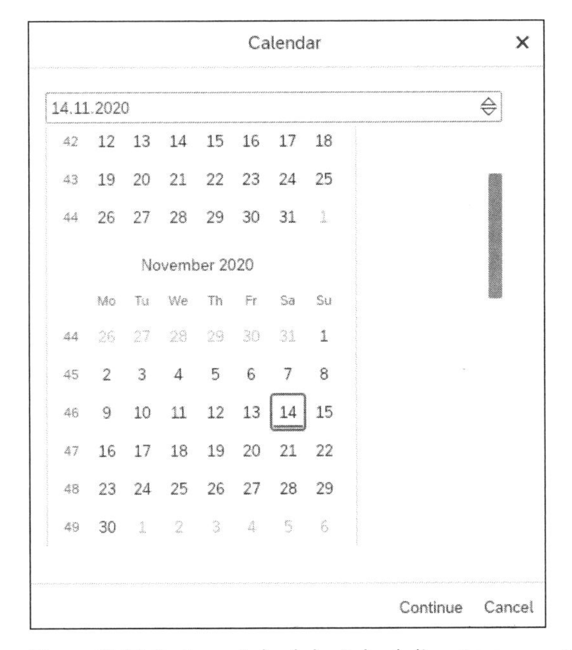

Figure 11.67 Delivery Schedule: Scheduling Agreement

5. After releases have been made on the scheduling agreement and transmitted to the Ariba Network, the supplier can fulfill these orders.
6. The supplier selects the scheduling agreement release for processing, as shown in Figure 11.68.

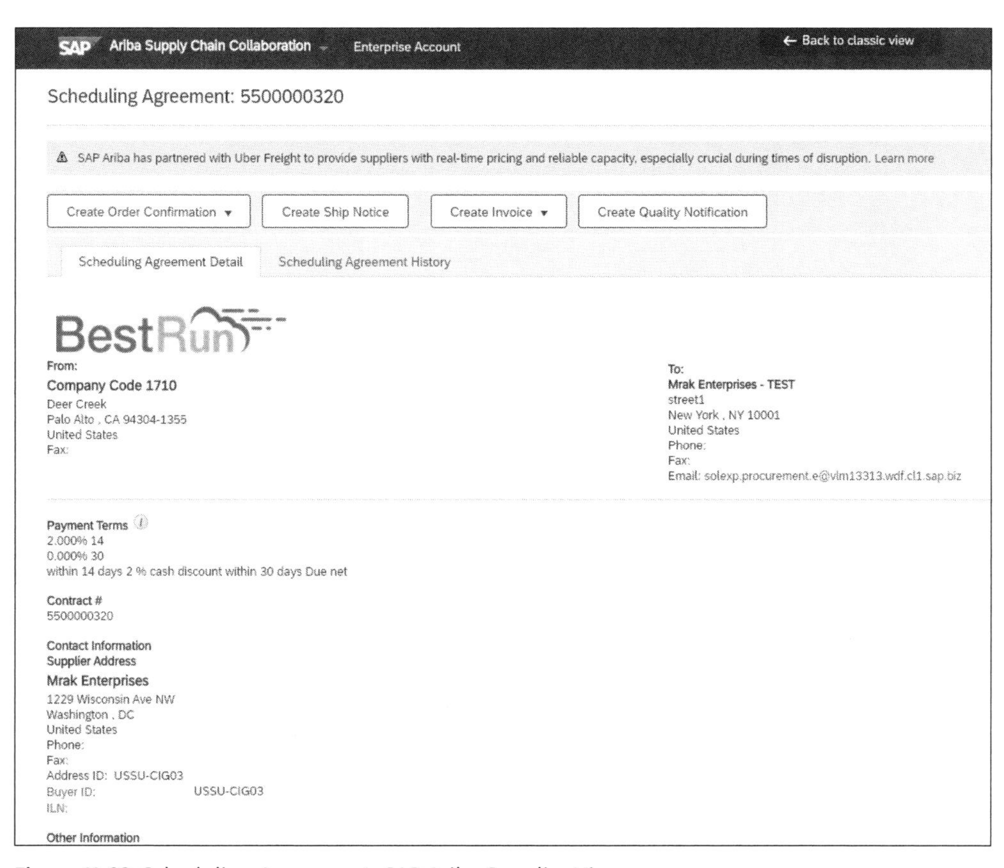

Figure 11.68 Scheduling Agreement: SAP Ariba Supplier View

7. The supplier selects the line item to process and proceeds, as shown in Figure 11.69, to create a ship notice, shown in Figure 11.70.

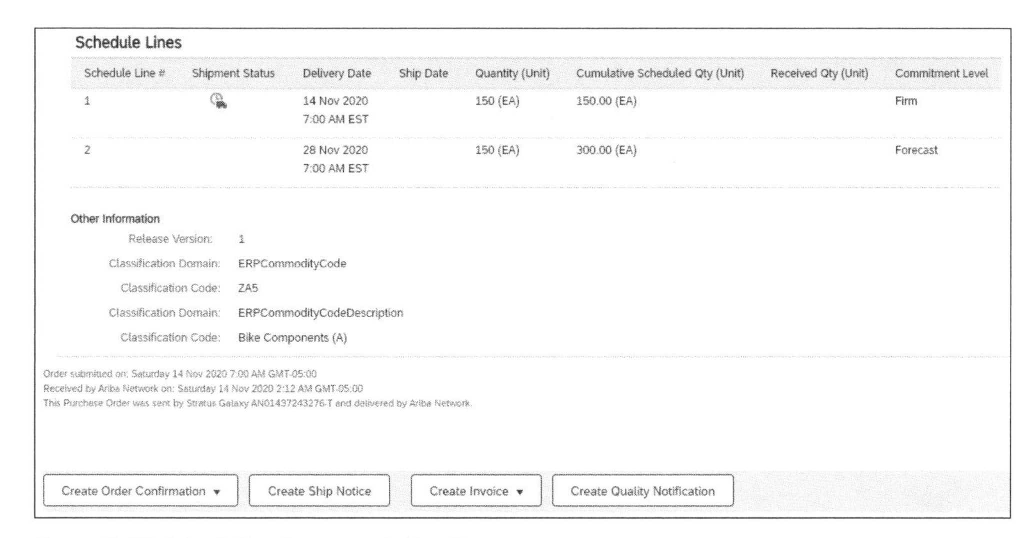

Figure 11.69 Scheduling Agreement: Line Item

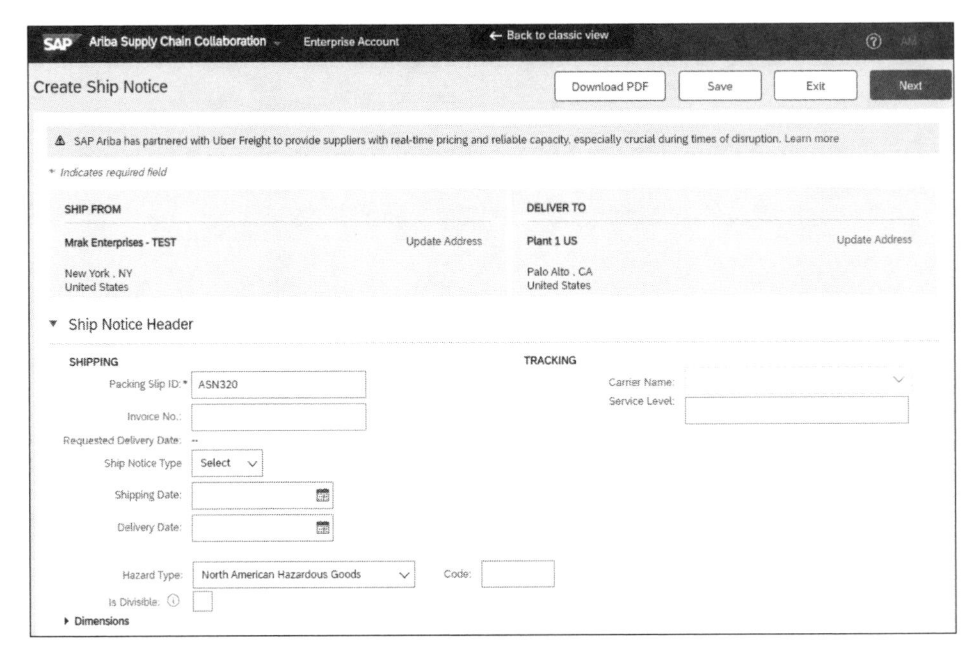

Figure 11.70 Creating an ASN

8. After the ASN is created and the buyer receives the item, the buyer creates a goods receipt in its SAP S/4HANA backend system, as shown in Figure 11.71, where the buyer reviews the **Supplier Confirmations** on the scheduling agreement.

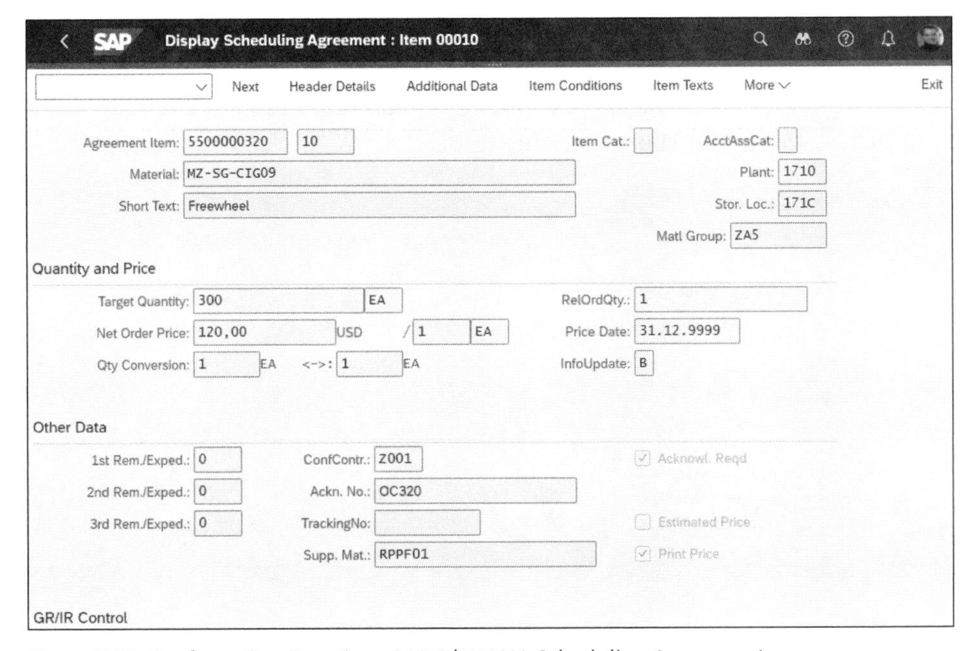

Figure 11.71 Confirmation Overview: SAP S/4HANA Scheduling Agreement

9. In the example shown in Figure 11.72, the buyer can see that an ASN has been created by the supplier (in the Ariba Network) and transmitted to the buyer's SAP S/4HANA system.

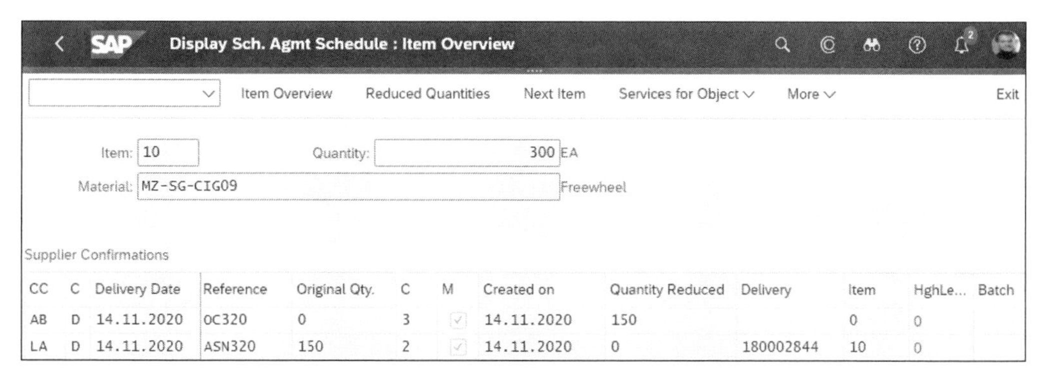

Figure 11.72 Advanced Shipping Notification in SAP S/4HANA Confirmation Overview

10. The buyer then clicks on the **Change Inbound Delivery** tile in SAP S/4HANA, as shown in Figure 11.73.

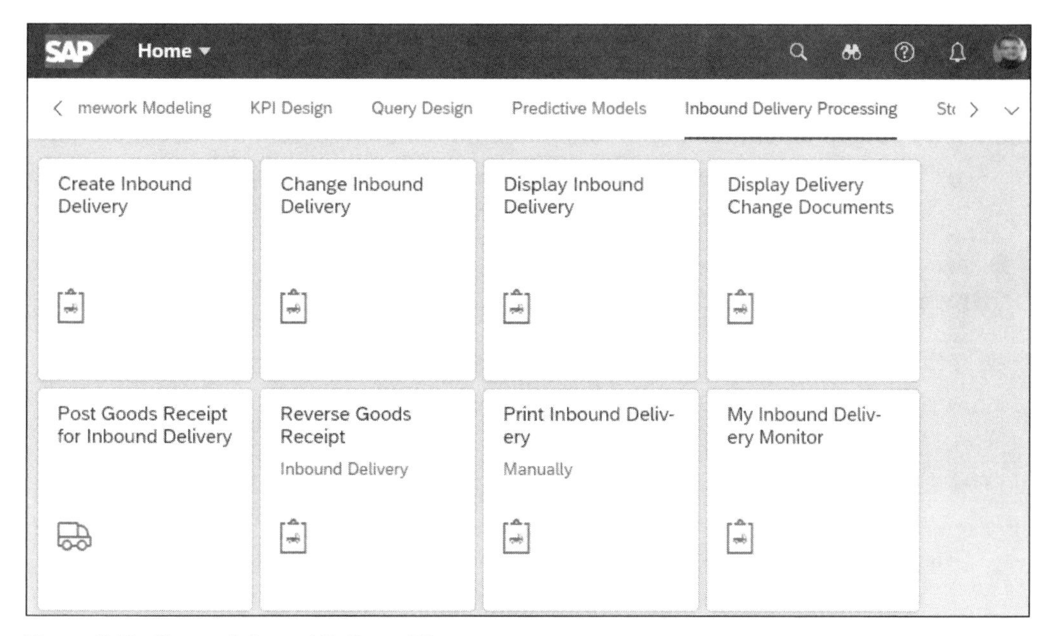

Figure 11.73 Change Inbound Delivery Tile

11. The buyer then creates entries, as shown in Figure 11.74.

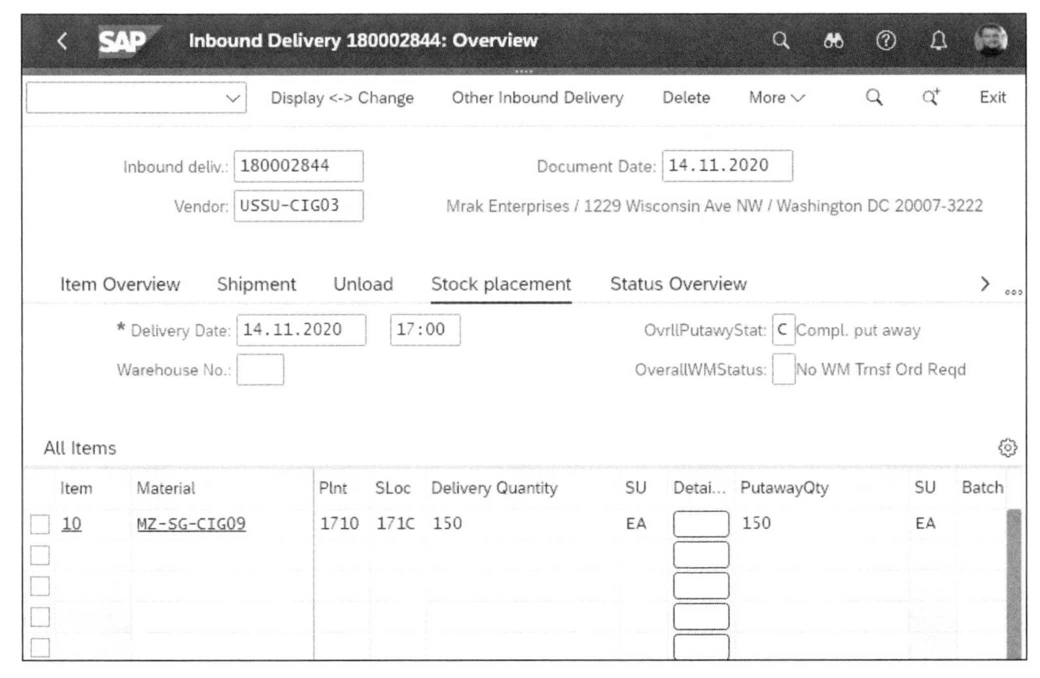

Figure 11.74 Posting a Goods Receipt against the Inbound Delivery

12. After the goods receipt has been created, the 150 items received will now appear in the SAP S/4HANA scheduling agreement, as shown in Figure 11.75.

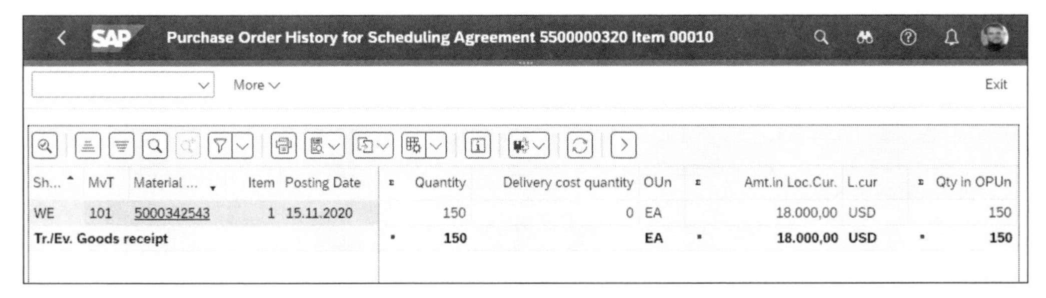

Figure 11.75 Display Scheduling Agreement with Goods Receipt Quantity 150

This order was an ERS order, which means that after the goods receipt has been entered in SAP S/4HANA, an invoice will be created automatically in the system and then shared with the supplier via the Ariba Network. The supplier can view the complete document flow in its area as can the buyer in the SAP S/4HANA system, as shown in Figure 11.76.

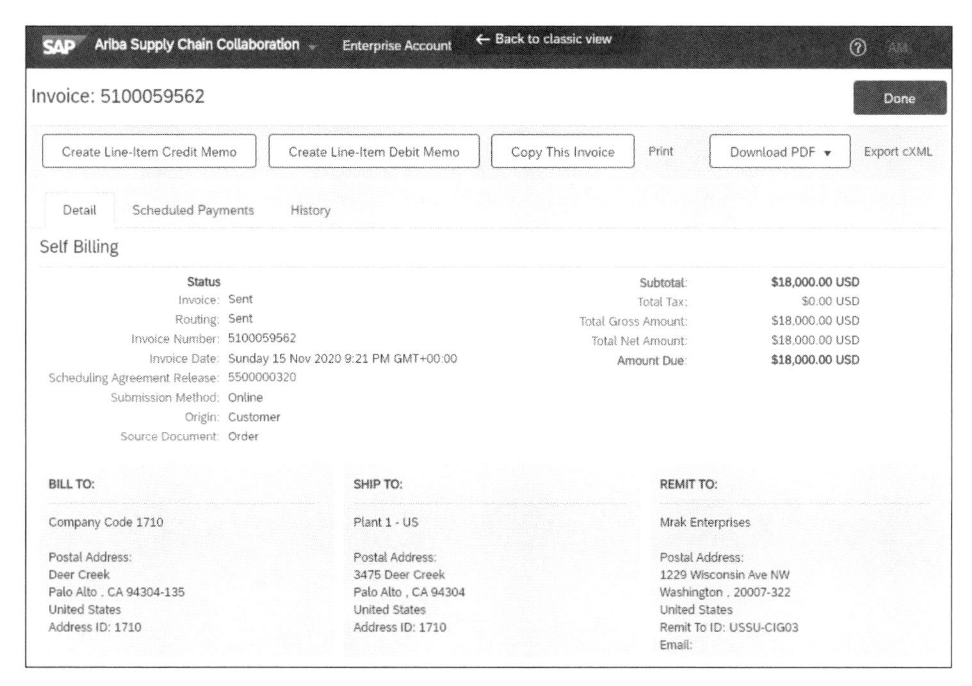

Figure 11.76 SAP Ariba Supply Chain Collaboration for Buyers: Supplier Invoice View

SAP Ariba Supply Chain Collaboration for Buyers Quality Management

The third type of planning activity supported in SAP Ariba Supply Chain Collaboration for Buyers is invoicing and quality management. Quality management is often conducted in a back-and-forth manner, where the buyer laboriously reviews the shipment and inspects the items, communicates the findings to the supplier, requests issues be rectified with the supplier over the phone, and so on. With SAP Ariba Supply Chain Collaboration for Buyers, issues logged in the buyer's SAP S/4HANA environment seamlessly update the supplier, the turnaround time for fixing issues can be significantly reduced, and certificates can be included with notifications, as shown in Figure 11.77.

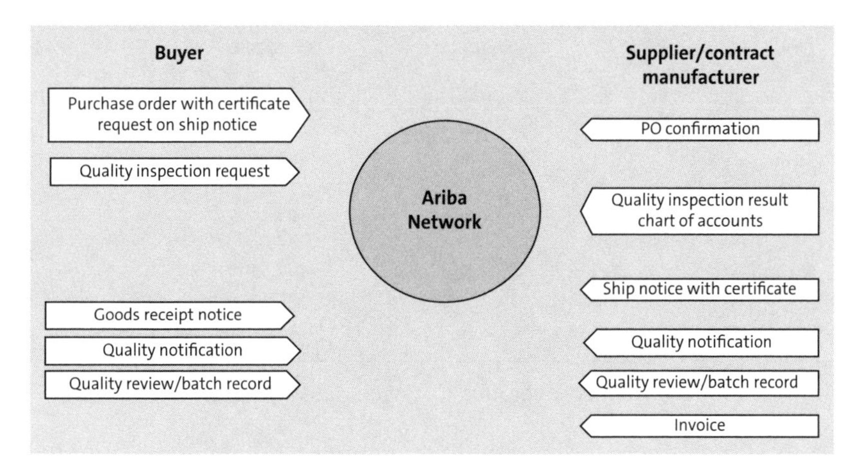

Figure 11.77 SAP Ariba Supply Chain Collaboration for Buyers: Quality Management Process

In this planning activity, the following steps are taken:

1. Logging in to SAP Ariba Supply Chain Collaboration for Buyers, the supplier can click on the **Quality Management** tab and view three submenu options: **Review**, **Inspection**, and **Notification**.
2. Under **Review**, the supplier can search a date range or a type, as shown in Figure 11.78.

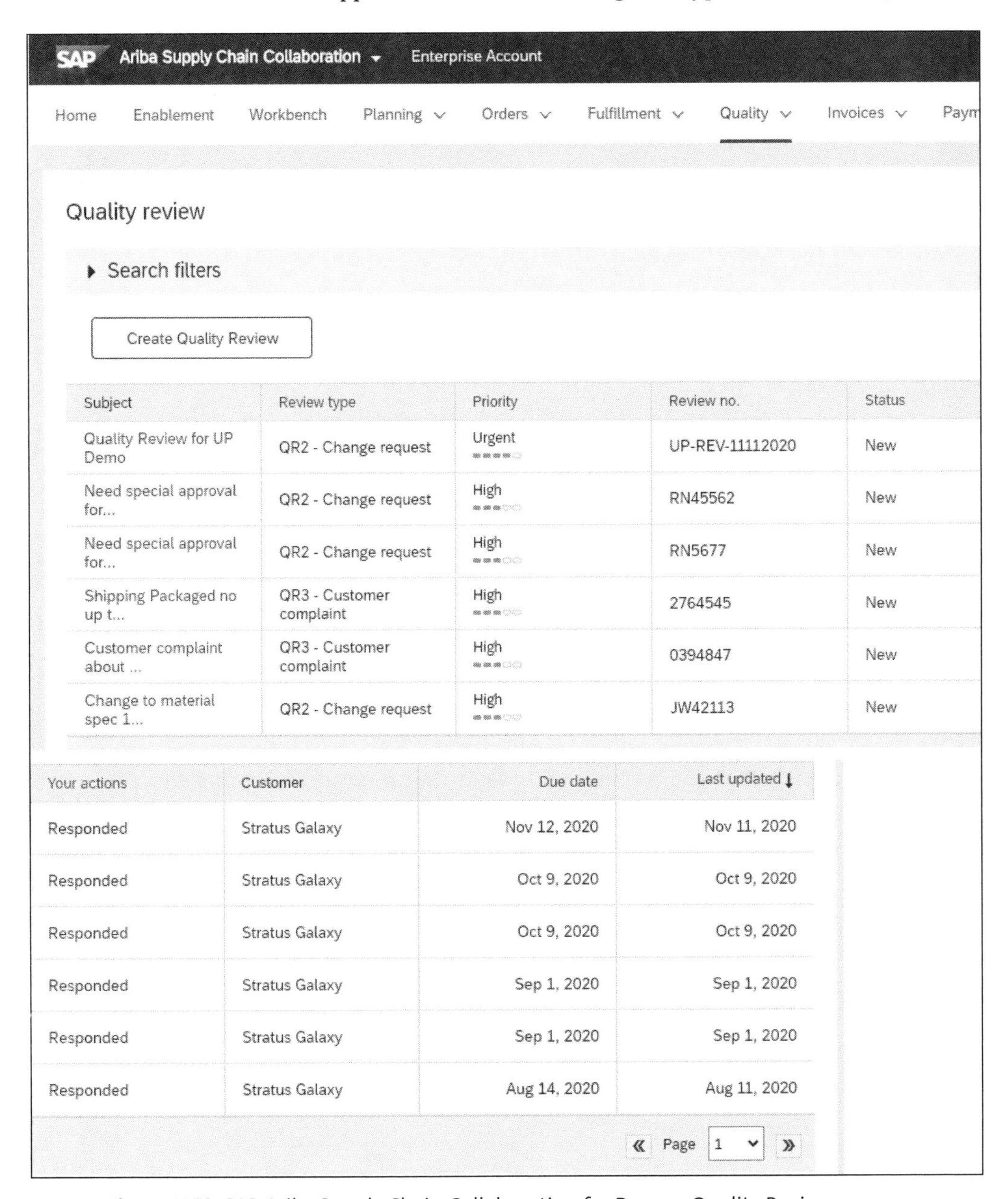

Figure 11.78 SAP Ariba Supply Chain Collaboration for Buyers: Quality Review

3. The supplier can further drill down into buyer change requests on component or quality audit notifications. A buyer can submit a change request and then review the results. In the example shown in Figure 11.79, the buyer has specified the component that is to be reviewed based on specific changes made to it via a change request.

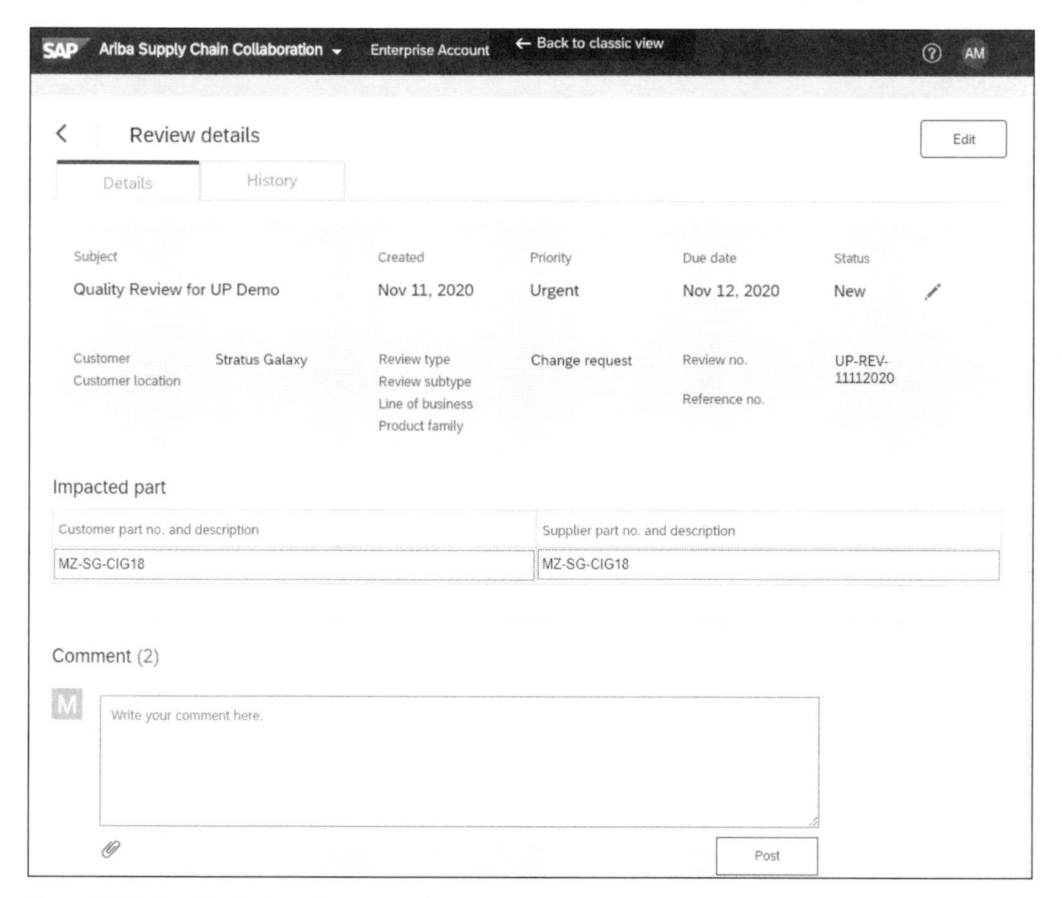

Figure 11.79 Quality Review: Component

4. During the quality management activities for inspection, the buyer completes the process of putting an item into quarantined stock, inspecting the item, and then accepting the item. These changes are reflected in SAP Ariba Supply Chain Collaboration for Buyers for the supplier to review, as shown in Figure 11.80.

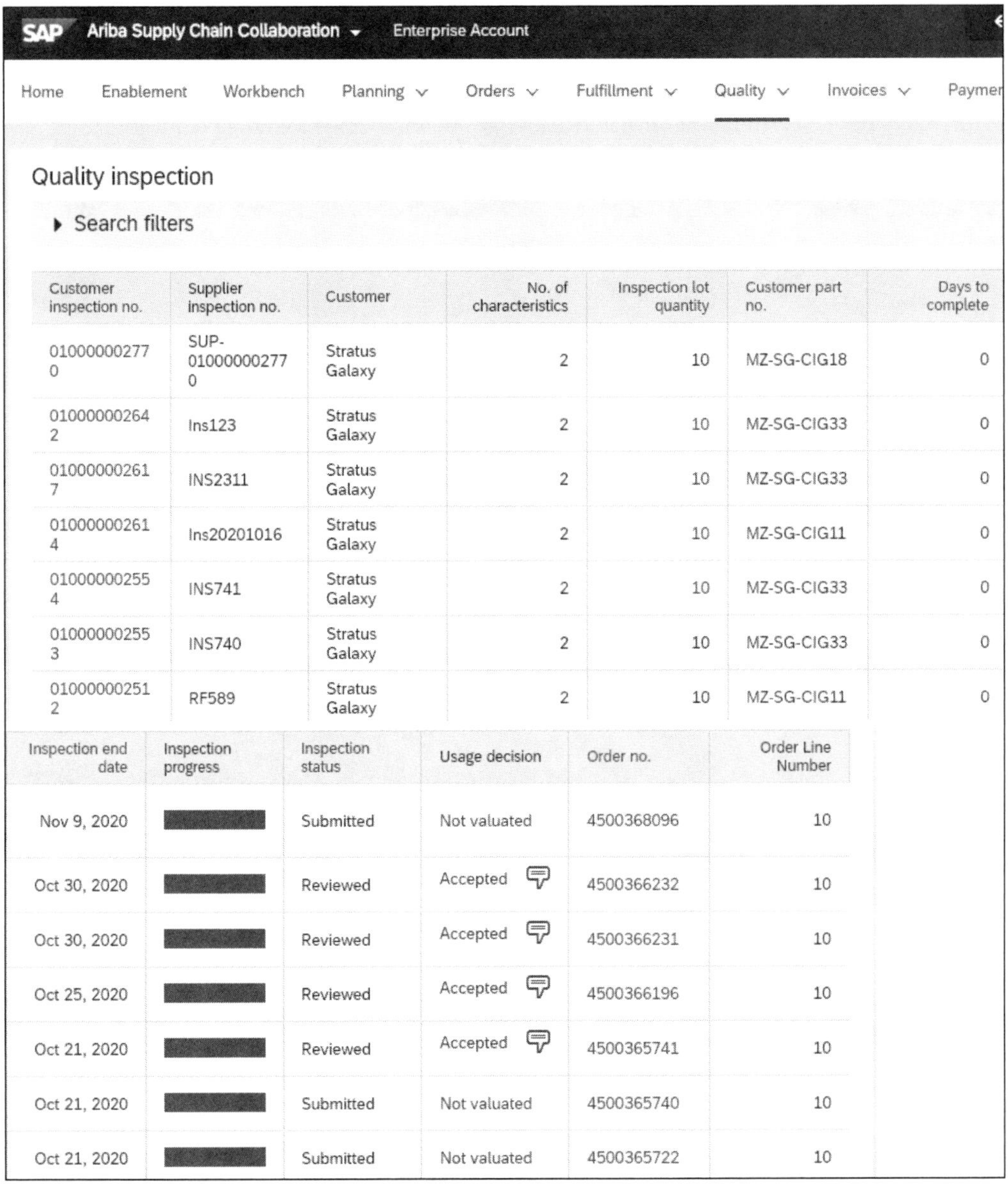

SAP Ariba Supply Chain Collaboration ▾ Enterprise Account

Home Enablement Workbench Planning ∨ Orders ∨ Fulfillment ∨ Quality ∨ Invoices ∨ Paymen

Quality inspection

▸ Search filters

Customer inspection no.	Supplier inspection no.	Customer	No. of characteristics	Inspection lot quantity	Customer part no.	Days to complete	Inspection end date	Inspection progress	Inspection status	Usage decision	Order no.	Order Line Number
010000002770	SUP-010000002770	Stratus Galaxy	2	10	MZ-SG-CIG18	0	Nov 9, 2020		Submitted	Not valuated	4500368096	10
010000002642	Ins123	Stratus Galaxy	2	10	MZ-SG-CIG33	0	Oct 30, 2020		Reviewed	Accepted	4500366232	10
010000002617	INS2311	Stratus Galaxy	2	10	MZ-SG-CIG33	0	Oct 30, 2020		Reviewed	Accepted	4500366231	10
010000002614	Ins20201016	Stratus Galaxy	2	10	MZ-SG-CIG11	0	Oct 25, 2020		Reviewed	Accepted	4500366196	10
010000002554	INS741	Stratus Galaxy	2	10	MZ-SG-CIG33	0	Oct 21, 2020		Reviewed	Accepted	4500365741	10
010000002553	INS740	Stratus Galaxy	2	10	MZ-SG-CIG33	0	Oct 21, 2020		Submitted	Not valuated	4500365740	10
010000002512	RF589	Stratus Galaxy	2	10	MZ-SG-CIG11	0	Oct 21, 2020		Submitted	Not valuated	4500365722	10

Figure 11.80 SAP Ariba Supply Chain Collaboration for Buyers: Quality Inspection

5. In the details area for the inspection line, the supplier can view the pass/fail rate and the defects found in the shipment, as shown in Figure 11.81.

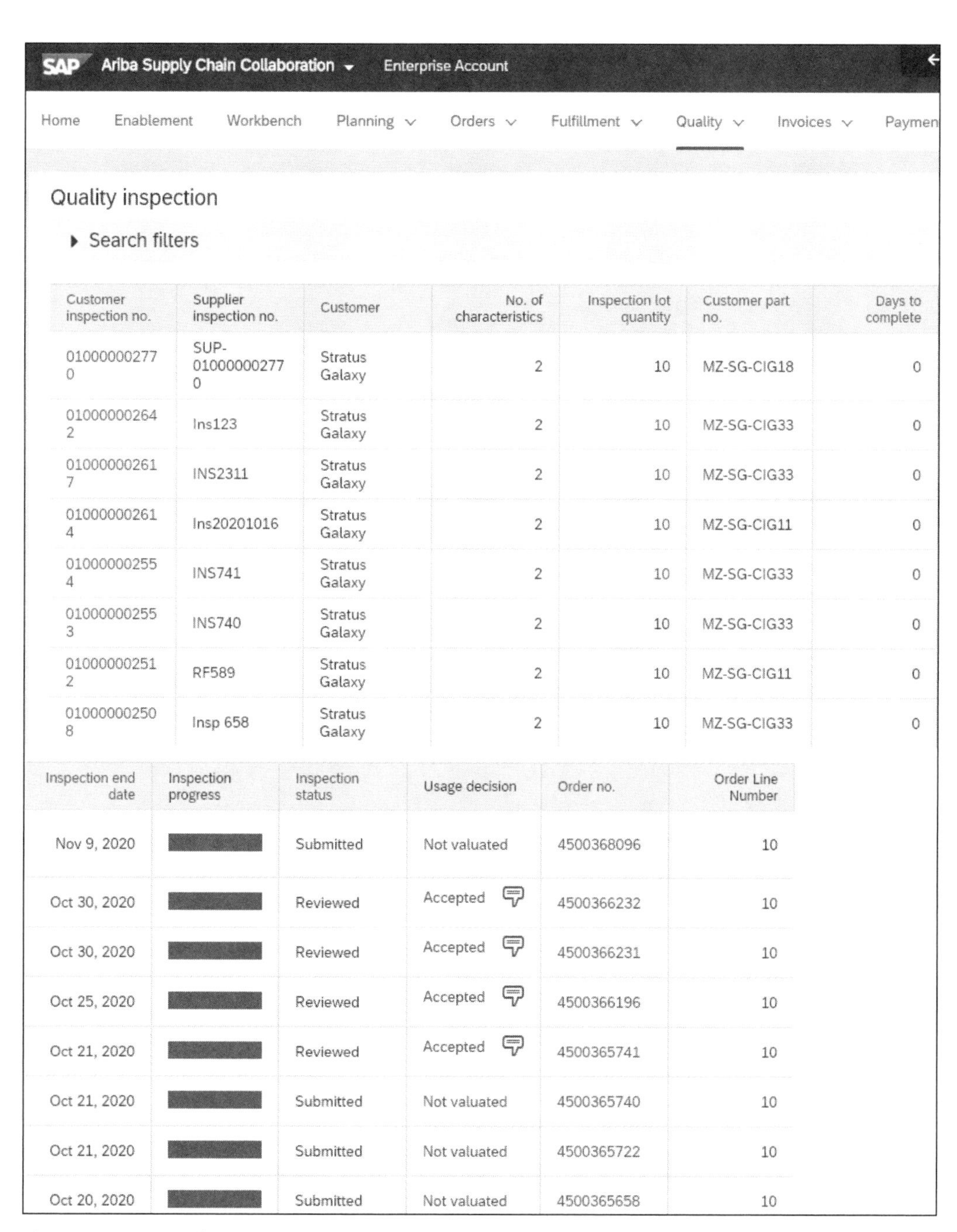

Figure 11.81 SAP Ariba Supply Chain Collaboration for Buyers: Inspection Details

6. Under the **Quality • Notifications** area in SAP Ariba Supply Chain Collaboration for Buyers, a supplier can see the notes directly assigned to specific orders, as shown in Figure 11.82.

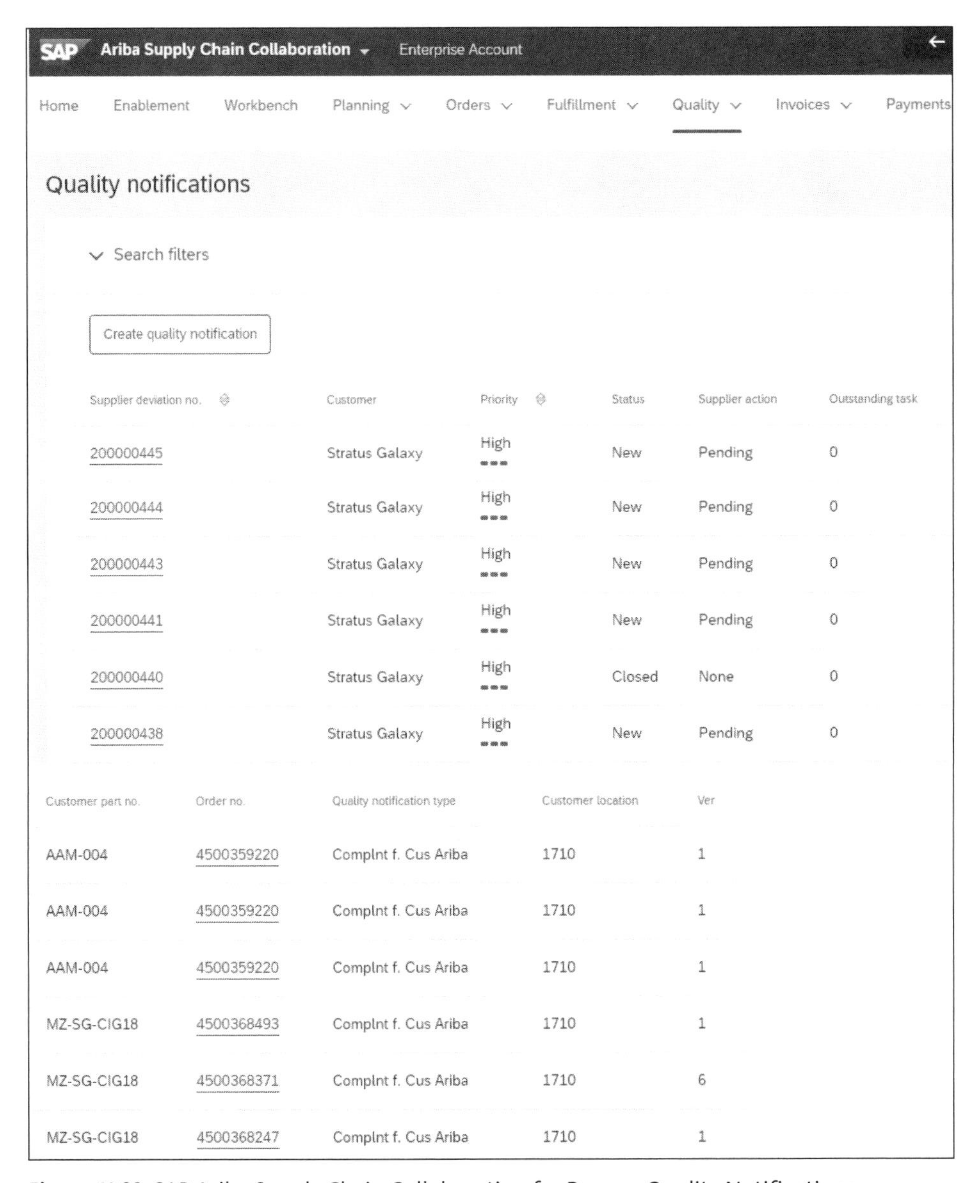

Figure 11.82 SAP Ariba Supply Chain Collaboration for Buyers: Quality Notifications

7. A supplier no longer has to sift through emails to find an order and its associated documents. These details are all captured in the **Quality Notifications** section in SAP Ariba Supply Chain Collaboration for Buyers, where the supplier can also respond directly to the buyer's concerns regarding the order, as shown in Figure 11.83.

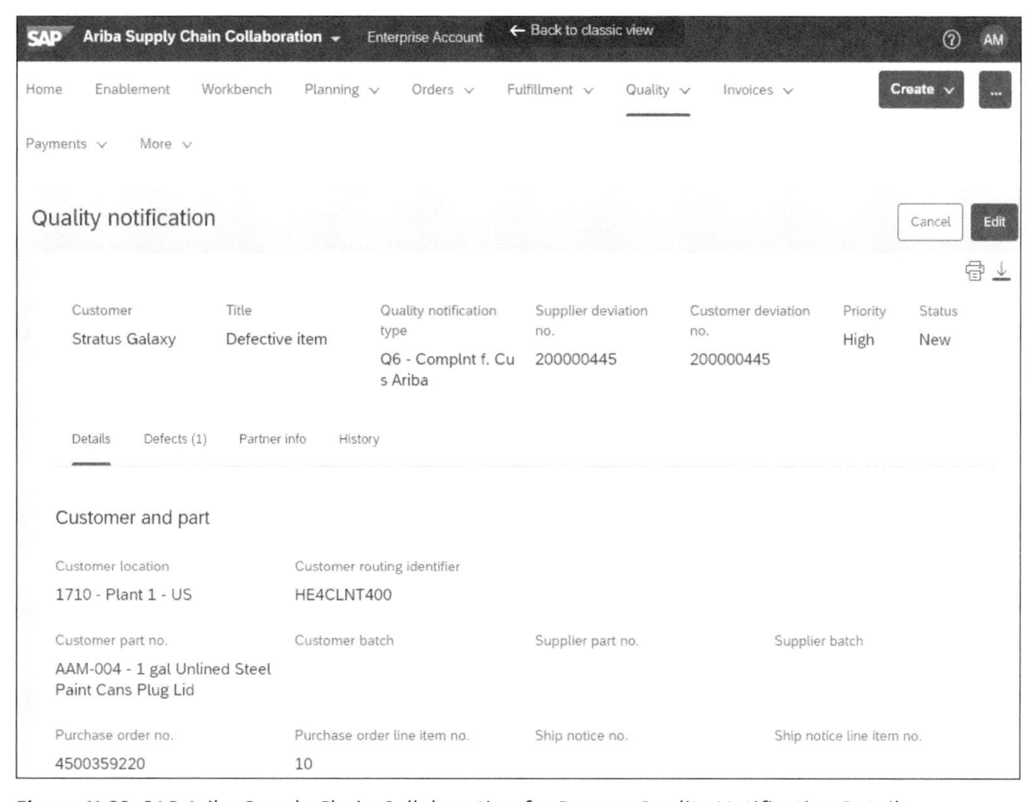

Figure 11.83 SAP Ariba Supply Chain Collaboration for Buyers: Quality Notification Details

SAP Ariba Supply Chain Collaboration for Buyers has gone through several iterations and much evolution at this point, all with the common goal of augmenting the Ariba Network's functionality and capabilities for supporting direct procurement. Many of these documents and processes require additional integration and design.

In the next section, we'll describe the latest, most comprehensive integration platform for connecting and integrating SAP Ariba to ERP in general.

11.5 SAP Ariba Cloud Integration Gateway

The latest and future go-to integration approach for SAP Ariba is the SAP Ariba Cloud Integration Gateway. Released in 2018, the SAP Ariba Cloud Integration Gateway unifies and standardizes several approaches and existing integrations between SAP ERP (or SAP S/4HANA) and SAP Ariba. This solution has a large base upon which to build, with more than 200 integrations between the SAP Ariba and SAP ERP systems already in existence. These integrations are split roughly down the middle between the Ariba Network and SAP Ariba Procurement and SAP Ariba Sourcing applications. The application integrations are listed in Table 11.10.

▪ Purchasing organization ▪ Purchasing groups ▪ Company code ▪ Payment terms ▪ Plants ▪ Incoterms ▪ Material master ▪ Material master/plant ▪ Material groups ▪ Bill of material (BOM) ▪ Manufacturer part number ▪ Purchase info record ▪ SAP Ariba Supplier Lifecycle and Performance outbound (pass-through) ▪ SAP Ariba Supplier Lifecycle and Performance inbound (pass-through) ▪ Article master ▪ Characteristics ▪ Profile (class) ▪ Merchandise category ▪ Display set ▪ Article/site ▪ Genetic article variant ▪ RFQ to quote request	▪ Sourcing award to PO ▪ Sourcing award to outline agreement ▪ Contract workspace to outline agreement ▪ Account categories ▪ Assets ▪ Asset class ▪ Company codes ▪ Cost centers ▪ Cost center languages ▪ ERP commodity codes ▪ General ledger ▪ Internal orders ▪ Plant ▪ Payment terms ▪ Purchase groups ▪ Tax codes ▪ WBS elements ▪ User data ▪ User group mapping ▪ Remittance locations ▪ Supplier data ▪ Supplier location	▪ Contracts ▪ Funds management derive ▪ Company code internal order mapping ▪ Company code work breakdown structure (WBS) element mapping ▪ Purchasing organization supplier combo ▪ Account category field status ▪ Plant to purchasing organization ▪ Currency conversion rates ▪ Fund management objects ▪ Purchasing organizations ▪ General ledger languages ▪ Fund ▪ Earmarked fund ▪ Functional area ▪ Funds center ▪ Commitment item ▪ Budget period ▪ FM area ▪ Grant ▪ Material PO ▪ Service PO ▪ Blanket/limit PO	▪ Change PO ▪ Cancel PO ▪ Close PO ▪ Goods receipt ▪ Service entry sheet ▪ Invoices (OK2Pay) ▪ Non-PO invoice (one-time) ▪ Expense report ▪ PO status ▪ Change PO status ▪ Receipt status ▪ Invoice status ▪ Remittances ▪ Catalog ▪ Advance payment requests ▪ Advance payments ▪ Cancel advance payment ▪ Requisition ▪ Change requisition ▪ Reservation ▪ Budget check ▪ Service entry sheet response ▪ Import requisitions ▪ Asset shell ▪ Inventory stock update ▪ FI invoice

Table 11.10 Integrations between SAP S/4HANA/SAP ERP and SAP Ariba Procurement and SAP Ariba Sourcing Applications

Ariba Network integrations can be further divided between buy-side or sell-side integrations, or between integrations applicable to buyers and those applicable to suppliers. The integrations that pertain to the buyer are document-centric, as shown in Table 11.11.

▪ Standard PO ▪ Change PO ▪ Cancel PO ▪ Order enquiry request ▪ Order confirmation ▪ Inbound ship notice ▪ Outbound ASN ▪ Outbound payment remittance ▪ Inbound remittance advice ▪ Remittance cancellation	▪ Inbound receipt ▪ Outbound receipt ▪ Service entry sheet ▪ Service entry status update ▪ Invoice (including credit memo) ▪ Invoice status ▪ Carbon copy (CC) invoice ▪ Order ▪ Change order ▪ Cancel order	▪ Remittance advice* ▪ Quote request ▪ Quote message ▪ Payment batch file ▪ Inbound payment proposal ▪ Outbound payment proposal ▪ Subcontract PO ▪ Consignment PO ▪ Component consumption ▪ Scheduling agreement (including forecast/delivery schedule and JIT)	▪ Product replenishment ▪ Transfer movement/product activity ▪ Quality notification ▪ Quality inspection ▪ Order confirmation ▪ Invoice ▪ Shipment notification ▪ Delivery schedule (forecast) ▪ Delivery schedule (JIT)

Table 11.11 Integrations between SAP S/4HANA/SAP ERP and the Ariba Network: Buy Side

The supplier-side integrations for the Ariba Network and SAP S/4HANA/SAP ERP pertain to various EDI protocols, as outlined in Table 11.12.

Table 11.13 provides further integrations.

X12 v4010	EDIFACT D01B	EDIFACT D96A	GS1 EANCOM 2002
▪ 810 ▪ 820 ▪ 204 ▪ 214 ▪ 824 (in and out) ▪ 830 (forecast) ▪ 830 (commit) ▪ 832 ▪ 842 (in and out) ▪ 846 (in and out) ▪ 850 ▪ 855 ▪ 856 ▪ 860 ▪ 861 ▪ 862 ▪ 864 ▪ 866 ▪ 997 (in and out)	▪ ORDERS ▪ ORDCHG ▪ DESADV ▪ INVOIC	▪ ORDERS ▪ ORDCHG ▪ ORDRSP ▪ INVOIC ▪ DESADV ▪ CONTRL ▪ RECADV ▪ INVRPT ▪ REMADV ▪ DELFOR ▪ DELJIT ▪ APERAK ▪ IFTMIN ▪ IFTSTA	▪ ORDERS ▪ ORDRSP ▪ ORDCHG ▪ INVOIC ▪ DESADV ▪ REMADV

Table 11.12 Integrations between SAP S/4HANA/SAP ERP and the Ariba Network: Sell Side (Part 1)

GS1 EANCOM 97	GS1 GUSI	OAGiS v9.2	PiDX v1.61
▪ ORDERS ▪ ORDCHG ▪ DESADV ▪ INVOIC ▪ REMADV	▪ MultiShipmentOrder ▪ DespatchAdvice ▪ ReplenishmentProposal ▪ ReplenishmentRequest ▪ ProductForecast ▪ GoodsRequirement ▪ ReceiptAdvice ▪ ComsumptionReport ▪ InventoryActivityorInventoryStatus	▪ ProcessPurchaseOrder (PO) ▪ AcknowledgePurchaseOrder ▪ NotifyShipment ▪ ProcessReceiveDelivery ▪ NotifyPlanningSchedule ▪ NotifyInventoryConsumption ▪ NotifyProductionOrder ▪ NotifyInventoryBalance ▪ ConfirmBOD	▪ OrderRequest ▪ OrderChange ▪ OrderResponse ▪ Invoice ▪ InvoiceResponse ▪ Receipt ▪ ReceiptAcknowledgement ▪ Exception ▪ InvoiceResponse-SESR ▪ AdvancedShipNotice

Table 11.13 Integrations between SAP S/4HANA/SAP ERP and the Ariba Network: Sell Side (Part 2)

11.5.1 SAP Cloud Platform Integration and SAP Ariba Cloud Integration Gateway

The SAP Ariba Cloud Integration Gateway underpinned by SAP Cloud Platform Integration provides a fast, flexible solution to integrate with backend systems, trading partners, and SAP Ariba solutions, as shown in Figure 11.84. The SAP Ariba Cloud Integration Gateway is the go-to integration platform for both Ariba Network and SAP Ariba solutions as well as SAP cloud and on-premise solutions. You can download the SAP Ariba Cloud Integration Gateway via the Software Downloads link on the SAP Support Portal as an add-on via *https://support.sap.com*. Standard documentation can be found on the SAP Help Portal via *https://help.sap.com/viewer/product/ARIBA_CIG/*. The SAP Ariba Cloud Integration Gateway provides self-service tools to quickly configure, test, and extend processes, and it incorporates automated upgrades via Software Update Manager (SUM) and monitoring with tools (e.g., SAP Solution Manager) to provide an integration solution that fits seamlessly with your existing SAP landscape to minimize TCO and ensure smooth transaction flow.

The SAP Ariba Cloud Integration Gateway will serve as the comprehensive lynchpin for preexisting and future integrations between SAP S/4HANA and SAP Ariba, as well as for non-SAP backend systems.

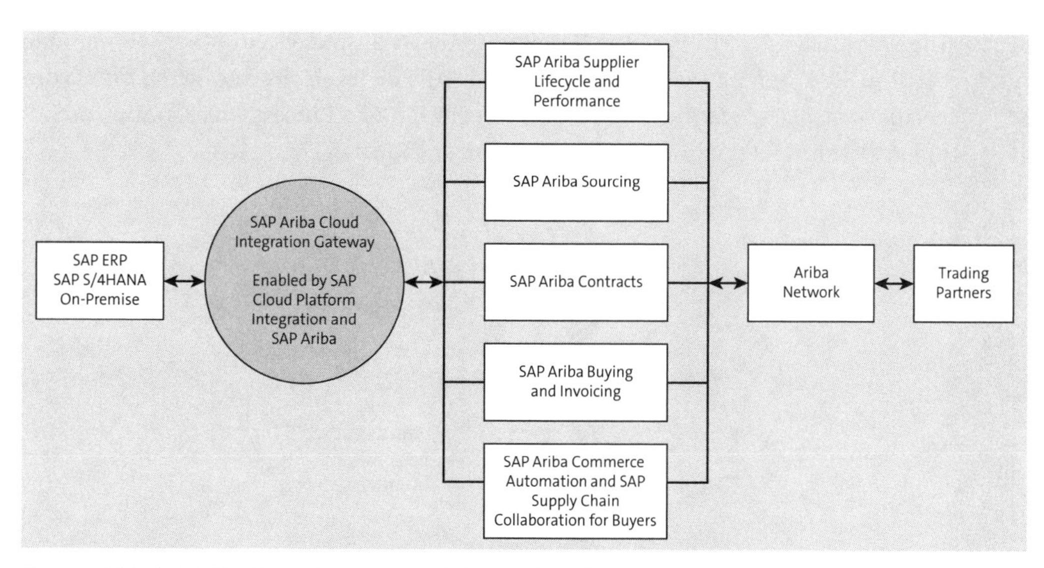

Figure 11.84 SAP Ariba Cloud Integration Gateway: Buy-Side Integration

Architecture

The SAP Ariba Cloud Integration Gateway provides the portal UI to unify the integration options for the administrator, as well as a configuration wizard, mapping repository, centralized testing, and monitoring. For the integration "heavy lifting" of mapping, transformation, and messaging, the SAP Ariba Cloud Integration Gateway relies on SAP Cloud Platform Integration, which connects to the SAP Ariba Cloud Integration Gateway add-on for SAP ERP or connects to SAP S/4HANA via HTTPS for incoming calls from and to the cloud connector Secure Sockets Layer (SSL) tunnel for outgoing calls to the ERP environment from the SAP Ariba Cloud Integration Gateway layer, as shown in Figure 11.85.

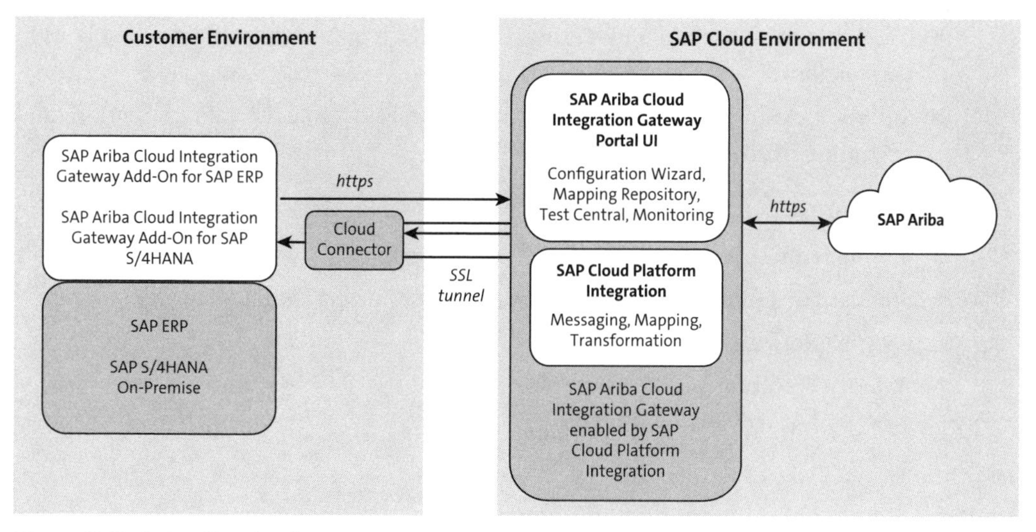

Figure 11.85 SAP Ariba Cloud Integration Gateway: Architecture

The SAP Ariba Cloud Integration Gateway connects a cloud environment to multiple backend ERPs, at both the SAP S/4HANA and SAP ERP level, directly via the cloud connector or mediated via your SAP Process Integration/SAP Process Orchestration or SAP Cloud Platform Integration middleware layer, as shown in Figure 11.86.

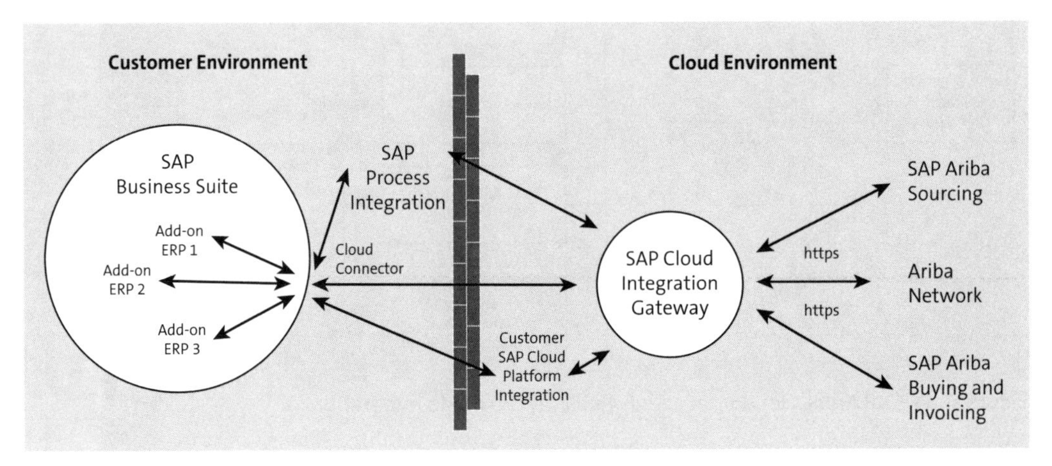

Figure 11.86 SAP Ariba Cloud Integration Gateway: Deployment with Multiple Backend ERPs

Setup

The SAP Ariba Cloud Integration Gateway has three main areas: the backend SAP ERP area, the integration layer of the SAP Ariba Cloud Integration Gateway, and the SAP Ariba Cloud solutions area. In the SAP ERP area, you'll enable the add-on to the SAP Ariba Cloud Integration Gateway. The add-on has the following features:

- Enriches outbound messages
- Extensible
- Configurable in the SAP ERP Transaction SPRO area (more detail provided later in this section)
- Supports cloud transactional integration scenarios, master data extraction, and migration from cloud integration

The SAP Ariba Cloud Integration Gateway area provides the following:

- Simple setup supported by configuration wizard
- Standard mappings for all integration scenarios, including the following:
  - Ariba Network
  - SAP Ariba Strategic Sourcing Suite
  - SAP Ariba Procurement portfolio

- Multi-ERP by system ID/realm ID, meaning multiple backend systems can be connected via the same SAP Ariba Cloud Integration Gateway instance to SAP Ariba solutions
- Support for conditional routing
- Test central with automation
- One-click deployment to production
- Transaction tracking

On the SAP Ariba Solutions side, after you've enabled the SAP Ariba Cloud Integration Gateway as an administration and endpoint, you can both push and pull data from the backend systems, as well as leverage asynchronous web services and enhanced message contents to avoid having to look up the message in SAP.

As shown in Figure 11.87, setting up the SAP Ariba Cloud Integration Gateway entails three steps. First, you'll configure the buy-side account for the SAP Ariba Cloud Integration Gateway and install and configure the SAP Ariba Cloud Integration add-on for SAP ERP and SAP S/4HANA. Next, you'll extend using the extension framework in the add-on interfaces in SAP, and the custom eXtensible Stylesheet Language (XSLT) can be used for additional data mapping. Finally, you'll test the integration using predefined scripts and any customized test cases required.

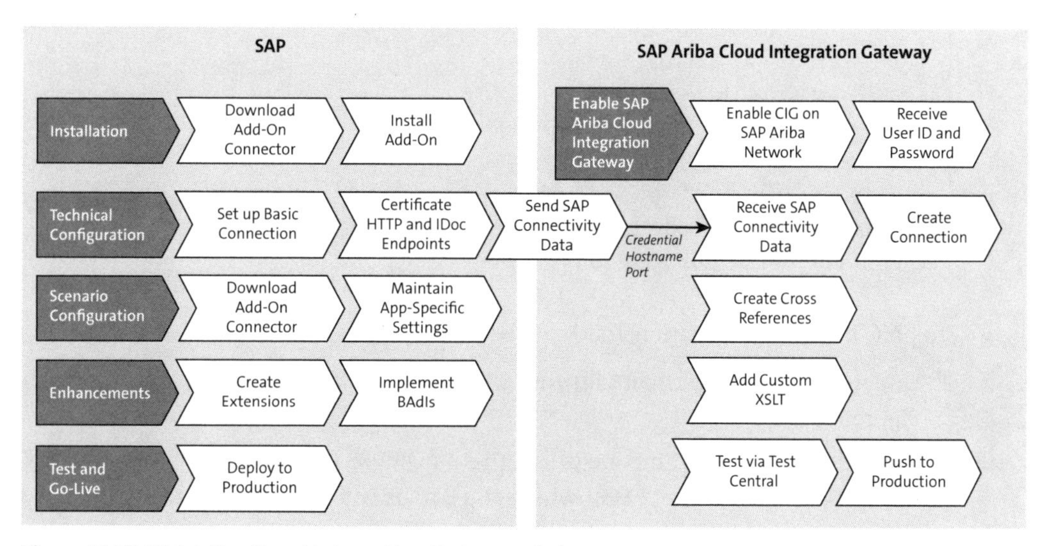

Figure 11.87 SAP Ariba Cloud Integration Gateway: Setup

Setting up the SAP Ariba Cloud Integration Gateway can entail five individual steps, where the installation, technical configuration, scenario configuration, extensions/enhancements, and test/go-live comprise individual steps.

11.5.2 Backend Configuration

You'll need to consider multiple systems and steps when integrating SAP ERP or SAP S/4HANA. The main systems are the backend, the SAP Ariba Cloud Integration Gateway, and the relevant solutions within the SAP Ariba Cloud. Given the robust integration layers available to the user, you can now perform a wide variety of integration tasks and permutations in the backend system, the SAP Ariba Cloud Integration Gateway, and the receiving SAP Ariba system. In this section, we'll begin with configuring the backend SAP S/4HANA and SAP ERP systems, moving then to configuring the SAP Ariba Cloud Integration Gateway and SAP Ariba solutions.

Beginning with the backend SAP ERP or SAP S/4HANA environment, the configuration is now organized in its own Transaction SPRO menu path. The technical and scenario configuration can be accessed by following the menu path **SPRO • Integration with Other mySAP Components • Ariba Cloud Integration • Global Settings** (for the technical configuration) and **SPRO • Integration with Other SAP Components • SAP Ariba Cloud Integration • Ariba Network Integration** (for the scenario configuration), as shown in Figure 11.88.

- Integration with Other SAP Components
  - SAP Ariba Cloud Integration Gateway
    - Global Settings
    - Ariba Network Integration
    - Master Data Integration
    - Master Data Native Integration
    - SAP Ariba Strategic Sourcing Suite Integration
    - SAP Ariba Procurement Integration
    - Configuration Activities to Transfer Data

Figure 11.88 SAP Ariba Cloud Integration Gateway: Technical and Scenario Configuration

To perform the configuration, follow these steps:

1. Under **Global Settings**, maintain your SSL certificates by adding details, as shown in Figure 11.89.
2. Maintain your remote function call (RFC) connections for SAP Ariba, as shown in Figure 11.90, via Transaction SM59, where you'll maintain RFCs in general.

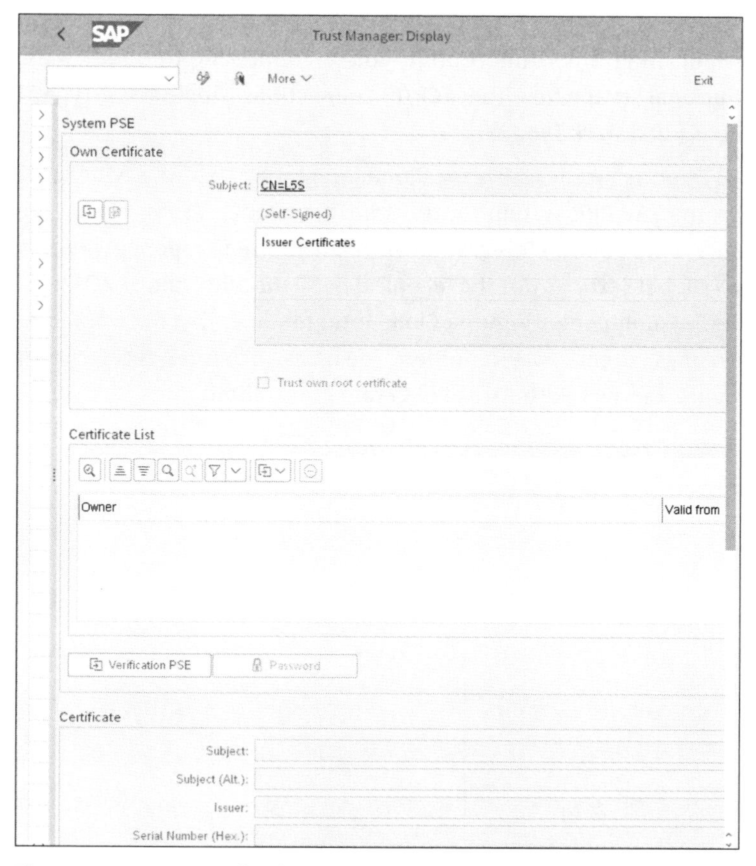

Figure 11.89 SAP ERP Cloud Integration Gateway Technical Settings: Maintain Trust Certificates

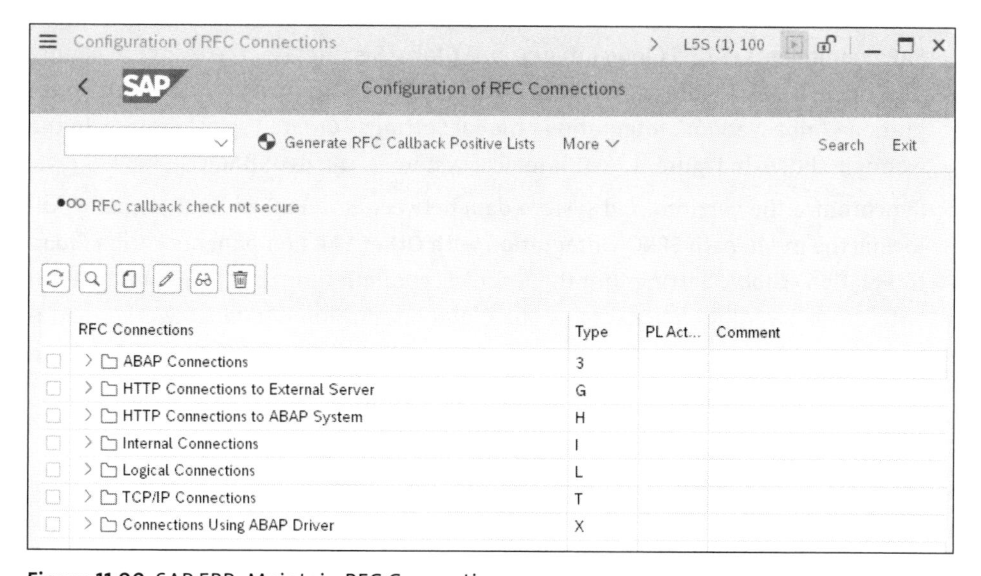

Figure 11.90 SAP ERP: Maintain RFC Connections

Make sure you have the right access credentials for the SAP Ariba Cloud Integration Gateway when establishing this connection. If you're connecting to middleware, you'll need an additional setting for either **TCP/IP Connections** (single stack, **Type T**) or **ABAP Connections** (dual stack **Type 3**).

3. Define the XML HTTP ports for the SAP Ariba Cloud Integration Gateway, which represent the ports in the SAP ERP system for transmitting IDocs, as shown in Figure 11.91. Follow the menu path **SPRO • Integration with Other SAP Components • Ariba Cloud Integration • Global Settings** (for the technical configuration) and **SPRO • Integration with Other SAP Components • Ariba Cloud Integration • Ariba Network Integration • Global Settings • Create Port Definition**. You can also define a receiver port by selecting **Configure Receiver Port** instead of **Create Port Definition**.

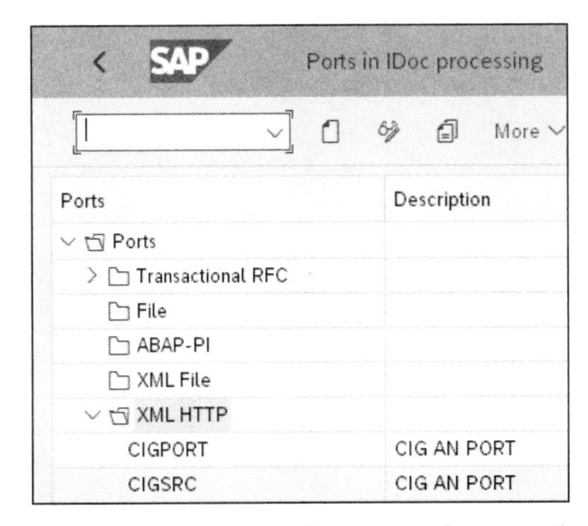

Figure 11.91 SAP ERP: Defining Ports for SAP Ariba Cloud Integration Gateway

4. Define the logical system by following the menu path **SPRO • Integration with Other SAP Components • Ariba Cloud Integration • Global Settings** (for the technical configuration) and **SPRO • Integration with Other mySAP Components • Ariba Cloud Integration • Ariba Network Integration • Global Settings • Create Logical System**. In the example shown in Figure 11.92, this logical system is called **ARIBACIG**.
5. Synchronize the versions and system data between SAP ERP and SAP Ariba by following the menu path **SPRO • Integration with Other SAP Components • Ariba Cloud Integration • Global Settings** (for the technical configuration) and **SPRO • Integration with Other mySAP Components • Ariba Cloud Integration • Ariba Network Integration • Global Settings • Send Information to the SAP Ariba Cloud Gateway • Synchronize SAP Information with the SAP Ariba Cloud Integration Gateway**.

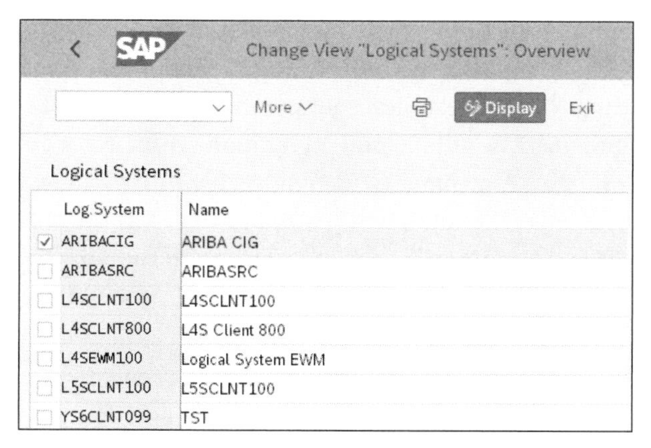

Figure 11.92 SAP ERP: Defining the Logical System for SAP Ariba Cloud Integration Gateway

6. Select the setups you want to migrate (**Sourcing, Procurement,** or **Both**), and enter your user name and password to trigger the job, as shown in Figure 11.93. You can view the status of this job in Transaction SLG1. You only need run this job once, during the initial configuration.

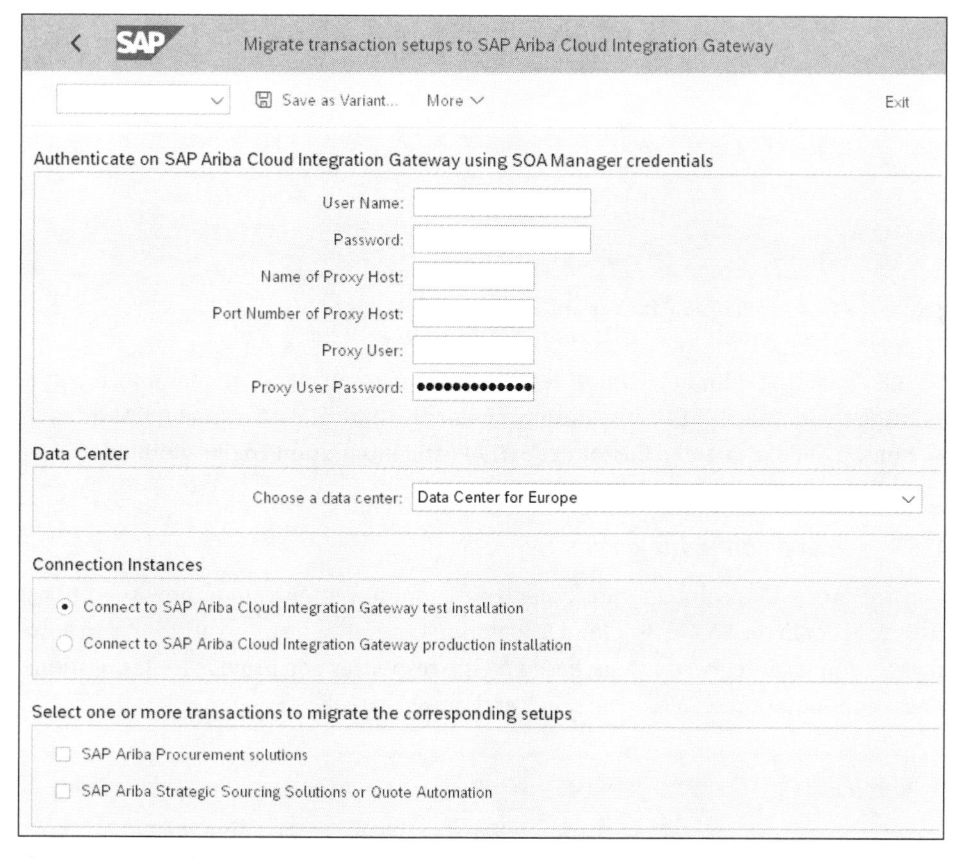

Figure 11.93 Synchronizing Versions between SAP ERP and SAP Ariba

7. If desired, support the comments and attachments in this integration by continuing with the steps provided in **SPRO • Integration with Other SAP Components • Ariba Cloud Integration • Global Settings** (for the technical configuration and **SPRO • Integration with Other mySAP Components • Ariba Cloud Integration • Ariba Network Integration • Global Settings • Support Attachments/Comments**. For further attachment support in areas not supported in the configuration, you can leverage the user exit business add-in (BAdI) ARBCIG_ES_ATTACHMENT_UTIL. Similarly, you can take the same approach to support comments, where the user exit is ARBCIG_ES_COMMENTS_UTIL.
8. Map your variant and partition to differentiate between SAP Ariba Sourcing data and SAP Ariba Buying data by following the menu path **SPRO • Integration with Other SAP Components • Ariba Cloud Integration • Global Settings** (for the technical configuration) and **SPRO • Integration with Other mySAP Components • Ariba Cloud Integration • Ariba Network Integration • Global Settings • Map Variant and Partition to Procurement and Sourcing**. These variants are maintained in a table, shown in Figure 11.94, to avoid confusion in transmitted messages.

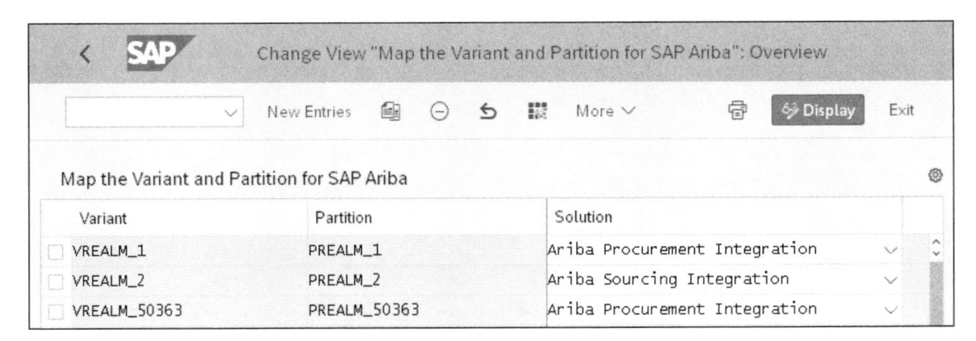

Figure 11.94 SAP Ariba Table for Variant and Partition Master

9. The remaining steps in technical settings pertain to setting up middleware. If you're using middleware in this integration, enable the middleware using **Enable Integration via Middleware** and **Customize SLD API and Integration Engine-Admin for PI**.

11.5.3 General Configuration

The SAP Ariba Cloud Integration Gateway integration wizard underpins the project functionality in the SAP Ariba Cloud Integration Gateway. The integration wizard automates complex integration steps, based on the responses you provide to its questions. These configurations can then be saved and reused as well.

Prerequisites

The prerequisites for starting the Ariba Network integration in SAP Ariba Cloud Integration Gateway are as follows:

- You must have an active buyer account and valid trading buyer/supplier relationship on the Ariba Network.
- Your buyer account must have access to the SAP Ariba Cloud Integration Gateway and be set up to send documents back and forth from the SAP Ariba Cloud Integration Gateway.
- You must have access to the SAP Support Portal to download the SAP Ariba Cloud Integration Gateway add-on for SAP ERP or SAP S/4HANA.
- You must configure certificates and add-ons in SAP ERP or SAP S/4HANA.
- The buyer needs to be added to the **Ariba Network Buyer Orgs** under **Feature Availability** by SAP Ariba customer support.
- An admin user with either **SAP Ariba CIG Configuration** or **SAP Ariba CIG Access** permissions assigned must be created for *https://integration.ariba.com* to access the SAP Ariba Cloud Integration Gateway configuration.

The integration wizard helps you go live in six simple steps for providing basic information, configuring connections, mapping, data entry for cross-referencing additional value maps, testing, and planning the cutover for deployment. We'll provide more detail for each of these steps in the following subsections.

Basic Information

Let's start with the basic information. Logging in as the admin user, select **Configure the Cloud Integration Gateway** and select the checkbox. Upon saving, you'll be issued a p-user to your email and be able to define a unique password. This p-user will be used to send information to the SAP system from SAP Ariba Cloud Integration Gateway and provide details while creating an integration project to connect to the Ariba Network. This connection is created when you select **Cloud Integration Gateway Setup**, as shown in Figure 11.95.

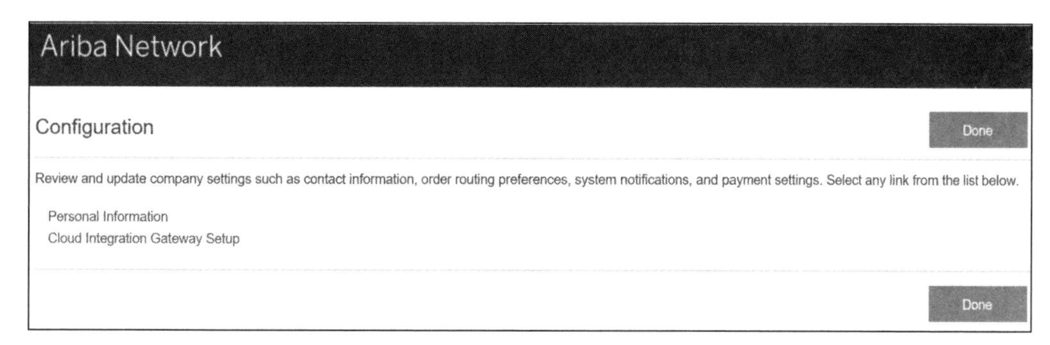

Figure 11.95 SAP Ariba Cloud Integration Gateway: Setup

Next, select **Create Project**, and enter the **Project Name** of your choice. Then, select the **Product** you want to integrate (i.e., **Ariba Network**, **Ariba Sourcing**, or **Ariba Procurement**), as shown in Figure 11.96.

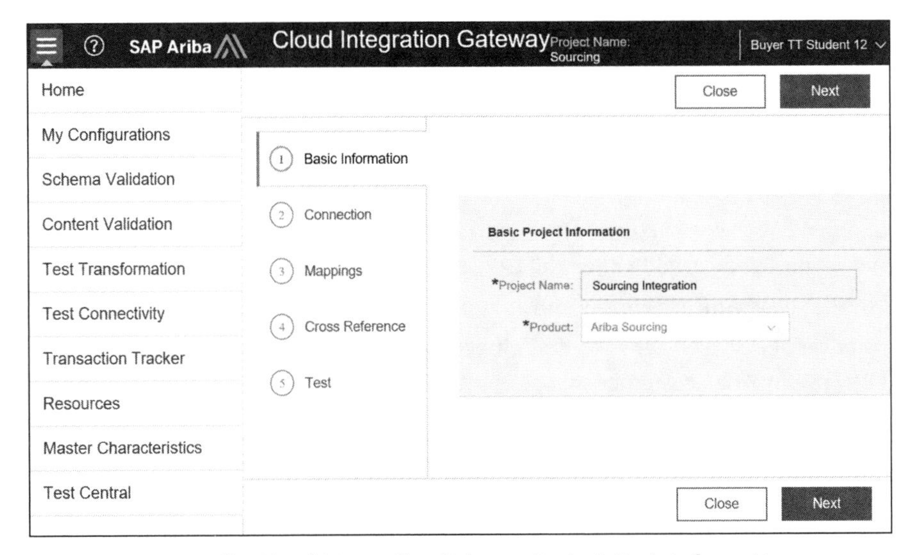

Figure 11.96 SAP Ariba Cloud Integration Gateway Project: Basic Information

Configuring Connections within a Project

After you've completed the **Basic Information** step, next you'll maintain the connections, as shown in Figure 11.97.

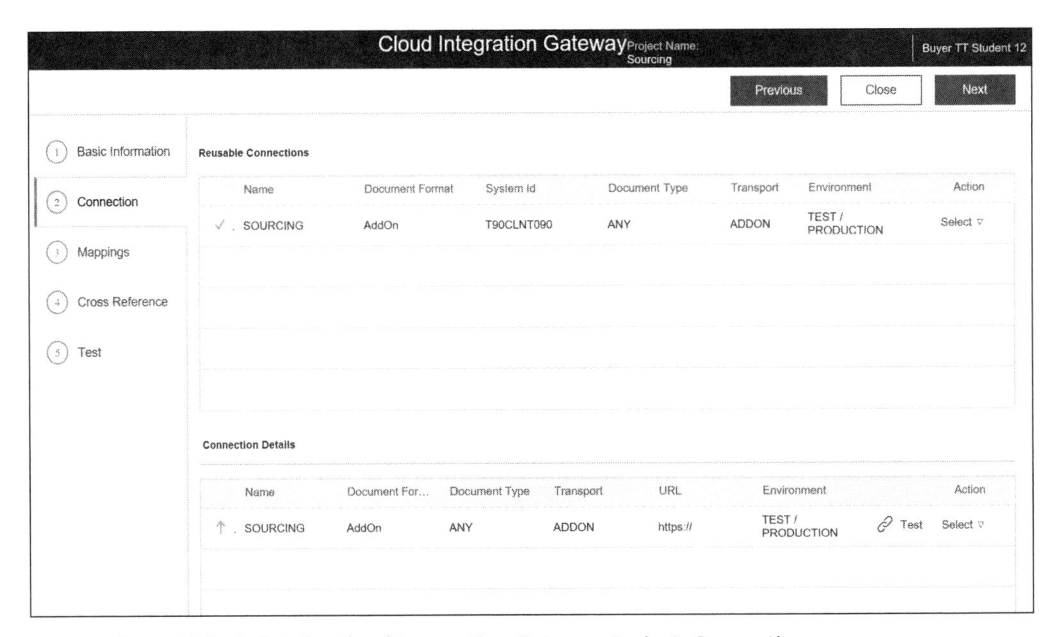

Figure 11.97 SAP Ariba Cloud Integration Gateway Project: Connections

The SAP backend information you enter while setting up the backend system connection with SAP Ariba should be available in the **Reusable Connections** dropdown list. You can also add a new connection by selecting **Add New Connection**. The connection details include the name, the transport type (if using an add-on and integrating via the cloud connector, you won't need this require mappings step) or HTTPS (if you're connecting via middleware such as SAP Process Integration), the backend system ID, and the environment (test or production).

Connection details will contain the system information for the backend connection as well as on the SAP Ariba side. After all the information is verified, click **Save** to continue.

Setting Up Mappings Using Data Maps

In the next steps, you'll establish the mappings for your documents and data. The first step only applies if you're connecting via HTTPS and middleware and isn't applicable for buyer-side integrations. In this step, you'll map the connection based on your chosen middleware. For the buy-side integration with SAP Ariba Cloud Integration Gateway, standard mappings are available for XSLT and stored in the database.

1. Enter data as cross-references for additional value maps. Cross-references are optional. Similar to the backend configuration options, you can define in SAP Ariba Cloud Integration Gateway cross-references for parameters, UOM mappings, and lookup values, as shown in Figure 11.98. With parameters, you can upload and download UOM sets as CSV files. The lookup table aligns SAP Ariba names with customer naming for documents.

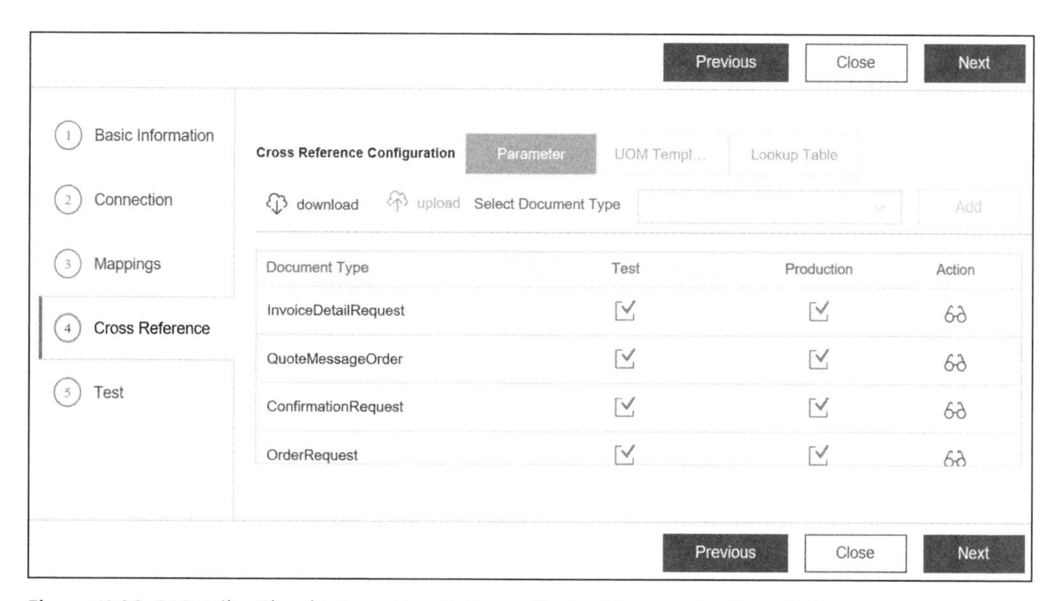

Figure 11.98 SAP Ariba Cloud Integration Gateway Project: Parameter Cross-References

2. Create a test script, and schedule the testing.
3. After you've defined any cross-references and mappings required and confirmed the project, the project status will change to **In Testing**." Now, you're ready for testing. In

the test section, you can define which tests should be run, as shown in Figure 11.99. After you've run the test scenarios successfully, the project is set to **Tested**.

4. Now let's get ready for deployment. This means successfully conducting testing, confirming your communications and organizational change management plans, and finalizing the go-live date.
5. After you've completed the test scenarios, the **Go Live** button will unlock. Click this button to deploy the integration.

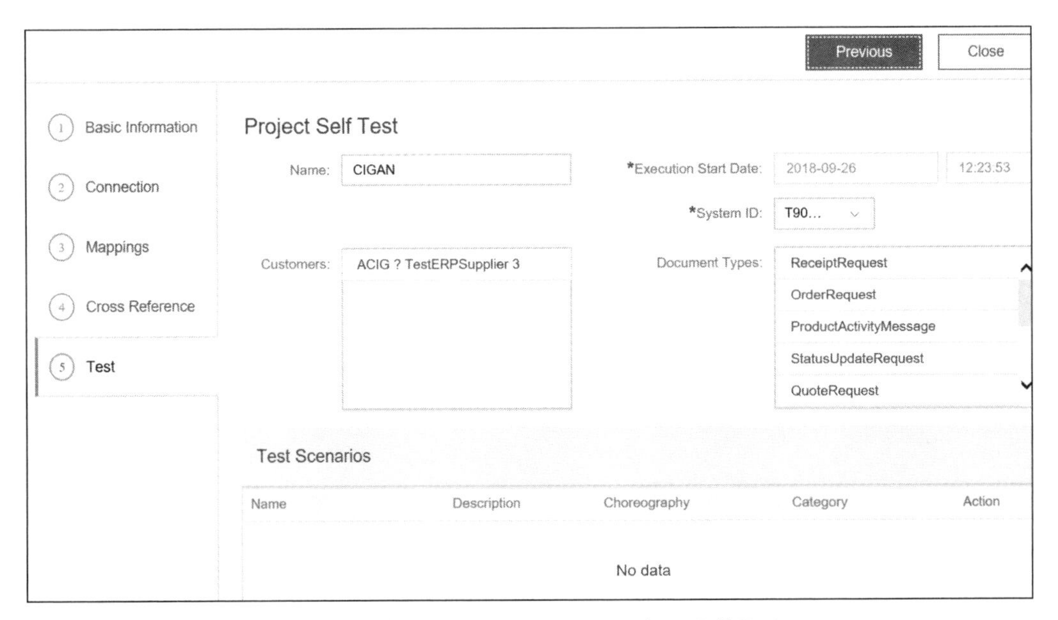

Figure 11.99 SAP Ariba Cloud Integration Gateway Project: Self-Test

11.5.4 Ariba Network Integration

After you've completed the configuration steps for SAP Ariba Cloud Integration Gateway, you can start integrating SAP Ariba solutions, including the Ariba Network. In this section, we'll outline the necessary configuration steps as well as take a look at integrating with older versions of SAP ERP.

Ariba Network Integration in SAP S/4HANA and SAP ERP

Ideally, you'll be integrating SAP Ariba with SAP S/4HANA, which has native integrations to many solutions in SAP Ariba and also has the most immediate integration with SAP Ariba Cloud Integration Gateway. With the configuration of SAP Ariba Cloud Integration Gateway completed, you can now configure SAP S/4HANA at the scenario level.

1. Follow the menu path **SPRO • Integration with Other SAP Components • SAP Ariba Cloud Integration Gateway • Ariba Cloud Integration • Global Settings** (for the technical configuration) and **SPRO • Integration with Other SAP Components • Ariba**

Cloud Integration • Ariba Network Integration • Ariba Network Integration • General Settings • Set Up Interface • Configure the Connections to Send Messages.

2. Set your receiver logical system, as shown in Figure 11.100, and the required types of communications and solutions. Then, run the transaction.

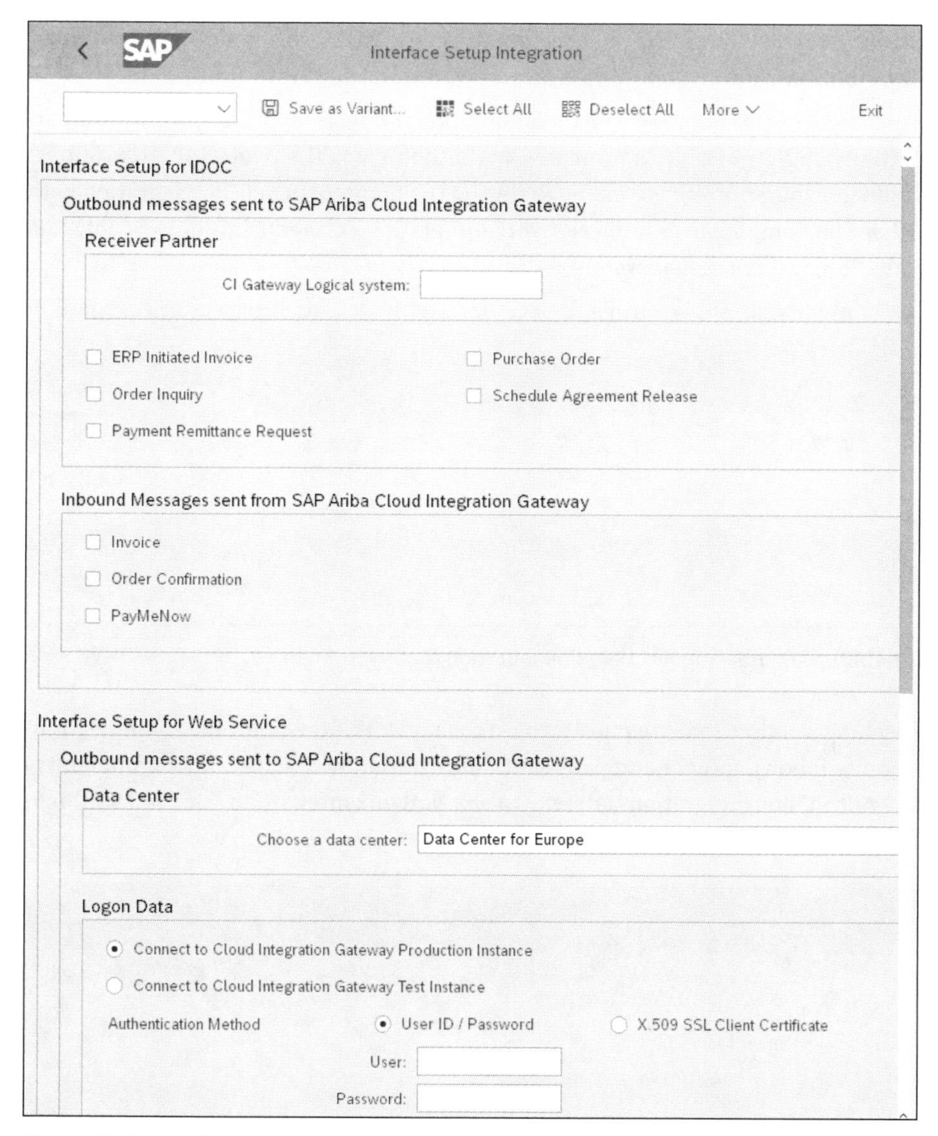

Figure 11.100 Configuring Connections to Send Messages: SAP S/4HANA to SAP Ariba Cloud Integration Gateway

3. You can also set up payment remittances by following the menu path **SPRO • Integration with Other SAP Components • SAP Ariba Cloud Integration Gateway • Ariba Cloud Integration • Create Partner Profile for House Bank**. In the following screen, establish a house bank with partner profile type B.

4. You can maintain parameters, such as the date/time for status updates and the batch job for sending out notifications, by following the menu path **SPRO • Integration with Other SAP Components • SAP Ariba Cloud Integration Gateway • Ariba Network Integration • General Settings • Document Status Update** under the **Document Status Update** section. If you use this setting, you'll need to schedule the batch job `ARBCIG_INBOUND_IDOC_SUR` in Transaction SM36 for **PayMeNow** and **Order Confirmation** status update.
5. Because SAP Ariba Supply Chain Collaboration for Buyers technically comprises part of the Ariba Network integration area, at this point, you'll activate SAP Ariba Supply Chain Collaboration for Buyers by following the menu path **SPRO • Integration with Other SAP Components • Ariba Network Integration • General Settings • Enable SAP Ariba Supply Chain Collaboration**.
6. Select **Modify**, as shown in Figure 11.101, to activate this integration on the SAP S/4HANA side.

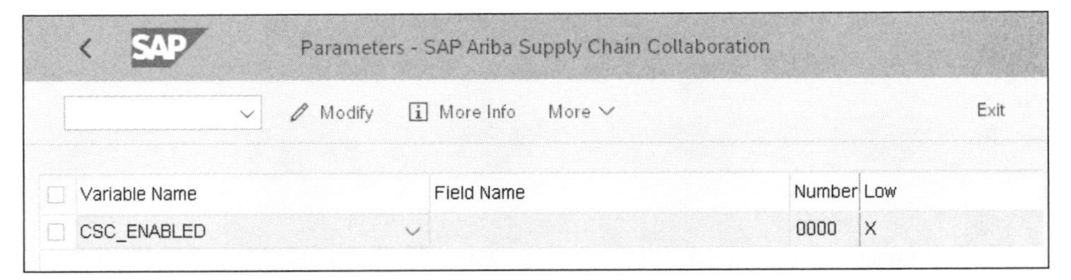

Figure 11.101 SAP Ariba Supply Chain Collaboration for Buyers Enabled

7. To enable vendors that are registered on the network, you'll now maintain relationships by following the menu path **SPRO • Integration with Other SAP Components • SAP Ariba Cloud Integration Gateway • Ariba Network Integration • General Settings • Enable Suppliers**, as shown in Figure 11.102.

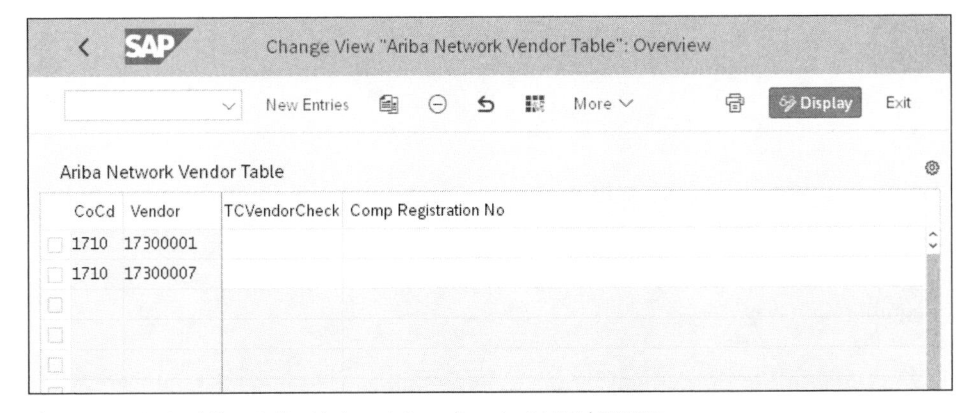

Figure 11.102 Enabling Ariba Network Suppliers in SAP S/4HANA

8. Next, you'll maintain document-specific settings for all in-scope documents for integration with the Ariba Network by following the menu path **SPRO • Integration with Other SAP Components • SAP Ariba Cloud Integration Gateway • Ariba Network Integration • Application Specific Settings**. Select the in-scope documents to maintain. Under this node, you'll find a further node for documents specific to SAP Ariba Supply Chain Collaboration for Buyers, allowing you to maintain these specific documents as well, as shown in Figure 11.103.

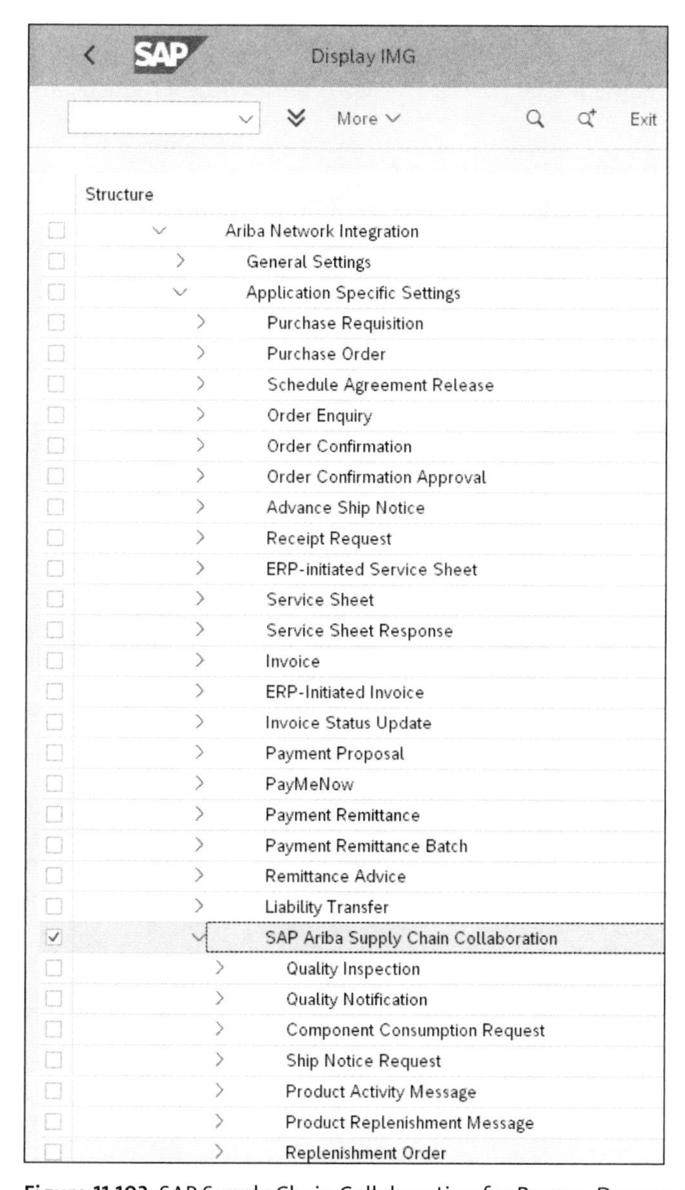

Figure 11.103 SAP Supply Chain Collaboration for Buyers: Document Maintenance

Ariba Network Integration in SAP Ariba Cloud Integration Gateway

To complete the Ariba Network integration in SAP Ariba Cloud Integration Gateway, follow these steps:

1. Ensure that you have user role authorization to access SAP Ariba Cloud Integration Gateway.
2. Log in to your Ariba Network buyer account, and access the **Administration • Configuration** link.
3. Select the **Cloud Integration Gateway Setup**.
4. Read the terms of the privacy statement, select the checkbox to accept these terms, and click **Save**.
5. Again, select the **Cloud Integration Gateway Setup**.
6. On the next page, select the **Go to the Ariba Cloud Integration Gateway** link.
7. If this iteration is the first integration iteration within SAP Ariba Cloud Integration Gateway, you'll receive a p-user and a password reset email from *notification@sapnetworkmail.com*. For subsequent access to SAP Ariba Cloud Integration Gateway, you won't receive further p-users. Each p-user is associated with a single Ariba Network buyer account and is used to send information from the SAP backend to SAP Ariba Cloud Integration Gateway and to provide details while creating a project to connect to the Ariba Network.
8. When you configure your SAP Ariba Cloud Integration Gateway account for the first time, you'll be prompted to send your backend system details. Complete this step prior to setting up your account or configuring SAP Ariba Cloud Integration Gateway further.

Now, you're ready to proceed with further document, master data, and solution integration between your SAP backend and SAP Ariba Cloud Integration Gateway.

11.5.5 Additional Integration Options for Older Versions of SAP ERP

In addition to native integration with SAP S/4HANA, three other major integration options are available for connecting SAP Ariba with SAP ERP 6.0, the previous SAP ERP platform:

- SAP Business Suite add-on direct connection
- SAP Business Suite add-on via SAP Process Orchestration "mediated" integration
- SAP Ariba Adapter integration via SAP Process Orchestration

The SAP Business Suite add-on can be connected directly to the Ariba Network, transmitting and converting documents in the SAP Ariba XML format—cXML. The SAP Business Suite add-on can also leverage a mediated approach, either via SAP Process Orchestration on-premise at the customer or cloud-based SAP Cloud Platform Integration middleware in SAP Ariba. The SAP Ariba Adapter, built prior to the acquisition of Ariba by SAP, leverages SAP Process Orchestration to convert IDocs into cXML, and vice versa.

The SAP Business Suite add-on is an SAP product that is installed on SAP ERP and facilitates the integration and translation of documents into cXML, as well as transmission to and from the Ariba Network. The SAP Business Suite add-on supports additional middleware and mediated approaches but can also communicate directly with the Ariba Network. Both buyers and suppliers can implement these integrations between SAP ERP and the Ariba Network to enable communications/documents back and forth.

The following prerequisites are mandatory for SAP ERP if implementing either SAP Business Suite add-on approach, depending on the selected connectivity technology options for POs, and invoice automation:

- For direct connectivity via web service or connectivity via SAP Cloud Platform Integration:
  - SAP ERP 6.0 SP Stack 15 or higher
  - SAP Business Suite add-on for Ariba Network integration 1.0 SP5
- For connectivity via web service and SAP Process Orchestration:
  - SAP ERP 6.0 SP Stack 15 or higher
  - SAP Business Suite add-on for Ariba Network integration 1.0 SP5
  - SAP Process Integration for SAP NetWeaver 7.3 or higher with Ariba Network adapter for Ariba Network integration CI-5 or higher installed. For versions lower than CI-5, the SAP Ariba integration toolkit is required to mediate the connection.
- For connectivity via IDocs and SAP Process Orchestration:
  - SAP Process Integration or higher with the Ariba Network adapter for Ariba Network integration CI-5 or higher installed
  - SAP ERP 6.0 or higher
  - (Optional) SAP Solution Manager

The first step in integrating and automating POs and invoices between SAP ERP and the Ariba Network is choosing which integration option to use. The SAP Business Suite add-on is the declared go-forward path on SAP's roadmap, so integrations leveraging the SAP Business Suite add-on should be prioritized over integration options using the SAP Ariba Adapter. However, in some instances, the SAP Ariba Adapter may still have a temporary advantage, for example, to support an additional document type or if installing the SAP Business Suite add-on isn't preferred, perhaps due to centralized communication requirements via IDocs from SAP ERP (where middleware has been declared the standard integration approach and IDocs need to be issued from SAP ERP). Even in these cases, evaluating the SAP Business Suite add-on with SAP Process Orchestration mediation as an option makes sense because no further development is planned on the SAP Ariba Adapter going forward.

If the SAP Business Suite add-on is selected as the integration path between SAP ERP and the Ariba Network, the next decision you'll need to make is whether to connect directly to the Ariba Network or to have SAP Process Integration or another middleware component mediate the connection. For the latter case, policies and internal IT standards may require that all communication between SAP ERP move through a middleware component, or other security concerns can play a factor in deciding to use a mediated approach. From a pure cost and simplicity standpoint, the direct connection is the least expensive and least complex.

11.5.6 Integration Pointers for SAP ERP and SAP Business Suite

SAP ERP and the SAP Business Suite also support integration options with the Ariba Network. Similar to the SAP Ariba Cloud Integration Gateway configuration, you can find the options by following the menu path **SPRO • Integration with Other SAP Components • SAP Business Suite Integration Component for Ariba**, as shown in Figure 11.104:

1. After maintaining your login information for the Ariba Network in the **Define Credentials and End Points for Ariba Network** area, you then define your message settings and mappings for cXML. Depending on the connectivity approach you've adopted, whether direct or mediated via SAP Process Integration, you'll then complete either the direct connection or the connection to SAP Process Integration.
2. If you're using UOMs in your SAP ERP or SAP Business Suite instance that are different from the UOM used in the Ariba Network, you can map these units in SAP by following the menu path **SPRO • Integration with Other mySAP Components • SAP Business Suite Integration Component for Ariba • Map Unit of Measure Codes cXML Messages**. You can configure both inbound and outbound messages, as shown in Figure 11.105.

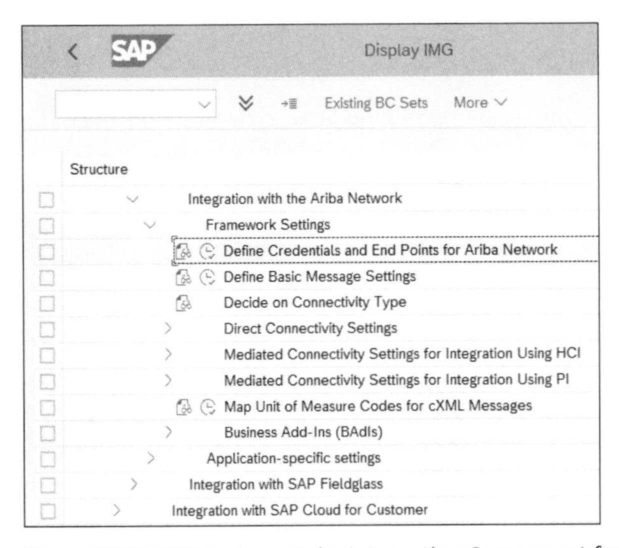

Figure 11.104 SAP Business Suite Integration Component for Ariba Network

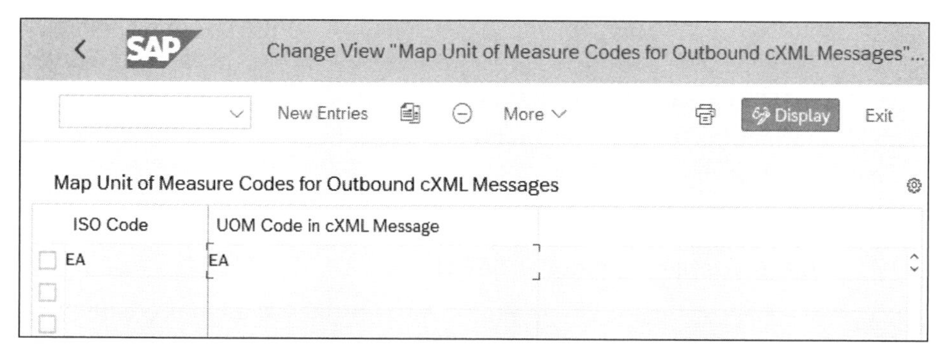

Figure 11.105 Map Unit of Measure Codes for Outbound cXML Messages

3. Next, complete the following configuration steps:
   - Assign an Ariba Network ID by SAP ERP company code. This ID allows the Ariba Network to identify the sending company code of a message. You can also use the same network ID for multiple company codes using the BAdI `Outbound Mapping`.
   - You can then define the message output control, which can apply to the documents listed in Table 11.14.

11

	Application	Standard Output Type	cXML Message Type
POs	EF	NEU	ORDR
Scheduling Agreement Release (FRC)	EL	LPH1	ORDR
Scheduling Agreement Release (JIT)	EL	LPJ1	ORDR
Updates to service entry sheets	ES	NEU	STAT
Invoices (ERS)	MR	ERS6	CCINVC
Shipping notifications	V2	LD00	SHIP
Billing documents (SD)	V3	RD00	INVC

Table 11.14 Defining Message Output Control: Applicable Documents

4. You can map both message texts and general texts from the documents by following the menu path **SPRO • Integration with Other SAP Components • SAP Business Suite Integration Component for Ariba • Application-Specific Settings • SAP Integration Component for Ariba • Define Document Specific Customizing/Map Texts of SAP ERP and Ariba Network.**
5. Adapt the SAP ERP Incoterm to cXML **TransportTerms** field in the event that nonstandard Incoterms or differences require this adaptation.
6. Specify whether you'll be connecting to the Ariba Network as a buyer or as a supplier. As a supplier, you're essentially selling, which activates sales and distribution

processes instead of procurement ones. Thus, you must maintain your mapping settings for the applicable documents in the correct section, either **SPRO • Integration with Other SAP Components • SAP Business Suite Integration Component for Ariba • Application-Specific Settings • SAP Integration Component for Ariba • Integration for Buyers** or **Integration for Vendors**.

11.5.7 SAP Master Data Integration in SAP ERP

To set up master data integration, follow the menu path **SPRO • Integration with Other SAP Components • SAP Ariba Cloud Integration Gateway • Master Data Integration**. You must first review and complete the applicable setup steps, beginning with the **Configure External Commands for Operating System** and the **Configure Ariba Incremental Extract Event** tasks on through **Maintain Parameters** and **Maintain Filters** tasks. After you've maintained your master data integration settings, you can schedule jobs to upload the required master data via Transaction SM36. The jobs you would schedule are `ARBCIG_MASTER_DATA_EXPORT` for general master data, `ARBCIG_BUYER_CATALOG_EXPORT` for catalogs, and `ARBCIG_BOM_MASTER_DATA_EXPORT` for BOMs.

If needed, you can resend a master data set or catalog by following the menu path **SPRO • Integration with Other SAP Components • SAP Ariba Cloud Integration Gateway • Master Data Integration • General Settings • Resend Master Data or Catalog Upload Request**. Note that this activity doesn't render BOMs. For each area and job, numerous enhancements are available in this menu section.

11.5.8 SAP Ariba Sourcing Integration

SAP Ariba Sourcing is a key area in SAP Ariba that is clearly superior to the tools on offer in SAP S/4HANA. By virtue of the Ariba Network and the general advantages of the cloud, sourcing activities and functionality are invariably more robust when conducted in SAP Ariba, than in attempting to use the sourcing cockpit in SAP ERP alone. Fortunately, the integration between the two areas is also fairly built out and robust. This next section will detail this further.

SAP Ariba Sourcing Integration in SAP ERP

To set up SAP Ariba Sourcing integration, follow the menu path **SPRO • Integration with Other SAP Components • SAP Ariba Cloud Integration Gateway • SAP Ariba Strategic Sourcing Suite Integration • General Settings • Setup the Interface • Configure the Connections to Send Messages**. First, you'll configure the interface itself, as shown in Figure 11.106, by maintaining the logical system, types of data to be transmitted outbound (e.g., quote requests) and inbound (e.g., SAP Ariba Sourcing RFPs), and login data and proxy information.

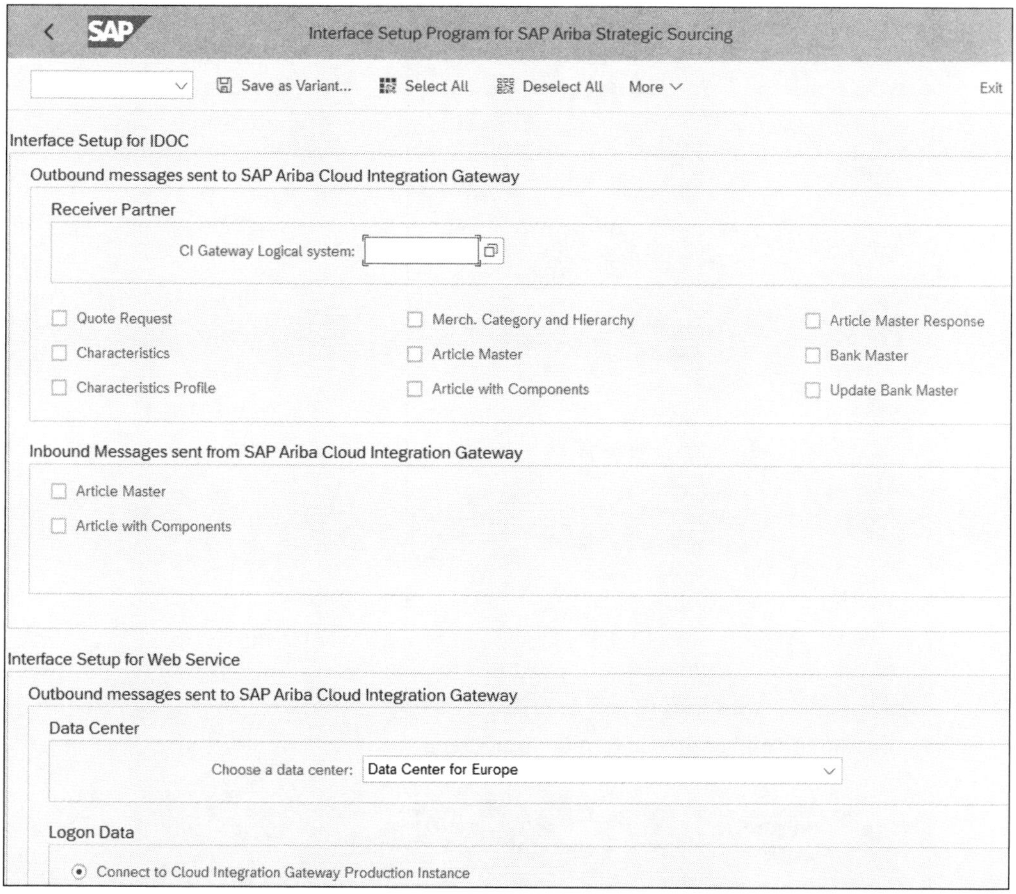

Figure 11.106 Interface Setup Program for SAP Ariba Strategic Sourcing

Note

You can also integrate article masters, instead of material masters. Article masters are part of SAP's Industry Solution for Retail (IS-Retail) which is supported within this integration framework. Industry solutions tailor SAP ERP functionality to a particular industry.

Next, you'll maintain the conditions for output control by following the menu path **SPRO • Integration with Other SAP Components • SAP Ariba Cloud Integration Gateway • SAP Ariba Strategic Sourcing Suite Integration • Application Specific Settings • Request for Quotation • Conditions for Output Control**. Highlight one of the rows, as shown in Figure 11.107, and then select **Condition Records**, **Procedures**, **Output Types**, or **Access Sequences** to maintain these different areas, as shown in Figure 11.108.

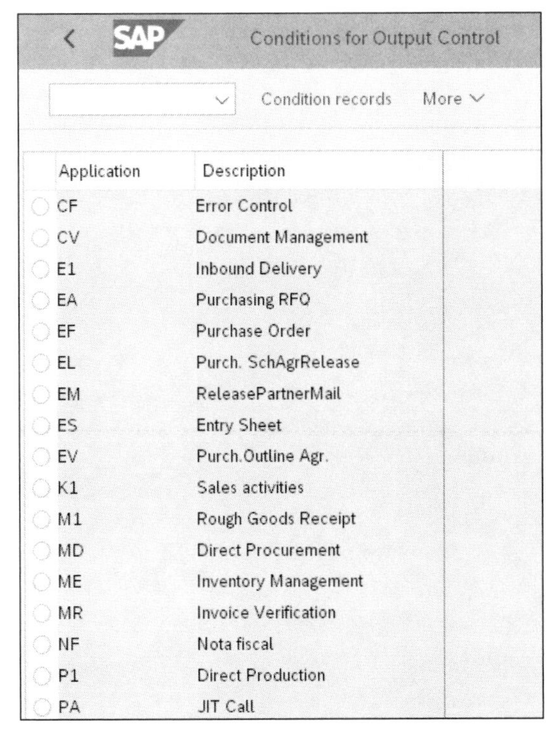

Figure 11.107 Conditions for Output Control Screen

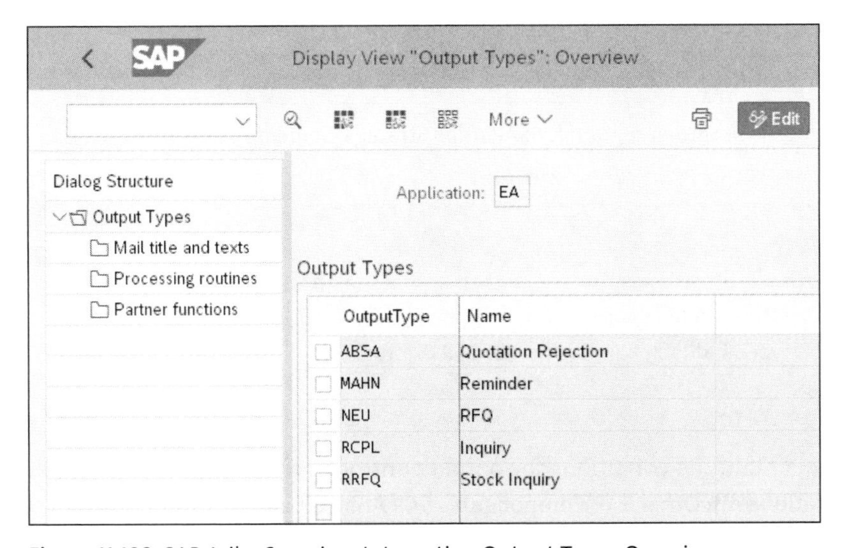

Figure 11.108 SAP Ariba Sourcing: Integration Output Types Overview

If desired, you can include custom messages with your RFP document exchanges by following the menu path **SPRO • Integration with Other SAP Components • SAP Ariba Cloud Integration Gateway • SAP Ariba Strategic Sourcing Suite Integration • Application Specific Settings • Request for Quotation • Maintain Parameters for Request for**

Quotation. You can add default content for both the internal and external messages that are included in the RFQ documents/steps, as shown in Figure 11.109.

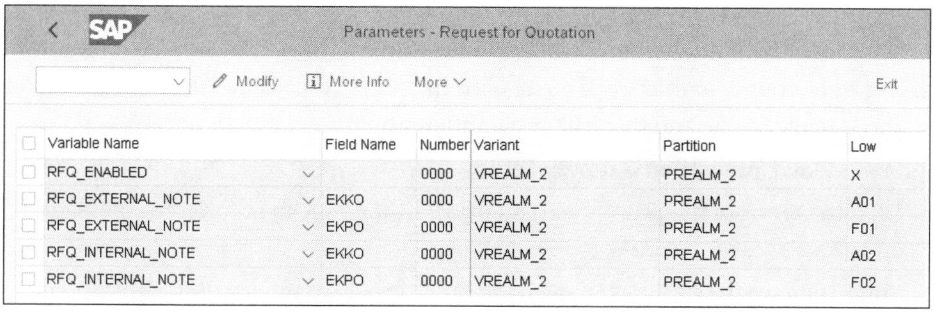

Parameters - Request for Quotation

Modify | More Info | More | Exit

Variable Name	Field Name	Number	Variant	Partition	Low
RFQ_ENABLED		0000	VREALM_2	PREALM_2	X
RFQ_EXTERNAL_NOTE	EKKO	0000	VREALM_2	PREALM_2	A01
RFQ_EXTERNAL_NOTE	EKPO	0000	VREALM_2	PREALM_2	F01
RFQ_INTERNAL_NOTE	EKKO	0000	VREALM_2	PREALM_2	A02
RFQ_INTERNAL_NOTE	EKPO	0000	VREALM_2	PREALM_2	F02

Figure 11.109 RFQ_EXTERNAL_NOTE and RFQ_INTERNAL_NOTE

For extensions, BAdI implementations are available for SAP Ariba Sourcing integration, including BAdI ARBCIG_SRC_CONTRACT_CREATE, which maps the contract fields between systems and is especially useful for RFx processes that result in contracts rather than POs. To access these user exits, follow the menu path **SPRO • Integration with Other SAP Components • SAP Ariba Cloud Integration Gateway • SAP Ariba Strategic Sourcing Suite Integration • Business Add-Ins (BAdIs) • SAP Ariba Sourcing and SAP Ariba Contracts Integration • Contract**. The same functionality is also available for POs in this area by using the BAdI ARBCIG_SRC_QUOTE_PO because POs play such a key role in SAP Ariba Procurement.

SAP Ariba Sourcing Integration in SAP Ariba Cloud Integration Gateway

The prerequisites for starting the SAP Ariba Sourcing integration in SAP Ariba Cloud Integration Gateway are as follows:

- You must have an enabled SAP Ariba Sourcing realm connected to an active buyer account and a valid trading buyer/supplier relationship on the Ariba Network.
- Your buyer account must have access to SAP Ariba Cloud Integration Gateway and be set up to send documents back and forth from SAP Ariba Cloud Integration Gateway.
- You must access to SAP Support Portal to download the SAP Ariba Cloud Integration Gateway add-on for SAP ERP or SAP S/4HANA.
- You must configure certificates and add-ons in SAP ERP or SAP S/4HANA.
- The sourcing realm must be added to your Ariba Network buyer account under **Feature Availability** by SAP Ariba customer support.

Note that, for SAP Ariba Sourcing integration, you only need to enable SAP Ariba Cloud Integration Gateway through the SAP Ariba Sourcing realm. For transactions such as Quote Request, Award, and Contract, you'll need to enable SAP Ariba Cloud Integration Gateway from the Ariba Network as well.

Similar to establishing the Ariba Network integration in SAP Ariba Cloud Integration Gateway, you'll now complete the following steps:

1. Ensure that you have user role authorization to access SAP Ariba Cloud Integration Gateway, as well as the customer administrator role.
2. Log in to your Ariba Network buyer account, access the **Administration** link under the **Manage** menu, and then click on **Configuration**.
3. Select **Cloud Integration Gateway Setup**.
4. Read the terms of the privacy statement, select the checkbox to accept these terms, and click **Save**.
5. You'll then see the status as **Enabled** for SAP Ariba Cloud Integration Gateway. Again, select **Cloud Integration Gateway Setup**.
6. On the next page, select the **Go to the Ariba Cloud Integration Gateway** link.
7. If this iteration is the first integration iteration within SAP Ariba Cloud Integration Gateway, you'll receive a p-user and a password reset email from *notification@sap-networkmail.com*. For subsequent accesses to SAP Ariba Cloud Integration Gateway, you won't receive further p-users. Each p-user is associated with a single Ariba Network buyer account and is used to send information from the SAP backend to SAP Ariba Cloud Integration Gateway, and to provide details while creating a project to connect to SAP Ariba Sourcing.
8. When you configure your SAP Ariba Cloud Integration Gateway account for the first time, you'll be prompted to send your backend system details. Complete this step prior to setting up your account or configuring SAP Ariba Cloud Integration Gateway further.

Now, you're ready to proceed with further document, master data, and solution integration between your SAP backend and SAP Ariba Cloud Integration Gateway.

SAP Ariba Buying, SAP Ariba Procurement, and P2O/P2P

SAP Ariba Procurement, which includes SAP Ariba Buying, previously was called SAP Ariba Procure-to-Pay and SAP Ariba Procure-to-Order, depending on how far the process extended before handing off to the backend ERP system. This integration is now covered by SAP Ariba Cloud Integration Gateway as well.

11.5.9 SAP Ariba Procurement Integration in SAP ERP

Similar to SAP Ariba Sourcing, a configuration area is available for SAP Ariba Procurement in SAP. As this area covers more possible integrations than did SAP Ariba Sourcing, the options are commensurately more extensive during configuration, as shown in Figure 11.110.

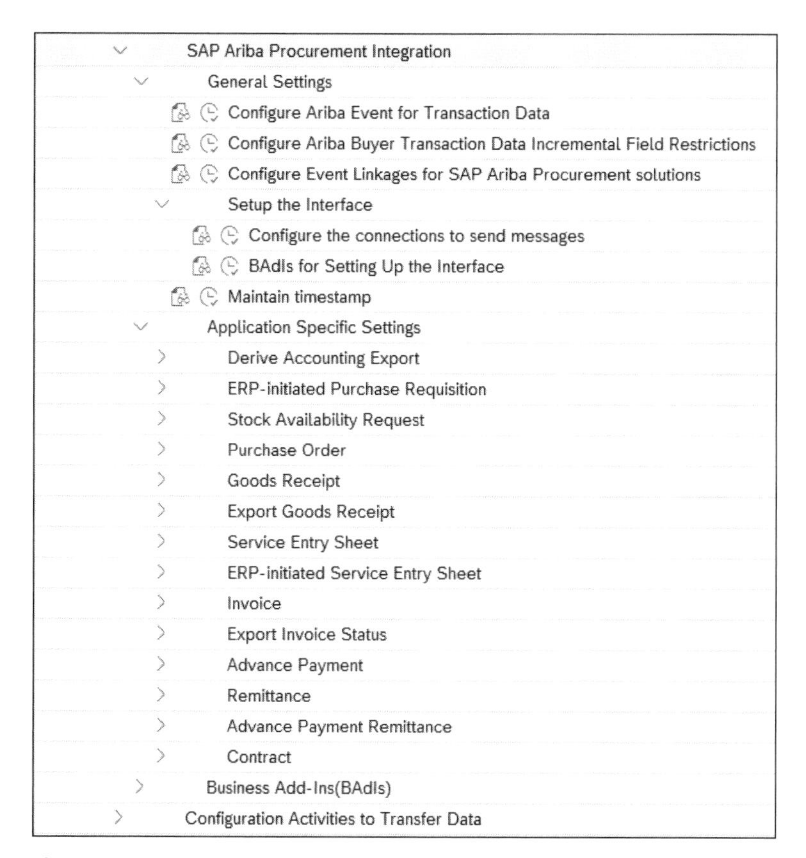

Figure 11.110 SAP Ariba Procurement Integration

Follow these steps:

1. As with SAP Ariba Sourcing, you'll first establish the baseline communication settings between systems by following the menu path **SPRO • Integration with Other SAP Components • SAP Ariba Cloud Integration Gateway • SAP Ariba Procurement Integration • General Settings • Setup Interface • Configure Connections and Send Messages**, as shown in Figure 11.111.
2. In the procurement setup, login data is required just like earlier in the SAP Ariba Sourcing setup, but, in this case, you'll also specify whether catalog upload requests and/or PO-specific inbound messages coming from SAP Ariba Cloud Integration Gateway should be included.
3. After you've finished setting up the interface, you can define the different document type integrations in the application-specific settings found by following the menu path **SPRO • Integration with Other SAP Components • SAP Ariba Cloud Integration Gateway • SAP Ariba Procurement Integration • Application-Specific Settings**. This area covers a number of document types used in procurement activities, including SAP ERP–initiated purchase requisitions (see Figure 11.112), PO header output types, and output control (see Figure 11.113).

Parameters - ERP Requisition Export

Modify | More Info | More | Exit

Variable Name	Field Name	Number	Variant	Partition	Low
BUY_PR_CHANGE_ENABLED		0000	VREALM_350014	PREALM_350014	X
BUY_PR_EXPORT_CHANGES		0000	VREALM_350014	PREALM_350014	X

Figure 11.111 Interface Setup Program for SAP Ariba Procurement

Parameters - ERP Requisition Export

Modify | More Info | More | Exit

Variable Name	Field Name	Number	Variant	Partition	Low
BUY_PR_CHANGE_ENABLED		0000	VREALM_350014	PREALM_350014	X
BUY_PR_EXPORT_CHANGES		0000	VREALM_350014	PREALM_350014	X

Figure 11.112 SAP Ariba Procurement: SAP ERP-Initiated Purchase Requisition Export

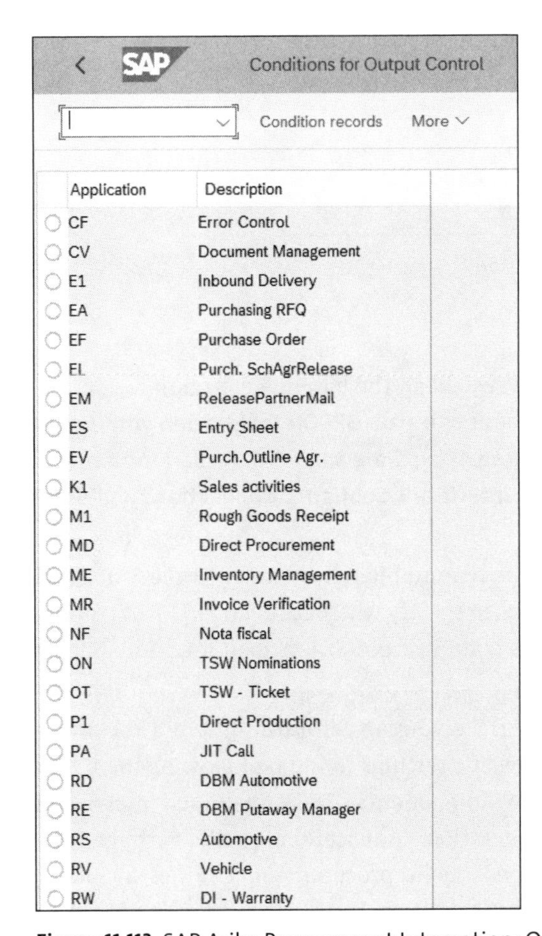

Conditions for Output Control

Condition records | More

Application	Description
CF	Error Control
CV	Document Management
E1	Inbound Delivery
EA	Purchasing RFQ
EF	Purchase Order
EL	Purch. SchAgrRelease
EM	ReleasePartnerMail
ES	Entry Sheet
EV	Purch.Outline Agr.
K1	Sales activities
M1	Rough Goods Receipt
MD	Direct Procurement
ME	Inventory Management
MR	Invoice Verification
NF	Nota fiscal
ON	TSW Nominations
OT	TSW - Ticket
P1	Direct Production
PA	JIT Call
RD	DBM Automotive
RE	DBM Putaway Manager
RS	Automotive
RV	Vehicle
RW	DI - Warranty

Figure 11.113 SAP Ariba Procurement Integration: Output Control

4. For receipts, both goods receipts, shown in Figure 11.114, and service entry sheets, shown in Figure 11.115, are covered. Further configuration of service entry sheet responses is shown in Figure 11.116.

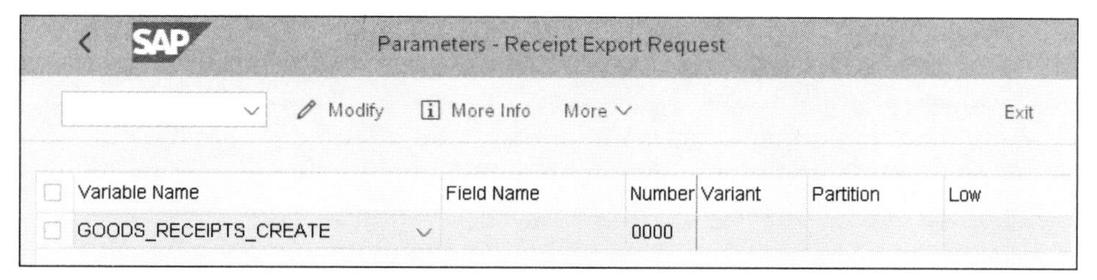

Figure 11.114 SAP Ariba Procurement: Goods Receipt Export

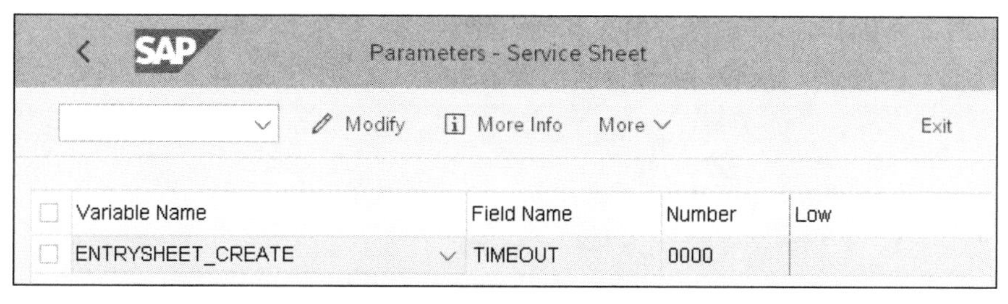

Figure 11.115 SAP Ariba Procurement: Service Entry Sheet Export

11

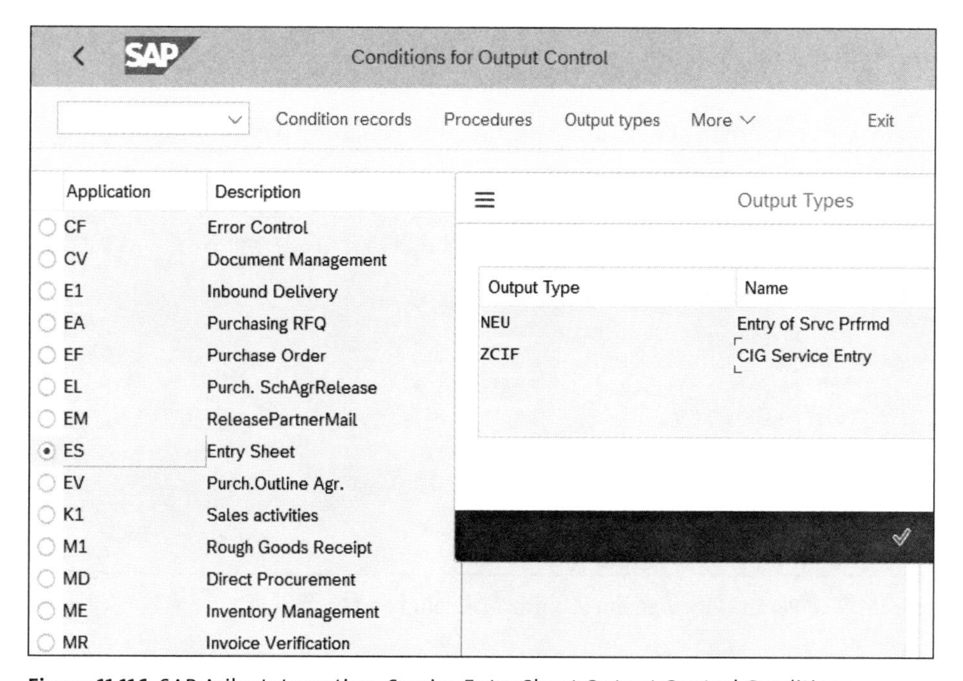

Figure 11.116 SAP Ariba Integration: Service Entry Sheet Output Control Conditions

5. You can also allow service entry sheets to be revoked in both systems by updating the function module `MS_UPDATE_SERVICE_ENTRY` in SAP. For invoicing, you can also set the invoice export settings, as shown in Figure 11.117, as well as advance payment settings, as shown in Figure 11.118.

Parameters - Payment Export Request

Modify | More Info | More | Exit

Variable Name	Field Name	Number	Variant	Partition	Low
ASSET_TR_TYP_FI_NONPO		0000			
INVOICE_PUSH		0000			
P2P_INVOICE_ENABLED		0000	VREALM_50363	PREALM_50363	X

Figure 11.117 SAP Ariba Procurement: Payment Export Request

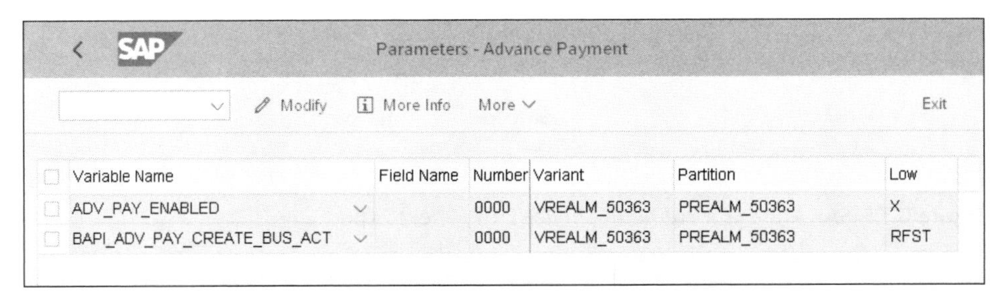

Parameters - Advance Payment

Modify | More Info | More | Exit

Variable Name	Field Name	Number	Variant	Partition	Low
ADV_PAY_ENABLED		0000	VREALM_50363	PREALM_50363	X
BAPI_ADV_PAY_CREATE_BUS_ACT		0000	VREALM_50363	PREALM_50363	RFST

Figure 11.118 SAP Ariba Procurement: Advance Payment

For contracts, you can configure discount and price condition parameters, as shown in Figure 11.119.

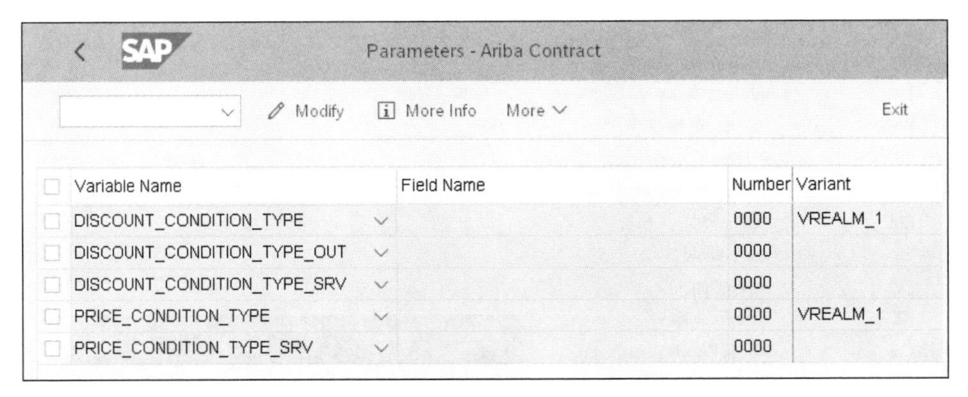

Parameters - Ariba Contract

Modify | More Info | More | Exit

Variable Name	Field Name	Number	Variant
DISCOUNT_CONDITION_TYPE		0000	VREALM_1
DISCOUNT_CONDITION_TYPE_OUT		0000	
DISCOUNT_CONDITION_TYPE_SRV		0000	
PRICE_CONDITION_TYPE		0000	VREALM_1
PRICE_CONDITION_TYPE_SRV		0000	

Figure 11.119 SAP Ariba Procurement: Contract Discount and Price

Now, follow the menu path **SPRO • Integration with Other SAP Components • SAP Ariba Cloud Integration Gateway • SAP Ariba Procurement Integration • Application-Specific**

Settings • Contracts • Maintain Field Map for Contracts. You can also map the individual contract fields between the two systems, as shown in Figure 11.120.

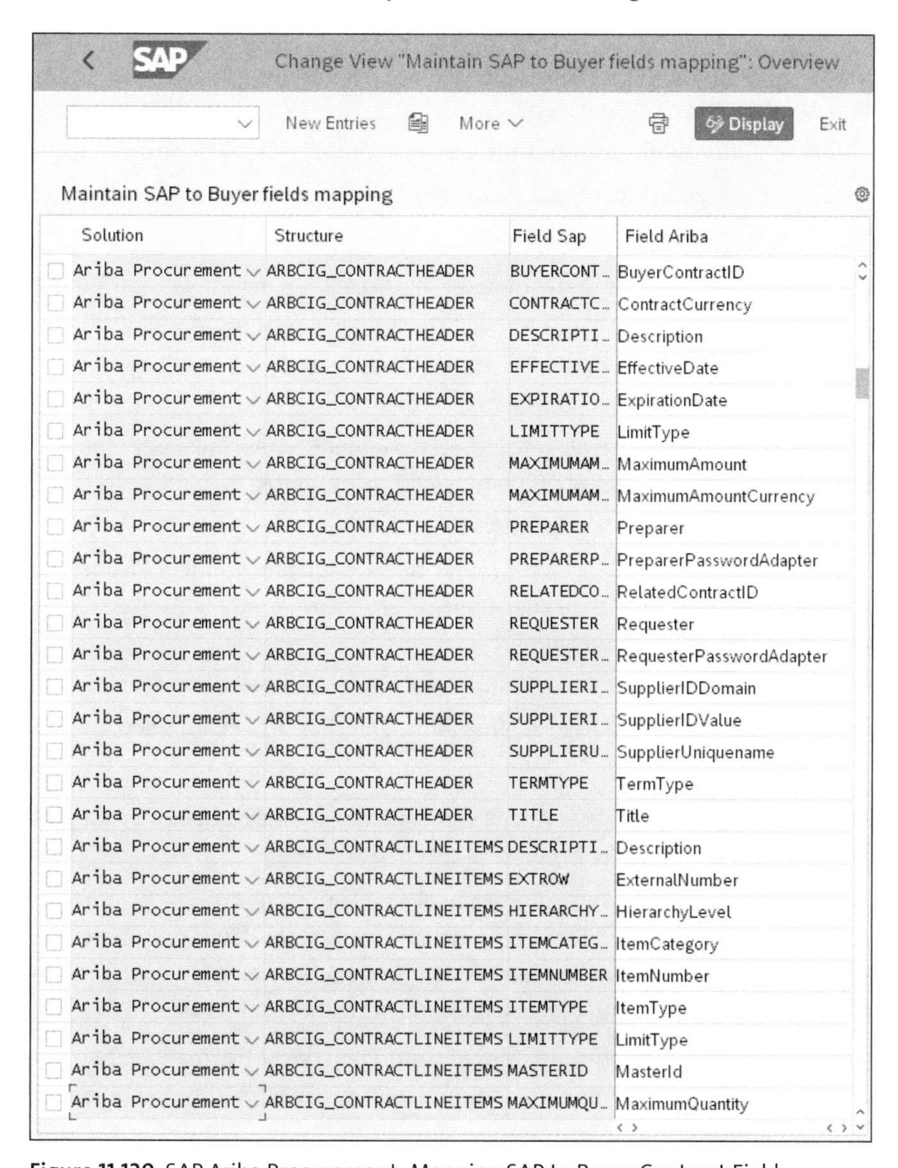

Figure 11.120 SAP Ariba Procurement: Mapping SAP to Buyer Contract Fields

SAP Ariba Procurement Integration Steps with SAP Ariba Cloud Integration Gateway

The prerequisites for starting the SAP Ariba Procurement integration steps in SAP Ariba Cloud Integration Gateway are as follows:

- You must have an enabled SAP Ariba procurement realm connected to an active buyer account and a valid trading buyer/supplier relationship on the Ariba Network.

- Your buyer account must have the following:
  - Access to SAP Ariba Cloud Integration Gateway and be set up to send documents back and forth from SAP Ariba Cloud Integration Gateway.
  - Access to SAP Support Portal to download the SAP Ariba Cloud Integration Gateway add-on for SAP ERP or SAP S/4HANA.
  - You must configure certificates and add-ons in SAP ERP or SAP S/4HANA.
  - The procurement realm must be added to your Ariba Network buyer account under **Feature Availability** by SAP Ariba customer support.

Similar to establishing the Ariba Network and SAP Ariba Sourcing integration in SAP Ariba Cloud Integration Gateway, you'll also need to complete the following steps:

1. Ensure that you have user role authorization to access SAP Ariba Cloud Integration Gateway, as well as the customer administrator role.
2. Log in to your Ariba Network buyer account, access **Administration** link under the **Feature Manager** menu, and then click on **Feature Availability Status**.
3. Select the **Cloud Integration Gateway Setup** and click **Edit**.
4. Click the **Add/Remove** button. Search on and select your procurement realm (the realm you're integrating), and then click **Done**. Your procurement realm should now be displayed under the **Enable for Sites** list.
5. Log in in as an SAP Ariba procurement admin, and choose **Customization • Parameters**.
6. Choose **Application • Common • Asynch Integration**, and click on **Details**. Select the **Yes** radio button and save.

Note that some of these steps may need to be performed by SAP Ariba customer support or by individuals with service manager–level access to the system.

Now, for SAP Ariba Procurement integration, you'll enable the SAP Ariba Cloud Integration Gateway procurement realm for both procurement master data and transaction integration. For enabling the invoice management and combo realms, you'll need to enable SAP Ariba Cloud Integration Gateway from the Ariba Network as well.

After you've completed the prerequisite steps for setting up the SAP Ariba Procurement realm integration in SAP Ariba Cloud Integration Gateway, you'll perform the following steps:

1. Ensure that you have user role authorization to access SAP Ariba Cloud Integration Gateway, as well as the customer administrator role.
2. Log in to your Ariba Network buyer account, access the **Administration** link under the **Manage** menu, and then click on **Core Administration**.
3. Select **Cloud Integration Gateway Setup**.
4. Read the terms of the privacy statement, and select the checkbox to accept these terms. In addition, select the checkboxes for each required task and integration event as needed, and click **Save**.

5. You'll then see the status as **Enabled** for SAP Ariba Cloud Integration Gateway. Again, select **Cloud Integration Gateway Setup**.
6. Select the **Go to the Ariba Cloud Integration Gateway** link on the next page.
7. If this iteration is the first integration iteration within SAP Ariba Cloud Integration Gateway, you'll receive a p-user and a password reset email from *notification@sapnetworkmail.com*. For subsequent accesses to SAP Ariba Cloud Integration Gateway, you won't receive further p-users. Each p-user is associated with a single Ariba Network buyer account and is used to send information from the SAP backend to the SAP Ariba Cloud Integration Gateway and to provide details while creating a project to connect to the SAP Ariba Procurement realm.
8. When you configure your SAP Ariba Cloud Integration Gateway account for the first time, you'll be prompted to send your backend system details. Complete this step before setting up your account or configuring SAP Ariba Cloud Integration Gateway further.
9. Now, you're ready to proceed with further document, master data, and solution integration between your SAP backend system and SAP Ariba Cloud Integration Gateway.
10. Repeat the earlier steps to activate the **Enable Asynchronous Integration Events for Webservice** feature. Read the terms of the privacy statement, select the checkbox to accept these terms, and click **Save**.
11. You'll then see the status as **Enabled** for SAP Ariba Cloud Integration Gateway. Again, select **Cloud Integration Gateway Setup**.
12. On the next page, select the **Go to the Ariba Cloud Integration Gateway** link.
13. If this iteration is the first integration iteration within the SAP Ariba Cloud Integration Gateway, you'll receive a p-user and a password reset email from *notification@sapnetworkmail.com*. For subsequent accesses to SAP Ariba Cloud Integration Gateway, you won't receive further p-users. Each p-user is associated with a single Ariba Network buyer account and is used to send information from the SAP backend to SAP Ariba Cloud Integration Gateway and to provide details while creating a project to connect to the SAP Ariba Procurement realm.
14. When you configure your SAP Ariba Cloud Integration Gateway account for the first time, you'll be prompted to send your backend system details. Complete this step before setting up your account or configuring SAP Ariba Cloud Integration Gateway further.
15. Now, you're ready to proceed with further document, master data, and solution integration between your SAP backend system and SAP Ariba Cloud Integration Gateway.

11.5.10 SAP Ariba Cloud Integration Gateway Extensions

For extensions, you can access the extension framework in the add-on, which includes an extension structure for all IDocs and proxies. BAdIs are available under the structure /ARBA/ARIBA/EXTENSION. In SAP, a BAdI code sample looks like this:

```
MOVE lw_int_edidd-sdata TO lw_e1edk01.
lw_extension-objkey = lw_e1edk01-belnr.
lw_extension-obname = 'PO'.
lw_extension-fieldname = 'ZZ_TEL_NUMBER'.
lw_extension-fieldvalue = '080-98765432'.
MOVE lw_extension TO lw_int_edidd-sdata.
lw_int_edidd-segnam = '/ARBA/HDEXTN'.
INSERT lw_int_edidd INTO int_edidd INDEX sy-tabix + 1.
```

When this BAdI is implemented in SAP, SAP Ariba will load the XLST mapping for the extended segment. The output in SAP Ariba should look like this:

```
<Extrinsic name="Objkey">1000065263</Extrinsic>
<Extrinsic name="ObjName">PO</Extrinsic>
<Extrinsic name="FieldName">ZZ_TEL_NUMBER</Extrinsic>
<Extrinsic name="FieldValue">080-987654432</Extrinsic>
```

After you've completed the installation, configuration, and extension of this functionality, you're ready for testing. In **Test Central**, you'll find a repository of test cases that can simulate a supplier "response" and propagate a test result.

11.5.11 Migrating to SAP Ariba Cloud Integration Gateway

SAP is supporting migration to SAP Ariba Cloud Integration Gateway with migration tools, and maintenance for cloud integration is still planned until further notice. The supported migration scenarios are as follows:

- SAP ERP with the SAP Ariba Cloud Integration Gateway add-on
- Migration for ERP cloud integration adapter
- Migration for ERP, SAP Business Suite add-on
- SAP S/4HANA with the SAP Ariba Cloud Integration Gateway add-on
- Migration for SAP S/4HANA cXML integration
- Migration for SAP S/4HANA Cloud Integration adapter

You can transition your integration platform from an existing integration up to SAP Ariba Cloud Integration Gateway in a variety of a ways. In a "big bang" approach, the entire architecture is transitioned at once. Alternatively, you can pick and choose specific areas to transfer and move them in a phased manner. SAP Ariba Cloud Integration Gateway can run alongside supported versions of Cloud Integration version 9 (CI-9).

Scenarios initiated in CI-9 can be completed in SAP Ariba Cloud Integration Gateway, allowing for hybrid integration layers, which are required because of namespace changes and the fact that the new namespace also contains more than 4,000 ABAP objects for use. Dozens of test cases are provided to support migration testing and consistency. Ultimately, the target end state for your integration layer should be SAP Ariba Cloud Integration Gateway, as shown in Figure 11.121.

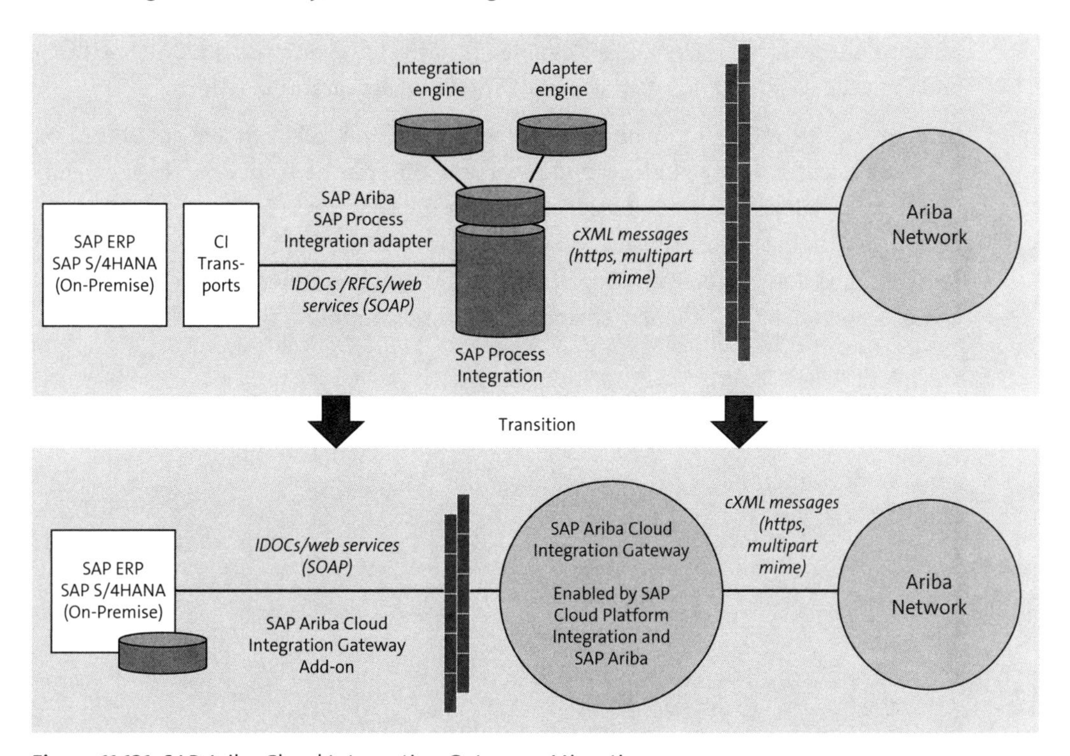

Figure 11.121 SAP Ariba Cloud Integration Gateway: Migration

The high-level steps for migrating to SAP Ariba Cloud Integration Gateway are as follows:

1. Download and install the applicable the SAP Ariba Cloud Integration Gateway add-on for your version of SAP ERP.
2. Migrate (copy) the database tables.
3. Transfer configurations from SAP to SAP Ariba Cloud Integration Gateway.
4. Migrate your updated BAdI code.

Note

For more information on migrating SAP Ariba Cloud Integration Gateway, see the SAP Ariba Cloud Integration Gateway migration guide at *https://help.sap.com*.

11.6 Summary

While cloud solutions offer unprecedented savings in terms of maintenance and ease of use, data and processes from on-premise ERP systems, such as SAP ERP and SAP S/4HANA, are often core to augmenting and completing processes in SAP Ariba. In this chapter, we outlined several integration options for pairing SAP's cloud portfolio solutions for procurement with SAP ERP environments. For SAP Ariba, many of these integrations are available today and accessible via native integrations with SAP S/4HANA and accessible comprehensively as part of SAP Cloud Integration Platform.

In terms of integration, more linkages are being built and will be native out of the box in future releases in all of SAP's cloud-based and on-premise solutions. However, just being an SAP cloud solution doesn't mean that the solution is completely integrated with SAP's digital core solutions today. For now, understanding your integration requirements and options for your project is key to successfully realizing the end-to-end processes in SAP's cloud-based and on-premise solutions.

Chapter 12
Conclusion

In this chapter, we'll review the numerous SAP Ariba solutions we've discussed in previous chapters and offer some predictions about the future of procurement.

SAP has invested billions of dollars in its procurement solution portfolio over the years, both in its acquisitions (SAP Ariba, SAP Fieldglass, and SAP Concur) and its in-house development (SAP S/4HANA sourcing and procurement). Over the last five years, SAP has invested more in cloud-based procurement solutions than in all previous years combined. As more transactions move to the cloud, the value of networks and flexible processes integrated with cloud networks will drive accelerated growth and efficiencies. SAP is uniquely positioned for this future, with the Ariba Network and its $1.5-trillion transaction run rate, the largest business-to-business network of its kind and as a recognized leader in vendor management systems with SAP Fieldglass and the travel and expense management with SAP Concur. From an install base standpoint, over 70% of digital transactions worldwide already touch an SAP system at some point.

While the technologies supporting procurement transactions, as well as the sheer volume of transactions, have evolved dramatically since the inception of computing (especially in the last five years) many of the processes and procurement scenarios have remained largely static. A company still needs to identify demand and then find, vet, and negotiate with suppliers. Once a supplier has been selected, the organization can create contracts and/or purchase order agreements, take receipt of goods/services provided, process the invoice, and, finally, analyze performance to make the next round of purchasing activities smarter. On the supplier side, a company needs to find the demand, negotiate terms with potential customers, obtain, produce the good or service to fulfil orders, and then submit an invoice. The technology underpinning these steps now includes real-time analytics of supplier/spend and automated processing of most orders once submitted, as well as massive networks to facilitate collaboration, document exchange, and new supplier identification.

12.1 Summary

Much has changed in SAP Ariba since the first edition of this book. The Ariba Network continues to be of paramount importance to the overall solution offering, as the Ariba

Network touches and underpins so many different SAP Ariba solutions. In this second edition, Chapter 2 on supplier collaboration chapter is followed by Chapter 3 on SAP Ariba Supplier Lifecycle and Performance and SAP Ariba Supplier Risk solutions, which weren't available in 2015 when SAP PRESS published the first edition of this book.

SAP Ariba Supplier Lifecycle and Performance helps SAP Ariba customers make sense of their supply base from a qualification and onboarding standpoint and thus helps ensure ongoing performance analysis based on both objective and subjective criteria. Having native integration with the Ariba Network means SAP Ariba Supplier Lifecycle and Performance can interface with suppliers and drive updates to supplier records in a "one-to-many" manner. Being integrated with SAP S/4HANA allows for the seamless transfer of supplier master records and updates to backend systems. SAP Ariba Supplier Lifecycle and Performance can essentially move your supplier management activities to the frontend, keeping things up-to-date in real time and distributing the relevant updates to backend systems to keep the supplier master data in sync.

SAP Ariba Sourcing also changed since the last edition, adding the direct procurement functionality in 2018 to support sourcing of bills of materials (BOMs) and other related types of direct materials and configurations. With SAP Ariba Discovery and SAP Ariba Spot Buy Catalog, this book concluded its tour of SAP Ariba solutions for what was formerly called "upstream" in the SAP Ariba world. In this process area, finding the right partner is paramount. Many of these suppliers, over 2 million of them, are already transacting on the Ariba Network today.

Another key part of SAP Ariba upstream solutions are contracts. SAP Ariba Contracts takes RFx awards or single sourcing events and turns them into objects that can be reused for transactions other than simple conversions to purchase orders. RFx awards don't need to result in a verbatim contract or require the creation of new terms and conditions. With clause libraries and negotiation steps, SAP Ariba Contracts makes a contract more than a follow-on RFx. With SAP Ariba Contracts, you can draw from a clause library, define terms and conditions negotiating steps, and operationalize your contracts for procurement via SAP Ariba Buying and SAP Ariba Catalog.

Once called "downstream" in SAP Ariba vernacular, the SAP Ariba guided buying capability, SAP Ariba Buying, and SAP Ariba Catalog turn upstream sourcing and contract creation efforts into actual consumption. The SAP Ariba guided buying capability and SAP Ariba Buying can integrate at both the system level and the process level with existing SAP ERP backend systems, and thus, invoice processing can occur in a different system for SAP Ariba procure-to-order scenarios in SAP Ariba Buying and for invoices to be processed by SAP Ariba in an SAP Ariba procure-to-pay (P2P) scenario.

SAP Ariba Buying also leverages a host of SAP Ariba tools to increase effectiveness, such as SAP Ariba Discovery for identifying and onboarding new suppliers, the guided buying capability in SAP Ariba for a consumer-grade buying experience, and SAP Ariba Spot Buy Catalog for obtaining one-off quotes in a straightforward manner during the sourcing of a requisition/shopping cart. SAP Ariba Catalog serves as the content repository for all catalog-related activities and items. SAP Ariba Catalog also functions as a basic shopping tool, allowing users to create shopping carts directly in the SAP Ariba Catalog, and drives the shopping process.

The Ariba Network plays a significant role in tying supplier updates together for SAP Ariba Supplier Lifecycle and Performance, and the Ariba Network plays a similar role in other process steps of the SAP Ariba solution portfolio. From SAP Ariba Buying processes that require sourcing via SAP Ariba Discovery, SAP Ariba Spot Buy Catalog, and SAP Ariba Catalog updates managed by suppliers, to RFx and contracting, the Ariba Network is the lynchpin supporting suppliers and spending.

In addition to analyzing supplier performance in SAP Ariba Supplier Lifecycle and Performance, spend performance and visibility requires analysis on the transactional side of the procurement equation. SAP Ariba Spend Analysis offers a comprehensive suite of reports and analytics engines for understanding trends in your spending. The core engines in these solutions include data validation, supplier matching, rationalization, business rule, inference, and machine learning. Custom commodity taxonomies, supplier diversity, and supplier risk management are optional add-ons.

A key takeaway in procurement in general is that decisions that may initially appear straightforward based on a piece of analysis may not always be so easy. Macro-level conditions can impact whether a negotiation proves successful. For example, a consolidated market may prevent significant discounts from being obtained where they otherwise would be. Cleansing data for analysis is also a key step that should be implemented prior to jumping to conclusions on the data. Supplier duplication and incorrect categorization of spend can create blind spots in analysis conducted too quickly.

SAP Ariba requires integration when interfacing with SAP ERP environments. Some cloud implementations are standalone, and in this scenario, you could skip Chapter 10 on integration. While many types of integrations with the Ariba Network now come standard out of the box, or are planned with SAP ERP systems, others require further fine-tuning or building out. The main SAP Ariba platform for integration is now the SAP Ariba Cloud Integration Gateway. Many SAP Ariba implementations still rely on older cloud integration approaches such as the Cloud Integration toolset (CI-X), the SAP Business Suite add-on, or the Ariba Network Adapter.

Cloud-based projects are different from traditional, onsite software implementations. Often requiring less formality and avoiding the phased timeline of an onsite implementation, the architecture undergirding the cloud-based application, as well as the application itself, are already in place. This difference enables businesses to focus more quickly on value realization from the project, rather than the technical aspects of implementation. SAP Activate is the latest project management methodology for managing cloud projects and will evolve with SAP solutions, much like the ASAP methodology did for on-premise implementations.

A key element of procurement implementations is looking beyond the internal boxes and systems involved and making supplier enablement a core part of the project. Many software-minded projects forget or downplay supplier enablement, which is not usually a must-have for getting a system up and running. However, once live, without supplier collaboration and participation, a system can quickly fail to realize its project benefits for efficiency and automation. Without supplier participation, a procurement system is less than half as effective.

The project methodologies reviewed in this book incorporate a supplier enablement approach for this reason, and any future methodologies for procurement will need to keep this area in focus. With the Ariba Network, the importance of onboarding suppliers becomes core to both the customer and the solution provider, since both the customer and SAP Ariba stand much to gain from having every supplier participating in the network.

12.2 The Future of Procurement Solutions

In the case of procurement solutions, while the underpinning technologies grow at exponential rates in computing, often the processes they support stay largely static, which creates both tension and opportunity. Similarly, technology in general evolves at a breakneck pace, while other areas, such as adjudication and the legal realm, move resoundingly slowly. The distance growing between these two areas creates immense pressures and the conditions for a tectonic-type earthquake when realignment does finally occur.

12.2.1 Procure-to-Pay and Order-to-Cash Processes

The view of a traditional supplier and customer relationship is also outdated. Customers and suppliers often interchange their roles during procurement or supply chain processes. Many suppliers are avid customers on the Ariba Network, leveraging the network as a significant channel to move their goods and services.

SAP Ariba and SAP S/4HANA have begun to integrate these shared areas for order-to-cash and procure-to-pay, first by consolidating customer and supplier master records into a single comprehensive record called the business partner. The business partner model of combining customer and supplier records will likely be the go-to model for all future cloud solutions that rely on master data from the digital core (ERP). For example, SAP Ariba Supplier Lifecycle and Performance requires the business partner model for integration with an SAP ERP backend, even if the ERP system is on an earlier release of SAP than SAP S/4HANA. As the Ariba Network continues to grow in importance and size, you should expect SAP's seller-side solutions in customer experience, SAP Customer Experience, and other areas to further extend their functionality and integration packages into the Ariba Network.

12.2.2 In-Memory Computing, Real-Time Analytics, Machine Learning, and Decision-Making

With real-time analytics now at the heart of many SAP solutions, external as well as internal data can be used to form actionable insights for procurement. Already, SAP Ariba and SAP S/4HANA are using machine learning to better reconcile invoices in an automated fashion. Both SAP Ariba and SAP S/4HANA environments offer dashboards to help users understand in real time what is occurring in their business areas and focus areas. Faster in-memory computing's impacts do not stop there. Insights generated within the system and from real datasets are coming at a speed that benefits from machine-based responses, so that the information can be acted upon in time and the full benefits of that information can be gained. The dashboard capability becomes a springboard for machine learning and the automation of decision-making. Many automated buying and selling scenarios already take place today, in stock trading and online ad-buying, for example. Systems running complex algorithms make purchasing decisions in an automated fashion. Material requirements planning (MRP) was a precursor, in a sense, to these sophisticated buying systems. MRP automates the decision-making process to reorder points and inventory replenishment. Next-level MRP enriches some reorder point decisions with external market condition data and/or seasonal influences on price and availability.

12.2.3 Big Data

Machine learning runs on data, and the more relevant data, the more precise decisions can be derived. Big data also will undoubtedly influence further innovations in procurement systems, as will the need for continued simplifications of processes. The Ariba Network generates reams of data every millisecond of operation, as do all of the SAP solutions running today. Expect this data to drive the enhancements to both existing and new solutions in the future.

12.2.4 Blockchain

At the heart of a transaction between two parties lies the issue of trust. For centuries, a person, an organization such as a bank or marketplace, or a digital clearing house has acted as the transaction broker, ensuring and insuring that the two parties transacting arrived at their mutually agreed to outcomes. However, this insurance and middleman comes at a cost. Blockchain concepts have been around decades but have matured to the point where in the near future the traditional broker model can be augmented or even replaced with a more cost-effective and efficient distributed ledger and notary-like service built on a blockchain model.

12.2.5 Consumer-Grade User Experiences

As consumers at home increasingly expect similar user experiences at work, including in their procurement activities, user interfaces and processes will need further adaptation and streamlining to keep up the pace.

12.2.6 SAP's Intelligent Enterprise

SAP's intelligent enterprise strategy consolidates these future-state topics into a comprehensive strategy and offering that customers begin deploying and consuming today. Currently, organizations face three significant challenges:

- To deliver a next-generation customer experience in a world of disruption
- To drive maximum cost synergies to fund innovation
- To better engage employees to attract and retain top talent

The intelligent enterprise strategy from SAP leverages artificial intelligence (AI), machine learning, Internet of Things (IoT), and analytics to:

- Redefine the end-to-end customer experience.
- Deliver a step change in productivity.
- Transform workforce engagement.

SAP Ariba and the other procurement solutions in the portfolio from SAP Concur and SAP Fieldglass buttress the second component in the intelligent enterprise of delivering a step change in productivity.

SAP Ariba is part of the network and spend management area in the intelligent enterprise strategy. This strategy drives off three main components:

- **Intelligent suite**
  This suite consists of line of business solutions in SAP's cloud and digital core portfolio.
- **Digital platform**
  This platform turns on both a holistic, real-time approach to data management with the SAP Data Hub and SAP HANA, and the SAP Cloud Platform and SAP Analytics Cloud for insight, integration, cross-solution coordination, and execution.
- **Intelligent technologies**
  SAP Leonardo technologies for the Internet of Things, machine learning, and artificial intelligence drive this next level awareness, agility, and capability throughout the individual applications and the intelligent enterprise in general.

12.2.7 Conclusion

SAP intends to remain the leader in procurement systems, both on-premise and in the cloud. With SAP Ariba and SAP S/4HANA Sourcing and Procurement as the digital core, as well as SAP Fieldglass for contingent labor and SAP Concur for travel and expenses, SAP provides a comprehensive portfolio of solutions for all procurement scenarios. Further, with the Ariba Network, SAP has a platform with limitless possibilities for growth and further efficiencies. SAP Ariba, as well as this entire procurement solution area, will continue to underpin SAP's slogan, "helping the world run better and improving people's lives."

The Authors

Rachith Srinivas has spent more than 15 years in technology with more than 12 years in SAP America, joining SAP's Professional Services organization in 2007, serving eventually as a platinum consultant in the SAP SRM practice. Since 2007, Rachith has been implementing and leading complex SAP procurement projects globally and has worked as a trusted advisor in this space on many strategic SAP accounts.

Rachith has a Master of Business Administration degree from McCombs School of Business at UT Austin and a Master of Science in Software Engineering. Prior to joining SAP as a functional consultant, Rachith worked as an architect and lead programmer, working on developing complex J2EE based solutions for verticals including health, insurance, and finance. In addition, Rachith is also a prior SAP PRESS author, having published a book on SAP Ariba. He resides in Dallas with one son.

Matthew Cauthen has spent over half of his more than 15-year career working at SAP America, joining SAP's Professional Services organization in 2011 as a materials management and supplier relationship management (SRM) consultant. He has extensive experience in implementing and supporting SAP procurement projects with a focus on SAP Ariba procurement applications, SAP S/4HANA, and systems integration.

Prior to joining SAP, Matthew held a number of positions in finance and IT fields. He has a bachelor's degree from the University of Georgia, and a master's degree in business administration from Emory University. He resides in Atlanta.

The Contributors

Justin Ashlock has spent over half of his 20-year-plus career in technology at SAP America, serving as the lead consultant for hundreds of global SAP customer projects and engagements supporting over $100 billion in procurement and logistics activities worldwide. Justin is a vice president of procurement services delivery within SAP's Intelligent Spend Management organization. In this role, he provides general management for the North American procurement services practice, while also providing global delivery strategy and consistency to MEE and CIS regions. Justin holds a bachelor's degree from the University of California, Berkeley, and a master's degree in business administration from the University of Notre Dame.

Juan Barrera started at SAP 17 years ago as a senior consultant, progressively taking on a series of consulting, services, and leadership roles, supporting sales and delivery. Currently, Juan serves as the global head of enablement for the cloud services delivery organization in the SAP procurement practice and is one of the services excellence leaders for the North America practice supporting the services delivery organization with enablement, learning, and training activities. Juan is an engineer based in San Jose, California, who for many years learned, mastered and implemented ERP and all the SAP Ariba solutions on the supply chain, procurement, and network areas for customers seeking real business results and procurement best practices in North America, Latin America, Europe and Asia, leveraging the SAP Ariba solutions with the world's largest enterprise business network.

Juan holds a bachelor's degree from the Pontificia Universidad Javeriana in Colombia, and he has a lifelong fascination for how technology constantly changes and improves processes and businesses in today's collaborative and innovation-driven economy.

Index

A

Accounting data ... 460
Accounts payable ... 43, 478
Ad hoc dynamic discount ... 497, 503
Ad hoc supplier-driven ... 498
Ad hoc user-driven ... 498
Administration console ... 420
Advanced shipping notification (ASN) ... 72, 447
Agreement release collaboration ... 608
Amending contracts ... 317
aml.analysis.HostedSpendExt.csv ... 553
aml.analysis.InvoiceAnalysis.csv ... 553
aml.analysis.PurchaseOrderAnalysis.csv ... 553
Analysis ... 33
Analysis data schema ... 539
Analytical reports ... 161
Application lifecycle management (ALM) ... 48
Approval flow ... 302
Approval process ... 378
Approval rules ... 419
Approval task ... 238
  adding approvers ... 240
  configure ... 240
Approval workflow ... 388
Approver
  adding ... 303
  matrix ... 304
Architecting ... 418
Ariba Network ... 29, 36–37, 39, 53–54, 60, 63, 77, 166, 379, 384, 419, 444, 480, 569, 574, 616–617
  buyer account ... 76
  buying ... 85
  customer and supplier interaction ... 578
  ID ... 461
  integration ... 650
  purchase order and invoice automation ... 86
  score ... 417
  supplier profile ... 82
Ariba Network Adapter ... 423, 447, 675
Ariba Network for sellers ... 55
Assigning tasks ... 69

B

Banking questions ... 158
Benchmark data ... 535
Best practices ... 416, 418
Bid processing ... 597
Bid transformation auction ... 214
Big data ... 677
Bill of materials (BOM) ... 383
Billing process ... 386
Blanket purchase order ... 384
Bookmarking ... 298
  format ... 298
  partial bookmarking ... 299
Budget data ... 379
Business network ... 53, 498
Business process management (BPM) ... 48
Business requirements ... 58
Business requirements workbook (BRW) ... 438
Business rule engine ... 533
Buyer task ... 70
Buying strategy ... 328

C

Capital management ... 497
Cash flow ... 496
Cash flow controls ... 505
Cash management ... 375
Catalog ... 34, 55, 57
Catalog interchange format (CIF) ... 420, 447
  file ... 379
Category change request (CCR) ... 556
Category strategy ... 328
  high-touch ... 329
  low-touch ... 329
  no-touch ... 328
  self-service ... 328
Change management ... 63, 421, 512
Change order ... 72
Check run ... 477
Clause library ... 674
Cleansing data ... 675
Cloud implementation model ... 44, 675
Cloud integration ... 675
Cloud Integration adapter ... 466
Cloud solution ... 28, 35–36, 38
Cloud Vendor Management System (VMS) ... 20
Collaboration request ... 381, 384
Collaborative shopping cart ... 419
Commerce ... 55

Commitment management ... 283
Communication plan ... 58
Complex sourcing ... 245
Compliance ... 463, 477
Compliance message ... 58
Compliance requirements ... 63
Component consumption message ... 613–614
Component inventory message ... 614
Component receipt ... 613
Component ship notice ... 613
Compound report ... 535
Configuration document ... 516
Consignment collaboration ... 615
Consignment movement ... 615
Consumption ... 32
Contingent labor ... 357, 461–462, 476
management ... 464
Continuous supplier registration ... 62
Contract ... 320, 510
compliance ... 419
data ... 565
management ... 55
management life cycle ... 56
manufacturing collaboration ... 612
Contract application programming interface ... 326
Contract authoring ... 297
Contract awareness report ... 557
Contract compliance ... 323
Contract consumption ... 313
Contract execution ... 313
Contract expiration report ... 310
Contract management ... 280, 283
Contract manufacturing ... 603
Contract planning ... 281
Contract request project ... 290
Contract term document ... 321
Contract terms attributes ... 314
Contract workspace ... 143
procurement ... 313
Contract workspaces ... 294, 316
Contracts ... 311
Core administration ... 420
Cross-functional collaboration ... 94
CSV file ... 486
Custom approval flow ... 150, 240, 303
Custom facts ... 566
Custom reporting ... 565
Customer project manager ... 537
Customer requirements ... 434
Customer spend ... 57
Cutover ... 47
Cutover management (COM) ... 48
Cutover planning ... 443
cXML ... 607, 609, 654–655
cXML document ... 486
cXML documents ... 67, 79

D

Dashboard ... 534
Data access control ... 551
Data archiving ... 48
Data collection ... 65
Data enrichment ... 542
Data extraction ... 538
Data migration ... 47–48
Data preparation ... 416
Data rationalization ... 460
Days payable outstanding ... 486, 496, 508, 510
Days sales outstanding ... 83, 508
Demand planning ... 32
Deployment ... 418, 510
Deployment project plan ... 431
Designated support contact ... 455
Digital signature authentication ... 479
Dimension tables ... 566
Direct materials ... 614
Direct procurement ... 41
Discount ... 486
Discount group ... 497, 503
Discounts ... 387
Discrepancies ... 481
Document ... 291
Document cleansing ... 298
Document properties ... 299
Document property type ... 300
Document signer ... 309
DocuSign ... 308
Dual user verification ... 505
Dutch auctions ... 214
Dynamic discounting ... 55, 497, 503
Dynamic filters ... 379, 384

E

Ease of exit ... 52
E-commerce ... 82
Electronic data interchange ... 31, 379, 616
Electronic invoice management ... 481
Electronic signature ... 308
Engagement risk ... 181
Engagement risk project ... 195
ERS invoicing ... 603

Evaluated receipt settlement (ERS) 615
Event messages 215
Extract, transform, and load (ETL) tool 541
Extrinsics 81

F

Face discount percent 497
Federated process control (FPC) 458
FI/CO 466, 487
File validation 544
  extended reports 544
Fiscal hierarchies 554
Flight plan 415–416, 418
Follow-on documents 599
Forecast 604
Forecast planning and collaboration 617
Forward auctions 214
Functional buyer 329
Fuzzy search 382

G

Generated subscription 385
Global compliance 483
Global invoicing 483
Go-live 425
Goods receipt 449, 582
Group
  adding 303
Guided buying 40, 327
  catalog 343
  category 342
  configuration 333
  forms 345
  group 341
  home page 340
  implementation 331
  preferred supplier 344
  procurement 357
  purchasing unit 342
  supplier 344
  supplier management 344
  tile 338
  user 341

H

Help community 349
  adding content 351
  configuring 352
  moderation 353
Historical data 559
Home page 340
Hybrid deployment 39
Hybrid solution 28, 39

I

IBM WebSphere DataStage 541
IDoc 655
Implementation 331, 415
Implementation methodology 28
Independent contractor 462
Index auction 214
Indirect procurement 40, 388
Inference engine 534
Integration 423, 569
Internal compliance 65
Internal invoices 381
Inventory 604
Inventory management 487
Inventory visibility 617
Invitation letter 66, 73
Invoice 33, 43, 72, 378–379, 444, 447, 481, 504, 565
  processes 477
  processing 375
Invoice automation 479
Invoice conversion 481
Invoice creation 583
Invoice enhancement 615
Invoice exception 78
Invoice management 477
Invoice order data 554
Invoice status 72
Issues management 181–182
Issues management project template 192

J

Japanese auctions 215

K

Key performance indicator (KPI) 49, 388
Kickoff 78, 511
Kit 383

L

Labor spend 461
Landing pages 339
Large-volume data set 565

License agreement ... 23
License agreements ... 283
Licensing ... 83
Local catalog ... 385
Local subscription ... 385

M

Machine learning engine ... 534
Manage benchmarking ... 557
Managed service provider (MSP) ... 468, 470
Manual risk scoring ... 171, 198
Master agreement ... 299
Master data ... 160, 458, 487
  import ... 161
Material classification ... 439
Material procurement ... 386–387
Material requirements planning ... 41, 445, 616, 677
Materials management (MM) ... 37
Maverick spend ... 388, 555
Mediated approach ... 655
Message Board ... 316
Migration path ... 64
Milestone ... 294, 425, 471, 511
Modular questionnaire ... 131
  approve ... 135
  deny ... 135
  request additional information ... 135
  templates ... 156
Modular questionnaires ... 134
Monitoring tasks ... 69
MRO ... 40
Multitenant cloud environment ... 49

N

Negotiation planning ... 281
Network catalog ... 385
Network growth ... 56, 62, 217
Noncompliant supplier ... 65
Notification ... 215

O

Off Contract Spend report ... 561
OK2Pay ... 378, 423, 444, 481
Onboarding supplier ... 66
Onboarding wave ... 417
Online analytical processing ... 27
Online transaction processing ... 27
On-premise project ... 424, 439
On-premise software ... 416
On-premise solution ... 28, 30, 36, 39, 51
On-time payment ... 486
Open Catalog Interface (OCI) ... 446
Open supplier registration report ... 162
Open supplier requests report ... 163
OpenText VIM ... 488
Operating cost ... 479
Opportunity search ... 545, 557
Order collaboration ... 614
Order confirmation ... 72
Order fragmentation ... 557
Order fulfillment ... 565, 579
Organization change management ... 48

P

Packing Slip ID ... 608
Parallel approval flow ... 240
Parametric refinement ... 382
Part numbers ... 382
Payment details ... 84
Payment terms ... 504
P-card reconciliation ... 379
Pilot approach model ... 425
Pilot users ... 512
Pipeline tracking ... 210
Planning ... 418
PO collaboration ... 602
PO vs. Non-PO Spend report ... 561
Post-award contract management ... 281
Preferred supplier management ... 128
  documents ... 154
  tasks ... 155
  team ... 155
  template ... 154
Prepackaged report ... 310
Price change ... 559
Price insensitive ... 559
Price variance data (PPV) ... 557
Price variation ... 557
Principal-agent problem ... 31
Private message ... 215
Procurement ... 28, 35, 376, 382, 419, 464
  Internet-based ... 20
Procure-to-pay ... 34, 281
Product sourcing ... 249
  dashboard ... 252
  manager parameters ... 261
  parameters ... 260
Production environment ... 62
Project
  template ... 287

Project approval activity report 164
Project message 215
Project notification letter 59
Project template 137
documents tab 290
overview tab 289
tasks tab 293
team tab 295
Prorated discount scale 497
Punch-in invoice 381
Punchout 379
catalog 385, 415
item 385
messages 385
process 419
Purchase agreement 32
Purchase order 61, 72, 365, 387, 447, 449, 463, 481, 565
data 554
reports 534
Purchase order and invoice automation ... 574, 608
Purchase order line item views 614
Purchase price alignment (PPA) 557
Purchase-to-pay channels 95
Purchasing integration 574
Purchasing policy 347

Q

Quality 605
Quality gate 48
Quality inspection 603
Quality notification 603
Quality review 603
Questions 141
Quick quote 244

R

Rationalization engine 533
Realm 512
Real-time analytics 677
Receipts 379
Registration 61
Relevance ranking 384
Remittance 381
Remittance address 510
Remittance advice 498
Remittance information 78
Remote shopping site 419
Replenishment 604
Reporting 65, 419
Request for information 213
Request for proposals 595
Requests for quotation 70, 591, 588
Requisition 379, 381, 565
process 378
Return order collaboration 603
Returns management 379
Reverse auction 214
RFx 463
RFx event template 233
RFx events for materials 257
Risk assessment 195
Risk assessment project template 191
Risk data 565
Risk management 95, 172
Risk mitigation 463
Risk scoring 189
Rollout process 536

S

Sales cycle 55
SAP Activate 45, 49, 424
SAP Ariba 27–28, 31, 327, 505
automatic reconciliation process 481
guided buying capability 37, 40, 330
integration 569
integration projects 569
mobile app 34
portfolio 95, 172
supply chain finance 508
SAP Ariba Adapter 655
SAP Ariba Analysis 541
SAP Ariba Buying 33–34, 37
SAP Ariba Buying and Invoicing
contracts 321
SAP Ariba Catalog 29, 33–34, 37, 376, 382, 419, 577
enablement 419
filter 383
hierarchy 383–384
refresh 383
SAP Ariba Cloud Integration Gateway 37, 635
extensions 670
integration wizard 646
migration 670
SAP Ariba Cloud Integration Gateway architecture 639
SAP Ariba Commerce Automation 388
SAP Ariba Contract Invoicing 482

SAP Ariba Contracts ... 23, 34, 66, 143, 146, 280–284, 293, 295, 308, 311, 315–316, 325, 419, 449, 554, 563–564
consuming contracts ... 317
implementation ... 284
master data ... 307
SAP Ariba Data Enrichment Services ... 537, 539
integration ... 547
SAP Ariba Discount Management ... 57, 498, 503
SAP Ariba Discount Management professional ... 34, 55
SAP Ariba Discovery ... 29, 34, 66, 242, 244, 283, 377, 381–382, 674
SAP Ariba Integration Toolkit ... 423
SAP Ariba invoice automation ... 34, 508, 574
SAP Ariba invoice conversion services (ICS) add-on
invoice threshold ... 74
SAP Ariba invoice conversion services (SAP Ariba ICS) add-on ... 480–481, 486
SAP Ariba Invoice Management ... 37, 57, 478–479, 482
SAP Ariba Invoice Management professional ... 555
SAP Ariba Invoice professional ... 438
SAP Ariba mobile ... 37
SAP Ariba on-demand ... 50
SAP Ariba P2X ... 39
SAP Ariba payment automation ... 34, 508
SAP Ariba payment management ... 57, 478, 502, 505
SAP Ariba portfolio ... 201
SAP Ariba procure to pay (P2P) ... 481, 508, 555
SAP Ariba Procurement
integration ... 667
SAP Ariba Procurement content ... 34, 39, 57, 375, 379, 382, 384–385, 415, 445–447
timestamp ... 421
SAP Ariba procure-to-pay (P2P) ... 84, 375, 414
SAP Ariba realm ... 49
SAP Ariba services procurement ... 386, 414
SAP Ariba Shared Services ... 50
SAP Ariba Sourcing ... 23, 29, 32, 34, 66, 201, 244, 282, 381, 554, 564
configure ... 221
dashboard ... 244
implementation ... 217
integration ... 658, 661
intergration ... 586
SAP Ariba Sourcing, savings and pipeline tracking add-on ... 282
SAP Ariba Spend Analysis ... 34, 283, 388, 529, 532, 539, 554, 564
data collection ... 541
data file manager ... 556, 563
enrichment ... 548
integration ... 547, 553
project manager ... 556
SAP Ariba Spend Analysis basic ... 533
SAP Ariba Spend Analysis professional ... 533
SAP Ariba spend management ... 563
SAP Ariba Spot Buy Catalog ... 327, 381, 674
SAP Ariba Strategic Sourcing ... 40, 249
SAP Ariba supplier enablement ... 67
SAP Ariba supplier enablement automation ... 73
SAP Ariba supplier enablement methodology ... 56
SAP Ariba Supplier Information and Performance Management ... 96
SAP Ariba Supplier Lifecycle and Performance ... 29, 32, 34, 37, 96, 282, 438
configuration ... 136
implementation ... 136
process flow ... 105
SAP Ariba Supplier Management ... 331, 564
SAP Ariba Supplier Risk ... 29, 37, 173
configure ... 185
implementation ... 183
SAP Ariba Supply Chain ... 35
SAP Ariba Supply Chain Collaboration for Buyers ... 569, 600, 617
quality management ... 629
SAP ASAP ... 47
SAP Business Suite ... 448
SAP Business Suite Add-On ... 447, 449
SAP Business Suite add-on ... 79, 654, 675
SAP Cloud Platform Integration ... 449
SAP Concur ... 31, 357
SAP ERP ... 37, 72, 78, 447, 481, 554, 655
SAP ERP materials management (MM) ... 382
SAP Fieldglass ... 20, 28
SAP Fieldglass vendor management system ... 37, 375
SAP Human Capital Management (SAP HCM) ... 471
SAP Integrated Business Planning ... 617
SAP Invoice Management by OpenText ... 37
SAP Master Data ... 658
SAP NetWeaver ... 79
SAP Pay ... 34, 37, 388, 478, 502–503, 511
payment ... 518

SAP Process Integration ... 448
SAP Process Orchestration ... 423, 640
SAP R/3 ... 28
SAP S/4HANA ... 27, 569
SAP Solution Manager ... 655
SAP SRM ... 37
SAP Supplier Lifecycle and Performance ... 564
SAP Supplier Lifecycle Management ... 37
SAP Supplier Relationship Management (SAP SRM) ... 382
SAP Supplier Relationship Management (SRM) ... 446, 574
SAP Supply Network Collaboration ... 605
Savings form ... 210
Savings target ... 561
Scheduling agreement ... 608–610, 622
Scheduling agreement release ... 602, 605
Scheduling agreement release collaboration ... 608
Security access ... 67
Self-billing ... 603, 615
Self-register ... 53
Self-service tools ... 486
Seller collaboration console ... 65
Seller value calculator ... 83
Serial approval flow ... 240
Service entry sheet ... 449
Service procurement ... 386
Service-level agreement (SLA) ... 468
Services procurement ... 40
Shared services ... 512
Ship notifications ... 379, 444
ShipNoticePortion ... 612
Signature request ... 309
Signature task ... 308–309
Site configuration data ... 460
SM questionnaire project response report ... 164
Solution design ... 46
Solution objective ... 426
Solution testing ... 47
Source system ... 563
Source type ... 564
Sourcing ... 34, 199
Sourcing event ... 56, 211, 244
Sourcing event content ... 213
Sourcing event type ... 213
Sourcing library ... 237
Sourcing project template ... 226
copy ... 226
Sourcing request ... 204
Sourcing request template ... 206, 221
copy ... 221
Sourcing scenario ... 244
Spend analysis ... 529
implementation ... 535
Spend categorization area ... 387
Spend category ... 462
Spend management ... 312, 530
Spend.aml.analysis.SpendAnalysis.csv ... 553
Spot Buy capability ... 37, 335
Spot buying ... 245
SSP-Psoft ... 564
SSP-SAP ... 564
Standalone implementation ... 444, 499
Star schema ... 539, 546, 566
Statement of work (SOW) ... 461
Strategic sourcing ... 39, 245
Style mapping ... 298
Subcontract manufacturing ... 603
Subcontracting purchase order ... 613
Subelement ... 610
Subject matter expert (SME) ... 421, 470, 512
Subscription ... 420
Subscription model ... 49, 467
Supplier
collaboration ... 53, 93, 171, 279
connectivity ... 480
country-based invoice rules ... 81
data ... 416, 460
deployment ... 60, 62
determination ... 379
enablement ... 53, 72, 415–416, 512
Enablement Status Report ... 77
Enablement Task Status Report ... 77
fees ... 82
fragmentation ... 557, 560
group ... 81
information portal ... 68
matching engine ... 533
membership program (SMP) ... 82
network ID ... 78
onboarding ... 49, 53
optimization cost (SOC) ... 558
parentage ... 559
parentage report ... 558
performance ... 675
profile ... 243
profile data ... 64
registration ... 85
relationship management ... 555
self-services ... 37
summit ... 59, 68
wave plan ... 63
Supplier 360° view ... 99
Supplier category
status request ... 129
Supplier certification management ... 135

Supplier collaboration
strategies ... 54
Supplier disqualification ... 125
template ... 152
Supplier enablement ... 72
services ... 484
status ... 71
Supplier footprint ... 560
Supplier integration ... 78
Supplier management activities ... 99
Supplier onboarding ... 60
Supplier qualification ... 122
approve ... 124
create new supplier ... 123
deny ... 124
request additional information ... 124
response ... 124
Supplier qualification template ... 150
Supplier record ... 606
Supplier registration ... 112, 116
approve ... 119
deny ... 119
request more information ... 119
update ... 114
Supplier registration template ... 145
Supplier request ... 106
approve ... 110
deny ... 110
manage ... 110
resubmit ... 112
Supplier request form ... 139
Supplier request process ... 105
Supplier request template
add content ... 140
configure ... 138
create ... 138
design ... 138
Supplier response ... 166
Supplier risk ... 194
Supplier risk users ... 187
Supplier self-registration ... 109
Suppliers ... 32
Supply base strategy ... 94
Supply risk
monitoring ... 194
Systems integrations leads ... 424

T

Tactical procurement ... 476
Tactical sourcing ... 353
Tasks ... 293
Tax information questions ... 159
Tax invoicing ... 487
Technical solution management ... 48
Technical workstream ... 415, 423
Template ... 288
Temporary labor ... 386
Terms and conditions (T&Cs) ... 280
Terms and conditions management ... 280
Testing ... 62
Timeline ... 63, 511–512
Time-to-value acceleration ... 49
Top-down hierarchy ... 566
Total cost auction ... 214
Total user experience ... 35
Track and trace functionality ... 505
Transaction ... 84, 553
Transaction type ... 72
Transparency ... 386

U

Unified supplier portal ... 243
UNSPSC ... 384
UNSPSC code ... 546
Uploading suppliers ... 68
User data ... 460
User exits ... 51
User strategy ... 328–329

V

Vendor consignment ... 602
Vendor data ... 66
Vendor data export report ... 77
Vendor invoice processing ... 603
Vendor management ... 78
Vendor management system ... 461
Vendor on-premise (VOP) ... 468
Vetting ... 464

W

Work order ... 604
Workbook ... 512
Worker quality ... 463
Workflow approval task ... 148, 300
Workflow rule ... 485
Workflow task ... 238, 301
Working capital ... 95, 486, 488, 504

- Configure purchasing, sourcing, invoicing, evaluation, and more
- Run your sourcing and procurement processes in SAP S/4HANA
- Analyze your procurement operations

Justin Ashlock

Sourcing and Procurement with SAP S/4HANA

Your comprehensive guide to SAP S/4HANA sourcing and procurement is here! Get step-by-step instructions to configure sourcing, invoicing, supplier management and evaluation, and centralized procurement. Learn how to integrate SAP S/4HANA with SAP Ariba, SAP Fieldglass, and more. Then, expertly run your system after go-live with predictive analysis and machine learning. See the future of sourcing and procurement!

716 pages, 2nd edition, pub. 02/2020
E-Book: $79.99 | **Print:** $89.95 | **Bundle:** $99.99

www.sap-press.com/5003

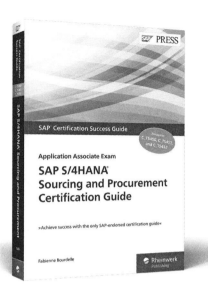

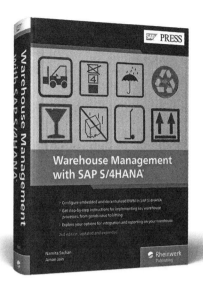

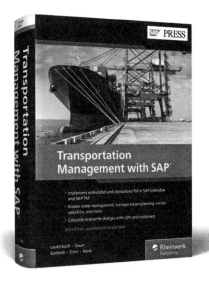

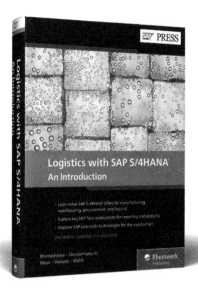